Dynamics of Coupled Systems in High-Speed Railways

Dynamics of Coupled Systems in High-Speed Railways

Theory and Practice

Weihua Zhang

ELSEVIER

Science Press
Beijing

Elsevier
Radarweg 29, PO Box 211, 1000 AE Amsterdam, Netherlands
The Boulevard, Langford Lane, Kidlington, Oxford OX5 1GB, United Kingdom
50 Hampshire Street, 5th Floor, Cambridge, MA 02139, United States

Library of Congress Cataloging-in-Publication Data
A catalog record for this book is available from the Library of Congress

British Library Cataloguing-in-Publication Data
A catalogue record for this book is available from the British Library

ISBN: 978-0-12-813375-0

For information on all Elsevier publications visit our website at
https://www.elsevier.com/books-and-journals

Publisher: Matthew Deans
Acquisition Editor: Glyn Jones
Editorial Project Manager: Naomi Robertson
Production Project Manager: Surya Narayanan Jayachandran
Cover Designer: Matthew Limbert

Typeset by TNQ Technologies

Contents

Preface

The author experienced the rapid growth of China's passenger railway system, has been engaged in research on rail vehicle dynamics, and in the construction of experimental platforms for rail vehicle dynamics over the past two decades. As chair of the Bogie research team, the author was involved in the development of China's first high-speed train, the *China Star*. Since 2006, the author has served as a member of a team of experts who constitute the innovation research team for China's high-speed trains and has been involved in the entire development process: introduction, digestion, absorption, and re-innovation. Furthermore, the author contributed to the development of the *Harmony* 380 series high-speed trains and the *Fuxing* 400 series China standard high-speed trains, which substantially advanced innovation in China's high-speed trains. Owing to these hugely significant contributions to the China high-speed railway, the author has been honored five times with the National Science and Technology Award.

The author developed the theoretical framework of the coupled system dynamics for the high-speed railway system. This theory treats the high-speed train as the primary component and considers associated sub-systems that could affect vehicle dynamics (railway tracks, airflow, power supply system, pantograph, and overhead catenary system) to achieve global optimization of the high-speed railway system. The coupled system dynamics of high-speed trains serves as an essential theoretical framework for the life-cycle dynamics research and test system for the design, manufacture, utilization, and maintenance of high-speed trains. Furthermore, the author developed a cyclic variable based integration method that accelerates modeling and calculation efficiency for a long-range train system. A sliding window method was proposed for the coupled vehicle/track dynamic model, enabling the coupled system dynamic model of high-speed trains to simulate a train operating on a track of infinite length. The arc due to contact loss and its associated arc force arising from the pantograph-catenary interaction under vibration conditions were also formulated. Based on the relaxation factor and equilibrium state, a numerical simulation method for fluid-structure coupling in a high-speed train system was developed; it substantially improves calculation stability and efficiency.

The author also headed the construction of a digital simulation platform, a basic research and experimentation platform, and a service performance research and experimentation platform for high-speed train systems. These research platforms have become the most important elements of the life-cycle dynamics research and test system for high-speed trains and serve as the foundation for the establishment and verification of the coupled system dynamic model of high-speed trains. This further facilitates the application of the coupled system dynamic model of high-speed trains in innovation research on the China high-speed railway and makes the coupled system dynamic model a significant novelty of the *Beijing-Shanghai high-speed railway project* that won the National Science & Technology Progress Award (special class).

This book is based on the author's previous book *Theory and Practice of the Dynamics of Coupled Systems of High-Speed Train* published by China Science Press in 2013, and the author's recent achievements. Throughout this journey, the author received considerable support and help. The author would like to express sincere appreciation to his doctoral supervisor, Professor Shen Zhiyun, an academician at the Chinese Academy of Sciences (CAS) and the Chinese Academy of Engineering (CAE), who consistently encouraged the author to explore the theory of dynamics of coupled systems of high-speed trains. The author also thanks colleagues at the State Key Laboratory of Traction Power for their active cooperation and support that ensured the full completion of the life-cycle dynamics research and test system for high-speed train design, manufacture, utilization, and maintenance processes. The author would also like to thank all teachers and graduate students who participated in his various scientific research projects; their research achievements greatly enrich this book. Finally, the author thanks his students, and teachers from the School of Foreign Languages of Southwest Jiaotong University, for their hard work on the English version of this book.

Because the theory of dynamics of coupled systems of high-speed trains has been proposed only recently, its theoretical framework for the life-cycle dynamics research and test system of high-speed train design, manufacture, utilization, and maintenance thus needs further improvement in practice. Owing to the author's limited knowledge on traction power supply and traction drive, and some related researches still being under investigation, this book, to some degree, is not quite systematic and its content is incomplete. Hopefully, it will be gradually improved in the coming years. Your suggestions and corrections will be highly appreciated.

Weihua Zhang
19th September 2018
Chengdu, Sichuan Province, China

CHAPTER 1

Introduction

Chapter Outline

Dynamics of Coupled Systems in High-Speed Railways. https://doi.org/10.1016/B978-0-12-813375-0.00001-7

The high-speed railway is the product of over half a century of continuous development, and more than 37,000 km of high-speed railway lines have been built around the world. Presently, Europe and Asia, characterized by German, French, Japanese, and Chinese technologies, have become the epicenter of the high-speed railway [1–27]. Although these technology systems have different design philosophies, each one *explores new design technology* to achieve faster, safer, and more comfortable high-speed trains. Thus, highly advanced and efficient manufacturing technologies together with a vast amount of operational experience have been employed to develop new high-speed railway systems.

As the core of the high-speed railway, the high-speed train system consists of not only the high-speed train but also the high-speed railway track, overhead catenary system (power supply system), communications system, and other subsystems, as shown in Fig. 1.1.

Considering the traction mode adopted, the high-speed train can be further classified as a concentrated power train or distributed power train. Owing to the unique advantages of these traction modes, both types of high-speed train are widely used across the world. The high-speed railway track technology also constitutes one of the essential components of the high-speed train system. There are two types of track systems: ballasted track and slab track. The ballasted track is characterized by low noise emission, and the slab track is well known for its relatively small track irregularities compared with the ballasted track. Designers of railway tracks must balance the trade-off between both track systems to achieve optimal dynamic performance for the vehicle and track systems. Depending on the suspension type, the high-speed overhead catenary system can be classified into simple catenary system, stitched catenary system, and compound catenary system. Considerable effort involving the optimization of catenary parameters has been made to further improve the current collection performance of different catenary systems.

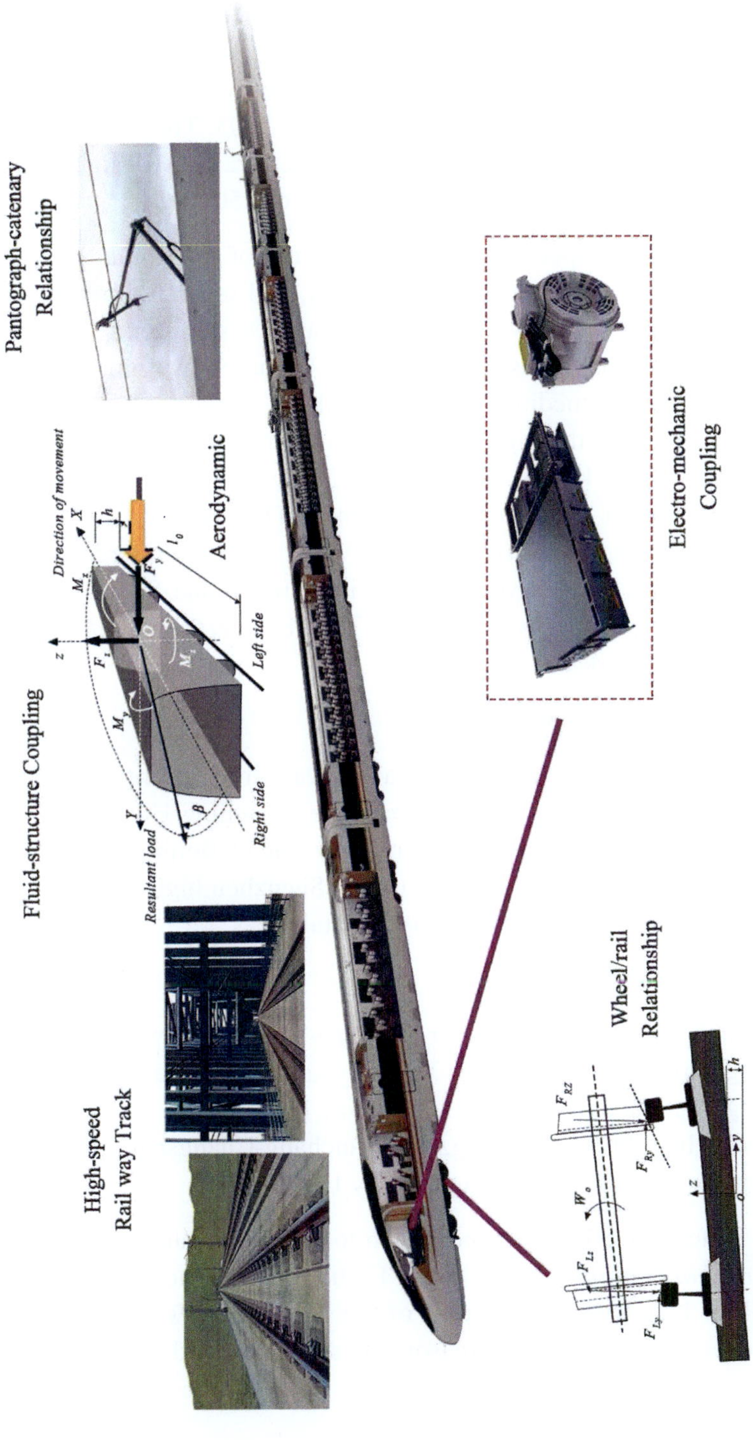

Figure 1.1
High-speed train system.

1.1 Development and technical features of the China high-speed railway

1.1.1 Development of the China high-speed railway

On August 16th, 1999 construction began on China's first high-speed railway line, and the 404 km Qinhuangdao-Shenyang line was completed on October 12th, 2003. This high-speed railway project was designed and constructed entirely by Chinese railway designers and engineers. The top speed was 200 km/h, with a reserve capacity of 250 km/h.

In the preceding decades, the low operating speed and limited rail lines of the China railway system, which was the main driver of the Chinese economy, had severely limited the development of China's economy and the growth of its mass transit system for a long time. In the twenty-first century, an agreement was reached by the Chinese government to develop a high-speed railway system, and a medium-long term railway network plan was published. The plan was for the rail lines of the high-speed railway with design speeds of 200 km/h or more to be longer than 12,000 km by 2020 [28]. In 2008, this goal was adjusted to 18,000 km with the approval of the China State Council. This included approval for the *Four verticals and Four horizontals* railway network and three intercity railway lines on the Yangtze River Delta, Pearl River Delta, and Circum-Bohai-Sea [29]. These together form an extensive high-speed railway network. The *Four verticals and Four horizontals* high-speed railway network consists of:

Four vertical railway lines running north to south:

Beijing-Shanghai high-speed railway line or Beijing-Shanghai railway line, Beijing-Wuhan-Guangzhou-Shenzhen (Hong Kong) high-speed railway line or Beijing-Hongkong railway line, Beijing-Shenyang-Harbin (Dalian) high-speed railway line or Beijing-Harbin high-speed railway line, and Shanghai-Hangzhou-Ningbo-Fuzhou-Shenzhen high-speed railway line — a passenger railway line in China's southeast coastal areas.

Four horizontal railway lines running east to west:

Qingdao-Shijiazhuang-Taiyuan high-speed railway line or Qingdao-Taiyuan railway line, Xuzhou-Zhengzhou-Lanzhou high-speed railway line or Xuzhou-Lanzhou railway line, Shanghai-Nanjing-Wuhan-Chongqing-Chengdu high-speed railway line or Shanghai-Wuhan-Chengdu railway line, Shanghai-Hangzhou-Nanchang-Changsha-Kunming high-speed railway line or Shanghai-Kunming railway line.

On June 29th, 2016, the China State Council announced a plan to expand the almost complete *Four verticals and Four horizontals* network to the Eight verticals and Eight horizontals network (shown in Fig. 1.2), based on the plan of the *Four verticals and Four horizontals*. The new plan considers passenger flows, reasonable standards, and other requirements for the expansion.

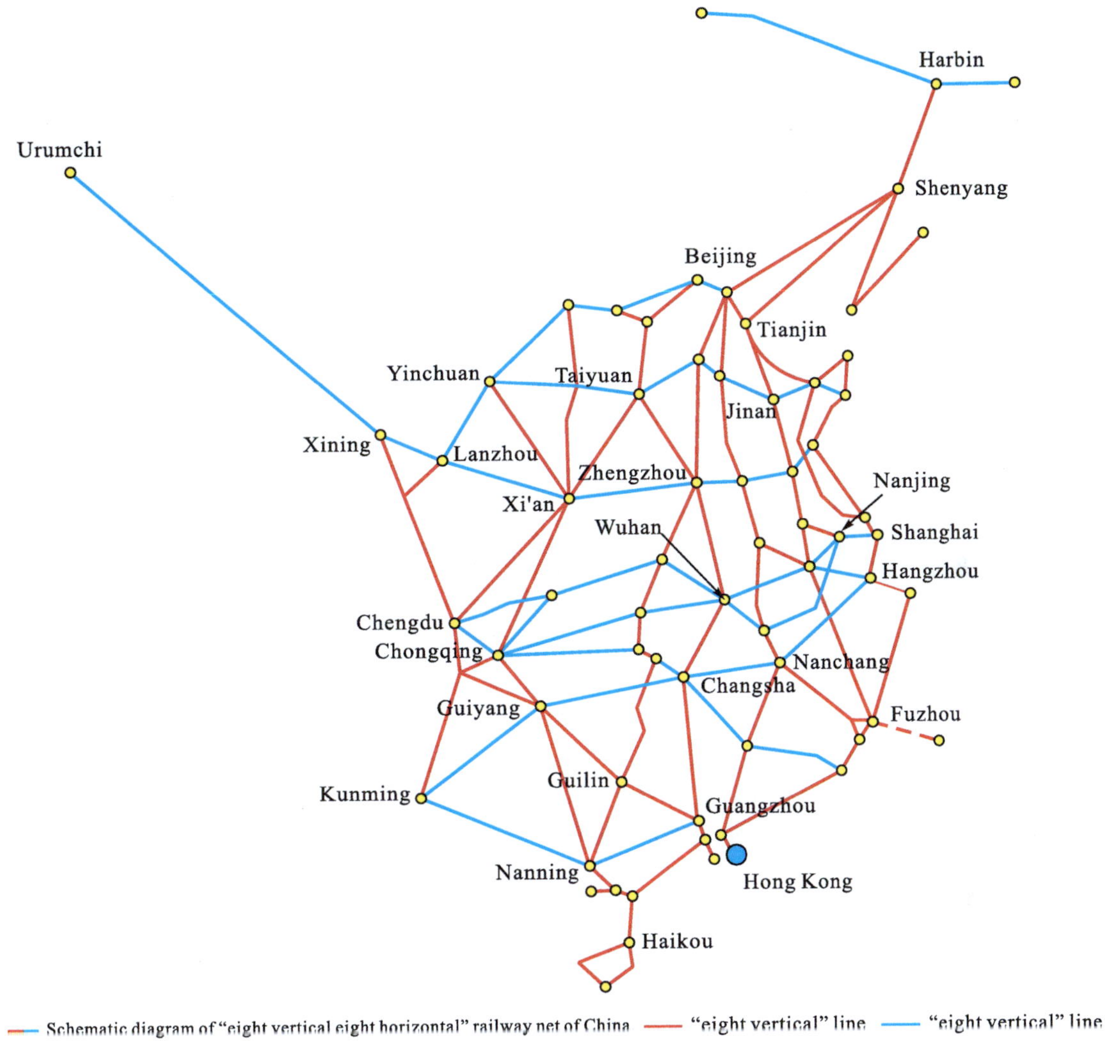

Figure 1.2
Medium-long term plan for the China high-speed railway network.

This plan aims to develop a higher standard high-speed railway network by 2030, and the *Eight verticals and Eight horizontals* railway network is the main corridor of this high-speed railway network while other regional and intercity railway lines supplement the main corridor. The plan also specifies the construction standards for the high-speed railway network.

The new high-speed railway line will use standards with design speeds of 250 km/h or higher (a lower standard may be used in some areas due to complex geological or climate conditions). The railway lines passing through areas with dense population, built-up areas, and big cities will use standards with design speeds of 350 km/h, while regional railway

Figure 1.3
A high-speed railway viaduct in China.

lines will use standards with speeds of 250 km/h or lower, and the design speed of intercity railway lines would not exceed 200 km/h.

The *Eight verticals* high-speed railway network consists of:

The Coastal line, Beijing-Shanghai line, Beijing-Hong Kong (Taiwan) line, Beijing-Harbin-Beijing-Hong Kong-Macau line, Hohhot-Nanning line, Beijing-Kunming line, Baotou (Yinchuan)-Haikou line, and Lanzhou (Xi'an) Guangzhou line.

The *Eight horizontals* high-speed railway network consists of:

The Sufenhe-Manzhouli line, Beijing-Lanzhou line, Qinghai-Yinchuan line, Europe-Asia transportation line, the railway line along the Yangtze River, the Shanghai-Kunming line, Xiamen-Chongqing line, and Guangzhou-Kunming line.

Based on the *Eight verticals and Eight horizontals* railway network, further expansion is planned for regional railway lines to increase the coverage of the high-speed railway system. Both high-speed and conventional railway systems are supported.

By 2017, 2,985 high-speed trains had been manufactured by the China Railway Rolling Stock Corporation (CRRC) with over 5,200 high-speed trains operating on the China high-speed railway network daily. The total route of the high-speed railway rail lines that were put into operation already exceeded 25,000 km and could be extended to 30,000 km by 2020. This makes China the country with the most high-speed trains and the greatest length of high-speed railway lines. The 1,318 km JinHu high-speed railway line is characterized by the highest design standards in the world. It has a typical slab track design that allows for an operating speed of 350 km/h, with 86.5% of the line constructed as a viaduct for superior track quality.

1.1.2 China high-speed railway technologies

1.1.2.1 Railway line

In China, most high-speed railways are constructed as double tracks. For high-speed railway lines with design speeds of 300 km/h and above, the track spacing is 5 m, the track gradient is less than 2%, and the minimum curve radius is 7,000 m. The railway viaduct, as shown in Figure 1.3, is widely used in high-speed railways to maintain track irregularities at a relatively low level to avoid track settlement, which could be caused by rivers and gullies. Furthermore, the railway design must also consider the effects of the horizontal and vertical alignment of the railway line to get rid of the vertical vibration arising from the track design.

1.1.2.2 Railway track

Because of the characteristics of low-level irregularities, the slab track is widely used in the China high-speed railway. The slab track technology, however, was initially imported from other countries due to lack of manufacturing technology for track slabs in China. In September 2004, a 13.16 km test slab track built using Chinese technology for the first time was applied to the Suining-Chongqing railway line. In 2007, a field test at a speed of 232 km/h demonstrated that the performance of this test slab track meets design standards.

To develop a 350 km/h high-speed railway, China imported additional slab tracks from German companies, such as Max Bögl, RAIL.ONE, and Züblin. Beijing-Tianjin and Beijing-Shanghai high-speed railway lines use the Max Bögl track slab, Wuhan-Guangzhou railway line uses the Rheda 2000 slab track provided by RAIL.ONE, while Zhengzhou-Xian railway line uses Züblin technology. The Harbin-Dalian railway line uses Japanese Shinkanshen slab track technology. In China, slab tracks are further classified into four types based on differentiating characteristics: China Railway Track System type-I (CRTS I), CRTS II, CRTS II dual-block type, and CRTS III. The slab track based on Japanese Shinkanshen technology is classified as CRTS I. CRTS II and CRTS II dual-block type slab tracks are slab tracks from Max Bögl and RAIL.ONE, respectively. New slab tracks designed by the Chinese railway company in November 2009 are designated CRTS III and have been successfully applied on the Chengdu-Dujiangyan high-speed railway line.

Based on over 20,000 km of operating experience and experimental data from different types of railway tracks, China is presently able to design, manufacture, construct, and maintain different railway tracks, and possesses an understanding of the properties of the different slab tracks, which has led to a systematic increase in the design standards of the China high-speed railway.

To reduce the influence of track irregularities under high-speed conditions, China further improved track construction standards to reduce track irregularities. In addition to seamless rail technology, significant effort has been made to reduce tolerances in the manufacturing and installation procedures. The manufacturing precision for track slabs is 0.1 mm, and allowable tolerance for track gauge is +/-1 mm. Tolerances in height, torsional direction, and track direction are less than 2 mm, while only a 1 mm deviation is allowed in the horizontal direction. The material, U71MnK, is used to manufacture rails with a length of 100 m using a rolling procedure. High-speed turnouts are also employed in high-speed railways to achieve the high speeds, which allow the train pass through the turnout at a speed of 350 km/h from the facing direction and at 120–250 km/h from the curve side.

1.1.2.3 Tunnel

In China, the tunnel cross-section of most pre-existing railway lines is about 64 m^2. Because the section area of tunnel directly impacts the vehicle's air resistance and air tightness, the tunnel section area of the China high-speed railway is much higher at 100 m^2. Furthermore, buffer-structured entrances and assistant tunnel wells are built to improve the aerodynamic performance of high-speed trains.

1.1.2.4 Train control system

The China high-speed railway network is characterized not only by its high speeds but also by the high-density regions it connects. This combination makes the high-speed train a viable top choice for public transportation. At the request of the China high-speed railway, the Chinese train control system (CTCS) was developed for better control of railway operations. In the CTCS-2 system, the transponder and railway signal transmission systems enable the driver to operate the train at speeds of 200–250 km/h. The CTCS-3 system is based on a wireless signal transmission system and permits the driver to operate the train at further speeds of 300–350 km/h.

From the viewpoint of compatibility, the CTCS-3 train control system can be superimposed onto the CTCS-2 train control system, and the CTCS-2 can act as a backup system for the CTCS-3 train control system. When the radio block center or wireless communication fails, the CTCS-2 train control system can take control of the train in place of the CTCS-3 control system. Due to the advanced performance of the CTCS train control system, the minimum operating interval of a China high-speed train is 3 min and is generally about 5 min in practice.

1.1.2.5 Power supply system

The power supply system of the China railway system is a single-phase alternating current of 27.5 kV/50 Hz, and the power supply mode is AT (auto-transformer) with a 50 km long supply arm. The traditional overhead catenary system is the simple catenary system, which is also employed in the Beijing-Tianjin intercity railway line, with a design speed of 350 km/h. To further improve the current collection performance of the pantograph-catenary system, especially that of the double-pantograph at a speed of 350 km/h, the China high-speed railway adopted the stitched catenary system with a tensile force of 30 kN in the contact wire.

1.1.3 Development of the China high-speed train

Research for the development of the China high-speed train began at the end of the previous century and reached fruition when the first independently developed high-speed train, *China Star* (See Fig. 1.4), was developed by the China Railway Corporation at the beginning of this century. The *China Star* is a power-concentrated train with a design speed of 270 km/h, which employs an AC drive system and 2M9T marshaling and has a capacity of 726 passengers. In a field test on November 27th, 2002, the *China Star* electric multiple unit (EMU) reached the highest operating speed on Chinese railway at 321.5 km/h.

For higher speeds and more advanced high-speed train technology, the China State Council proposed a guideline "introducing advanced technology, joint design and manufacturing, and building the China brand." Accordingly, the Ministry of Railway implemented a project for high-speed train "introduction, mastering, assimilation, and reinvention." In 2004, China Railway imported 200 km/h high-speed trains from Alstom, Bombardier, and Kawasaki, and a 300 km/h high-speed train from Siemens in 2006.

Figure 1.4
The *China Star* high-speed train.

Figure 1.5
"CR" China standard high-speed train.

The requirements for high-speed railways, such as long distance, large capacity, high density, and short travel time, present significant challenges to the design, manufacture, operation, and maintenance of the high-speed train. To localize the introduced high-speed trains and develop China's own high-speed train technology, most of the research institutes were convened to investigate the critical technologies in high-speed trains: system integration, car body, bogie, traction transformer, traction converter, traction motor, traction control, train control network system, and braking system. Furthermore, ten supplementary technologies were also considered: air condition system, dejectax collecting device, door, window, windshield, coupler and buffer, pantograph, power supply system, interior decoration, and seat.

Through innovative research and the mastering and assimilation of imported technologies, the high-speed train has been successfully localized and subjected to further self-sustaining innovation. Currently, two speed series, two brands, and ten types of high-speed trains have been developed. The two speed series are the 300–350 km/h series and the 200–250 km/h series. Eight of the ten types of high-speed trains are of the China Railway High-speed (CRH) brand while two types are of the China Railway (CR) brand. Under the CRH brand are the CRH1 produced jointly with Bombardier in China, CRH2 built with Kawasaki technology, the CHR3 (Siemens technology), CRH5 (Alstom technology), CRH6 (independent innovation), and the new generation of high-speed trains CRH380A, CRH380B, and CRH380D (jointly produced with Bombardier in China). The CR brand represents China standard high-speed train CR400AF and CR400BF, as shown Figure 1.5.

The following illustrates some of the technical features and parameters of a high-speed test train for higher speeds. This train was developed independently by CRRC Qingdao

Sifang Co. Ltd. drawing on forward-looking, fundamental, and theoretical research with a view on safety and reliability under much higher speed conditions. The primary purpose focuses on three objectives:

(1) To provide a technical platform for high-speed train safety studies because running safety is a determining factor in the development of high-speed train technology. Thus, the technical platform is expected to improve the safety margin through the exploration of running stability, structural strength, train-track network matching relationships, and other safety systems for high-speed trains under higher speed conditions, to create improved guidelines for high-speed train applications.
(2) To provide a test platform for fundamental research. With a high-speed test train for higher speeds, forward-looking, fundamental, and theoretical research on high-speed trains can be conducted within a broader and higher speed range that can reveal train dynamic behaviors under high-speed conditions. Furthermore, such studies will strengthen the industry-university research collaboration, thereby creating a sustainable environment for innovation.
(3) To provide an application and verification platform for new materials and new technologies. The high-speed train technology system can be improved through research on various new materials and technologies. The purpose of this platform is to ensure the sustainable development of China's high-speed train technology, accelerate the training of design teams, and improve the standardizing system.

The 500 km/h test train for higher speeds is based on independent, innovative products in the new generation CRH high-speed train. Improving the critical speed and traction of the vehicle, in combination with lower resistance and dedicated research effort on system integration, head form, the car body, bogie, traction, braking, and other systems have contributed to the achievement of operational safety and reliability at higher speeds. The main technical parameters of this train are summarized in Table 1.1.

Main technical features:

1.1.3.1 Different head forms in the lead and tail cars

To achieve 500 km/h test speed, the aerodynamic resistance of the lead car and the aerodynamic lift force of the tail car must be reduced. Thus, different head forms are designed for the lead and tail cars. The head form of the lead car is shaped like a *sword*, while the tail car is shaped more like a *rocket*. The head form of the tail car was designed based on a pre-existing CRH train characterized by good aerodynamic performance, while the sword-shaped lead car is a completely new design with lower air resistance and lateral force. Both lead and tail cars are illustrated in Fig. 1.6 [30].

Table 1.1: Main technical parameters of the 500 km/h CRH high-speed train.

Parameters	Value
Marshalling type	six motor vehicles
Operating speed	400 km/h
Test speed on track	$\geq$ 500 km/h
Control device	IGBT water-cooled VVVF
Braking method	Regeneration braking, direct-through type electric pneumatic braking
Train network	Industrial Ethernet
Power supply mode	25 kv/50 Hz AC, power supply of aerial cable
Car length	153.5 m
Head car length	27/26.5 m (different head forms)
Middle car length	25 m
Width of car body	2950 mm
Car height	3650 mm (new wheel, excluding pantograph)
Length between two truck centers	17,500 mm
Bogie	Bolsterless bogie with air spring
Wheelbase	2,800 mm
Wheel diameter	920/860 mm
Track gauge	1,435 mm
Maximum axle load	15 t
Traction power	22,800 kW (test)
Shaft power	950 kW (test)

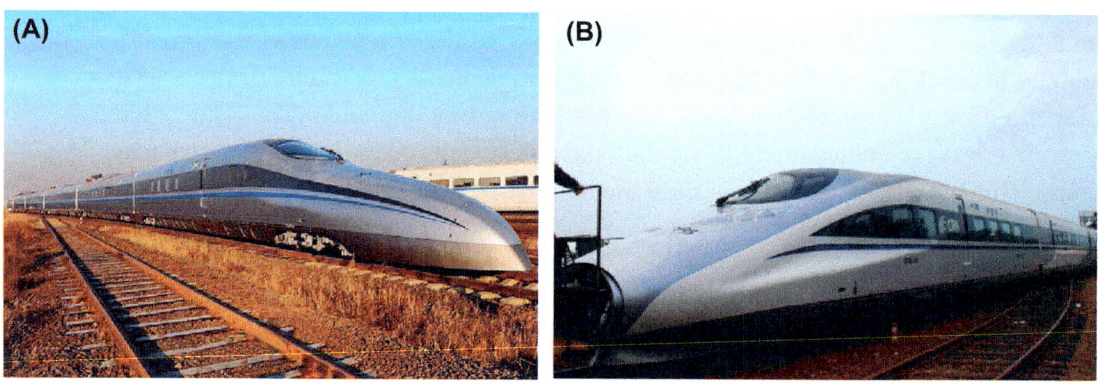

Lead car of the test train
Tail car of the test train

Figure 1.6
Test train for higher speeds. (A) Lead car of the test train (B) Tail car of the test train.

1.1.3.2 Small aerodynamic resistance on the car body

To reduce air resistance and lateral aerodynamic force, the section of the car body is further optimized. The width of the car body is reduced to 2,950 mm from 3,380 mm used in CRH380A, and the height of the car body is decreased to 3,650 mm from 3,700 mm.

Furthermore, the radius of the roof of the car body is increased to increase the stiffness of the car body and reduce the lateral aerodynamic force on it.

1.1.3.3 Larger wheelbase in bogie

To increase the critical speed of the train and its compatibility with a high-power motor, the wheelbase is increased to 2,800 mm from 2,500 mm used in the CRH 380A. The bogie is of a bolsterless design and is equipped with an anti-roll bar and an air spring system. A yaw damper employs a redundancy design to enhance stability and damping in the bogie. This results in a very high critical speed of 750 km/h for the high-speed test train for higher speeds, which can effectively ensure dynamic performance of this train at 500 km/h.

1.1.3.4 High traction power

To overcome air resistance under high-speed conditions, the traction power of the train is enhanced, in addition to reduce the section area of the car body and head form optimization. The traction power of each axle of the test train reaches 950 kW, which is the highest traction power of all high-speed trains.

1.1.3.5 Development of a new network

The test train employs a new network system developed by CRRC Zhuzhou Institute Co. This network system is based on industrial Ethernet technology, which permits a very high information transmission capacity for the train's operating, control, and safety information. Meanwhile, a network system for the test is also installed on the train to synchronously acquire the experimental data.

1.1.3.6 Application of new materials

Apart from the new technologies in the network and the air resistance braking system, several new materials, such as carbon fiber, magnesium alloy, and novel nanometer soundproofing have been employed in the design of test train.

Through a review of critical technologies in French, German, Chinese, and Japanese high-speed railways, it can be concluded that these technical systems have their own unique characteristics in the high-speed train or the high-speed railway line that makes the railway system more diverse and also culminates in extensive design experience. High-speed railway technology consists of not only high-speed train technology, high-speed track technology, and high-speed overhead catenary technology, but also the coupling technologies among the subsystems. The higher operating speed is expected to introduce stronger coupling effects on the train-track-tunnel-catenary-aerodynamic system environment. The design of the subsystems thus must consider the coupling loads arising from these coupling systems. The low-cost tunnels in Japan face significant challenges to high-speed operation because the section area is small — even with the buffer section near the tunnel entrance taken into consideration. This is evident from the longitudinal impact arising from air resistance and air pressure fluctuation when the train enters the tunnel. The ballasted track used in France is characterized by low construction cost and can effectively absorb wheel-rail noise (2—3 dB). However, from a long-term viewpoint, ballast impact and track settlement may affect the train's dynamic performance and the reliability of equipment under the car body.

In the last half-century, railway technologies related to vehicle dynamics, track dynamics, pantograph dynamics, and aerodynamics have advanced significantly. Under high-speed conditions, however, the coupling effects among these elements become much more critical and inevitably constitutes a tremendous challenge to high-speed railway design.

1.2 Literature review of railway dynamics

Apart from the traditional vibration problems in the mechanical system, the high-speed train is also subject to the stability problem. The mechanical system in the high-speed train has much more complex than the traditional mechanical system. When the train operates on the track, loss of stability could occur due to disturbances arising from the track, airflow, and other external excitations. The instability could also affect other subsystems. Therefore, the dynamics or vibration problems in the subsystems of high-speed trains are also quite important, although the dynamics of the high-speed train is at the core of railway dynamics. The dynamics of the railway system mainly includes vehicle system dynamics or train system dynamics, track system dynamics, vehicle-track coupling dynamics, train aerodynamics, and pantograph-catenary system dynamics. On the dynamics of the high-speed train, a coupling system dynamics of the high-speed train based on the dynamics of each subsystem was proposed by the author [31].

1.2.1 Vehicle system dynamics

Vehicle system dynamics, taking the rail vehicle as the research object, is mainly dedicated to studying the relationship between the vehicle and external excitations. The primary purpose is thus to understand the dynamic behaviors of the rail vehicle and some special dynamic phenomena during operation, such as hunting stability, curving dynamic performance, and ride comfort.

1.2.1.1 Hunting stability

The earliest definition of stability was given by Torricelli in the 17th century, while a more precise definition was given by Lyapunov in 1892 [32]. Stability can be further classified into static stability and dynamic stability. In the rail vehicle, the anti-overturning stability and anti-derailment stability of the vehicle determined by force equilibrium are referred to as static stability. The hunting stability of the vehicle, as an inherent feature of the rail vehicle system determining the dynamic performance of the high-speed train under high-speed conditions, is referred to as dynamic stability [33].

The hunting motion of the wheelset was initially observed in the early stages of the railway [34], while the early studies were primarily focused on the wheel-rail contact relationship rather than the mechanism of hunting stability. In 1916, Carter [35] researched the mechanism of hunting stability and concluded that hunting instability is primarily related to the cone-shaped wheel tread and the wheel-rail creep forces at the wheel-rail interface. These constitute the theoretical foundation for hunting stability and facilitates research on hunting stability of the rail vehicle. Using stability theory, De Pater researched the hunting stability of railway vehicles. Matsudaira [36, 37] researched the linear stability of hunting motion taking into consideration the cone-shaped wheel tread in a bogie model. Wickens [38] studied bogie hunting stability using a linear wheel-rail creep force model and concluded that wheel-rail creep forces and the conicity of the wheel tread are the main causal factors of hunting motion, and a more detailed dynamic model should be used in the calculation of vehicle critical speed.

With the development of wheel-rail contact mechanics, a number of nonlinear wheel-rail creep force models were developed to evaluate wheel-rail creep forces at the wheel-rail interface [39−41]. In the early stages, linear models were widely used in research on vehicle hunting stability due to limited computational efficiency. The studies suggest that the linear model yields results with relatively low accuracy compared to those obtained using a nonlinear model — which facilitates the development of the nonlinear vehicle dynamic model. Considering the nonlinearities in lateral and longitudinal creep force, Cooperider [42] researched the nonlinear dynamic behaviors of rail vehicles. Wickens [43] studied the effects of the worn wheel profile on the nonlinear stability of the bogie. Using the Poincaré section method, Knudsen *et al.* [44] researched the bifurcation and chaos phenomena of a single wheelset. Considerable effort was also invested into research on bifurcation and chaos of rail vehicles by Petersen [45, 46] and Hans True *et al.* [47−50] based on the Vermeulen-Johnson nonlinear wheel-rail creep force model. Consequently, a methodology for calculating the nonlinear critical speed of a vehicle system was developed. Zboinski *et al.* [51] studied hunting stability of rail vehicles in a curve. Zeng [52] researched the characteristic value of a vehicle system using the QR algorithm and linear stability theory. The shooting method was also employed to identify stable and unstable limit cycles in the vehicle system. Zhang [53] numerically researched the characteristics of bifurcation and limit cycles in a nonlinear vehicle system using stability and bifurcation theories, which led to a criterion for determining the characteristics of the limit cycle and bifurcation in the vehicle system. These were validated through vehicle roller rig tests at the State Key Laboratory of Traction Power, Southwest Jiaotong University.

Active control has also been employed to control hunting stability of the bogie. Goodall *et al.* [54] derived the motion of equations for both an unconstrained wheelset and an independent wheelset and formulated the state equation block diagram to illustrate the feedback effect of wheel-rail creep force and equivalent conicity. The research proposes several active control methods for wheelset hunting stability. Pearson and Goodall *et al.* [55] pointed out that the yaw damper could transmit high-frequency vibrations to the car body and thus cause some undesirable effects. To investigate the yaw feedback-based active control method, a well-validated active control model coupled with a rail vehicle model was developed. The results suggest that co-simulation technology can be used to design the active control. Bruni *et al.* [56] designed and manufactured an active yaw damper based on brushless motor technology. The laboratory test shows that the controllable motor-based damper can dissipate much more energy than the traditional yaw damper without increasing force transmissibility to the car body. Shen [57] also conducted some valuable research on stability control.

1.2.1.2 Curving dynamic performance

Due to geographical conditions, the movement of the rail vehicle is inevitably subjected to curves and not just tangents alone. Considerable effort has thus been invested in research on curving dynamics to better understand wheel-rail interaction and reduce dynamic forces in the vehicle system, thus preventing overturning and derailment [58–59].

The study of the curving dynamics of rail vehicles experienced four phases [33]. In the 1930s, Heumann and Poter studied curving negotiation using the friction center method. Due to breakthrough progress in wheel-rail creep theory in the 1960s, Newland and Boocock [231] pointed out that wheel-rail creep force can enhance the steering capability of the wheelset in a curve. Because the wheel-rail creep forces and wheel-rail contact relationship, as well as the suspensions in the rail vehicle, are considered linear characteristics, such vehicle motion in a curve is thus referred to as steady-state motion. Steady-state motion is defined as a vehicle operating on a track (with constant curvature and super-elevation) at a constant speed. Under such conditions, the centrifugal force is balanced by gravity in the lateral direction, and the curving dynamics can be converted into the static problem. In a tight curve, large lateral displacement of the wheelset, however, can yield large creepage at the wheel-rail interface and introduce strong nonlinearities in the wheel-rail creep forces and wheel-rail contact relationship. In the late 1970s, Elkins and Gostling *et* al. [232] developed a new calculation method for the nonlinear curve passing by introducing kinematic constraint to different curve sections. In each curve segment, the motion of the rail vehicle is considered as steady-state motion. In the 1980s, Nagurka, Hedrick, and Wormley *e*t al. [233] conducted more comprehensive research on curve passing considering the dynamic responses of the vehicle arising from the transition section and circular curve section. Consequently, more complete curving dynamics can be obtained. This proposed method yields more realistic responses and is thus referred to as a dynamic curve passing.

To study dynamic curve passing and improve the curving dynamic performance of railway vehicles, considerable effort has been invested in both numerical simulations and experimental research. Chi [61, 62] employed a magneto-rheological coupled wheelset in the front wheelset together with independent rotation wheels in the rear wheelset to improve curving dynamic performance. Furthermore, a variety of radial bogies have also been developed to improve curving dynamics [63–65]. In recent years, research into control technologies on curving performance has become much more popular among railways [60]. Matsumoto *et al.* [234] from Japan's National Traffic Safety and Environment Laboratory (NTSEL) developed a well-validated active steering bogie. The bogie design incorporates an active steering mechanism between the car body and the bogie frame, which forces the bogie into a radial position as the vehicle negotiates a curve. The numerical and experimental results show that the active steering bogie can substantially reduce lateral force in both inner and outer wheels, thus improving curving dynamic performance. Shen *et al.* conducted many fundamental studies on bogie active control strategies aimed at improving curving dynamic performance, including yaw motion control for the wheelset and bogie frame, wheelset lateral motion active control, primary suspension control, and relative motion control for wheelsets. Using the 1/10 scale vehicle model, field test, as well as the numerical simulation, Suda *et al.* [235] developed a wheel-rail friction controller for improved curving dynamic performance.

1.2.1.3 Ride comfort

Apart from hunting stability and curving dynamic performance of railway vehicles, ride comfort is another important concern for vehicle system dynamics. Research on ride comfort could be utilized in parameter optimization of passive suspension, control strategies for active and semi-active suspension, and rigid-flexible coupling-based high-frequency vibration isolation and absorption.

To improve the ride comfort of rail vehicles, tuning of bogie suspension parameters is the primary method for mitigating vibration [66–69]. Chi [70–71] studied the influence of hunting motion on vehicle lateral ride comfort and pointed out that the natural frequencies of the rail vehicle should be distant from the excitation frequencies. Wu [72] researched the effects of track gauge, curvature, lateral direction, and horizontal super-elevation on lateral acceleration and displacement of a car body, and discussed track maintenance-associated problems. To investigate the ride comfort problem induced by tail car waggle in the Shinkanson EMU, Teng *et al.* [73] studied the influence of stiffness and damping between two car bodies and concluded that the longitudinal damper between two car bodies can effectively eliminate tail car waggle. This conclusion was further validated by experimental results. Using frequency response analysis, Diedrichs *et al.* [236] performed a sensitivity analysis of ride comfort with respect to vehicle parameters, to understand ride comfort in a high-speed light passenger car in a tunnel. The results show that the car body mass can significantly influence lateral vibration due to aerodynamic forces. A single-car equivalent model considering the lateral dampers between car-bodies was formulated by Kagami Takashi *et al.* [74] to investigate the effects of lateral dampers on lateral vibration in the rail vehicle arising from air pressure fluctuation in the tunnel.

In the 1960s, active suspension technology was proposed to further improve vehicle ride comfort. The active suspension system equipment, however, includes a sensor, actuator, and control device as well as an external energy source, which could substantially increase the cost. The semi-active suspension technology is more realistic in comparison to the active suspension. In 1974, Karnopp *et al.* [237], of the University of California, developed a semi-active suspension system using a controllable damper that can adjust the damping force based on the damping control law. Based on this protocol, the control strategies of the semi-active suspension, the structure of the damper, damping control law, and parameter sensitivity were also well researched. Valasek *et al.* [238] proposed a skyhook semi-active control strategy, which has been widely employed in semi-active suspensions. Sasaki [239] also conducted multiple studies on the semi-active suspension. A semi-active suspension was developed and completed, and its validity demonstrated through a roller rig test at Southwest Jiaotong University. The results suggest that the lateral semi-active suspension can effectively lower the peak acceleration of the car body. Furthermore, the on-off control strategies for the semi-active suspension have also been well researched using numerical simulation [240–243]. Yao [244] and Dong [245], of the Academy of Railway Sciences, theoretically and experimentally studied the semi-active suspension of rail vehicles.

Elastic deformation in the car body could also contribute considerably to passenger ride comfort, resulting in the development of rigid-flexible coupled dynamics in the railway. The frequency of the first bending vibration mode of the car body is around 8~12 Hz near the sensitivity frequency range of the human body, and can thus have a significant influence on the vertical ride comfort in a vehicle. It is thus expedient to study related countermeasures for lowering vibration arising from the car body's elastic vibration. Tomioka *et al.* [246] suggested that bending vibration is primarily related to the traction rod and anti-yaw damper between the car body and bogie. The linkages between the car body and bogie must be optimized to reduce the bending vibration of the car body, thereby improving the ride comfort of the rail vehicles. To suppress car body vibration induced by the first bending vibration mode, Foo *et al.* [247] installed two electro-hydrostatic actuators at both the front and rear ends, and one electromagnetic actuator at the medium position of the car body. An active control algorithm is used to control these active actuators. Tamai *et al.* [248] studied the LQG algorithm-based active system for controlling elastic vibration of railway vehicles. Sugahara *et al.* [249] researched the effects of damping of the primary suspension and air spring on the elastic vibration of the car body. Using a finite element model of a soft-seat car body, Zhu *et al.* [250] formulated a rigid-flexible coupled vehicle dynamic model to investigate the feasibility of fuzzy control strategy in reducing the elastic vibration of a car body. Suzuki *et al.* [251] applied viscoelasticity material and a constraint layer to the car body to improve the damping ratio of the car body without changing its structure. The results suggest that this method can also lower the elastic vibration of the car body. Takigami *et al.* used external electricity circuits together with piezoelectric ceramics to dissipate the bending vibration of the car body. Using a small-scale simulation model, Benatzky and Kozek *et al.* [252] studied the active control ability of a piezoelectric stack actuator on elastic vibration of a car body. Lu *et al.* [253] researched active control methods for reducing the elastic vibration of a high-speed passenger car body. Zhou *et al.* [254] researched the influence of a vibration isolator on car body vibration. Zeng *et al.* [255] also theoretically analyzed the elastic vibration of railway vehicles, established the concept of resonant speed of the rail vehicle, and discussed the application of semi-active control in the rail vehicle. Huang *et al.* [256] researched the effects of suspension parameters on the elastic vibration of the car body, e.g., the effects of track rod stiffness on elastic vibration.

These studies suggest that the development of vehicle system dynamics experienced three stages: from a linear model to a nonlinear model, passive suspension to active suspension, and low frequency to high frequency.

1.2.2 Train system dynamics

In traditional vehicle system dynamics, the single vehicle neglecting the longitudinal degree of freedom and the longitudinal constraints between two car bodies is usually taken as the research object. However, each car shows an inconsistent response to the time delay caused by traction and braking. The study of train system dynamics is originally aimed at investigating longitudinal dynamic behavior of trains and longitudinal force transmissibility. The development of the study of train dynamics was driven by the demands of heavy haul transportation in America, Canada, Australia, and the former Soviet Union. These countries have conducted a vast amount of research on train dynamics to improve the operating performance of heavy haul vehicles. Japan has also performed several studies on train dynamics, with their research focused primarily on the passenger train because of car body vibrations in the tail car and other associated problems [75, 76].

With the application of the heavy haul train in America in the 1970s, problems associated with train dynamics became extremely obvious due to increased train length and loading. Thus, in 1971, a research program on train-track dynamics was undertaken in America, with its primary focus on the influence of train operation, topographic conditions, and climate on train dynamic performance. This research plan was implemented primarily by the Association of American Railroads (AAR). The Federal Railroad Administration (FRA) and the Ground Transportation Technology Center of the National Research Council Canada (NRC), as joint participants, also participated in this research program. Through more than 10 years of research funded by this research program, train system dynamics achieved tremendous progress in the numerical simulation model, which gave rise to many train dynamic models, such as the longitudinal model (quasi-static model [77, 78] and dynamic model [79]), quasi-static lateral model [80–82], and vertical dynamic model [83]. Meanwhile, considerable effort has been made through field tests to improve the railway track, vehicle design, and railway operation [84–86]. These mainly include:

- Longitudinal dynamic behaviors of trains: To investigate longitudinal dynamic behaviors of the vehicle and coupler force, the vehicle is usually modeled as a concentrated mass neglecting the wheel-rail contact relationship. Each vehicle is inter-connected through a nonlinear coupler model to account for longitudinal impacts between two car bodies, especially during traction and braking procedures and when subjected to the vertical railway line condition.
- Traction calculation for trains: The train dynamic model is employed to investigate the traction and braking procedure as well as normal operating conditions when the train operates between two railway stations, to obtain the relationships between vehicle speed, traction force, and traction power with operational rail lines.
- Quasi-static lateral dynamics of the train: Curving dynamics of trains was researched in terms of wheel-rail contact force, coupler force, and derailment coefficient in order to study running safety and design standards for the railway track.
- Train operation simulator: The train dynamic model is also used to provide input for the train operation simulator to simulate the operating behaviors of the train. Because the driving simulator is a typical semi-physical (manipulating system) and semi-virtual (train dynamic model) hybrid simulation system, the longitudinal train dynamic model must be simplified to improve integration speed.

From the 1980s, the train system dynamic models, represented by the America train-track research program, can be classified as: longitudinal dynamic model, lateral dynamic model, and vertical dynamic model.

1.2.2.1 Longitudinal dynamic model

The primary purpose of the longitudinal dynamic model is to study the longitudinal dynamics of trains. Based on different purposes, the longitudinal train dynamic model can be further classified into the quasi-static model and dynamic model.

(1) Quasi-static model

The quasi-static model is generally used to study braking distance, signal space, and traction power of the locomotive, as well as the quasi-static force between the train and track. Because the internal interaction and transient forces of the vehicle are not the focus, numerical simulation is usually implemented using a simplified model. The following lists some assumptions employed in the model:

a) Neglect suspensions in the bogie.
b) Neglect control of the track center.
c) Two adjacent cars are connected by a rigid coupler.
d) Within the considered speed range, traction and braking properties are modeled as a linear or parabolic curve.
e) Only the longitudinal degree of freedom is taken into consideration for each car.
f) Transient force of the coupler is calculated using the quasi-static method.
g) Each car of the train is subjected to identical acceleration.

(2) Longitudinal dynamic model [79]

The longitudinal dynamic model is used to study the longitudinal dynamic forces between two adjacent cars during operation, and it facilitates the coupler design. In this model, each vehicle is modeled as a concentrated mass and coupled with couplers. The following lists some assumptions used in the modeling:

a) The goods in the vehicle are in relative motion with respect to the car body. The vehicle and goods are thus considered as a concentrated mass point in the dynamic model.
b) Only longitudinal motion is considered, and other motions arising from the curving and vertical gradient are not considered in the model.
c) Track irregularity is ignored.
d) The underframe of the car body is modeled as rigid, although longitudinal impact could lead to elastic deformation in the underframe.
e) The gravity center of the vehicle is located on the center line of the coupler, and the pitch motion of the vehicle is thus ignored.

1.2.2.2 Lateral dynamic model

The lateral dynamic model is mainly used to study the curving dynamic performance of a train and the associated wheel-rail lateral forces and coupler forces arising from a curved track. This model can also be employed to investigate train buckling caused by longitudinal instability of train. In the train-track dynamics research program, AAR employed this model to investigate track geometry and curving dynamics of the train [81, 82]. Owing to limitations in computational efficiency, the proposed model, however, neglects the influence of braking, acceleration, and track irregularities, which makes the model a quasi-static lateral dynamic model. In this model, the modeling of the coupler is key, and it also assumes that:

a) Coordinates of the track center is known.
b) The coupler between two cars is modeled as a rigid bar without bending deformation.
c) There is no limit to the roll angle of the coupler.
d) The distance between two cars remains constant regardless of the angles of the coupler.

1.2.2.3 Vertical dynamic model

In the vertical direction, the vertical dynamic model is generally employed to investigate the vertical dynamic response of the vehicle system instead of the quasi-static model. Such vertical dynamic response is mainly caused by wheel-rail vertical contact force at the wheel-rail interface arising from track irregularities [83]. In the train-track dynamics research program, the vertical dynamic model is modeled with three degrees of freedom (DOF), including longitudinal, vertical, and pitch. In the heavy-haul vehicle dynamic model, the goods are usually modeled with a longitudinal degree of freedom. This model is thus able to investigate dynamic responses of the vehicle and goods. There are some assumptions employed in the vertical dynamic model:

a) Only vertical forces and motions are considered.
b) The car body is modeled as rigid, and each car is connected through a spring element representing the stiffness of the underframe and coupler.
c) The goods are modeled as a rigid body fixed on the car body through the spring element.
d) The bogie of each car is modeled as rigid and subjected to dynamic forces from both primary and secondary suspensions.
e) Resistances arising from wind, ramp, and curve are considered as negligible.
f) The coupler represented by the spring element can affect the vertical and pitch motions of the car body. When the vertical force applied on the coupler is greater than the vertical friction force arising from the longitudinal force at the coupler interface, relative sliding could occur between couplers.
g) All motions are very small.

The large number of degrees of freedom and strong nonlinearities in the train system presents significant challenges to solve train dynamic problems using analytical methods, and it is not easy to obtain an analytical result even if it is based on a simplified model. In the early stages, there were two types of simplified models used to obtain analytical solutions: continuous mass system and discretized mass system. In the continuous mass system, the vehicle is considered as an elastic bar, and the elastic constraint and mass in the vehicle are distributed along the vehicle length direction, while the locomotive is modeled as a concentrated mass located on one end of an elastic bar. In the discretized mass system, the vehicle is modeled as a concentrated mass interconnected through a linear spring element. Based on these simplified models, the analytical solution, however, still can only be obtained in the case of less complex vehicles and relatively simple interaction forces between two vehicles and the results obtained can only be used for qualitative analysis owing to low accuracy.

Because the train dynamic system is a complex dynamic system, some simplifications have to be made based on requirements for research, e.g., the wheel-rail contact relationship and creep forces at the wheel-rail interface as well as the axle load transfer are usually ignored in the model. To reduce the degrees of freedom and improve computational efficiency, longitudinal, lateral, and vertical motions are usually decoupled — resulting in the development of the longitudinal, lateral, and vertical dynamic model. However, these simplifications invariably introduce significant calculation errors into the results.

The high requirements for calculation accuracy led to the development of a more precise and more complex vehicle dynamic model, and thus, more systematic simulation platforms such as SIMPACK, NUCARS, VAMPIRE, ADAMS/rail, have been developed. These simulation platforms can establish more comprehensive dynamic model considering a variety of degrees of freedom for each component, interaction forces between each component as well as the wheel-rail creep forces at the wheel-rail interface. These dynamic models thus can yield more precise results, although much more effort is needed to model every detail in the model. However, as the number of vehicles in the train increases, the degrees of freedom in the train system will be multiplied and further increase the computational cost. Moreover, as long as the number of vehicles is changed, the dynamic model and associated equations of motion have to be modified, which suggests poor flexibility and commonality in the modeling.

At the end of the 1980s and in the early 1990s, considerable effort went into research on train dynamics to support the operation of a 10,000-ton heavy haul train on the Datong-Qinhuangdao specific line for coal transportation in China [75]. Train dynamic simulation software [87] and a train operations driving simulator [88] with proprietary intellectual property rights were developed. Furthermore, a NEWMARK-based explicit integration method [89] was proposed to improve the calculation speed of the numerical model. The train dynamic models and the research methods employed in China's studies are similar to those used by the AAR in its train-track dynamics research program.

Compared with freight train dynamics, studies on passenger train dynamics is relatively recent, as railway transportation was strongly affected by road transportation and the aviation industry in the late 1980s. To revitalize railway transportation, increasing the number of vehicles in the train and vehicle speed became a strategy for improving railway transportation capacity. However, owing to limitations in coupling and braking devices, the

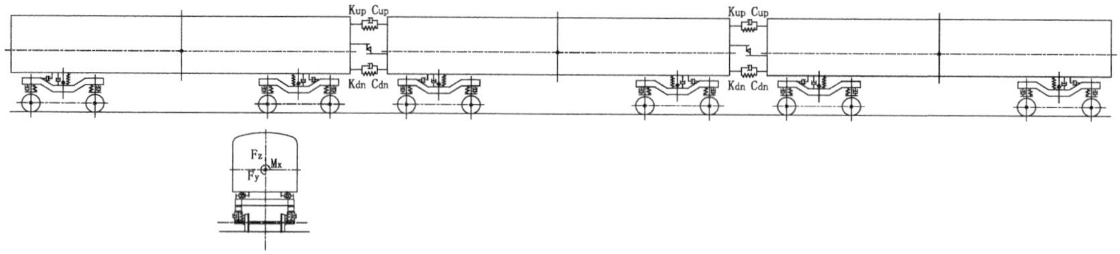

Figure 1.7
Three-car train dynamic model.

train is subjected to substantial longitudinal impact during the acceleration procedure. Until the mid-late 1990s, dramatic advances were achieved in railway by increasing the operating speed of the rail vehicle, and the high-speed EMU serves as the most important role model in China's railway transportation [90].

In 2003, Liu [75] researched the longitudinal dynamics of the high-speed train as his Ph.D. thesis and developed a three-dimensional train dynamic model based on the conclusion that the three-car train dynamic model can yield results with accuracy as good as those obtained using the full train model [91]. In addition, the three-car train dynamic model can also reduce the complexity of numerical simulation arising from the nonlinearities in the train system. This model, consisting of the lead car, middle car, and tail car, is thus taken as the basic research objective for studies on train dynamics. Because the primary purpose of the train dynamic model is to investigate hunting stability, curving dynamics, and nonlinear responses of the train, the mass of the track is thus neglected, and only the equivalent stiffness and damping in the vertical and lateral directions are considered in the model. Figure 1.7 illustrates the train dynamic model in which the model consists of three cars, and each adjacent car is connected using a foldable windshield. In addition, a lateral damper is installed on top of the car body, and two longitudinal dampers are installed on both sides of the car body. Due to the absence of a traction model of the locomotive, this train dynamic model is just a lateral-vertical coupled dynamic model and cannot consider the longitudinal dynamics of the train.

Using a 5-car train dynamic model, Zeng [92] researched hunting stability of the train on both tangent and curved tracks. Considering traction and braking procedure, and various running resistances, a lateral-vertical-longitudinal coupled dynamic model was developed with 34 degrees of freedom. Table 1.2 summarizes degrees of freedom (DOF) considered in the train dynamic model. In the model, the nonlinearities arising from the suspensions, wheel-rail contact relationship, wheel-rail creep force, and the coupler have also been taken into consideration.

Figure 1.8 demonstrates longitudinal forces acting on a single vehicle, where

Table 1.2: Degrees of freedom considered in the model

Rigid	Degree of freedom						Remarks
	Longitudinal	Lateral	Vertical	Roll	Pitch	Yaw	
Car body	x_c	y_c	z_c	f_c	β_c	y_c	-
Bogie frame	x_{fi}	y_{fi}	z_{fi}	f_{fi}	β_{fi}	y_{fi}	$i = 1 \sim 2$
Wheelset	x_{wi}	y_{wi}	(z_{wi})	(f_{wi})	β_{wi}	y_{wi}	$i = 1 \sim 4$

Note: Values in bracket are non-independent degrees of freedom

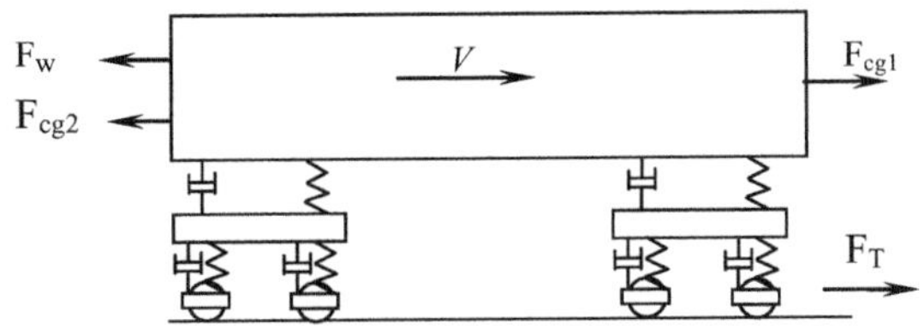

Figure 1.8
Free-body diagrams of a rail vehicle.

Fcg1: longitudinal forces caused by the lead car through the coupler-buffer system and dampers.

Fcg2: longitudinal forces caused by the tail car through the coupler-buffer system and dampers.

Fw: operating resistance, including resistances arising from the mechanical system and wind, which can be determined with the Davis formula:

$$F_w = \left(a + bv + cv^2\right)G \tag{1.1}$$

where *a*, *b*, and *c* represent different constants that vary with locomotive type, *v* is train running speed, and *G* is the weight of the train.

FT is defined as the traction force being applied on the wheelset, which is determined by the power of the traction motor. In line with train operating speed requirements and running condition, the train resistance can be calculated, and then the traction power *PT* for normal operation can be determined. It is also assumed that train power remains constant during the entire operation procedure and that it is equally distributed to each wheelset in the motor vehicle. The traction force of the *i*th wheelset at a given moment *t* is thus obtained as:

$$F_{Ti}(t) = P_T/(N_{car} \times 4 \times V_{wi}) \tag{1.2}$$

where *Ncar* is the number of motor vehicles in the train and *Vwi* is the longitudinal speed of the *i*th wheelset.

Three kinds of train marshaling types, including M-T-T-T-M, M-T-M-T-M, and T-M-T-M-T were considered in this study, which led to 170 DOFs in the 5-car train dynamic model.

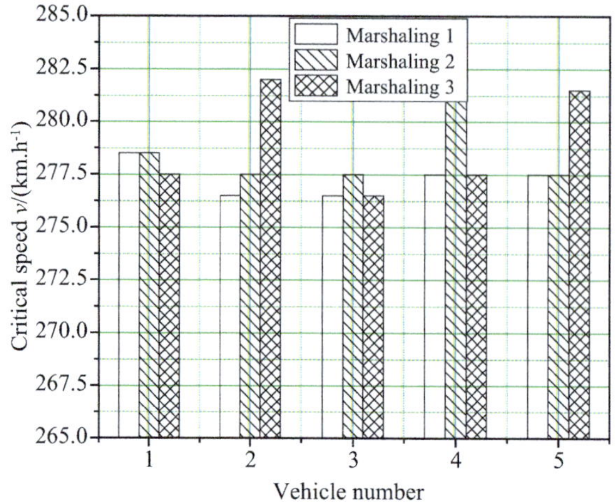

Figure 1.9
Nonlinear critical speed of every vehicle in terms of marshaling.

"M" and "T" represent the motor and trailer vehicle, respectively. While, "-" denotes the internal linkages between adjacent cars.

The study concluded that the linear critical speed of a single motor car and trailer car are 431.1 km/h and 431.8 km/h, respectively. The linear critical speed of the train is obtained through an analysis of the characteristic value of the train model in the absence of damping and stiffness in the lateral damper located between two cars. The results suggest that the linear critical speeds of marshaling type 1 ~ 3 are 431.0 km/h, 431.3 km/h, and 431.1 km/h, respectively. Fig. 1.9 illustrates the nonlinear critical speed of the train and shows that the single vehicle model and the train model yield similar nonlinear critical speeds. This implies that the nonlinear critical speed of the train can be acquired with a single vehicle model.

Although the 5-car train dynamic model considers both the motor and trailer vehicles as well as aerodynamic forces and mechanical resistances, there were differences in the aerodynamic forces caused by different vehicles in the train, such as aerodynamic-induced resistance and pressure for the lead car, and lifting force and lateral disturbance acting on the tail car. These differences were not considered in the model and could lead to discrepancies between the model and actual conditions.

1.2.3 Track system dynamics

Track system dynamics constitutes one of the essential components of the coupled system dynamics of the high-speed train. The track is mainly used to absorb vibration arising from wheel-rail interaction to ensure running safety and stability of the high-speed train. Owing to the wheel-rail interaction, the track system could be subjected to rail corrugation, fatigue, fracture, damage in sleepers, hardening of ballast, ballast bed smashing, track slab fracture, and track deformation, which could affect its service performance. Furthermore, the characteristics of the track system also contribute considerably to railway noise emission. Surface roughness and local irregularities can yield strong oscillations at the wheel-rail interface, which is transmitted to both the wheelset and track structure. These also generate radiated air-noise and structural vibration noise from the car body and track sub-structure.

Track dynamics, as the fundamental dynamic issue in the railway system, can be dated back to 1867. Winkler proposed an elastic foundation beam theory in which the rail is represented as a beam continuously supported on an elastic foundation, and deformation of the elastic foundation is proportional to the pressure under the beam. This theory further established the theoretical foundation for track dynamics. In 1926, Timoshenko developed an elastic foundation beam model by considering the rail as an infinite long beam continuously supported on an elastic foundation, and researched vertical deflection and stress fluctuation in the rail under a moving load. However, the effects of wheel-rail contact force and the inertial effects of the vehicle and sleeper were not considered in the study [93]. In 1943, Dorr proposed a more comprehensive track dynamic model to investigate problems arising from increased vehicle speed. In 1958, Mathews *et al.* analyzed the dynamic responses of rail subjected to a moving harmonic loading in which the rail was modeled as a Bernoulli-Euler beam. Bogacz [95] modeled the rail using the Timoshenko beam theory instead of the Bernoulli-Euler beam theory. In 1989, Grassie and Cox [96] established a continuous foundation beam model together with a flexible sleeper model to investigate high-frequency vibration in the track system. Sato *et al.* [97] expanded the single layer track beam model to a three-layered discrete supported mass-spring-damping model. By 1990s, the high-speed and heavy-haul rail vehicles had been well developed, and this facilitated the development of track dynamics. To investigate dynamic responses induced by wheel and rail defects, Cai and Raymond [98] established a two-layer rail-sleeper track model in which 40 sleepers were taken into consideration. Knothe [99] and Zhai *et al.* [100] introduced a three-layer discrete supported mass-spring-damping model into the vehicle-track coupled dynamic model. Diana and Cheli *et al.* [101] studied the interactions between the upper structures of the railway track and the vehicle. Li and Selig [102] analyzed dynamic responses of the track foundation under wheel-rail interaction. Bogacz [103], Popp [104], and Belotserkovskiy [105] established an infinite long lumped parameter track model to investigate dynamic responses of the vehicle. Using the coupled vehicle-track dynamic model, Ripke and Knothe [106] studied wheel-rail interaction in the high frequency range in 1995. Oscarsson and Dahlberg *et al.* [107] modeled the rail as a Timoshenko beam and formulated a train-track-ballast coupled dynamic model. Frequency responses of the rail obtained through field tests were employed to identify parameters used in the model. Auersch [108] considered the dynamic properties of the roadbed in the coupled vehicle-track dynamic model and concluded that roadbed stiffness affects only the low-frequency response of the system. Drozdzie [109] and Wilson [110] experimentally and theoretically researched vehicle-track dynamics in the turnout zone in 1999. Oscarsson [111] introduced the random properties of the track system (rail support stiffness, ballast stiffness, mass of the ballast and roadbed, and sleeper bay) into the coupled vehicle-track dynamic model. Sun and Dhanasekar [112] established a five-layer track model and researched vertical force between the vehicle and track. Jin *et al.* [113, 114] formulated a three-dimensional vehicle-track coupled model to study the influence of the flexibility of the track on the formation of rail corrugation and the associated dynamic responses. Xiao *et al.* [115, 116] developed a three-layer vehicle-track dynamic model to study the influence of track structural failure on the derailment of vehicles. In the model, the rail is modeled as a Timoshenko beam, while the sleeper is represented by an Euler beam. These studies employed the beam model to describe flexibility in the rail and sleeper, which invariably introduces discrepancy owing to simplification in the beam model. The three-dimensional solid finite element is thus considered a more realistic modeling method for the track dynamic system [117]. Using the finite element method, Ripke [106] and Andersson [118] researched high-frequency response at the wheel-rail interface. Wu and Thompson [119] analyzed wheel-rail vibration noise using the vehicle-track dynamic model coupled with a flexible wheelset model. Several studies employed the finite element method to model the vehicle-track dynamic model while researching dynamic responses of trains under earthquake excitations [120–123].

Research on track dynamics in China is relatively recent when compared with similar research in other countries. In 1993, Zhai [124] proposed the concept of vehicle-track coupled dynamics and established a vertical vehicle-track dynamic model and three-dimensional vehicle-track coupled dynamic model, which are used to study wheel-rail interaction, vehicle-track-bridge coupled vibration, and random vibration at the wheel-rail interaction. Representing the rail as a Timoshenko beam, Hu *et al.* [125] established a vertical coupled dynamic model for the truck and track system. Li and Zeng [126] formulated a vehicle-track coupled dynamic model in which the track was described as a set of discrete track segments with 30 degrees of freedom. Using the finite element method, Jin [127] developed a locomotive-track coupled dynamic model and analyzed stress and deformation in the wheelset and track. Wang and Ren *et al.* [128, 129] developed a turnout-vehicle coupled dynamic model and studied the vehicle-turnout interaction. This study facilitated the improvement of the high-speed turnout. Cai *et al.* [130, 131] considered the effects of roadbed structure while researching vehicle-track-roadbed vertical coupled dynamics. Using the vehicle-track dynamic model, the dynamic issues arising when the vehicle passes through the transition section between the road and bridge were also studied, and consequently, the theoretical foundation for design and maintenance of the transition area was established [132–136]. Jin and Xiao [137–140] established a vehicle-track dynamic model for the ballasted track and slab track systems, and researched wheel-rail related dynamic problems, such as train derailment, rail failure, wear, and wheel-rail contact.

These studies suggest that the development of the track dynamic model experienced four phases: lumped parameter model → single layer continuously elastic supported beam model, in the absence of track sub-structures → multi-layer continuously elastic supported beam model, considering vibration in track sub-structures → discrete supported beam model or finite element model considering rail, sleeper, ballast, and roadbed. After developments over decades, improved track dynamics became an essential part of railway system dynamics. Furthermore, the application of track dynamics has also been substantially extended. It has been widely used in research on vibration and strength analysis for track structures and dynamic problems arising from turnout and the transition section between the road and bridge, rather than wheel flat and rail joint-induced dynamic responses alone.

1.2.4 Train aerodynamics

With vehicle operating speed increasing continuously, the effects of airflow are significantly intensified. Train aerodynamics, which is usually neglected in the low speed range, could thus present significant challenges to high-speed operation and limit the speed increase for the high-speed train. For example, crosswinds would pose a significant threat to running safety of a high-speed train, and the transient air pressure fluctuation caused by the train crossing could significantly affect ride comfort of passengers as well as the running safety of the train. The micro-pressure wave near the tunnel exit can also strongly affect the environment around the tunnel. Furthermore, aerodynamic noise could also substantially increase with increased vehicle speed, which is usually expressed as the sixth or eighth power of the train operating speed. Aerodynamic-induced resistance is a small proportion of train resistance when the train is operated at low speeds but accounts for over 70% when train speed is more than 200 km/h.

1.2.4.1 Train aerodynamics in the presence of environmental wind

In the presence of environmental wind, the flow field around the train becomes much more complex. This deteriorates train aerodynamics while intensifying aerodynamic forces, and thus impacts train running safety. It could lead to the occurrence of overturning and derailment. Considerable effort has been invested into investigating aerodynamic problems in the presence of environmental wind, such as the characteristics of airflow fields. To secure the operating safety of the high-speed train under environmental wind conditions, some European railway companies launched a joint research project, TRANSAERO [141], in 1996. In 2010, UNIFE published some corresponding standards specifying evaluation methods for crosswind running safety and its associated test procedures [142].

Using the wind tunnel test results of the APT-P train, Cooper [143] concluded that the possibility of overturning under environmental wind conditions could increase by around 12% if vehicle speed increases by 5%. Khier *et al.* [144] theoretically researched the aerodynamics as a high-speed train operates on an embankment under environmental wind conditions. Using an ICE3 high-speed train model, Diedrichs *et al.* [145, 146] studied the aerodynamics of a train on both flat ground and on an embankment. They concluded that in the presence of environmental wind, the running safety index of the high-speed train on the embankment is much higher than that obtained on flat ground. Using wind tunnel tests and numerical simulations, Baker and his research team [147−151] also conducted many studies on train aerodynamics under the action of environmental wind. In the winter of 2001 and 2002, Baker *et al.* [151] conducted a crosswind test on a real train at the Eskmeals coastline in which two suspension conditions — released and locked, and two car body postures — 6° inclination and no inclination, were taken into consideration. The test results indicate that both the suspension state and car body posture pose significant influence on the aerodynamic lift forces on the train. The lift forces obtained in the case of the inclined car body together with a locked suspension are much greater than those obtained with no car body inclination combined with a released suspension. Furthermore, both suspension states also yield discrepancies in the side force coefficient when the drift angle is relatively small.

In China, Tian [152], of Central South University, also conducted many studies on train aerodynamics, including train aerodynamic performance under environmental wind conditions and train crossing aerodynamic performance. Zhou [153] and Gao *et al.* [154, 155] researched the aerodynamics of the passenger train and freight train in terms of pressure, flow field distribution, and train aerodynamic forces due to environmental wind, as well as the influence of vehicle speed, wind speed, and road conditions. An overturning model subjected to aerodynamic forces was formulated to study the operating safety margin of the train. Zhang [151−161] of Southwest Jiaotong University and his research team proposed an evaluation method for operating safety of the high-speed train under wind-induced conditions. In this method, fluid equation and a turbulence model of the train are used to calculate the flow field and aerodynamic forces, which are then applied to a multi-body train dynamic model to compute the running safety index of the train. The associated standards for running safety are then employed to evaluate the running safety of the train in the presence of environmental wind. Yu [156−158] researched the effects of environmental wind on the flow field of the high-speed train, including the relationship between operating speed, wind speed, wind direction angle, and aerodynamic forces. The evaluation method for operating safety under wind-induced conditions is also used to investigate the operating safety of a high-speed train on flat ground, embankments, and bridges under the action of environmental wind. These studies led to operating safety margins for the high-speed train on flat ground, embankments, and bridges, respectively.

1.2.4.2 Train crossing aerodynamics

When two trains cross each other, the airflow around both the lead and tail cars is fiercely disturbed, and consequently there are huge fluctuations in the air pressure on the car body. This transient pressure wave can yield both negative and positive peaks in the air pressure over a short period, which is regarded as a train crossing transient air pressure wave [162, 163]. This pressure wave can adversely affect the running safety and ride comfort of a high-speed train.

The train crossing pressure wave can be simulated using a scaling model test. Neppert *et al.* [164] researched the train crossing pressure wave using three 1/25 scaled train models with different shapes. The results obtained are closely comparable to those obtained using a full-scale train model. These models were also employed to study the relationship between track spacing and the train crossing pressure wave. Using a scaled high-speed magnetic train, Howell [165] performed several wind tunnel tests under the action of environmental wind. Comparing results from tests conducted in the presence or absence of environmental wind, the research concluded that crosswinds can considerably affect train crossing aerodynamic forces.

Regarding the numerical simulation of a train crossing, Fuji [166] calculated the three-dimensional compressible Euler/Navier-Stokes equation via the finite differential method and studied the air pressure and aerodynamic forces during the train crossing process. Jaeho *et al.* [167, 168] employed the FDS method together with the LU-SGS method to solve the Euler/Navier-Stokes equation and concluded that this method can better simulate the air pressure fluctuation and aerodynamic force fluctuations arising from a train crossing. Tian [152] researched the relationship between track spacing and the train crossing pressure wave using numerical simulation. Zhang [151−161] also researched the aerodynamic characteristics and associated vehicle dynamics during the high-speed train crossing. Li [162] researched the aerodynamics and vehicle dynamic performance during the crossing procedure of high-speed trains using FLUENT and SIMPACK commercial software. On the basis of Li's work, Zhao [163] studied the train crossing under the crosswind environment, considering different running speeds, track spacing, road condition, crosswind speeds, and wind directions, which culminated in a running safety margin for train crossing in the presence of environmental wind.

1.2.4.3 Train tunnel aerodynamics

When the train passes through a tunnel or crosses in the tunnel, the flow field around the train is completely different from those obtained when outside of the tunnel due to tunnel constraints, which could lead to aerodynamic problems for the high-speed train [169].

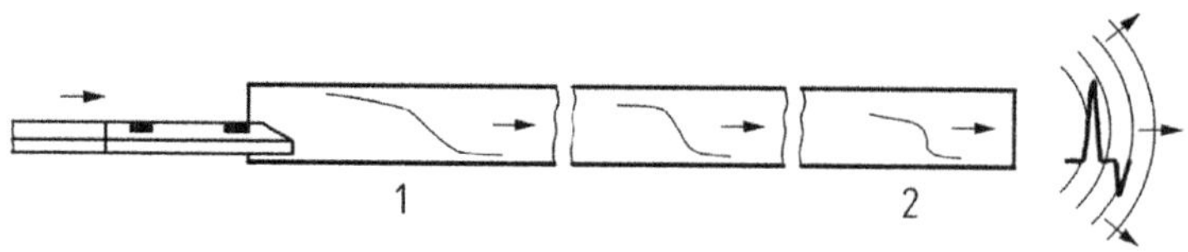

Figure 1.10
Wave generation, transmission, and radiation.

When the head part of the train enters the tunnel, a pressure wave is generated owing to extruded air in front of the train. Similarly, an expansion wave is formed when the tail part of the train enters the tunnel, owing to the sudden increase in space behind the train. The pressure wave and expansion wave are transmitted and radiated in the tunnel. When the pressure wave is transmitted to the exit, some of the pressure wave is reflected near the exit to form an expansion wave that is then transmitted back to the entrance, while the rest of the pressure wave continues to radiate outwards to form a micro-air pressure wave, as shown in Fig. 1.10. The round-trip transformation of the wave can form a series of tunnel aerodynamics problems.

The numerical simulation of tunnel aerodynamics requires significant computational resources, and early research on tunnel aerodynamics was limited mainly to the method of characteristics. In the 1960s, Hara *et al.* [170] established a one-dimensional flow theory for the tunnel pressure wave propagation law, using the method of characteristics. Based on previous research, Wood and Pope *et al.* [171, 172] improved the method of characteristics for the one-dimensional problem and developed a numerical program for the tunnel pressure wave. With advancements in computer hardware, the two-dimensional or three-dimensional numerical methods were further developed to investigate tunnel aerodynamics. Gregorie *et al.* [173] theoretically and experimentally researched aerodynamic problems occurring when the train enters a tunnel. A numerical simulation was used to solve the non-stationary three-dimensional Euler equations. The results suggest that the fluctuations of air pressure on the car body obtained from the numerical simulation show good agreement with those obtained through tests. Shin *et al.* [174] numerically solved a three-dimensional Reynolds average Navier-Stokes equation and studied the characteristics of flow and pressure when the train enters and exits a tunnel.

Research on tunnel aerodynamics in China is relatively recent. In the 1990s, Mei and Yu *et al.* [175, 176] researched aerodynamic problems arising from the vehicle passing through a tunnel and a crossing in the tunnel, based on the one-dimensional model and using the method of characteristics. Using commercial software, Gao et al. conducted a comprehensive analysis of three-dimensional tunnel aerodynamics [177–179] in which the influence of the tunnel buffer structure and tunnel vertical well were studied. Using a fluid software, FLUENT, combined with a multi-body dynamic software, SIMPACK, Feng *et al.*

[180] studied tunnel aerodynamics and the related vehicle dynamics, concluding that train aerodynamic forces and surface pressure are in a square relation with tunnel blockage ratio, and that the average resistance of the train in the tunnel is higher by about 12% than when it is operating on an open line.

These studies are related primarily to train aerodynamics, and few studies are focused on the running safety problems induced by train aerodynamics. The so-called off-line simulation method is usually employed to investigate running safety of the high-speed train under the action of aerodynamic forces [181]. In this method, the fluid control equation is primarily used to obtain the time history of the aerodynamic forces. Then, those aerodynamic forces are applied on a dynamic model of the high-speed train to calculate dynamic responses of the train. The associated standards are then employed to evaluate the running safety of the high-speed train under the action of aerodynamic forces. This method neglects the coupling effects between the aerodynamic model and the multi-body dynamic model of the train [181, 182]. The aerodynamic forces caused by airflow around the train can alter the posture of the car body, thereby changing the airflow near the train, and the altered airflow could consequently contribute to further change in the car body's posture. Research on fluid-structure coupled dynamics for the high-speed train is thus expedient for a better representation of the relationship between the high-speed train and airflow [175, 176].

1.2.5 Pantograph-catenary system dynamics

The pantograph-catenary system, as one of the most critical technologies for ensuring the stability of current collection and the running safety of the high-speed train, has become much more popular in academic research. A vast number of studies have been conducted on the pantograph-catenary system to further increase the operating speed of the high-speed train. It has been established that the dynamics of the pantograph-catenary system under high-speed conditions is the most import factor for current collection performance and the reliability of the pantograph-catenary system.

Research on pantograph-catenary dynamics is at the core of unraveling the pantograph-catenary relationship. Thus, significant endeavors related to pantograph-catenary dynamics have been undertaken on a global scale, resulting in many valuable achievements in modeling and numerical simulation of pantograph-catenary dynamics. On the modeling of overhead catenary systems, in Japan, the contact wire and messenger wire are usually modeled as a concentrated point mass, while the dropper is considered as a massless rigid rod. This method can simulate the relaxation phenomenon in the dropper. Bending and tension deformation in the overhead contact wire and messenger wire, however, is not taken into consideration. In America, the factors above are all considered in modeling the catenary system, and the Lagrange method is used to formulate the equations of motion for the catenary system, which are then solved using the modal method. Due to the complexity of this model, it can only be employed in the simple catenary system, instead of the model with the relaxation phenomenon in the dropper.

There are two methods used to model the catenary system in Germany. In the first method, the finite element method is used to model the node, dropper, and spring in the catenary system, and is solved using the modal superposition method to calculate the total reaction forces in the catenary. This method can be used to simulate the elastic catenary system in the absence of bending stiffness in the beam. In the second method, the contact wire and messenger wire are discretized into a set of small segments, and the finite differential method is used to formulate the differential equations for the system. This method can substantially increase computational speed using a parallel computation algorithm. The research mentioned earlier suggest that the catenary models are mainly represented by the lumped mass model, Euler beam model, and finite element model.

In China, Zhang [183–186] proposed a modal method for modeling the catenary system in which the mode shape functions of the catenary system obtained through modal analyses are treated as generalized variables to build the equations of motion for the catenary system. Cai *et al.* [187–189] represented the overhead contact line equipment as a Bernoulli-Euler beam, while the messenger wire is modeled as an Euler beam-lumped mass model. Li [190] established a mathematical model for the catenary system. Using the finite element model, Wu *et al.* [191] developed a numerical model for the catenary system considering pre-tension in the dropper. This model was then employed to study both the dynamic and static characteristics of the catenary system.

Regarding modeling of the pantograph, the pantograph models can be classified into lumped parameter model, multi-body dynamic model, and finite element model. Depending on the number of components considered in the catenary model, the lumped parameter model can be further divided into single mass model, two-mass model, and three-mass model as well as a multi-mass model. Because only one degree of freedom is considered in the single-mass model, this model is limited to the low frequency range of the pantograph. An [192] modeled the pantograph as a single mass model and used the Mathieu equation to describe the equations of motion for the pantograph. Using the single mass model of the pantograph, Wu *et al.* [193–195] derived the boundary of steady current collection and then researched the effect of the stiffness variation of the contact wire, and propagation of the vibration wave. The two-mass model and three-mass model, representing the pantograph system as a mass-spring-damping model, are also widely employed in research on pantograph dynamics [196–206]. Apart from the lumped parameter models of the pantograph, some new modeling methods have been developed that offer a better representation of the pantograph. Based on the three-mass model, Cho *et al.* [207] considered the pitch motion of the pantograph head. Mei *et al.* [208]

established a nonlinear pantograph dynamic model, and the Tayler series is used to linearize the model. Based on this model, a vertical coupled pantograph-catenary dynamic model, considering the effects of vehicle and track irregularities, was then developed. Considering the planar motions of sliding panel and brackets as well as the pitching motions of two sliding panels, a more comprehensive pantograph dynamic model was developed by Li [209, 210]. Cai [187] also modeled a multi-body dynamic model for the pantograph and catenary system.

Much research has also been made to deepen our understanding of the dynamics and the current collection performance of the pantograph-catenary system. Metrikine *et al.* [211] modeled the pantograph-catenary interaction as a point load moving along the contact wire at a constant speed, which is used to investigate the dynamic behaviors of the pantograph-catenary system when subjected to moving load. Garcia *et al.* [212] studied the initial equilibrium condition of the pantograph-catenary dynamic model. Based on the lumped parameter model of the pantograph, Park *et al.* [213] formulated the equations of motion for the pantograph-catenary coupled system and researched the dynamic responses of the pantograph-catenary system. The current collection performance of the pantograph-catenary system, in terms of dynamic characteristics, the propagation speed of contact wire, and contact loss rate were also researched by Yu [214] and Mitsuo *et al.* [215]. Cho *et al.* [207] studied the influence of pre-relaxation of the contact wire on contact pressure and contact loss rate. Mei *et al.* [216] researched stiffness and dynamic behaviors of the simple catenary and stitched catenary system. Liu [217] optimized the geometrical parameters of the high-speed pantograph to improve the dynamic performance of the pantograph. Using virtual prototype technology, Huang [218] researched the strength of the pantograph and evaluated its geometry parameters. Using the modal method, Liu *et al.* [219] calculated the pantograph-catenary interaction and the associated dynamic responses in the catenary.

Under high-speed conditions, the influence of high speed airflow, vehicle vibration, and irregularities on the contact wire tend to be more significant, thus bringing tremendous challenges to pantograph-catenary dynamic performance. A more comprehensive dynamic model for the pantograph-catenary system, considering the full-degree of freedom of the pantograph, high-speed airflow, elastic deformation of structure, vehicle vibration, and the irregularities on the contact wire need to be taken into consideration for future research.

The aerodynamic characteristics of the pantograph also have a significant influence on the current collection performance of the pantograph. Studies show that aerodynamic resistance of the pantograph accounts for 7–14% of the total aerodynamic resistance of the train [220]. High-speed airflow around the pantograph not only causes noise emission but also structural vibration due to the associated aerodynamic lift forces and the lateral forces. These could further intensify the vibration in the pantograph and consequently impact current collection performance under high-speed conditions. There have been a few studies on the aerodynamics of the pantograph aimed at optimizing the aerodynamic performance of the pantograph and fluid-structure vibration when subjected to a simple loading. In Japan, a tunnel test and numerical simulation involving measurements of aerodynamic forces and airflow around the pantograph were conducted to optimize pantograph aerodynamic performance and to reduce noise emission in the pantograph. This study facilitates further research on the pantograph system [221, 222]. In China, noise emission from the pantograph was initially researched using tunnel tests and field tests [223–225]. Recently, numerical simulation is also employed in research on the aerodynamics of the pantograph, thanks to the development of computational technologies [226, 227].

Apart from the airflow-induced disturbance in the pantograph-catenary system, the vehicle and track system can also affect the dynamic performance of the pantograph-catenary system under high-speed operating conditions. Regarding the influence of the train on the pantograph-catenary system, Zhai *et al.* [228] developed a pantograph-catenary coupled model combined with a coupled vehicle-track dynamic model to study the influence of locomotive-track interaction on the pantograph-catenary system. The results conclude that the vehicle-track coupled dynamics significantly influence pantograph-catenary dynamics under high-speed conditions when the vehicle operates on a track with numerous irregularities. Mei *et al.* [208] compared the linear and nonlinear models of the pantograph and concluded that the vehicle-track interaction-induced oscillation is more significant in the nonlinear pantograph model when compared with the linear model, and the resulting oscillation can also increase contact pressure on the catenary. These further suggest that the nonlinear model is more realistic compared with the linear model.

The studies mentioned earlier suggest that the pantograph-catenary system, vehicle-track interaction, and airflow around the train constitute a coupled dynamic system for the high-speed train. The vehicle-track interaction and the interaction between the airflow and pantograph-catenary system tend to be more evident with increased vehicle speed. In current research, pantograph-catenary dynamics is the critical component of the pantograph-catenary relationship, while most research is limited primarily to the pantograph-catenary system rather than the entire train system, and the influence of airflow and the vehicle system are considered using only a simple coupling model or an external loading. Under high-speed conditions, the pantograph-catenary system must be considered within the context of the entire coupling system of the high-speed train to characterize the excitations arising from strong airflow and the vehicle/track system. These facilitate the enhancement of the pantograph-catenary system to satisfy the requirements of current collection under high-speed conditions.

1.3 The necessity of studying the high-speed train coupling system

1.3.1 Particularity of the railway system

The railway has been progressively improved upon since its advent 180 years ago. Although the operating speed of trains has improved from the initial 10 km/h to 350 km/h and loading capacity has reached ten thousand tons, wheel-rail interaction still cannot be characterized accurately using the numerical model. The railway system is inevitably subjected to derailment, rail corrugation, and rolling contact fatigue problems [31, 229]. From the perspective of academic research, there are three basic problems that need to be researched in the railway system:

(1) Scale effect

The length of a train ranges from a hundred to a thousand meters, while the contact patch of the wheel-rail interaction giving rise to traction and braking forces is only in millimeters. The dimension ratio of the train length over the contact patch is in millions. This brings significant challenges to numerical simulation and experimental research.

(2) Time effect

The service life of a rail vehicle is about 25 ~ 30 years, and its performance changes over time. Apart from the deterioration caused by structural failure, the time-variant characteristics in the vehicle parameters is another contributor to performance deterioration. Such time-variant characteristics also lead to significant difficulty in characterizing the railway system using either the numerical model or experiments.

(3) Spatial effect

Trains are operated throughout the vast China mainland. The spatial span causes changes in the track structures in addition to environmental change during a single day of service. These variations can substantially alter train operating performance and thus must be taken into consideration in the numerical model.

Additionally, the strong nonlinearities in the wheel-rail interaction and track structure together with uncertainties in track structure and track irregularity also present significant challenges to accurately characterize the railway system. It is thus expedient to investigate the modeling methodology for the coupling system and the running stability of the nonlinear system as well as its random vibration.

1.3.2 Dynamic problems in the high-speed railway

To develop China's own high-speed train technologies, some key technologies must be mastered, such as high-speed bogie technologies, lightweight car body technologies, traction technologies, and braking technologies. These technologies are invariably related to the vehicle system dynamics. Bogie technology, as one of the essential components in the vehicle, strongly affects the dynamic performance of the vehicle, including running stability, ride comfort, and running safety. Apart from the manufacturing technologies, the stiffness and modal design of the car body are mainly based on the vehicle system dynamics, while shape design of the car body is highly dependent on the aerodynamics. The traction and braking computation of the train is usually established based on train longitudinal dynamics. The longitudinal impact and longitudinal acceleration limit during traction and braking are also part of the basis for evaluating vehicle dynamic performance. The running control of the train also should be based upon system dynamics, considering the characteristics of the vehicle, line, and power supply system. Therefore, system dynamics can be treated as the primary research area for overcoming dynamics-induced challenges to the development of the China high-speed railway [30].

1.3.2.1 Hunting stability

In vehicle system dynamics, the vehicle is modeled as a deterministic dynamic system consisting of the wheelset, bogie frame, and car body, and each component is connected through either the primary or secondary suspension. This dynamic system is different from the traditional mechanic system owing to the wheel-rail contact relationship. Although the wheel-rail relationship can theoretically be treated as a deterministic constraint between the wheel and rail, the wheel-rail contact relationship and wheel-rail creep forces introduce strong nonlinearities into the vehicle system. The creep forces at the wheel-rail interface act as the steering capability for the wheelset and enable the self-centering capability of the wheelset. Furthermore, wheel-rail creep forces also provide the damping effect in the system. The damping effect generated by the creep force, however, decreases with increased vehicle speed and consequently becomes insufficient to suppress vehicle oscillation. Thus, resulting in instability. When the vehicle loses stability, the wheelset starts hunting. This can further intensify vehicle oscillation and lead to derailment. The speed at which the high-speed train begins to lose stability is also regarded as the critical speed of the high-speed train.

1.3.2.2 System vibration

Due to track irregularities, each component of the vehicle system is subjected to oscillations, and this affects passenger ride comfort. Furthermore, intensified oscillations could also induce greater dynamic loads on the vehicle components, resulting in fatigue failure and reduced reliability.

The track-subgrade system is a continuous system in which a vibration wave is propagated in the form of body waves (compression wave, shear wave) and surface waves (Rayleigh wave). The propagation speed of such waves is usually regarded as the critical speed of the track. When the operating speed of the train is close to or greater than the critical speed of the track, the track-subgrade system tends to lose stability. Experimental results suggest that as the operating speed of train exceeds the critical speed of the track, the high-speed train-induced track oscillation can reach 10 times that of the oscillation under normal conditions. This can severely damage the track and increase maintenance cost. The deteriorated track also affects the running safety of the high-speed train. The critical speed of the track should be 1.5–2 times that of the operating speed of the train. In high-speed railway design, the critical speed of the track can be improved using a ballastless track. For soft foundations, a bridge is usually employed to improve foundation stiffness, and consequently, the critical speed of the track. This is described as "replacing the road with bridges." In the Jin-Hu high-speed railway line, the length of the bridge accounts for 80.5% of the total length.

Furthermore, track irregularities and stiffness differences along the track also give rise to oscillations in the vehicle system. These oscillations increase with increased vehicle speed and limit further increase in vehicle speed. Therefore, corresponding vibration isolation methods must be adapted to isolate the vibration caused by track irregularities. Meanwhile, maintaining good ride comfort under high-speed conditions and different track conditions presents another challenge to high-speed train dynamics.

1.3.2.3 Pantograph-catenary vibration

The overhead catenary system is usually employed to provide electricity for high-speed railway. The pantograph-catenary interaction also produces vibration in the high-speed train system that substantially affects the current collection performance of the pantograph, and consequently, the operating speed of the high-speed train. Furthermore, pantograph-catenary interaction can also cause wear on the pantograph and catenary system, which further contributes to the noise problem.

In pantograph-catenary system dynamics, the vertical stiffness difference and irregularities along the overhead contact wire significantly influence current collection performance. These influences may be weakened by optimization of structural and suspension parameters. The propagation speed of the overhead catenary system is defined as the transmission speed of the wave on the catenary system. When the vehicle speed approaches the propagation speed of the catenary system, the interaction force between the pantograph and catenary is significantly increased and consequently wear and fatigue problems on the contact wire are intensified. These can also lead to the failure of current collection. Therefore, the propagation speed of the overhead contact wire is the most critical factor limiting further increase in vehicle speed. For the overhead catenary system of the high-speed railway, a catenary system characterized by light weight and high strength should be developed to improve the propagation speed of the catenary system.

1.3.2.4 Aerodynamic disturbance

The aerodynamics of the high-speed train is another primary research area for vehicle dynamics. The high-speed train is usually designed with a streamlined shape to reduce air resistance. However, the air resistance on a high-speed train increases with increased vehicle speed. The air resistance is thus another crucial factor limiting the increase of vehicle speed, and the balance point of traction force and resistance is usually the limit of the train operating speed.

The high-speed airflow not only gives rise to resistance on the vehicle but also affects the dynamics of the vehicle. Especially in the presence of crosswinds, the risk of overturning for the high-speed train is increased, and overturning could occur under extreme conditions. Under high-speed conditions, the overturning safety of the high-speed train is related to both crosswind speed and vehicle speed. Therefore, different wind speeds lead to different limits on vehicle speed. This presents another challenge for the design of the high-speed train.

The foregoing limits and challenges that the high-speed train encounters suggest that the optimal system for the high-speed train must be achieved by improving its dynamic performance as well as better consideration of the coupling effects of the high-speed train on other sub-systems.

The operation of the high-speed train is influenced by the railway track, overhead catenary system, traction power supply, and train operation control system, as well as restrictions arising from air disturbance, resistance, and noise. The high-speed train and its associated sub-systems thus constitute an extensive coupling system, which includes wheel-rail coupling relationship, pantograph-catenary coupling relationship, fluid-structure coupling relationship, electro-mechanical coupling relationship, and associated environmental coupling relationships, as shown in Fig.1.11. It is expected that the coupling effects of the high-speed train on associated sub-systems will tend to be more significant under high-

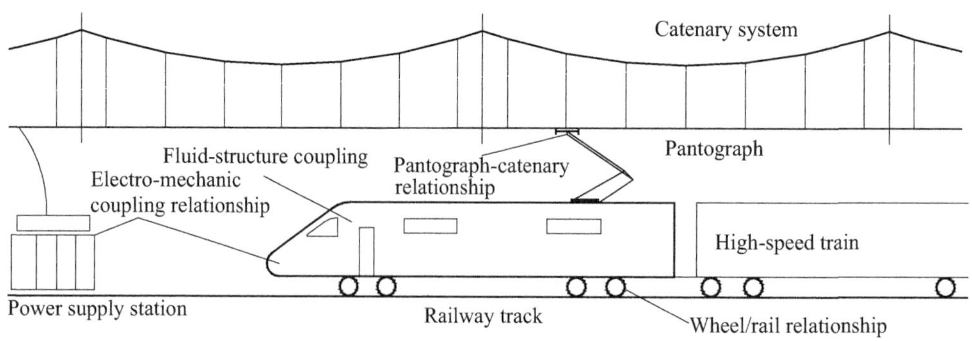

Figure 1.11
Coupling system of the high-speed train.

speed operating conditions. All coupling effects must be taken into consideration to improve the operating quality of the high-speed train. High-speed operation of the high-speed train would intensify the wheel-rail interaction and pantograph-catenary interaction, which would significantly increase wear generation, current collection performance, and consequently, the operating performance of the vehicle. The high-speed airflow could also affect the dynamic performance of the vehicle owing to induced air resistance. Furthermore, high power supply and high-density traveling also place high requirements on the power supply system, and the matching relationship between train and power supply system can significantly affect the stability of the power supply system. Therefore, technological innovation with respect to the high-speed train must address both the dynamics of the train and the coupling relationship of the high-speed train with other sub-systems, to achieve optimal performance in the high-speed train and its coupling systems. To achieve this goal, the operating performance of the high-speed train must be researched using the coupling system dynamics of the high-speed train while considering the railway track, overhead catenary system, airflow, and other associated sub-systems.

1.4 Research on coupling system dynamics of the high-speed train

Research on traditional vehicle system dynamics usually consists of hunting stability, ride comfort, and safety. For train dynamics, traction and braking procedures also have to be taken into consideration to better represent the longitudinal motion of the train. Apart from those considered in the vehicle or train system dynamics, the coupling system dynamics of the high-speed train also considers the influence of the track, overhead catenary system, airflow, and power supply, to investigate coupling effects between the train and its associated systems [31, 229, 230].

1.4.1 Research on vehicle system dynamics

1.4.1.1 Hunting stability

Hunting stability is at the core of vehicle system dynamics, and can severely affect the dynamic performance of the vehicle. In train dynamics, owing to differences in airflow and constraints between two cars, the dynamic performance of each car is different. It is thus necessary to investigate the hunting stability of the vehicle using coupling dynamics of the high-speed train, which could include the following:

- Critical speed of each car, considering different marshaling types. Defining the critical speed of the train via the lowest critical speed of vehicles in the train.
- Hunting stability of the train, considering different line conditions, such as tangent, curve, and ramp.
- Hunting stability of train, considering different operational conditions, including acceleration, constant speed, and deceleration.

- Hunting stability of the train in the presence of failure scenarios.
- Influence of structural and suspension parameters on train stability, to further optimize suspension parameters of the vehicle.

1.4.1.2 Running safety

Research on running safety of the high-speed train is most closely related to derailment. Thus, some safety indexes of derailment, including derailment coefficient, wheel unloading rate, lateral wheelset force, and overturning coefficient, are used to quantify the running safety of the train. Research on the running safety of the high-speed train mainly includes:

- Influence of marshaling types on the running safety of each car in the train, including the derailment coefficient, wheel unloading rate, lateral wheelset force, and overturning coefficient.
- The running safety of the train under different line conditions, such as tangent, curve, and ramp.
- Influence of operating conditions on running safety, including acceleration, constant speed, and deceleration.
- Running safety under different failure scenarios.
- The running safety of a high-speed train subjected to crosswinds and earthquake.

1.4.1.3 Ride comfort

Oscillation of the vehicle arising from wheel-rail interaction not only causes vibration in vehicle components but also affects the ride comfort of passengers. Therefore, research on ride comfort is not limited to car body vibration, and the characteristics of the vehicle system also need to be researched further to attenuate vibration in each component of the vehicle system. Research on ride comfort of the high-speed train could include:

- Vibration of the main components in each car, including acceleration and ride comfort index.
- Vibration of the vehicle under different operating conditions, including acceleration, constant speed, and deceleration.
- Vibration of the vehicle under different failure scenarios.
- Influence of the suspension mode (rigid mode) and structural mode on train vibration.
- The influence of suspension parameters on ride comfort.

1.4.2 Research on coupling relationship

Apart from vehicle system dynamics, the coupling system dynamics of the high-speed train also accounts for the influence of wheel-rail contact, pantograph-catenary interaction, and the interaction between the train and airflow.

1.4.2.1 Wheel-rail contact relationship

Wheel-rail contact relationship constitutes the most important coupling relationship in the coupling dynamics of the high-speed train, which strongly affects the dynamic performance of the vehicle and track systems. In traditional vehicle dynamics, research on wheel-rail contact relationship is limited to the matching relationship in the contact patch, including wheel-rail contact geometry, rolling contact mechanics, material, and hardness that affect the wheel-rail contact performance. However, it is expected that the structure and parameters of the track can also affect the dynamic performance of the train through the wheel-rail contact relationship, especially under high-speed conditions. It is thus expedient to investigate the coupling between vehicle and track, and not just wheel-rail contact alone. Research on wheel-rail contact relationship could include:

- Wheel-rail profile matching.
- Wheel-rail material matching.
- Wheel-rail hardness matching.
- Track (including transition section) stiffness optimization.
- Track alignment.
- Limits of track irregularity.

1.4.2.2 Pantograph-catenary coupling relationship

The pantograph-catenary relationship strongly affects the power supply and consequently, the operating performance of the train, which is thus also a critical component of coupling system dynamics of the high-speed train. Pantograph-catenary interaction, however, has not yet been well researched compared with wheel-rail contact relationship. For the Chinese high-speed train, research on the high-speed pantograph-catenary relationship could include:

- Wave behavior and propagation speed of the catenary system.
- Tracking properties and parameter optimization of the pantograph-catenary system.
- Aerodynamic forces on the pantograph and oscillations in the pantograph-catenary system arising from the airflow.
- Contact wire irregularity and associated vibration in the pantograph-catenary system.
- Current collection performance of the double-pantograph.

1.4.2.3 Fluid-structure coupling relationship

With a continuously increasing train speed, fluid-structure interaction becomes significantly more critical in coupling system dynamics of the high-speed train. With respect to the high-speed train, the fluid-structure coupling relationship is defined as the interaction forces between airflow and the train. The fluid-structure coupling relationship should thus consider the influence of airflow on the dynamic performance of the high-speed train, such as:

- Air resistance and aerodynamic lift forces in the track and vehicle systems under different conditions.
- Aerodynamic behavior, characteristics of the airflow field, and air pressure distribution on the train surface.
- Influence of airflow on the motion of the train.
- Influence of airflow on running safety and the running safety margin in the presence of crosswinds.
- Influence of train wind on the track and nearby objects as well as people along the track.

1.4.2.4 Electro-mechanical coupling relationship

In the coupling system dynamics of the high-speed train, the electromechanical coupling relationship is mainly used to investigate the relationship between train dynamics and the power supply system, focusing primarily on the dynamic responses of the power supply system during acceleration and braking procedures, as well as the influence of dynamic states of the power supply system on train operation. Related research could include:

- Coupling relationship between operating conditions of the train (acceleration, braking, and constant speed) and the power supply system.
- Influence of track conditions on dynamic responses of the power supply system.
- Influence of environmental conditions on dynamic responses of the power supply system.
- The dynamic responses of the power supply system when multiple trains with different operating condition are located on one power supply arm.
- The characteristics of resonance and harmonic vibration in the power supply system under different train operating conditions.

References

[1] Qian L. Summary of the technical development of high-speed train with the speed of 350 km.h-1 in the world. China Academy of Railway Sciences 2007;28(4):66–72.

[2] Wu G, Li B. Development history and technical features of French TGV. Foreign Rolling Stock 2007;44(1):1–4.

[3] Zhu J. Enlightenment of development of high-speed railway in German, French and Japan. Urban and Regional Planning 2011;(3):60–9.

[4] Li J, Xu J, Zhao G, Li J, He Z. Travel notes of investigation of urban railway system in Europe. Tianjin Construction Science and Technology 2007;(2):53–6.

[5] Zheng T. Successful experiences of French high-speed railway. Railway Knowledge 2006;(6):14–7.

[6] Lacote F. TGV: High-speed train in French. Impact of Science on Society 1992;(2):133–42.

[7] Xie X. French high-speed train at 574.8km/h creates world record again. Railway Knowledge 2007;(5):32−7.
[8] Baidu Library: TGV high-speed railway in French operation. http://wenku.baidu.com/view/8296ac04cc1755270722086f.
[9] Alstom Transport. The AGV–a cutting-edge technology integrator. Innotrans-Berlin; Sep 2008.
[10] Wikipedia TGV 001.http://en.wikipedia.org/wiki/TGV_001.
[11] Cathelin K. TGV POS high-speed electrical train. Foreign Locomotive &Rolling Stock Technology 2012;(3):7−13.
[12] TGV maintenance. Website: http://www.trainweb.org/tgvpages/images/maintain/.
[13] Baidu Library Ballastless track technology. http://baike.baidu.com/view/3077415.htm.
[14] Li C. Study on new construction technology and engineering cost of ballastless track in Suining-Chongqing railway line. Tianjin: Master's thesis of Tianjin University; 2007.
[15] Luo Z. Analysis of structure stress and wheel-rail dynamic action of ballastless track of high-speed railway. Chengdu: Doctor Degree Dissertation of Southwest Jiaotong University; 2008.
[16] Yu Z, Zhang H, Wu H, Chang Y. Study on catenary-pantograph system of high-speed electrical railway. China Academy of Railway Sciences 1999;20(1):59−72.
[17] Li R, Wang M. Summary of German High-speed Trains. Foreign Rolling Stock 2005;42(6):1−6.
[18] Lu N, Liu J. Research status of thin plate welding of aluminum alloy car body of high-speed train. Technology Innovation and Application 2012;(5):20.
[19] Bombardier Inc. M-Size Bogies–MD523, MD524 and MD530 for long distance and high-speed coaches. 2004.
[20] Siemens transportation systems. First class bogies-the complete programme for high-quality railway transportation. 2008.
[21] Rui D, Zhao L. Brief introduction of track technology of high-speed railway in German. Railway Standard Design 2006;S1:144−6.
[22] Zhu F. Design, Construction and Detection Point of Catenary Elastic Chain-shaped Suspension. Railway Standard Design 2008;(1):79−81. 87.
[23] Takatsu T, Gan L. History and future of Japan's high-speed rail. Urban Planning International 2011;26(6):6−15.
[24] Webpage Technical features of Shinkansen high-speed trains, http://www.863p.com/transportation/transporttransit/200609/7464.html.
[25] Huang J. Application and task in future of aluminum alloy car body. Electric Locomotives & Mass Transit Vehicles 1989;(2):45−9.
[26] Luo J. Exploration of post-construction zero-settlement construction theory of underline engineering for special passenger line. Subgrade Engineering 2008;(3):73−4.
[27] Li X. Technology of high-speed railway. Beijing: China Railway Publishing House; 2008.
[28] Zhang W, Wang B. Innovative Development of China High-speed Trains. Electric Drive for Locomotives 2010;(1):8−12. +69.
[29] Website Development Planning Division of Ministry of Railway, brief introduction of national *Middle-Long-Term Railway Network Plan*. http://www.china-mor.gov.cn/tljs/tlt/201101/W020120130553870286334.jpg.
[30] Zhang W, Zeng J. Development of China's high-speed trains and the study on VSD. In: 6th International Symposium on Speed-up, Safety and Service Technology for Railway and Maglev Systems (STECH'12), Seoul Korea; 2012.
[31] Zhang W, Zhang S. Study on dynamics of coupled systems in high-speed trains. Journal of Southwest Jiaotong University 2008;43(2):147−52.
[32] Liao X. Theory, method and application of stability. Wu Han: Huazhong University of Science and Technology Press; 2010.
[33] Wang F. Vehicle system dynamics. Beijing: China Railway Publishing House; 1994.

[34] Marshall CFDA. History of British Railways down to the Year 1830. London: Oxford University Press; 1938. p. 147−8.

[35] Carter FW. On the stability of running of locomotives. Proceedings of the Royal Society of London. Series A, Containing Papers of a Mathematical and Physical Character 1928;121:585−611.

[36] Kenzo I. Comparison of hunting motion characteristics of vehicles on test board and actual track. Foreign Rolling Stock 1994;1(3):40−5.

[37] Matsudaira, T. Dynamics of High-speed Rolling Stock. Quartely Reports of RTRI, Special Issue, 1960, 57-65; 1961, 20-26; 1962, 13-22; 1963, 21-27; 1964, 21-26.

[38] Wickens AH. The dynamic stability of a simplified four-wheeled railway vehicle having profiled wheels. International Journal of Solids and Structures 1965;l(4):385−406.

[39] Vermeulen PJ, Johnson KL. Contact of non-spherical bodies transmitting Tangential forces. J Appl Mech 1964;31:338−40.

[40] Kalker JJ. A fast algorithm for the simplified theory of rolling contact. Vehicle Styst Dyn 1982;11(1):1−13.

[41] Shen ZY, Hedrick JK, Elkins JA. A comparison of alternative creep force models for rail vehicle dynamic analysis. In: Proceeding of the 8th IAVSD on Dynamics of Vehicles on Roads and Tracks. Cambridge: Swets and Zeitlinger; 1984. p. 591−605.

[42] Cooperrider NK. The hunting behavior of conventional railway trucks. ASME Journal of Engineering and Industry 1972;94:752−62.

[43] Wickens AH. The dynamic stability of railway vehicle wheelsets and bogies having Profiled wheels. Int J Solids Sturct 1965;1(3):319−41.

[44] Knudsen C, Feldberg R, Jaschinski A. Nonlinear dynamic phenomena in the behavior of a railway wheelset model. Nonlinear Dynam 1991;2:389−404.

[45] Petersen KC, True H. Periodic, bi-periodic and chaotic dynamic behaviour of railway vehicles. Proceedings of the 9th IAVSD Symposium, Linköping University 1986;15:208−21.

[46] Petersen KC. Chaos in a railway bogie. Acta Mech 1986;61:89−107.

[47] True H. Dynamics of a rolling wheelset. Appl Mech Rev 1993;46:438−44.

[48] True H, Kass-Petersen C. A bifurcation analysis of nonlinear oscillations in railway vehicle. Veicle Styst Dyn 1983;12:5−6.

[49] True H, Kass-Petersen C. A bifurcation analysis of nonlinear oscillations in railway vehicles. In: Proceedings of the 8th IAVSD Symposium, vol 13. Cambridge: Swets and Zeitlinger; 1984. p. 655−65.

[50] True H. Railway vehicle chaos and asymmetric hunting. Vehicle System Dynamics 1983;6:225−637.

[51] Zboinski K, Dusza M. Self-exciting vibrations and Hopf's bifurcation in non-linear stability analysis of rail vehicles in a curved track. Eur J Mech A-Solid 2010;29:190−203.

[52] Zeng J. Numerical calculation of hunting motion turnout and extreme ring of vehicle system. Journal of Railway 1996;18(3):13−9.

[53] Zhang W. Dynamic simulation of locomotive. China Railway Publishing House; 2005.

[54] Goodall R, Li H. Solid axle and independently-rotating railway wheelsets-a control engineering assessment of stability. Vehicle System Dynamics 2000;33(1):57−67.

[55] Pearson JT, Goodall R. An active stability system for a high-speed railway vehicle. UK: Electronic system and control division research, Loughborough University; 2003. p. 11−4.

[56] Bruni S, Resta F. Active control of railway vehicles to avoid hunting instability. IEEE/ASME International Conference on Advanced Intelligent Mechatronics Proceedings, Como, Italy 2001.

[57] Shen G, Goodall RM. Active yaw relaxation for improved bogie performance. Vehicle System Dynamics 1997;28:273−89.

[58] Schielen W. Multibody Systems Handbook. Springer-Verlag; 1990.

[59] Kortüm, W. Review of Multibody Computer Codes for Vehicle System Dynamics [J]. Vehicle System Dynamics,1744-5159, V22(2), (Supplement), 1993: 3-31.

[60] Zhai W. New Progress of Study on Vehicle System Dynamics of International Railway. Railway vehicle 2004;42(1):1−5.

[61] Chi M, Wang K, Fu M, Ni W, Zhang W. Study on curving performance for bogie with magnetic fluid coupled wheelsets. Journal of the China Railway Society 2002;24(4):28–33.
[62] Chi M, Wang K, Fu M, Ni W, Zhang W. Analysis on wheel-rail lateral force of the bogie with independently rotating wheels for rear wheelsets. Journal of Traffic and Transportation Engineering 2002;2(2):32–5.
[63] Shen ZY, Zhang XF, Dai HY, et al. Curving behavior and lateral stability of a three-axle forced-steering freight bogie for heavy haul railway transportation. In: Proc. of the 11th IAVSD Symposium, Canada; 1989.
[64] Chen J, Wang K, Jiang J. Simulation Research on Small Radius Curve-passing Dynamics Performances of Tilting Trains with Self-steering Radial Bogies. Railway Locomotive 2006;26(5):16–20.
[65] Chi M, Zhang W, Zeng J, Dai H, Wu P. New Flexible Coupled Radial Bogie with Independently Rotating Wheels. Chinese Journal of Mechanical Engineering 2008;44(3):9–15.
[66] Wang F, Zhu C, Chen J. Theory calculation and test analysis of lateral vibration of passenger car bogie. Journal of the China Railway Society 1981;3(1):14–26.
[67] Liu J, Wang K, Feng B. Influence of lateral damper on locomotive riding quality. Journal of Traffic and Transportation Engineering 2006;6(3):1–4.
[68] Chi M, Zhang W, Zeng J, Dai H, Wu P. Study of Suspension Parameter of High-speed Passenger Car Bogie. Journal of Dalian Jiaotong University 2007;28(3):13–9.
[69] Wu H. Study on Coupling Vibration of High-speed EMU Between Car body and Underframe Equipment. Chengdu: Doctor Degree Dissertation of Southwest Jiaotong University; 2012.
[70] Chi M, Zhang W, Zeng J, Jin X, Zhu M. Impact of hunting motion on railway vehicle stability. Journal of Vibration Engineering 2008;21(6):639–43.
[71] Chi M, Zhang W, Zeng J, Dai H, Wu P. Vibration response characteristics of railway vehicles. Journal of Traffic and Transportation Engineering 2007;7(5):6–11.
[72] Wu G. Geological status of track and vehicle lateral vibration. Journal of Railway 1985;(4):11–7.
[73] Fujimoto Y, Ding S. Lateral vibration of Shinkansen motor train unit and its vibration reduction measures. Power Traction Express 1995;7:21–7.
[74] Toshimitsu T, Cai Q. Improvement of comfort level of high-speed train by cross damper of between car bodies. Foreign Rolling Stock 2009;46(1):8–15.
[75] Liu H. Study on the Key of Dynamic Problems in High-speed Train. Chengdu: Doctor Degree Dissertation of Southwest Jiaotong University; 2003.
[76] Garg VK, Dukkipati RV. Dynamics of Railway Vehicle Systems. California: Academic Press; 1984.
[77] Luttrell NW, Gupta RK, Low EM, et al. Train Operation Simulator-User's Manual. Research Report R-198. Association of American Railroads; April 1976.
[78] Low EM, Garg VK. Train Operation Simulator-Validation Report. Research Report R-235. Association of American Railroads; November 1978.
[79] Martin GC, Tideman H. Detailed Longitudinal Train Action Model-Technical Documentation. Research Report R-221. Association of American Railroads; February 1977.
[80] Thomas LR, MacMillan RD, Martin GC. Quasi-Static Lateral Train Stability Model. R-209. Association of American Railroads; April 1976.
[81] Chang EH, Shum KL, Abbott RA, et al. Effects of spiral length on reverse curves with minimum tangent length for slow speed operation. Research Report R-309. Association of American Railroads; 1978.
[82] Shum KL, Singh SP, Chang EH. Study of effects of spiral length on lateral stability on simple curve negotiation. Research Report R-355. Association of American Railroads; 1979.
[83] Raidt JB, Shum KL, Martin GC, et al. Detailed vertical train stability model-technical documentation. Research Report R-267. Association of American Railroads; March 1977.
[84] Gogineni BR, Garg VK. A Dynamic Model for Longitudinal Train Action. Research Report R-297. Association of American Railroads; April 1978.

[85] Low EM, Garg VK. Detailed Longitudinal Train Action Model-Computation Program Documentation. Research Report R-296. Association of American Railroads; July 1978.
[86] Se YH, Martin GC. Lateral/Vertical Model-User's Manual. Computation Program Documentation. R-237. Association of American Railroads; February 1977.
[87] Ma Dawei. Heavy Haul Train, Its Test and Research (the 4th Part continued)-Research of Simulation on Heavy Haul Trains (1st half). Rolling Stock 1999;37(2):11–5.
[88] Zhao Xiaoling, Qian Xuejun. SS4 improved locomotive driving simulator. Electric Drive for Locomotives 1996;(2):16–8. +24.
[89] Zhai WM. Two simple fast integration methods for large-scale dynamic problems in engineering. International Journal for Numerical Methods in Engineering 1996;39(24):4199–214.
[90] Li Xuefeng. Development and Comprehensive Performance Test of EMU in China. China Railway 2002;(4):30–2.
[91] Eikhoff BM, Evans JR, Minnis AJ. A review of modeling methods for railway vehicle suspension components. Vehicle System Dynamics 1995;24:469–96.
[92] Luo Ren, Zeng Jing. Hunting Stability Analysis of Train System and Comparison with Single Vehicle Model. Chinese Journal of Mechanical Engineering 2008;44(4):184–8.
[93] Timoshenko S. Method of analysis static and dynamical stresses in rail. In: Proc. of the 2nd International Congress of Applied Mechanics, Zurich; 1927.
[94] Mathews PM. Vibrations of a beam on elastic foundation. Zeitschrift fur Angewandte Mathematik und Mechanik 1958;(38):105–15.
[95] Bogacz R, Krzyzynski T, Popp K. On the generalisation of Mathews' problem of the vibrations of a beam on elastic foundation. Zeitschrift fur Angewandte Mathematik und Mechanik 1989;(69):243–52.
[96] Grassie SL, Cox SJ. The dynamic response of railway track with flexible sleepers to high frequency vertical excitation. In: Proc. of Institution of Mechanical Engineers, London; 1984.
[97] Sato Y, Odaka T, Takai H. Theoretical analyses on vibration of ballasted track. Quarterly Report of Railway Technology Institution (Jpn.) 1988;29(1):30–2.
[98] Cai A, Raymond GP. Modeling the dynamic response of railway track to wheel-rail impact loading. Structural Engineering and Mechanics 1994;2(1):95–112.
[99] Knothe K, Grassie SL. Modeling of railway track and vehicle/track interaction at high frequencies. Vehicle System Dynamics 1993;(22):209–62.
[100] Zhai WM, Sun X. A detailed model for investigating vertical interact ion between railway vehicle and track. Vehicle System Dynamics 1994;23(Supplement):603–15.
[101] Diana G, Cheli F, Bruni S, et al. Interaction between railroad superstructure and railway vehicles. Vehicle System Dynamics Vehicle System Dynamics 1994;23(Supplement):75–86.
[102] Li D, Selig ET. Wheel/track dynamic interaction: track substructure perspective. Vehicle System Dynamics 1995;24(Supplement):183–96.
[103] Bogacz R, Krzyzynski T, Popp K. On dynamics of systems modeling continuous and periodic guide ways. Archives of Mechanics 1993;45:575–93.
[104] Kruse H, Popp K, Krzyzynski T. On steady state dynamics of railway tracks modeled as continuous periodic structures. Machine Dynamics Problems 1998;20:149–66.
[105] Belotserkovskiy PM. On the oscillations of infinite periodic beams subjected to a moving concentrated force. Journal of Sound and Vibration 1996;193:705–12.
[106] Ripke B, Knothe K. Simulation of high frequency vehicle-track interactions. Vehicle System Dynamics 1995;24(Supplement):72–85.
[107] Osearsson J, Dahlberg T. Dynamic train/track/ballast interaction computer model: and full-scale experiments. Vehicle System Dynamics 1998;28(Supplement):73–84.
[108] Auerseh L. Vehicle-track interaction and soil Dynamics. Vehicle System Dynamics 1998;28(Supplement):553–8.
[109] Drozdzie J, Sowinski B, Groll W. The effect of railway vehicle-track system geometric deviation on its dynamics in the turnout zone. Vehicle System Dynamics 1999;33(Supplement):641–52.

[110] Gurule S, Wilson N. Simulation of wheel-rail interaction in turnouts and special track work. Vehicle System Dynamics 1999;33(Supplement):143–54.
[111] Osearsson J. Dynamic train-track-ballast interaction with unevenly distributed track properties. In: 17th IAVSD Symposium on Dynamics of Vehicles on Roads and on Tracks, Lyngby, Denmark; 2001.
[112] Sun YQ, Dhanasekar M. A dynamic model for the vertical interaction of the rail track and wagon system. Int. Journal of of Solids Struct. 2002;39:1337–59.
[113] Jin XS, Wen ZF, Wang KY, et al. Three-dimensional train/track model for study of rail corrugation. Journal of Sound Vibration 2006;293(3–5):830–55.
[114] Jin XS, Wen ZF, Xiao XB, et al. A numerical method for prediction of curved rail wear. Multibody System Dynamics 2007;(18):531–57.
[115] Xiao XB, Wen ZF, Jin XS, et al. Effects of track support failures on dynamic response of high-speed tracks. International Journal of Nonlinear Sciences and Numerical Simulation 2007;8(4):615–30.
[116] Xiao XB, Jin XS, Wen ZF, et al. Effect of tangent track buckle on vehicle derailment. Multi-Body Dynamic. 2011;25(1):1–41.
[117] Iwnicki S. Handbook of Railway Vehicle Dynamics. London: Taylor & Francis; 2006.
[118] Andersson C, Abrahamsson T. Simulation of interaction between a train in general motion and a track. Vehicle System Dynamics 2002;38(6):433–55.
[119] Wu TX, Thompson DJ. Vibration analysis of railway track with multiple wheels on the rail. Journal of Sound Vibration 2001;239(1):69–97.
[120] Yang Y, Wu Y. Dynamic stability of trains moving over bridges shaken by earthquakes. Journal of Sound Vibration 2002;258(1):65–94.
[121] Tanabe M, Matsumoto N, Wakui H, et al. A simple and efficient numerical method for dynamic interaction analysis of a high-speed train and railway structure during an earthquake. Trans. ASME Journal of Computer Nonlinear Dynamics 2008;3(2):1–8.
[122] Ju SH, Li HC. Dynamic interaction analysis of trains moving on embankments during earthquakes. Journal of Sound Vibration 2011;330:5322–32.
[123] Xiao XB, Liang L, Jin XS. A study of the derailment mechanism of a high-speed train due to an earthquake. Vehicle System Dynamic 2011;50(3):449∼470.
[124] Wickens AH. Fundamentals of rail vehicle dynamics. Netherland: Swets & Zeitlinger Publishers; 2003.
[125] Hu Yongsheng, Tan Fuxing, Lu Zhenggang. Simulation verification of TBDS wheel-rail coupling model and its application. Journal of Railway 1996;18(3).
[126] Li Dejian, Zeng Qiangyuan. Dynamic Analysis of Train-Tangent-Track Space-Coupling Time-Varying System. Journal of Railway 1997;19(1):101–7.
[127] Jin Yuyun. The Finite Element Analysis on the Vertical Coupling Dynamics of vehicle/track system. Beijing: Doctor Degree Dissertation of China Academy of Railway Sciences; 2000.
[128] Wang Ping. Study on wheel-rail system dynamics in turnout zone. Chengdu: Doctor Degree Dissertation of Southwest Jiaotong University; 1997.
[129] Ren Zunsong. Dynamic study on vehicle-turnout system. Chengdu: Doctor Degree Dissertation of Southwest Jiaotong University; 2000.
[130] Liang Bo, Cai Ying, Zhu Dongsheng. Dynamic analysis of vehicle-road vertical coupling system. Journal of Railway 2000;22(5):65–71.
[131] Su Qian, Cai Ying. Analyses of space-time varying system coupling system of subgrade structure of high-speed railway. Journal of Southwest Jiaotong University 2001;36(5):509–13.
[132] Wang Qichang, Cai Chengbiao, Luo Qiang, et al. Allowable values of track deflection angles on high-speed railway bridge-subgrade transition sections. Journal of Railway 1998;20(3):109–13.
[133] Luo Qiang, Cai Ying, Zhai Wanming. Dynamic performance analyses of road-bridge transition section of high-speed railway. Engineering Mechanics 1999;16(5):65–70.
[134] Mao Dejun, Mao Xiaoyan, Du Houzhi. Dynamic Response Analyses on Track Transition in High-speed Railway. Journal of East China Jiaotong University 2001;18(1):35–40.

[135] Cai Chengbiao. Theory and Application of Train-Track-Bridge Coupling Vibration in High-speed Railways. Chengdu: Doctor Degree Dissertation of Southwest Jiaotong University; 2004.
[136] Wang Yu, Zhai Wanming, Wang Qichang, et al. A new method for confirming length of rail transition section. Journal of Railway Engineering Society 1999;(4):25–8.
[137] Jin Xuesong, Wang Kaiyun, Wen Zefeng, et al. Impact of rail lateral non-uniform supporting stiffness on rail wave wear. Acta Mechanica Sinica 2005;37(6):737–49.
[138] Xiao Xinbiao, Jin Xuesong, Wen Zefeng. Effect of hard spot on dynamic response of railway sleeper. Journal of Traffic and Transportation Engineering 2007;7(6):28–35.
[139] Xiao Xinbiao, Jin Xuesong, Wen Zefeng. Impact of sub-rail supporting failure on dynamic response of linear track. Acta Mechanica Sinica 2008;40(1):67–78.
[140] Jin Xuesong, Guo Jun, Xiao Xinbiao, Wen Zefeng, Zhou Zhongrong. Key Scientific Problems in the Study On Running Safety of High-speed Trains. Engineering Mechanics 2009;26(supplement):8–22.
[141] Burkhard SW. TRANSAERO: a European initiative on transient aerodynamics for railway system optimization. 2002. Sperling.
[142] European Standard. Railway applications - Aerodynamics - Part 6: Requirements and test procedures for cross wind assessment. 2010. 1 (BS EN 14067-6:2010).
[143] Cooper RK. Wind tunnel tests on a 1/25th scale APT-P model. British Rail Research Division Technical Memorandum TMAERO 1982;43.
[144] Khier W, Breuer M, Durst F. Flow structure around trains under side wind conditions: a numerical study. Computers&Fluids 2000;29(2):179–95.
[145] Dierichs B. On computational fluid dynamics modeling of crosswind effects for high-speed rolling stock. Institution of Mechanical Engineers Part F: Journal of Rail of Rapid Transit 2003;217:203–26.
[146] Dierichs B, Sima M, Orellano A, et al. Crosswind stability of a high-speed train on a high embankment. Institution of Mechanical Engineers Part F: Journal of Rail of Rapid Transit 2007;221:205–25.
[147] Baker CJ. Ground vehicles in high cross winds. Part I: steady aerodynamic forces. Journal of Fluids and Structures 1991;5:69–90.
[148] Baker CJ. Ground vehicles in high cross winds. Part II: unsteady aerodynamic forces. Journal of Fluids and Structures 1991;5:91–111.
[149] Baker CJ. Ground vehicles in high cross winds. Part III: The interaction of aerodynamic forces and the vehicle system. Journal of Fluids and Structures 1991;5:221–41.
[150] Baker CJ. The simulation of unsteady aerodynamic crosswind forces on trains. Journal of Wind Engineering and Industrial Aerodynamics 2010;98:88–99.
[151] Baker CJ, Jones J, Lopez-Calleja F, et al. Measurements of the crosswind forces on trains. Journal of Wind Engineering and Industrial Aerodynamics 2004;92:547–63.
[152] Tian Hongqi. Train aerodynamics. Beijing: China Railway Publishing House; 2007.
[153] Zhou Dan, Tian Hongqi, Yang Mingzhi, et al. Comparison of Aerodynamic Performance of Different Kinds of Wagon Running on Embankment of the Qinghai Tibet Railway under Strong Cross wind. Journal of Railway 2007;29(5):32–6.
[154] Gao Guangjun, Tian Hongqi, Zhang Jian. Cross-wind affection on double container train. Journal of Traffic and Transportation Engineering 2004;4(2):45–8.
[155] Gao Guangjun, Tian Hongqi, Yao Song, et al. Effect of Strong Cross-wind on the stability of Trains Running on the Lanzhou-Xinjiang Railway Line. Journal of Railway 2004;26(04):36–40.
[156] Yu Mengge. Study on the wind-induced security for high-speed trains. Chengdu: Doctor Degree Dissertation of Southwest Jiaotong University; 2010.
[157] Yu Mengge, Zhang Jiye, Zhang Weihua. Running attitudes of car body and wheelset for high-speed train under cross wind. Journal of Traffic and Transportation Engineering 2011;11(4):48–55.
[158] Yu Mengge, Zhang Jiye, Zhang Weihua. Wind-induced Security of High-speed Trains on the Ground. Journal of Southwest Jiaotong University 2011;46:989–95.
[159] Liu Jiali, Yu Mengge, Zhang Jiye, et al. Study on Running Safety of High-speed Train under Crosswind by Large Eddy Simulation. Journal of Railway 2011;33(4):13–21.

[160] Yu Mengge, Zhang Jiye, Zhang Weihua. Study on running safety of high-speed train at 350km/h under crosswind. Machinery Design and Manufacture 2011;5:174−6.
[161] Tian Li, Zhang Jiye, Zhang Weihua. Coupling Dynamics Performance of Vehicle-track Under Crosswind. Journal of Traffic and Transportation Engineering 2011;11(5):55−60.
[162] Li Xuebing. The Study on Fluid-Induced Vibration of High-speed in Passing Events. Chengdu: Master's thesis of Southwest Jiaotong University; 2010.
[163] Zhao Yinghui. Dynamic Performances of High-speed Trains Passing Each Other. Chengdu: Master's thesis of Southwest Jiaotong University; 2012.
[164] Nepper H, Sanderson R. Unersuchungenzurzugbegnung, bauwerkund personer passage in wasser kanal, vortrag während des seminars "schnellbahn aerodynamik", am 7, end 8. In: DFVLR-Forschungszentrum AVA-Gottingen; Juni 1978.
[165] Howell JP. Aerodynamic response of maglev train models to a crosswind gust. Journal of Wind Engineering and Industrial Aerodynamics 1986;22(2):205−13.
[166] Fujj K, Ogawa T. Aerodynamics of High-speed trains passing by each other. Computer & Fluids 1995;24(8):897−908.
[167] Hwang J, Lee D. Numerical Simulation of flow field around high-speed trains passing by each other. American Institute of Aeronautics & Astronautics 1999:422−32.
[168] Hwang J, Yoon TS, Lee DH, et al. Numerical study of unsteady flow field around high-speed trains passing by each other. JSME International Journal 2001;44(3):451−64.
[169] Zhang Lei. Study on Aerodynamic Performance of High-speed Trains Through the Tunnel. Chengdu: Master's thesis of Southwest Jiaotong University; 2012.
[170] Hara T. Aerodynamical problems when train is running into with large velocity. Railway Technical Research Report No153. 1960.
[171] Pope CW. Tunnel aerodynamic s. In: Proc. of Railway Aerodynamics Symposium. University of Nottingham. Department of Civil Engineering; 1984.
[172] Wood WA. Wave action associated with a train entering a tunnel. In: Paper on 1st ISAVVT; 1973.
[173] Gregoire R, Rety JM, Masbermat F. Experimental study (scale 1/70) and numerical simulations of the generation of pressure wave and micro-pressure waves due to high-speed train-tunnel entry. In: Paper on 4th ISAVVT; 1982.
[174] Shin CH, Park WG. Numerical study of flow characteristics of the high-speed train entering into a tunnel. Mechanics Research Communication 2003;30(4):287−96.
[175] Mei Yuangui. Study on numerical simulation of tunnel pressure wave of high-speed railway. Chengdu: Doctor Degree Dissertation of Southwest Jiaotong University; 1997.
[176] Yu Nanyang, Mei Yuangui. Study on Main Parameters Effecting Pressure Transients While Train Passing Through Tunnel. China Academy of Railway Sciences 2003;24(6):67−9.
[177] Luo Jianjun, Gao Bo, Wang Yingxue. Numerical simulation of two-dimensional unsteady flow when high-speed train passing a tunnel. Journal of Railway 2003;25(2):68−73.
[178] Luo Jianjun, Gao Bo, Wang Yingxue, et al. Numerical study on pressure change when high-speed train passing a tunnel. China Academy of Railway Sciences 2003;24(4):82−6.
[179] Luo Jianjun, Gao Bo, Wang Yingxue. Numerical study on tunnel-hood when high-speed train passing a tunnel. Acta Aerodynamica Sinica 2003;21(3):376−81.
[180] Feng ZP, Zhang JY, Zhang WH. The Vehicle Dynamical Analysis of a High-speed Train Passing through a Tunnel. Applied Mechanics and Materials 2010;29-32:835−40.
[181] Tian Li. Study on Calculation Method and Dynamic Performance of Fluid-structure Interaction of High-speed Train. Chengdu: Doctor Degree Dissertation of Southwest Jiaotong University; 2012.
[182] Cui Tao. Study on Fluid-Solid Coupling Vibration and Running Safety of High-speed Trains. Chengdu: Doctor Degree Dissertation of Southwest Jiaotong University; 2011.
[183] Zhang Weihua. Study on Dynamic Performance of Pantograph-Catenary System. Chengdu: Master's thesis of Southwest Jiaotong University; 1988.

[184] Zhang Weihua, Shen Zhiyun. Dynamics Study of High-speed Pantograph/Catenary System. Journal of Southwest Jiaotong University 1991;26(1):105–11.
[185] Zhang Weihua, Shen Zhiyun. Study on Pantograph Dynamics. Journal of Railway 1993;15(1):23–30.
[186] Zhang Weihua. On Dynamic Behavior of the Quasi-High-Speed Catenary. Journal of Southwest Jiaotong University 1997;32(2):187–92.
[187] Cai Chengbiao, Zhai Wanming. Study on Simulation of Dynamic Performance of Pantograph-Catenary System at High-speed Railway. Journal of Railway 1997;19(5):38–43.
[188] Cai Chengbiao, Zhai Wanming. Static stiffness analysis of catenary of high-speed railway. Journal of Southwest Jiaotong University 1996;31(2):185–90.
[189] Cai Chengbiao, Zhai Wanming. Vibration characteristic analysis of catenary of high-speed railway. Journal of Southwest Jiaotong University 1997;32(5):497–501.
[190] Li Fengliang, Suqian, Sun Yan. The Dynamic Model and Differential Equations of the Catenary. Journal of Changsha Railway University 1996;14(2):90–3.
[191] Wu Tianxing. Analysis and calculation of catenary by FEM. Journal of Railway 1996;18(3):44–9.
[192] An Xiaolian. Current collector. Beijing: China Railway Publishing House; 1984. p. 34–44.
[193] Wu Tianxing. Current collection simulation analysis on pantograph-catenary dynamic system 弓网高速动态受流仿真研究. Journal of Railway 1996;18(4):55–61.
[194] Wu TX, Brennan MJ. Basic analytical study of pantograph–catenary system dynamics. Vehicle System Dynamics 1998;30(6):443–56.
[195] Wu TX, Brennan MJ. Dynamic stiffness of a railway overhead wire system and its effect on pantograph–catenary system dynamics. Journal of Sound and Vibration 1999;219(3):483–502.
[196] Zhang WH, Mei GM, Zeng J. A Study of Pantograph/Catenary System Dynamics with Influence of Presag and Irregularity of Contact Wire. Proc. Vehicle System Dynamics 2002;37(Supplement):593–604.
[197] Zhang WH, Liu Y, Mei GM. Evaluation of the coupled dynamical response of a pantograph–catenary system contact force and stresses. Vehicle System Dynamics 2006;44(8):645–58.
[198] Lee K. Analysis of dynamic contact between overhead wire and pantograph of a high-speed electric train. Proc. IMechE. Part F: Journal of Rail and Rapid Transit 2007;221:157–66.
[199] Lopez-Garcia O, Carnicero A, Maroño JL. Influence of stiffness and contact modelling on catenary-pantograph system dynamics. Journal of Sound and Vibration 2007;299:806–21.
[200] Kim JW, Chae HC, Park BS, et al. State sensitivity analysis of the pantograph system for a high-speed rail vehicle considering span length and static uplift force. Journal of Sound and Vibration 2007;303(3-5):405–27.
[201] Cho YH. Numerical simulation of the dynamic responses of railway overhead contact lines to a moving pantograph, considering a nonlinear dropper. Journal of Sound and Vibration 2008;315(3):433–54.
[202] Zhou Ning, Zhang Weihua. Study of catenary coupling system dynamics based on direct integration method. China Academy of Railway Sciences 2008;29(6):71–6.
[203] Zhou Ning, Zhang Weihua, Wang Dong. Lumped Mass Model for Dynamic Performance Simulation of Pantograph. Journal of Southwest Jiaotong University 2011;46(3):398–403.
[204] Zhang WH, Zhou N, Li RP, et al. Pantograph and catenary system with double pantographs for high-speed trains at 350 km/h or higher. Journal of Modern Transportation 2011;19(1):7–11.
[205] Bucca G, Collina A, Manigrasso R, et al. Analysis of electrical interferences related to the current collection quality in pantograph–catenary interaction. Proc IMechE Part F: Journal of Rail and Rapid Transit. 2011;225(5):483–500.
[206] Zhou N, Zhang WH. Investigation on Dynamic Performance and Parameter Optimization Design of Pantograph and Catenary System. Finite element in analysis and design 2011;47(3):288–95.
[207] Cho YH, Lee K, Park Y, et al. Influence of contact wire pre-sag on the dynamics of pantograph-railway catenary. International Journal of Mechanical Sciences 2010;52(11):1471–90.
[208] Mei Guiming, Zhang Weihua. Dynamic model and behavior of pantograph/catenary system. Journal of Traffic and Transportation Engineering 2002;2(1):20–5.

[209] Li Fengliang, Sun Yan, Su Qian. Reduction quality of TSG3 pantograph. Journal of Railway 1998;20(2):55–8.
[210] Li Fengliang, Li Min, Tang Jianxiang. Modelling and parameter test of pantograph. Journal of Central South University (JCR Science Edition) 2006;37(1):194–9.
[211] Metrikine AV, Bosch AL. Dynamic response of a two-level catenary to a moving load. Journal of Sound and Vibration 2006;292(3-5):676–93.
[212] Lopez-Garcia O, Carniceroa A, Torresb V. Computation of the initial equilibrium of railway overheads based on the catenary equation. Engineering Structures 2006;28(10):1387–94.
[213] Park TJ, Han CS, Jang JH. Dynamic sensitivity analysis for the pantograph of a high-speed rail vehicle. Journal of Sound and Vibration 2003;266(2):235–60.
[214] Yu D, Lo KL. Analysis on catenary current-collection theory. In: Proc. of the Universities Power Engineering Conference; 2000. p. 44–5.
[215] Mitsuo A, Issei N, Hiorkazu K. Current collection characteristics and improvement methods of high-tension overhead catenary systems. Electrical Engineering in Japan 1998;123(4):67–76.
[216] Mei Guiming, Zhang Weihua. Dynamic performances of different structural catenary. Journal of Traffic and Transportation Engineering 2002;2(2):27–31.
[217] Liu Hongjiao. The optimization of pantograph structure parameters and the research on active control of pantograph. Chengdu: Master's thesis of Southwest Jiaotong University; 2002.
[218] Huang Biao. Study on the system of pantograph/catenary based on virtual prototyping. Chengdu: Master's thesis of Southwest Jiaotong University; 2004.
[219] Liu Yi, Zhang Weihua, Mei Jiaming. Study of dynamic stress of the catenary in the pantograph/catenary vertical coupling movement. Journal of Railway 2003;25(4):23–6.
[220] Yan Junmao. Vehicle engineering (the 2nd edition). Beijing: China Railway Publishing House; 2003.
[221] Ikeda M, Yoshida K, Suzuki M. A flow control technique utilizing air blowing to modify the aerodynamic characteristics of pantograph for high-speed train. Journal of Mechanical Systems for Transportation and Logistics 2008;1(3):264–71.
[222] Suzuki M, Ikeda M, Koyama T. Flow control on pantograph with air intake and outlet. Quarterly Report of RTRI 2007;48(4):236–9.
[223] Zhang Hong, Yu Zhengping, Wu Hongbiao. Aerodynamic force model and wind tunnel test of pantograph. China Academy of Railway Sciences 1995;16(1):37–49.
[224] Cai Guohua. Measurement of pantograph aerodynamic force of high-speed train. Experiments and Measurements in Fluid Mechanics 2004;18(1):53–6.
[225] Zhang Jianhui, Yang Jiong, Yao Yong. Wind Tunnel test investigation on pantograph drag reduction of high-speed train. Journal of Railway Science and Engineering 2010;7(6):116–21.
[226] Song Honglei, Wu Junyong, Wu Yan, et al. Influence of Aerodynamic to High-speed Pantograph Current Collection Characteristics. Electric Railway 2010;(1):28–32.
[227] Wu Yan, Wu Junyong, Zheng Jihao. Simulation of High-speed Pantograph Dynamic Performance Based on Finite Element Model and Aerodynamic Pantograph Model. Journal of Southwest Jiaotong University 2009;44(6):855–9.
[228] Zhai Wanming, Cai Chengbiao. Effect of Locomotive-Track Coupling Vibrations on Pantograph-Catenary System Dynamics. Journal of Railway 1998;20(1):32–8.
[229] Zhang Weihua E. MU Integration and Bogie. Beijing: China Railway Publishing House; 2011.
[230] Zhang Shuguang. Study on Design Method of High-speed Train. Beijing: China Railway Publishing House; 2009.
[231] Boocock D. Steady-state motion of railway vehicles on curved track. Journal of Mechanical Engineering Science 1969;11(6):556–66.
[232] Elkins JA, Gostling RJ. A general quasi-static curving theory for railway vehicles. Vehicle System Dynamics 1977;6(2-3):100–6.
[233] Wormley DN, Hedrick JK, Nagurka ML. Analysis of rail transit vehicle dynamic curving performance. United States: Urban Mass Transportation Administration; 1984.

[234] Matsumoto A. Multibody Dynamics Simulation and Experimental Evaluation for Active-Bogie-Steering Bogie. Foreign Rolling Stock 2010;4:014.
[235] Suda Y, Iwasa T, Komine H, et al. Development of onboard friction control. Wear 2005;258(7-8):1109–14.
[236] Diedrichs B, Berg M, Stichel S, et al. Vehicle dynamics of a high-speed passenger car due to aerodynamics inside tunnels. Proceedings of the Institution of Mechanical Engineers, Part F: Journal of Rail and Rapid Transit 2007;221(4):527–45.
[237] Karnopp D, Crosby MJ, Harwood RA. Vibration control using semi-active force generators. Journal of engineering for industry 1974;96(2):619–26.
[238] Valášek M, Novak M, Šika Z, et al. Extended ground-hook-new concept of semi-active control of truck's suspension. Vehicle system dynamics 1997;27(5-6):289–303.
[239] Sasaki K, Nagai MA. A semi-active suspension of tilting train. In: The Proceedings of International Symposium on Seed-up and Service Technology for Railway and Maglev Systems: STECH 2003. The Japan Society of Mechanical Engineers; 2003. p. 214–9.
[240] Margolis DL. The response of active and semi-active suspension of feedback signals. Vehicle System Dynamics 1982;11(4):267–82.
[241] Wang Y. Research on characteristics of on-off control for semi-active suspensions of vehicle. Chinese of Mechanical Engineering 2002;38(6):148–51.
[242] Zeng J, Dai H, Wu P. Dynamics performance study of railway passenger car system based on On/Off damping control. China Railway Science 2004;25(6):27–31.
[243] Ding W, Bu J. Research on lateral on-off semi-active suspension of railway vehicle. Chinese of Mechanical Engineering 2004;40(9):161–70.
[244] Yao J, Sun Q. Theoretical research & development on the semi-active damper of locomotive &car. Railway Locomotive & Car 2004;24(S):6–9.
[245] Dong X. Investigation of lateral semi-active damper for the high-speed train. Master thesis. Beijing Jiaotong University; 2003.
[246] Tomioka T, Takigami T. Reduction of bending vibration in railway vehicle carbodies using carbody–bogie dynamic interaction. Vehicle System Dynamics 2010;48(S1):467–86.
[247] Foo E, Goodall RM. Active suspension control of flexible-bodied railway vehicles using electro-hydraulic and electro-magnetic actuators. Control Engineering Practice 2000;8(5):507–18.
[248] Tamai EH, Sotelo J. LQG-control of active suspension considering vehicle body flexibility. In: Control Applications, 1995., Proceedings of the 4th IEEE Conference on. IEEE; 1995. p. 143–7.
[249] Sugahara Y, Kazato A, Koganei R, et al. Suppression of vertical bending and rigid-body-mode vibration in railway vehicle car body by primary and secondary suspension control: results of simulations and running tests using Shinkansen vehicle. Proceedings of the Institution of Mechanical Engineers, Part F: Journal of Rail and Rapid Transit 2009;223(6):517–31.
[250] Zhu H, Liu S. Study on active suspension of rail vehicles based on flexible car body with Fuzzy control strategy. Chinese of Mechanical Engineering 2006;17(18):1883–7.
[251] Suzuki Y, Chonan S. Theoretical analysis for flexural vibration damping method of rolling stock carbody. Transactions of the Japan Society of Mechanical Engineers, Part C 1996;62(598):2132–9.
[252] Benatzky C, Kozek M. An identification procedure for a scaled metro vehicle-flexible structure experiment. In: Control Conference (ECC), 2007 European. IEEE; 2007. p. 4639–44.
[253] Lu Z. Flexible vibration active control and ride quality improvement. Chinese of Mechanical Engineering 2005;17(10):1026–30.
[254] Zhou J. Simulation on passive vibration control of carbody of railway passenger vehicles. Journal of Tongji University (Natural Science) 2009;37(8):1085–9.
[255] Zeng J, Luo R. Vibration analysis of railway passenger car system by considering flexible carbody effect. Journal of the China Railway Society 2007;29(6):19–25.
[256] Huang C, Zeng J, Wu P, Luo R. Study on carbody flexible vibration reduction for railway passenger carriage. Engineering Mechanics 2010;27(12):250–6.

CHAPTER 2

Dynamic modeling of coupled systems in the high-speed train

Chapter Outline

Dynamics of Coupled Systems in High-Speed Railways. https://doi.org/10.1016/B978-0-12-813375-0.00002-9

2.1 Basic definitions

The dynamics of coupled systems in the high-speed train is a train-oriented science concerned with an intricate system consisting of the high-speed train itself and several subsystems, such as the track, airflow, power supply, and catenary. It deals with the system dynamics to realize simulation, optimization, and control of coupled systems in the high-speed train.

Figure 1.11 shows the coupled systems of the high-speed train: vehicle subsystem, track subsystem, pantograph-catenary subsystem, and power-supply subsystem, as well as the airflow subsystem that affects the train dynamics. Different subsystems can interact and respond to related actions. Distinct coupled relationships exist between several subsystems, including the traditional coupled relationship between train vehicles, the wheel-rail coupled relationship, the pantograph-catenary coupled relationship, the electromechanical coupled relationship between the power supply subsystem and the train, and the fluid-solid coupled relationship between the train and the environment. These are all illustrated in Figure 2.1.

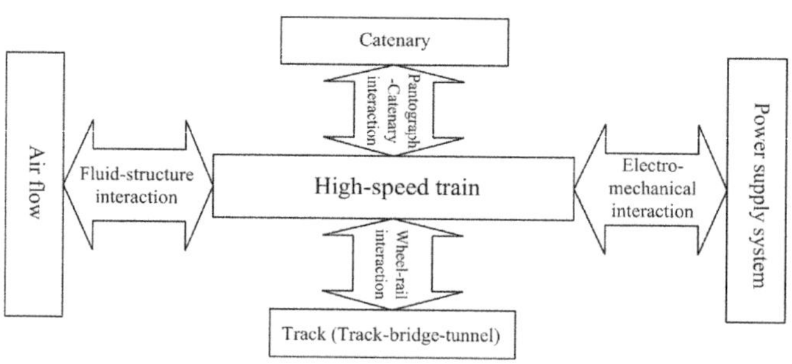

Figure 2.1
Coupled relationships between subsystems in the high-speed train.

The key concerns for the dynamics of coupled systems in the high-speed trains are identified as follows:

(1) High-speed train,
(2) Dynamics,
(3) Dynamic subsystems related to the train,
(4) Coupled relationships between the train and subsystems.

This book highlights key points regarding the high-speed train, coupled systems, coupled relationships, and dynamics. From this perspective, it not only addresses the extension of the traditional vehicle system but emphasizes the integrity and relevance of the dynamic system considered: it focuses on subsystems but pays more attention to the coupled relationships between them; it contains traditional vehicle dynamics but extends to the generalized dynamics of coupled systems.

Figure 2.2 shows a more specific relations diagram of the dynamics of the coupled systems of the high-speed trains. First, the train is fundamentally an assembly of motor cars and trailer cars. All cars are research objects, being vehicle units in traditional vehicle dynamics. The coupled forces between two vehicle units should be included in the dynamic model of the train system. Secondly, the pantograph is installed on one vehicle unit and thus moves with the train. The sliding contact of the pantograph strips with the contact wires results in the continuous interaction between pantograph and catenary through which electric transmission is achieved. Next, tracks are dynamically coupled to the train system by the wheel-rail interaction. The track is composed of rails, ballastless track plates, embankments, and viaducts (if any). These components are also coupled in the force way. Finally, aerodynamic force is the resistance against the train's motion

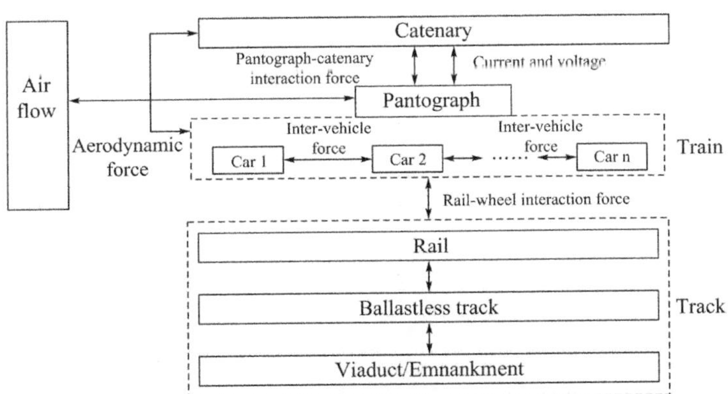

Figure 2.2
Relations diagram for the dynamics of coupled systems in the high-speed train.

caused by airflow. The aerodynamic force exerted on the train, the pantograph, and the catenary can lead to vibrations in relevant subsystems. It should be pointed out that except for the electromechanical subsystem, the subsystems directly influence the coupled system in the force way. The electromechanical subsystem is coupled by means of the conversion process between electrical and mechanical energies. Electric motors deliver torque that drives or brakes the wheels of the train. The wheel-rail interaction can yield traction or braking forces, causing the acceleration, deceleration, or constant speed of the train.

Based on train dynamics, the dynamic models of the vehicle subsystem, track subsystem, pantograph and catenary subsystem, power supply subsystem, traction and transmission subsystem, and airflow subsystem are presented first. These are followed by the coupled models of the vehicles, pantograph and catenary, and the wheel-rail interaction.

2.2 Dynamic modeling for subsystems

2.2.1 Vehicle subsystem

The general mechanics method, typically represented by Newtonian mechanics, is often used to model traditional vehicle dynamics, where the vehicle is dynamically equivalent to different mass blocks representing wheelsets, frame and vehicle body, and other components. Each mass block is a research object. The use of Newton's law of mechanics yields the motion equations of each object. A dynamic model [2~6] of the vehicle subsystem is then established by assembling the motion equations of all objects. However, the Newtonian mechanics method is not suitable for large systems with many components. It has become common practice that the dynamics of the vehicle system is modeled and simulated using computational multibody system dynamics formulation.

A multibody system consists of rigid and flexible bodies that are interconnected with each other by various constraints and forces. From this perspective, mechanical systems, robots, space structures, and biomechanical systems are all typical multibody systems. Figure 2.3 presents a picture of a real vehicle system. The real vehicle system is a collection of a

Figure 2.3
The picture of a real train vehicle.

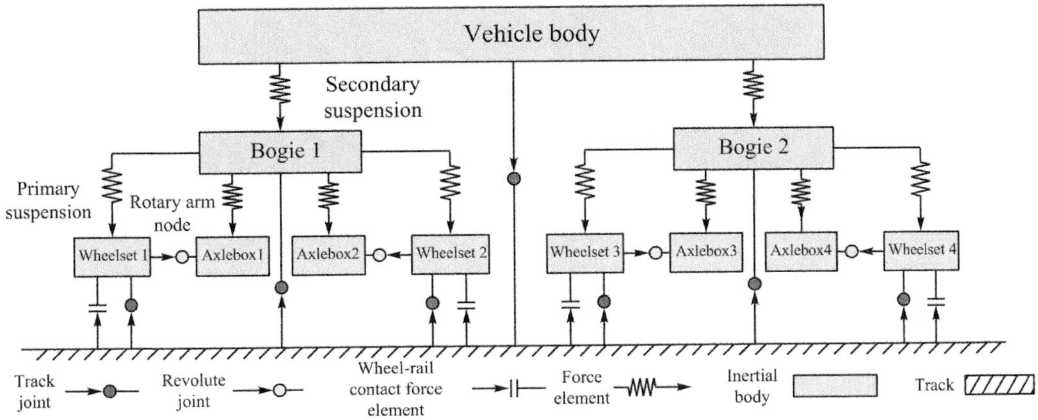

Figure 2.4
Topological graph of the vehicle system.

vehicle body, frame, wheelsets, motors, axle box, and other components connected by steel spring, rubber stack, hydraulic damper, air spring, and bearings. The vehicle system is also a multibody system, as defined earlier. In multibody dynamics system formulation, the vehicle body, frame, wheelsets, and axel box are modeled using rigid bodies. The steel spring, rubber stack, hydraulic damper, and air spring are treated with force elements, while bearings are idealized by hinge constraints. Using the topological relations shown in Figure 2.4, the multibody model of the vehicle system is obtained.

Compared with a general multibody system, the vehicle system has two distinct aspects: (1) the vehicle system runs along special orbits, and (2) there are complicated rolling contacts between the wheels and the rails. Consequently, based on the orbit coordinate system, a modeling method is developed into which the characteristics of the vehicle system fit.

2.2.1.1 Multi-rigid-body modeling of the vehicle system

(1) Multi-rigid-body dynamics theory

Multibody system dynamics formulation involves rigid and flexible bodies, which are distinguished by elastic deformations. The rigid body is a solid body in which the distance between two arbitrary points of the body remains invariant. In vehicle dynamics analysis, the main components of the vehicle system, such as the vehicle body, frames, and wheelsets, can be considered as rigid bodies because they have higher stiffness. Their deformations are thus ignored.

Without loss of generality, an absolute coordinate system *OXYZ* is defined for the vehicle dynamics. It is assumed that this coordinate system provides a unique standard for all bodies in the vehicle system. In this sense, this coordinate system can also be referred to

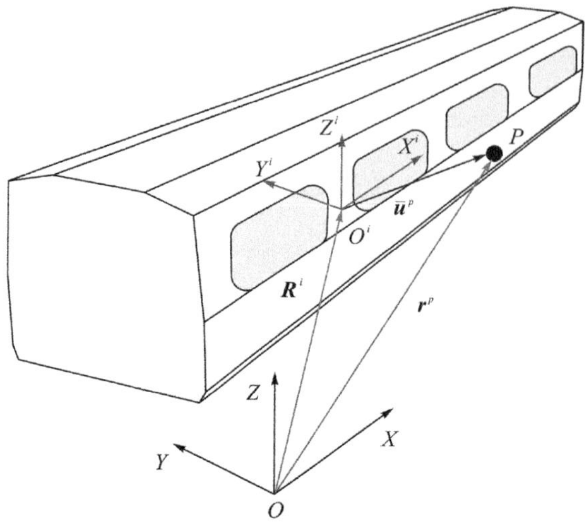

Figure 2.5
Configuration description of an arbitrary point on the rigid body.

as the global and inertial frame of reference. For reference purposes, each rigid body is assigned a body-fixed coordinate system. For example, the coordinate system $O^iX^iY^iZ^i$ is the body system of the rigid body i, as shown in Figure 2.5. Consider an arbitrary point P on one rigid body. In the multi-rigid-body theory, the absolute position vector $\mathbf{r}^p$ of point P can be expressed as:

$$\mathbf{r}^p = \mathbf{R} + \mathbf{u}^p = \mathbf{R} + \mathbf{A}\bar{\mathbf{u}}^p \tag{2.1}$$

where $\mathbf{R}$ denotes the global position vector of the origin of the body coordinate system, $\bar{\mathbf{u}}^p$denotes the local position vector of point P in the body coordinate system, $\mathbf{u}^p$ is the global vector of $\bar{\mathbf{u}}^p$, and $\mathbf{A}$ is the rotation matrix of the body with respect to the global coordinate system.

The derivative of Equation (2.1) with respect to time t leads to the velocity vector of point P:

$$\dot{\boldsymbol{r}}^p = \dot{\boldsymbol{R}} + \dot{\mathbf{A}}\bar{\boldsymbol{u}}^p = \dot{\boldsymbol{R}} + \boldsymbol{\omega} \times \boldsymbol{u}^p = \dot{\boldsymbol{R}} + \mathbf{A}\tilde{\bar{\boldsymbol{u}}}^p\bar{\boldsymbol{\omega}}^i \tag{2.2}$$

where $\boldsymbol{\omega} = \begin{bmatrix} \omega_x & \omega_y & \omega_z \end{bmatrix}^{\mathrm{T}}$ is the angular velocity of the body in the global coordinate system, and $\bar{\boldsymbol{\omega}} = \begin{bmatrix} \bar{\omega}_x & \bar{\omega}_y & \bar{\omega}_z \end{bmatrix}^{\mathrm{T}}$is the angular velocity vector of the body in the body coordinate system. Similarly, the acceleration vector of point P can be obtained as:

$$\ddot{\boldsymbol{r}}^p = \ddot{\boldsymbol{R}} + \ddot{\mathbf{A}}\bar{\boldsymbol{u}}^p = \ddot{\boldsymbol{R}} + \mathbf{A}\left\{\tilde{\bar{\boldsymbol{\alpha}}}\bar{\boldsymbol{u}}^p + \tilde{\bar{\boldsymbol{\omega}}}\tilde{\bar{\boldsymbol{\omega}}}\bar{\boldsymbol{u}}^p\right\} \tag{2.3}$$

where $\bar{\mathbf{a}}$ denotes the angular acceleration vector defined in the body coordinate system. The overbar '$\sim$' of a vector $\boldsymbol{x}$represents the calculation of the skew matrix as:

$$\tilde{\boldsymbol{x}} = \begin{bmatrix} 0 & -x_3 & x_2 \\ x_3 & 0 & -x_1 \\ -x_2 & x_1 & 0 \end{bmatrix} \tag{2.4}$$

Skew matrices $\tilde{\bar{\mathbf{u}}}$, $\tilde{\bar{\mathbf{a}}}$, and $\tilde{\bar{\boldsymbol{\omega}}}$ in Equation (2.2) and Equation (2.3) can be understood in a similar manner.

Various methods for describing the orientation of a rigid body in three dimensions are available in the literature, such as the direction cosine matrix, Euler angles, and Euler parameters. The direction cosine matrix has nine elements, but only three are independent. Six constraint equations are therefore required to determine the direction cosine matrix. The method of direction cosine matrix is rarely used to orientate the rigid body. The method of Euler angles provides three independent variables to represent three successive rotations of the rigid body in a given sequence. It makes the description of the rotation of the rigid body with as few coordinates as possible. However, singularities exist in this method.

The rotation of a rigid body can also be described using the Euler's rotation theorem. Euler's rotation theorem states that any rotation can be obtained as a single rotation around a fixed axis $\boldsymbol{v} = \begin{bmatrix} v_1 & v_2 & v_3 \end{bmatrix}^{\mathrm{T}}$ in space with finite angle θ. According to the Euler rotation theorem, four dependent variables, also called Euler parameters, are given as follows:

$\lambda_0 = \cos\frac{\theta}{2}$, $\lambda_i = v_i \sin\frac{\theta}{2}$, $(i = 1, 2, 3)$. Thus, the vector of Euler parameters to orientate the rigid body can be defined as:

$$\boldsymbol{\Lambda} = \begin{bmatrix} \lambda_0 & \lambda_1 & \lambda_2 & \lambda_3 \end{bmatrix}^{\mathrm{T}} \tag{2.5}$$

Using Equation (2.5), the rotation matrix A can be written explicitly in terms of the Euler parameters as:

$$\mathbf{A} = \begin{bmatrix} 2(\lambda_0^2 + \lambda_1^2) - 1 & 2(\lambda_1\lambda_2 - \lambda_0\lambda_3) & 2(\lambda_1\lambda_3 + \lambda_0\lambda_2) \\ 2(\lambda_1\lambda_2 + \lambda_0\lambda_3) & 2(\lambda_0^2 + \lambda_2^2) - 1 & 2(\lambda_2\lambda_3 - \lambda_0\lambda_1) \\ 2(\lambda_1\lambda_3 - \lambda_0\lambda_2) & 2(\lambda_2\lambda_3 + \lambda_0\lambda_1) & 2(\lambda_0^2 + \lambda_3^2) - 1 \end{bmatrix} \tag{2.6}$$

where the four Euler parameters should satisfy the following equation:

$$\boldsymbol{\Lambda}^{\mathrm{T}}\boldsymbol{\Lambda} = 1 \tag{2.7}$$

The Euler parameters have an advantage and are thus used in common dynamic software. Considering Equation (2.1) and Equation (2.5), the vector of generalized coordinates to describe general motion and rotation of a rigid body can be defined as:

$$q = [\boldsymbol{R}^{\mathrm{T}} \quad \Lambda^{\mathrm{T}}]^{\mathrm{T}} \tag{2.8}$$

Introduce the following matrix and vector:

$$\mathbf{B} = \begin{bmatrix} \mathbf{I} & -2\mathbf{A}\tilde{\bar{\boldsymbol{u}}}^{p}\overline{\mathbf{G}} \end{bmatrix} \tag{2.9}$$

$$\boldsymbol{a}_{\mathrm{v}} = \mathbf{A}\tilde{\bar{\omega}}\tilde{\bar{\omega}}\bar{\boldsymbol{u}}^{p} \tag{2.10}$$

where the three by four matrix $\overline{G}$ is given as:

$$\overline{\mathbf{G}} = \begin{bmatrix} -\lambda_1 & \lambda_0 & \lambda_3 & -\lambda_2 \\ -\lambda_2 & -\lambda_3 & \lambda_0 & \lambda_1 \\ -\lambda_3 & \lambda_2 & -\lambda_1 & \lambda_0 \end{bmatrix}$$

Using Equation (2.5) and considering the definition of the matrix $\overline{G}$, the angular velocity vector $\overline{\boldsymbol{\omega}}$ can be expressed as:

$$\overline{\omega} = 2\overline{\mathbf{G}}\dot{\Lambda}$$

Similarly, the acceleration vector $\overline{\boldsymbol{\alpha}}$ can be expressed as:

$$\overline{\alpha} = 2\overline{\mathbf{G}}\ddot{\Lambda}$$

Using Equation (2.9) and Equation (2.10), and considering Equation (2.2) and Equation (2.3), the following equations can be obtained:

$$\dot{\mathbf{r}}^{p} = \mathbf{B}\dot{\mathbf{q}}, \quad \ddot{\mathbf{r}}^{p} = \mathbf{B}\ddot{\mathbf{q}} + \boldsymbol{\alpha}_{v} \tag{2.11}$$

Consider an unconstrained multibody system that consists of N rigid bodies. The generalized coordinate vector $\mathbf{q}$ of the system considered has the form:

$$\boldsymbol{q} = \begin{bmatrix} \boldsymbol{q}^{1\mathrm{T}} & \boldsymbol{q}^{2\mathrm{T}} & \cdots & \boldsymbol{q}^{N\mathrm{T}} \end{bmatrix}^{\mathrm{T}}$$

The dynamic equations of the system considered can be described using Lagrange's equations of the second kind, as:

$$\mathbf{M}\ddot{q} = \boldsymbol{Q}_{\mathrm{e}} + \boldsymbol{Q}_{\mathrm{v}} \tag{2.12}$$

where $\mathbf{M}$ is the mass matrix of the system; $\mathbf{Q}_v$ is the vector of centrifugal and Coriolis inertia forces; and $\mathbf{Q}_e$ is the vector of externally applied forces, including gravity, spring, damper, and actuator forces. The system mass matrix $\mathbf{M}$ is a diagonal matrix expressed as:

$$\mathbf{M} = diag\left(\mathbf{M}^1, \mathbf{M}^2, \cdots \mathbf{M}^i, \cdots \mathbf{M}^N\right)$$

where $\mathbf{M}^i$ is the mass matrix of the ith rigid body, and can be expressed as:

$$\mathbf{M}^i = \int_{V^i} \rho^i \mathbf{B}^{\mathrm{T}} \mathbf{B} \mathrm{d}V^i \tag{2.13}$$

in which ρ^i denotes the density of the ith rigid body. Similarly, the vector $\mathbf{Q}_v$ is given as:

$$\mathbf{Q}_v = diag\left(\mathbf{Q}_v^1, \mathbf{Q}_v^2, \cdots \mathbf{Q}_v^i, \cdots \mathbf{Q}_v^N\right)$$

where the vector $\mathbf{Q}_v^i$ can be expressed as:

$$\mathbf{Q}_{\mathrm{v}}^i = -\int_{V^i} \rho^i \mathbf{B}^{\mathrm{T}} \boldsymbol{a}_{\mathrm{v}} \mathrm{d}V^i \tag{2.14}$$

The constrained system is now considered. It is assumed that the system constraints are ideal. The use of D'Alembert's principle and the virtual work principle leads to dynamic equations of the constrained system in differential-algebraic form, with Lagrange multipliers:

$$\begin{cases} \mathbf{M}\ddot{q} + \mathbf{C}_q^{\mathrm{T}} \boldsymbol{\lambda} = \boldsymbol{Q} \\ \boldsymbol{C}(\boldsymbol{q}, t) = \boldsymbol{0} \end{cases} \tag{2.15}$$

where $\boldsymbol{C}(\boldsymbol{q}, t)$ is the constraint equation of the system, including the constraint condition in Equation (2.7); $\mathbf{C}_q$ is the Jacobian matrix of constraint equations; $\boldsymbol{\lambda}$ is the vector of unknown Lagrange multipliers associated with the system constraints; and $\boldsymbol{Q} = \boldsymbol{Q}_{\mathrm{e}} + \boldsymbol{Q}_{\mathrm{v}}$ is the generalized force vector of the system.

(2) Vehicle system modeling based on the orbital coordinate system

One of the distinct differences between the vehicle system and the general multibody system is that the vehicle moves along a special orbit. The orbital coordinate-based method is thus used here to model the vehicle system. In this method, each rigid body is attached to an orbital coordinate system. For example, the orbital coordinate system $O^tX^tY^tZ^t$ attached to body i is shown in Figure 2.6. As shown in Figure 2.6, the origin of the attached coordinate system can be uniquely determined using the arc length s.

In this moving frame: the X^t axis indicates the heading direction of the vehicle system and is tangent to the orbit, the Z^t axis is perpendicular to the orbital plane and points vertically upwards, and the Y^t axis is determined by the right-hand rule. In the orbital coordinate system, the generalized coordinates of the ith rigid body can be defined as:

$$\boldsymbol{p}^i = \begin{bmatrix} s^i & y^i & z^i & \boldsymbol{\Lambda}^{i\mathrm{T}} \end{bmatrix}^{\mathrm{T}} \tag{2.16}$$

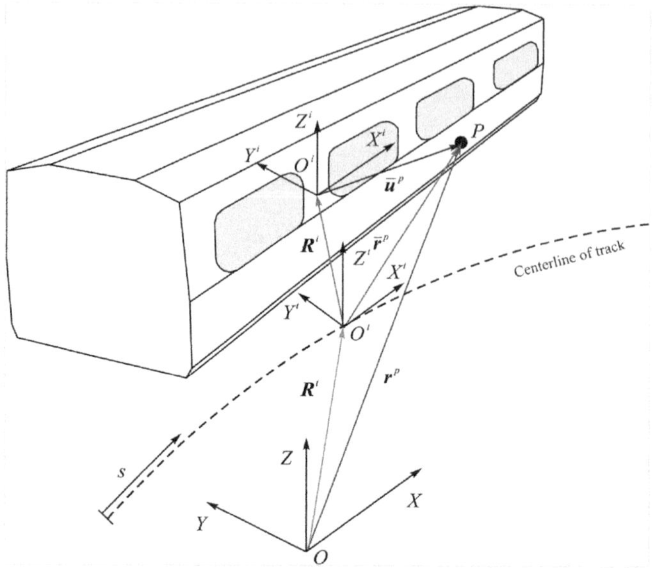

Figure 2.6
Configuration description of the rigid body in orbit coordinates.

where s^i represents the arc length coordinate of the origin of the current coordinate system, while y^i and z^i are respectively the transverse and vertical position components of an arbitrary point on the ith rigid body defined in its orbital coordinate system, and Λ^i represents four Euler parameters used to orientate the ith rigid body.

Let $\mathbf{A}^i$ be the transformation matrix of the ith rigid body relative to the orbital coordinate system. In vehicle dynamics analysis, the orientation of the train body is usually determined by roll angle, pitch angle, and yaw angle. The roll angle is the angle of the rigid body rotating with the X^t axis of its orbital coordinate system. The pitch angle is the angle of the rigid body rotating with the Y^t axis of its orbital coordinate system. The yaw angle is the angle of the rigid body rotating with the Z^t axis of its orbital coordinate system. The roll, pitch, and yaw angles of the rigid body can be obtained from four Euler parameters.

Thus, the position vector $\mathbf{r}^p$of an arbitrary point P in the ith rigid body can be expressed in orbital coordinates as:

$$\boldsymbol{r}^p = \boldsymbol{R}^t + \mathbf{A}^t\bar{\boldsymbol{r}}^p = \boldsymbol{R}^t + \mathbf{A}^t\boldsymbol{R}^i + \mathbf{A}^t\mathbf{A}^i\bar{\boldsymbol{u}}^p \tag{2.17}$$

where $\mathbf{R}^t$ is the arc length s-dependent position vector of the origin of the orbital coordinate system; $\mathbf{A}^t$ is the transformation matrix of the orbital coordinate system relative to the global coordinate system; $\bar{\mathbf{r}}^p$ defined in the orbital coordinate system, is the local position vector of point P with respect to the origin of the orbital coordinate system; $\bar{\mathbf{u}}^p$ defined in the vehicle coordinate system is the local position vector of point P with respect to the origin of the vehicle coordinate system; and $\mathbf{A}^i$ is the transformation matrix of the vehicle coordinate system relative to the orbital coordinate system. Note that the transformation matrix $\mathbf{A}^t$ can be determined with three Euler angles $\psi(s^i)$, $\theta(s^i)$, and $\phi(s^i)$ using the 3-2-1 sequence.

The derivative of Equation (2.17) with respect to time t yields the velocity vector of point P:

$$\begin{aligned}\dot{\boldsymbol{r}}^p &= \dot{\boldsymbol{R}}^t + \dot{\mathbf{A}}^t\boldsymbol{R}^i + \mathbf{A}^t\dot{\boldsymbol{R}}^i + \dot{\mathbf{A}}^t\mathbf{A}^i\overline{\boldsymbol{u}}^p + \mathbf{A}^t\dot{\mathbf{A}}^i\overline{\boldsymbol{u}}^p \\ &= \left(\frac{\partial \boldsymbol{R}^t}{\partial s}\right)\dot{s} + \left(\frac{\partial \mathbf{A}^t}{\partial s}\right)\dot{s}\boldsymbol{R}^i + \mathbf{A}^t\dot{\boldsymbol{R}}^i + \left(\frac{\partial \mathbf{A}^t}{\partial s}\right)\dot{s}\mathbf{A}^i\overline{\boldsymbol{u}}^p + \mathbf{A}^t\dot{\mathbf{A}}^i\overline{\boldsymbol{u}}^p\end{aligned} \tag{2.18}$$

Let $\boldsymbol{i}^t, \boldsymbol{j}^t, \boldsymbol{k}^t$ be unit vectors along the positive directions of the X^t-, Y^t-, and Z^t- axes of the orbital coordinate system, respectively. Using the geometry shown in Figure 2.6,

$$\frac{\partial \boldsymbol{R}^t}{\partial s} = \boldsymbol{i}^t$$

Using the chain rule, the following relationship can be obtained:

$$\frac{\partial \mathbf{A}^t}{\partial s} = \frac{\partial \mathbf{A}^t}{\partial \psi}\frac{\partial \psi}{\partial s} + \frac{\partial \mathbf{A}^t}{\partial \theta}\frac{\partial \theta}{\partial s} + \frac{\partial \mathbf{A}^t}{\partial \phi}\frac{\partial \phi}{\partial s}$$

Considering the notations defined above, the following equation can be obtained:

$$\mathbf{L} = \left[\boldsymbol{i}^t + \left(\frac{\partial \mathbf{A}^t}{\partial s}\right)\boldsymbol{R}^i + \left(\frac{\partial \mathbf{A}^t}{\partial s}\right)\mathbf{A}^i\overline{\boldsymbol{u}}^p \quad \boldsymbol{j}^t \quad \boldsymbol{k}^t \quad -2\mathbf{A}^t\mathbf{A}^{t}\widetilde{\overline{\boldsymbol{u}}}^p\overline{\mathbf{G}}\right] \tag{2.19}$$

Using Equation (2.19), Equation (2.18) can be rewritten as:

$$\dot{\boldsymbol{r}}^p = \mathbf{L}\dot{\boldsymbol{p}} \tag{2.20}$$

Using Equation (2.20), the acceleration vector of point P can be expressed as:

$$\ddot{\boldsymbol{r}}^p = \mathbf{L}\ddot{\boldsymbol{p}} + \boldsymbol{\gamma}_{\mathrm{v}} \tag{2.21}$$

where the dynamic term γ_v is given as:

$$\begin{aligned}\boldsymbol{\gamma}_{\mathrm{v}} &= \dot{\mathbf{L}}\dot{\boldsymbol{p}} \\ &= \left(\frac{\partial \boldsymbol{i}^t}{\partial s} + \frac{\partial^2 \mathbf{A}^t}{\partial s^2}\boldsymbol{R}^i\right)\dot{s}^2 + 2\frac{\partial \mathbf{A}^t}{\partial s}\left(\dot{\boldsymbol{R}}^i - \mathbf{A}^i\widetilde{\overline{\boldsymbol{u}}}^p\overline{\boldsymbol{G}}\right)\dot{s} - 2\mathbf{A}^t\left(\dot{\mathbf{A}}^i\widetilde{\overline{\boldsymbol{u}}}^p\overline{\boldsymbol{G}} + \mathbf{A}^i\widetilde{\overline{\boldsymbol{u}}}^p\dot{\overline{\boldsymbol{G}}}\right)\dot{\Lambda}^i\end{aligned} \tag{2.22}$$

Using the same derivation presented above, dynamic equations of the vehicle system can be expressed in orbital coordinates as:

$$\begin{cases}\mathbf{M}\ddot{\boldsymbol{p}} + \mathbf{C}_p^{\mathrm{T}}\boldsymbol{\lambda} = \boldsymbol{Q}_{\mathrm{e}} + \boldsymbol{Q}_{\mathrm{v}} + \boldsymbol{Q}_{\mathrm{wr}} \\ \boldsymbol{C}(\boldsymbol{p}, \boldsymbol{t}) = \boldsymbol{0}\end{cases} \tag{2.23}$$

where the overall mass matrix M is the assembly of block diagonal matrices M^i, for i = 1,2, ..., N. The mass matrix Mi of the *i*th rigid body can be calculated as:

$$\mathbf{M}^i = \int_{V^i} \rho^i \mathbf{L}^{\mathrm{T}} \mathbf{L} \mathrm{d} V^i \tag{2.24}$$

The generalized force vectors $\mathbf{Q}_{\mathrm{e}}$ and $\mathbf{Q}_{\mathrm{v}}$ are defined in Equation (2.23) and $\mathbf{Q}_{\mathrm{wr}}$ is the generalized force vector related to wheel-rail contact forces. Similarly, these three vectors can be obtained using the assembling technique. The ith block vector of $\mathbf{Q}_{\mathrm{v}}$ is given as:

$$\mathbf{Q}_{\mathrm{v}}^i = -\int_{V^i} \rho^i \mathbf{L}^{\mathrm{T}} \gamma_{\mathrm{v}} \mathrm{d} V^i \tag{2.25}$$

The overall mass matrix of the system is now a state-depended matrix, not a constant one, owing to the movement of the orbital coordinate system. Furthermore, dynamic terms associated with higher derivatives of orbital geometry are included. The orbital curve is differentiable to at least second order, for accuracy of the results obtained.

(3) Force element library for the vehicle system

Several components of the vehicle system can be treated as force elements, such as the steel spring, rubber stack, hydraulic shock absorber, rocker node, air spring, anti-rolling torsion bar, traction bar, lateral bump stop, car end coupling device, friction pair. Typical force elements include the linear or nonlinear axial force element, three direction force element, torsional force element, spring-damping serial force element, friction force element, and air spring force element.

1 Linear spring-damper parallel axial force element

The axial force element refers to the element of the force acting along the elemental direction in length. An example illustrating this is the linear spring-damper parallel axial force element shown in Figure 2.7. Let $O^iX^iY^iZ^i$ and $O^jX^jY^jZ^j$ be two vehicle coordinate systems fixed in rigid bodies i and j, respectively, while $O^{ti}X^{ti}Y^{ti}Z^{ti}$ and $O^{tj}X^{tj}Y^{tj}Z^{tj}$ are the orbital coordinate systems of corresponding rigid bodies. Two connection points of the force element are located at points P^i and P^j on two rigid bodies. Let K, C, F_0, and l_0 be stiffness, damping, initial force, and initial length of the force element, respectively. With the definitions and variables above, the force element can be analyzed. To begin, obtain the position vector of the connect points of the force element in the global coordinate system. The length and time derivative of the force element can be calculated next, and the force of the force element is obtained. Finally, the generalized force of this force is calculated. More details are presented below.

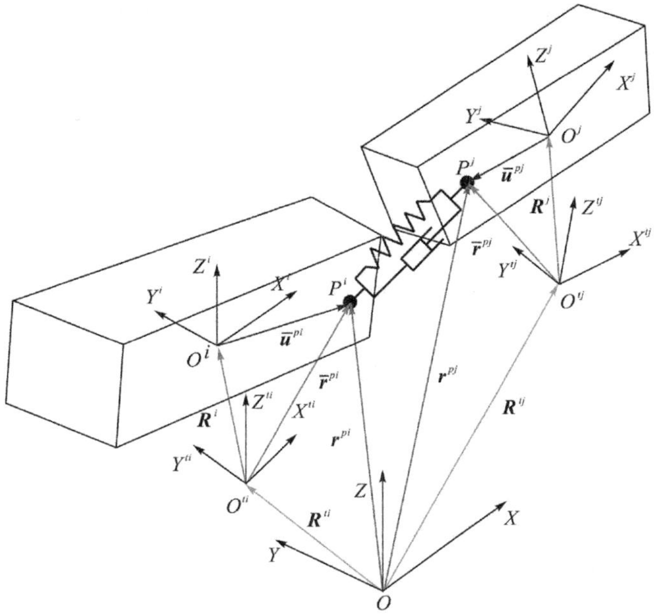

Figure 2.7
Spring-damper parallel force element.

The position vector of connection points P^i and P^j in the global coordinate system can be expressed as:

$$\boldsymbol{r}^{ij} = \boldsymbol{r}^{pi} - \boldsymbol{r}^{pj} = \boldsymbol{R}^{ti} + \mathbf{A}^{ti}\boldsymbol{R}^i + \mathbf{A}^{ti}\mathbf{A}^i\bar{\boldsymbol{u}}^{pi} - \left(\boldsymbol{R}^{tj} + \mathbf{A}^{tj}\boldsymbol{R}^j + \mathbf{A}^{tj}\mathbf{A}^j\bar{\boldsymbol{u}}^{pj}\right) \tag{2.26}$$

Applying Equation (2.26) yields:

$$\dot{\boldsymbol{r}}^{ij} = \dot{\boldsymbol{r}}^{pi} - \dot{\boldsymbol{r}}^{pj} = \mathbf{L}^i\dot{\boldsymbol{p}}^i - \mathbf{L}^j\dot{\boldsymbol{p}}^j \tag{2.27}$$

Therefore, the magnitude of the element force can be calculated as follows:

$$F_s = K(l - l_0) + C\dot{l} + F_0 \tag{2.28}$$

where l is the length of the force element and $\dot{l}$ is the rate of change of the length l. In this equation, the variables l and $\dot{l}$ can be calculated as:

$$l = \sqrt{\boldsymbol{r}^{ij\mathrm{T}}\boldsymbol{r}^{ij}} \tag{2.29}$$

$$\dot{l} = \hat{\boldsymbol{r}}^{ij\mathrm{T}}\dot{\boldsymbol{r}}^{ij} \tag{2.30}$$

where $\hat{\boldsymbol{r}}^{ij} = \boldsymbol{r}^{ij}/l$ is the unit direction vector of the force element. Using the principle of virtual work and considering Equation (2.26), the generalized force vectors associated with the two bodies are given as:

$$\begin{aligned} \boldsymbol{Q}_{\mathrm{e}}^i &= -F_s\mathbf{L}^{i\mathrm{T}}\hat{\boldsymbol{r}}^{ij} \\ \boldsymbol{Q}_{\mathrm{e}}^j &= -F_s\mathbf{L}^{j\mathrm{T}}\hat{\boldsymbol{r}}^{ij} \end{aligned} \tag{2.31}$$

Similarly, the generalized forces of other force elements can be calculated once the positions of the connection points and orientations of the relevant rigid bodies are known.

2 Spring-damper serial force element

Spring-damper serial force elements are widely used in vehicle systems to simulate the hydraulic shock absorber and rubber joints installed at its tip, as shown in Figure 2.8.

Let l be the distance between connection points of this force element, which can be calculated using Equation (2.29). The spring has a deformed length l_1, initial length l_0, and stiffness K. The damper is of length l_2 and damping coefficient C. Ignoring the inertia effects, the spring force F_s is equal to the damper force F_d. This force element has the following relationships:

$$l = l_1 + l_2 \tag{2.32}$$

$$F_s = F_d \tag{2.33}$$

where $F_s = K(l_1 - l_0)$ and $F_d = C\dot{l}_2$. The first derivative of Equation (2.32) with respect to time leads to:

$$\dot{l} = \dot{l}_1 + \dot{l}_2 \tag{2.34}$$

Considering the equations above, the following equation can be obtained:

$$\dot{l}_2 + \frac{K}{C} l_2 = \frac{K}{C}(l - l_0) \tag{2.35}$$

Note that Equation (2.35) should be solved together with the motion equations of the vehicle system in Equation (2.15) or Equation (2.23).

3 Air spring model

Air springs are important secondary suspension components in high-speed vehicle systems. The air spring is a complex system that comprises of an air spring body, a throttle hole, an

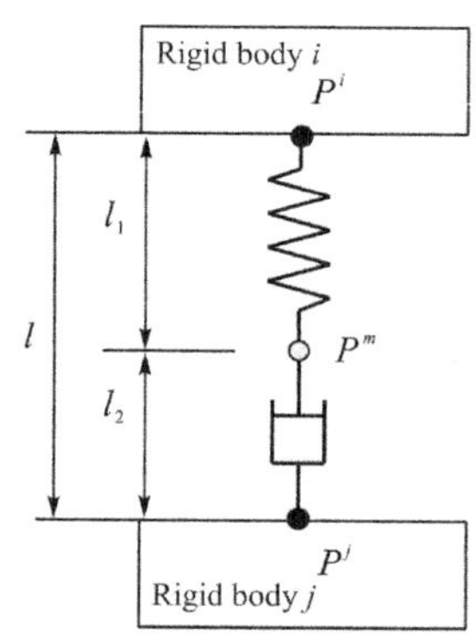

Figure 2.8
Spring-damper serial force element.

additional air chamber, a pipe, a differential pressure valve, and a height control valve, as shown in Figure 2.9. The air spring has good linear property — static stiffness in the working range — and can be simulated using a linear three-dimensional force element. In general, the dynamic characteristics of the air spring have some effects on vehicle dynamic performance as the train runs at high speed. A more accurate air spring model is therefore required for analysis.

Ignoring minor factors, considering the air spring body, the additional gas chamber, the throttle, differential pressure valve, and height control valve, and assuming that the air in the spring body is ideal, the vertical dynamic model [7] of the air spring can be established according to state equations of ideal gas. The state equations of the air in the spring body are:

$$\dot{P}_1 = \left(\dot{m}_1 R T_1 + m_1 R \dot{T}_1 - n P_1 V_1^{n-1} \dot{V}_1\right) / V_1^n \tag{2.36}$$

$$C_v m_1 \dot{T}_1 = -\beta \dot{T}_1 - P_1 \dot{V} + C_p \dot{m}_1 T_p \tag{2.37}$$

In Equation (2.36) and Equation (2.37), P_1 and V_1 represent the internal pressure of the air spring body and the volume of the air spring body, respectively; m_1 and T_1 represent the air mass and the absolute temperature in the air spring body, respectively; R is a gas constant; n is a polytropic exponent; C_v and C_p are specific heat at constant volume and constant pressure, respectively; β is the heat transfer coefficient; and T_p is the temperature of the gas flowing into the body.

Let A_e be the effective area of the air spring, which is associated with pressure P_1. The vertical force F_{sz} provided by the air spring is then given as:

$$F_{sz} = P_1 A_e(P_1) \tag{2.38}$$

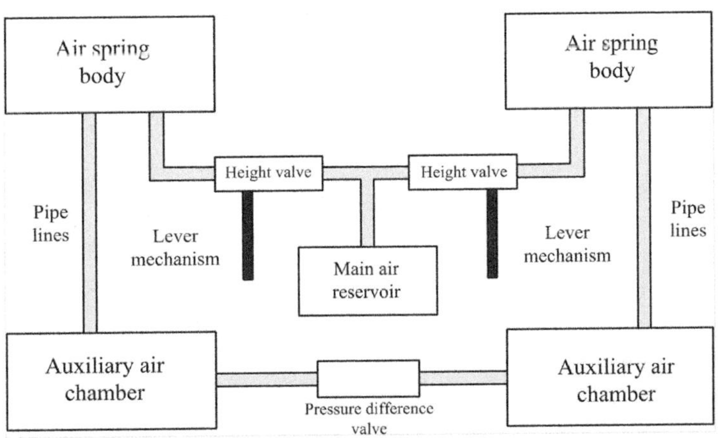

Figure 2.9
Air spring.

Like the spring body, the dynamic equations of the additional chamber can be expressed as:

$$\dot{P}_2 = \left(\dot{m}_2 RT_2 + m_2 R\dot{T}_2\right)/V_2^n \tag{2.39}$$

$$C_v m_2 \dot{T}_2 = -\beta_2 \dot{T}_2 + C_p \dot{m}_2 T_p \tag{2.40}$$

In Equations (2-39) and Equation (2.40), P_2 and V_2 are the pressure and volume of additional air in the chamber, respectively; m_2 and T_2 are the mass and absolute temperature of additional air in the chamber, respectively.

The gas exchange in the throttle hole can be treated as an adiabatic process. In thermodynamics theory, the flow rate of the gas exchange is related to its pressure ratio P_D/P_U. The mass flow q can be written as:

$$q = \begin{cases} C_d A_s \sqrt{P_U \rho_U \dfrac{2\kappa}{\kappa - 1}\left(\left(\dfrac{P_D}{P_U}\right)^{\frac{2}{\kappa}} - \left(\dfrac{P_D}{P_U}\right)^{\frac{\kappa+1}{\kappa}}\right)} & \dfrac{P_D}{P_U} > 0.5282 \\ C_d A_s \sqrt{P_U \rho_U \kappa \left(\dfrac{2}{\kappa - 1}\right)^{\frac{\kappa+1}{\kappa-1}}} & \dfrac{P_D}{P_U} \leq 0.5282 \end{cases} \tag{2.41}$$

where P_D is the maximum gas pressure on the body, P_U is the minimum gas pressure in the air chamber, C_d is the flow coefficient of the throttle orifice, A_s is the area of the throttle orifice, ρ_U is the density of the gas from the throttle, and k is the adiabatic index.

Consider a height valve model with a delay mechanism in the current air spring, which has an insensitive band of about 10 mm and a delay of about 3 s. According to the working principle of the differential pressure valve, when the pressure difference between the left and right air chambers is higher than the prescribed pressure, the differential pressure valve is open. This process can be simplified as a step function. Figure 2.10 shows nonlinear variations for the opening of the height valve and differential pressure valve with respect to the actuator displacement S of the height valve and pressure difference ΔP.

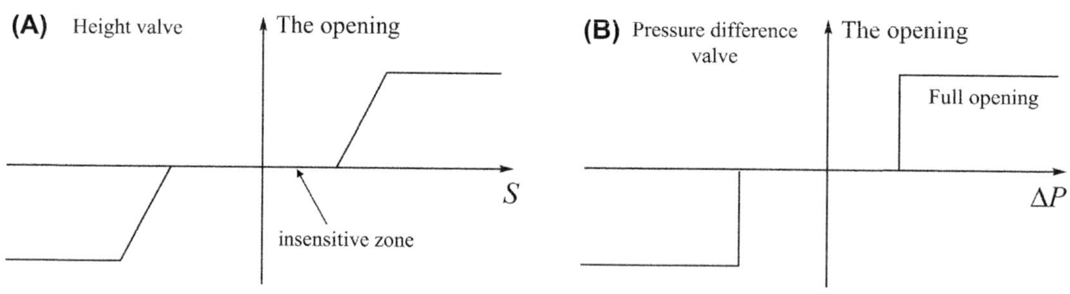

Figure 2.10
Nonlinear characteristics of the height valve and differential pressure valve.

4 External Load Elements

The vehicle system in motion is always subjected to some external forces, such as wind loads, forces between interacted vehicles, actuator forces, and coupled forces between the vehicle and other railway subsystems. These forces are addressed in the global coordinate system and included in the dynamic model of the vehicle system as external load elements. Without loss of generality, any external load exerted on the rigid body can be reduced to a resultant force Fe and a resultant moment Te in space:

$$\boldsymbol{F}_{\mathrm{e}} = [F_x \quad F_y \quad F_z]^{\mathrm{T}}, \boldsymbol{T}_{\mathrm{e}}[T_x \quad T_y \quad T_z]^{\mathrm{T}} \tag{2.42}$$

Using the principle of virtual work, the generalized forces of external force associated with the resultant force Fe and resultant moment Te can be expressed as:

$$\boldsymbol{Q}_{\mathrm{e}} = \mathbf{L}^T \boldsymbol{F}_{\mathrm{e}} + \overline{\mathbf{H}}^{\mathrm{T}} \boldsymbol{T}_{\mathrm{e}} \tag{2.43}$$

where:

$$\overline{\mathbf{H}} = \begin{bmatrix} \mathbf{A}^i \overline{\mathbf{G}}^t \dfrac{\partial \boldsymbol{\theta}^t}{\partial s} & \mathit{0} & \mathit{0} & \mathit{0} & \overline{\mathbf{G}}^i \end{bmatrix} \tag{2.44}$$

in which

$\boldsymbol{\theta}^t = \begin{bmatrix} \psi^t & \theta^t & \phi^t \end{bmatrix}^{\mathrm{T}}$ is the vector of Euler angles for the orientation of the corresponding orbital coordinate system and $\mathbf{0}$ is the 3×1 null matrix.

2.2.1.2 Rigid-flexible coupled model of the vehicle system

In traditional vehicle dynamics, the inertial components of the system are usually treated as rigid bodies. It is reasonable and economical for most analysis, but with the development of the high-speed train, the flexible effects of some components of the vehicle system should be considered, especially for lightweight design of the vehicle system. The main components, such as the vehicle body, frame, and wheelsets, have small deformations owing to higher stiffness. Hence, the modal superposition method can be used to describe the elastic deformations of these components. For this purpose, modal vibration information of components that need to be considered as elastic bodies can be obtained using finite element analysis software. Thus, the motions of these components, e.g., the displacement and velocity vectors of points on them, can be re-described with the obtained modal information. Using the same method described earlier, the rigid-flexible coupled model of the vehicle system can be established. Figure 2.11 shows a general flow process of the rigid-flexible coupled modeling and simulation analysis of the vehicle system. More details are provided below.

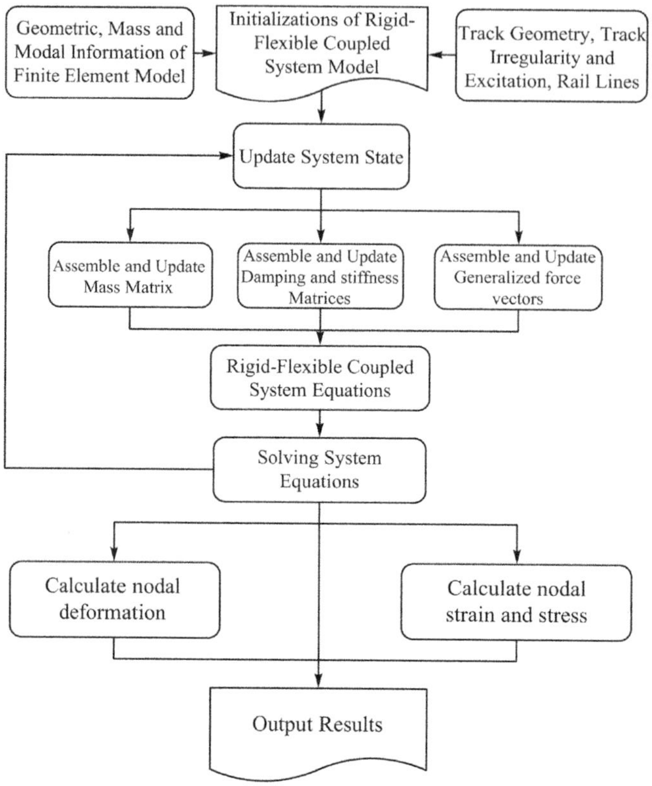

Figure 2.11
Flow process diagram for rigid-flexible coupled modeling of the vehicle system.

(1) Finite Element Modal Extraction

In general, an elastic body has infinite degrees of freedom. To model an elastic body, the finite element method can be used, and many degrees of freedom are required for the sake of precision. In this regard, millions of degrees of freedom are usually required to model complex structures such as the vehicle body and the frame. Fortunately, the components mentioned earlier under the assumption of small deformations can be modeled by the modal superposition method with at most dozens of degrees of freedom. The purpose of using the modal superposition method is to obtain a reasonable and complete structure modal set, which directly affects the accuracy and precision of the displacement analysis. The modal set refers to an assembly of the displacements of all nodes. The free modal set and the Craig-Bampton modal set [8] (C-B modal set) are mainly used in flexible dynamics analysis. The C-B modal set contains non-orthogonal constraint modes and orthogonal fixed interface normal modes. Besides some lower characteristic modes that contribute largely towards structural deformation, constraint modes (static responses induced by static loads or displacements) are also included in the C-B modal set to compensate for truncation errors of the higher modes and improve convergence and precision of the results obtained.

Here, all degrees of freedom of the structure are divided into two sets. One set is the set $\boldsymbol{u}_{\mathrm{B}}$ of degrees of freedom that describe boundary conditions between the structure and other interacted structures. The other set is the set $\boldsymbol{u}_{\mathrm{I}}$ of inner degrees of freedom of the structure. N_{o} and N_{i} are the numbers of degrees of freedom of sets $\boldsymbol{u}_{\mathrm{B}}$ and $\boldsymbol{u}_{\mathrm{I}}$, respectively. Thus, the equation of motion of the structure can be written in block form as:

$$\begin{bmatrix} \mathbf{M}_{\mathrm{BB}} & \mathbf{M}_{\mathrm{BI}} \\ \mathbf{M}_{\mathrm{IB}} & \mathbf{M}_{\mathrm{II}} \end{bmatrix} \begin{bmatrix} \ddot{u}_{\mathrm{B}} \\ \ddot{u}_{\mathrm{I}} \end{bmatrix} + \begin{bmatrix} \mathbf{K}_{\mathrm{BB}} & \mathbf{K}_{\mathrm{BI}} \\ \mathbf{K}_{\mathrm{IB}} & \mathbf{K}_{\mathrm{II}} \end{bmatrix} \begin{bmatrix} u_{\mathrm{B}} \\ u_{\mathrm{I}} \end{bmatrix} = \begin{bmatrix} F_{\mathrm{fB}} \\ 0 \end{bmatrix} \tag{2.45}$$

where $\mathrm{M_{BB}}$, $\mathrm{M_{BI}}$, $\mathrm{M_{IB}}$, and $\mathrm{M_{II}}$ are block mass matrices of the structure; $\mathrm{K_{BB}}$, $\mathrm{K_{BI}}$, $\mathrm{K_{IB}}$, and $\mathrm{K_{II}}$ are block mass matrices of the structure; and $\boldsymbol{F}_{\mathrm{fB}}$ are the applied forces at boundaries associated with set $\boldsymbol{u}_{\mathrm{B}}$. Using the Craig-Bampton technique, $\boldsymbol{u}_{\mathrm{B}}$ and $\boldsymbol{u}_{\mathrm{I}}$ can be transformed to $\boldsymbol{u}_{\mathrm{B}}$ and $\boldsymbol{q}_{\mathrm{m}}$ as:

$$\boldsymbol{u} = \begin{bmatrix} u_{\mathrm{B}} \\ u_{\mathrm{I}} \end{bmatrix} = \begin{bmatrix} \mathbf{I} & \mathbf{0} \\ \phi_B & \phi_I \end{bmatrix} \begin{bmatrix} u_{\mathrm{B}} \\ q_{\mathrm{m}} \end{bmatrix} = [\boldsymbol{\Phi}_{\mathrm{C}} \quad \boldsymbol{\Phi}_{\mathrm{N}}] \begin{bmatrix} u_{\mathrm{B}} \\ q_{\mathrm{m}} \end{bmatrix} = \boldsymbol{\Phi}_{\mathrm{CB}}\boldsymbol{q} \tag{2.46}$$

where $\boldsymbol{q}_{\mathrm{m}}$ is the vector of mode coordinates associated with inner degrees of freedom of the structure, $\mathbf{I}$is the $N_O \times N_O$ identity matrix, $\mathbf{0}$ is the $N_O \times N_i$ zero matrix, ϕ_{B} is the transformation matrix between boundary degrees of freedom for rigid displacement and internal degrees of freedom for elastic deformation of the structure, ϕ_{I} is the transformation matrix between modal response and elastic deformation of internal degrees of freedom, Φ_{C} is the constraint modal set, and Φ_{N} is the modal set that describes the fixed boundary characteristics. The Craig-Bampton set Φ_{CB} is a non-orthogonal set. Considering the orthogonalization of

$$\mathbf{M}_{\mathrm{CB}} = \boldsymbol{\Phi}_{\mathrm{CB}}^{\mathrm{T}}\mathbf{M}_{\mathrm{f}}\boldsymbol{\Phi}_{\mathrm{CB}}, \quad \mathbf{K}_{\mathrm{CB}} = \boldsymbol{\Phi}_{\mathrm{CB}}^{\mathrm{T}}\mathbf{K}_{\mathrm{f}}\boldsymbol{\Phi}_{\mathrm{CB}} \tag{2.47}$$

where $\mathrm{M_{CB}}$ and $\mathrm{K_{CB}}$ are normalized matrices associated with matrices $\mathrm{M_f}$ and $\mathrm{K_f}$, the frequency ω and eigenvector ϕ can be calculated as:

$$\left\{\mathbf{K}_{\mathrm{CB}} - \omega^2\mathbf{M}_{\mathrm{CB}}\right\}\phi = \mathbf{0} \tag{2.48}$$

The frequency ω and the eigenvector ϕ are nominal. By using the transformation matrix $\mathbf{T} = \begin{bmatrix} \phi_1 & \phi_2 & \cdots & \phi_{N_m} \end{bmatrix}$, in which N_m denotes the total number of constraint modes and fixed boundary characteristics modes, the orthogonalized C-B modal set is then obtained as:

$$\boldsymbol{\Phi} = \boldsymbol{\Phi}_{\mathrm{CB}}\mathbf{T} \tag{2.49}$$

In terms of translational and rotational degrees of freedom, the orthogonalized C-B modal set $\mathbf{\Phi}$ consists of the modal set $\mathbf{\Phi}$t of translational degrees of freedom and the modal set $\mathbf{\Phi}$r of rotational degrees of freedom, which can be shown as:

$$\mathbf{\Phi} = [\mathbf{\Phi}_{\mathrm{t}} \quad \mathbf{\Phi}_{\mathrm{r}}] \tag{2.50}$$

(2) Rigid-flexible coupled modeling theory for vehicle systems

Consider the elastic deformation of the inertial body in the orbital coordinate system, as shown in Figure 2.12. The position vector of point P in the elastic body can be expressed as:

$$\boldsymbol{r}^P = \boldsymbol{R}^t + \mathbf{A}^t\boldsymbol{R}^i + \mathbf{A}^t\mathbf{A}^i\overline{\boldsymbol{u}}^P = \boldsymbol{R}^t + \mathbf{A}^t\boldsymbol{R}^i + \mathbf{A}^t\mathbf{A}^i\left(\overline{\boldsymbol{u}}_0^P + \overline{\boldsymbol{u}}_{\mathrm{f}}^P\right) \tag{2.51}$$

where $\overline{\mathbf{u}}_0^P$ is the initial position vector of point P in the elastic body and $\overline{\mathbf{u}}_f^P$ is the local vector induced by elastic deformation, which can be expressed using the mode superposition method as:

$$\overline{\boldsymbol{u}}_{\mathrm{f}}^P = \boldsymbol{\Phi}_{\mathrm{t}}^P \boldsymbol{q}_{\mathrm{f}} \tag{2.52}$$

in which, $\mathbf{q}_{\mathrm{f}}$ is the vector of modal coordinates that describe elastic deformation of the body and Φ_t^P is the modal shape matrix.

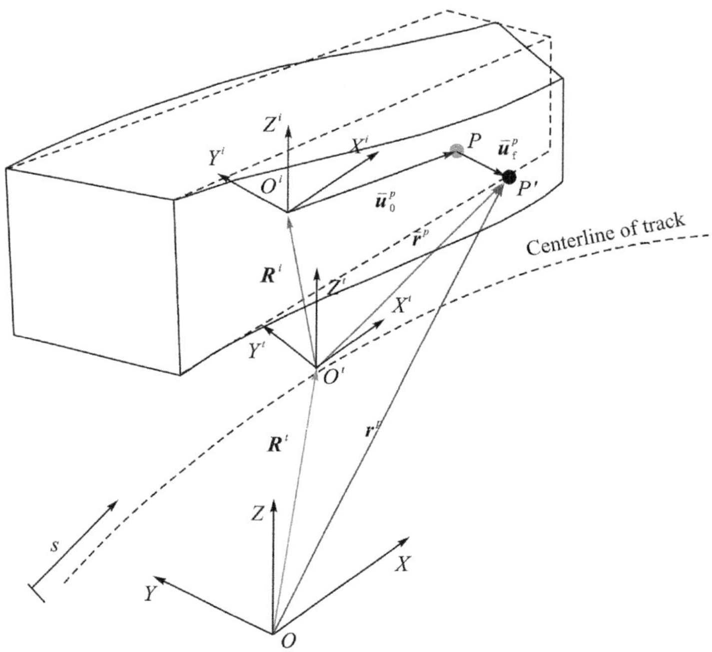

Figure 2.12
Configuration description of a flexible body.

Using Equation (2.51), the velocity vector of point P can be obtained:

$$\begin{aligned} \dot{r}^p &= \dot{R}^t + \dot{\mathbf{A}}^t R^i + \mathbf{A}^t \dot{R}^i + \dot{\mathbf{A}}^t \mathbf{A}^i \overline{u}^p + \mathbf{A}^t \dot{\mathbf{A}}^i \overline{u}^p + \mathbf{A}^t \mathbf{A}^i \mathbf{\Phi}_t^p \dot{q}_f \\ &= \left(\frac{\partial R^t}{\partial s}\right)\dot{s} + \left(\frac{\partial \mathbf{A}^t}{\partial s}\right)\dot{s} R^i + \mathbf{A}^t \dot{R}^i + \left(\frac{\partial \mathbf{A}^t}{\partial s}\right)\dot{s}\mathbf{A}^i \overline{u}^p + \mathbf{A}^t \dot{\mathbf{A}}^i \overline{u}^p + \mathbf{A}^t \mathbf{A}^i \mathbf{\Phi}_t^p \dot{q}_f \end{aligned} \tag{2.53}$$

Considering Equation (2.53), the following equation can be obtained:

$$\mathbf{L} = \begin{bmatrix} i^t + \left(\frac{\partial \mathbf{A}^t}{\partial s}\right) R^i + \left(\frac{\partial \mathbf{A}^t}{\partial s}\right) \mathbf{A}^i \overline{u}^p & j^t & k^t & -2\mathbf{A}^t \mathbf{A}^i \widetilde{\overline{u}}^p \mathbf{G} & \mathbf{A}^t \mathbf{A}^i \mathbf{\Phi}_t^p \end{bmatrix} \tag{2.54}$$

Introduce the following matrices:

$$\mathbb{R} = \begin{bmatrix} \mathbf{i}^t + \left(\frac{\partial \mathbf{A}^t}{\partial s}\right) \mathbf{R}^i + \left(\frac{\partial \mathbf{A}^t}{\partial s}\right) \mathbf{A}^i \overline{\mathbf{u}}^p & \mathbf{j}^t & \mathbf{k}^t \end{bmatrix}, \ \mathbf{A} = \mathbf{A}^t \mathbf{A}^i, \text{ and } \mathbf{B} = -2\mathbf{A}^t \mathbf{A}^i \widetilde{\overline{u}}^p \mathbf{G}$$

Using these matrices, Equation (2.54) is then reduced to:

$$\mathbf{L} = \begin{bmatrix} \mathbb{R} & \mathbf{B} & \mathbf{A}\Phi_t^p \end{bmatrix} \tag{2.55}$$

The vector of generalized coordinates of flexible body i, including modal coordinates, can be now defined as:

$$p = \begin{bmatrix} s & y & z & \Lambda^{\mathrm{T}} & q_f^{\mathrm{T}} \end{bmatrix}^{\mathrm{T}} \tag{2.56}$$

Using Equation (2.56), Equation (2.53) can be rewritten as:

$$\dot{r}^p = \mathbf{L}\dot{p} \tag{2.57}$$

Note that $\overline{u}^p$ in the term $\mathbf{L}$ contains an initial position and deformation of an elastic body, and is therefore not a constant vector in contrast with its equivalent that is related to the rigid body. See Equation (2.19) and Equation (2.54).

With the same method used above, the mass matrix of a flexible system can be obtained, and is expressed as:

$$\mathbf{M}^i = \int_{V^i} \rho^i \mathbf{L}^{\mathrm{T}} \mathbf{L} \mathrm{d}V^i = \int_V \rho \begin{bmatrix} \mathbb{R}^{\mathrm{T}}\mathbb{R} & \mathbb{R}^{\mathrm{T}}\mathbf{B} & \mathbb{R}^{\mathrm{T}}\mathbf{A}\mathbf{\Phi}_t^p \\ & \mathbf{B}^{\mathrm{T}}\mathbf{B} & \mathbf{B}^{\mathrm{T}}\mathbf{A}\mathbf{\Phi}_t^p \\ \text{symm} & & \mathbf{\Phi}_t^{p\mathrm{T}}\mathbf{\Phi}_t^p \end{bmatrix} \mathrm{d}V = \begin{bmatrix} \mathbf{m}_{RR} & \mathbf{m}_{R\theta} & \mathbf{m}_{Rf} \\ & \mathbf{m}_{\theta\theta} & \mathbf{m}_{\theta f} \\ \text{symm} & & \mathbf{m}_{ff} \end{bmatrix} \tag{2.58}$$

where submatrices $\mathbf{M_{RR}}$ and $\mathbf{M_{ff}}$ are constant and are associated with translation reference and elastic coordinate, respectively. Submatrix $\mathbf{M_{\theta\theta}}$ is associated with the rotational inertia tensor. Submatrix $\mathbf{M_{R\theta}}$ is associated with inertia coupling between translation and rotation. Submatrices $\mathbf{M}_{\mathbf{Rf}}$ and $\mathbf{M}_{\theta\mathbf{f}}$ are associated with inertia coupling between reference motion and elastic deformation. These submatrices can be calculated as:

$$\begin{cases} \mathbf{m}_{RR} = \int_V \rho \mathbb{R}^{\mathrm{T}} \mathbb{R} \mathrm{d}V \cong \mathbb{R}^{\mathrm{T}} \mathbb{R} \sum_{p=1}^{N_n} m^p = \mathbb{R}^{\mathrm{T}} \mathbb{R} m \\ \mathbf{m}_{R\theta} = \int_V \rho \mathbb{R}^{\mathrm{T}} \mathbf{B} \ \mathrm{d}V \cong -2\mathbb{R}^{\mathrm{T}} \mathbf{A} \sum_{p=1}^{N_n} m^p \tilde{\overline{\boldsymbol{u}}}^p \overline{\mathbf{G}} \\ \mathbf{m}_{Rf} = \int_V \rho \mathbb{R}^{\mathrm{T}} \mathbf{A} \boldsymbol{\Phi}_{\mathrm{t}}^p \mathrm{d}V \cong \mathbb{R}^{\mathrm{T}} \mathbf{A} \sum_{p=1}^{N_n} m^p \boldsymbol{\Phi}_{\mathrm{t}}^p \\ \mathbf{m}_{\theta\theta} = \int_V \rho \mathbf{B}^{\mathrm{T}} \mathbf{B} \ \mathrm{d}V \cong 4\overline{\mathbf{G}}^T \sum_{p=1}^{N_n} m^p \tilde{\overline{\boldsymbol{u}}}^{p\mathrm{T}} \tilde{\overline{\boldsymbol{u}}}^p \overline{\mathbf{G}} \\ \mathbf{m}_{\theta f} = \int_V \rho \mathbf{B}^{\mathrm{T}} \mathbf{A} \boldsymbol{\Phi}_{\mathrm{t}}^p \ \mathrm{d}V \cong -2\overline{\mathbf{G}}^{\mathrm{T}} \mathbf{A} \sum_{p=1}^{N_n} m^p \tilde{\overline{\boldsymbol{u}}}^{p\mathrm{T}} \boldsymbol{\Phi}_{\mathrm{t}}^p \\ \mathbf{m}_{ff} = \int_V \rho \boldsymbol{\Phi}_{\mathrm{t}}^{p\mathrm{T}} \boldsymbol{\Phi}_{\mathrm{t}}^p \ \mathrm{d}V \cong \sum_{p=1}^{N_n} m^p \boldsymbol{\Phi}_{\mathrm{t}}^{p\mathrm{T}} \boldsymbol{\Phi}_{\mathrm{t}}^p \end{cases} \tag{2.59}$$

where m is the mass of the flexible body.

In the multibody system dynamics formulation, equations of motion of the flexible body have the form:

$$\begin{cases} \mathbf{M}\ddot{\boldsymbol{p}} + \mathbf{D}\dot{\boldsymbol{p}} + \mathbf{K}\boldsymbol{p} + \boldsymbol{C}_{\boldsymbol{p}}^{\mathrm{T}} \boldsymbol{\lambda} = \boldsymbol{Q}_{\mathrm{e}} + \boldsymbol{Q}_{\mathrm{v}} + \boldsymbol{Q}_{\mathrm{wr}} \\ \boldsymbol{C}(\boldsymbol{p}, t) = \boldsymbol{0} \end{cases} \tag{2.60}$$

where $\mathbf{D}$ is the damping matrix and $\mathbf{K}$ is the stiffness matrix, which are given as:

$$\mathbf{K} = \begin{bmatrix} \mathbf{0} & \mathbf{0} & \mathbf{0} \\ \mathbf{0} & \mathbf{0} & \mathbf{0} \\ \mathbf{0} & \mathbf{0} & \mathbf{K}_{\mathrm{ff}} \end{bmatrix}, \quad \mathbf{D} = \begin{bmatrix} \mathbf{0} & \mathbf{0} & \mathbf{0} \\ \mathbf{0} & \mathbf{0} & \mathbf{0} \\ \mathbf{0} & \mathbf{0} & \mathbf{D}_{\mathrm{ff}} \end{bmatrix} \tag{2.61}$$

in which non-zero submatrices K_{ff} and D_{ff} are:

$$\mathbf{K}_{\mathrm{ff}} = \mathrm{diag}\left(\omega_1^2 \quad \omega_2^2 \quad \cdots \quad \omega_{N_m}^2 \right), \ \mathbf{D}_{\mathrm{ff}} = \mathrm{diag}\left(2\zeta_1\omega_1 \quad 2\zeta_2\omega_2 \quad \cdots \quad 2\zeta_{N_m}\omega_{N_m} \right) \tag{2.62}$$

where ω_i is the ith natural frequency of the flexible body for index $i = 1, 2, \ldots, N_m$, and ζ_i is the corresponding equivalent viscous damping ratio.

(3) Dynamic Stress Calculation

Based on rigid-flexible coupled dynamics [9], there are three methods for calculating the dynamic stress of the structure: (a) In the first method, the structure is discretized by finite elements and nodes. After the responses of the nodes of all elements are obtained using numerical methods, the strain and stress of an arbitrary point on the structure can be calculated based on strain and stress transformation matrices and the responses of all nodes. (b) In the second method, the strain and stress of the structure under a unit dynamic load can be obtained *from* a finite element software, and then the stress and strain of the structure under arbitrary dynamic loads can be calculated using linear superposition methods. (c) The third method is to directly analyze the transient responses of the structure with a finite element software. This method is computationally expensive and is not used often.

This book introduces a stress and strain superposition-based method for dynamic stress analysis, which is similar to the classical displacement mode superposition method. It is assumed that the structure has small deformations and prestress is not considered. According to finite element theory, the stress and strain of the structure can be calculated as:

$$\boldsymbol{\varepsilon} = \mathbf{B}\boldsymbol{u} \tag{2.63}$$

$$\boldsymbol{\sigma} = \mathbf{E}\mathbf{B}\boldsymbol{u} \tag{2.64}$$

In Equation (2.63) and Equation (2.64), $\mathbf{B}$ is the strain matrix, $\mathbf{E}$ is the elastic matrix, and $\boldsymbol{u}$ is the displacement vector of the nodes. Considering Equation (2.52), Equation (2.63) and Equation (2.64) are rewritten as:

$$\varepsilon = \mathbf{B}\boldsymbol{\Phi}\boldsymbol{q}_f = \boldsymbol{\Phi}_\varepsilon \boldsymbol{q}_f \tag{2.65}$$

$$\boldsymbol{\sigma} = \mathbf{E}\mathbf{B}\boldsymbol{\Phi}\boldsymbol{q}_f = \boldsymbol{\Phi}_\sigma \boldsymbol{q}_f \tag{2.66}$$

In Equation (2.65) and Equation (2.66), $\boldsymbol{\Phi}_\varepsilon$ and $\boldsymbol{\Phi}_\sigma$ are strain modal and stress modal matrices, respectively.

2.2.1.3 Vehicle system dynamics modeling extension

Traditional vehicle system dynamics primarily pays attention to running stability, ride comfort, and curve passing performance of the vehicle. For the development of the high-speed train, more relevant problems deserve more in-depth research. These problems enrich and extend traditional vehicle system dynamics in various aspects, e.g., advanced modeling theories and methods, such as parameter time-varying system modeling; topological time-varying system modeling; and the integration of multi-body dynamics modeling with other modeling methods. For further illustration, some common problems in modern high-speed train dynamics are modeled below.

(1) Fault dynamics modeling [10, 11]

Fault simulation and analysis in traditional vehicle dynamics are usually based on parameters and data obtained from the failed system. To some extent, the simulation is unable to account for transient dynamics of the system at the moment the faults arise. However, the failing vehicle system demonstrates critical behaviors owing to time-varying parameters. Fault dynamics of the vehicle system is modeled and studied thereafter.

Take the rupture of the rubber capsule in the air spring as an example. The rupture of the rubber capsule can lead to the rapid release of gas from the air spring, and consequently, the secondary suspension system would suddenly fail. The emergency spring would then kick in. The entire process is short. The overall stiffness of the secondary suspension system changes drastically, and the corresponding suspension forces change accordingly. This causes a strong impact on the vehicle body and frame and even leads to safety problems. The failing of the air spring in the vehicle system is modeled to simulate this process. The failing model includes three parts: the air spring model that acts normally, the model that accounts for the decay of secondary suspension stiffness during the breakdown, and the failed model of the air spring after the breakdown. The linear three direction force element can be simulated by the air spring model that acts normally. The other two models are discussed below.

1 Modeling the decay of secondary suspension stiffness The degassing of the air spring is the process of the secondary suspension stiffness decaying. The secondary suspension stiffness during this process usually decays in a complex way. Let K_a be the stiffness of the air spring body and K_1 be the stiffness of the emergency rubber. As the air spring acts normally, the equivalent stiffness of the secondary suspension system can be determined as:

$$K_0 = K_a K_1 / (K_a + K_1) \tag{2.67}$$

However, during the breakdown, the equivalent suspension stiffness cannot easily be calculated using Equation (2.67). In physics, the release of gas will lead to contact of the upper wear plate of the air spring with the wear plate of the emergency rubber, and the equivalent suspension stiffness would suddenly change to K_1. Here, it is assumed that the equivalent suspension stiffness during the breakdown varies from K_0 to zero in a nonlinear manner, which depends on two critical time constants t_0 and t_1, as shown in Figure 2.13. Thus, the equivalent stiffness of the secondary suspension system during this process has the following relations:

$$K_s = \begin{cases} K_0 & t \le t_0 \\ K_0\left(1 - \sqrt{1 - \left(\dfrac{t - t_0 - t_e}{t_e}\right)^2}\right) & t_0 < t \le t_0 + t_e \\ K_1 & t > t_0 + t_e \end{cases} \tag{2.68}$$

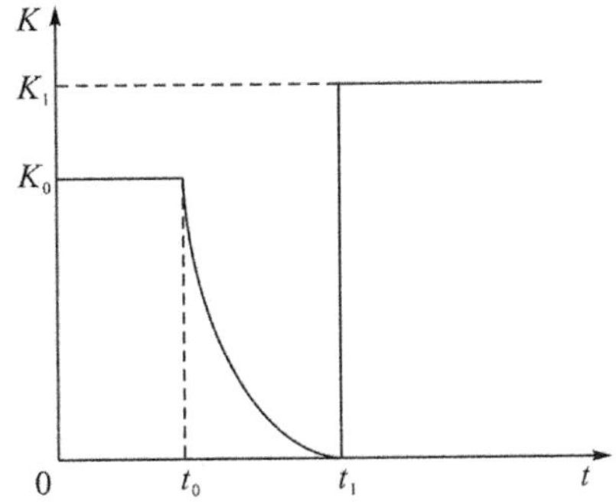

Figure 2.13
Time-varying curve of the equivalent stiffness of the secondary suspension system during the breakdown of the air spring.

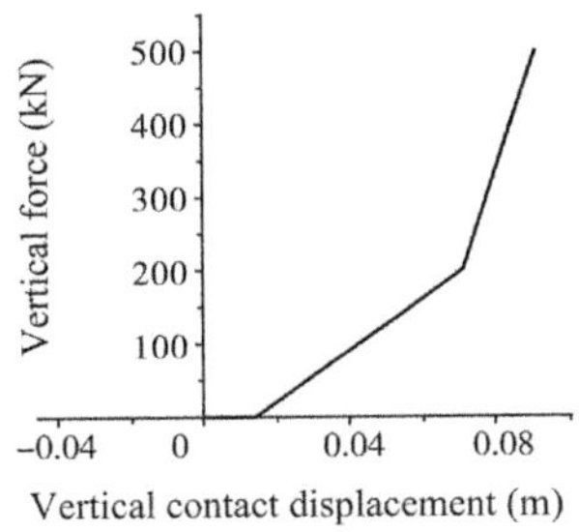

Figure 2.14
Variations of the vertical force of the emergency rubber with respect to contact displacement.

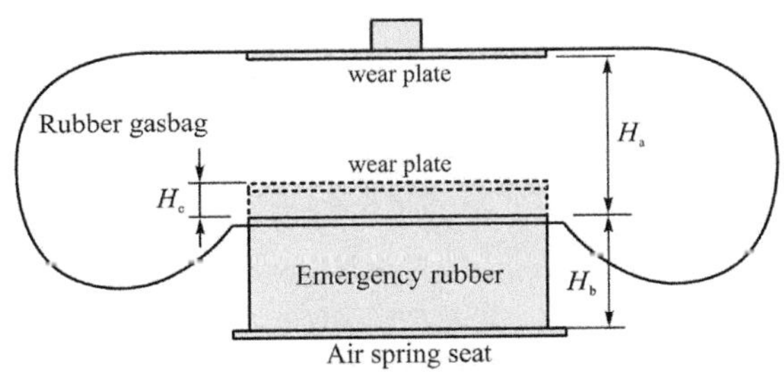

Figure 2.15
Schematic of the air spring.

In the process above, the vertical stiffness of the emergency rubber can be obtained from the curves plotted in Figure 2.14.

2 Stick-slip contact modeling Figure 2.15 shows the schematic of the air spring. H_a is the working height of the air spring, H_b is the working height of the emergency rubber spring, and H_c is the compression of the emergency rubber spring while in operation. As the upper wear plate comes into contact with the wear plate of the emergency spring, the secondary suspension system can be simplified as a stick-slip contact model.

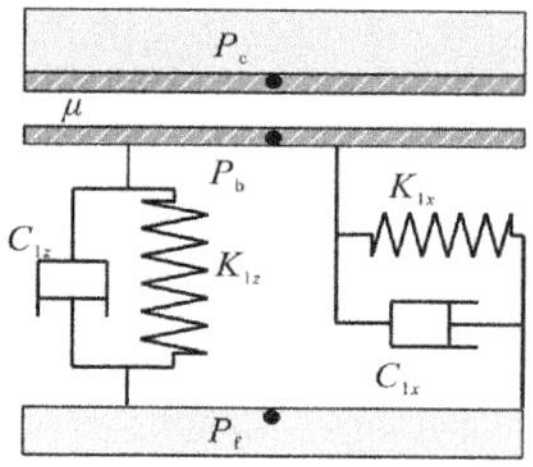

Figure 2.16
Schematic of the stick-slip contact model.

In this model, the longitudinal and lateral relative slides between components of the secondary suspension are considered, and the vertical separation of contacted components is permitted. Therefore, the stick-slip contact model has complex dynamic behaviors. The vertical force of this model is treated as a nonlinear contact force element, while the longitudinal and lateral forces are both equivalent to the spring-damper models, as shown in Figure 2.16. The model parameters are as follows: K_{1z} is the vertical stiffness of the emergency rubber spring.$C_{1z}$is the vertical damping of the emergency rubber spring. K_{1x} is the lateral stiffness of the emergency rubber spring. C_{1x} is the lateral damping of the emergency rubber spring. The emergency rubber spring has the same stiffness and damping in the longitudinal direction as those in the lateral direction. Let μ be the friction coefficient between the down wear plate and the emergency spring. For the calculation, the positions of contact point P_c in the upper wear plate and P_f in the air spring seat are obtained using vehicle dynamic analysis, and the position P_b of the emergency rubber wear plate can be determined from the stick-slip contact model.

Three states exist in the current model, including stick, slip, and separation. The calculations of the vertical contact Fz and transverse force F_x are illustrated in Figure 2.17. As shown in Figure 2.17, Δx is the transverse deformation of the emergency rubber, Δv is the transverse deformation rate of the emergency rubber, v is the transverse

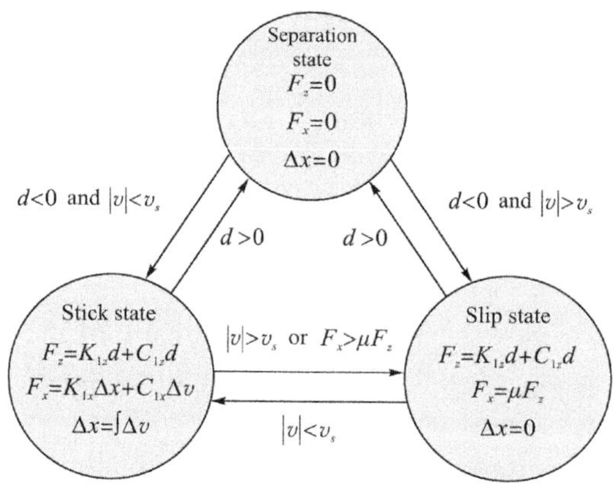

Figure 2.17
Stick-slip contact force element model.

relative velocity between the upper and lower wear plates, and v_s is the critical speed. The critical speed v_s is often small and can be taken as 0.001 m/s. When v is lower than v_s, the force element changes from slip state to stick state. Let d be the vertical compression of the emergency rubber. When $d > 0$, the upper and lower wear plates detach. Assuming that the force element is initially in a stuck state. At every time step of numerical integration, the vertical and transverse forces of the force element can be calculated. See Figure 2.17.

3 Simulation results Fault dynamics, which is caused by the air spring, is simulated and analyzed below. It is assumed that the air spring should be deflated at t = 1 s. The main results obtained are presented in Figure 2.18 and Figure 2.19. It can be concluded from Figure 2.18 and Figure 2.19 that the deflation of the air spring causes the vehicle body to sink suddenly. Curves plotted in the two figures show that the forces of the secondary suspension system change sharply at t = 1 s, vertical contact forces between wheels and rails decrease obviously, and the rate of wheel load reduction suddenly increases and even exceeds the standard limit.

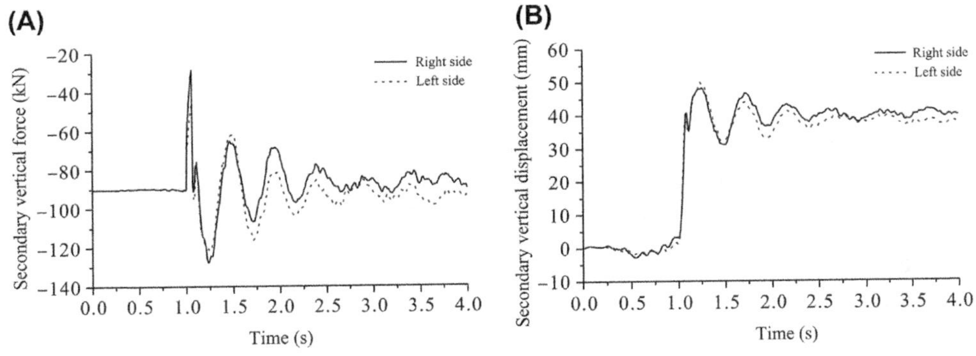

Figure 2.18
Time history for vertical force and displacement of the secondary suspension system.

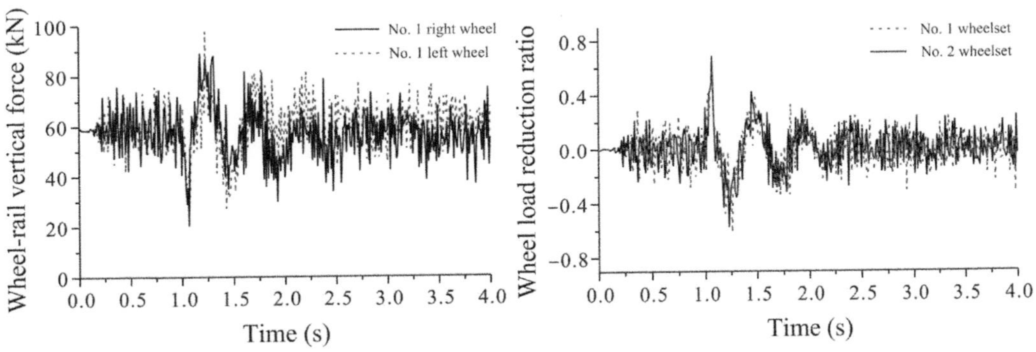

Figure 2.19
Time history for vertical wheel-rail contact force and the rate of wheel load reduction.

(2) Compact modeling using vehicle system dynamics and the discrete element method

Landslides unexpectedly occur anywhere along the railway. The rolling soil, rocks, and earth fall onto and obscure the track, which greatly hampers the running vehicles. It thus constitutes a new research topic on the impact dynamics between the vehicle and obstacles. In the current research for this book, co-modeling is adopted to deal with such complex impact problems. Take a train car rolling accident as an example. The vehicle rests on a gradient track with a mortar rubble line terminal stopper. For unknown reasons, the vehicle slides and eventually impacts the stopper at a speed of 16 km/h. The impact dynamics in this process is analyzed below using a combination of the discrete element method and the multibody system dynamics formulation.

The mortar rubble stopper is made up of rubble and soil. The rubble and soil are treated as spherical particles in the discrete element method. In contrast, the vehicle system is treated as the rigid body. There are contacts, slips, and cohesion between the particles. These effects should be considered in the analysis. Furthermore, the dissipation of the impact energy between the particles should also be included in the current research, because of viscous damping. For these reasons, the constitutive model shown in Figure 2.20 is used to describe the interaction between the particles. In Figure 2.20, F_{nt} is the non-tension connection, F_c is the contact force, F_r is the friction force, and F_{pb} is the parallel constraint force. K_n, D_n, K_s, and D_s are the normal stiffness, normal damping, tangential stiffness, and tangential damping of the contact force element, respectively. μ is the friction coefficient. K_{pbn} and K_{pbs} are the normal and tangential stiffness of the parallel constraint element, respectively.

The front of the head vehicle is much more rigid than the terminal stopper. It can be seen from the fact that the accident vehicle is usually minimally dented in the front. Hence, it can be assumed that the head vehicle would be equivalent to a rigid wall. The analysis is then carried out at discrete time steps by considering the equivalent impact between a rigid moving wall and the particles representing the mortar rubble stopper. The motion of the

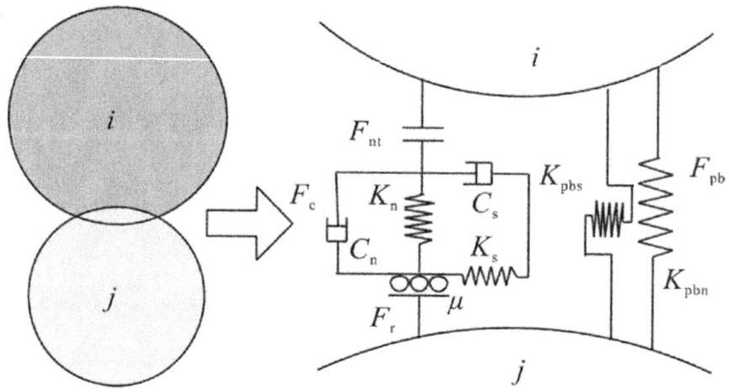

Figure 2.20
The constitutive relation of soil particles.

rigid wall can be obtained using a simulation of the dynamic model of the vehicle system, which has been established earlier in the multibody system dynamics formulation. Once the motion of the rigid wall is known, the impact forces between the rigid wall and the particles can be determined based on the contact theory. The impact forces are transformed into corresponding generalized forces and included in the model of the vehicle system, as well as the models of the particles. The impact analysis is performed at the current time step. Then, the simulation continues to the next time step and ceases at a designated step. The coupled relationship between the models of the vehicle system and the mortar rubble stopper in the analysis is shown in Figure 2.21.

Simulation results are obtained and presented in Figure 2.22 and Figure 2.23. Figure 2.22 shows the impact-induced dynamical processes of the train system and the terminal stopper at certain characterization times. Figure 2.23(a) shows variations in the impact forces between the vehicle system and the terminal stopper in the time domain. Figure 2.23(b) shows variations in the total kinetic energy of the vehicle in the time domain. As shown in Figure 2.23, the impact force is a maximum of 891.55 kN, and the dissipated energy is about 634.7 kJ. Based on the simulation results, further analysis shows that the acceleration of the head vehicle reaches its maximum at 1.614 g, and the vehicle system can travel about 20 m and then cease. Furthermore, the impact forces at the current level cannot produce the damage to the main structure and only a small dent in the front part of the head vehicle, which is in accordance with the measured data in the field.

(3) Derailment dynamics modeling

Recently, significant focus has been placed on derailment criteria and mechanisms. In contrast, the dynamics of the train after derailment seldom formed part of previous studies. Moreover, a two-dimensional reduced model assembled using mass blocks is used in the common analysis and is unable to account for the dynamic interactions between the components of the complex vehicle system and the tracks. To better simulate and

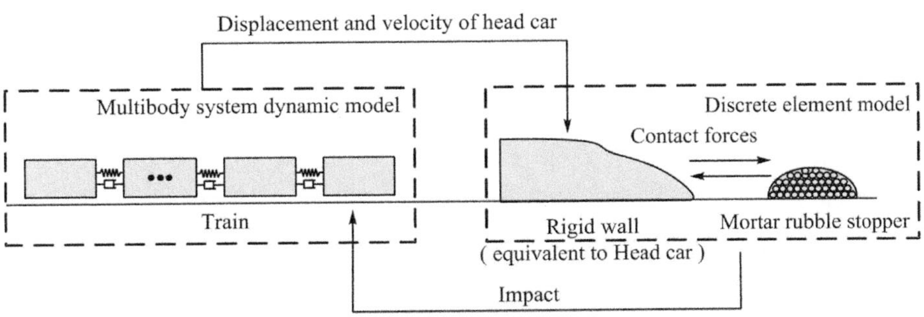

Figure 2.21
Coupled model of the vehicle and discrete particle elements representing the mortar rubble stopper.

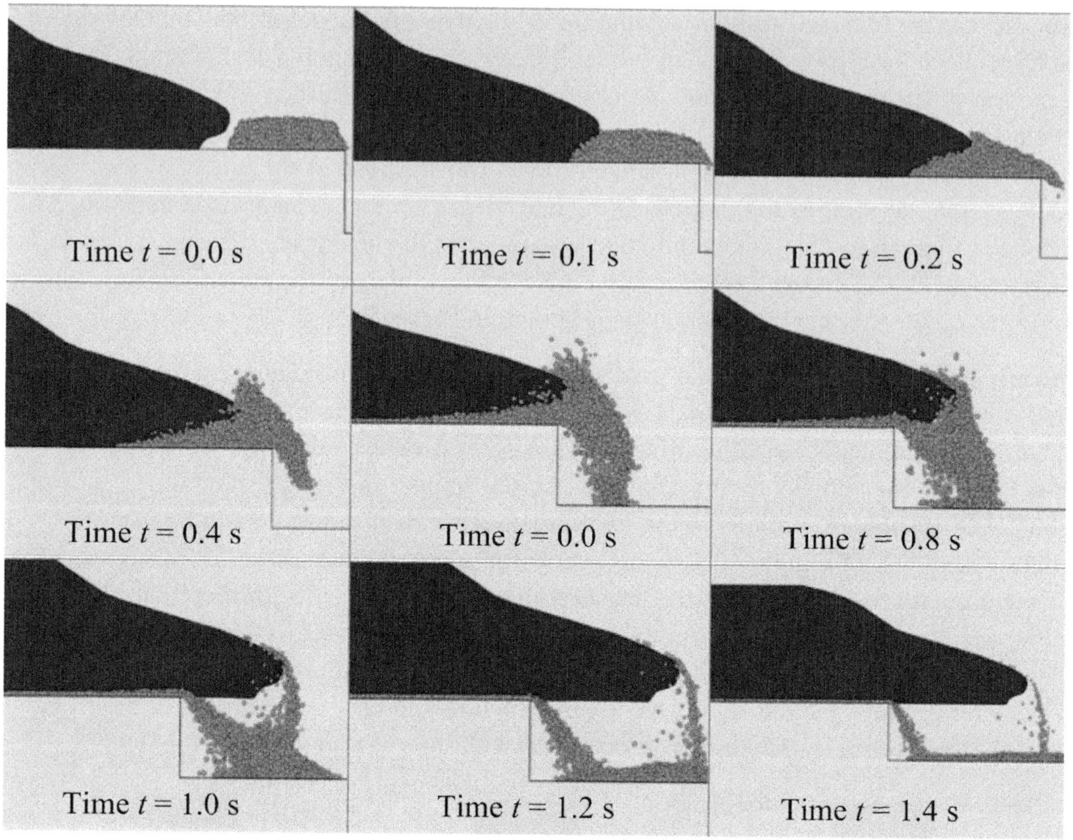

Figure 2.22
Dynamical processes of the train system and the terminal stopper at characterization times.

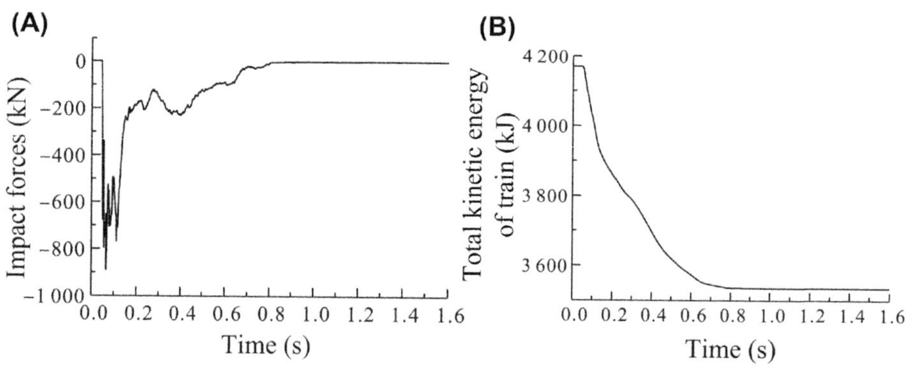

Figure 2.23
Time histories of the impact forces between the train and the terminal stopper, and the total kinetic energy of the train.

understand the dynamics of the system after the derailment, a more accurate model is required and is discussed below.

Derailment is a complex process with variations in topological properties and system parameters. In this process, the wheelsets contact with the rails before derailment, and contact with the slabs after derailment. In general, the wheelsets, the rails, and the slabs have irregular geometric shapes. These conditions result in three-dimensional complex contact behaviors. To facilitate the solution of the contact problems, two sets of the wheelset and the rail are considered.

(1) A wheel-rail contact element is established for the calculation of wheel-rail force using the traditional wheel-rail tread and rail profile. It is irrelevant to the three-dimensional geometry of the wheel-rail surface.
(2) Three-dimensional geometry of the wheelset and the rail is used to calculate the contact force between the wheelset and the rail, and between the wheelset and the slab. For this purpose, contact points are set on these components. With the aid of these points, the contacts are tested and found within the traditional vehicle dynamic simulation. Only the profile of the rail head is needed. However, in the derailment dynamics simulation, the profile needs to be extended so that the contact point can still be found after the derailment, as shown in Figure 2.24. The extension part is lower than the real three-dimensional surface of the track slab, such that contact computations do not cease.

After derailment, the main components of the vehicle come into contact with the track, e.g., contact of the wheel and track plate, wheel and rail, electrical machine and rail, brake disc and rail. Knowing and using the three-dimensional geometry of the wheelsets, motors, rails, gearboxes, motors, and brake discs, the contact forces between these components can be calculated. The derailment criteria are that the tread shifts 70 mm transversely, and the vertical displacement of the tread is 30 mm. At this point, the wheel rim is located on the rail top, and the critical state of the derailment is reached.

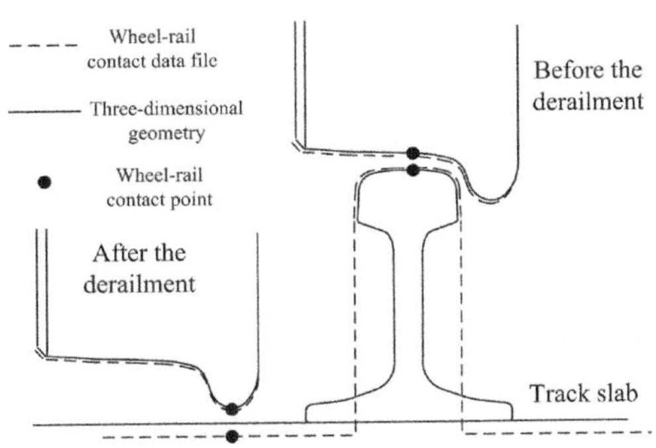

Figure 2.24
Contact model.

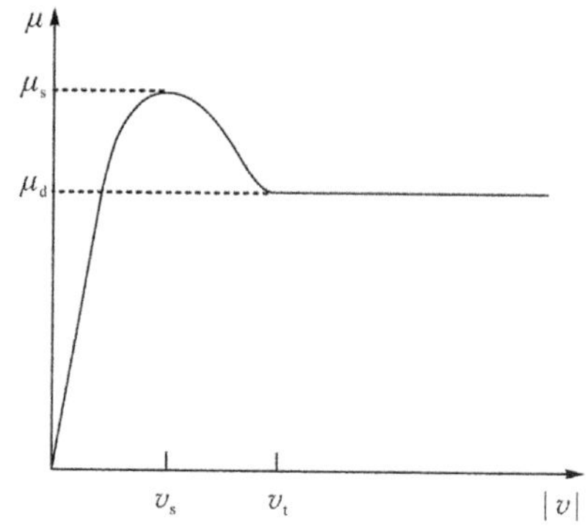

Figure 2.25
Variations of the friction coefficient against the relative velocity between two contact bodies.

During and after the derailment, the contact problems can be addressed with the classical theory of Hertz contact, with consideration given to the hysteresis damping effect. In the Hertz contact theory, the normal force F_N can thus be expressed as:

$$F_N = \begin{cases} K_{Hz}\delta^{3/2}\left(1 + \dfrac{3\left(1-\varepsilon^2\right)}{4}\dfrac{\dot{\delta}}{\dot{\delta}^{(-)}}\right) & \delta > 0 \\ 0 & \delta \le 0 \end{cases} \tag{2.69}$$

where K_{Hz} is the Hertz contact stiffness, δ is the penetration, ε is the restoring coefficient, $\dot{\delta}$ is the approaching velocity of two contact bodies and $\dot{\delta}^{(-)}$ is the approaching velocity of two bodies at the moment that the contact occurs. The tangential force is mainly caused by the relative slide between contact bodies and can be expressed as:

$$F_T = -F_N\mu(|v_s|\widehat{v})_s \tag{2.70}$$

where μ is the friction coefficient, $\mathbf{v}_s$is the relative sliding velocity, and the unit vector $\widehat{\mathbf{v}}_s$ indicates the tangential direction. The friction coefficient μ may vary with the sliding velocity$\mathbf{v}_s$, as shown in Figure 2.25. In this figure, μ_s and μ_d represent the static friction coefficient and sliding friction coefficient, respectively; and V_s and V_t represent the critical velocity associated with the static friction and critical sliding velocity associated with the dynamic friction, respectively.

Using the current model and other relevant models in this book, the dynamics of the train system during and after derailment is analyzed. Figure 2.26 shows the dynamic states of the train system at characterization times. It is pointed out that derailment can lead to variations in the system parameters and even damage to the system. These aspects should be included for the sake of enhanced simulation.

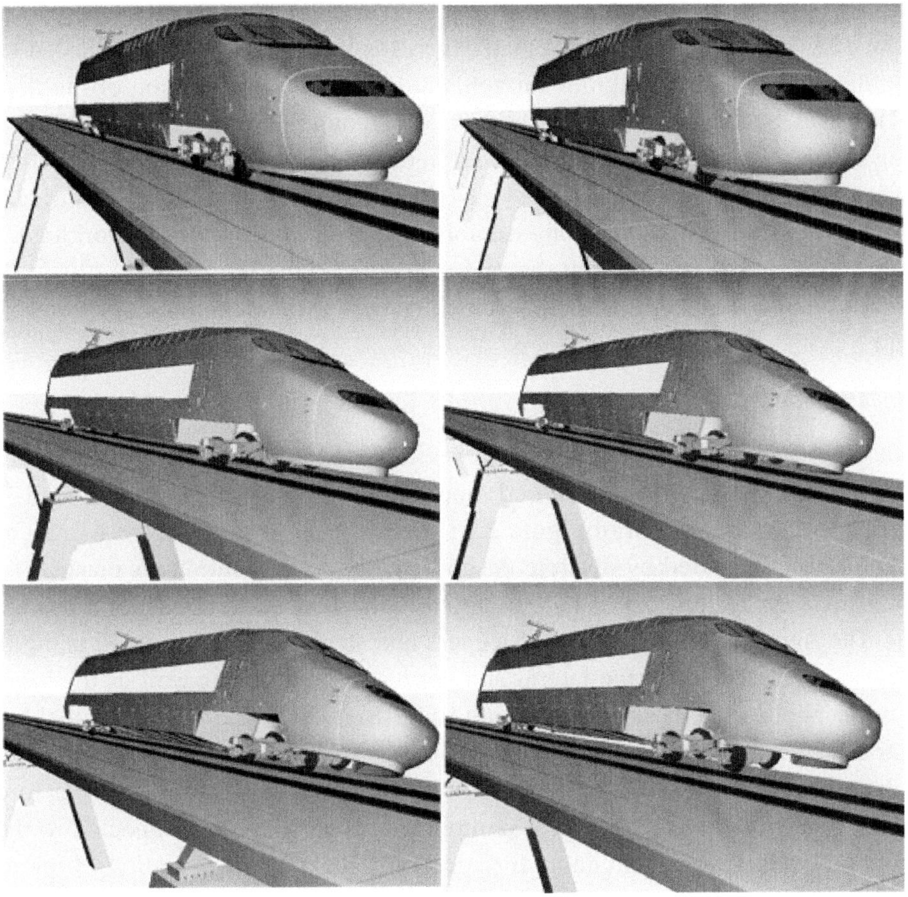

Figure 2.26
Dynamic behaviors of the vehicle during and after derailment at characterization times.

2.2.2 Track system modeling

There are more than a million kilometers of common railway that use ballasted track beds around the world, and the existing major railway lines in China mostly have this same form, which permits a maximum speed of over 250 km/h. In the past, the high-speed track structures in many countries used ballast beds, including parts of the Shinkansen lines in Japan, the main sections of the European high-speed railway, and the Seoul-Busan high-speed line built by South Korea a few years ago. The Beijing-Shanghai high-speed railway in China also has more than 50 kilometers of ballast beds. The ballasted track structures have advantages of good elasticity, good energy absorption performance, and low noise. When used in long tunnels, the ballasted track structures can absorb the energy of the micro-barometer wave and effectively restrain or eliminate the sonic boom phenomenon caused by the high-speed train passing through the tunnel. Furthermore, the construction cost is about 30% lower than that of the ballastless tracks. Therefore, the ballasted track plays a significant role and has significant applications in the construction of high-speed railways. However, according to Japan Railway Maintenance Work statistics, the cost of maintaining ballasted tracks on the Shinkansen line was lower than that for ballastless tracks in the first nine years, but the reverse is true nine years later. Ballastless track excels in good stability, uniform rigidity, strong durability, and low maintenance workload. Thus, ballastless tracks are currently used in the main sections of the high-speed rails. In this book, the ballasted track dynamics model and ballastless track dynamics model are introduced.

2.2.2.1 Ballasted tracks

In the ballasted track model, the track system is simplified as a vibration system that consists of two mass (sleeper and track bed) and a three-layer spring damping (rail-sleeper-track-roadbed), as shown in Figure 2.27. The left and right rails are treated as Timoshenko beams supported by discrete dashpot springs on a continuous elastic foundation. The vertical, transverse, and torsional motions of the Timoshenko beam are considered. The interactions between the sleepers and steel rails are modeled using the damping linear-spring forces in the lateral and vertical directions, as well as the sleepers and track beds. The sleeper is simplified as an Euler-Bernoulli beam. The vertical bending vibration, transverse rigid motion, and torsion motions of the sleepers are considered. The track bed is discretized into rigid mass blocks, which interact with the shear forces characterized by shear stiffness and shear damping. The track bed is connected to the subgrade by the linear springs and dampings, and only the vertical vibration of the track bed is considered.

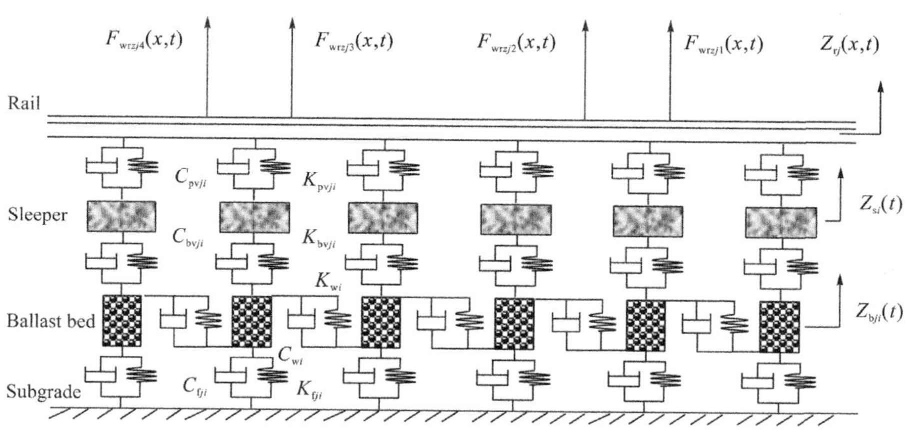

Figure 2.27
Ballasted track model.

From an engineering perspective, the rails have long lengths and are assumed to be infinitely long Timoshenko beams. In the actual analysis, they are often treated as simply supported Timoshenko beams of finite lengths. The effects of the length and boundary condition of the truncated beam on the dynamics of the system are discussed in [12]. As suggested in [12], for the ballast tracks, the length of the truncated beam is no less than 52.8 m, (larger than 88 times the distance between two adjacent sleepers), and the truncated beam is simply supported at two ends. It is assumed that the vehicle travels the tracks at a running speed v. Let $F_{wrzjk}(t)$ denote the wheel-rail interaction force, in which index $j = L$ and R, and index $k = 1, 2, \ldots, Nw$. Here, subscript "L" and "R" are the left and right tracts, respectively. Nw is the number of wheelsets. Let $Rs(t)$ be the support forces, in which index $s = 1, 2, \ldots, Ns$, and Ns is the number of supports over the truncated beam. From this perspective, the equations of motion of the rails are obtained as follows, based on the Timoshenko beam theory [13]:

The equations for the lateral bending motion of the rail can be expressed as:

$$\begin{cases} m\dfrac{\partial^2 y(x,t)}{\partial t^2} + \kappa_y GA\left[\dfrac{\partial \psi_y(x,t)}{\partial x} - \dfrac{\partial^2 y(x,t)}{\partial x^2}\right] = -\sum\limits_{i=1}^{N_s} R_{yi}(t)\delta(x - x_{si}) + \sum\limits_{j=1}^{N_w} F_{wryj}(t)\delta(x - x_{w_j}) \\ \rho I_z \dfrac{\partial^2 \psi_y(x,t)}{\partial t^2} + \kappa_y GA\left[\psi_y(x,t) - \dfrac{\partial y(x,t)}{\partial x}\right] - EI_z\dfrac{\partial^2 \psi_y(x,t)}{\partial x^2} = 0 \end{cases} \tag{2.71}$$

where y is the lateral deflection of the rail, Ψ_y is the slope of the deflection curve of the rail with respect to the y-axis, m is the mass per unit length of the rail, ρ is the density of the rail, A is the cross-sectional area of the rail, I_z is the second moment of the area in relation to the z-axis, G is the shear modulus of the rail, E is Young's modulus of the rail, κ_y is the lateral shear factor, $\delta(x)$ is the Dirac delta function, R_{yi} is the lateral component of the support force R_i, F_{wryj} is the lateral component of the wheel-rail force, x_{si} is the coordinate of the fastener I, and x_{wj} is the coordinate of the wheel j. Similarly, the equations for the vertical bending motion of the rail can be expressed as:

$$\begin{cases} m\dfrac{\partial^2 z(x,t)}{\partial t^2} + \kappa_z GA\left[\dfrac{\partial \psi_z(x,t)}{\partial x} - \dfrac{\partial^2 z(x,t)}{\partial x^2}\right] = -\sum\limits_{i=1}^{N_s} R_{zi}(t)\delta(x - x_{si}) + \\ \qquad \sum\limits_{j=1}^{N_w} F_{wrzj}(t)\delta(x - x_{w_j}) \\ \rho I_y \dfrac{\partial^2 \psi_z(x,t)}{\partial t^2} + \kappa_z GA\left[\psi_z(x,t) - \dfrac{\partial z(x,t)}{\partial x}\right] - EI_y\dfrac{\partial^2 \psi_z(x,t)}{\partial x^2} = 0 \end{cases} \tag{2.72}$$

where z is the vertical deflection of the rail, Ψ_z is the slope of the deflection curve of the rail with respect to the z-axis, κ_z is the vertical shear factor, I_y is the second moment of the area in relation to the y-axis, R_{zi} is the vertical component of the support force R_i, and F_{wrzj} is the vertical component of the wheel-rail force. Finally, the equations for the torsional motion of the rail can be expressed as:

$$\rho I_0 \frac{\partial^2 \phi(x,t)}{\partial t^2} - GK \frac{\partial^2 \phi(x,t)}{\partial x^2} = -\sum_{i=1}^{N_s} M_{si}(t)\delta(x - x_{si}) + \sum_{j=1}^{N_w} M_{Gj}(t)\delta(x - x_{w_j}) \tag{2.73}$$

where ϕ_r is the torsional deflection of the rail, I_0 is the polar moment of inertia of the rail cross-section, M_{si} and M_{Gj} are the equivalent moments on the rail, and K is a constant. The shear factors κ_y and κ_z used in the analysis are 0.4507 and 0.5329, respectively.

Using the method of separation of variables and considering the normalized shape functions, the following equations can be obtained for the lateral, vertical, and torsional motions:

$$y(x,t) = \sum_{k=1}^{NMY} Y_k(x) q_{yk}(t) \tag{2.74}$$

$$\psi_y(x,t) = \sum_{k=1}^{NMY} \psi_{yk}(x) w_{yk}(t) \tag{2.75}$$

$$z(x,t) = \sum_{k=1}^{NMZ} Z_k(x) q_{zk}(t) \tag{2.76}$$

$$\psi_z(x,t) = \sum_{k=1}^{NMZ} \psi_{zk}(x) w_{zk}(t) \tag{2.77}$$

$$\phi(x,t) = \sum_{k=1}^{NMT} \Phi_k(x) q_{Tk}(t) \tag{2.78}$$

In the preceding equations, $q_{yk}(t)$, $q_{zk}(t)$, and $q_{Tk}(t)$ are the generalized coordinates in relation to the lateral, vertical, and torsional motions of the rail, respectively. $w_{yk}(t)$ and $w_{zk}(t)$ are the generalized coordinates in relation to the rail cross-sectional rotations about the y and z axes, respectively. $Y_k(x)$, $Z_k(x)$, and $\Phi_k(x)$ are the kth normalized modal shape functions of lateral bending, vertical bending, and torsion of the rail, respectively. $\psi_{yk}(x)$ and $\psi_{zk}(x)$are the kth normalized modal shape functions in relation to the rail cross-sectional rotations about the y and z axes, respectively. NMY, NMZ, and NMT are the total number of mode coordinates used in the analysis. In Equation (2.74) to Equation (2.74), the normalized shape functions can be given as [14]:

$$Y_k(x) = \sqrt{\frac{2}{m l_{\text{tim}}}} \sin\left(\frac{k\pi}{l_{\text{tim}}} x\right) \tag{2.79}$$

$$\Psi_{yk}(x) = \sqrt{\frac{2}{\rho I_y l_{\text{tim}}}} \cos\left(\frac{k\pi}{l_{\text{tim}}} x\right) \tag{2.80}$$

$$Z_k(x) = \sqrt{\frac{2}{ml_{\text{tim}}}}\sin\left(\frac{k\pi}{l_{\text{tim}}}x\right) \tag{2.81}$$

$$\Psi_{zk}(x) = \sqrt{\frac{2}{\rho I_z l_{\text{tim}}}}\cos\left(\frac{k\pi}{l_{\text{tim}}}x\right) \tag{2.82}$$

$$\Phi_k(x) = \sqrt{\frac{2}{\rho I_0 l_{\text{tim}}}}\sin\left(\frac{k\pi}{l_{\text{tim}}}x\right) \tag{2.83}$$

where l_{tim} is the length of the beam used in the analysis.

Using the orthogonality of normalized modes and the properties of the Dirac delta function, the lateral, vertical, and torsional motions of the Timoshenko beam can be governed by ordinary second-order differential equations in normalized modal coordinates.

In the lateral direction, motion equations of the beam can be expressed as:

$$\begin{cases} \ddot{q}_{yk}(t) + \dfrac{\kappa_y GA}{m}\left(\dfrac{i\pi}{l_{\text{tim}}}\right)^2 q_{yk}(t) + \kappa_y GA\dfrac{i\pi}{l_{\text{tim}}}\sqrt{\dfrac{1}{m\rho I_z}}w_{yk}(t) = \\ \qquad -\displaystyle\sum_{i=1}^{N_s} R_{yi}(t)Y_k(x_{si}) + \sum_{j=1}^{N_w} F_{wryj}(t)Y_k(x_{w_j}) \\ \ddot{w}_{yk}(t) + \left[\dfrac{\kappa_y GA}{\rho I_z} + \dfrac{EI_z}{\rho I_z}\left(\dfrac{i\pi}{l_{\text{tim}}}\right)^2\right]w_{yk}(t) - \kappa_y GA\dfrac{i\pi}{l_{\text{tim}}}\sqrt{\dfrac{1}{m\rho I_z}}q_{yk}(t) = 0 \quad (k = 1 \sim NMY) \end{cases} \tag{2.84}$$

In the torsional direction, motion equations of the beam can be expressed as:

$$\ddot{q}_{Tk}(t) + \frac{GK}{\rho I_0}\left(\frac{i\pi}{l_{\text{tim}}}\right)^2 q_{Tk}(t) = -\sum_{i=1}^{N_s} M_{si}(t)\Phi_K(x_{si}) + \sum_{j=1}^{N_w} M_{Gj}(t)\Phi_k(x_{w_j}) \quad (k - 1 \sim NMT) \tag{2.85}$$

In the vertical direction, motion equations of the beam can be expressed as:

$$\begin{cases} \ddot{q}_{zk}(t) + \dfrac{\kappa_z GA}{m}\left(\dfrac{i\pi}{l_{\text{tim}}}\right)^2 q_{zk}(t) + \kappa_z GA\dfrac{i\pi}{l_{\text{tim}}}\sqrt{\dfrac{1}{m\rho I_y}}w_{zk}(t) = \\ \qquad -\displaystyle\sum_{i=1}^{N_s} R_{zi}(t)Z_k(x_{si}) + \sum_{j=1}^{N_w} F_{wrzj}(t)Z_k(x_{w_j}) \\ \ddot{w}_{zk}(t) + \left[\dfrac{\kappa_z GA}{\rho I_y} + \dfrac{EI_y}{\rho I_y}\left(\dfrac{i\pi}{l_{\text{tim}}}\right)^2\right]w_{zk}(t) - \kappa_z GA\dfrac{i\pi}{l_{\text{tim}}}\sqrt{\dfrac{1}{m\rho I_y}}q_{zk}(t) = 0 \quad (k = 1 \sim NMZ) \end{cases} \tag{2.86}$$

In previous track models, the sleeper is treated as a rigid mass with lateral, vertical, and torsional degrees of freedom. These track models can account for the frequency response characteristics of the sleeper in the lateral and vertical directions. To consider the effect of higher frequencies of the sleeper, it is modeled using the Euler-Bernoulli beam. The vertical bending, lateral, and torsional motions of the sleeper are considered. The vertical motion equation of the sleeper is:

$$E_s I_s \frac{\partial^4 z_s(y,t)}{\partial y^4} + \frac{M_s}{l_s}\frac{\partial^2 z_s(y,t)}{\partial t^2} = -F_{bzLi}\delta(y+d_{sLeq}) - F_{bzRi}\delta(y-d_{sReq}) + R_{zLi}(t)\delta(y+d_r) + R_{zRi}(t)\delta(y-d_r) \tag{2.87}$$

where z_s is the vertical displacement of the sleeper; $E_s I_{s,}$ is the bending stiffness of the sleeper; $M_{s,}$ is the total mass of the sleeper; l_s is the length of the sleeper; dr is half the distance between the left and right rails; R_{zLi} and R_{zRi} are the forces of the left and right rails, respectively, acting on the sleeper I; F_{bzLi} and F_{bzRi} are the interaction forces between the sleeper i and the left and right track beds, respectively; d_{sLeq} and d_{sReq} are the equivalent distances between the center of the sleeper to the applied points of the forces F_{bzLi} and F_{bzRi}, respectively. F_{bzLi} and F_{bzRi} are expressed as:

$$F_{bzLi} = \int_{-l_s/2}^{0} k_{beq}[z_{si}(y,t) - Z_{bLi}(t)]dy + \int_{-l_s/2}^{0} c_{beq}\left[\dot{z}_{si}(y,t) - \dot{Z}_{bLi}(t)\right]dy \tag{2.88}$$

$$F_{bzRi} = \int_{0}^{l_s/2} k_{beq}[z_{si}(y,t) - Z_{bRi}(t)]dy + \int_{0}^{l_s/2} c_{beq}\left[\dot{z}_{si}(y,t) - \dot{Z}_{bRi}(t)\right]dy \tag{2.89}$$

where Z_{bLi} and Z_{bRi} are the vertical displacements of the left and right track beds, respectively, and k_{beq} and c_{beq} are the equivalent damping and stiffness per unit area of the track beds, as shown in Figure 2.28.

The vertical response of the sleeper consists of two vibrating components, including the rigid motion modes and flexible bending modes of the high frequencies. Thus, the vertical vibration response of the sleeper can be expressed using the mode superposition method [5] as:

$$z_{si}(y,t) = \sum_{k=1}^{NMSR} Z_{sRk}(y)T_{sRk}(t) + \sum_{k=1}^{NMSE} Z_{sEk}(y)T_{sEk}(t) \tag{2.90}$$

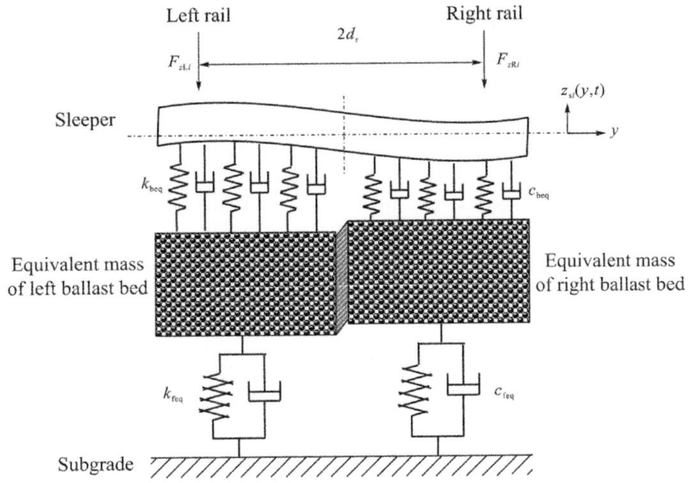

Figure 2.28
Euler-Bernoulli beam model of the sleeper.

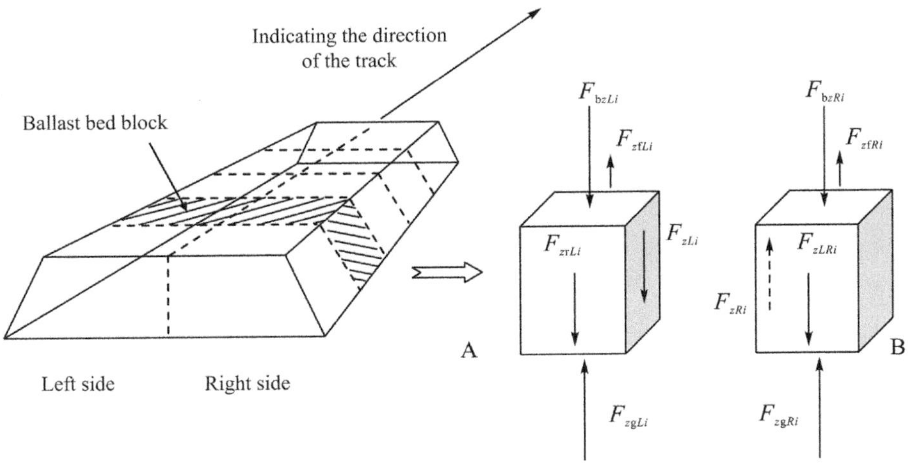

Figure 2.29
Force balance diagram of the track bed.

where *NMSR* and *NMSE*, respectively, are the numbers of rigid and flexible bending modes of the sleeper, $T_{sRk}(t)$ and $T_{sEk}(t)$ are the corresponding normalized coordinates, and $Z_{sRk}(y)$ and $Z_{sEk}(y)$ are the kth modal shape functions given as:

$$\begin{cases} Z_{sRk}(y) = \sqrt{\dfrac{1}{M_s}} \quad (k=1) \\ Z_{sRk}(y) = \sqrt{\dfrac{3}{M_s}}\left(1-\dfrac{2y}{l_s}\right) \quad (k=2) \\ Z_{sEk}(y) = \sqrt{\dfrac{1}{M_s}}[(\cosh \alpha_k y + \cos \alpha_k y) - C_k(\sinh \alpha_k y + \sin \alpha_k y)] \quad (k=1,2,\cdots,NMSE) \end{cases} \tag{2.91}$$

where α_k and C_k are two coefficients associated with the vibration frequency response characteristics and modal shapes of the sleeper at free boundaries [5]. The lateral motion equation of the sleeper i is expressed as:

$$M_s \ddot{Y}_{si} = (F_{yLi} + F_{yRi}) - F_{ysbi} \tag{2.92}$$

where F_{yLi} is the lateral force between the sleeper i and the left rail, F_{yRi} is the lateral force between the sleeper i and the right rail, and F_{ysbi} is the equivalent lateral support force of the track bed. The yaw, pitch, and longitudinal motions of the sleeper are not considered.

It is noted that the vibration responses of the track structures with frequencies below 100 Hz can be better simulated using the rigid mass model of the sleeper. In contrast, the use of the Euler-Bernoulli beam model of the sleeper extends the simulation to a frequency range below 600 Hz. For the wheel-rail noise, the components of frequencies below 500 Hz are produced primarily by rail-sleeper interaction [15]. Furthermore, the torsion and shear deformation of the sleeper contribute to little towards its dynamics. Therefore, the sleepers are best modeled using Euler-Bernoulli beams instead of Timoshenko beams.

It remains difficult to fully model the track bed. For the sake of simplicity and addressing the main dynamic behaviors, the track bed is treated as a rigid mass with a vertical degree of freedom, as shown in Figure 2.30. Thus, the vertical motion equation of the left track bed can be expressed as:

$$M_{bL}\ddot{Z}_{bLi} = F_{bzLi} + F_{zrLi} + F_{zLRi} - F_{zgLi} - F_{zfLi} + M_{bL}g \tag{2.93}$$

where Z_{bLi} is the vertical displacement of the ith left track bed; M_{bL} is the mass of each left track bed; F_{bzLi}, F_{zrLi}, F_{zLRi}, and F_{zfLi} are the vertical shear forces between the adjacent left track beds; F_{zgLi} is the interaction force between the subgrade and the left

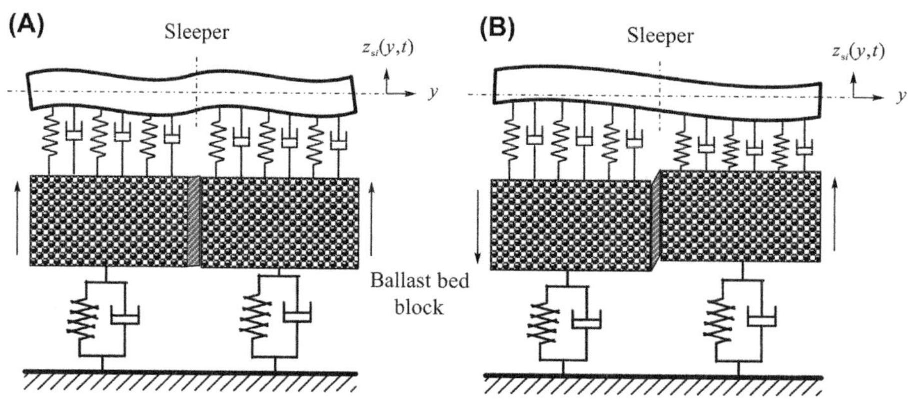

Figure 2.30
The track bed model.

track bed; and g is gravitational acceleration. Similarly, the vertical motion equation of the right track bed is expressed as:

$$M_{\mathrm{bR}}\ddot{Z}_{\mathrm{bR}i} = F_{\mathrm{bzR}i} + F_{\mathrm{zrR}i} + F_{\mathrm{zLR}i} - F_{\mathrm{zgR}i} - F_{\mathrm{zfR}i} + M_{\mathrm{bR}}g \tag{2.94}$$

where $Z_{\mathrm{bR}i}$ is the vertical displacement of the ith right track bed; M_{bR} is the mass of each right track bed; $F_{b\mathrm{zR}i}$, $F_{z\mathrm{rR}i}$, and $F_{z\mathrm{fR}i}$ are the vertical shear forces between the adjacent right track beds; $F_{z\mathrm{gR}i}$ is the interaction force between the subgrade and the right track bed. The current model of the track bed can account for the torsional motion, as shown in Figure 2.30.

2.2.2.2 Ballastless track on the embankment

The ballastless track has advantages of good stability, uniform stiffness, strong durability, and low maintenance. It is thus used in modern high-speed rails. The literature [6] establishes a three-dimensional dynamic model of the ballastless track slab based on plate theory. The vertical responses of the slab traversed by the train are calculated using the modal superposition method while its lateral vibration is considered as rigid motion. In this book, the slab is modeled using three-dimensional solid finite elements, which can simultaneously take into consideration the vertical, lateral, and coupled vibrations. The model of the slab used is shown in Figure 2.31. The track slab model used is

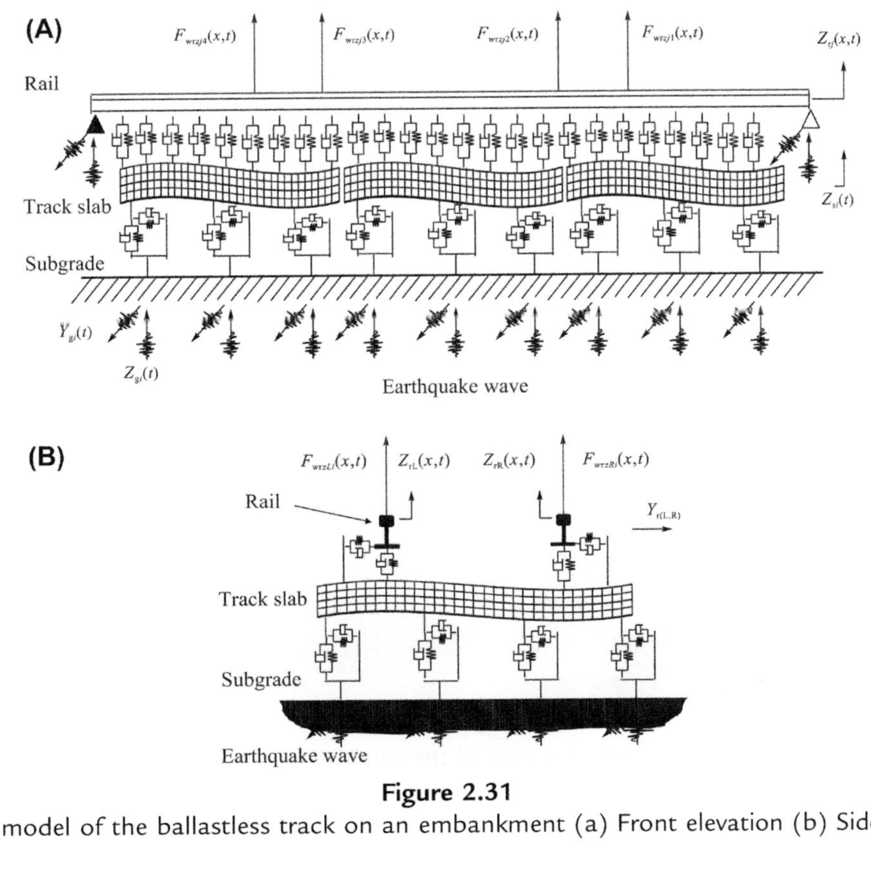

Figure 2.31
Dynamic model of the ballastless track on an embankment (a) Front elevation (b) Side elevation.

shown in Figure 2.31. Using this model, the dynamic behavior of the high-speed vehicle and track, and the running safety of high-speed trains under earthquake conditions can be studied [14].

For the ballastless track system, the track is mainly composed of the rail, the fastener, the track slab, and the cement–emulsified asphalt mortar (CA mortar) layer. In a computational model of the slab on the Chinese high-speed railway, as shown in Figure 2.32, the left and right rails are treated as Timoshenko beams supported by discrete fasteners on an elastic foundation, which include the vertical, lateral, and torsional vibrations of the rails. The slabs are modeled using three-dimensional solid finite elements. The fasteners and the CA mortar layer are considered as uniformly viscoelastic elements.

In the dynamic analysis of the ballastless track system, the suggested length of the rail is 65.0 m [12], covering 100 sleepers and 10 track slabs. The truncated rail is simply supported at two ends. It is assumed that the analyzed structure is uniformly exposed to earthquake excitation. The motion equations of the rail under earthquake excitation can be given as follows: The lateral motion equations of the rail under earthquake excitation are expressed as:

$$\begin{cases} \rho A\dfrac{\partial^2 y}{\partial t^2}+\kappa_y GA\left(\dfrac{\partial\psi_y}{\partial x}-\dfrac{\partial^2 y}{\partial x^2}\right)=-\displaystyle\sum_{i=1}^{N_s}R_{yi}(t)\delta(x-x_{si}) \\ \qquad\qquad\qquad\qquad\qquad +\displaystyle\sum_{j=1}^{N_w}F_{wryi}(t)\delta(x-x_{pj})-\rho A\dfrac{d^2Y_g(t)}{dt^2} \\ \rho I_z\dfrac{\partial^2\psi_y}{\partial t^2}-EI_z\dfrac{\partial^2\psi_y}{\partial x^2}+\kappa_y GA\left(\psi_y-\dfrac{\partial y}{\partial x}\right)=0 \end{cases} \tag{2.95}$$

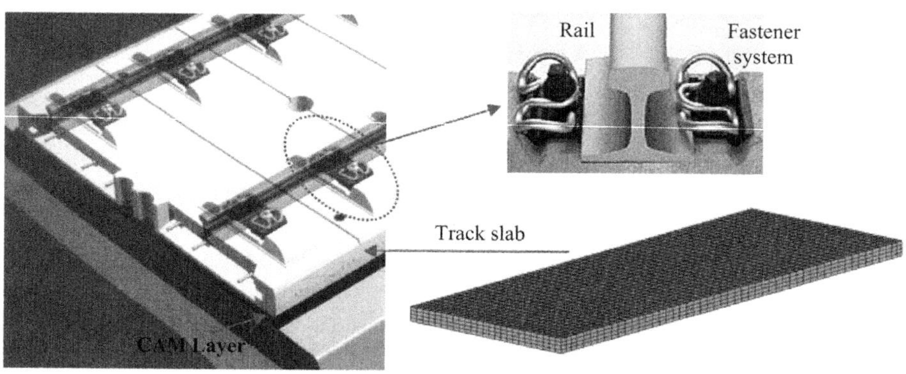

Figure 2.32
Common track slab structure of the Chinese high-speed railway.

The vertical motion equations of the rail under earthquake excitation are expressed as:

$$\begin{cases} \rho A\dfrac{\partial^2 z}{\partial t^2} + \kappa_z GA\left(\dfrac{\partial\psi_z}{\partial x} - \dfrac{\partial^2 z}{\partial x^2}\right) = -\sum_{i=1}^{N_s} R_{zi}(t)\delta(x-x_{si}) \\ \qquad + \sum_{j=1}^{N_w} F_{wrzj}(t)\delta(x-x_{p_j}) - \rho A\dfrac{d^2 Z_g(t)}{dt^2} \\ \rho I_y\dfrac{\partial^2\psi_z}{\partial t^2} - EI_y\dfrac{\partial^2\psi_z}{\partial x^2} + \kappa_z GA\left(\psi_z - \dfrac{\partial z}{\partial x}\right) = 0 \end{cases} \tag{2.96}$$

The torsional motion equations of the rail under earthquake excitation are expressed as:

$$\rho I_0\frac{\partial^2\phi}{\partial t^2} - GK\frac{\partial^2\phi}{\partial x^2} = -\sum_{i=1}^{N_s} M_{si}(t)\delta(x-x_{si}) + \sum_{j=1}^{N_w} M_{Gj}(t)\delta(x-x_{p_j}) \tag{2.97}$$

In the Equation (2.95), (2-96), and (2-97), $Y_g(t)$ and $Z_g(t)$ represent the lateral and vertical components, respectively, of the excitations and other variables and parameters defined earlier. Using the mode superposition method, the three equations can be rewritten in generalized coordinates as follows:

The lateral motion equations of the rail are:

$$\begin{cases} \ddot{q}_{yk}(t) + \dfrac{\kappa_y GA}{m}\left(\dfrac{k\pi}{l_{\text{tim}}}\right)^2 q_{yk}(t) - \kappa_y GA\dfrac{k\pi}{l_{\text{tim}}}\sqrt{\dfrac{1}{m\rho I_z}}w_{yk}(t) = -\sum_{i=1}^{N_s} R_{yi}(t)Y_k(x_{si}) \\ \qquad + \sum_{j=1}^{N_w} F_{wryi}(t)Y_k(x_{p_j}) + \rho A\ddot{Y}_g(t)\int_0^L Y_k(x)dx \\ \ddot{w}_{yk}(t) + \left[\dfrac{\kappa_y GA}{\rho I_z} + \dfrac{EI_z}{\rho I_z}\left(\dfrac{k\pi}{l_{\text{tim}}}\right)^2\right]w_{yk}(t) - \kappa_y GA\dfrac{k\pi}{l_{\text{tim}}}\sqrt{\dfrac{1}{m\rho I_z}}q_{yk}(t) = 0 \quad (k = 1\sim\text{NMY}) \end{cases} \tag{2.98}$$

The vertical motion equations of the rail are:

$$\begin{cases} \ddot{q}_{zk}(t) + \dfrac{\kappa_z GA}{m}\left(\dfrac{k\pi}{l_{\text{tim}}}\right)^2 q_{zk}(t) - \kappa_z GA\dfrac{k\pi}{l_{\text{tim}}}\sqrt{\dfrac{1}{m\rho I_y}}w_{zk}(t) = \\ -\sum_{i=1}^{N_s} R_{zi}(t)Z_k(x_{si}) + \sum_{j=1}^{N_w} F_{wrzj}(t)Z_k(x_{p_j}) + \rho A\ddot{Z}_g(t)\int_0^L Z_k(x)dx \\ \ddot{w}_{zk}(t) + \left[\dfrac{\kappa_y GA}{\rho I_y} + \dfrac{EI_y}{\rho I_y}\left(\dfrac{k\pi}{l_{\text{tim}}}\right)^2\right]w_{zk}(t) - \kappa_z GA\dfrac{k\pi}{l_{\text{tim}}}\sqrt{\dfrac{1}{m\rho I_y}}q_{zk}(t) = 0 \quad (k = 1\sim\text{NMZ}) \end{cases} \tag{2.99}$$

The torsional motion equations of the rail are:

$$\ddot{q}_{Tk}(t)+\frac{GK}{\rho I_0}\left(\frac{k\pi}{l_{\text{tm}}}\right)^2 q_{Tk}(t)=-\sum_{i=1}^{N_s} M_{\text{s}i}(t)\Phi_k(x_{\text{s}i})+\sum_{j=1}^{N_w} M_{\text{G}j}(t)\Phi_k\left(x_{\text{p}_j}\right)\quad (k=1\sim\text{NMT}) \tag{2.100}$$

The finite element model of the track slab is shown in Figure 2.32. The CA mortar layer is treated as planar continuous spring damping. The motion equations of the track slab in global coordinates can be expressed as:

$$[M]\{\ddot{x}\}_i+[C]\{\dot{x}\}_i+[K]\{x\}_i=\{F^{\text{rs}}\}_i+\{F^{\text{g}}\}_i\quad (i=1\sim N_{\text{slab}}) \tag{2.101}$$

where $[M]$, $[C]$, and $[K]$ are the mass, damping, and stiffness matrices of the track slabs. The subscript i is an index of the track slab. The total number N_{slab} of track slabs is 10 or 11 in the computational analysis. $\{x\}_i$ is the displacement vector of the ith track slab, $\{x^{\text{g}}\}_i$ is the seismically excited displacement vector under the track slab, $\{F^{\text{rs}}\}_i$ is the force vector between the rail and the track slab, and $\{F^{\text{g}}\}_i$ is the force vector between the foundation and the track slab. Considering the foregoing variables, the following equations can be obtained:

$$\{F^{\text{rs}}\}_i=K_{\text{p}}[\{x^{r}\}-\{x\}_i]+C_{\text{p}}[\{\dot{x}^{r}\}-\{\dot{x}\}_i]\quad (i=1\sim N_{\text{slab}}) \tag{2.102}$$

$$\{F^{\text{g}}\}_i=K_{\text{CAM}}[\{x\}_i-\{x^{\text{g}}\}_i]+C_{\text{CAM}}[\{\dot{x}\}_i-\{\dot{x}^{\text{g}}\}_i]\quad (i=1\sim N_{\text{slab}}) \tag{2.103}$$

In Equation (2.102) and (2.103), K_{p} and C_{p} are stiffness and damping of the fastener, respectively; K_{CAM} and C_{CAM} are stiffness and damping, respectively, of the CA mortar layer. $\{F^{\text{rs}}\}_i$ includes the forces R_{yi} and R_{zi} between the sleeper and rail. Using mode superposition method and the modal shape functions of the track slab, the vibration equations of the track slab have the reduced form:

$$M_n\ddot{X}_{in}+C_n\dot{X}_{in}+K_nX_{in}=P_n\quad (i=1\sim N_{\text{slab}},\ n=1\sim N_{\text{mode}}) \tag{2.104}$$

where $\{X\}_i=[X_{1i}, X_{2i}, \ldots, X_{N\text{slab}i}]^T$, and is the vector of the normalized mode coordinates of the ith track slab; M_n, C_n, and K_n are the generalized mass, damping, and stiffness matrices, respectively; and P_n is the generalized force vector of the external loads on the ith track slab. They can be expressed as:

$$\begin{cases} M_n=\{\phi\}_n^{\text{T}}[M]\{\phi\}_n,\quad C_n=\{\phi\}_n^{\text{T}}[C]\{\phi\}_n,\quad K_n=\{\phi\}_n^{\text{T}}[K]\{\phi\}_n \\ P_n=\{\phi\}_n^{\text{T}}\{F^{\text{rs}}\}_i\{\phi\}_n^{\text{T}}\{F^{\text{g}}\}_i\quad (i=1\sim N_{\text{slab}},\ n=1\sim N_{\text{mode}}) \end{cases} \tag{2.105}$$

where $\{\Phi\}_n$ is the nth modal vector, which can be obtained through modal analysis of the track slab using ANSYS. This book considers the first 20 modes ($N_{\text{mode}}=20$), including 6 rigid body modes and 14 elastic modes. Figure 2.33 shows the first three modal

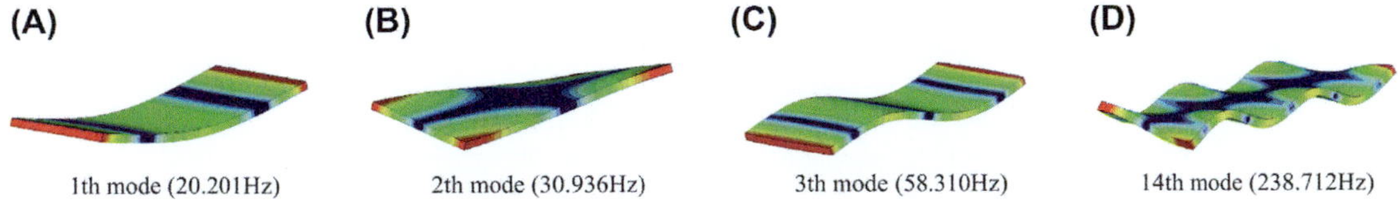

Figure 2.33
Modal deformations of the track slab.

deformations of the slab. Based on vibration theory, the displacement of the track slab i can be expressed as:

$$\{x\}_i = \sum_{n=1}^{N_{\text{mode}}} \{\phi\}_n X_{in} = \{\Phi\}\{X\}_i \quad (i = 1 \sim N_{\text{slab}}) \tag{2.106}$$

2.2.2.3 Ballastless track on a bridge

The modeling of the ballastless track on a bridge is similar to that on an embankment. The only difference between the two structure types is that the former is on a bridge while the latter is on an embankment. The bridge structure under the rail can be modeled using the finite element method and the analytical method.

In the finite element method, the modeling and simulation analysis are performed with consideration given to the actual size, and material properties of the structure, which are similar to those of the track slab modeled using three-dimensional solid elements. The interaction of the track slab with the bridge is addressed by the support forces of the track slab. Figure 2.34 shows a dynamic model of the track slab on the bridge.

The vibration equation of bridge structure (including piers) can be expressed as:

$$M_{bn}\ddot{X}_{bin} + C_{bn}\dot{X}_{bin} + K_{bn}X_{bin} = P_{bn} \quad (i = 1 \sim N_{\text{bridge}},\ n = 1 \sim N_{\text{mode}}) \tag{2.107}$$

where $\{X_b\}_i$ is the displacement vector of the ith bridge structure, M_{bn}, C_{bn}, and K_{bn} are the mass, damping, and stiffness matrices, respectively, and P_{bn} is the vector of the external loads on the ith bridge structure. The mass, damping and stiffness matrices M_{bn}, C_{bn}, and K_{bn}, and the vector P_{bn} are:

$$\begin{cases} M_{bn} = \{\phi_b\}_n^{\mathrm{T}}[M_b]\{\phi_b\}_n, \quad C_{bn} = \{\phi_b\}_n^{\mathrm{T}}[C_b]\{\phi_b\}_n, \quad K_{bn} = \{\phi_b\}_n^{\mathrm{T}}[K_b]\{\phi_b\}_n \\ P_{bn} = \{\phi_b\}_n^{\mathrm{T}}\{F_b^{\mathrm{rs}}\}_i\{\phi_b\}_n^{\mathrm{T}}\left\{F_b^{\mathrm{g}}\right\}_i \quad (i = 1 \sim N_{\text{bridge}},\ n = 1 \sim N_{\text{mode}}) \end{cases} \tag{2.108}$$

where $\{\Phi_b\}_n$ is the nth modal vector and can be obtained using finite element analysis software, e.g., ANSYS. Using Equation (2.95) to Equation (2.18) yields the dynamic responses of the ballastless track on the bridge.

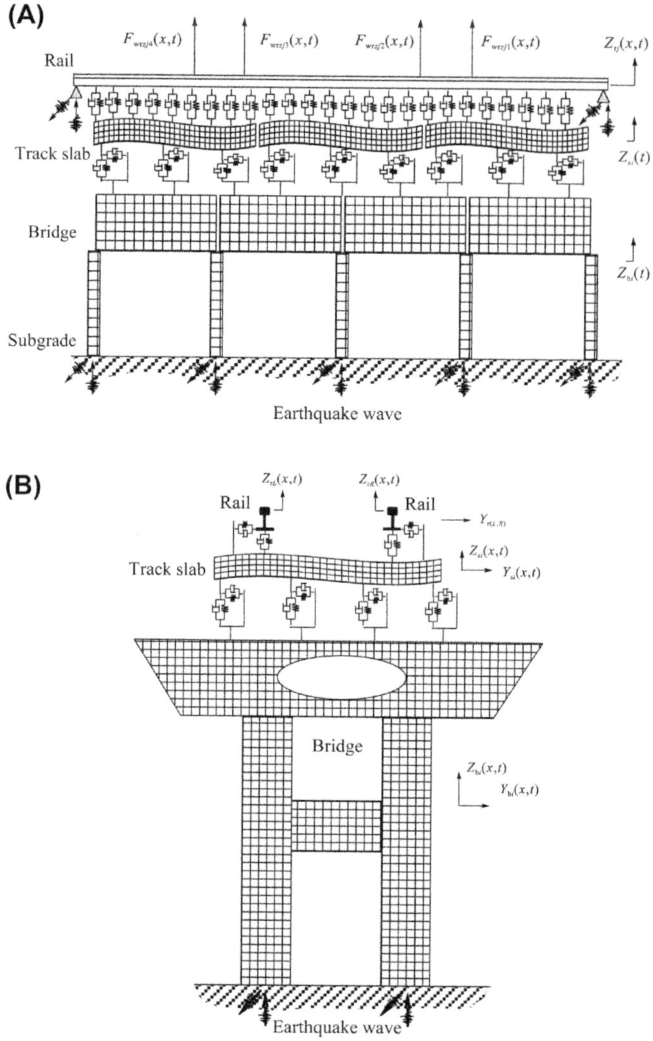

Figure 2.34
Dynamic model of the ballastless track on the bridge.

In contrast, the analytical method uses the Euler-Bernoulli beam to model the bending vibration of the bridge. The interaction of the track slab with the bridge is considered as the uniform distributed loads caused by the supports and is included in the dynamic model of the bridge. Figure 2.35 shows the dynamic model of the ballastless track model on a bridge.

The lateral motion equation of the bridge can be expressed as:

$$E_b I_{by} \frac{\partial^4 y_b(x,t)}{\partial x^4} + m_b \frac{\partial^2 y_b(x,t)}{\partial t^2} = p_y(x,t) - \rho_b A_b \frac{d^2 Y_g(t)}{dt^2} \tag{2.109}$$

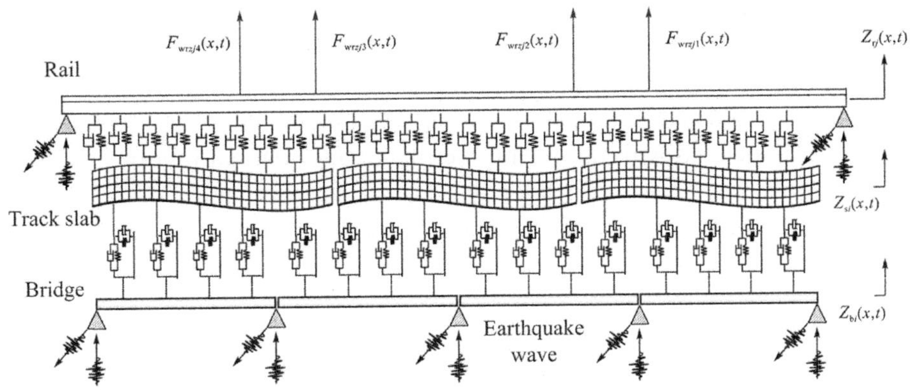

Figure 2.35
Dynamic model of the ballastless track on a bridge section.

where $y_b(x,t)$ is the lateral deflection of the bridge, m_b is the mass of the bridge per unit length, ρ_b is the density of the bridge, A_b is the cross-sectional area of the bridge, E_b is Young's modulus of the bridge, I_{by} is the second moment of area of the bridge cross-section about the y-axis, p_y is the lateral force between the bridge and the slabs, and Y_g is the lateral component of the earthquake excitation. The vertical motion equation of the bridge can be expressed as:

$$E_b I_{bz} \frac{\partial^4 z_b(x,t)}{\partial x^4} + m_b \frac{\partial^2 z_b(x,t)}{\partial t^2} = p_z(x,t) - \rho_b A_b \frac{d^2 Z_g(t)}{dt^2} \tag{2.110}$$

where $z_b(x,t)$ is the vertical deflection of the bridge, I_{bz} is the second moment of area of the bridge cross-section about the z-axis, p_z is the vertical force between the bridge and the slabs, and Z_g is the vertical component of the earthquake excitation. The torsional motion equation of the bridge can be expressed as:

$$\rho_b I_{b0} \frac{\partial^2 \phi_b(x,t)}{\partial t^2} - G_b K_b \frac{\partial^2 \phi_b(x,t)}{\partial x^2} = M_b(x,t) \tag{2.111}$$

where $\varphi_b(x,t)$ is the torsional deflection of the bridge, I_{b0} is the polar moment of area of the bridge cross-section, $G_b K_b$ is the torsional stiffness of the bridge, and M_b is the torsional moment between the bridge and the slabs.

Using the method of separation of variables and considering the normalized modal shapes, the following equations can be obtained:

$$y_b(x,t) = \sum_{k=1}^{NY} Y_{bk}(x) q_{byk}(t) \tag{2.112}$$

$$z_b(x,t) = \sum_{k=1}^{NZ} Z_{bk}(x) q_{bzk}(t) \tag{2.113}$$

$$\phi_b(x,t) = \sum_{k=1}^{NT} \Phi_{bk}(x) q_{bTk}(t) \tag{2.114}$$

In Equation (2.112) to Equation (2.114), $q_{byk}(t)$, $q_{bzk}(t)$, and $q_{bTk}(t)$ are the normalized coordinate vectors associated with the lateral, vertical, and torsional motions of the bridge; $Y_{bk}(x)$, $Z_{bk}(x)$, and $\Phi_{bk}(x)$ are the corresponding lateral, vertical, and torsional modal shape functions; *NY*, *NZ*, and *NT* are the numbers of modes used in the analysis. Considering the simply supported boundary conditions, the normalized modal shape functions are [14]:

$$Y_{bk}(x) = \sqrt{\frac{2}{m_b l_b}} \sin\left(\frac{k\pi}{l_b} x\right) \tag{2.115}$$

$$Z_{bk}(x) = \sqrt{\frac{2}{m_b l_b}} \sin\left(\frac{k\pi}{l_b} x\right) \tag{2.116}$$

$$\Phi_{bk}(x) = \sqrt{\frac{2}{\rho_b I_{b0} l_b}} \sin\left(\frac{k\pi}{l_b} x\right) \tag{2.117}$$

where l_b is the length of the bridge.

Using the orthogonality of the modes and considering Equations (2.109), (2.110), and (2.111), the lateral motion equation of the kth mode of the bridge is given as:

$$\ddot{q}_{byk}(t) + \frac{E_b I_{by}}{m_b}\left(\frac{k\pi}{l_b}\right)^4 q_{byk}(t) = \int_0^{l_b} p_y(x,t)\sin\left(\frac{k\pi}{l_b}x\right)dx + \frac{1-(-1)^k}{k\pi}\rho A l_b \ddot{Y}_g(t) \tag{2.118}$$

Similarly, the vertical motion equation of the kth mode of the bridge is given as:

$$\ddot{q}_{bzk}(t) + \frac{E_b I_{bz}}{m_b}\left(\frac{k\pi}{l_b}\right)^4 q_{bzk}(t) = \int_0^{l_b} p_z(x,t)\sin\left(\frac{k\pi}{l_b}x\right)dx + \frac{1-(-1)^k}{k\pi}\rho A l_b \ddot{Z}_g(t) \tag{2.119}$$

and the torsional motion equation of the kth mode of the bridge is given as:

$$\ddot{q}_{bTk}(t) + \frac{G_b K_b}{\rho_b I_{b0}}\left(\frac{i\pi}{l_b}\right)^2 q_{bTk}(t) = \int_0^{l_b} M_b(x,t)\sin\left(\frac{k\pi}{l_b}x\right)dx \tag{2.120}$$

Note that the uniformly distributed loads p_y and p_z of the track slab acting on the bridge through the CA mortar are calculated as:

$$p_y(x,t) = k_{y\mathrm{CAM}}[Y_s(x,t) - y_b(x,t)] + c_{y\mathrm{CAM}}\left[\dot{Y}_s(x,t) - \dot{y}_b(x,t)\right] \tag{2.121}$$

$$p_z(x,t) = k_{z\mathrm{CAM}}[Z_s(x,t) - z_b(x,t)] + c_{z\mathrm{CAM}}\left[\dot{Z}_s(x,t) - \dot{z}_b(x,t)\right] \tag{2.122}$$

In Equation (2.121) and Equation (2.122), $k_{y\mathrm{CAM}}$, $k_{z\mathrm{CAM}}$, $c_{y\mathrm{CAM}}$, and $c_{z\mathrm{CAM}}$ are the stiffness and damping of the CA mortar layers in the horizontal and vertical directions, respectively.

Using Equations (2-95) ~ (2-106) and Equations (2-109) ~ (2-122), the dynamic responses of the track system can be numerically obtained.

2.2.3 Pantograph modeling

In recent years, considerable research has been done on the dynamics of the pantograph-catenary system, and several standard mathematical models have been developed. There are different pantograph models, including the multi-rigid body model, lumped mass model, the rigid-flexible coupled model, and the full flexible body model. Based on the different pantograph models, the apparent difference in the dynamic performance of the pantograph and catenary system can be found. The TSG 19 pantograph shown in Figure 2.36 is modeled using the four methods mentioned earlier. Essential details are provided below.

2.2.3.1 Multi-rigid body modeling

The pan-head of the pantograph can be modeled using the multi-rigid system dynamic theory [17]. In the multi-rigid body model, the dynamic model of the pan-head is usually simple and easily established, while a nonlinear model of the frame must be considered based on its geometric shape and physical parameters, as shown in Figure 2.36(a). The pantograph considered is equipped with double slide plates. The pan-head is treated as two simple equivalent masses of the vibration system. By ignoring the translational motions of the pan-head in the other directions, the vertical motion equations of the pan-head are governed by:

$$m_h\ddot{y}_{h1} + F_{c1} - F_{h1} = 0 \tag{2.123}$$

$$m_h\ddot{y}_{h2} + F_{c2} - F_{h2} = 0 \tag{2.124}$$

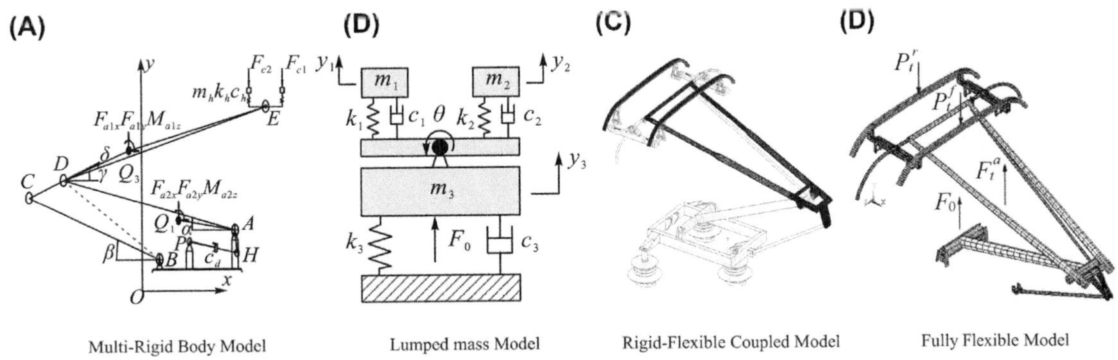

Figure 2.36
Different pantograph models.

In Equation (2.123) and Equation (2.124), m_h is the mass of the front (rear) slide plate: $\ddot{y}_{h1}$ and $\ddot{y}_{h2}$ are the vertical accelerations of the front and rear slide plates, respectively, F_{c1} and F_{c2} are the contact pressure between the catenary and the front and rear slide plates, respectively; and F_{h1} and F_{h2} are the spring forces of the front and rear slide plates, respectively, which can be expressed as:

$$F_{h1} = -k_h(y_{h1} - y_E + l_d \sin\theta_h) - c_h\left(\dot{y}_{h1} - \dot{y}_E + l_d \cos\theta_h \cdot \dot{\theta}_h\right) \tag{2.125}$$

$$F_{h2} = -k_h(y_{h2} - y_E - l_d \sin\theta_h) - c_h\left(\dot{y}_{h2} - \dot{y}_E - l_d \cos\theta_h \cdot \dot{\theta}_h\right) \tag{2.126}$$

In Equation (2.125) and Equation (2.126), k_h and c_h, respectively, are the stiffness and damping between the pan-head and frame; l_d is half the distance between the front and rear slide plates; and θ_h is the angle of the pan-head frame in the horizontal direction.

Furthermore, the rotational motion equation of the pan-head can be expressed as:

$$J\ddot{\theta}_h + (F_{h1} - F_{h2})l_d \cos\theta_h = 0 \tag{2.127}$$

where J is the moment of inertia of the pan-head.

The pantograph frame can be treated as a rod structure with only one degree of freedom represented by the angle α of the lower arm. Thus, the dynamic equilibrium equation of the pantograph frame can be given as:

$$\begin{aligned} &f_1(\alpha)\ddot{\alpha} + f_2(\alpha)\dot{\alpha}^2 + f_3(\alpha)\dot{\alpha} + f_4(\alpha)\mathrm{sgn}(\dot{\alpha}) \\ &+ f_5\left(\alpha, \dot{\alpha}, y_{h_1}, y_{h_2}, \theta_h, \dot{y}_{h_1}, \dot{y}_{h_2}, \dot{\theta}_h\right) = 0 \end{aligned} \tag{2.128}$$

where the variables $f_1(\alpha)$, $f_2(\alpha)$, $f_3(\alpha)$, $f_4(\alpha)$, and $f_5(\alpha)$ are expressed as:

$$f_1(\alpha) = J_1 + m_1 l_{AQ_1}^2 + k_2^2 J_2 + m_2 k_2^2 l_{BQ_2}^2 + k_3^2 J_3$$

$$+ m_3\left[l_{AD}^2 + k_3^2 l_{DQ_3}^2 + 2k_3 l_{AD} l_{DQ_3} \cos(\alpha + \gamma + \angle EDQ_3)\right]$$

$$f_2(\alpha) = k_2 k_{12} J_2 + m_2 k_2 k_{12} l_{BQ_2}^2 + k_3 k_{13} J_3 + m_3\left[k_3 k_{13} l_{DQ_3}^2 + k_{13} l_{AD} l_{DQ_3} \cos(\alpha + \gamma\right.$$

$$\left. + \angle EDQ_3) - k_3(1 + k_3) l_{AD} l_{DQ_3} \sin(\alpha + \gamma + \angle EDQ_3)\right] f_3(\alpha) = c_d \frac{dl_{PH}}{d\alpha} l_{AH} \sin\angle AHP$$

$$+ U_A + k_2^2 U_B + (k_2 + k_3)^2 U_C + (1 + k_3)^2 U_D$$

$$f_4(\alpha) = B_A + |k_2| B_B + |k_2 + k_3| B_C + |1 + k_3| B_D$$

$$\begin{aligned} f_5\left(\alpha, \dot{\alpha}, y_{h_1}, y_{h_2}, \theta_h, \dot{y}_{h_1}, \dot{y}_{h_2}, \dot{\theta}_h\right) &= -M_z + F_{h_1}(l_{AD} \cos\alpha + k_3 l_{DE} \cos\gamma - k_5 l_d \cos\theta_h) \\ &+ F_{h_2}(l_{AD} \cos\alpha + k_3 l_{DE} \cos\gamma + k_5 l_d \cos\theta_h) \\ &+ m_1 g l_{AQ_1} \cos(\alpha - \angle DAQ_1) + k_2 m_2 g l_{BQ_2} \cos\beta \\ &+ m_3 g\left[l_{AD} \cos\alpha - k_3 l_{DQ_3} \cos(\gamma + \angle EDQ_3)\right] \end{aligned}$$

In these expressions of five variables, m_1, m_2, and m_3 are the mass of the low frame, the pull rod, and the upper frame, respectively; J_1, J_2, and J_3 are the rotational inertia around the center of the mass of the low frame, the pull rod, and the upper frame, respectively; U_A, B_A, U_B, B_B, U_C, B_C, U_D, and B_D are the damping and dry friction of joints A, B, C, and D, respectively; c_d is the damping coefficient of the damper; M_z is the uplift moment;

$$k_2 = \frac{d\beta}{d\alpha},\ k_3 = \frac{d\gamma}{d\alpha},\ k_{12} = \frac{dk_2}{d\alpha},\ \text{and } k_{13} = \frac{dk_3}{d\alpha}$$

Derived from the above, the nonlinear multi-rigid body model of the pantograph is thus established. The other pantographs can be modeled using a similar method for the various uses. For different pantographs, minor differences may exist in their differential equations of motion. For example, some models take the mass of the balance rod into account while other models consider the effects of dry friction of the system. Sometimes, the pantograph may be equipped with a pan-head. In this case, the pan-head is considered as a reduced vibration system with a single degree of freedom.

The pantograph is installed in the high-speed train and moves with the train. On the one hand, the vibrations of the vehicle body can negatively affect the pantograph, resulting in large fluctuation in the displacement of contact points, inducing strong structural vibrations, and aggravating the current collection quality of the pantograph-catenary system. On the other hand, the vibrations of the pantograph—catenary system may also be transmitted to the vehicle body along the structure and have dynamic effects on the performance of the vehicle system.

The effects above can be included in the dynamic model of the high-speed train system in Equation (2.23) or Equation (2.60) as a form of generalized forces $\boldsymbol{Q}_p$. To show this, the multi-rigid body model of the pantograph established earlier is reconsidered here. Let $\ddot{y}_A$ be the base excitation of the pantograph caused by the movement of the train. Then, Equation (2.128) can be rewritten as:

$$\begin{aligned} & f_1(\alpha)\ddot{\alpha} + f_2(\alpha)\dot{\alpha}^2 + f_3(\alpha)\dot{\alpha} + f_4(\alpha)\mathrm{sgn}(\dot{\alpha}) \\ & + f_5\left(\alpha, \dot{\alpha}, y_{h_1}, y_{h_2}, \theta_h, \dot{y}_{h_1}, \dot{y}_{h_2}, \dot{\theta}_h, \ddot{y}_A\right) = 0 \end{aligned} \tag{2.129}$$

where $f_i(\alpha)$, for the index $i = 1, 2, 3, 4$, is the same as in Equation (2.128) and $f_5(\alpha)$ is given by:

$$\begin{aligned} f_5\left(\alpha, \dot{\alpha}, y_{h_1}, y_{h_2}, \theta_h, \dot{y}_{h_1}, \dot{y}_{h_2}, \dot{\theta}_h, \ddot{y}_A\right) = & -M_y + F_{h_1}(l_{AD}\cos\alpha + k_3 l_{DE}\cos\gamma - k_5 l_d \cos\theta_h) \\ & + F_{h_2}(l_{AD}\cos\alpha + k_3 l_{DE}\cos\gamma - k_5 l_d\cos\theta_h) \\ & + m_1 g l_{AQ_1}\cos(\alpha - \angle DAQ_1) + k_2 m_2 g l_{BQ_2}\cos\beta \\ & + m_3 g\left[l_{AD}\cos\alpha + k_3 l_{DQ_3}\cos(\gamma + \angle EDQ_3)\right] \\ & + \left[m_1 l_{AQ_1}\cos(\alpha - \angle DAQ_1) + m_2 k_2 l_{BQ_2}\cos\beta \right. \\ & \left. + m_3 l_{AD}\cos\alpha + m_3 k_3 l_{DQ_3}\cos(\gamma + \angle EDQ_3)\right]\ddot{y}_A \end{aligned} \tag{2.130}$$

Considering Equation (2.19), the interaction between the pantograph and the vehicle can be investigated by co-modeling the vehicle system and the pantograph-catenary system. Compared with the vehicle, the pantograph has such a small mass that its contribution to the mass of the vehicle can be ignored. Previous studies and experimental data also validate this conclusion. Consequently, the vehicle can be considered as the base excitation for the analysis of the pantograph-catenary system, while the corresponding effects are not considered in the analysis of the vehicle system.

High-speed trains exposed to the air. When the train is running at a low speed, the aerodynamics of the pantograph has a weak effect on current collection and is often ignored accordingly. However, when the train speed reaches 200km/h or above, the effect of the pantograph aerodynamics on current collection becomes significant. In China, the high-speed pantographs, such as the DSA250, SSS400+, and DSA380, are all equipped with guide plates to improve aerodynamic performance and to meet the requirements for high-speed current collection. With these pantographs, it is necessary to study their aerodynamics.

During the motion of the train, the high-speed airflow acts on the pantograph and causes aerodynamic resistance and the aerodynamic lift force and moment of its rod components. It eventually forms the resultant vertical aerodynamic uplift forces [18]. The aerodynamic uplift force is the key factor in influencing the quality of the current collection, and it can be adjusted using the guide plate. Currently, offline simulation and fluid-solid coupled simulation are the main methods for studying pantograph-catenary dynamics under air turbulence. The offline simulation is described as follows: the time domain response of the aerodynamic uplift force is obtained through an experiment or other simulation. Then the aerodynamic uplift force at each time step can be calculated using an interpolation technique based on discrete response data, and the simulation can be performed. The effects of the change in aerodynamic force, which is caused by the movement of the pantograph, are not addressed in the offline simulation. In contrast, the coupled system model is considered in the fluid-solid coupled simulation, which can address the aerodynamics of the pantograph as well as the pantograph-catenary interaction.

Again, consider the multi-rigid body model of the pantograph to show the fluid-solid coupled simulation. The aerodynamic simulation of the pantograph is performed, and the aerodynamic resistances and the aerodynamic lift forces of the front and rear slide plates are calculated based on the simulation results. After the corresponding aerodynamic uplift forces $F_{\mathrm{a}1}$ and $F_{\mathrm{a}2}$ are calculated, the motion equations of the pan-head are expressed as:

$$m_h\ddot{y}_{h1} + F_{c1} - F_{h1} - F_{a1} = 0 \tag{2.131}$$

$$m_h\ddot{y}_{h2} + F_{c2} - F_{h2} - F_{a2} = 0 \tag{2.132}$$

The aerodynamic resistances and the aerodynamic lift forces and moments on the upper frame and lower arm of the pantograph are reduced to their mass centers. The resultant forces and moments are calculated. With the corresponding derivations above, the motion equation of the frame under air turbulence can be established, which has the same form as Equation (2.218). In the established equations, the variable f_5 is expressed as:

$$\begin{aligned}
f_5\left(\alpha,\dot{\alpha},y_{h_1},y_{h_2},\theta_h,\dot{y}_{h_1},\dot{y}_{h_2},\dot{\theta}_h\right) &= -M_y + F_{h_1}(l_{AD}\cos\alpha + k_3 l_{DE}\cos\gamma - k_5 l_d\cos\theta_h) \\
&+F_{h_2}(l_{AD}\cos\alpha + k_3 l_{DE}\cos\gamma - k_5 l_d\cos\theta_h) \\
&-F_{a1x}\left[l_{AD}\sin\alpha - k_3 l_{DQ_3}\sin(\gamma + \angle EDQ_3)\right] \\
&-F_{a1y}\left[l_{AD}\cos\alpha - k_3 l_{DQ_3}\cos(\gamma + \angle EDQ_3)\right] - k_3 M_{a1z} \\
&-F_{a2x} l_{AQ_1}\sin(\alpha - \angle DAQ_1) - F_{a2y} l_{AQ_1}\cos(\alpha - \angle DAQ_1) \\
&+M_{a2z} + m_1 g l_{AQ_1}\cos(\alpha - \angle DAQ_1) + k_2 m_2 g l_{BQ_2}\cos\beta \\
&+m_3 g\left[l_{AD}\cos\alpha - k_3 l_{DQ_3}\cos(\gamma + \angle EDQ_3)\right]
\end{aligned} \tag{2.133}$$

where F_{a1x}, F_{a1y}, and M_{a1z} are the aerodynamic resistance, lift force, and moment, respectively, of the upper frame at its mass center; and F_{a2x}, F_{a2y}, and M_{a2z} are the aerodynamic resistance, lift force, and moment, respectively, of the lower arm bar at its mass center.

2.2.3.2 Lumped mass modeling

The lumped mass model of the pantograph is also in current use. Based on the multi-rigid body model, it is composed of several equivalent masses representing the frame, as shown in Figure 2.36(b). In general, the lumped mass of the frame can be obtained through linearization of its nonlinear motion equation at a certain height. Considering this, Equation (2.218) is processed, and a reduced linear model of the pantograph can be obtained, which consists of two masses: the damper and the spring. The motion equation of the pan-head can be found in Equations (2.123) ~ (2.127), while the motion equation of the frame is given as:

$$m_3\ddot{y}_3 + c_3\dot{y}_3 + k_3 y_3 + F_1 + F_2 = F_0 \tag{2.134}$$

Similarly, when the frame is equivalent to two lumped masses, four lumped mass models of the pantograph are then obtained, see Figure 2.37. The pan-head is modeled by two masses m_1 and m_2, and the motion equations are identical with those derived above. The mass m_3 is subjected to three forces, as shown in Figure 2.37, and its motion equation can

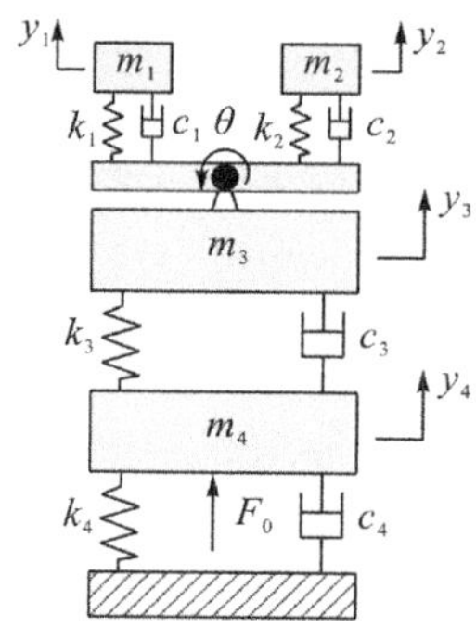

Figure 2.37
Four lumped mass models of the pantograph.

be easily obtained by applying $F_0 = 0$ in Equation (2.134). The motion equation of mass m_4 can be written as:

$$m_4\ddot{y}_4 + (c_4 + c_3)\dot{y}_4 + (k_4 + k_3)y_4 - c_3\dot{y}_3 - k_3y_3 = F_0 \tag{2.135}$$

2.2.3.3 Rigid-flexible coupled modeling

As the pantograph running speed increases, the interaction frequency of the pantograph and the catenary coupled system also increases continually. Increasingly significant contributions of the higher frequency band to dynamic performance has been identified, and the flow-induced vibration caused by strong airflow makes flexible deformation of the critical components (pan-head and frame) noticeable. To meet the requirements of the analysis mentioned earlier, the pantograph is treated as the equivalent of a rigid-flexible hybrid body, while regarding the pan-head or frame as a flexible body and the other parts of the pantograph as rigid bodies, as shown in Figure 2.36(c). With the finite element method, the flexible body model of these critical components is established. Using the fixed joints, the flexible body is connected to the other rigid bodies. The rigid-flexible coupled model of the pantograph is then obtained as:

$$[M_p]\{\ddot{u}_p\} + [C_p]\{\dot{u}_p\} + [K_p]\{u_p\} = \{f(t)\} \tag{2.136}$$

where $[M_p]$ is the mass matrix of the pantograph, $[C_p]$ is the damping matrix of the pantograph, $[K_p]$ is the stiffness matrix of the pantograph, $\{u_p\}$ is the displacement vector of the coordinates used, and $\{f(t)\}$ is the pantograph load vector.

2.2.3.4 Fully flexible modeling

All components of the pantograph are treated as flexible bodies. Each flexible body is discretized by solid or shell elements and is connected to the other flexible body by joint constraints. In this regard, all flexible bodies dynamically interact by further exerting the

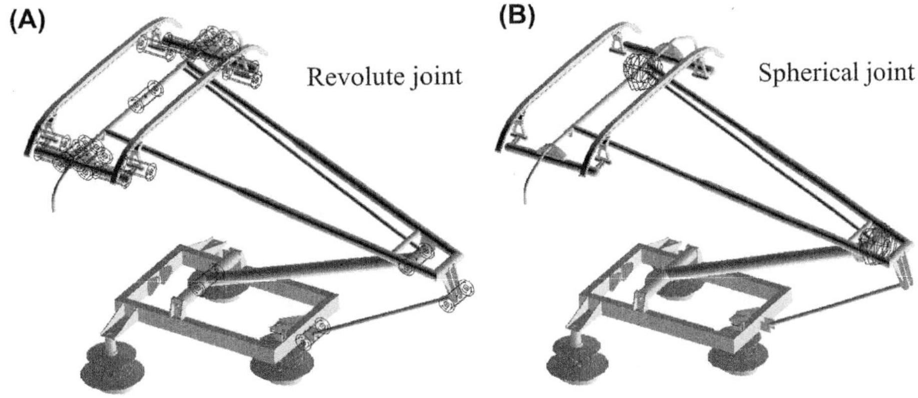

Figure 2.38
Joints used in the fully flexible model of the pantograph.

loads and a fully flexible model of the pantograph is established, as shown in Figure 2.36(d). In general, the components of the pantograph can be modeled using shell elements, such as the pan-head frame, the upper arm, the lower arm, the crossbar, and the base frame. The integrally shaped components are discretized using solid elements. For example, the slide plate of the pan-head is welded by two parts. Each part of the pan-head is discretized by the solid element with a different density.

There are three kinds of joints used in the fully flexible model: fixed joints, revolution joints, and spherical joints. The fixed joint is used to model the interaction of two components whose six relative motions are fully restricted. The revolution joint is used to simulate the interactions between the pan-head suspension systems, between the upper arm and the lower arm, and between the lower arm and the base frame. The spherical joint is used primarily to simulate the interaction between the balance rod and the upper frame, and between the balance rod and the lower frame. For example, as far as the RSG19 pantograph is concerned, the relevant joints used in the fully flexible model are shown in Figure 2.38. In this model, there are 21 revolution joints and 2 spherical joints. The fully flexible model of the pantograph has many degrees of freedom, and the motion equations of the pantograph can be governed in the same form as Equation (2.136).

2.2.4 Catenary modeling

The catenary model plays an essential role in the study of the dynamic behaviors of the pantograph-catenary system. Many available methods can be used to model the catenary with different purposes. In general, there are two kinds of modeling methods. For the first method, structural modes are used as generalized coordinates to establish the motion equations of the catenary. This method is quite convenient for catenary modeling with good precision and in broad applications. However, it addresses complex catenary systems poorly and is thus primarily used for simple catenaries. The second method simplifies the structure, and a reduced model of the catenary can be established using analytical or numerical methods. The second method is straightforward and can produce valuable results.

2.2.4.1 modal-based modelling method [2]

This book introduces a modeling method of the catenary based on the Fourier expansion and modal techniques. Figure 2.39 shows a simple catenary model, which is composed of messenger wires, contact wires, droppers, supporters, and limiters. Consider the flexible catenary in the tension length L. As shown in Figure 2.39, k_T is the stiffness of the supporter, m_{AT} is the mass of the supporter, N_T is the number of supporters, m_{BT} is the mass of the registration arm, N_D is the number of droppers, k_{Di} is the stiffness of the dropper, m_D is half the mass of the dropper, S_A is the tension of the messenger wire, S_B is the tension of the contact wire, E_{IA} is the bending stiffness of the messenger wire, E_{IB} is the bending stiffness of the contact wire, ρ_A is the density of the messenger wire, and ρ_B is the density of the contact wire.

Let Z_A and Z_B be the vertical displacement of the messenger and contact wires, respectively. As shown in Figure 2.39, the contact and messenger wires can be equivalently jointed at the ends of the anchor section. This means the corresponding displacement and stiffness at the ends of the anchor section are zeros. The conditions are:

$$z|_0 = z|_L = 0 \tag{2.137}$$

$$\left.\frac{\partial^2 z}{\partial x^2}\right|_0 = \left.\frac{\partial^2 z}{\partial x^2}\right|_L = 0 \tag{2.138}$$

where $z = Z_\mathrm{A}$ or $z = Z_\mathrm{B}$. Then, the displacements Z_A and Z_B of the wires can be expanded using the Fourier expansion technique and are expressed as:

$$z_A(x,t) = \sum_i^N A_i(t)\sin\left(\frac{i\pi x}{L}\right) \tag{2.139}$$

$$z_B(x,t) = \sum_i^N B_i(t)\sin\left(\frac{i\pi x}{L}\right) \tag{2.140}$$

In Equation (2.139) and Equation (2.140), A_i and B_i are the Fourier coefficients of the ith expansion term associated with the displacements of the messenger and contact wires, the coordinate x is the distance measured from the starting end of the anchor session, and N is the number of expansion terms.

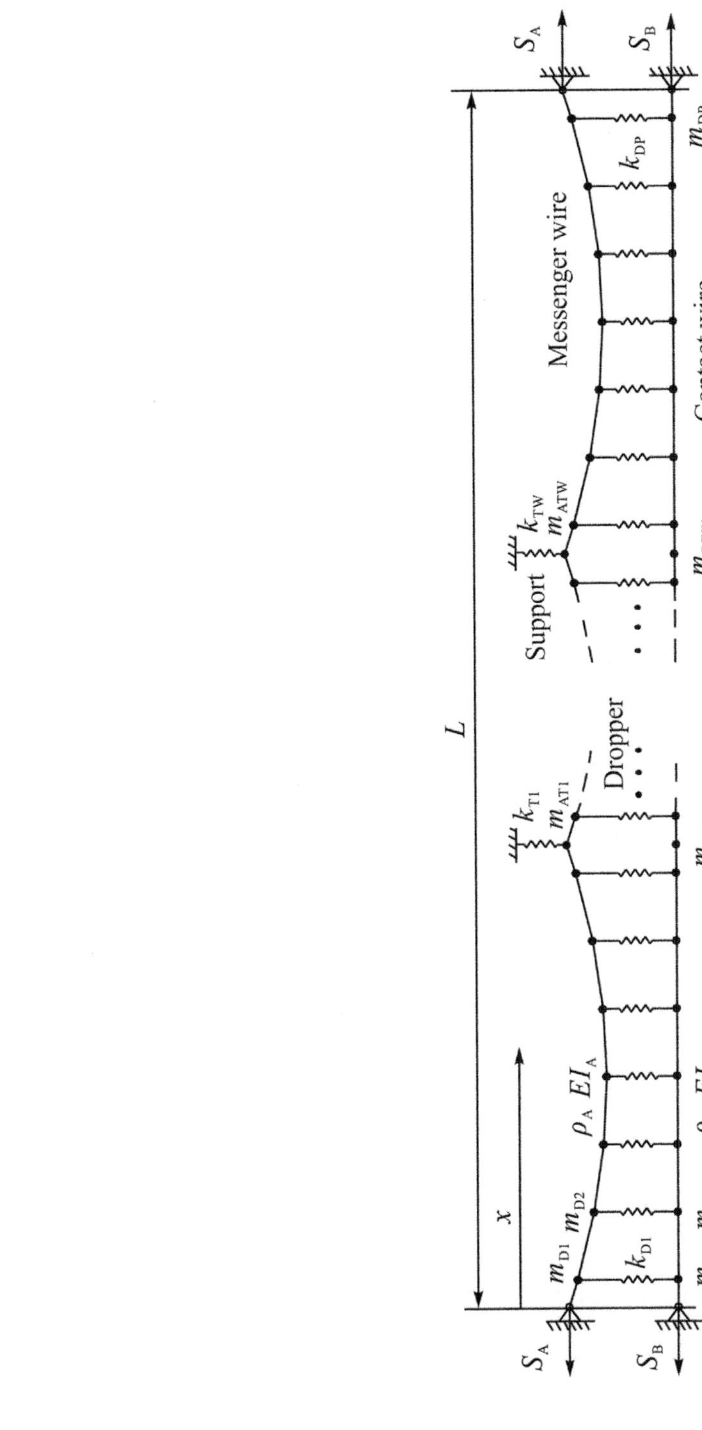

Figure 2.39
Schematic of a simple catenary model.

The Lagrange's equations of the second kind are used here to derive motion equations of the catenary. For this purpose, the kinetic energy T and the potential energy V of the system are calculated first. The total kinetic energy of the system can be expressed as:

$$T = T_m + T_c + T_d + T_s + T_l \tag{2.141}$$

where T_m, T_c, T_d, T_s, and T_l are the kinetic energy of the messenger wire, contact wire, dropper, supporter, and limiter, respectively. Considering the tensile and bending deformations primarily, the total potential energy of the system can be expressed as:

$$V = V_T + V_b + V_d + V_s \tag{2.142}$$

where V_T is the potential energy caused by tensile deformation, V_b is the potential energy caused by bending deformation, V_d is the potential energy of the dropper, and V_s is the potential energy of the supporter. Thus, the Lagrangian can be expressed as:

$$\Gamma = T - V \tag{2.143}$$

Using the Lagrangian Γ, the following equations can be obtained:

$$\frac{d}{dt}\left(\frac{\partial \Gamma}{\partial \dot{A}_i}\right) - \frac{\partial \Gamma}{\partial A_i} = Q_{Ai} \tag{2.144}$$

$$\frac{d}{dt}\left(\frac{\partial \Gamma}{\partial \dot{B}_i}\right) - \frac{\partial \Gamma}{\partial B_i} = Q_{Bi} \tag{2.145}$$

where A_i and B_i are the generalized coordinates, and Q_{Ai} and Q_{Bi} are the corresponding generalized forces. Consider the free vibrations of the system. This means $Q_{Ai} = 0$ and $Q_{Bi} = 0$. Using the conditions $Q_{Ai} = 0$ and $Q_{Bi} = 0$, Equation (2.144) and Equation (2.143) can be simplified and rewritten in the matrix form as:

$$[M]_{2N\times 2N}\begin{pmatrix}\ddot{A}\\ \ddot{B}\end{pmatrix}_{2N\times 1} + [K]_{2N\times 2N}\begin{pmatrix}A\\ B\end{pmatrix}_{2N\times 1} = (0) \tag{2.146}$$

where **[M]** is the mass matrix and **[K]** is the stiffness matrix. The eigenvalue problem of Equation (2.146) yields:

$$\det\left([K] - \omega^2[M]\right) = 0 \tag{2.147}$$

Equation (2.147) may lead to $2N$ eigenfrequencies and eigenvectors of the system under free vibration. Let ω_i be the ith eigenfrequency. Its corresponding eigenvector can be expressed as:

$$\{A, B\}_i^T = \{a_{i1}, a_{i2}, \cdots, a_{iN}, b_{i1}, b_{i2}, \cdots, b_{iN}\}^T \tag{2.148}$$

where a_{ij} and $b_{ij,}$ for $j = 1 \sim N$ are the components of the eigenvectors A_i and B_i. Therefore, the ith modal shape function can be expressed as:

$$\Phi_{Ai} = \sum_{j=1}^{N} a_{ij} \sin\frac{j\pi x}{L}$$
$$\Phi_{Bi} = \sum_{j=1}^{N} b_{ij} \sin\frac{j\pi x}{L} \tag{2.149}$$

Thus, the motion equation of the catenary is written in modal coordinates as:

$$m_i\ddot{q}_i + 2m_i\omega_{ni}\zeta_i\dot{q}_i + m_i\omega_{ni}^2 q_i = Q_i(t).(i = 1, 2, \cdots, N) \tag{2.150}$$

where m_i is the ith modal mass, ω_i the ith modal frequency, ζ_i the ith modal damping, and $Q_i(t)$ is the ith generalized modal force. Considering the external force $f(x,t)$ applied on the catenary, the ith generalized force can be expressed as:

$$Q_i = \int_0^L f(x,t)\Phi_i(x)dx \tag{2.151}$$

2.2.4.2 Direct modeling methods

With information on the geometric positions, topological relations, and forces on the structural components, the catenary can be directly modeled by means of analytical or numerical methods. The two methods are introduced below.

First, the analytic method [19] is demonstrated. It is assumed that the contact wire can be treated as a Bernoulli-Euler beam. The tension of the contact wire is T_c, the line density of the contact wire is ρ_c, and the bending stiffness is EI_c. Based on the theory of the Bernoulli-Euler beam, the motion equation of the contact wire is then expressed as:

$$\rho_c\frac{\partial^2 y_c(x,t)}{\partial t^2} + EI_c\frac{\partial^4 y_c(x,t)}{\partial x^4} - T_c\frac{\partial^2 y_c(x,t)}{\partial x^2} + k_d(y_m - y_c)\delta(x - x_n) = P\delta(x - vt) \tag{2.152}$$

where $y_c(x,t)$ is the vibration displacement of the contact wire, k_d is the stiffness of the dropper, P is the contact pressure between the pantograph and the catenary, and v is the speed of the pantograph.

In a similar way, the motion equation of the messenger wire is then expressed as:

$$\rho_m\frac{\partial^2 y_m(x,t)}{\partial t^2} + EI_m\frac{\partial^4 y_m(x,t)}{\partial x^4} - T_m\frac{\partial^2 y_m(x,t)}{\partial x^2} + k_d(y_m - y_c)\delta(x - x_n) + k_s y_m\delta(x - x_s) = 0 \tag{2.153}$$

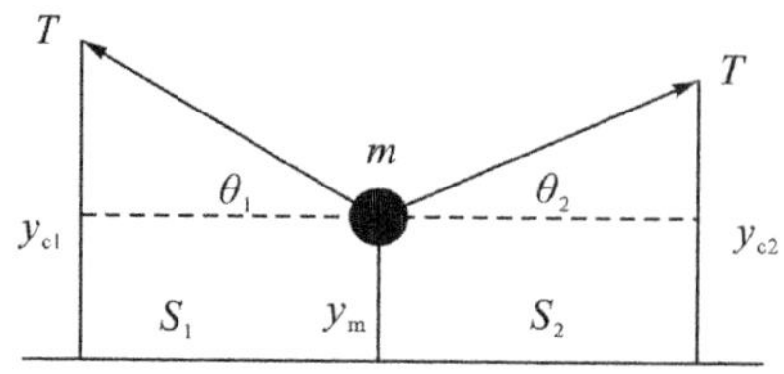

Figure 2.40
Force balance analysis of the lumped mass model of the messenger wire.

where ym(x,t) is the vibration displacement of the messenger wire, ρ_m is the line density, EI_m is the bending stiffness, T_m is the tension, and k_s is the stiffness of the supporter. The messenger wire can also be characterized using a lumped mass model, as shown in Figure 2.40. Considering that the angles θ_1 and θ_2 are both small, the lumped mass model of the messenger wire can be expressed as [20]:

$$T\left[\frac{y_{c1}-y_m}{S_1}+\frac{y_{c2}-y_m}{S_2}\right]=m\ddot{y}_m \tag{2.154}$$

where T is the tension.

Apart from this analytic method of establishing the catenary model, a numerical method can also be used, e.g., the finite element method. The numerical method is introduced next. With this method, the geometric model of the catenary should first be established, and then the contact wire and messenger wire are considered as either beam elements or cable elements. However, due to the relatively small line density of the dropper compared to the others, the dropper is modeled as a spring element with a mass at both ends. Furthermore, the clamp of the dropper and some additional components are simulated by mass element [21]. Thus, the finite element model of the catenary can be established directly by meshing the geometric model. The mass matrix, stiffness matrix, and damping matrix of the catenary model and the load vector may be obtained by assembling the element matrix into global matrices. Therefore, the motion differential equations of the catenary can be derived as follows:

$$[M_c]\{\ddot{u}_c\}+[C_c]\{\dot{u}_c\}+[K_c]\{u_c\}=\{f(t)\} \tag{2.155}$$

where, $[\boldsymbol{M_c}]$ is the mass matrix, $[\boldsymbol{C_c}]$ is the damping matrix, $[\boldsymbol{K_c}]$ is the stiffness matrix, $\{\boldsymbol{u_c}\}$ is the displacement vector of the coordinates used, and $\{\boldsymbol{f(t)}\}$ is the load vector.

The finite element-based numerical method has the advantages of flexibility, convenience, and higher accuracy. It thus is widely used for this analysis. However, the results yielded by this method can be affected by certain factors, such as structure configurations, element types, and computational techniques. Previous studies [22] suggest that it is reasonable to model the catenary with considerations given to the structural configuration at the equilibrium state, the pulling out of the catenary, and the axial stiffness of the element.

Figure 2.41
Real picture of the Pantograph-catenary system on a viaduct.

For many high-speed passenger lines, the catenary systems are usually built in the viaduct in China, as shown in Figure 2.41, to diminish the effects of track irregularity on the vibrations of the vehicle system and the current collection of the pantograph. This way, the vibrations of the bridge can be transferred to the catenary system through tower supports and affects the dynamics of the pantograph—catenary system, as well as the quality of the current collection. Thus, a method is introduced below, which can account for the effects of the train-bridge interaction on the dynamics of the pantograph-catenary system.

Figure 2.42 shows a coupled model, including the viaduct, high-speed train, and pantograph- catenary system. The modeling of the vehicle, the track, and the pantograph has been presented earlier and is not repeated here. The modeling of the catenary with consideration on the bridges is introduced below.

The messenger wire is assumed to be a Bernoulli-Euler beam. Let y_s be the vertical displacement of the bridge. Using Equation (2.153), the dynamic equations of the messenger wire can be rewritten as:

$$\rho_m \frac{\partial^2 y_m(x,t)}{\partial t^2} + EI_m \frac{\partial^4 y_m(x,t)}{\partial x^4} - T_m \frac{\partial^2 y_m(x,t)}{\partial x^2} + k_d(y_m - y_c)\delta(x - x_n)$$
$$+k_s(y_m - y_s)\delta(x - x_s) = 0 \qquad (2.156)$$

Thus, the influence of the viaduct on the pantograph-catenary dynamics can be analyzed using an updated model of the messenger wire, together with dynamic models of the relevant systems derived above.

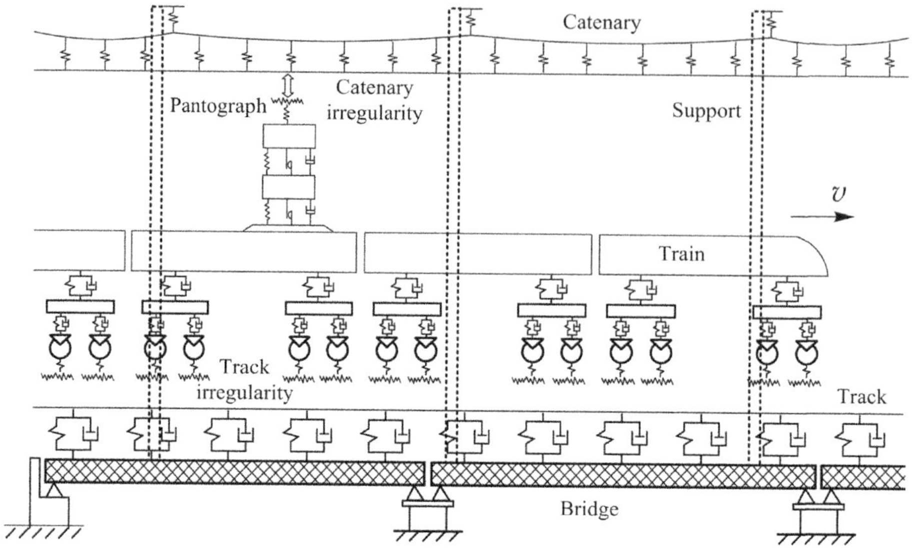

Figure 2.42
Schematic of the dynamic model of the pantograph-catenary system on a viaduct.

The catenary is designed in line with wind speeds. Two different wind speeds are considered. They are the operational wind speed and design wind speed. The operational wind speed is used for wind deviation of the catenary under normal conditions. The design wind speed is used for structural strength. The natural wind consists of average wind and fluctuating wind. However, in the traditional design, the maximum span is determined by checking static wind deflection, which ignores the influences of the fluctuating wind on span design and the quality of the current collection. For the engineering structures with a period of primary self-excited vibration less than 0.25 s, the influence of fluctuating wind is not considered. The catenary is a flexible structure with lightweight and low natural frequency and is sensitive to wind loads. Under wind loads, it is prone to large deformations and vibrations and exhibits strong nonlinear characteristics. Therefore, it is necessary to analyze the dynamic responses of the catenary with consideration given to fluctuating wind so that the design wind speed of the catenary is more accurate [23].

The fluctuating wind is treated as the dynamic load. Let $V(x_i, y_i, z_i, t)$ be the wind speed on the structure at the point (x_i, y_i, z_i). The wind speed V can be expressed as:

$$V(x_i, y_i, z_i, t) = \overline{V}(z_i) + v(x_i, y_i, z_i, t) \tag{2.157}$$

where $\overline{V}(z_i)$ is the average wind velocity and $v(x_i, y_i, z_i, t)$ is the fluctuating wind speed.

The average wind velocity $\overline{V}(z_i)$ can exponentially vary in height. The average wind velocity $\overline{V}(z)$ at the height z can be calculated using the mean wind velocity $\overline{V}_{10}$ at a height of 10 m as:

$$\overline{V}(z) = \overline{V}_{10}\left(\frac{z}{10}\right)^{\alpha} \tag{2.158}$$

where α is the surface roughness.

The autoregressive (AR) model in the linear filtering technique is used here to simulate the catenary in the wind field. The height of the catenary structure is usually about 1.5 m. The wind speed difference caused by variation in the catenary height is small, so the Davenport spectrum is adopted. It is independent of height and expressed as:

$$S_u(n) = 4K\overline{V}_{10}^2 \frac{x^2}{n(1+x^2)^{4/3}} \tag{2.159}$$

where n is the frequency of the fluctuating wind in Hertz, K is the coefficient that reflects the surface roughness, and $x = \frac{1200n}{\overline{V}_{10}}$.

The average span of the high-speed railway catenary is about 50 m. Considering the spatial correlation of the wind speed, the simulation point of the wind speed is chosen at every 25 m interval along the contact wire, as shown in Figure 2.43.

Figure 2.44 shows the time history of the fluctuating wind speed of the contact wire at point 10 when the average wind speed is 30 m/s. It can be seen in Figure 2.44 that the wind speeds at the simulation points are very different at different times. The fluctuation can exert dynamic effects on the catenary. To verify the accuracy of the simulation results, they are transformed in the frequency domain. Compared with the target spectrum in the double logarithmic coordinate system, the simulated spectrum is in good agreement with the target spectrum, as shown in Figure 2.44(b).

The finite element method is applied to a three-dimensional dynamic model of the catenary. Using modal analysis, the modal shapes of the catenary are plotted in Figure 2.45.

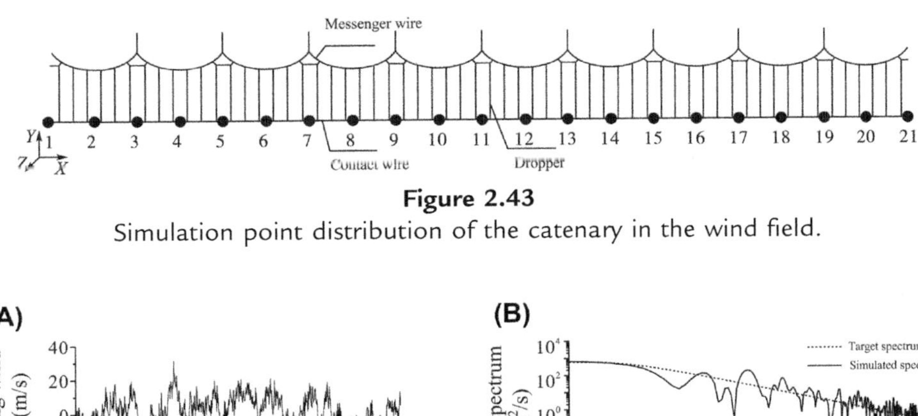

Figure 2.43
Simulation point distribution of the catenary in the wind field.

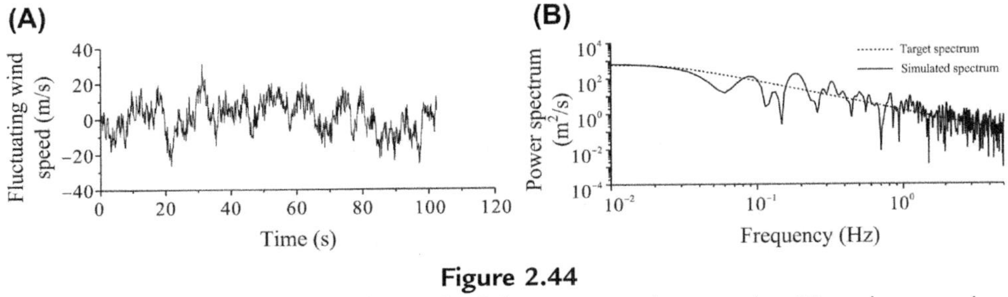

Figure 2.44
Time history of the fluctuating wind speed of the contact wire at point 10 and comparisons of the simulated spectrum with the target spectrum.

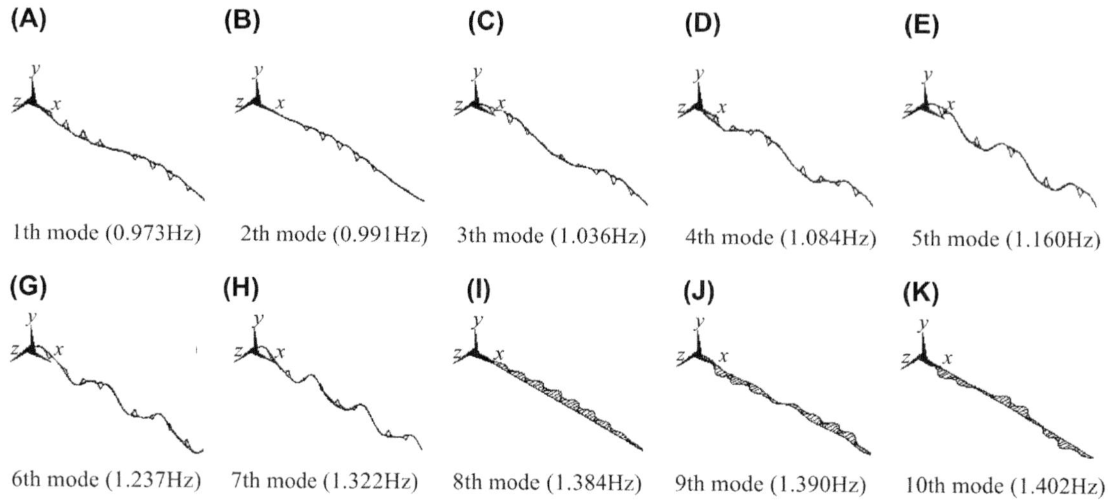

Figure 2.45
Modal shapes of the catenary.

As shown in Figure 2.45, the natural frequencies of the catenary are closely distributed. The first seven modal shapes are the vertical vibrations, the 8th and 10th modal shapes are the lateral vibrations, and the 9th modal shape is the coupled lateral and vertical vibration. Thus, the conclusion is that the catenary in the wind field has complex vibrations.

In summary, the three-dimensional model of the catenary in the wind field can finally be expressed in matrix form as:

$$\mathbf{M}_{ci}\{\ddot{\mathbf{u}}_{ci}\} + \mathbf{C}_{ci}\{\ddot{\mathbf{u}}_{ci}\} + \mathbf{K}_{ci}\{\mathbf{u}_{ci}\} = \{\mathbf{F}_{ci}(t)\} \tag{2.160}$$

where $\mathbf{M}_{ci}$, $\mathbf{C}_{ci}$, and $\mathbf{K}_{ci}$ are the mass, damping and stiffness matrices, respectively; $\{\ddot{\mathbf{u}}_{ci}\}$, $\{\dot{\mathbf{u}}_{ci}\}$, and $\{\mathbf{u}_{ci}\}$ are the acceleration, velocity, and displacement vectors of the nodes in the catenary; and $\{\mathbf{F}_{ci}(t)\}$ is the wind load vector applied to the nodes. Using Equation (2.160), simulation of the catenary dynamics under crosswind conditions can be performed.

2.2.5 Airflow modeling

The aerodynamic computational model includes a mathematical model and a geometric model. The mathematical model is the theoretical basis of the solution. The geometric model primarily includes the train shape, line environment, and boundary conditions.

2.2.5.1 Mathematical model

The fluid control equations consist of three main equations: the continuity equation, the momentum conservation equation, and the energy conservation equation. They are essentially the mass conservation law, Newton's motion law, and energy conservation law, respectively. When the train is running at high speeds, the number of Reynolds in the flow field around the train is usually much larger than 10^7, and the flow field is in a completely turbulent state.

By introducing the fluxφ, the fluid control equations have the general form [24, 25]:

$$\frac{\partial(\rho\boldsymbol{\varphi})}{\partial t}+div(\rho\mathbf{v}\boldsymbol{\varphi})=div(\Gamma grad\boldsymbol{\varphi})+\mathbf{S} \tag{2.161}$$

where ρ is the fluid density, t is time, the operator div is the divergence of a vector, $\boldsymbol{v}$ is the velocity vector of the fluid, the operator $grad$ is the gradient, $\Gamma$is the diffusion coefficient, and $\boldsymbol{S}$ is the source term. Replacing the flux φ in Equation (2.161) with the proper fluid variables, and considering the corresponding diffusion coefficients Γ and source terms $\boldsymbol{S}$, the continuity equation, the momentum conservation equation, the energy conservation equation, the turbulent kinetic energy equation, and the turbulent energy dissipation rate equation can be obtained, as shown in Table 2.1.

In any finite region in the fluid field, integrating the general transport equation leads to

$$\frac{\partial}{\partial t}\int_V \rho\boldsymbol{\varphi}dV+\int_A n\cdot(\rho\mathbf{v}\boldsymbol{\varphi})dA=\int_A n\cdot(\Gamma grad\boldsymbol{\varphi})dA+\int_V \mathbf{S}dV \tag{2.162}$$

where V is the volume of the finite region, and A is the surface area of the control volume. Let $\bar{\boldsymbol{v}}$ be the velocity of the outer control surface A, then the integral expression of the transport equation is rewritten as [24,25]:

$$\frac{\partial}{\partial t}\int_V \rho\boldsymbol{\varphi}dV+\int_A \mathbf{n}\cdot(\rho(\mathbf{v}-\bar{\mathbf{v}})\boldsymbol{\varphi})dA=\int_A \mathbf{n}\cdot(\Gamma grad\boldsymbol{\varphi})dA+\int_V \mathbf{S}dV \tag{2.163}$$

Table 2.1: Fluid variables and the corresponding diffusion coefficients and source terms.

	φ	Γ	S
Continuity equation	1	0	0
Momentum conservation equation	$u_i(i=1,2,3)$	$\mu+\mu_t$	$\frac{\partial}{\partial x_i}(-p)+\frac{\partial}{\partial x_j}(\tau_{ij}+\tau'_{ij})$
Turbulent kinetic energy equation	k	$\mu+\frac{\mu_t}{\sigma_k}$	$\mu_t s_{ij}\frac{\partial u_i}{\partial x_j}-\rho\varepsilon$
Turbulent energy dissipation rate equation	ε	$\mu+\frac{\mu_t}{\sigma_\varepsilon}$	$\frac{\varepsilon}{k}\left(C_{1\varepsilon}\mu_t s_{ij}\frac{\partial u_i}{\partial x_j}-C_{2\varepsilon}\rho\varepsilon\right)$

Table 2.2: Pantograph Contact Stiffness.

Loading stiffness (N/mm)	Uploading stiffness (N/mm)
3384	3484.4

If $\bar{v} = 0$, Equation (2.163) is the Euler form of the fluid equation. If the surface $\bar{v} = v$, Equation (2.163) then is the Lagrangian form of the fluid equation. Equation (2.163) is an arbitrary Lagrangian-Eulerian (ALE) form of the fluid equation if t $\bar{v} \neq v$ and $\bar{v} \neq 0$.

2.2.5.2 Geometric model

In recent years, numerical simulation has been widely applied to train aerodynamics with the development of computer hardware and improvements in computational speed. Though the results obtained from the numerical simulation may be different from the experimental results, the numerical simulation can provide a more global analysis of the flow field, compared with experimental tests. Numerical simulation will play an increasingly important role in train aerodynamics studies. Being the key aspect of train aerodynamic simulation, geometric modeling is presented below.

To obtain satisfactory simulation results, the following basic principles should be addressed in geometric modeling:

(1) In the model of the vehicle system, at least the head, middle, and rear vehicles should be included with consideration for their original shapes.
(2) The computation domain in the fluid field is chosen in such a way that the boundaries do not interfere with the airflow around the train. Meanwhile, it is necessary to further verify the independence of the computation domain from the results obtained for different problems.
(3) The boundary conditions are carefully taken so that the main phenomena are captured, and the effects of truncation errors are minimized.

There are two kinds of typical geometric modeling problems in train aerodynamics, which correspond to the aerodynamics of the train under ambient winds and the aerodynamics of trains passing by each other. Figure 2.46 shows the computation domain and the boundary

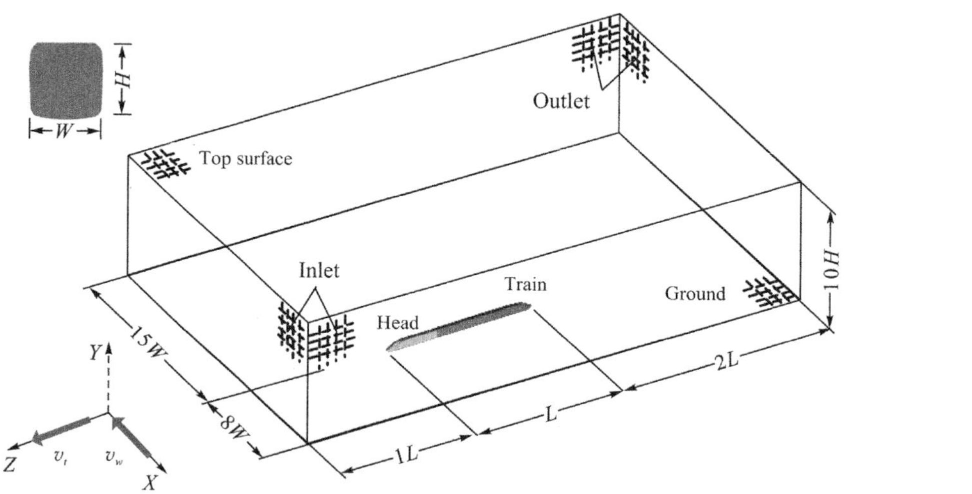

Figure 2.46
Computational region of the high-speed train under crosswind.

conditions [25] in a model of the train aerodynamics under ambient winds. For this problem, the parameters and boundary conditions of the computational model can be determined using wind tunnel tests, and the numerical simulation is then performed. As shown in Figure 2.46, the computation domain can be regarded as a cuboid around the train. The boundary conditions of the flow field are taken as follows: The front head and left side surfaces of the head train are the conditions of the velocity entry, while the back/behind and right side surfaces of the rear train are the conditions of the pressure outlet, that is, $P_{\text{outlet}} = 0$. The top surface of the train is considered as a symmetrical boundary condition. Let L, W, and H be the length, weight, and height of the train, respectively. The computational region of the flow field represented by the cuboid has at least the length 4L, the weight 23W, and the height 10H. The longitudinal distance between the entrance end of the train and the tip of the nose of the train head is at least L. The longitudinal distance between the exit end of the train and the tip of the nose of the train head is at least 2L. The height from the top of the region to the ground is at least 10H. The lateral distance between the windward side of the region and the centerline of the track is at least 8W. The leeward side of the region is at least 15W from the centerline of the track.

Figure 2.47 and Figure 2.48 show the computation domain and boundary conditions [26] for the aerodynamics of regular trains passing by each other. The slip grid method is often used to numerically simulate this problem. The boundary conditions of the flow field in simulation are considered as follows: the end surfaces of the upper part and two sides of the train are treated as the symmetrical boundary conditions; the end surfaces in the front and the rear of the train are treated as the pressure outlet conditions, that is, $P_{\text{outlet}} = 0$; the interface of the two grid regions is treated as the interacted boundary condition, that is, the interface flow field information of the two areas is mutually transmitted; the grid region containing the train moves with the train; the height from the top of the region to the ground is at least 10H; the distance between the left and right sides of the computation domain is at least 10W from the center of the line; the line spacing is D0; the longitudinal distance between the tips of two train heads is D1. Experiences show that the length of D1 should be more than 50 m.

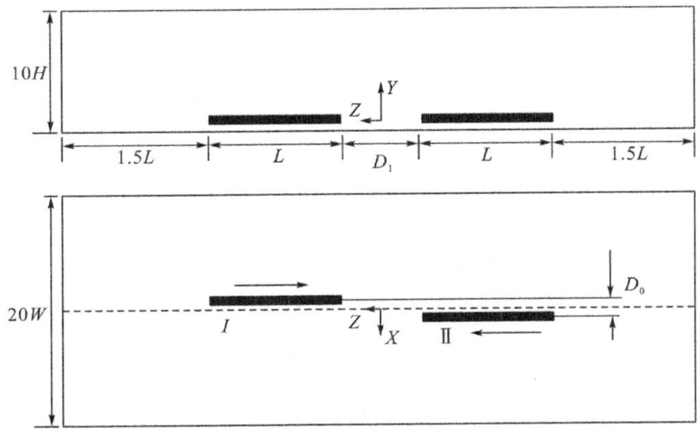

Figure 2.47
Computation domain of two trains passing by each other.

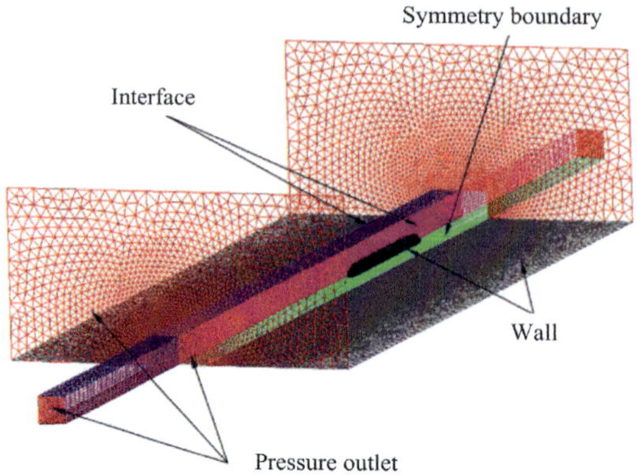

Figure 2.48
The boundary conditions of two trains passing by each other.

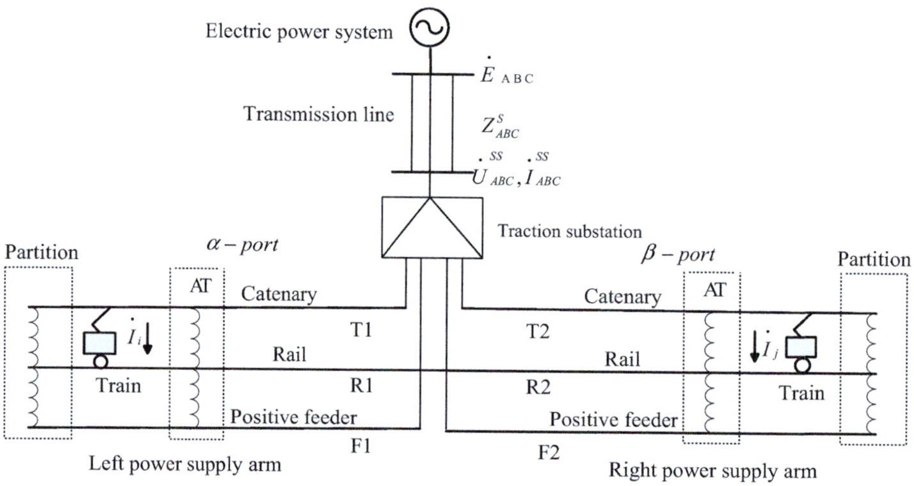

Figure 2.49
Traction power supply system of the high-speed railway.

2.2.6 Power System Modeling

Figure 2.49 presents the traction power supply system of the high-speed railway. The power supply system powered by the traction substation is usually referred to as an external power supply or primary system. It consists of a traction network and a traction substation [27, 28].

The electric power system and transmission line provide power for electrified railways at a voltage level of 110 kV or 220 kV. The traction load of the electrified railway is the first level load.

Traction substation: the role of the traction substation is to convert the electrical energy supplied by the electric power system into a form that provides electric traction and is compatible with its power supply mode. The core component of the traction substation is the traction transformer. Common traction transformer wiring methods include YNd11, Scott, Vv, and Vx. To balance the negative sequence, the traction transformer also adopts a commutation connection.

Traction power supply mode: the power supply mode of the traction network can be divided into direct power supply mode (DF), direct power supply mode with return line (DN), Booster-Transformer (BT) power supply mode, Auto-Transformer (AT) power supply mode and Coaxial Cable (CC) power supply mode, etc. Spatial distribution of the traction network varies with different power supply modes. The high-speed railway generally adopts the AT power supply mode. The corresponding model is presented in Figure 2.49.

2.2.6.1 Traction substation model

Traction substation equivalent circuit: in addition to the wiring form of the traction transformer and the phase sequence, the mode of operation of the electric power system and the impact of voltage loss on the power transmission line should also be considered. In Figure 2.49, the three-phase voltage of the electric power matrix is $\dot{E}_{ABC} = \left[\dot{E}_A, \dot{E}_B, \dot{E}_C\right]^T$, and the three-phase impedance matrix of the electric power matrix is $Z^S_{ABC} = \begin{bmatrix} z_{aa} & z_{ab} & z_{ac} \\ z_{ba} & z_{bb} & z_{bc} \\ z_{ca} & z_{cb} & z_{cc} \end{bmatrix}$. Considering system impedance and transmission line impedance, the traction transformer leakage resistance matrix is $Z^{TT}_{ABC} = \begin{bmatrix} z_{TA} & z_{TAB} & z_{TAC} \\ z_{TBA} & z_{TB} & z_{TBC} \\ z_{TCA} & z_{TCB} & z_{TC} \end{bmatrix}$, three-phase voltage is $\dot{U}^{SS}_{ABC}$, and $\dot{I}_{ABC}$ on the incoming line of the traction substation is equivalent to the three-phase current on the primary side of the traction load. The traction power supply system structure model [29] is shown in Figure 2.50.

2.2.6.2 Simulation model of the traction power supply system

Generally, the structure of the traction network is a parallel multi-conductor transmission line system consisting of conductors and backflow network conductors, regardless of whether the current traction power supply mode is single-line or double-line [30 , 31]. Based on the chain structure of the entire system, a general mathematical model of the

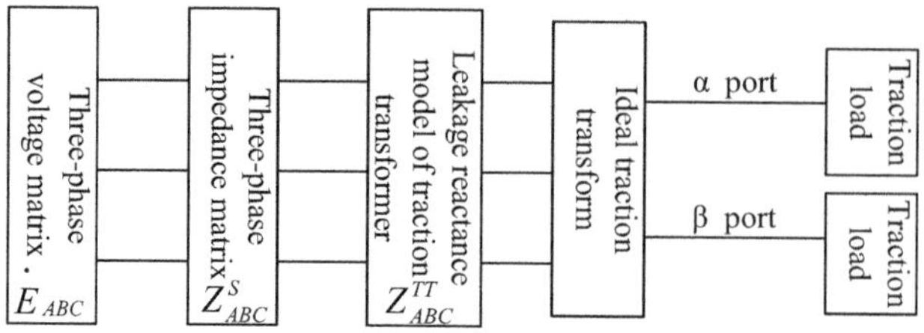

Figure 2.50
Structure model of the traction power supply system.

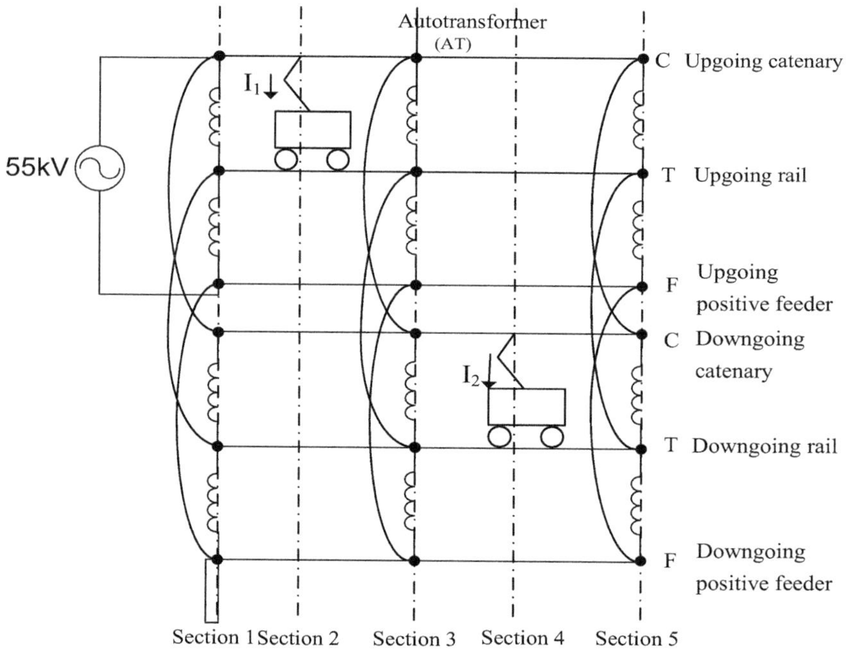

Figure 2.51
Chain network of the traction network.

traction power supply system can be established by combining the traction substation, traction load, and traction network. The model can be adapted to the multi-train power flow calculation under different traction network power supply modes.

There is a locomotive on the upgoing and downgoing lines of the traction network, as shown in Figure 2.51. The traction network can be divided into five parts: traction substation, locomotive 1, AT, locomotive 2, and power supply arm, according to the structure characteristics of the traction power supply system and the actual position of the train. This chain network consists of longitudinal series elements and parallel elements. The series element is the parallel multi-conductor line between adjacent sections, and the parallel element is the transverse element on the sections.

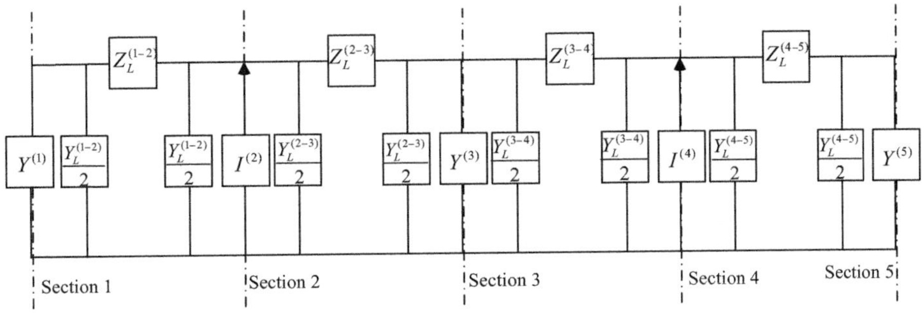

Figure 2.52
Chain Network Equivalent Circuit.

The equivalent circuit of the chain network of the traction network is shown in Figure 2.52. $Y^{(1)}$ on section 1 represents the traction substation model, $I^{(2)}$ and $I^{(4)}$ on sections 2 and 4 represent the load current, $Y^{(3)}$ on section 3 represents the AT model, $Y^{(5)}$ on section 5 represents the model of the end of the traction network. The equivalent circuit of the parallel multi-wire between the sections is modeled as a Π-type circuit, and $Y^{(1\text{-}2)}$ is the traction network admittance of the ground from 1 to 2. For the AT complex line shown in Figure 2.52, the number of traction mesh conductors is six. Therefore, the order of each impedance matrix and that of the admittance matrix is6×6.

For any type of traction network, the number of parallel conductors can be set as m, and then the order of each impedance matrix and admittance matrix can be set as$m \times m$. The power supply arm has a total of n sections. Taking each section as a node, the admittance matrix of the power supply arm can be obtained by modeling the series-parallel elements on a section and merging the admittance matrix of the parallel elements, as shown in Equation (2.164).

$$\mathbf{Y} = \begin{bmatrix} Y_1 + Z_1^{-1} & -Z_1^{-1} & & & \\ -Z_1^{-1} & Z_1^{-1} + Y_2 + Z_2^{-1} & -Z_2^{-1} & & \\ & -Z_2^{-1} & Z_2^{-1} + Y_3 + Z_3^{-1} & & \\ & \ddots & \ddots & \ddots & \\ & & & Z_{n-2}^{-1} + Y_{n-1} + Z_{n-1}^{-1} & -Z_{n-1}^{-1} \\ & & & -Z_{n-1}^{-1} & Z_{n-1}^{-1} + Y_n \end{bmatrix} \tag{2.164}$$

Taking the traction substation as a unit, a model of the traction power supply system centered on the traction substation can be established by combining the chain structures of two power supply arms, a and b, in the substation, as shown in Figure 2.53, considering that Z_N is an electrical phase separation matrix.

The node voltage equation of the traction substation is presented as Equation (2.165).

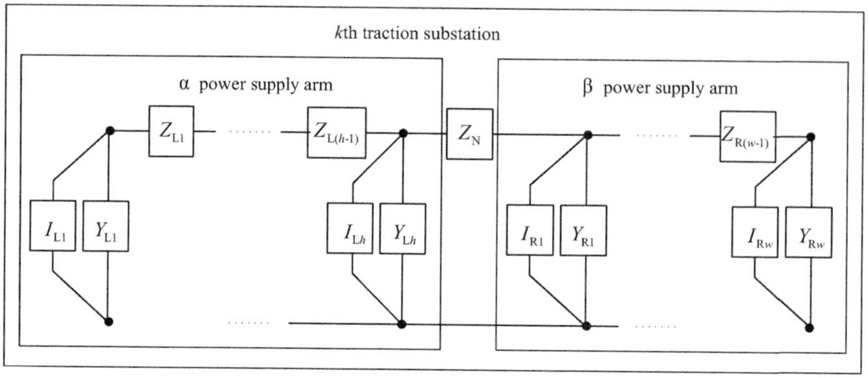

Figure 2.53
Chain network structure centered on the traction substation.

$$\begin{bmatrix} I_{L1} \\ \vdots \\ I_{Lh} \\ I_{R1} \\ \vdots \\ I_{Rw} \end{bmatrix} = \begin{bmatrix} Y_{L1}+Z_{L1}^{-1} & -Z_{L1}^{-1} & \cdots & 0 & 0 & \cdots & 0 \\ \vdots & \vdots & \cdots & \vdots & \vdots & \cdots & \vdots \\ 0 & 0 & \cdots & Z_{L(h-1)}^{-1}+Y_{Lh}+Z_N^{-1} & -Z_N^{-1} & \cdots & 0 \\ 0 & 0 & \cdots & -Z_N^{-1} & Z_N^{-1}+Y_{R1}+Z_{R1}^{-1} & \cdots & 0 \\ \vdots & \vdots & \cdots & \vdots & \vdots & \cdots & \vdots \\ 0 & 0 & \cdots & 0 & 0 & \cdots & Z_{R(w-1)}^{-1}+Y_{Rw} \end{bmatrix} \begin{bmatrix} U_{L1} \\ \vdots \\ U_{Lh} \\ U_{R1} \\ \vdots \\ U_{Rw} \end{bmatrix} \tag{2.165}$$

2.2.7 Modeling of the drive system [32]

2.2.7.1 High-speed train transmission system topology

The function of the traction drive system of the high-speed train is to convert electric energy into mechanical energy to drive the train and to convert mechanical energy into electric energy that can be used to feed electric power back into the power grid. When the train is running, the traction drive system converts a 25 kV, 50 Hz, single-phase alternating current on the catenary into a three-phase alternating current to supply the traction motor. In order to meet the traction requirements of locomotives, it is necessary to control the moment speed of the motor during operation. In train braking, priority is given to dynamic braking. During this process, the traction motor operates as a generator and converts the three-phase AC output from the traction motor into a 25 kV, 50 Hz single-phase AC that is fed back to the power grid, ultimately realizing regenerative braking.

The traction drive system primarily consists of the pantograph (including its high voltage electrical equipment), traction transformer, four-quadrant pulse rectifier, intermediate link, traction inverter, traction motor, and other electrical equipment as shown in Figure 2.54.

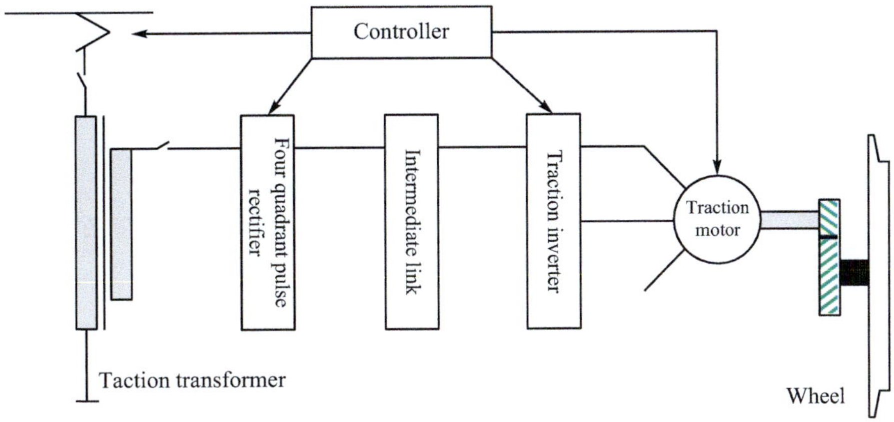

Figure 2.54
Traction drive system of the high-speed train.

The pantograph supplies the 25 kV single-phase power frequency alternating current of the catenary to the traction transformer, and the single-phase AC power that is stepped down by the transformer is supplied to the pulse rectifier. The pulse rectifier then converts the single-phase alternating current into direct current and outputs the direct current to the traction inverter through the intermediate DC circuit and draws the output voltage and current of the inverter. The three-phase alternating current with controllable frequency is supplied to the three-phase asynchronous traction motor. The torque and speed outputs at the shaft end of the traction motor are transmitted to the wheelset through the gear transmission and are converted into the tractive effort and linear velocity at the wheel rim.

Presently, an AC-DC-AC main circuit structure with a DC voltage intermediate circuit is used for the domestically produced three-phase AC drive high-speed train. Figure 2.55 shows the three-level main circuit of a high-speed train of the China High-Speed Railway. Taking this type of high-speed train as an example, the mathematical model of the input voltage from the pantograph to the output torque of the traction motor is given below.

2.2.7.2 Mathematical model of the traction drive system of a type of EMU

(1) Three-level pulse rectifier

The main circuit structure of an NPC three-level pulse rectifier is shown in Figure 2.56. In this figure, u_1 is the voltage on the DC side support capacitanceC_1, and u_2 is the voltage on the DC side supporting capacitorC_2. The ideal switching functions, S_A and S_B, are defined as Equations (2.166) and (2.167), respectively. Using the ideal switching function and ignoring the winding resistance, the main circuit of the three-level pulse rectifier in Figure 2.56(a) can be equivalent to the circuit in Figure 2.56(b).

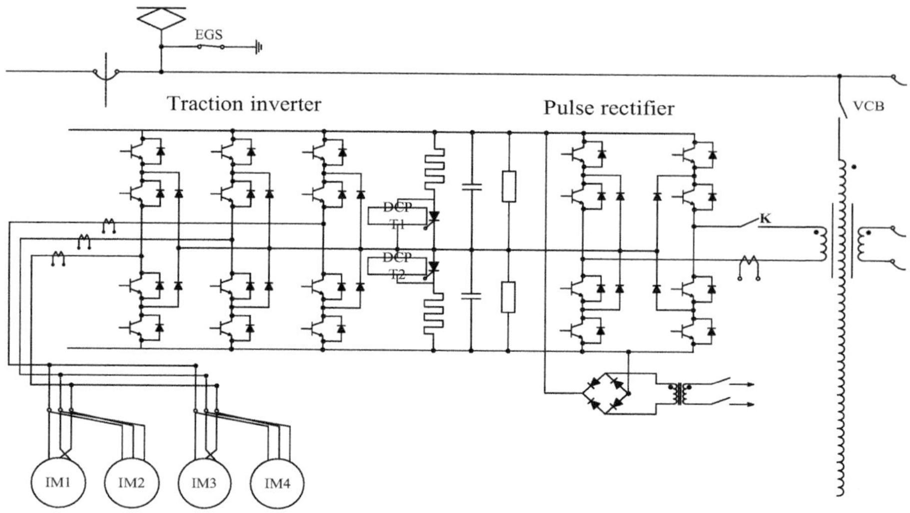

Figure 2.55
Main circuit of the CRH2 EMU.

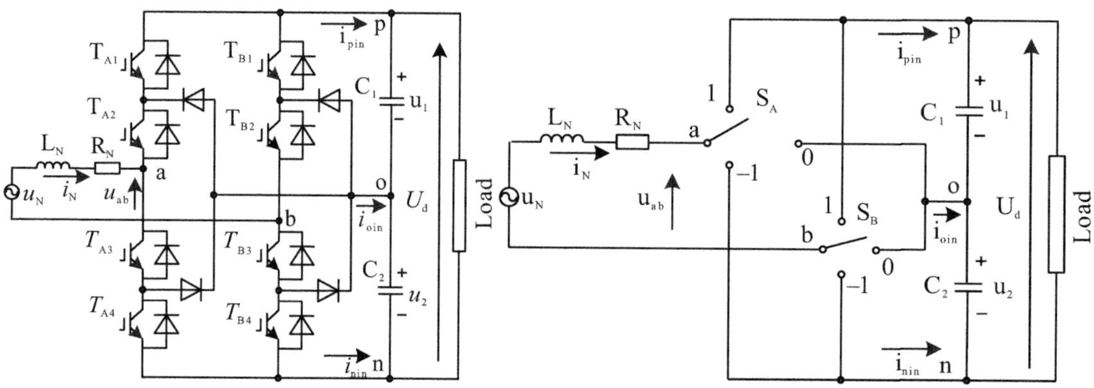

Figure 2.56
Two-level pulse rectifier topology.

$$S_A = \begin{cases} 1(\mathrm{p}) & T_{a1} \text{ and } T_{a2} \text{ conduct} \\ 0(\mathrm{o}) & T_{a2} \text{ and } T_{a3} \text{ conduct} \\ -1(\mathrm{n}) & T_{a3} \text{ and } T_{a4} \text{ conduct} \end{cases} \tag{2.166}$$

$$S_B = \begin{cases} 1(\mathrm{p}) & T_{b1} \text{ and } T_{b2} \text{ conduct} \\ 0(\mathrm{o}) & T_{a2} \text{ and } T_{a3} \text{ conduct} \\ -1(\mathrm{n}) & T_{b3} \text{ and } T_{b4} \text{ conduct} \end{cases} \tag{2.167}$$

The voltage and current at the input of the grid side are analyzed using Kirchhoff's law, and the voltage at the input of the grid side can be expressed as:

$$u_{ao} = \frac{S_A(S_A+1)}{2}u_1 - \frac{S_A(S_A-1)}{2}u_2 \tag{2.168}$$

$$u_{bo} = \frac{S_B(S_B+1)}{2}u_1 - \frac{S_B(S_B-1)}{2}u_2 \tag{2.169}$$

$$u_{ab} = u_{ao} - u_{bo} = \left[\frac{S_A(S_A+1)}{2} - \frac{S_B(S_B+1)}{2}\right]u_1 - \left[\frac{S_A(S_A-1)}{2} - \frac{S_B(S_B-1)}{2}\right]u_2 \tag{2.170}$$

Assuming that the switching tube is an ideal model (there is no power loss or energy storage in the commutation process), the instantaneous power on the AC side and the DC side would be equal, such that:

$$u_{ab}i_s = u_1 i_{pin} - u_2 i_{nin} \tag{2.171}$$

Substituting Equation (2.170) into Equation (2.171):

$$i_{pin} = \frac{S_A(S_A+1) - S_B(S_B+1)}{2}i_s \tag{2.172}$$

$$i_{nin} = \frac{S_A(S_A-1) - S_B(S_B-1)}{2}i_s \tag{2.173}$$

(2) Three-level traction inverter

The main circuit of the system when the three-level inverter supplies power to the asynchronous traction motor is presented in Figure 2.57(a). The switch model of an ideal three-level inverter circuit is shown in Figure 2.57(b). The circuit structure of each phase

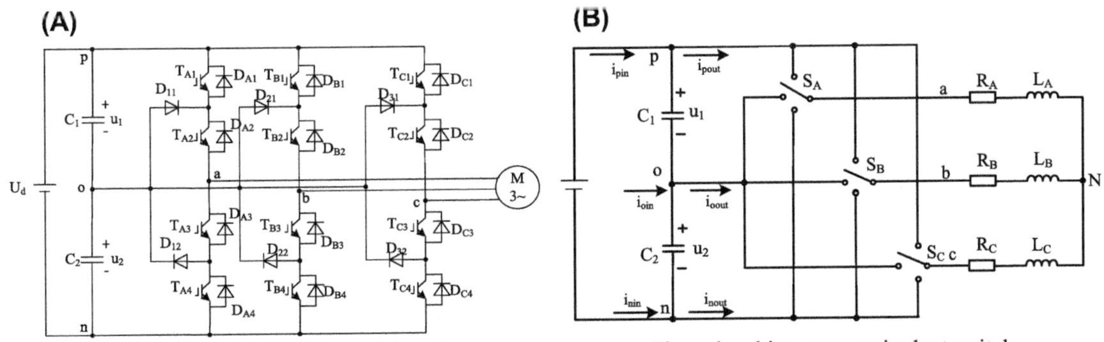

Three-level inverter main circuit topology Three-level inverter equivalent switch

Figure 2.57

Three-level inverter-motor topology. (A) Three-level inverter main circuit topology (B) Three-level inverter equivalent switch.

of the bridge arm can be simplified into a single-pole three-throw switch S that communicates with the DC side. Each bridge arm has three states: state P (upper bridge arm switch device turned on), state O (auxiliary switch device turned on), and state N (lower bridge arm switch device turned on).

Defining S_i ($i = A, B, C$) as an ideal switching function, the states P, O, and N are represented by 1, 0, and -1, respectively. The output phase voltage of each bridge arm of the inverter can be expressed as:

$$\begin{cases} u_{ao} = \dfrac{S_A(S_A+1)}{2}u_1 - \dfrac{S_A(S_A-1)}{2}u_2 \\ u_{bo} = \dfrac{S_B(S_B+1)}{2}u_1 - \dfrac{S_B(S_B-1)}{2}u_2 \\ u_{co} = \dfrac{S_C(S_C+1)}{2}u_1 - \dfrac{S_C(S_C-1)}{2}u_2 \end{cases} \tag{2.174}$$

The capacitive voltage of the intermediate DC link can be expressed as:

$$\begin{cases} C_1\dfrac{du_1}{dt} = i_{pin} - i_{pout} \\ C_2\dfrac{du_2}{dt} = -i_{nin} + i_{nout} \\ C_1\dfrac{du_1}{dt} - C_2\dfrac{du_2}{dt} = -i_{oin} + i_{oout} \end{cases} \tag{2.175}$$

The output current of the intermediate DC link can be expressed as:

$$\begin{cases} i_{\text{pout}} = \dfrac{S_A(S_A+1)i_a + S_B(S_B+1)i_b + S_C(S_C+1)i_c}{2} \\ i_{nout} = \dfrac{S_A(S_A-1)i_a + S_B(S_B-1)i_b + S_C(S_C-1)i_c}{2} \\ i_{oout} = -S_A^2 i_a - S_B^2 i_b - S_C^2 i_c = (S_C^2 - S_A^2)i_a + (S_C^2 - S_B^2)i_b \end{cases} \tag{2.176}$$

(3) Mathematical model of the traction motor

The AC asynchronous traction motor is a major execution link in the traction drive system. It is a multivariable system of a high order, nonlinear, strong coupling, and its static characteristics, dynamic characteristics, and control technology are far more complex than those in DC motors. The traditional mathematical model of the asynchronous motor in a three-phase coordinate system is complex. Here, we discuss only the ω_r-ψ_r-i_s state equation of the two-phase synchronous rotation dq coordinate system. By changing this model a little, the other types of two-phase coordinate systems can also be obtained. Torque formula:

$$\begin{aligned} T_e &= \frac{n_p L_m}{L_r}\left(i_{sq}\psi_{rd} - L_m i_{sd} i_{sq} - i_{sd}\psi_{rq} + L_m i_{sd} i_{sq}\right) \\ &= \frac{n_p L_m}{L_r}\left(i_{sq}\psi_{rd} - i_{sd}\psi_{rq}\right) \end{aligned} \tag{2.177}$$

State equations of ω_r-ψ_r-i_s are set as follows:

$$\frac{d\omega_r}{dt} = \frac{n_p^2 L_m}{JL_r}\left(i_{sq}\psi_{rd} - i_{sd}\psi_{rq}\right) - \frac{n_p}{J}T_L \tag{2.178}$$

$$\frac{d\psi_{rd}}{dt} = -\frac{1}{T_r}\psi_{rd} + (\omega_s - \omega_r)\psi_{rq} + \frac{L_m}{T_r}i_{sd} \tag{2.179}$$

$$\frac{d\psi_{rq}}{dt} = -\frac{1}{T_r}\psi_{rq} - (\omega_s - \omega_r)\psi_{rd} + \frac{L_m}{T_r}i_{sq} \tag{2.180}$$

$$\frac{di_{sd}}{dt} = \frac{L_m}{\sigma L_s L_r T_r}\psi_{rd} + \frac{L_m}{\sigma L_s L_r}\omega_r\psi_{rq} - \frac{R_s L_r^2 + R_r L_m^2}{\sigma L_s L_r^2}i_{sd} + \omega_s i_{sq} + \frac{u_{sd}}{\sigma L_s} \tag{2.181}$$

$$\frac{di_{sq}}{dt} = \frac{L_m}{\sigma L_s L_r T_r}\psi_{rq} - \frac{L_m}{\sigma L_s L_r}\omega_r\psi_{rd} - \frac{R_s L_r^2 + R_r L_m^2}{\sigma L_s L_r^2}i_{sq} - \omega_s i_{sd} + \frac{u_{sq}}{\sigma L_s} \tag{2.182}$$

where T_e is motor torque; n_p is the number of pole pairs; $R_s$is stator resistance; $R_r$is rotor resistance; $L_m$is excitation inductance; $L_s$is self-inductance of the stator; $L_r$is self-inductance of the rotor; i_{sd} and i_{sq} are the stator current d and q axis components of the motor; u_{sd} and u_{sq} are the stator voltage d and q axis components of the motor; ψ_{rd} and ψ_{rq} are the rotor flux d and q axis components of the motor rotor, respectively; $\omega_s$is the motor power frequency; $\omega_r$is the rotation speed of the motor; $J$is moment of inertia; T_L is load; T_L is motor electromagnetic time constant, and $T_r = \frac{L_r}{R_r}$; σ is leakage coefficient of the motor, and $\sigma = 1 - \frac{L_m^2}{L_s L_r}$.

2.3 Coupling models

Based on the vehicle system, rail system, pantograph-catenary system, power supply system, and the traction drive system, the carriage coupling, wheel-rail coupling, pantograph-catenaries coupling, and electromechanical coupling form a large high-speed train coupling system dynamics model.

2.3.1 Coupling model

2.3.1.1 Coupling model between vehicles

The connection between the vehicles in the high-speed train is an essential aspect of the train design because it consists of and completes not only the mechanical connection between the vehicles but also the electrical connection. It can affect vehicle boundaries, aerodynamics, and vehicle dynamics. Furthermore, the design of the vehicle connections should also consider the provision of safe and comfortable passages for passengers. The vehicle end connection device of the high-speed train includes mainly a close-coupled coupler buffer device, a shock absorber between the vehicles, an internal and external windshield, and electrical connection equipment. Among all these equipment, the coupler buffer device, shock absorber between vehicles, and the closed windshield device have the most significant influence on the dynamic performance of the train. To establish a high-speed train and track coupling dynamics model in this book, detailed dynamic modeling has been carried out for the close-coupled coupler cushioning device, shock absorber between vehicles, and close-coupled windshield device commonly used in high-speed trains in China, as shown in Figure 2.58.

(1) Coupler buffer device model

The coupler and draft gear is the most important device in the suspension system and is one of the most important components among train equipment. First, it primarily realizes various connections between the vehicles, such as the mechanical connection, windshield

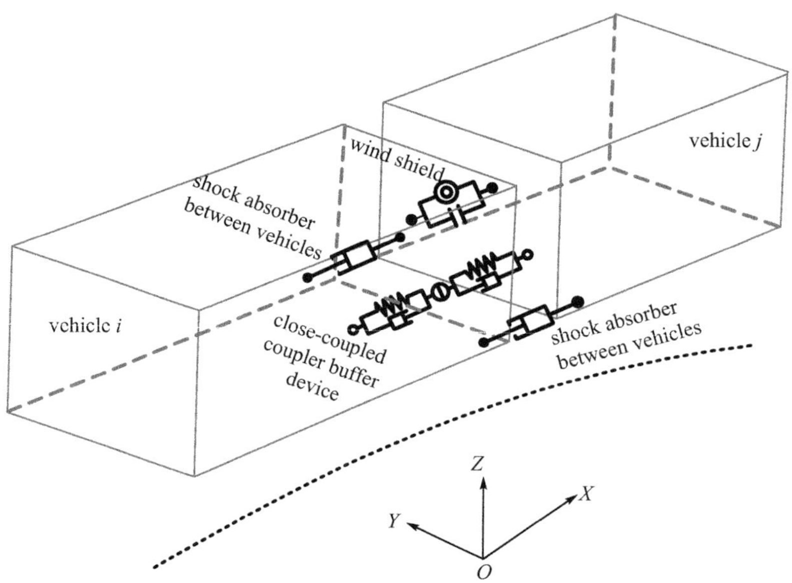

Figure 2.58
Connection element of the train end.

connection, and electrical connection. Secondly, it realizes the buffering effect between the vehicles and prevents damage to the connected car body. Furthermore, it also ensures the stationary state between vehicles and improves the ride comfort of the running vehicle. The coupler buffer system of China Railway High-speed (CRH) trains primarily consists of an automatic tightlock coupler at both ends of the vehicles, a semi-permanent coupler between the vehicles, and an emergency transition coupler for emergency rescue. The automatic tightlock coupler and the emergency transition coupler are located at the front and rear ends of the vehicles for reconnection to the train or connection between the vehicles. The semi-permanent coupler is located at the middle of the train for connection of adjacent vehicles in the train. To establish a high-speed train dynamic model for multi-vehicles, numerical modeling is required for only the semi-permanent coupler and draft gears for vehicle connections in the middle of the train. The left plane of Figure 2.59 shows the components of the semi-series semi-automatic tightlock coupler used in high-speed trains in China [33]. It shows that the coupler between the two trains is fixed by bolts and cannot be displaced in the horizontal plane. The entire coupler buffer device can freely rotate through a certain angle around the hook pin in horizontal and vertical directions to ensure that the train passes flat vertical curves. The link body of the hook body and the two hooks are virtually rigid parts, and the buffer can be extended and contracted axially to provide a buffer force between vehicles that relieves the longitudinal impulse of the train.

To establish a dynamic model of the coupler and draft gear, in this book we assume that the hook buffer system is hinged at the center of the base of the two vehicles bodies, and can only be axially stretched and deformed. When the train is running, the horizontal and vertical corners of the coupler are within the maximum free-angle of each design and do not have rigid contact with the vehicle body. The quality of the coupler system is negligible when considering the inertial contribution of the entire train. Thus, the coupler cushioning device is simplified as a two-force bar that can only expand and contract in the axial direction. The space coordinate of the coupler hinge point of any instantaneous front

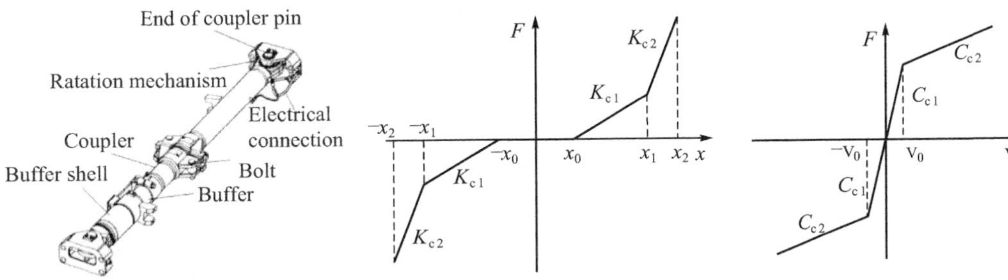

Figure 2.59

Diagram of a coupler and the stiffness damping nonlinear model of the coupler cushioning device.

and rear body hooks can be used to determine the force of the coupler cushioning device during train operation, and the component forces (including vertical, horizontal, and vertical forces) acting on the end of the vehicle body. Tightlock couplers are tightly connected by hooks and have no relative longitudinal displacement and can thus be considered as a force element. The end of the coupler pin facilitates the hooking of the coupler, but the angle of the coupler nod is relatively small. The coupler buffer is a series connection of two couplers, and the anti-dive stiffness is set at the tail pin of the coupler. The following basic assumptions are made for the establishment of the coupler force element model: (1) the coupler buffer device connected to each other between trains and vehicles is a rigid two-force straight rod, which is not bent and can only achieve expansion deformation axially; (2) during running of the train, the horizontal and vertical corners of the coupler are within the maximum free-angle angle of each design, and have no rigid contact with the vehicle body; (3) axial deformation of the coupler buffer system is negligible for the distance between the centers of the two adjacent couplers. Figures 2.60 and 2.61 are top and side views, respectively, of the tightlock coupler force elements.

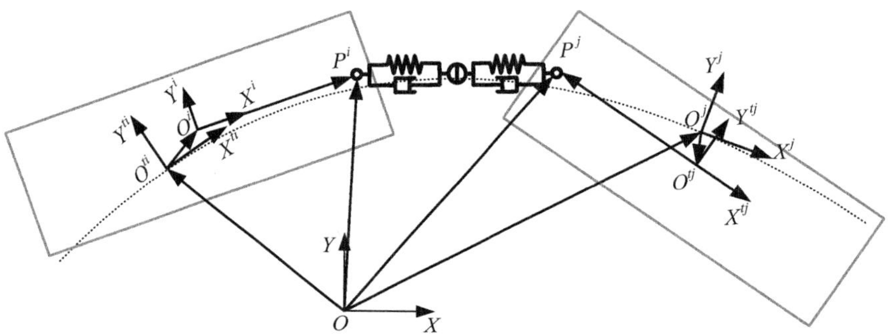

Figure 2.60
Tightlock coupler force element (top view).

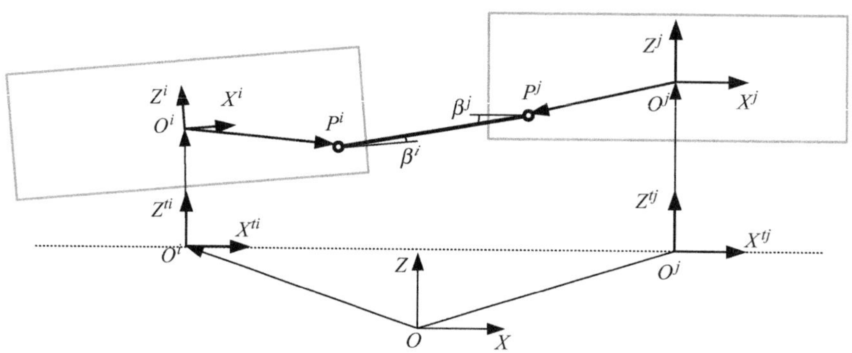

Figure 2.61
Tightlock coupler force element (side view).

The longitudinal force of the coupler is similar to the linear spring-damping parallel connection force in Section 2.2.2.1. The coupler is a strong nonlinear force element with complex tension and compression characteristics:

$$F_c = f_c(l - l_0) + g_c(\dot{l}) + F_0 \tag{2.183}$$

f_c, g_c are the nonlinear stiffness and damping of the train coupler, respectively. Because the angle of the coupler nod is small, a torsion spring that is resistant to the coupler is placed at the end of the coupler to calculate the nod moment. Under the global coordinate system, converting the length $\boldsymbol{r}^{ij}$of the coupler system into the train body coordinate system yields:

$$\bar{\boldsymbol{r}}^i = \mathbf{A}^{i\mathrm{T}}\mathbf{A}^{ti\mathrm{T}}\boldsymbol{r}^{ij}, \; \bar{\boldsymbol{r}}^j = \mathbf{A}^{j\mathrm{T}}\mathbf{A}^{tj\mathrm{T}}\boldsymbol{r}^{ij} \tag{2.184}$$

When the vehicle nods or passes a vertical curve, a nod angle is formed between the coupler and the vehicle body, as shown in Figure 2.61. The nod direction between the coupler and the front and rear body coordinate systems is calculated as:

$$\beta^i = \operatorname{asin}\left(\frac{\bar{r}_3^i}{|\bar{\boldsymbol{r}}^i|}\right), \; \beta^j = \operatorname{asin}\left(\frac{\bar{r}_3^j}{|\bar{\boldsymbol{r}}^j|}\right), \tag{2.185}$$

Based on the nodding angle, the nodding moment generated by the front of the vehicle can be calculated. The coupler force is determined using the dynamic characteristic curve of the coupler buffer system. Considering that it is difficult to obtain the dynamic characteristics of the coupler, only the nonlinear characteristics of the coupler buffer device with a bilinear model is stimulated, as shown in Figure 2.59. Although the tightlock coupler is also used in the high-speed train, there is still a 1~2 mm coupler clearance. This feature is also considered in the coupler buffer system model in this book. Thus, f_c and g_c can be expressed as:

$$f_c(x) = \begin{cases} 0 & |x| < x_0 \\ K_{c1}(x - x_0) & x_0 \le |x| \le x_1 \\ K_{c1}(x_1 - x_0) + K_{c2}(x - x_1) & |x| > x_1 \end{cases} \tag{2.186}$$

$$g_c(\dot{x}) = \begin{cases} C_1(\dot{x}) & |\dot{x}| \le v_1 \\ C_1(v_1) + C_2(\dot{x} - v_1) & |\dot{x}| > v_1 \end{cases} \tag{2.187}$$

where x_0 represents the coupler free clearance, $x = l$ - l_0 represents the telescopic length of the coupler cushioning device axially, K_1 and K_2 represent the coupler stiffness at different stages, and l_0 represents the initial equivalent length of the hook buffer device.

When the high-speed train is running in a straight line, the coupler force is represented primarily by the longitudinal component, and the lateral and vertical components of the coupler force are so small as to be negligible. When the train passes a flat vertical curve, the influence of the horizontal and vertical components of the coupler force on the train dynamics cannot be ignored. As the train passes the curve, the geometric centerline of the adjacent vehicle will rotate relative to the vehicle body, and the center line of the coupler will also rotate relative to the vehicle body. The resultant force of the coupler force is obtained using Equation (2.183). In the train dynamics model discussed in this book, the coupler is applied to the vehicle body as an external load.

(2) Shock absorber model between vehicles

For the high-speed train with a maximum running speed of over 250km/h, it is often necessary to install longitudinal or lateral dampers between the vehicles. The installation of a shock absorber system can effectively couple adjacent vehicles in the train, which improves the overall dynamic performance of the train system. Figure 2.62 shows the longitudinal coupling shock absorber installed on a high-speed train in China. Line tests and numerical studies have shown [34] that the longitudinal shock absorber not only reduces the longitudinal impact of the train but can also suppress the moving head and nodding movement of the vehicle. It can improve the critical velocity of the train system, and the lateral stability and ride comfort of the train. Thereby improving the overall dynamic performance of the high-speed train.

In this book, the longitudinal shock absorber between vehicles is simplified into a nonlinear spring-damping series force element using the spring-damped series force element described in Section 2.2.1.1. However, the damping characteristic is nonlinear, as shown in Figure 2.62. Thus, Equation (2.35) can be rewritten as:

$$\dot{l}_2 + \frac{K}{g_c\left(\dot{l}_2\right)} l_2 = \frac{K}{g_c\left(\dot{l}_2\right)}(l - l_0) \tag{2.188}$$

where$K$is the damper joint stiffness, and $g_c(x)$ is the nonlinear damping function.

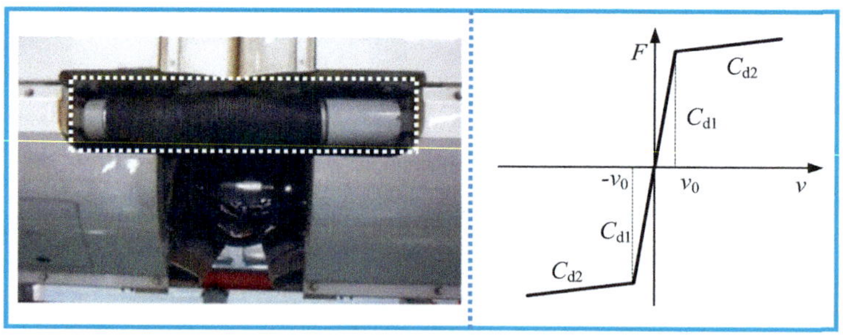

Figure 2.62
Picture of a longitudinal shock absorber and diagram of the nonlinear model of the longitudinal shock absorber between adjacent vehicles of the high-speed train.

(3) Model of the vestibule diaphragm device

The vestibule diaphragm device is a flexible connection between trains that provides passengers with a secure inter-vehicle through-passage to ensure a comfortable environment. The coupler force is determined by the dynamic characteristic curve of the coupler buffer system. Considering that it is difficult to obtain the dynamic characteristics of the coupler, in this book, only the nonlinear characteristics of the coupler buffer device with a bilinear model is stimulated, as shown in Figure 2.59. The tightlock vestibule diaphragm device is widely used as an internal windshield between vehicles in the high-speed train. A windshield with good tightness can effectively isolate the noise outside the vehicle, and it also has good performance regarding air tightness, heat preservation, and insulation. Figure 2.63 shows the tightlock windshield used by the CRH3 train. As the train running speed increases, the damping force of the train reduces. Thus, external

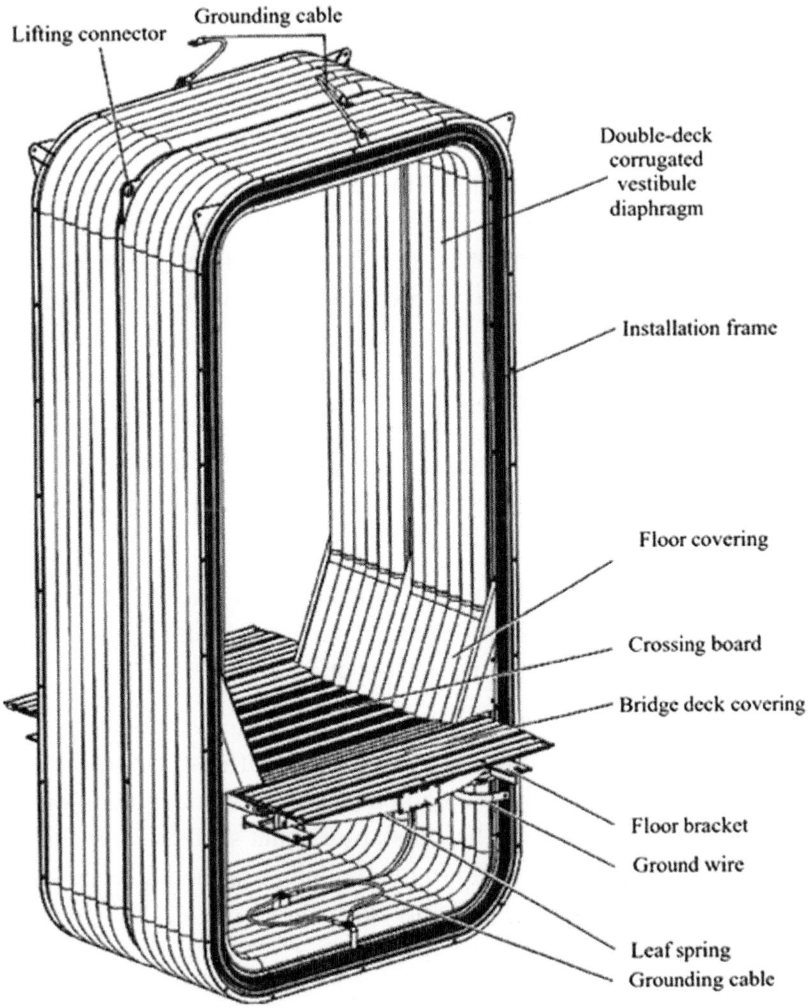

Figure 2.63
Tightlock windshield of the CRH3 EMU.

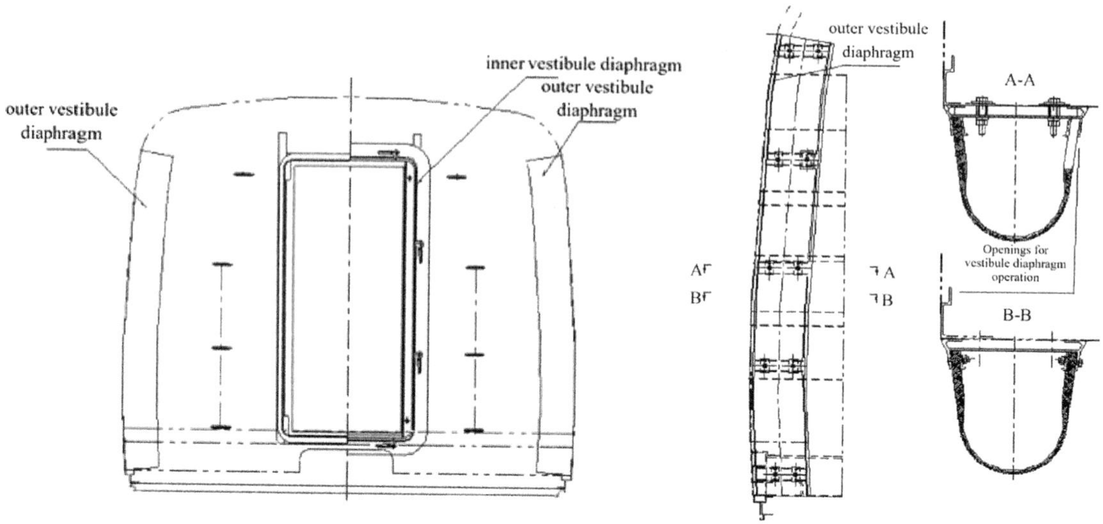

Figure 2.64
Outer windshield of the CRH2.

windshields are set between vehicles to make the train surface smooth and consequently, reduce wind resistance. Figure 2.64 shows the rubber outer windshield used in the CRH2. It is an intermittent windshield with its primary effect being drag reduction. The outer windshield shown in Figure 2.64 needs to be a compressed windshield because the outer windshield of the front and rear vehicles contact when the train passes a curve. The compressive stiffness of this type of windshield gives it the same shock absorption efficiency as a conventional vehicle end bumper.

Some external windshield products are totally enclosed, i.e., the whole outer windshield is installed at one end of the body with its structure somewhat similar to that of the inner windshield for noise isolation, and it is also sealed.

The primary effect of the internal and external windshields at the end of the vehicle is an interaction force caused by the relative position change of the vehicle ends. A simple windshield model can be regarded as a universal-friction element model.

Universal force element refers to a force element with three translational stiffness (longitudinal stiffness K_x, lateral stiffness K_y, and vertical stiffness K_z), translational damping, (longitudinal damping C_x, lateral damping C_y, and vertical damping C_z), torsional rigidity (roll torsional stiffness $K\varphi$, nodal torsional stiffness $K\theta$, and torsional stiffness $K\psi$), and torsional damping (roll torsion damping C_φ, nodal torsional damping C_θ, and torsional damping C_ψ) that can generate three directions of force and moment. The windshield equivalence model is shown in Figure 2.65. P^i and P^j belong to the marker

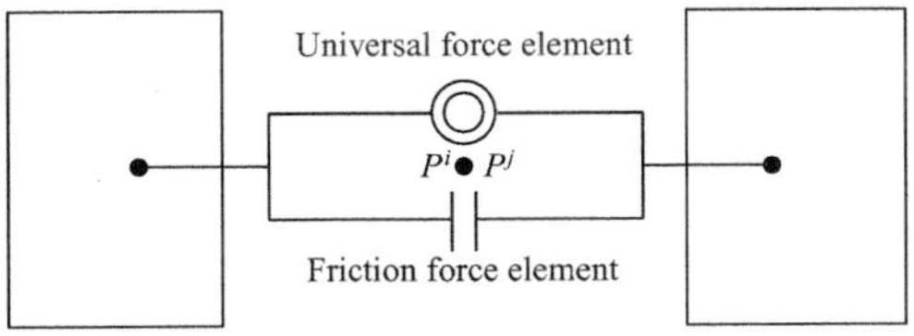

Figure 2.65
Windshield equivalence model.

point of the two bodies, respectively. In the windshield model, the two points coincide at the initial moment, which is located at the geometric center of the windshield. The marker points define the coordinates of the marker points parallel to the body coordinate system. The position vector and the transition matrix between the two coordinate systems are used to calculate the pulling quantum and rotation angle of the windshield for calculation of the windshield force. The force of the universal force element is given as:

$$F_c = K_s(Z_H - Z_c) \tag{2.189}$$

where F_{nx}, F_{ny}, F_{nz}, $T_{n\varphi}$, $T_{n\theta}$, and $T_{n\psi}$ are the initial force and initial moment in each direction; Δx, Δy, Δz, $\Delta\phi$, $\Delta\theta$, and $\Delta\psi$ are the displacements in three directions and the rotation angles in three directions, respectively. Furthermore, due to hysteresis in the vestibule diaphragm between the inner windshields, there is also a frictional behavior between the windshields. Thus, the internal and external windshields should also be considered using a friction model, which could be simplified into lateral and vertical shear friction forces and roll friction moments at the geometric center of the windshield. Assuming that the friction coefficient is correlated to the sliding velocity, as shown in Figure 2.66, the friction and friction moment can be calculated as:

$$\boldsymbol{F}_{\mathrm{f}} = \begin{bmatrix} \mu_x(\Delta\dot{x})F_{\mathrm{sz}} \\ \mu_y(\Delta\dot{y})F_{\mathrm{sx}} \\ \mu_z(\Delta\dot{z})F_{\mathrm{sx}} \end{bmatrix}, \quad \boldsymbol{T}_{\mathrm{f}} = \begin{bmatrix} \mu_\phi(\Delta\dot{\phi})F_{\mathrm{sx}}r_\phi \\ 0 \\ 0 \end{bmatrix} \tag{2.190}$$

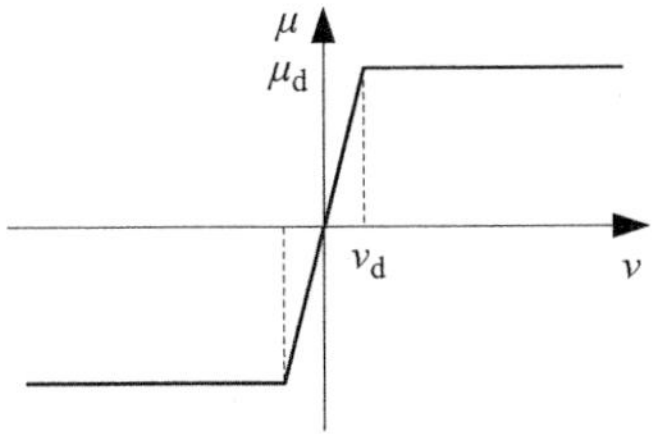

Figure 2.66
Friction coefficient curve.

where μ_y, μ and, μ_ϕ represent the transverse friction coefficient, the vertical friction coefficient, and side friction coefficient, respectively; and r_ϕ is the equivalent friction radius. The total force on the vehicle end connected windshield is:

$$\boldsymbol{F}_{\mathrm{fd}}=\boldsymbol{F}_{\mathrm{s}}+\boldsymbol{F}_{\mathrm{f}},\ \boldsymbol{T}_{\mathrm{fd}}=\boldsymbol{T}_{\mathrm{s}}+\boldsymbol{T}_{\mathrm{f}} \tag{2.191}$$

2.3.1.2 Coupling calculation method for the train

Traditional vehicle system dynamics generally only considers one vehicle and neglects longitudinal operation. Generally, only the simplified multi-mass model is used to study the train's operational behavior and braking performance. As kinetic research progresses, increasingly more accurate models are needed. High-speed trains are generally classified as a train of scattered power, and the bogie parameters of the motor car and the trail car are different, and each body has different quality parameters due to differences in structural design, the number of employees, and suspended equipment. Figure 2.67 shows the high-speed train in different marshalling forms.

Traditional train dynamics modeling and integration methods require the calculation of a large degree-of-freedom matrix determined by differential equations of motion of all the vehicles. However, the cyclic variable method treats a vehicle as a basic integration unit and does not require the calculation of the motion state of all variables in the train system at each time step, as in the traditional integration method. Instead, the first vehicle is calculated first, and then the second vehicle is calculated. In this case, all the vehicles are calculated one after the other until the state of motion of the tail car is calculated, and then the next step commences. In the new integral calculation method, because equations and the variables are reused, the new integration method can be called a cyclic variable method. Because the calculation is begun from the first vehicle in the integration, the calculation of the rear vehicle uses a new motion state that has been previously calculated, and a recursive effect is achieved. Thus, it is also called recursive integration method [2].

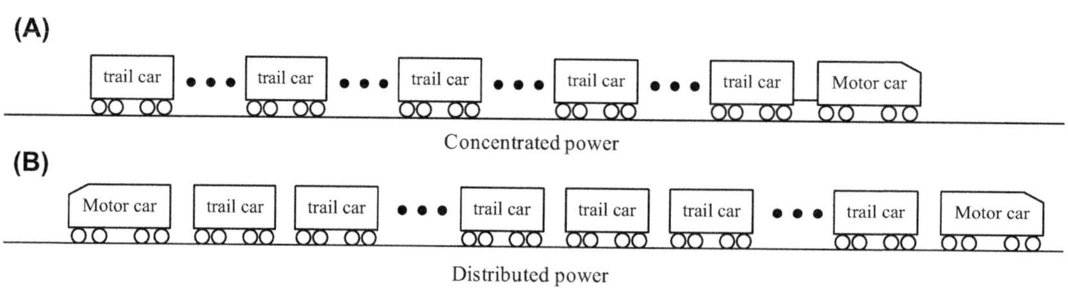

Figure 2.67
Train marshalling forms.

The determination of the integral initial of each vehicle in each time step and the calculation of the coupling force between the vehicles become the key issues to be solved by the cyclic variable method. During the integration process, in each cycle, the calculation results of each vehicle are stored in their respective intermediate variables for the next integration step. The initial value of each vehicle in each integration step and the calculation of the coupling force between vehicles can be scheduled directly using the intermediate variables, as shown in Figure 2.68.

In the new method, the parameter matrix corresponding to the vehicle type is based on the vehicle model determined by the vehicle location, and the integration is calculated based on the variable states in the temporary matrix. The new cyclic variable integration method needs only the different vehicle models and the different types of vehicle marshalling conditions. There is no need to calculate the matrix of all vehicle degrees of freedoms; consequently, it can effectively reduce the scale of the solution.

Therefore, this modeling process is convenient when working with a few workloads. Especially when the train marshalling changes, it only needs to modify the definition of the group, instead of remodeling and modifying the program like the traditional methods would. This improves the versatility and flexibility of the train dynamics simulation software. Due to limited original computing resources, it is necessary to recycle memory and processor resources. As computer technology has improved significantly, based on the basic principles of the cyclic variable method mentioned earlier, each vehicle can be regarded as a complete system. The coupler force is applied as an external force to the vehicle system to ensure the independence of the vehicle system, and parallel calculation is used to solve the entire train system. Based on the actual train formation, a complete train dynamics model is finally formed. The dynamic equation of the entire train system is:

$$\mathbf{M}^i \ddot{\boldsymbol{p}}^i = \boldsymbol{Q}_{\mathrm{c}}^i + \boldsymbol{Q}_{\mathrm{e}}^i + \boldsymbol{Q}_{\mathrm{v}}^i + \boldsymbol{Q}_{\mathrm{wr}}^i + j\boldsymbol{Q}_{\mathrm{cf}}^i + k\boldsymbol{Q}_{\mathrm{cr}}^i \quad (i = 1 \sim n) \tag{2.192}$$

where $\boldsymbol{Q}_{cf}$ is the generalized force between the front vehicles, $\boldsymbol{Q}_{cr}$ is the generalized force between the rear vehicles, and j and k are coefficients. When $i = 1, j = -1, k = 0$; when $i = n, j = 0, k = 1$; in other cases, $j = -1, k = 1$. Train dynamics can be solved using distributed computing or parallel computing. The process of train dynamics simulation is shown in Figure 2.69.

First, based on the train system configuration file, the dynamic model of each vehicle and the dynamic model of the coupling parts at the ends of the cars are generated, and each vehicle is solved independently using parallel calculation. Then, based on the state of the vehicle, the force of the coupling parts at the ends of the cars is calculated. If the calculation is not completed, the vehicle end force is applied to the vehicle system, the vehicle system state is updated, and the next step is applied until the simulation ends.

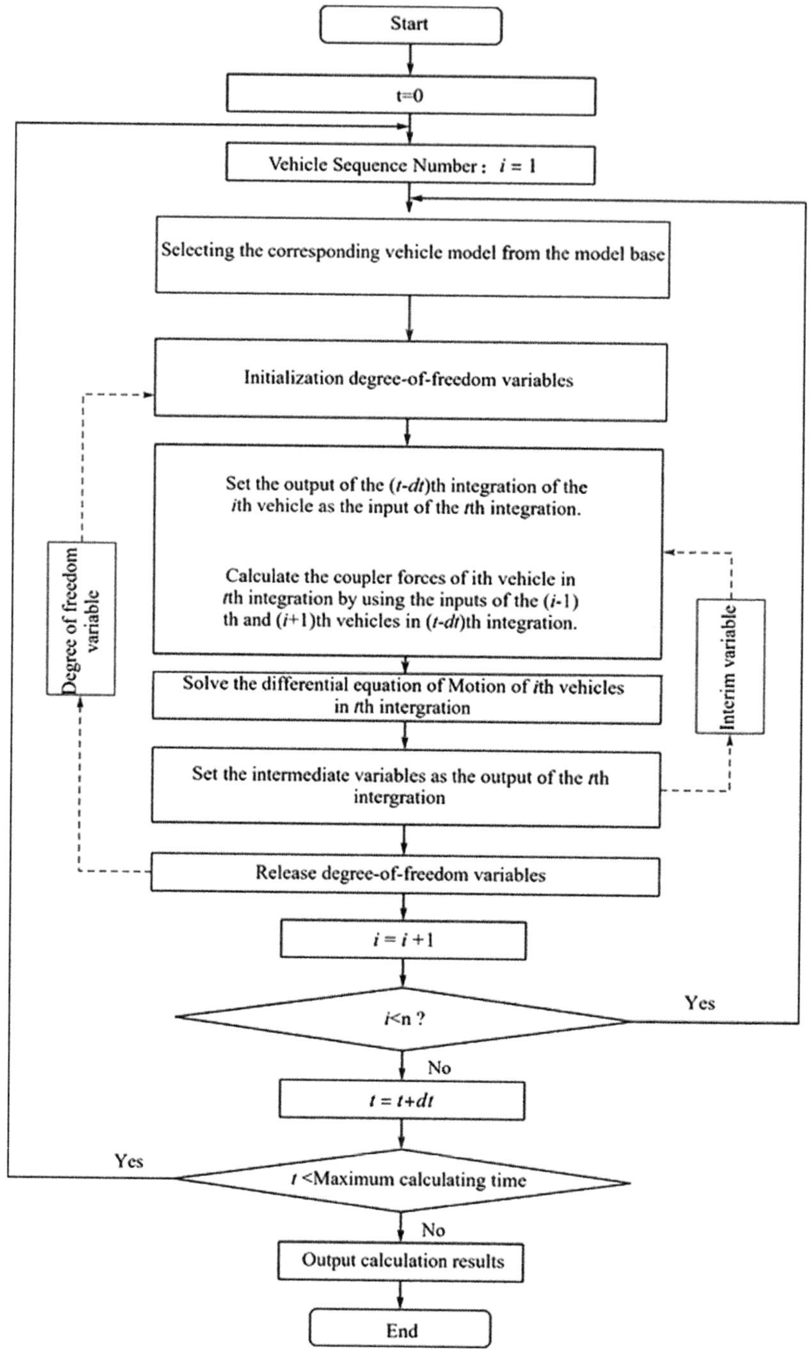

Figure 2.68
Loop variable method.

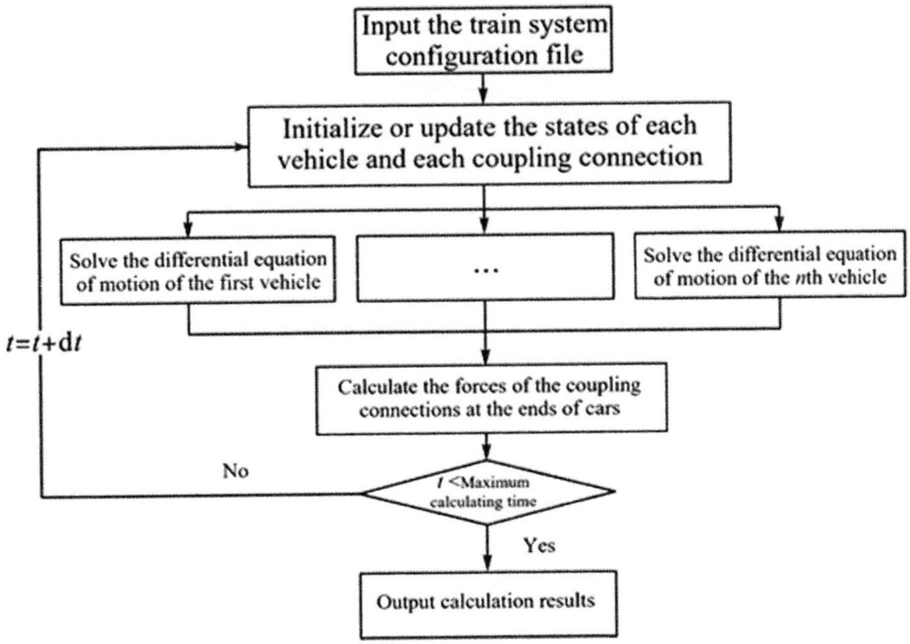

Figure 2.69
The train simulation process.

2.3.1.3 Pantograph catenary coupling model

The traditional contact force model of the pantograph-catenary system considers only the point contact state, regardless of the contact morphology. Considering the linear contact stiffness-based contact deformation, this point contact model uses only contact stiffness and contact deformation to calculate the pantograph-catenary contact force. With the development of pantograph-catenary modeling technology, the contact modeling method considering the contact shape of the pantograph needs to be improved.

The pantograph-contact model considering the shape of the contact point is shown in Figure 2.70, where the coordinate oxyz is the coordinate system of the pantograph system. The pantograph contact point is at point c, and cx'y'z' is the local coordinate system of the contact point. The origin of cx'y'z' is the contact point c, and the pantograph contact angle θ_c is rotated around the y' axis. The positive pressure of the contact point of the pantograph is F_c, and the calculation of the positive pressure adopts the traditional nonlinear contact stiffness calculation method. The normal pressure of the contact line on the pantograph slide is:

$$\begin{cases} F_c = -k_c(z'_{cph} - z'_{ccw}) \\ F_c = 0, \ if \ z'_{cph} - z'_{ccw} \le 0, \quad (off\text{-}line) \end{cases} \tag{2.193}$$

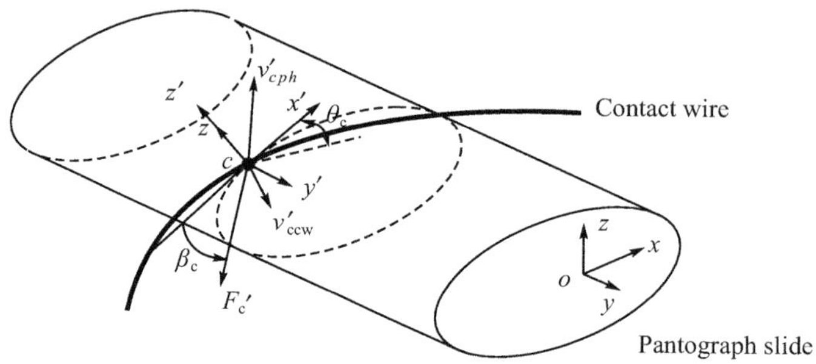

Figure 2.70
Pantograph contact model.

where, $z'_{cph} - z'_{ccw}$is the relative displacement of the pan head and the contact wire at the contact point along the normal point of the contact point, and k_c is the contact stiffness associated with the position of the contact point on the slider. It can be converted from the contact angle θ_c based on the vertical displacement of the pantograph at the contact point. Figure 2.71 shows the contact stiffness signal measured from the interaction of the CTMH150 contact wire with the metalized carbon contact strip material. It shows that the contact force and contact deformation are linearly correlated with the relative displacement. Table (2.2) lists the contact stiffness. Furthermore, the contact stiffness is different at different contact angles θ_c.

Because the frictional force of the relative sliding of the pantograph is considered, it is assumed that the friction coefficient of the contact point is correlated to the position c of the contact point of the slider and the running speed v of the contact strip, and the friction coefficient of the contact point is μ_{cv}. There is friction between the pantograph and catenary. Considering the directionality of the friction between the pantograph and the contact point:

$$F_\mu = \mu_{cv} F_c sign(v'_{cph} - v'_{ccw}) \tag{2.194}$$

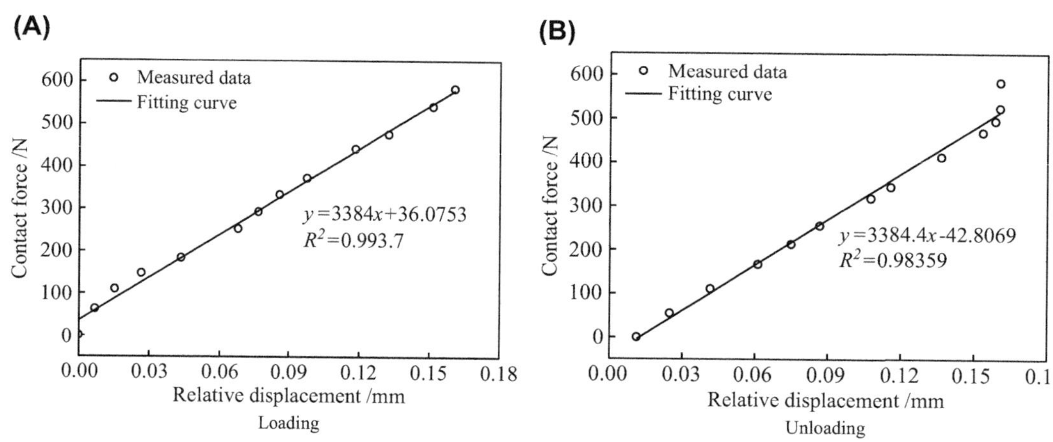

Figure 2.71
Data on loading and unloading pantograph contact stiffness. (A) Loading (B) Unloading.

where $v'_{cph} - v'_{ccw}$is the relative speed of the contact strip and the contact line at the tangent plane of the contact point. The frictional force of the pantograph $F_{\mu}$is the friction coefficient at the contact point multiplied by the positive pressure ($\mu_{cv}F_c$) with the direction on the tangent plane of the contact point and the reversed direction of the relative velocity of the pantograph.

It is assumed that the angle of the relative movement of the pantograph and the negative axis of x is β_c. Thus, the coupling force of the pantograph is the contact force F_c and pantograph frictionF_{μ}. Converting it to the oxyz of the coordinate system of the pantograph system, the interaction force is:

$$\begin{aligned} F_x &= -F_{\mu} \cos\beta_c \sin\theta_c - F_c \cos\theta \\ F_z &= -F_{\mu} \cos\beta_c \cos\theta_c + F_c \sin\theta \\ F_y &= F_{\mu} \sin\beta_c \end{aligned} \tag{2.195}$$

In the actual calculation, the pantograph also deviates from the train coordinate system, as the oxyz coordinates show in Figure 2.70, and the line coordinate system has a certain rotation angle. Therefore, it is necessary to correct Equation (2.195), and thus obtain the mutual force at the contact point of the pantograph.

Thus, under the premise of knowing the cross-sectional shape of the pantograph slide and the curved shape of the contact line, the coupling force between the pantograph and catenary can be obtained based on the state of motion of the slide plate and the catenary.

2.3.1.4 Wheel-rail coupling model

The rail contact can be treated as a special force element in a vehicle system model based on multibody dynamics. However, compared with the general force element, the rail contact force element is much more complicated. This is the main difference between the vehicle system and the general multibody system. The rail force is divided primarily into rail contact normal force and wheel-rail creep force. Calculation of the wheel-rail normal force includes the constraint reaction method and the nonlinear hertz theory. The calculation of the wheel-rail creep force has many relatively mature wheel-rail creep theories, such as the Vermeulen-Johnson theory, Kalker simplification theory using the future automotive systems technology simulator (FASTSim), Shen's theory, Polak nonlinear contact theory, and Kalker's three-dimensional precision theory (CONTACT).

The calculation procedure for wheel-rail normal force, wheel-rail creep force, and wheel-rail force is as follows. First, the position of the wheel-rail contact point is determined based on the positions of the wheel and rail, and the wheel-rail vertical force is calculated based on the contact permeability. Secondly, the wheel-rail creep rate is calculated based on the wheel pair and rail moving speed, and wheel-rail geometry information while the wheel-rail creeping force is calculated using wheel-rail creep theory. Finally, the wheel-rail normal force and the wheel-rail creep force are combined to give the wheel-rail force.

(1) Rail contact point calculation

Figure 2.72 is the basic rail contact system. The rail contact geometry is determined by the basic parameters of the rail contact state, referred to as the five elements of wheel-rail correlation.

1. Wheel and rail profile
2. Gauge $2d_T$
3. Rail cant θ_T
4. The distance between backs of the wheel flanges $2b$ or the distance from the nominal rolling circle to wheelset center l_0
5. The nominal radius of the wheel R_0

After the five elements of wheel-rail correlation are determined, the parameters of the rail contact geometric parameters mentioned above are the functions of the wheelset traverse Y_{Gw} and the wheelset yaw ψ. For a simple-arc rail profile or a straight rail profile, the accurate rail contact geometric parameters can be derived directly. However, for complex profiles, these parameters can only be solved numerically. Different wheel-rail profile matches have different wheel-to-rail contact geometry parameters. Once the basic parameters of the rail contact are adjusted, the rail contact geometry parameters also change as well.

There are many numerical methods for calculating the rail contact geometric parameters. In a test rig study, the profiles of the wheel and rail can be measured directly using a digital profiler, which is a discrete numerical correlation. Therefore, the rail contact geometric correlation of any tread profile can be derived using the trace line method in

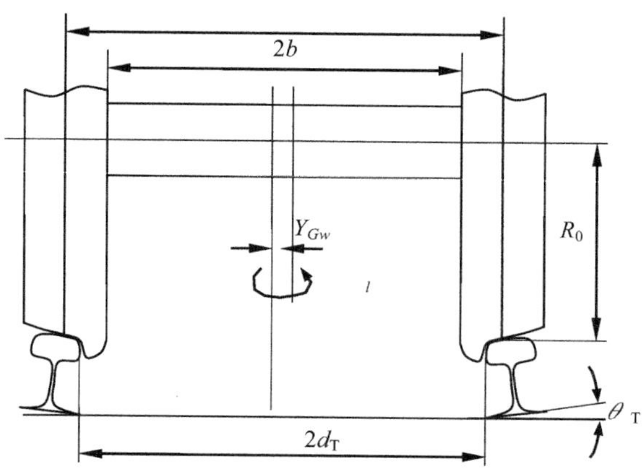

Figure 2.72
Geometric correlation in rail contact.

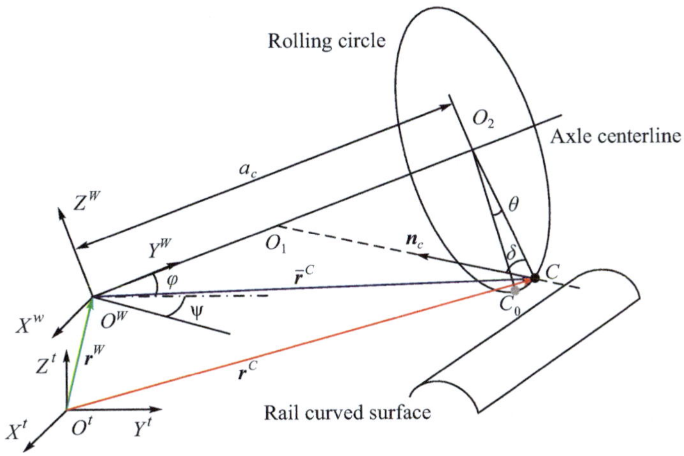

Figure 2.73
The trace method.

Reference [35]. To calculate the rail contact point using the trace method, the first step is to calculate the contact trace. The wheelset is considered to consist of a series of rolling circles, each of which has a possible contact point. These contact points must satisfy certain conditions such that the normal vector at the contact point must point to the axle centerline and be parallel to the *YOZ* plane of the orbital coordinate system. The spatial curve consisting of these possible contact points is the contact trace. The trace method is shown in Figure 2.73, where $O^tX^tY^tZ^t$ is the corresponding rail coordinate system of the wheel pair, and $O^WX^WY^WZ^W$ is the wheelset auxiliary coordinate system. The coordinate system coincides with the origin of the wheel-to-body coordinate system and the OY axis. However, it does not follow the wheel pair rotation. φ and ψ represent the roll and shake heads of the wheel pair, a_c is the lateral distance of the rolling circle to the center of the wheel pair and δ is the contact angle. When there is a shaking angle in the wheel pair, the contact point may no longer be the lowest point C_0 of the rolling circle but will lead or lag at point C. This phenomenon is called contact point lead/lag, and the angle between the center of the rolling circle and the two points is called the lead/lag angle θ.

The vector of possible contact points in the orbital coordinate system can be expressed as:

$$\boldsymbol{r}^C = \boldsymbol{r}^W + \mathrm{A}^W \bar{\boldsymbol{r}}^C \tag{2.196}$$

where

$$\boldsymbol{r}^W = \begin{bmatrix} 0 \\ y_W \\ z_W + R_0 \end{bmatrix}, \quad \mathbf{A}^W = \begin{bmatrix} \cos\psi & -\sin\psi\cos\varphi & \sin\psi\sin\varphi \\ \sin\psi & \cos\psi\cos\varphi & -\cos\psi\sin\varphi \\ 0 & \sin\varphi & \cos\varphi \end{bmatrix}, \quad \bar{\boldsymbol{r}}^C = \begin{bmatrix} R(a_c)\sin\theta \\ a_c \\ R(a_c)\cos\theta \end{bmatrix}$$

Based on the conditions that the normal vector $\boldsymbol{n}_c$ at the contact point needs to satisfy, the lead/lag angleθcan be solved as follows:

$$\theta = \operatorname{asin}\left(\frac{-\cos\varphi\sin\psi\tan\delta(a_c)}{\sqrt{1-\cos^2\varphi\sin^2\psi}}\right) - \operatorname{atan}(\sin\varphi\tan\psi) \tag{2.197}$$

where$R(a_c)$and$\delta(a_c)$are the rolling circle radius and contact angle, respectively. The possible contact points on the selected scroll circle can be determined by substituting θ into Equation (2.196). By updating the scrolling circles, the entire contact trace can be calculated by modifying the value of the scroll circle to the center-to-center lateral distance $a_{c.}$ The trace obtained is shown in Figure 2.74.

The maximum penetration δ0 between the contact trace projection and the rail's cross-section curve can be calculated by projecting the contact trace into the plane of the rail section, as shown in Figure 2.75. When the penetration is less than 0, it indicates that the wheel and rail have been disengaged and no contact has occurred.

(2) Rail contact normal force

To calculate the wheel-rail coupling force (the wheel-rail creep force *F*, the creep moment *M*, and the normal force *N*), the normal force of the rail contact (or positive pressure) should first be calculated using an equivalent nonlinear single-constraint spring model.

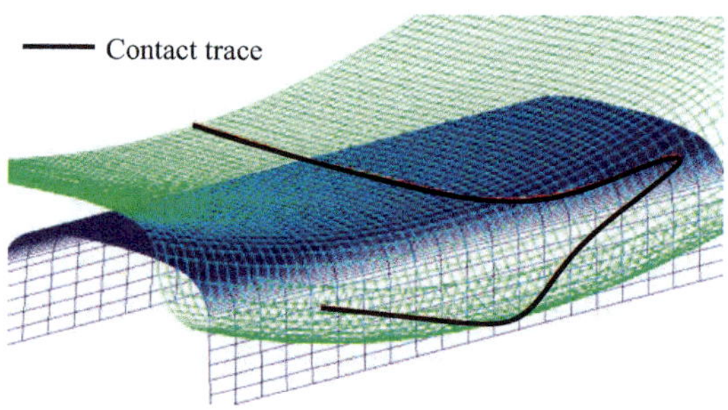

Figure 2.74
Contact Trace.

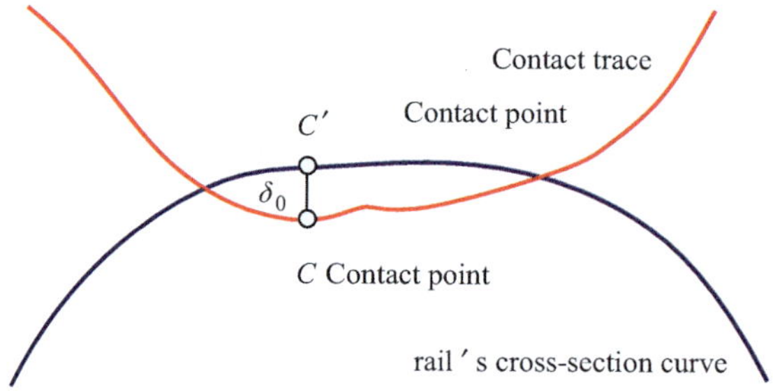

Figure 2.75
Contact point determination.

In most cases, the wheel-rail coupling force can be calculated using Hertz contact theory, which assumes that the initial surface gap function $h\ (x,\ y)$ of the two surfaces is a quadratic function, and the contact patch is an ellipse. The contact force distribution is an ellipsoid function:

$$h(x,y) = Ax^2 + By^2 \tag{2.198}$$

$$p(x,y) = p_0\sqrt{1-\left(\frac{x}{a}\right)^2-\left(\frac{y}{b}\right)^2} \tag{2.199}$$

where A and B are constants associated with contact geometry parameters, p_0 is the maximum contact pressure, and a and b are half of the short axis of the contact spot length. Let $\frac{1}{R_1}$ and $\frac{1}{R_1'}$ be the principal curvatures of contact body 1 at the contact point, and $\frac{1}{R_2}$ and $\frac{1}{R_2'}$ be the principal curvatures of contact body 2 at the contact point, such that:

$$B+A = \frac{1}{2}\left(\frac{1}{R_1}+\frac{1}{R_1'}+\frac{1}{R_2}+\frac{1}{R_2'}\right) \tag{2.200}$$

$$B-A = \frac{1}{2}\left[\left(\frac{1}{R_1}-\frac{1}{R_2}\right)^2+\left(\frac{1}{R_1'}-\frac{1}{R_2'}\right)^2+2\left(\frac{1}{R_1}-\frac{1}{R_2}\right)\left(\frac{1}{R_1'}-\frac{1}{R_2'}\right)\cos 2\alpha\right]^{1/2} \tag{2.201}$$

where α is the angle between the main directions corresponding to the principal curvatures. It is assumed that the total contact normal force is P, the contact area is an ellipse, and the long axis and short axis can be calculated as:

$$a = m\left[\frac{3PG^*}{4(B+A)}\right]^{1/3}, b = n\left[\frac{3PG^*}{4(B+A)}\right]^{1/3} \tag{2.202}$$

Let η be the angle between the normal contact planes, $\cos\eta = \frac{B-A}{B+A}$. After η is obtained, m and n can be obtained. $G^* = G_1 + G_2 = \left(1-v_1^2\right)/E_1 + \left(1-v_2^2\right)/E_2$ is the sum of the equivalent shear moduli of the two contact bodies, v_1、 v_2 and E_1、 E_2 are the Poisson ratio and the elastic modulus of the two contact objects, respectively. The calculated normal distance d_c of the contact point is taken as the geometric penetration amount δ_0, and the actual elastic deformation $\delta < \delta_0$. In this book, $\delta = 0.65\delta 0$ is taken, and the normal contact force can be calculated using the following formula:

$$P = \frac{4\beta}{3G^*\sqrt{A+B}}\delta^{3/2} = K_{Hz}\delta^{3/2} \quad (\delta > 0) \tag{2.203}$$

where $\beta = \left(\frac{\pi m}{2K(e)}\right)^{3/2}$. K_{Hz} can be obtained based on the correlation between β and A/B as shown in Table 2.3.

The major axis a and minor axis b of the rail contact ellipse can be determined using the normal force of the wheel-rail.

Table 2.3: Hertz coefficients.

A/B	1.0	0.7041	0.4903	0.3333	0.2174	0.1325	0.0718	0.0311	0.00765
β	0.318	0.3215	0.3322	0.3505	0.3819	0.4300	0.5132	0.6662	1.1450

(3) Wheel-rail creep force

The wheel-slip creep rate is defined in the contact patch coordinate system, as shown in the following equation:

$$\begin{cases} \xi_x = \dfrac{v_{w1} - v_{r1}}{V} \\ \xi_y = \dfrac{v_{w2} - v_{r2}}{V} \\ \xi_\psi = \dfrac{\Omega_{w3} - \Omega_{r3}}{V} \end{cases} \tag{2.204}$$

where ξ_x, ξ_y, ξ_ψ represent longitudinal creep rate, lateral creep rate, and spin creep rate, respectively. V is the nominal speed; $v_{w1}, v_{w2}, \Omega_{w3}$ are the longitudinal velocity, lateral velocity, and the spin angular velocity, respectively, defined in the rail contact locating coordinate system.

$$\begin{aligned} \xi_{r(l)x} &= \frac{V + \psi\dot{y} - R_{r(l)}\Omega + R_{r(l)}\psi\dot{\phi} \pm a_{r(l)}\dot{\psi}}{V} \\ \xi_{r(l)y} &= \frac{\left(\dot{y} - \psi V + \phi\dot{z} + R_{r(l)}\dot{\phi}\right)\cos\delta_{r(l)} + \left(\dot{z} - \dot{y}\phi \mp a_{r(l)}\dot{\phi}\right)\sin\delta_{r(l)}}{V} \\ \xi_{r(l)\psi} &= \frac{\pm(\Omega - \psi\dot{\phi})\sin\delta_{r(l)} + \dot{\psi}\cos\delta_{r(l)}}{V} \end{aligned} \tag{2.205}$$

where the subscript $r(l)$ corresponds to the right (left) wheel, and the upper and lower symbol correspond to the right and left wheel; y, z, ϕ, ψ are the traverse, vertical displacement, roll angle, and sway angle, respectively, of the opposite pairs of the wheel pair; R is the rolling circle radius at the contact point of the wheel pair; a is the lateral distance from the rolling circle to the center of the wheel pair; δ is the contact angle at the contact point; and Ω is the rolling acceleration of the wheel pair. To calculate the wheel-rail creep force under Shen's theory, the lateral and longitudinal creep forces should first be calculated using Kalker's linear theory:

$$\begin{pmatrix} T'_x \\ T'_y \\ M'_\psi \end{pmatrix} = Eab \begin{bmatrix} c_{11} & 0 & 0 \\ 0 & c_{22} & \sqrt{ab}c_{23} \\ 0 & -\sqrt{ab}c_{23} & abc_{33} \end{bmatrix} \begin{pmatrix} \xi_x \\ \xi_y \\ \xi_\psi \end{pmatrix} \tag{2.206}$$

where E is the elastic modulus; a and b are the major and minor semi-axes, respectively, of the contact spot ellipse; c_{11}, c_{22}, c_{23}, and c_{33} are Kalker coefficients. The composite creep force calculated using the Johnson and Vermeulen theory can be modified using Shen's theory:

$$T = \begin{cases} \mu N\left[\left(\dfrac{\overline{T}}{\mu N}\right) - \dfrac{1}{3}\left(\dfrac{\overline{T}}{\mu N}\right)^2 + \dfrac{1}{27}\left(\dfrac{\overline{T}}{\mu N}\right)^3\right] & \overline{T} \leq 3\mu N \\ \mu N & \overline{T} \leq 3\mu N \end{cases} \tag{2.207}$$

where $\overline{T}$ is the pre-synthesis creep force, and $\overline{T} = \sqrt{T'^2_x + T'^2_y}$. The modified longitudinal creep force and lateral creep force are:

$$\begin{pmatrix} T_x \\ T_y \end{pmatrix} = \varepsilon \begin{pmatrix} T'_x \\ T'_y \end{pmatrix} \tag{2.208}$$

where $\varepsilon = T/\overline{T}$. Shen's theory is close to the FASTSim result, with a simple algorithm and a high-efficiency solution. Therefore, it is widely used in vehicle system dynamics simulation.

2.3.1.5 Vehicle-track coupling excitation model

Considering track excitation in the large coupling system of the high-speed train, several excitation models include the fixed-point load model, moving-load model, moving-irregularity model, and the moving-vehicle model.

(1) Fixed-point load model

The fixed-point load model [36], shown in Figure 2.76, is the most simplified vehicle-track coupling excitation model in which the train system is simplified to a time-varying load excitation that is stationary relative to the line system in the longitudinal direction. This model is best suited for calculating the natural frequency, mode of vibration, and admittance characteristics of line systems, and the calculation results are closely comparable with the test data of the hammer tap or vibration exciter excitation method. Therefore, the fixed-point load model is often used to calculate the dynamic characteristics of line systems, especially in the frequency domain.

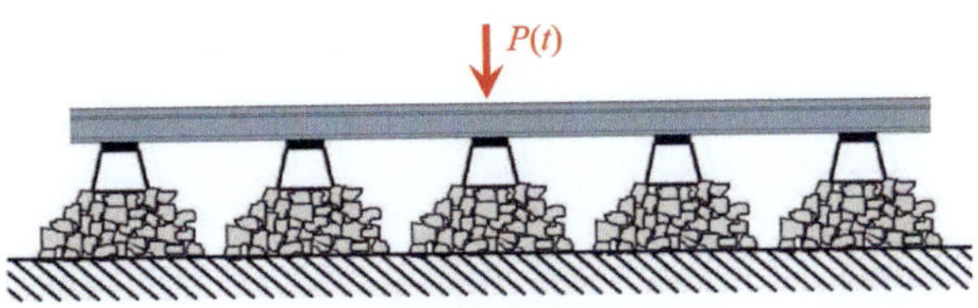

Figure 2.76
Fixed-point load model.

For the line system, the load it is subjected to can be expressed as:

$$F(x,t) = P(t)\cdot\delta(x - x_0) \tag{2.209}$$

where $F(x,t)$ is the load excitation of the train system on the line system, $P(t)$ is the magnitude of the time-varying load, $\delta(x)$ is the Dirac function, x_0 is the position of the load action point, and t is a time variable.

For the fixed-point load model, it should be noted that the dynamic response of the line system is correlated to the position of the load excitation point. For example, when the fixed-point load action position is above the sleeper support because the load action point is at the node of the rail pinned-pinned resonance mode, the result of this calculation is insufficient for determining the rail response for this frequency. Especially when this model is used to calculate the mode characteristics of the line system, if the load action point is located at the system modal node, the system dynamic response of the corresponding mode frequency may not be effectively excited, and the mode information may be missed. To solve this problem, we need to slightly change the longitudinal position of the load and pay attention to the phase frequency response characteristics of the system.

(2) Moving-load model

The moving-load model [36] is an extension of the fixed-point load model, as shown in Figure 2.77. The most significant difference between the moving-load model and the fixed-point model is that the load will move in the direction of the running train with v as its speed. This model is primarily used in the early theoretical study of railway system dynamics.

For the line system, the load it is subjected to can be expressed as:

$$F(x,t) = P(t)\bullet\delta(x - vt) \tag{2.210}$$

(3) Moving irregularity model

The moving irregularity model [36], shown in Figure 2.78, is another extension of the fixed-point load model, as, where the train system is simplified to have no relative motion

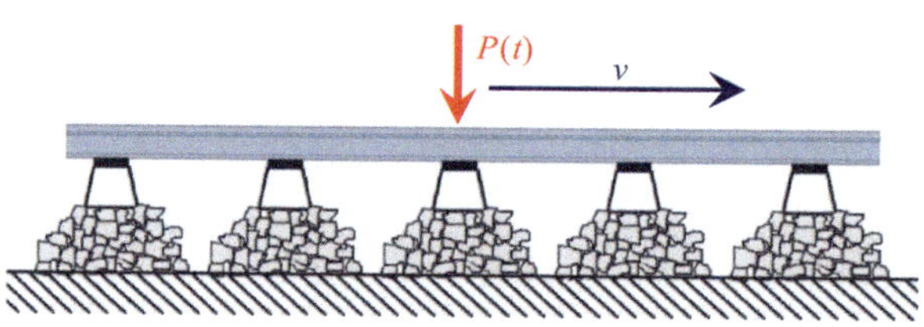

Figure 2.77
Moving-load model.

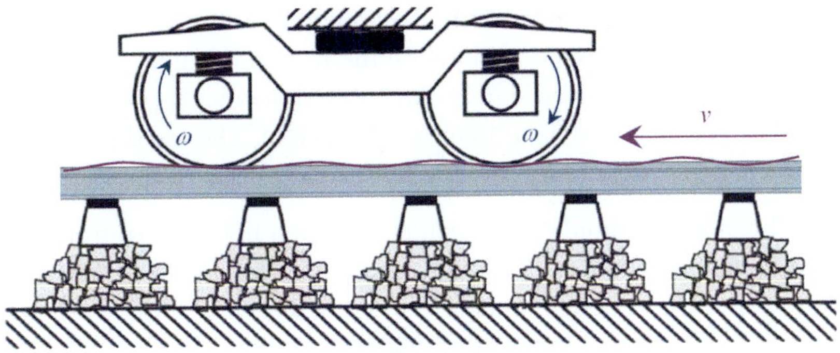

Figure 2.78
Moving irregularity model.

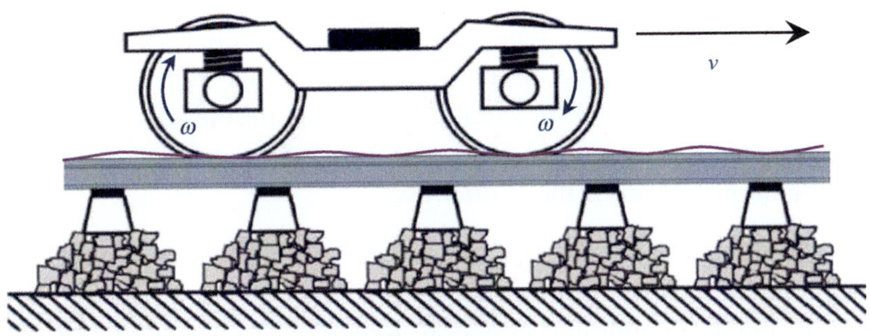

Figure 2.79
Moving vehicle model.

to the line system in the longitudinal direction. The irregularity of the wheel/rail is reversed in the direction of the running train with v as its speed.

This model considers the effects of vertical distribution on the dynamics of the train/line systems. Compared to the mobile vehicle model, it has the significant advantages of simplicity and high efficiency, especially in solving the frequency domain.

(4) Moving vehicle model

The moving vehicle model [36], shown in Figure 2.79, is the vehicle-line coupled excitation model that best matches actual train operation. Furthermore, this model is also the most complex of the existing vehicle-line coupled excitation models and is commonly used in time domain surveys for train/track system dynamics.

The preceding four vehicle-line coupled excitation models can be summarized into two categories from the perspective of train longitudinal motion: vehicle movement models and vehicle stationary models. In the vehicle movement model, the vehicle is considered as running on the rail, and the irregularity is regarded as a position-based variable. The mobile load model and moving vehicle model both belong to this category. In the vehicle static model, the vehicle is considered as fixed on the rail; in essence, the longitudinal position of the vehicle is always fixed during the entire dynamic response of the system. The system irregularity is regarded as a time-based variable. The fixed-point load model and the moving irregularity model both belong to this category.

The moving model can better simulate the spatial vibration behavior of trains and line systems in the horizontal and vertical direction, and the calculation — though large and complicated — is closer to the actual situation. In particular, in the line system, the influence of the medium-high frequency vibration characteristics on the dynamic response of the system needs to be considered, or the wave characteristics of the understructure of the line need to be captured (for example, the foundation soil layer), or finite element modeling analysis is needed. When it is necessary to study the dynamic behavior of a large system high-speed train under a long section line operation, because the line system is too large (beyond the current range of computing and storage capabilities), it is impossible to use the moving model in actual calculations. Furthermore, because the rails of the beam model adopted are too long, the accuracy of the rail mode is lowered, and it is difficult to describe the vibration of the short wave.

To overcome the shortcomings of the moving model, the longitudinal motion of the train is often simplified, and a fixed-point model is adopted — with the advantage that the train system is stationary in the longitudinal direction. According to the Saint-Venant principle, it is necessary to select only a certain length of the line segment for the calculation to ensure the accuracy of the system dynamic response result. Consequently, the amount of calculation is significantly reduced, and the numerical process is simplified.

However, when the fixed-point model is used, the number of under-rail supports included in the calculation line section is fixed. The longitudinal positions of these under-rail supports do not change during the entire calculation process, and consequently, variations in the longitudinal stiffness of the under-rail structure cannot be considered in this model. In essence, the model cannot handle irregular or nonuniform under-rail supports (fasteners, sleepers, track beds, and subgrades) resulting from train operation, and make incentives on high-speed train/line large systems and the system dynamics characteristics. It cannot reflect the effect of the wheel over rail cross-frequency and periodic discrete support frequency in the high-speed train/line coupling dynamics. Furthermore, the model cannot represent the dynamic response of the under-rail structure, which is the most obvious shortcoming of the fixed-point model [37].

Combining the strengths of the two preceding models, the *sliding window model* can be used to more efficiently simulate the coupled excitation of the vehicle line. The basic principles and ideas can be found in the *tracking window* in Reference [37] (shown in Figure 2.80). The sliding window model, as its name implies, is a train system that is

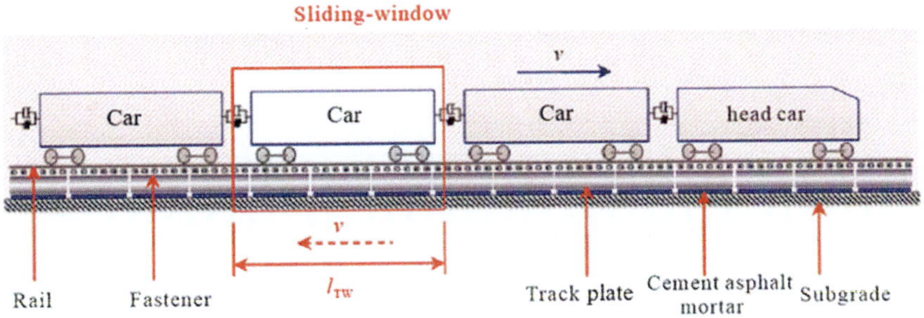

Figure 2.80
Sliding Window Model [37].

stationary. The line system, including rails, fasteners, rail cushion layers, track plates, sleepers, track beds, ballast, and the wheel-rail surface irregularity move in the opposite direction of the running train, at the same speed as the train.

(5) Sliding window model

The sliding window model takes into full account the effects of discrete under-rail support and track irregularities, especially the support characteristics of a line varying unevenly in the longitudinal direction. This model is closer to the actual situation of the train moving between the sleepers. At each calculation time step, there is an assessment to calculate the under-rail support within the length of the rail. According to the Saint-Venant principle, only the under-rail support within the calculated length of the rail is considered to have an excitation effect on the train-line system at that moment. The under-rail support outside the calculation area is thus negligible.

In the sliding window model, in order to select the high-speed railway sleeper, track plate, and bridge structure calculation window, the setting of the sliding window length l_{TW} is time-varying, as shown in Figure 2.81. The basic principles of the sliding window setting are: (1) different calculation windows are set for the sleepers, track plates, and bridges; (2) the window is selected to ensure that there is sufficient length, and the window can be guaranteed for reflecting the basic information of the vibration; (3) the length of each window should be set in consideration of the boundary point, and the length of the sleeper the rail) window(must be on the sleeper, not in the middle of the sleeper. Otherwise, the rail is suspended, and the boundary point cannot be calculated. The window boundary of the track plate must be at the edge of the track plate, and the window of the bridge must also be at the end of the bridge, not on the track plate and the bridge; (4) coordination between the windows should be maintained, i.e., the track panel window moves within the bridge window, and the sleeper (rail) window is within the window of the track panel. According to the basic principle of the sliding window setting, the length of the sliding window corresponding to different line types can be determined.

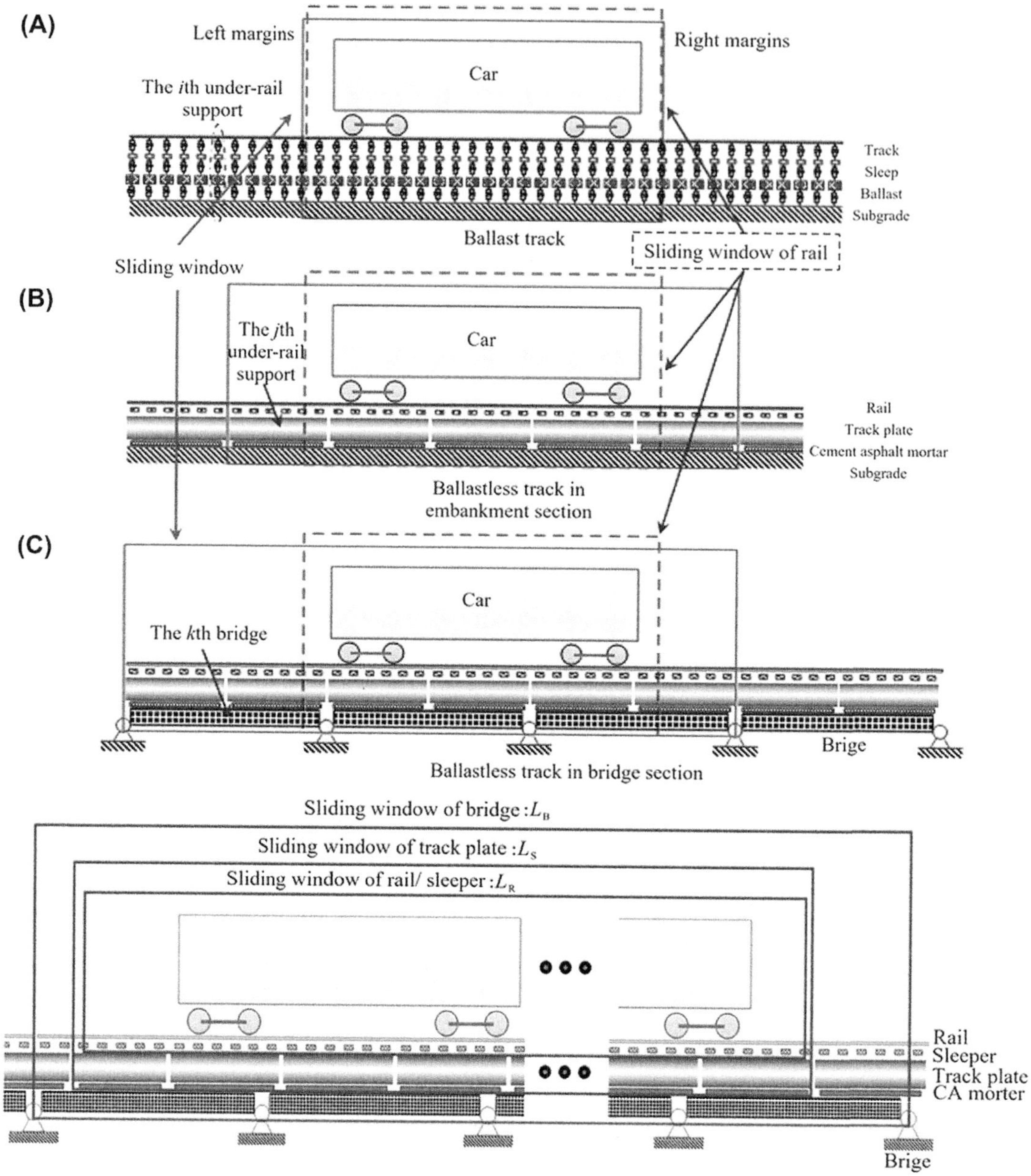

Figure 2.81
Sliding window length calculation model.

The sliding window jumps with the running train, and the jumping distance of the sleeper window is the spacing of the sleepers (generally 0.6 m); the jump distance of the track plate window is the length of the track plate (generally 6.5 m); and the jump distance of the elevated bridge window is the length of the bridge (generally 32 m). Therefore, the movement of each window continually recurs and alternates; thus, the calculation becomes very complicated. To make it simpler, the track and sleeper (rail), and the bridge windows move with the movement of the bridge window. However, because the sleeper track plates may be inconsistent with respect to each span bridge, the relative position between the three windows changes. Furthermore, the span of the bridge itself may change at different locations. Therefore, the three windows exist at the same time, and movement is coordinated according to the rules outlined above.

Considering the vibration attenuations of bridges, track plates and sleepers, and rails, the vibration outside the calculated window can be ignored. Therefore, the correlation among the length of the sliding window of the entire train/line l_{TW}, the length of the rail moving window l_R, the length of the sleeper moving window l_S, and the length of the bridge moving window l_B is $l_R \le l_S \le l_B \le l_{TW}$. If there is a ballastless track on the embankment section, there is no bridge window l_B.

For problems like car-line-bridge coupling, the choice of bridge window is crucial and is also the basis for the selection of track panels and sleeper windows. When the bridge window moves with the train step, the bridge window should, as much as possible, meet the train in the middle of the window. The distance from the front and rear of the window to the train should be greater than the area that guarantees the accuracy of the calculation.

Based on the above sliding window technology, not only can the calculation amount of the long line simulation be effectively reduced, but the simulation calculation of the train forward mode can be realized. Meanwhile, the composite structure lines, including the embankment section, transition section, and bridge section, can be handled more flexibly, making it possible to simulate the train operation under continuous long line conditions.

(2) Sliding window rail calculation method

When the sliding window is used for the calculation of the vehicle line coupling, because the sleeper, track board, and track bridge can be regarded as discrete objects, the moving in and out calculation window handles well. However, the rail is a continuum, even if a section of the rail is used as the beam unit for calculation. So far, no scholar has experimented with allowing the rail pass through the sliding window and continuously change according to the train position to realize the simulation of the train running on any of the long rails.

Assume that at t_0, the original calculation window of the rail is AB, and the window for the next calculation time$t_0 + \Delta t$ is A'B'. This is the effective area of the vibration displacement (deformation) of the rail in the virtual frame between A'B, and the vibration displacement of the rail outside the virtual frame can be neglected, as shown in Figure 2.82. At window AB, the rail is assumed to be a simply supported beam. Point A is the local coordinate origin of the beam, and the moving direction along the track is the

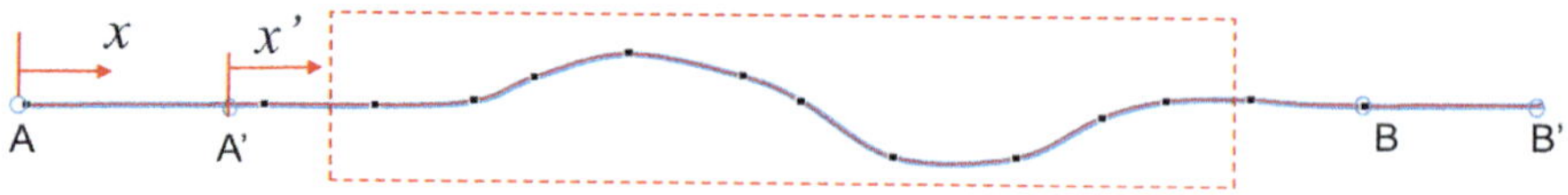

Figure 2.82
Sliding window.

x-axis. Then the vibration displacement of the window AB beam can be expressed as follows:

$$z_r(x, t_0) = \sum_{k=1}^{nz} Z_{rk}(x) q_{rk}(t_0) \tag{2.211}$$

Where $Z_{rk}(x)$ is the kth-order mode shape of the rail AB of the window region, $q_{rk}(t)$ is the kth-order mode amplitude, and nz is the modal order.

At the window A'B' of the next calculation time, the rail is assumed to be a simply supported beam. A' point is the local coordinate origin of the beam and the moving direction along the track is the x' axis. Then the vibration displacement z'_r of the window AB beam can be expressed as the following formula:

$$z'_r(x', t_0 + \Delta t) = \sum_{k=1}^{nz} Z_{rk}'(x') q'_{rk}(t_0 + \Delta t) \tag{2.212}$$

where $Z_{rk}'(x')$is the kth mode shape of the rail AB of the window zone, and $q'_{rk}(t)$ is the kth mode amplitude of the rail.

A'B shown in Figure 2.83 is a common section of the sliding windows AB and A'B at the two calculation timings before and after and it contains the effective area of the vibration displacement of the rail. nz sections are set in the common section in the public section, thus for two different windows, at the same section k, the coordinates of the rail in the two window coordinate systems are (x_k, z_k) and (x'_k, z'_k). Because they are the same point, $z_k = z'_k$.

$$\sum_{k=1}^{nz} Z_{rk}'(x_k') q'_{rk}(t_0 + \Delta t) = z_k(x_k, t_0) \quad k = 1, 2, \cdots, nz \tag{2.213}$$

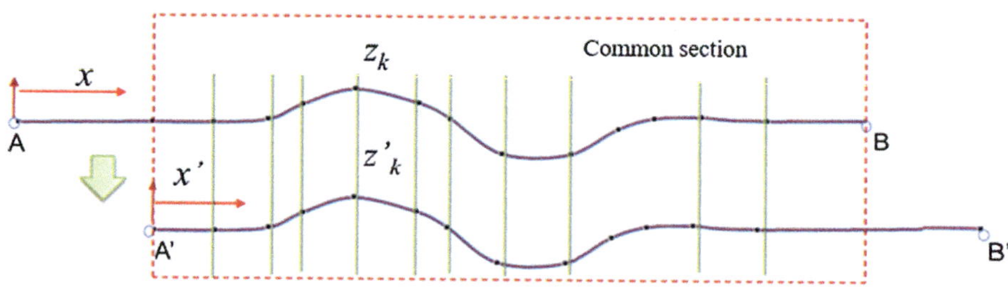

Figure 2.83
Sliding window section settings.

Because the vibration displacement of the rail $z_k$and the vibration mode of the window A'B', Z_{rk}', are known, Equation (2.213) becomes a linear equation of the modal amplitude q'_{rk}. After the modal amplitude is obtained, the window slip is realized by transmitting the rail vibration from the original window AB to the new window A'B'.

2.3.1.6 Fluid-solid coupling model

Due to the complexity of fluid-structure coupling calculations, the following calculation models are generally used in fluid-structure coupling calculations.

(1) Offline simulation model

The traditional aerodynamic model does not consider the state change of the train under aerodynamic forces and uses fluid dynamics software to calculate the steady-state aerodynamic torque, reloading this into the vehicle dynamics model to calculate the dynamic response. That is the offline simulation method.

Combined with the train dynamics equation (2.23) and fluid control equation (2.161), the formula for the offline simulation of the high-speed train is as follows:

$$\begin{cases} \dfrac{\partial}{\partial t}\displaystyle\int_V \rho\boldsymbol{\varphi} dV + \int_A \mathbf{n}\cdot(\rho(\mathbf{v}-\overline{\mathbf{v}})\boldsymbol{\varphi})dA = \int_A \mathbf{n}\cdot(\Gamma grad\boldsymbol{\varphi})dA + \int_V \mathbf{S}dV \\ \overline{\mathbf{v}} = \mathbf{V}_c \\ \mathbf{F}_{\mathrm{a}} = H(\boldsymbol{\varphi}) \\ \mathbf{M}\ddot{\mathbf{p}} = \mathbf{Q}_{\mathrm{c}} + \mathbf{Q}_{\mathrm{e}} + \mathbf{Q}_{\mathrm{v}} + \mathbf{Q}_{\mathrm{wr}} + \mathbf{F}_a \end{cases} \tag{2.214}$$

where $\mathbf{V}_c$is the velocity vector of the train and the aerodynamic force (moment) $\mathbf{F}_{\mathrm{a}}$can be expressed as an integral function H of the flow flux $\boldsymbol{\varphi}$.

The offline simulation method is a traditional train aerodynamic calculation method. It is only applicable to train aerodynamic calculation of steady-state airflow under conditions of low calculation accuracy.

(2) United simulation model

By loading aerodynamic force into the calculation model of train system dynamics in time, the aerodynamic changes caused by the adjustment of train running state can be acquired in time, and the real fluid-solid coupling calculation can be realized. Because co-simulation requires a new train state to calculate the aerodynamic force at each step, the calculation efficiency is very low. However, this model is the only choice for dynamic processes such as train crossing, passing through tunnels, and fluctuating wind.

Combined with the train dynamics equation (2.23) and the fluid control equation (2.161), the formulas for the united simulation of the high-speed train are as follows:

$$\begin{cases} \frac{\partial}{\partial t}\int_V \rho\boldsymbol{\varphi} dV + \int_A \mathbf{n}\cdot(\rho(\mathbf{v}-\overline{\mathbf{v}})\boldsymbol{\varphi})dA = \int_A \mathbf{n}\cdot(\Gamma grad\boldsymbol{\varphi})dA + \int_V \mathbf{S}dV \\ \mathbf{F}_{\mathrm{a}} = H(\boldsymbol{\varphi}) \\ \overline{\mathbf{v}} = \dot{\mathbf{R}} + \boldsymbol{\omega}\times\mathbf{u} \\ \left[\dot{\mathbf{R}}\ \boldsymbol{\omega}\right]^T = \dot{\mathbf{p}} \\ \mathbf{M}\ddot{\mathbf{p}} = \mathbf{Q}_{\mathrm{c}} + \mathbf{Q}_{\mathrm{e}} + \mathbf{Q}_{\mathrm{v}} + \mathbf{Q}_{\mathrm{wr}} + \mathbf{F}_{\mathrm{a}} \end{cases} \tag{2.215}$$

The fluid-solid coupling united simulation calculation process of the high-speed train is shown in Figure 2.84 [38,39]. Each coupling time interval is iterated as follows: first, update the calculation grid of the flow field according to the posture information of the train; secondly, calculate the fluid control equation based on the new grid, and obtain the flow field information in the fluid calculation area and the aerodynamic force of the train; finally, transmit the train aerodynamic information to the vehicle dynamics model, and calculate the vehicle dynamic response under aerodynamic forces. The spring analogy method and the remeshing method are used in the grid update of the fluid calculation.

In general, fluid-solid coupling calculations are inconsistent in terms of fluid and vehicle time scale requirements. The time iteration step of the vehicle dynamics solver is often order of magnitudes smaller than the time iteration step required by the train aerodynamic solver. If the time iteration step of the train aerodynamic calculation is too small, it will greatly affect the calculation efficiency. If the time iteration step of the vehicle dynamics calculation is too large, it easily causes oscillation or divergence in the calculation result. In the fluid-solid coupling calculation method for the high-speed train, the time iteration steps of the train aerodynamics and vehicle dynamics are selected to suit their respective magnitudes. Thus, the problem of inconsistent requirements for fluid-solid coupling calculation on the fluid and vehicle time scale can be fittingly solved [38].

In the fluid-solid coupling calculation for the high-speed train, the time iteration steps of the train aerodynamics and vehicle dynamics calculations are inconsistent. If aerodynamic

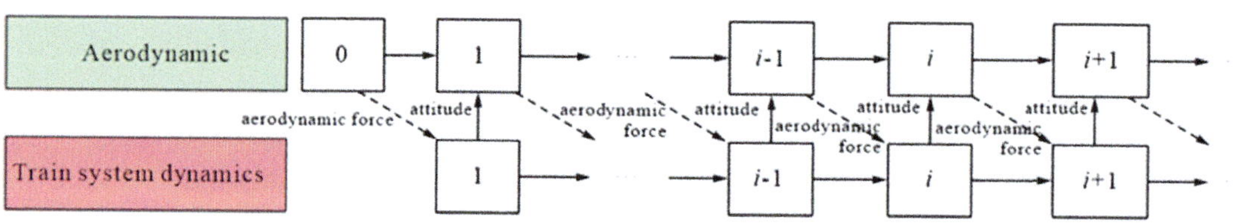

Figure 2.84
Fluid-solid coupling joint simulation calculation method for the high-speed train.

forces are transmitted to the vehicle dynamics solver as a constant force during coupling iteration, at the corresponding moments of two adjacent coupling iteration steps, the aerodynamic force transmitted to the vehicle dynamics solver may be abrupt. This may easily cause oscillation or divergence in the vehicle dynamics calculation. To avoid this phenomenon and make the aerodynamic force transmitted to the vehicle dynamics solver continuous, the aerodynamic force at the current moment and the previous moment can be interpolated. Thus, the aerodynamic force $\boldsymbol{F}_a$ [38] passed to the vehicle dynamics solver at time t (between t_n and t_{n-1}) is:

$$\mathbf{F}_a = \mathbf{F}_{n-1} + \frac{t - t_{n-1}}{t_n - t_{n-1}} (\mathbf{F}_n - \mathbf{F}_{n-1}) \tag{2.216}$$

where $\boldsymbol{F}_n$ and $\boldsymbol{F}_n$ are the aerodynamic force solved by the train aerodynamic solver at time $t_n$and t_{n-1}, respectively.

(3) Co-simulation model based on relaxation factor

A key issue in the fluid-solid coupling dynamics of the high-speed train is the iterative dissipation between train aerodynamics and train system dynamics: how to load the train aerodynamics to calculate the vehicle system dynamics response. To reduce this dissipation as much as possible, a flow-solid coupling simulation method based on the relaxation factor is proposed [25], as shown in Figure 2.85. Compared with the co-simulation method shown in Figure 2.84, the slack factor-based fluid-solid coupling joint simulation method has distinct differences: at the $(i + 1)$th time, i.e., the $(i + 1)$th coupling iteration step, the aerodynamics force transmitted from the aerodynamic solver to the vehicle system dynamics solver is not the aerodynamic force at the ith time, but obtained by predicting the aerodynamic information at the ith time and the $(i - 1)$th time.

Assuming f_{i-1} and f_i are the train aerodynamics calculated by the train aerodynamic solver at the $(i - 1)$th and ith times, respectively, and let Δtbe the time step, then the first-order derivative approximation of the train aerodynamics at the ith moment can be expressed as:

$$\dot{f}_i = \frac{f_i - f_{i-1}}{\Delta t} \tag{2.217}$$

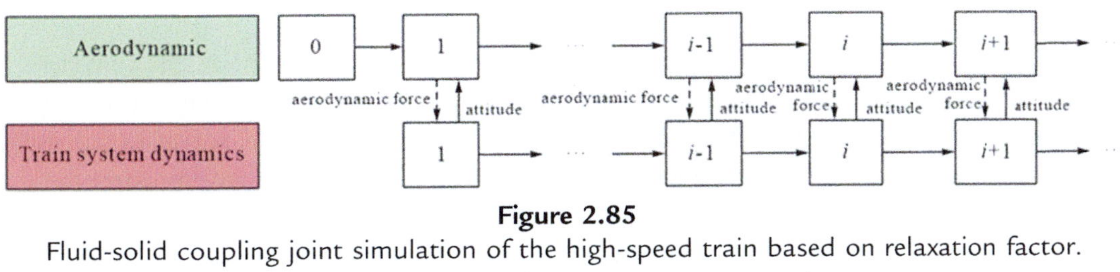

Figure 2.85
Fluid-solid coupling joint simulation of the high-speed train based on relaxation factor.

Because the time iteration step size $\Delta t$is relatively small, the aerodynamic force of the train at the $(i+1)$th time is estimated to be approximately:

$$f_{i+1} = f_i + \Delta t \times \dot{f}_i = 2f_i - f_{i-1} \tag{2.218}$$

Introducing the relaxation factor λ to correct the aerodynamics of the train at the $(i+1)$th time, the aerodynamic force transmitted to the vehicle system dynamics can be expressed as:

$$\widetilde{f}_{i+1} = (1-\lambda)f_i + \lambda f_{i+1} = (1+\lambda)f_i - \lambda f_{i-1} \tag{2.219}$$

If the parameter $\lambda$is equal to 0, the aerodynamic force transmitted from the aerodynamic solver to the vehicle system dynamics solver $\widetilde{f}_{i+1}$ at the $(i+1)$th time is $\widetilde{f}_{i+1}$, as described in the co-simulation method above.

The results show that [40]: if the relaxation factor λ is set to 0.5, the iteration dissipation between train aerodynamics and the train system dynamics is minimal.

(4) Equilibrium state model

Although ambient wind changes rapidly, there are many regional ambient winds that are stable over time. Ignoring the influence of external excitation such as track irregularity, if the speed of the ambient wind is relatively stable, the aerodynamics and the state of the train will stay in a relatively balanced state. In essence, the fluctuations in aerodynamics and state will be smaller after a certain period. Considering the change in train state under the action of aerodynamic force, the fluid-solid coupling calculation is first used to calculate the train state change under steady airflow. After the state is stabilized, the steady-state aerodynamic force is loaded into the train system dynamics model, and the offline fluid-solid coupling calculation is performed. That is the equilibrium state method [25]. Compared with the traditional aerodynamic model, this model can further improve calculation accuracy, calculation speed, and efficiency of the high-speed train fluid-solid coupling under a steady-state ambient wind. The equilibrium state method is as follows.

As shown in Figure 2.86, the calculation process is primarily:

1) Use the train aerodynamic solver to calculate the flow field information of the high-speed train in the initial state, and to calculate the aerodynamics.
2) Update the train aerodynamic force (moment) and assess whether the result converges according to convergence criteria: if it converges, skip to step 6, otherwise go to step 3.
3) Ignoring the influence of track irregularity, use the vehicle dynamics solver to calculate the vehicle dynamic response under the current aerodynamic force, and transmit the information on the stable state of the vehicle body to the interface code.
4) The interface code updates the grid of the flow field calculation area using the mesh update technology based on the returned information on the train. The train aerodynamic solver is also used to calculate the current state of the following vehicle flow field until there is less fluctuation in the amplitude of the train's aerodynamic force.

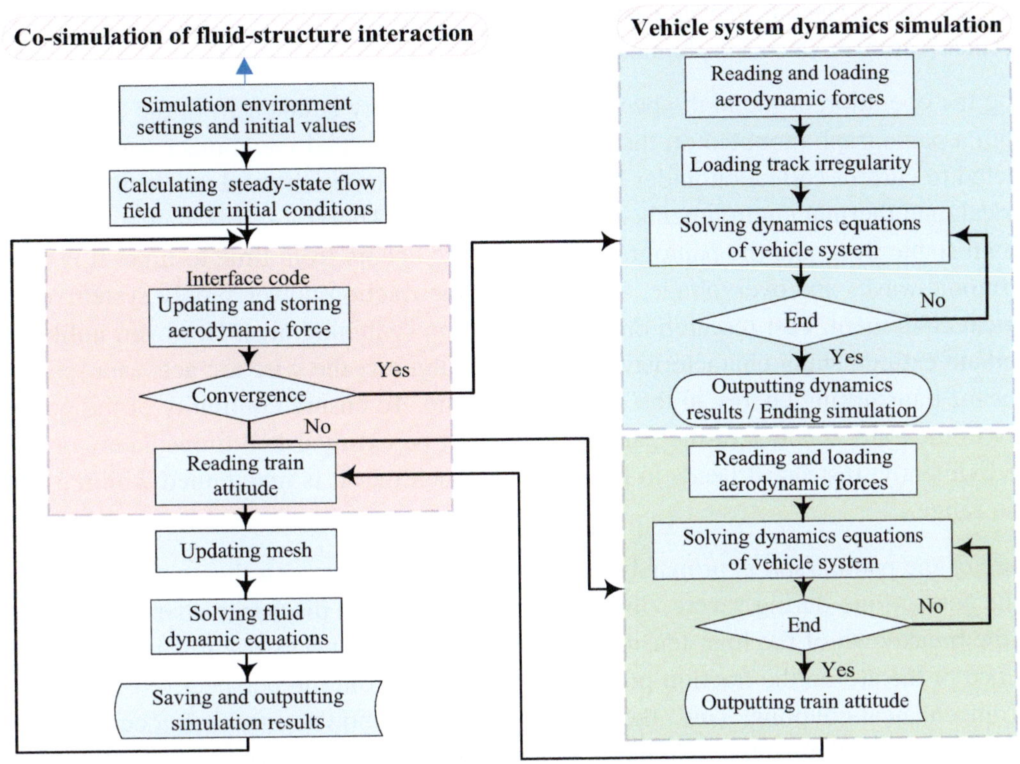

Figure 2.86
The fluid-solid coupling equilibrium state method for high-speed train.

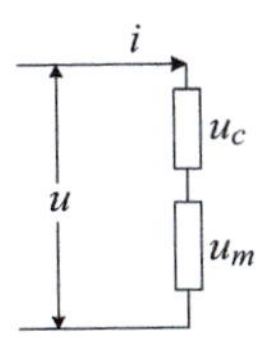

Figure 2.87
Cassie-Mayr series model.

5) Store the fluid calculation output and skip to step 2.
6) Load the track irregularity and calculate the vehicle dynamic response of the aerodynamic action under equilibrium conditions.

Assessing the aerodynamic convergence is required to establish a convergence criterion. The convergence criterion is expressed as:

$$\left| \frac{x_{i+1} - x_i}{x_i} \right| < \varepsilon \tag{2.220}$$

where x_i and x_{i+1} are the index magnitudes of the ith and $(i+1)$th time iteration steps and ε represents the convergence error.

2.3.1.7 Electromechanical coupling model [42]

(1) Vehicle-catenary electromechanical coupling

During the operation of the high-speed train, electric energy is taken from the catenary through a pantograph mounted on the roof. The pantograph and the contact wire are subjected to current collection under high-speed sliding conditions and mechanical, electrical, and thermal environments of high complexity. Therefore, it is challenging to maintain stable flow, and the pantograph-catenary arc occurs from time to time. It is prone to harmonic waves and overvoltage, which affects the traction power supply system and electrical equipment. Past research on arcs focused on switching arcs. However, unlike the immediate extinguishing characteristic of the switching arc, the wire arc acts as a special carrier for transmitting energy in the contact loss state. To ensure continuity of the power supply to the train, the arcs of the pantograph cannot be extinguished immediately or burned violently. Instead, it needs to be controlled such that it is maintained within a certain range.

The arc of the pantograph is primarily due to the clearance between the pantograph pan and the contact line during severe vibration, and a plasma arc discharge channel is formed after the breakdown of the high-tension field. Therefore, it is necessary to study the influence of the arc on the traction power supply system from the perspective of electromechanical coupling. Thus, the current key issue is how to establish a correlation between the arc of the pantograph and the vibration of the pantograph.

1) Pantograph arc model Theory and practice show that an arc is a complex plasma form with high nonlinearity between its voltage and current. If the arc is not burning, the gas between the two ends of the electrode does not have electrical conductivity and is in a high resistance state. As the gas breaks down, the arc continues to burn, and the arc resistance exhibits a negative resistance state while the arc voltage exhibits a relatively stable value. When the energy provided by the arc is insufficient, the arc resistance exhibits a positive resistance state, and the arc stability is weakened, then quickly extinguished. Therefore, the arc load can be simulated by a nonlinear resistance. The classic black-box models, including the Cassie model and Mayr arc model, can be applied to the circuit simulation analysis, as shown in equations (2.221) and (2.222).

Cassie arc model:

$$\frac{1}{g}\frac{dg}{dt} = \frac{d\ln g}{dt} = \frac{1}{\tau}\left(\frac{u^2}{U_c^2} - 1\right) \tag{2.221}$$

Mayr arc model:

$$\frac{1}{g}\frac{dg}{dt} = \frac{d\ln g}{dt} = \frac{1}{\tau}\left(\frac{ui}{P} - 1\right) \tag{2.222}$$

where $g$is arc conductance, u is arc voltage, U_c is arc voltage constant, $t$is arc duration, τ is the free parameter of the arc model, $i$is arc current; and $P$is dissipated power.

Based on studies on the Cassie and Mayr arc models, the Cassie arc model is suitable for high-current arc stable arcing, and the Mayr model is suitable for small current arcs. However, in the arcing-combustion-extinguishing process, the current of the pantograph arc increases from small to large and then decreases.

Using a combination of the Cassie and Meyer arc models (shown in Figure 2.87), a new in-series arc model can be established, as shown in equation (2.223).

$$\begin{aligned} \frac{dg_c}{dt} &= \frac{1}{\tau_c}\left(\frac{u^2g^2}{U_c^2 g_c} - g_c\right) \\ \frac{dg_m}{dt} &= \frac{1}{\tau_m}\left(\frac{u^2g^2}{p_0} - g_m\right) \\ \frac{1}{g} &= \frac{1}{g_c} + \frac{1}{g_m} \end{aligned} \tag{2.223}$$

We use MATLAB to establish a simulation module of the Cassie-Mayr series model and simulate the corresponding electrical characteristics using Simulink. The arc model consists of a voltage controlled current source, a differential equation editor, a fixed value detection, a step signal, and a voltage measuring module. The simulation results are shown in Figure 2.88 and Figure 2.89. The arc voltage shown in Figure 2.88 is obtained from the Cassie-Mayr series model, and the voltage shown in Figure 2.89 is obtained by measuring the pantograph arc. Figure 2.90 shows the continuous arc voltage. It shows that the simulated arc voltage waveform and the measured arc voltage waveform are similar and that the Cassie-Mayr series model can effectively reflect the arc characteristics. Thus, corresponding results can more accurately reflect the pantograph arc.

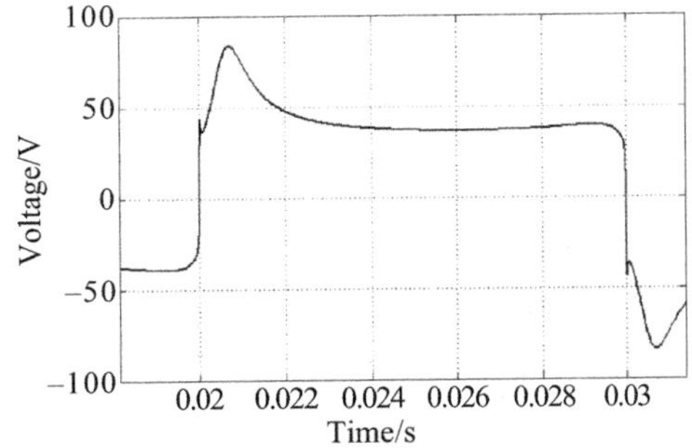

Figure 2.88
Cassie-Mayr arc voltage in series.

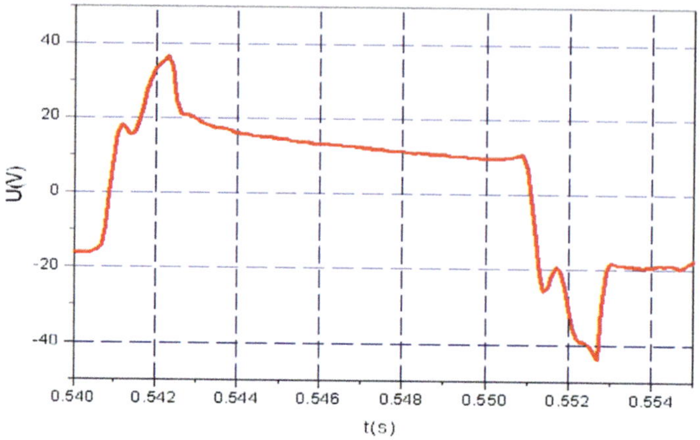

Figure 2.89
Arc voltage measured by arc voltage.

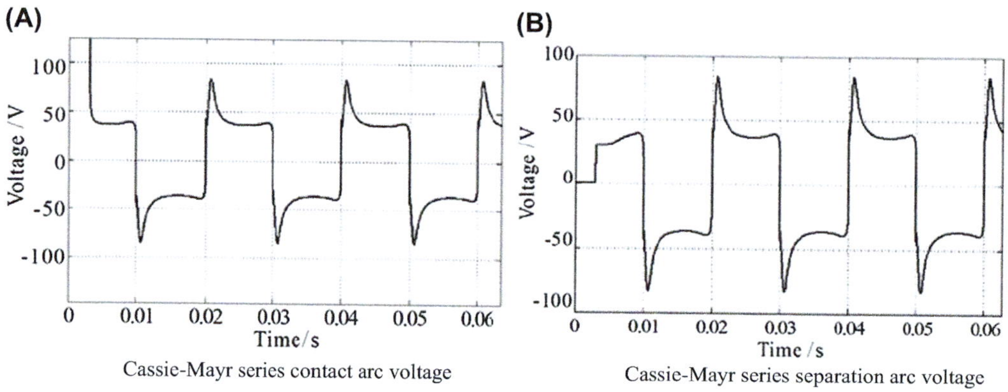

Figure 2.90
Cassie-Mayr series continuous arc voltage. (A) Cassie-Mayr series contact arc voltage (B) Cassie-Mayr series separation arc voltage.

Figure 2.90 shows that regardless of the contact process or separation process, the arcing voltage, the steady arc voltage, and the arc extinguishing voltage of the Cassie-Mayr series arc model are all constant. This is inconsistent with the actual situation of the offline arc of the pantograph because the gap between the pantograph and catenary changes continuously during the pantograph-catenary contact loss process, and the arcing voltage, stable arc voltage, and arc extinguishing voltage all change with changes in the gap. Therefore, the arc model obtained by directly connecting the Cassie and Mayr arc models can only reflect the arc shape of a certain period and lacks the ability to reflect the dynamic variation characteristics of the entire arc. The dynamic characteristics of the actual pantograph arc U_c and P_0 are not constant and should be correlated to the gap in the flow field.

The Cassie-Mayr arc model in series has been modified to establish a mathematical model suitable for the pantograph arc:

$$\frac{dg_c}{dt} = \frac{1}{\tau_c}\left(\frac{u^2 g^2}{U_c(d)g_c} - g_c\right)$$
$$\frac{dg_m}{dt} = \frac{1}{\tau_m}\left(\frac{u^2 g^2}{Cpo(d)} - g_m\right) \quad (2.224)$$
$$\frac{1}{g} = \frac{1}{g_c} + \frac{1}{g_m}$$

where C is the heat transfer coefficient of the arc, which is correlated to the airflow field around the arc and the heat dissipation coefficient of the pantograph-catenary system; and d is the length of the arc, also known as the pantograph vibration gap.

Equation (2.224) gives the simulation model of the pantograph arc. Figures 2.91 and 2.92 display the simulated arcing voltage and the measured arc voltage, respectively. As the gap (arc length) decreases, the arcing voltage, stable arc voltage, and extinction voltage all reduce. The simulated results are closely similar to the measured results. Therefore, Equation (2.224) can be used as a mathematical model of the pantograph arc.

(2) Vehicle-catenary electromechanical coupling

The pantograph-catenary contact force can be formulated as:

$$F_c = K_s(Z_H - Z_c) \quad (2.225)$$

whereF_c is pantograph-catenary contact force, $K_s$is contact stiffness, $Z_H$is the vertical displacement of pantograph head, and $Z_c$is the vertical displacement of the contact wire at the contact point.

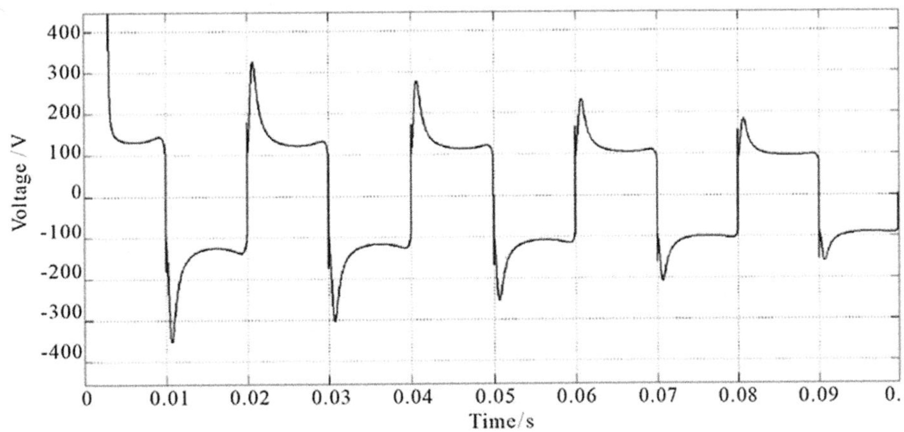

Figure 2.91
Arcing voltage.

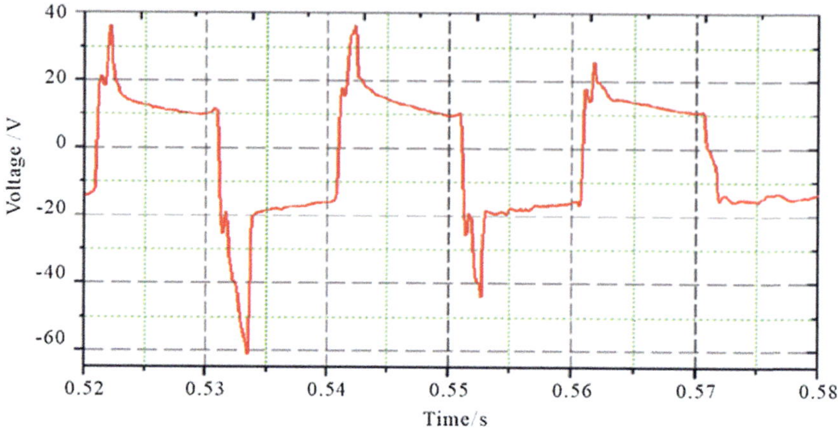

Figure 2.92
Real-time measurement of reignition arc voltage.

When the contact force F_c is less than zero, it indicates that the pantograph is offline. Therefore, the pantograph gap, mentioned earlier, can be expressed as:

$$d = Z_c - Z_H \qquad (d > 0) \tag{2.226}$$

Where Z_c and Z_H can be solved according to the pantograph coupling dynamic equations.

Therefore, the electromechanical coupling correlation of the vehicle-category can be established based on the pantograph vibration clearance and the pantograph arc model, as shown in the following formulas:

$$\begin{cases} \dfrac{dg_c}{dt} = \dfrac{1}{\tau_c}\left(\dfrac{u^2g^2}{U_c(d)g_c} - g_c\right) \\ \dfrac{dg_m}{dt} = \dfrac{1}{\tau_m}\left(\dfrac{u^2g^2}{Cp_0(d)} - g_m\right) \\ \dfrac{1}{g} = \dfrac{1}{g_c} + \dfrac{1}{g_m} \\ d = Z_c - Z_H \quad (d > 0) \end{cases} \tag{2.227}$$

According to the vehicle-catenary electromechanical coupling correlation established in Equation (2.227), the circuit analysis model can be used to study the influence of the offline pantograph on the traction power supply system under the conditions of electromechanical coupling, as shown in Figure 2.93. Um is the output voltage of the traction substation; Rc and Lc are the resistance and inductance of the catenary, respectively; Rr and Lr are the resistance and inductance of the rail, respectively; Cc is the

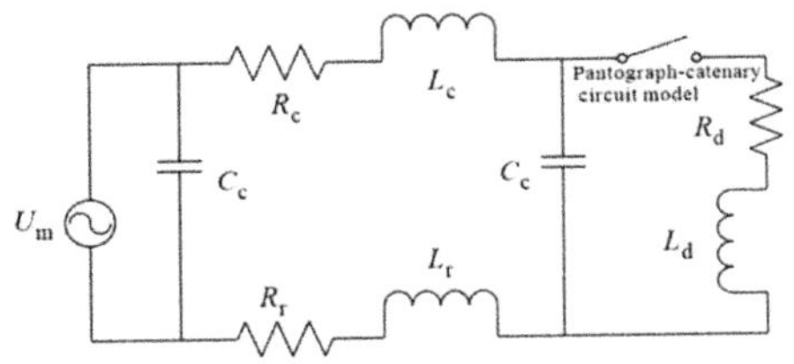

Figure 2.93
Vehicle-catenary electromechanical coupling simulation.

distributed capacitance between the catenary and rail; Rd and Ld characterize the equivalent resistance and inductance of the train, respectively.

To achieve the vehicle-catenary electromechanical coupling simulation. first, we need to calculate the pantograph vibration clearance *d* using the pantograph-catenary coupling dynamics model. Secondly, the vibration gap *d* is introduced into the pantograph arc model, and finally, the model is introduced into the traction power supply simulation model for calculation and analysis.

(2) Motor-wheel coupling

The model of the electric drive system only considers the power that transmits from the pantograph to the motor output torque. In fact, wheel, gearbox, and wheelset are also on the transmission path that converts the electrical energy into mechanical energy to drive or brake the train. Most of the high-speed train drive devices have a suspension mode and are usually connected to the transmission system by a flexible floating gear coupling. The traction motor is completely fixed to the track transom with a bolt connection. The output torque of the traction motor passes through a pinion and gear in turn, and finally drives the wheelset rotation, as shown in Figure 2.94. The driven large gear is press-fitted onto the axle. One end of the gearbox is suspended on the axle, and the other end is suspended on the frame beam by an elastic boom. Therefore, both the pinion and the gear need to be considered in the wheelset model in the vehicle system dynamics model.

In the general train dynamics study and the calculation of train traction, the mathematical model from the traction motor to the wheelset is relatively simple, considering only the transmission ratio of the gearboxes. In practice, the flexible floating gear coupling is a quasi-static mesh of multiple gears with substantially no disturbance to the output torque of the motor. However, when the dynamic behavior of the transmission gear is accurately considered, the gear meshing dynamics equation can be established in the transmission model, considering the vibration of the gear.

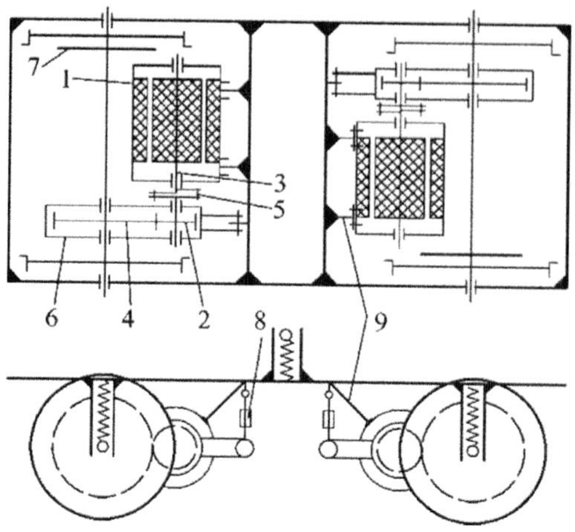

Figure 2.94
Flexible floating gear (WN) coupling frame suspension drive. 1: traction motor, 2: pinion, 3: drive shaft, 4: gear, 5: flexible coupling, 6: reduction gearbox, 7: brake disc, 8: gearbox hanging device, 9: motor hanging device.

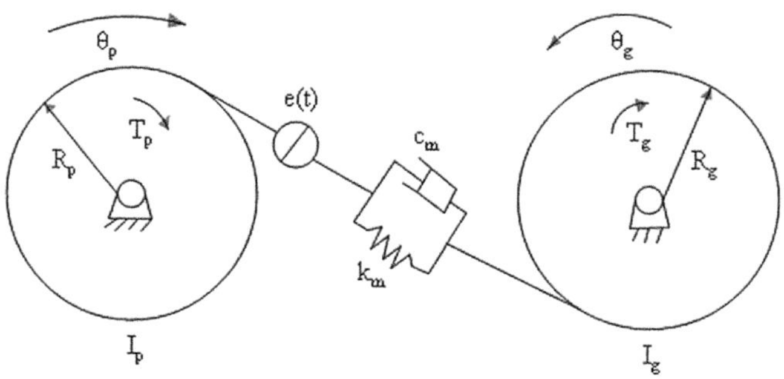

Figure 2.95
Gear vibration model.

When the transmission shaft and the elastic deformation of the support system are not considered, the gear system can be simplified as a gear pair torsional vibration system, as shown in Figure 2.95. θ_p and θ_g are the angular displacements of the active gear and passive gear, respectively; I_p and I_g are the moment of inertia of the active gear and passive gear, respectively; R_p and R_g are the base radius of the active gear and passive gear, respectively; i is the transmission ratio; $e(t)$ is the gear tooth meshing integrated error; k_m is the meshing joint stiffness; c_m is meshing damping; T_p and T_g are the external load moment acting on the active gear and passive gear, respectively. Then the kinetic equation can be expressed as [41]:

$$\begin{cases} I_p\ddot{\theta}_p + c_mR_p\left[R_p\dot{\theta}_p - R_g\dot{\theta}_g - \dot{e}(t)\right] + k_mR_p[R_p\theta_p - R_g\theta_g - e(t)] = T_p \\ I_g\ddot{\theta}_g - c_mR_g\left[R_p\dot{\theta}_p - R_g\dot{\theta}_g - \dot{e}(t)\right] - k_mR_g[R_p\theta_p - R_g\theta_g - e(t)] = -T_g \end{cases} \tag{2.228}$$

Define Δx as the relative displacement of the two gears on the meshing line:

$$\Delta x = R_p \theta_p - R_g \theta_g \tag{2.229}$$

Combining Equations (2.228) and (2.229) results in:

$$\begin{aligned} & I_p I_g \Delta \ddot{x} + \left(I_g c_m R_P^2 + I_p c_m R_g^2\right)(\Delta \dot{x} - \dot{e}(t)) + \left(I_g k_m R_P^2 + I_p k_m R_g^2\right)(\Delta x - e(t)) \\ & \quad = I_g R_p T_p + I_p R_g T_g \end{aligned} \tag{2.230}$$

This equation contains the relative rotation angle $\Delta x_0$of the system under the moment balance state. Let $\Delta \ddot{x}_0 = 0$, $\Delta \dot{x}_0 = 0$, and the relative rotation angle of the static deformation in the equilibrium state can be obtained. Here, $T_p = T_{p0}, T_g = T_{g0}$.

Substituting that into Equation (2.230) yields:

$$\left(I_g c_m R_P^2 + I_p c_m R_g^2\right)(- \dot{e}(t)) + \left(I_g k_m R_P^2 + I_p k_m R_g^2\right)(\Delta x_0 - e(t)) = I_g R_p T_{p0} + I_p R_g T_{g0} \tag{2.231}$$

Let $x = \Delta x - \Delta x_0$, and the formula in Equation (2.230) can be expressed as:

$$\frac{I_g I_p}{R_P^2 (I_g + i I_p)} \ddot{x} + c_m(\dot{x} - \dot{e}(t)) + k_m(x + \Delta x_0 + e(t)) = \frac{I_g T_p + i I_p T_g}{R_p (I_g + i I_p)} \tag{2.232}$$

Time-varying meshing stiffness is expressed as: $k_m = \overline{k} - \Delta k(t)$. Introducing the total equivalent excitation error and omitting a small amount, equation (2.232) can be updated to:

$$M\ddot{x} + C\dot{x} + \overline{k}x = \Delta k(t) e(t) \tag{2.233}$$

where $\overline{k}$ is the average stiffness, and $\Delta k(t)$ is the variable stiffness of the gear mesh stiffness. The system is a nonlinear time-varying stiffness system, and the gear could cause the change of stiffness during the meshing process, resulting in dynamic meshing force between the gear teeth. Even if the external excitation is zero, this internal dynamic excitation can still affect the vibration of the gear.

2.3.2 High-speed train coupling large system dynamics

2.3.2.1 High-speed train coupling large system dynamics model

Based on the mathematical models of the subsystems and the models of the coupling correlations, the high-speed train (coupling) large-system dynamic model can be constructed. Taking the high-speed train as the core, Figure 2.96 presents the high-speed train coupled large-system dynamics model using the vehicle-vehicle, wheel-rail, pantograph-catenary, solid-flow, and electromechanical coupling correlations. This model considers the subsystems, such as the line system, power supply system, pantograph-catenary system, transmission system, airflow, and other systems. The mathematical

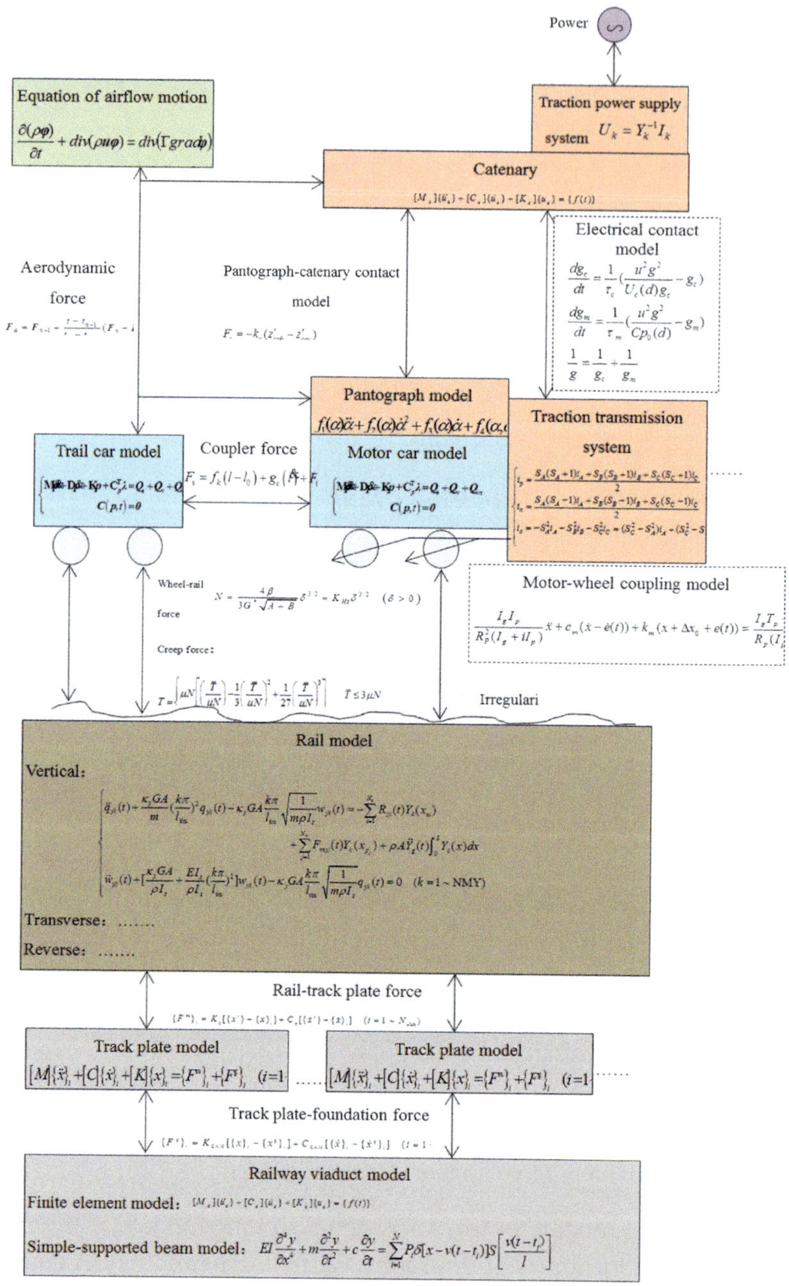

Figure 2.96
High-speed train coupled large-system dynamics model.

formulas shown in Figure 2.96 are just demonstrations, and there are often multiple modeling methods. The large system dynamics model reflects the complexities of multibody and polymorphic systems. The vehicle, line, and pantograph models can be a multi-rigid-body model or a rigid-flex hybrid model. The power supply system and traction drive system are simulated using electrical models, and the airflow is simulated using a hydrodynamic model. Although the train coupling large system dynamics model of the high-speed train is complex, its advantages are obvious. On the one hand, when considering the coupling correlation between subsystems, we can rely on different researchers to build different models. On the other hand, in simulation calculation, the coupling between different subsystems can be flexibly realized based on research needs, which will be introduced in the next chapter of the simulation platform.

2.3.2.2 Traction control in train operation

Based on the high-speed train coupling large system dynamics model, the simulation calculation can be performed under virtual operating conditions. Considering the electromechanical coupling, it is necessary to study the transient process of the traction transmission system and the power supply system during traction or electric braking. Currently, it is necessary to consider not only the running state of each train under the power supply arm but also the traction control process for more accurate simulation.

The so-called traction control is to control the rotation speed and torque of the traction motor. Asynchronous AC motors are primarily used in high-speed trains, but more advanced permanent magnet synchronous motors have not been widely used. The asynchronous traction motor control method has undergone three development processes: rotation speed difference-current control, field-oriented control, and direct torque control. The early slip-current control method was based on the steady-state mathematical model of the asynchronous motor, and its dynamic performance was of far less quality than that of the DC speed control system. In the 1970s, the theory of field-oriented control, also known as vector control, was introduced based on the control idea of the DC speed control system, and consequently, flux linkage and torque independent adjustments were achieved. Finally, the same dynamic response performance as that of the DC speed control system was achieved. The direct torque control introduced in the mid-1980s was based on the stator flux orientation, which has a simple mathematical model and better dynamic and static performance. Currently, the representative control systems of the traction control system are the rectifier current control and inverter magnetic field-oriented vector control.

Based on the working principle of the rectifier, the rectifier can be controlled by a transient direct current control scheme. The specific mathematical formulas are:

$$\begin{cases} I_{N1} = K_p(U_d^* - U_d) + 1\Big/T_i \int (U_d^* - U_d)dt \\ I_{N2} = I_d U_d / U_N \\ I_N^* = I_{N1} + I_{N2} \\ u_{ab}(t) = u_N(t) - \omega L_N I_N^* \cos \omega t - G_2\left[I_N^* \sin \omega t - i_N(t)\right] \end{cases} \tag{2.234}$$

where $G_2$is the proportional parameter, $I_d$is the intermediate DC link current, and $\omega$is the angular frequency of the grid side voltage. Figure 2.97 presents the control block of a two-level impulse rectifier in which the control system needs to feedback u_N, i_N, and U_d semaphores. The second-side output voltage $u_N$of the transformer at no-load can be obtained using the magnitude of the net voltage and the transformation ratio of the transformer. The 50 Hz BPF module shown in Figure 2.97 is a bandpass filter with a frequency band of 50 Hz. Its purpose is to extract the fundamental component of the sampled voltage on the line side signal. The module (1) is a voltage outer loop controller, and the difference between the set value of the support capacitor voltage and the actual value measured by the voltage sensor is passed through a constant voltage controller. Subsequently, the magnitude $I_{N1}$of the current on the line side $i_N$is the output. A PI regulator is usually used in the constant voltage controller. To reduce the load of the intermediate DC link voltage PI regulator and improve the dynamic response of the

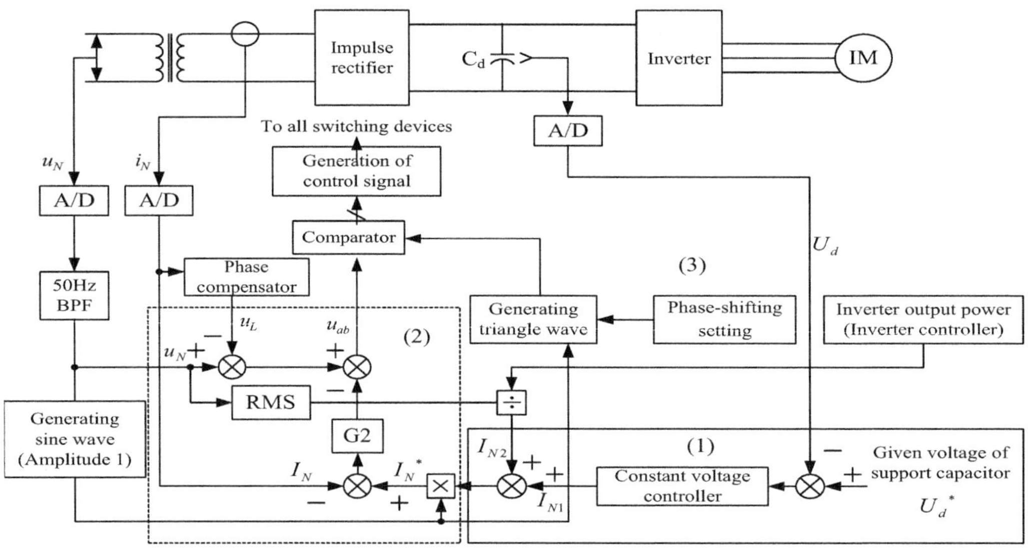

Figure 2.97

Two-level pulse rectifier transient current control block. (1) Voltage outer loop controller, (2) current inner loop controller, (3) SPWM signal generator.

PI regulator, the DC side output power is divided by the effective value of the second side of the traction winding voltage U_N to calculate the effective component of the given current I_{N2}, then added to $I_{N1}$of the mesh side to give the value of the grid side current I_N^*.

Module (2) is a transient current inner-loop controller. The modulating signal $u_{ab}(t)$ is generated using Equation (2.234), and the transient current control needs to feedback the inductor current on the line side. There is a link $G_2\left[I_N^* \sin \omega t - i_N(t)\right]$in the system, which provides the system with good dynamic response and quick adjustment of system parameter changes.

Module (3) is a pulse width modulated (PWM) signal generator, and the sine pulse width modulated (SPWM) control signal is generated by comparing a triangular carrier wave with a modulation wave $u_{ab}(t)$.

Transient direct current control is currently used in electric locomotives and high-speed trains. It has the advantages of simple implementation, effective suppression of the current harmonics of the secondary side traction winding, small voltage ripple on the DC side, and good dynamic response.

Field-oriented vector control is very promising for the future because it has certain advantages in preventing the wheelset from idling, and a similar system has been used in CRH2 series high-speed trains. The field-oriented vector control method is more complicated than the rectifier transient current control, which will not be covered here.

2.3.2.3 Service simulation of the high-speed train

In the past, vehicle system dynamics research, especially simulation calculations, primarily involved simulating the operational state to obtain responses under the operating boundary conditions. With the development of high-speed train coupling large-system dynamics, the concept of service simulation is increasingly recognized, and the basic concepts are presented in Reference [1]. The difference between the service model and the operation simulation is that the latter focuses only on the state of operation, especially the specific parameter matching the design phase and dynamic performance under line conditions, and the former is concerned with the evolution of dynamic performance during long-term service. The basis for running the simulation is the service simulation.

(1) Operation simulation

The running simulation is the simulation calculation of the high-speed train coupling large system under specific operation requirements, boundary conditions, and environmental constraints. Its primary aim is to grasp the dynamic responses of high-speed trains, and to perform high-speed train parameters optimization and its matching optimization with other coupling subsystems, as shown in Figure 2.98. Here, the so-called operational demand is the simulation running speed. The main line conditions of the boundary conditions include

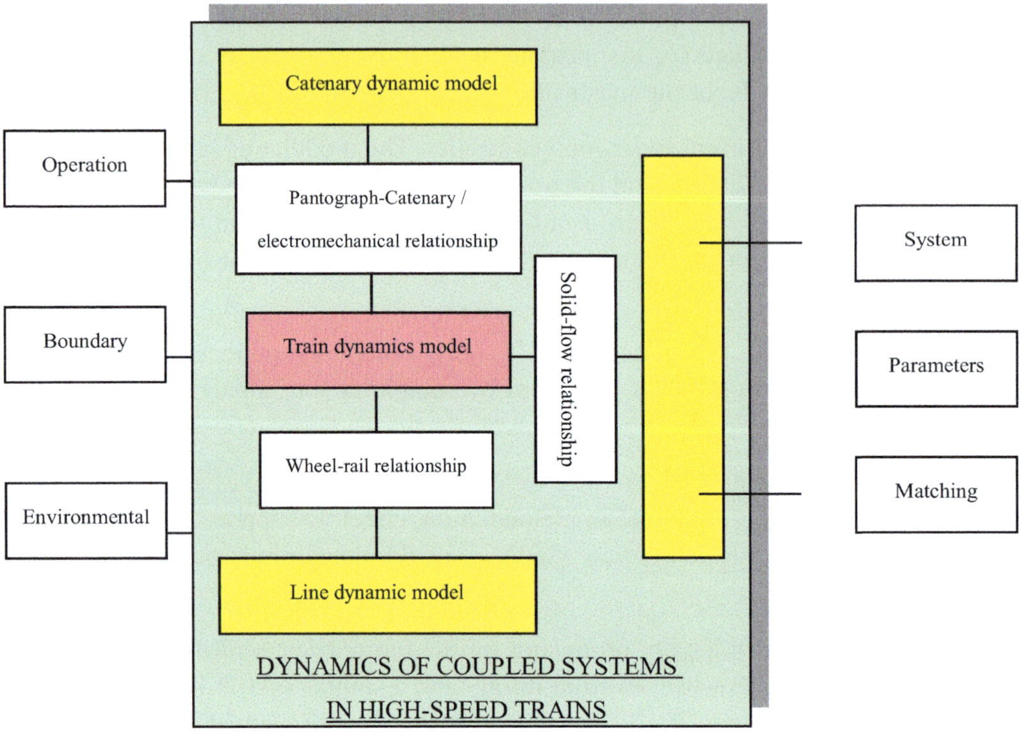

Figure 2.98
Running simulation.

the planar and vertical section, line irregularity, line spacing, catenary irregularity, and voltage fluctuations of the mains grid. The environmental constraints primarily include airflow conditions (especially crosswind conditions), earthquakes that seriously affect operational safety, and the impact of climate change on the dynamic performance of the high-speed train coupling large system during operation. The boundary and environmental models generally include:

(a) Orbital and catenary irregularities: Track irregularity is the main influence on vehicle vibration. In the vehicle system dynamics simulation calculation, track irregularity is generally treated as the boundary. As an important input, this irregularity is directly loaded on the rail surface, and the wheel is excited by the additional displacement. The contact wire irregularity has the same processing method as track irregularity, where the contact line irregularity is directly loaded onto the contact line, and the pantograph head is excited by the additional displacement.
(b) Airflow condition or wind load: Although the fluid-solid coupling model is established in the large-system model, the real wind load is still used as an external input, especially the fluctuating wind.

(c) Earthquake waves: As an external input, an earthquake wave acts on the foundation of the line or the pier of the elevated line. Its characterization includes the direction, intensity, and dynamic process of the wave (the earthquake spectrum). Generally, the earthquake wave acts on the train through the wheel-rail correlation.
(d) Environmental changes: The operating conditions affect parts of the train, especially the properties of polymer materials such as rubber and plastic. Sometimes, ice and snow can directly affect the quality of the catenary and contribute to failure of the vehicle suspension. Therefore, the impact of the operating conditions should be considered as an input to the system. Because the operating conditions primarily affect the components, only changes in the parameters of the components caused by the operating conditions should be considered in the running simulation or service simulation.

(2) Service simulation

Traditional vehicle system dynamics research is primarily used as a tool of computer-aided design (CAD) to guide structural design and determine the theoretical value of parameter design. However, the structural parameters of the high-speed train and its subsystems would change during long-term service, which causes changes in system performance and even induces safety misadventure. Therefore, the study of high-speed train coupling large system dynamics is not only the study of system conditions and performances but more importantly, service simulation from the perspective of service process characterization. In essence, it is the study of the evolution law of high-speed train parameters and performance during long-term service.

① *Calculation block diagram* The calculation block of the service simulation of the high-speed train coupled large system is shown in Figure 2.99. The high-speed train (coupling) large system model is described in the system model, including models of the pantograph system, power supply system, train system, traction drive system, and rail system. The boundaries and operating conditions that affect the performances of the high-speed train are used as system inputs. The system model is simulated, and the responses of the system can be obtained. System responses, in turn, can affect the performance of materials and structures, triggering material and structural failure and damage, which in turn leads to parametric changes in parameters that make certain parameters time-varying. Considering the variation of these time-varying parameters, the variations of structures, parameters, and performance of the high-speed train can finally be obtained in the simulation calculation. The service model can predict the dynamic performance of high-speed trains and their coupling systems, master the changes in high-speed train performance, and provide a basis for high-speed train operation, repair and, redesign.

This shows that the difference between the service simulation and the traditional dynamics simulation is that the simulation process ends when the latter receives the response results.

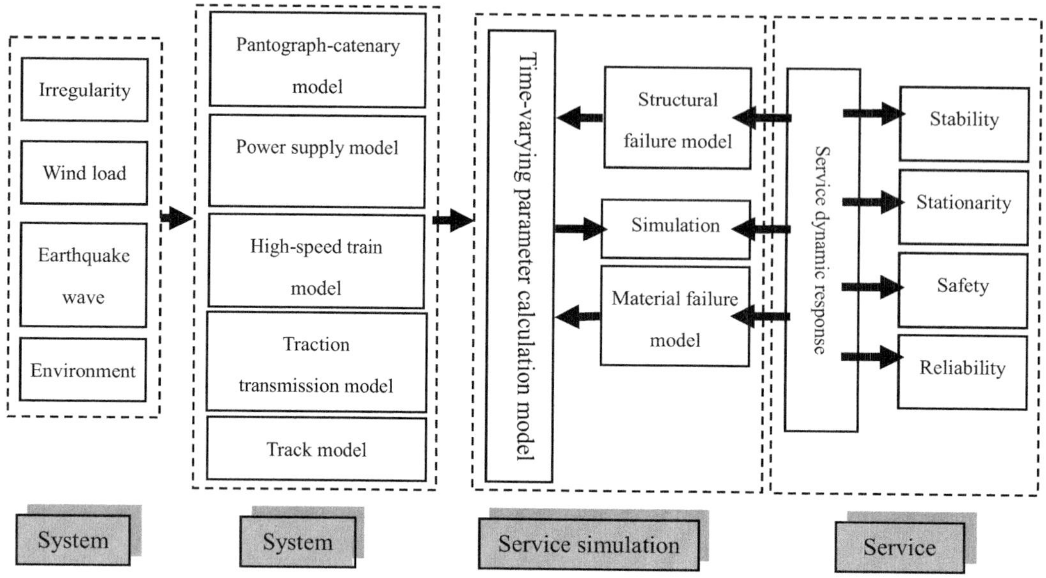

Figure 2.99
The simulation calculation of the high-speed train coupled large system.

In the service simulation, a time-varying calculation parameter model is added to the system model to form a service simulation model. Therefore, in the calculation, the service model and simulation calculation are used for iterations between the system response and the time-varying parameter model to realize the simulation of the service process of the high-speed train. Then the correlation between the system response and the generalized failure can be obtained to reveal the evolution of the dynamic behavior and performance of the high-speed train.

② *Failure model* To carry out the service simulation, the failure models involving material failure, structural damage, and parameter change should be established first, and loaded into the dynamic models of the high-speed train coupling large system to form the service model of the high-speed train coupling large system.

High-speed train service simulation is theoretically feasible, primarily by establishing a variety of accurate and comprehensive failure models, which is obviously difficult but unnecessary. What is necessary is to grasp the structure and parameters that are ineffective or variable and directly affect the operation performances of the high-speed train. The important parameters are as follows:

1) Parameter failure: The parameters primarily refer to geometric parameters such as suspension parameters and tread. Suspension parameters for high-speed trains include primary and secondary suspension stiffness and damping as well as dry friction. Changes in these parameters can directly affect the performance of the high-speed

train. Due to the influence of operating conditions and suspension components (especially rubber parts, friction pieces, and dampers), suspension parameter changes are unavoidable and should be taken into consideration in the service simulation. The geometrical parameters of the high-speed train are mainly concentrated in bogies, such as the wheel diameter, tread profile, and wheel weight Changes in the geometric parameters are often caused by frictional wear, plastic deformation, etc.

2) Structural failure: The structure determines the function of the system, and its failure affects the performance of the system. In high-speed trains and the corresponding coupling systems, the main structural failure modes are fatigue failure and friction wear. For the rail, due to the high-speed rolling, there are wheel-rail geometrical parameter changes and corrugation caused by the frictional wear of the wheel-rail. There are also contact spalling failures due to surface spalling, cracking, and knocking. For the vehicle, especially the bogie structure, fatigue of the bogie causes mainly failure accidents such as structural breaks. For the pantograph system, due to the high-speed sliding under current-carrying conditions, the failure mode is not only wear but also arc ablation.
3) Material failure: Material failure refers to changes in structural properties and parameters. It is primarily caused by the coupling of force, heat, electricity, and magnetism in an application, as well as environmental issues. Although material failure is often a slow and concealed process, its impact and damage cannot be underestimated.

References

[1] Zhang Weihua, Zhang Shuguang. Dynamics and Service Simulation for General Coupling of High-Speed Trains. Southwest Jiao Tong University 2008;43(2):147–52.

[2] Zhang Weihua. Vehicle dynamic simulation. Beijing: China Railway Publishing House; 2006.

[3] Zhai Wanming. Vehicle- track coupling system dynamics. Second Edition. Beijing: China Railway Publishing House; 2003.

[4] Wang Futian. Vehicle system dynamics. Beijing: China Railway Publishing House; 1994.

[5] Garg VK, Dukkipati RV. Dynamics of Railway Vehicle Systems. New York: Academic Press; 1984.

[6] Iwnicki S. Handbook of Railway Vehicle Dynamics. Taylor & Francis, CRC Press; 2006.

[7] Luo Ren, Zeng Jing, Wu Ping-bo. Influence of air spring on curve negotiation property of vehicle. Journal of traffic and Transportation Engineering 2007;7(5):15–8.

[8] Craig RR, Bampton MC. Coupling of Substructures for Dynamic Analysis. American Institute of Aeronautics and Astronautics Journal 1968;6:1313–9.

[9] Wallrapp O. Flexible Bodies in Multibody System Codes. Vehicle System Dynamics 1998;30:237–56.

[10] Gao Hao, Luo Ren, Mao-ru Chi, Huan-yun Dai. Safety analysis of railway vehicle in leakage process of air spring. Journal of traffic and Transportation Engineering 2012;12(3):60–6.

[11] Gao Hao, Huan-yun Dai. Dynamic Simulation Aanlysis of Train Crash to Mortar Rubble Line Terminal Stopper. China Railway Science 2012;33(3):61–6.

[12] Xiao XinBiao, Jin XueSong, Wen ZeFeng, Zhu MinHao, Zhang WeiHua. Effect of tangent track buckle on vehicle derailment. Multibody System Dynamics 2011;25:1–41.
[13] Timoshenko S, Young DH, JR WW. Vibration Problems in Engineering. 4th Edition. USA: John Wiley & Sons, Inc.; 1974.
[14] Xiao Xinbiao, Jin Xuesong, Wen Zefeng. Effect of earthquake on high speed railway vehicle running safety. Proceedings of the IAVSD'09. Stockholm, Sweden: KTH; 2009. p. 16–20.
[15] Thompson DJ, Jones JCO. A review of the modeling of wheel/rail noise generation. Journal of Sound and Vibration 2000;231:519–36.
[16] Cai ZQ. Modeling of rail track dynamics and wheel/rail interaction. Ph.D thesis. Kingston, Ontario, Canada: Queen's University; 1992.
[17] Mei Guiming. Study on Pantograph catenary system dynamics. Chengdu: Doctoral Dissertation of Southwest Jiao Tong University; 2010.
[18] Li Ruiping, Zhou Ning, Zhag Weihua, Mei Guiming, Zhen Zenbao. Calculation and Aanlysis of Pantograph Aerodynamic Uplift Force. Railway Transaction 2012;34(8):26–32.
[19] Park Tong-Jin, Chang-Soo Han, Jin-Hee Jang. Dynamic sensitivity analysis for the pantograph of a high-speed rail vehicle. Journal of Sound and Vibration 2003;266(2):235–60.
[20] Cai Chenbiao, Zhai Wanming. Study on Simulaition of Dynamic Performance of Pantograph-Catenary System at High-Speed Railway. Railway Transaction 1997;19(5):38–43.
[21] Zhou Ning, Li Ruiping, Zhang Weihua. Modeling and simulation of catenary based on negative sag method. Journal of traffic and transportation engineering 2009;9(4):28–32.
[22] Li Ruiping, Zhou Ning, Mei Guiming, Zhang Weihua. Finite element model for catenary in initial equilibrium state. Journal of Southwest Jiao Tong University 2009;44(5):732–7.
[23] Li Ruiping, Zhou Ning, Zhang Weihua, et al. Wind Load Simulation and Wind–induced Vibration Analysis of Catenary. In: International Symposium on Speed-up and Service Technology for Railway and Maglev Systems; 2012. Seoul,Korea.
[24] Cui Tao, Zhang Weihua. Study on Safety of Train in Side Wind with Changing States. Railway Transaction 2010;32(5):25–9.
[25] Tian Li. Approaches and Dynamic Performances of High-Speed Train Fluid-Structure. Doctoral Dissertation of Southwest Jiao Tong University; 2012.
[26] Li Xuebing. The Study on Flow-Induced Vibration of High-Speed in Passing Events. Master Thesis of Southwest Jiao Tong University; 2010.
[27] Cao Jianyou. Traction power supply system of electrified railway. Beijing: China Railway Publishing House; 1983.
[28] Li Qunzhan, He Jianmin. Analysis of traction power supply system. Chengdu: Southwest Jiao Tong University press; 2007.
[29] Li Qunzhan. Power supply analysis of traction substation and comprehensive compensation technology. Chengdu: Southwest Jiao Tong University press; 2006.
[30] Zhang Jinsi. On the Method of Calculating the Harmonic Distribution in Power Systems Caused by Electrical Traction. Journal of Southwest Jiao Tong University 1984;4:82–92.
[31] Wu Mingli. Study on Electrical Parameters and Mathematical Model of Traction Power Supply System. Doctoral Dissertation of Beijing Jiaotong University; 2006.
[32] Xiaoyun Feng. Electric traction AC drive and its control system. Beijing: Higher Education Press; 2009.
[33] Liang Li. Study on the Dynamic Behaviour of High-Speed Train/Track Coupling System Composed of Multiple Vehicles. Master Thesis of Southwest Jiao Tong University; 2012.
[34] Zhang Shuguang. Research on Design Method of High-Speed Train. Beijing: China Railway Publishing House; 2009.
[35] Wang Kaiwen. Study on Geometric corcorrelation between wheel and rail in arbitrary shape. Master Thesis of Southwest Jiao Tong University; 1982.

[36] Knothe KL, Grassie SL. Modeling of railway track and vehicle/track interaction at high frequencies. Vehicle System Dynamics 1993;22:209–62.
[37] Xiao Xinbiao, Jin Xuesong, Wen Zhenfeng. Effect of Track Support Failure on Dynamic Response of Tangent Track. Journal of mechanics 2008;40(1):67–78.
[38] Tian Li, Zhang Jiye, Zhang Weihua. Co-simulation of High-speed Train Fluid-Structure Interaction Dynamic in Crosswinds. Journal of Vibration Engineering 2012;25(2):138–45.
[39] Tian Li, Zhang Jiye, Li Zhongji, Zhang Weihua. Co-simulation of Fluid-Structure Interaction of High-Speed Train Based on Fluent and Simpack. Journal of Computational Mechanics 2012;29(5):675–80.
[40] Tian Li, Zhang Jiye, Zhang Weihua. An Improved algorithm for fluid-structure interaction of high-speed trains under crosswind. Journal of modern transportation 2011;19(2):75–81.
[41] Li Runfang, Wang Jianju. Gear System Dynamics. Beijing: Science Press; 1997.
[42] Han Yan, Xia He, Guo Weiwei. Dynamic Response of Cable-Stayed Bridge to Running Trains and Earthquakes. Engineering Mechanics 2006;23(1):93–8. 68.

CHAPTER 3

The simulation platform for the dynamics of coupled systems in high-speed trains

Chapter Outline

Dynamics of Coupled Systems in High-Speed Railways. https://doi.org/10.1016/B978-0-12-813375-0.00003-0

3.1 The framework of the simulation platform for the dynamics of coupled systems in high-speed trains

3.1.1 The function of the simulation platform for the dynamics of coupled systems in high-speed trains

The kernel function of this platform is to realize the coupling simulation of a high-speed train with the track, catenary, airflow, power supply system, etc.

It should also function to assist in the following:

(1) research on the coupling correlation between high-speed trains and other subsystems;
(2) design assessment of high-speed trains and their coupling systems;
(3) optimization analysis of high-speed trains and their coupling systems;
(4) decisions regarding the operation and maintenance of high-speed trains and their coupled systems.

To provide better support for the innovative design of, optimization of, performance evaluation of, and operation and maintenance decisions regarding high-speed trains and their coupled systems, some platform application requirements should be taken into account. For instance, subsystems may run in a heterogeneous environment with serial and parallel computing modes. And the flexibility organization, efficiency, and stability needs of the simulation process should be considered. The requirements that the platform be easy to use, easy to expand, and easy to maintain also should be satisfied. Therefore, the following requirements shall be met:

(1) In platform architecture: First, the platform has an open architecture, of which the interface design is based on normality and versatility so as to provide the system with good extendability. Second, it supports distributed simulation, where different computing modes and computing environments are allowed for each subsystem. Third, the efficiency, stability, and security of the computing network are guaranteed.
(2) In computing resource management: On one hand, the heterogeneous computing resources of the platform can be identified and uniquely managed. On the other hand, the computing network can be self-diagnosed and its topology can be generated automatically, etc.

(3) In the user interface: First of all, the platform needs a friendly human–computer interaction interface with clear functional areas and easily understood function icons and conforms to the general simulation users' operating habits. Second, it has an efficient simulation modeling function that can be visualized and parameterized. Third, the users can flexibly and quickly set the coupled simulation conditions composed of the different subsystems. Fourth, the interface prompts and operation sequence help the users control the simulation process smoothly. Fifth, the simulation data can be managed hierarchically, which includes simulation conditions data, simulation process data, simulation results data, user data, etc. Last, it has a rich display function for expressing various simulation results, including 2D charts, 3D scenes, and other display methods to help the users understand the system intuitively.

(4) In simulation condition scheduling: There are several common scheduling strategies in the platform, such as sequential scheduling algorithm, targeted scheduling algorithm, priority-based preemptive scheduling algorithm, etc. The new scheduling algorithms can be extended and managed easily.

(5) In calculation of simulation conditions: In the first place, under the premise of sufficient computing resources, the platform supports multiple simulation conditions performing simultaneously and balances the load of the computing network. In the second place, it provides a coupling data processing interface for subsystems with different data processing methods. In the third place, it can monitor the calculation process visually and present timely warnings of abnormal information during simulation.

3.1.2 Software architecture of the simulation platform for the dynamics of coupled systems in high-speed trains

In accordance with the functional requirements of a simulated platform for the dynamics of coupled systems of high-speed train, this software system adopts a distributed simulation system based on the client/server (C/S) mode with main function modules: simulating condition management module, preprocessing module, coupling simulation module, postprocessing module, and system management module. As shown in Fig. 3.1, the data between modules are transmitted and managed through the database system and files. The arrows represent the direction of the data flow.

The functions of each module of simulation software for the dynamics of coupled systems in high-speed trains are described in the following:

(1) Simulation condition management module
The main function of this module is to establish various simulation conditions, validate the simulation data for subsystems, distribute the simulation task, and manage the simulation conditions and results data.

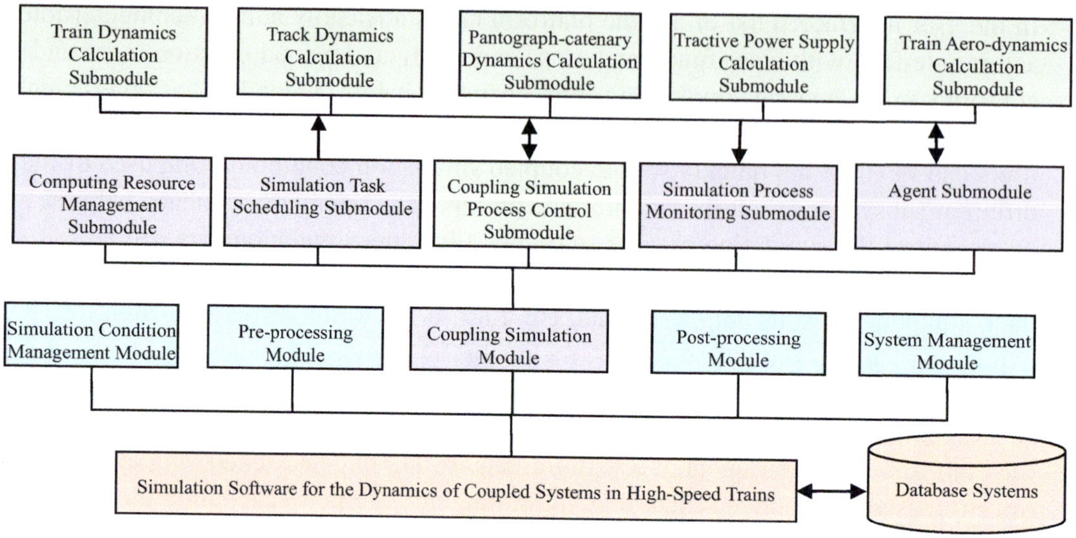

Figure 3.1
Modules of the simulation software for the dynamics of coupled systems in high-speed trains.

(2) Preprocessing module
The main function of this module is to map the actual physical models of each coupled simulation subsystem to a simulation model, thereby creating the preprocessing simulation models for all subsystems, which include preprocessing simulation models of high-speed train dynamics, track dynamics, pantograph—catenary dynamics, train aerodynamics, and tractive power supply and traction drive subsystem.

(3) Postprocessing module
The main function of this module is to process and analyze the coupling simulation results of each subsystem to obtain time—history curves and evolution rules, and then evaluate the performance of the subsystems, providing the basis for performance optimization. The module includes a 2D chart postprocessing submodule, a 3D visual operation simulation submodule, a tractive power supply postprocessing submodule, and an aerodynamics postprocessing submodule.

(4) System management module
The main function of this module is to maintain and manage the whole platform, such as user management, role management, rights management, etc.

(5) Coupling simulation module
The main function of this module is to realize the coupling calculation among the subsystems that have a coupling correlation with the high-speed train. It is the core of the simulation platform, including the computing resource management submodule, task scheduling submodule, coupling simulation process control submodule, agent

submodule, train dynamics coupling calculation submodule, track dynamics coupling calculation submodule, pantograph–catenary dynamics coupling calculation submodule, tractive power supply coupling calculation submodule, train aerodynamics coupling calculation submodule, and simulation process monitoring submodule. The functions of each submodule are described next.

1) Computing resource management submodule
The main function of this submodule is to obtain the usage information of the computing resources in the network via the agent submodule, monitor the available computing resources and their usage state, implement unified management of heterogeneous computing resources, and provide the computing resources for the simulation task. Also it can create the topology and realize the diagnosis of the computing network automatically, etc.
2) Task scheduling submodule
The main function of this submodule is to implement the scheduling of the tasks submitted by the user for simulation. The first step is to analyze the calculation requirements, priorities, and network resources, etc., of the task, and then to apply the scheduling strategy to schedule the calculation task to the appropriate computing resources. If computing resources are insufficient, computed tasks are queued for priority. The submodule can support automatic scheduling of multiple jobs and also allows users with administrative rights to extend the scheduling policy.
3) Coupling simulation process control submodule
The main function of this submodule is to perform coupling calculations on the submitted simulation task that is under the control of the coupler. The coupler controls the start and stop for one-step simulations and coordinates the simulation step size of subsystems such as train dynamics, track dynamics, pantograph–catenary dynamics, train aerodynamics, and tractive power supply and traction drive to keep the simulation process of the subsystems in a spatiotemporal synchronized form. It also controls the data reception, processing, and sending of each subsystem to realize the coupling simulation on every integral step.
4) Simulation process monitoring submodule
The main function of this submodule is to visually monitor the coupling simulation process, prompt or alarm for abnormal conditions, and make the calculation process transparent.
5) Agent submodule
The main function of this submodule is to obtain information on computing resources, such as the performance and usage rate of the CPU, the capacity and usage of the memory, the type of the operating system, etc. Continuous communication with the computing resource management module and the coupling control submodule enables

the management of heterogeneous computing resources, implementation of coupling simulation, fault diagnosis, and recovery. The agent submodule includes the coupler agent submodule and executor agent submodule. The coupler agent submodule is used to automatically download and launch the coupling simulation process control module and register related information to the task scheduling submodule. The executor agent submodule is used to automatically download and launch each subsystem's calculation module, obtain the operation status information of each subsystem calculation module, and send it to the coupling control submodule.

6) Train dynamics coupling calculation submodule
This submodule is used to control the single-step simulation of the train dynamics subsystem, output the simulation results of this step, and send the necessary data to the coupling control submodule; then it receives the needed data from the coupling control submodule to prepare the next step in the simulation.
7) Track dynamics coupling calculation submodule
This submodule is used to control the single-step simulation of the track dynamics subsystem, output the simulation results of this step, and send the necessary data to the coupling control submodule, and then receives the needed data from the coupling control submodule to prepare the next step in the simulation.
8) Pantograph–catenary dynamics coupling calculation submodule
This submodule is used to control the single-step simulation of the pantograph–catenary dynamics, output the simulation results of this step, and send the necessary data to the coupling control submodule, and then receives the needed data from the coupling control submodule to prepare the next step in the simulation.
9) Tractive power supply and traction drive coupling calculation submodule
This submodule is used to control the single-step simulation of the tractive power supply and traction drive subsystem, output the simulation results of this step, and send the necessary data to the coupling control submodule, and then receives the needed data from the coupling control submodule to prepare the next step in the simulation.
10) Train aerodynamics coupling calculation submodule
This submodule is used to control the single-step simulation of the train aerodynamics subsystem, output the simulation results of this step, and send the necessary data to the coupling control submodule, and then receives the needed data from the coupling control submodule to prepare the next step in the simulation.

The software architecture of the distributed simulation platform for the dynamics of coupled systems in high-speed trains is shown in Fig. 3.2, which includes the presentation layer, simulation condition scheduling layer, coupling simulation control layer, and simulation task execution layer.

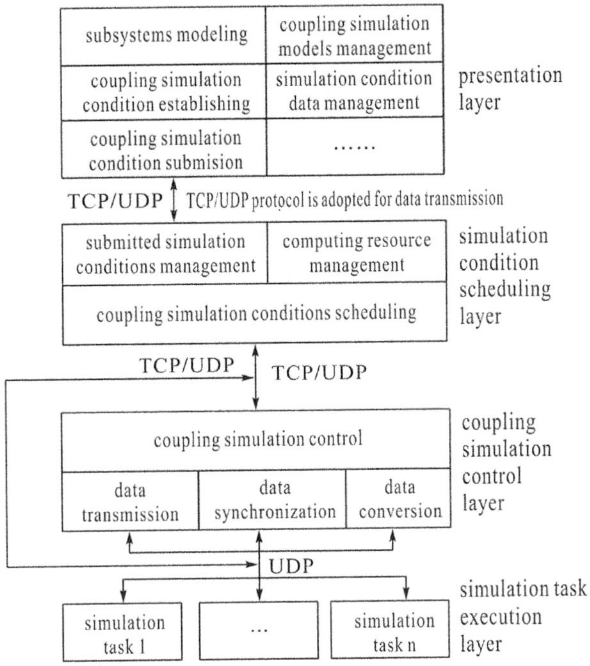

Figure 3.2

Software architecture of the simulation platform for the dynamics of coupled systems in high-speed trains.

The presentation layer is correlated to the user's operations on the foreground. It includes a simulation condition management module, a preprocessing module, a postprocessing module, a system management module, and a simulation process monitoring submodule.

The other three layers are correlated to the coupling calculation on the background, and the submodules except for the simulation process monitoring submodule in the coupling simulation process module are included. For example, the simulation task scheduling layer includes a simulation task scheduling submodule and a computing resource management submodule. The coupling simulation control layer includes an agent submodule and a coupling simulation process control submodule. The simulation task execution layer includes an agent submodule, a train dynamics coupling calculation submodule, a track dynamics coupling calculation submodule, a pantograph–catenary dynamics coupling calculation submodule, a tractive power supply and traction drive coupling calculation submodule, and a train aerodynamics coupling calculation submodule.

The three-layer architecture consisting of a scheduler (simulation task scheduling servers) –coupler (coupled simulation process control servers)–executor (subsystem simulation execution machines) has good versatility and extendability. When the scale of calculations is expanded, just increasing the numbers of couplers and executors can meet the requirements. The functions of each layer in the simulation platform are described as follows:

(1) Presentation layer
The main task of the presentation layer is to construct the simulation model of the subsystems involved in the coupled simulation, set the coupling simulation conditions and submit them, and manage the models, states, and results of the submitted simulation conditions.

(2) Simulation condition scheduling layer
The simulation condition scheduling layer queues the submitted simulation conditions and schedules them according to the usage state of the computing resources. The kernel module of this layer is named the scheduler and its functions are as follows:
1) dispatching the higher priority conditions to the light-load coupler in accordance with the load state of the coupling simulation control layer;
2) dispatching the submitted simulation task to the unoccupied computing resources in accordance with the load state of the simulation task execution layer and the resources demand of each simulation task;
3) instructing the coupler to get the simulation condition data, such as the correlation among subsystems, input and output of each subsystem, and coupling control parameters, etc.;
4) instructing the simulation task execution layer to download the necessary programs and launch them with the initialized input data and configuration information.

(3) Coupling simulation control layer
The coupling simulation control layer completes the process control and advancement of the coupling simulation for the submitted conditions. The kernel module of this layer is named the coupler, and its tasks are as follows:
1) receiving the simulation condition configuration file sent by the simulation condition scheduling layer, then extracting the key information from the file, and generating a coupling relation table and a parameter transfer table among the coupled subsystems in the simulation tasks; after that, starting the coupling simulation;
2) receiving and handling the interface data from each subsystem, and then transmitting them to the related subsystems by searching the coupling relation table;
3) checking the input data for each subsystem on every step, and sending a restart instruction to the subsystem when its input data are prepared; in this way, the synchronization of the simulation process among subsystems with different integral step is realized;
4) sending a state message to the simulation condition scheduling layer when the submitted simulation task is completed;

5) sending a warning message to the simulation condition scheduling layer when something wrong appears in the submitted simulation task;
6) deleting the distributed simulation data in the memory to prevent memory leaks.

(4) Simulation task execution layer

The simulation task execution layer performs the simulation tasks of the subsystems on the allocated computing resources. The kernel module of this layer is named the executor, whose main tasks are described as follows:

1) downloading the simulation program, input file, and configuration file of each subsystem involved in the coupling simulation task;
2) starting a one-step simulation of each subsystem when it receives the restart instruction from the coupler;
3) uploading the coupling interface data to the coupler when a one-step simulation of the subsystem is finished and waiting for the next simulation;
4) sending a request to the coupler to obtain the coupling interface input data for the next simulation;
5) receiving the notification information from the coupler;
6) sending local resource information to the scheduler periodically;
7) checking the process status of each simulation task, and sending the warning information to the scheduler in time when an exception occurs and then recovering the corresponding computing resources;
8) uploading the simulation results of each subsystem to the database or file server, and recovering the corresponding computing resources in time after completing the simulation.

3.1.3 The hardware architecture of the simulation platform for the dynamics of coupled systems in high-speed trains

Each subsystem has different requirements for computing resources and operating environment. To ensure the overall simulation efficiency of coupling simulations among multiple subsystems, the simulated platform for the dynamics of coupled systems in high-speed trains is deployed in a distributed computing environment connected by a high-speed network. The hardware components are shown in Fig. 3.3, which mainly includes subsystem simulation execution machines, in short called executors; simulation client computers; coupled simulation process control servers, in short called couplers; simulation task scheduling servers, in short called schedulers; database and file servers; simulation process monitoring computers; high-performance parallel computing clusters; and related network devices.

The simulation client computer is a client computer in which the user directly operates the platform. It is used for simulation preprocessing, setup and submission of simulation

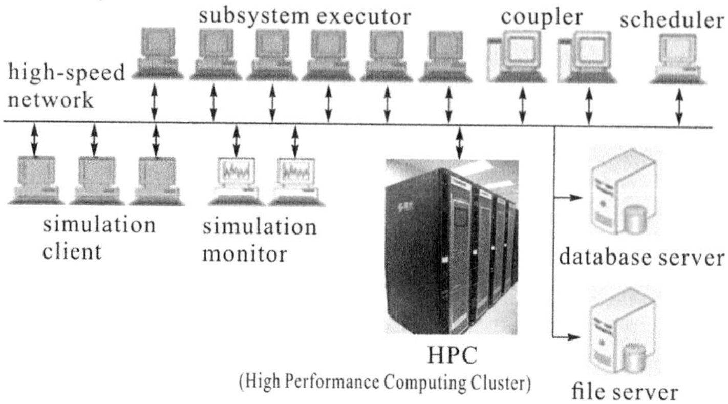

Figure 3.3
Hardware components of the simulation platform for the dynamics of coupled systems in high-speed trains. *HPC*, High Performance Computing Cluster.

conditions, postprocessing for simulation results, and system management. The main operating system is the Windows operating system and some of them use the Linux/Unix operating system. It is composed of a graphics workstation or ordinary PC of high performance.

The executor adopts the Windows operating system, which can run simulation modules of train dynamics, track dynamics, pantograph–catenary dynamics, train aerodynamics, and tractive power supply and traction drive subsystem. It is composed of a series of high-performance PCs and can be dynamically added or deleted according to the application requirements at any time.

The high-performance computer cluster adopts the Linux/Unix operating system, some nodes of which can adopt the Windows operating system. It has many computational cores and high computational efficiency, so it is preferentially used in the aerodynamic subsystem simulation.

The coupler is a center for data receiving, processing, and transmitting. All the coupling interface data of all simulation tasks are collected and distributed to each subsystem. In cases of multiple simulation conditions, it is very busy, because of the large data throughput, so a high processing capability of the CPU and excellent data throughput performance for the network card are required. Generally, the Windows operating system is run on it.

The scheduler is used to distribute the simulation tasks stably to the free computing resources. Redundant deployment of the scheduler should be considered while the computing resources are sufficient. The operating system for the scheduler can be Windows, Linux, or Unix.

The simulation process monitor is mainly used to monitor the process data, and its main functions are to obtain the process data of the simulation calculation from the database and display them through an independent simulation process monitoring system, so as to find the problems in the simulation in time and improve the application effectiveness. It is composed of a graphics workstation or ordinary or high-performance PC with the Windows operating system.

The database server stores data of various models, conditions, and simulation results. For calculations with a large storage capacity and high storage frequency, the memory database server is preferentially used, while for others, general servers with Windows, Linux, or Unix operating systems can be used.

Generally, the data reading and storing frequency of the file server is not too high, requires a large storing capacity, and runs on the Windows, Linux, or Unix operating system.

The main network of the simulation platform of coupled systems in high-speed trains is a gigabit network, of which hot standby is performed by two 10-gigabit core routing switches, and each subsystem access-layer switch uses gigabit access. The InfiniBand network is adopted in the performance computer cluster (HPC) as the computing network and the Gigabit network is adopted as the management network.

By analyzing the functional requirements, software architecture, and hardware components of the simulation platform for the dynamics of coupled systems in high-speed trains, the distributed hardware architecture with the C/S mode is proposed as shown in Fig. 3.4. The graphical client interface is developed to facilitate the user's interactive operation, and the simulating calculations are performed on the server with a focus on its performance.

The server adopts a three-layer architecture composed of "scheduler—coupler—executor" with high stability and scalability. The coupler and the executor are free to join or exit the platform's computing resource pool in accordance with the requirements of the simulation task and the usage situation of the computing resources. If one of the couplers or executors is abnormal, the other computing resource can play a backup role. And the multicoupler can also effectively cope with the load balance of the computing network caused by multiple simulation conditions.

3.2 Parametric and graphical modeling of high-speed trains

With the development of 3D graphics technology, detailed 3D computer-aided design (CAD) models have been established in the design of high-speed trains. Therefore, in the dynamics modeling of high-speed trains, the existing 3D CAD model information should be fully reused. Utilizing the function of the seamless integration of the high-speed train design platform with its dynamics analysis platform, high-speed train dynamics models

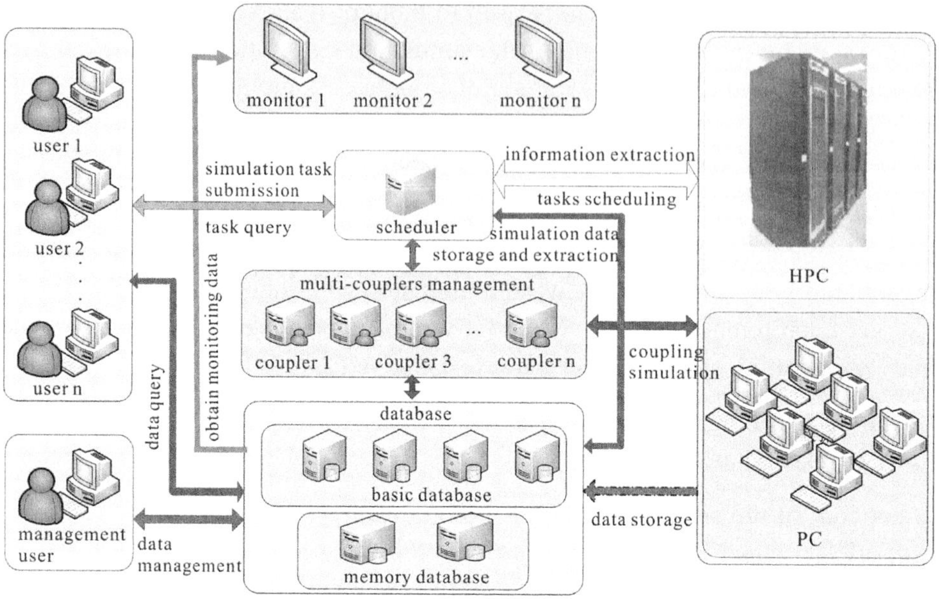

Figure 3.4
Hardware architecture of the simulation platform for the dynamics of coupled systems in high-speed trains. *HPC*, High Performance Computing Cluster; *PC*, personal computer.

can be built from CAD models directly. This enhances the direct reliance of the dynamics model on existing design models and also provides the basis for the optimization design of high-speed trains driven by dynamics simulation results. The dynamics attribute extraction technology oriented to CAD is used to extract the properties required for dynamics simulation from the existing high-speed train CAD model, which ensures the consistency of the design model and the dynamics model. The use of parameterized expressions of various properties in the dynamics model facilitates the updating and reuse of the model, and the use of graphical modeling techniques in the modeling process enhances the user's intuitive understanding of the model, reduces the difficulty of modeling, and allows the user to check and verify the model.

3.2.1 Dynamics property extraction techniques for computer-aided design models

The high-speed train CAD model contains information such as assembly, geometry, pose, mass, inertia, and mass center among its components. The parameters required for it are system topology, position, mass, inertia and centroid information, stiffness and damping characteristics, etc. Therefore, most of the dynamics property information of a high-speed train can be automatically extracted from its CAD model, while other parts are completed through user inputs. The extraction of the dynamics attributes of a high-speed train is based on its CAD model and feature tree. The system topology, overall symmetry center, and overall center position are extracted from the assembly model, and the center of gravity, moments of inertia and mass, etc., are all extracted from the part model [1]. Specific attributes are described as follows:

(1) Topological attributes:
- list of rigid bodies; list of primary/secondary suspension devices, antiroll devices, traction rods, hinged joints, universal joints, etc.;
- list of topology connections of rigid bodies.

(2) Geometric attributes:
- the essential parameters of the vehicle, such as the center distance of the bogies, the wheelbase, the center of gravity of the body and the bogie, etc.;
- the connections and the positions of the connection points among the primary and secondary suspension devices, antiroll devices, traction linkages, etc., and other related components.

(3) Physical attributes:
- the moments of mass and inertia of the body, bogie, wheelset, etc.

The relation between dynamics property extraction and dynamics modeling of high-speed train is shown in Fig. 3.5.

3.2.1.1 Definition of coordinate system for computer-aided design system and multibody dynamics system

The coordinate system in CAD software can generally be divided into two major categories: the world coordinate system, in short called WCS, and the user coordinate system, in short called UCS. For the convenience of description, this book makes the

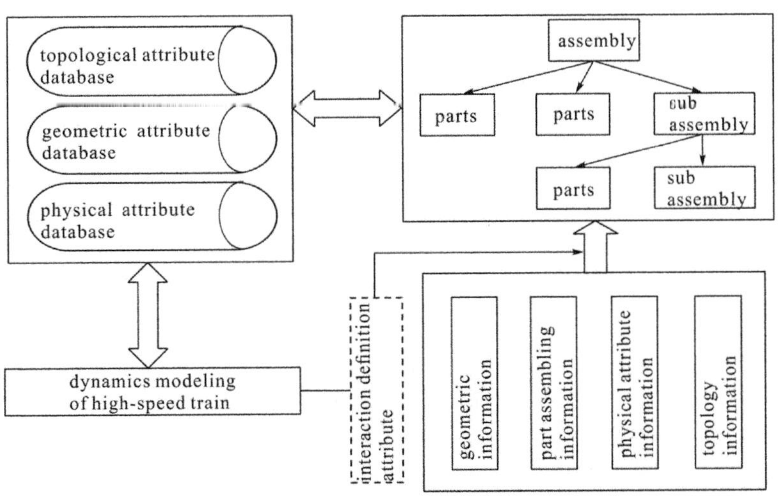

Figure 3.5
Dynamics property extraction and modeling technique based on a computer-aided design model.

following conventions for the UCS: the reference coordinate system for part modeling is called the part local coordinate system, abbreviated PLCS. The reference coordinate system for the parts assembled into the assembly is called the assembly local coordinate system, abbreviated ALCS. The parent–child correlation is built among different coordinate systems. For example, an assembly and its constituent parts can constitute a parent–child correlation.

In the multibody dynamics system, the mechanical abstraction of each part in the system is called component or body, and the coordinate system is divided into a WCS and a body-fixed coordinate system. The WCS is the absolute reference coordinate system of all the bodies. The body-fixed coordinate is an ortho-dimensional coordinate system that is fixed with a certain point of the body and describes the spatial position and orientation of the body.

When the CAD model is converted to a multibody dynamics model, the correspondence between various coordinate systems is shown in Table 3.1.

There is little difference in the definitions of different 3D CAD softwares, but the transformation correlation between coordinate systems is similar. Fig. 3.6 shows the transformation correlation between the coordinate systems in the CAD system. This model has only single-level assembly (multilevel assemblies are sequentially recursive). Assume that the direction cosine matrix of the ALCS in the WCS is defined as A^a, the direction cosine matrix of the PLCS in the WCS is defined as A^p, and the direction cosine matrix of the PLCS in the ALCS is defined as A^{ap}, then according to the nature of the direction cosine matrix [2], the conclusion can be drawn as $\mathrm{A}^p = \mathrm{A}^{ap}\mathrm{A}^a$, and $\mathrm{A}^{ap} = (\mathrm{A}^a)^{-1}\mathrm{A}^p = (\mathrm{A}^a)^T\mathrm{A}^p$.

3.2.1.2 Topological attribute extraction

The CAD assembly contains the information for each part. The main components that the vehicle system focuses on, such as the carbody, the bogie, the axle box, the wheelset, the motor, and other parts, are extracted and form the rigid body list, which is shown in Table 3.2.

A suspension parts list is shown in Table 3.3, including primary springs, primary vertical dampers, tumbler node, air springs, secondary vertical damper, secondary lateral damper, antihunting damper, antirolling torsion bars, traction bars, etc. The same force element type simulation can be used for different suspension components.

Table 3.1: Coordinate system correspondence between computer-aided design model and dynamic model.

Computer-aided design model	Dynamic model
Parts or assembly local coordinate system World coordinate system	Body-fixed coordinate system World coordinate system

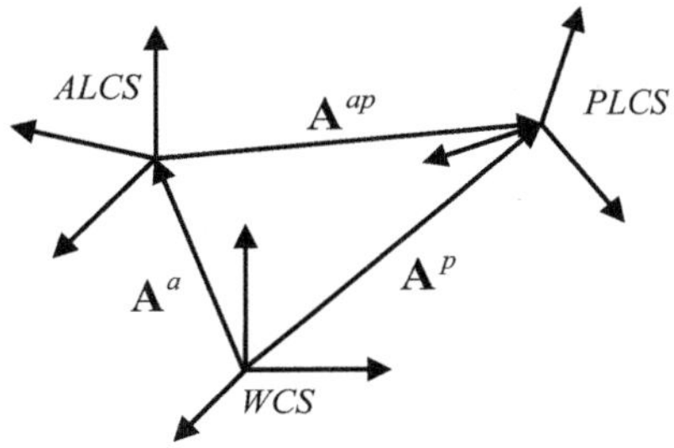

Figure 3.6
Coordinate system transformation in computer-aided design system. *ALCS*, assembly local coordinate system; *PLCS*, part local coordinate system; *WCS*, world coordinate system.

The list of joints is shown in Table 3.4. The types of joints include mainly rotating joints, translation joints, universal joints, spherical joints, etc.

The topology correlation is shown in Table 3.5, which contains the topology connection correlation between any two components. The first two columns are the numbers of the two components; the third column is the type of joints between the components, 0 indicates no connection; the fourth column is the number of force elements; the fifth column is followed by the corresponding force element number.

Table 3.2: List of rigid bodies.

ID	Type of rigid body	No.	Identification (name)
1	Vehicle body	1	Carbody
2	Bogie	2	Bogie_front
3	Bogie	2	Bogie_rear
...	...		...
15	Axel box	3	Box_rear_left

Table 3.3: List of force elements.

ID	Suspension type	No.	Force element type	Identification (name)
1	Primary spring	1	1	PS1_front_right
2	Primary spring	1	1	PS1_front_left
3	Air spring	2	2	AS_front_right
4	Air spring	2	2	AS_front_left
5	Antihunting damper	3	3	AHD_front_right
...	...	...	...	...
61	Traction bar	8	2	Tracbar_front
62	Traction bar	8	2	Tracbar _left

Table 3.4: List of joints.

ID	Type	Number	Identification (name)
1	Rotating joint	1	Gear_b1_front_right
2	Rotating joint	1	Gear_b1_front_left
3	Rotating joint	1	Gear_b1_rear_right
...	...	...	...
8	Rotating joint	1	Gear_b2_rear_left

Table 3.5: List of topology connections.

From_ID	To_ID	Joint type	Number of force elements	Force element number		
1	2	0	3	1		
1	3	0	3	3		
2	4	0	4	4	5	...
2	5	0	4	4	5	...
...	...	...	...	...	...	
15	3	0	2	10	11	...
15	7	1	0			

3.2.1.3 *Extraction of geometry properties*

The extraction of multibody dynamics geometric properties from the CAD environment includes mainly shaping dimension and locating dimension parameters for multibody dynamics analysis. The extraction of the shaping dimension is relatively simple, and the internal data of the CAD model can be directly accessed through the secondary development interface of the CAD system, from which the shaping dimension of the component features can be extracted. The process of extracting the shaping dimension from the CAD system is shown in Fig. 3.7. The shaping dimensions are recorded by recursively traversing the subassemblies, the parts in the subassemblies, and the features that make up the parts.

In the CAD system, the locating constraint dimensions of the feature on the component are usually defined based on the geometric elements of the feature (such as corner points, edges, faces, center lines, axes, etc.) [3]. When extracting the locating dimension between components, the mass center of the component is generally used as a reference, and the shaping dimension of the component and the locating constraint dimensions of the feature on the component are combined. Through spatial geometry operations, the final locating dimensions between components are obtained. In the selection of features, human–computer interaction is usually used.

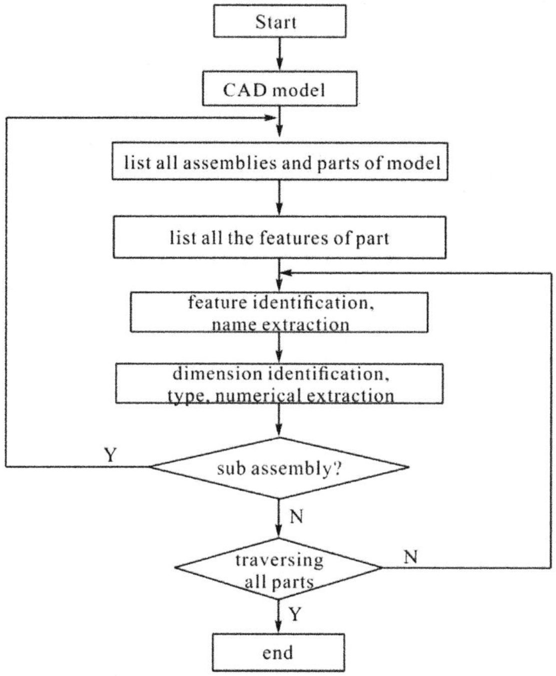

Figure 3.7
Shaping dimension extraction process [3]. *CAD*, computer-aided design.

Take the example of extracting the locating dimension of a certain type of train, which is shown in Fig. 3.8. The first step is to extract the location information of the components, such as body mass center M_c, mass center of the front bogie M_{f1}, mass center of the back bogie M_{f2}, mass center of the air spring M_{k1}, distance from the mass center of the air spring to its upper surface H_{ksc1}, distance from the mass center of the air spring to its lower surface H_{ksc2}, etc. Then through the calculation of spatial geometric relations between components (based on the rail surface and the vertical coordinates of the rail surface z $= 0$, the distance from the mass center of the vehicle body to the rail surface H_c, the bogie center distance L_c, the height from the upper surface of the air spring to the rail surface H_{ss1}, the height from the lower surface of the air spring to the rail surface H_{ss2}, and other locating dimensions can all be extracted.

The bogie center distance $L_c = X_{Mf1} - X_{Mf2}$ (X_{Mf1} represents the x-direction value of the mass center of the front bogie, the same below).

The height from the upper surface of the air spring to the rail surface $H_{ss1} = Z_{Mk1} + H_{ksc1}$.

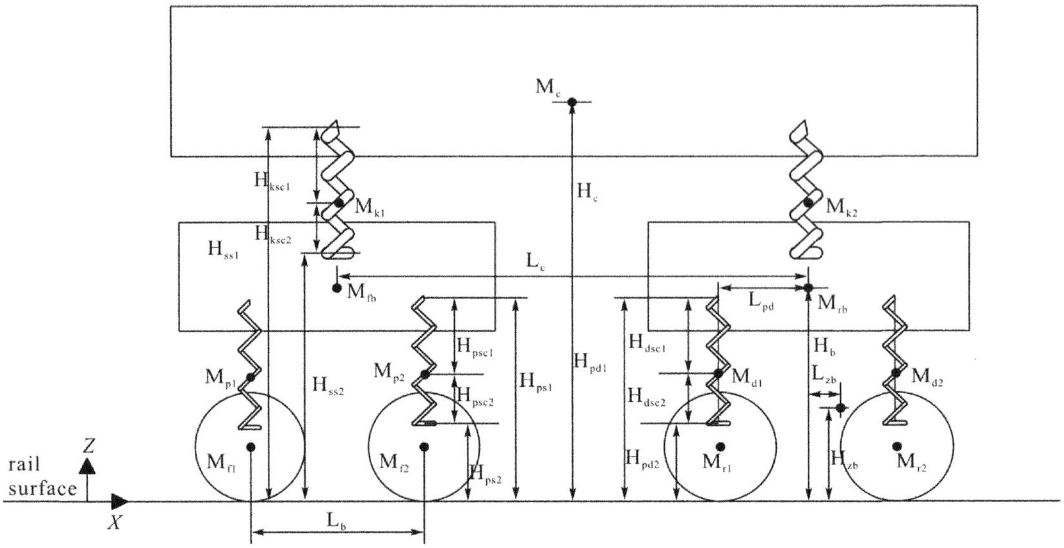

Figure 3.8
Parts of the locating dimension extraction of a vehicle [3].

The height from the lower surface of the air spring to the rail surface
$H_{ss2} = Z_{Mk1} - H_{ksc2}$.

The other sizes are calculated in the same way.

3.2.1.4 Extraction of physical attributes

Most CAD systems support secondary development. In view of this, the extraction of physical properties such as the mass center and the mass and inertia moments of the parts or assembly is relatively straightforward; they can be directly extracted from a CAD solid model with the secondary development interface of CAD software. Take CATIA as an example; its interface functions are:

Get Inertia (& density, & mass, position, matrix, ...)

After executing the function, the mass, barycentric coordinates, and inertia matrix of the component are assigned to the variables, including mass, position, and matrix, respectively.

Using the aforementioned principle, a CATIA-based inertial body dynamics property extraction module is developed, as shown in Fig. 3.9. After clicking on the "extraction attributes" on the interface, the carbody, bogie, wheelset, or assembly can be selected to extract the dynamics attribute of the part or assembly.

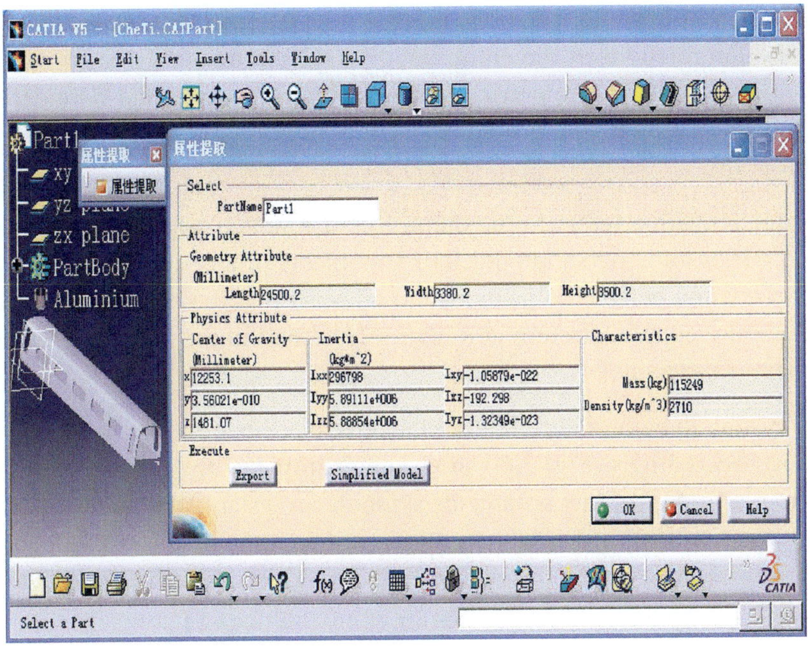

Figure 3.9

Inertial body dynamics property extraction for secondary development of CATIA.

3.2.2 Parametric dynamics modeling of high-speed trains

In the modeling of the dynamics of coupled systems in high-speed trains, the parametric design concept is mainly embodied in four ways: geometric model parameterization, design variable parameterization, behavioral correlation parameterization, and expression parameterization.

Parameterization of geometric models is mainly used for parametric modeling of various disciplines. Modelers use geometrically constrained sets of parameter predefined methods for design variables. Predefined parameter-based correlation functions are used in each subsystem module to pass specific values. The parameter is associated with a set of geometric constraints to drive the correlation function to automate the modeling of the structure. This modeling method matches the top-down design concept, which helps to make model changes quickly and reduce the workload of the modelers.

One of the main purposes of coupled system dynamics simulation is to perform system optimization and various design verifications. The parameterization of design variables is to preset the design variables concerned in a parametric form in the system during system modeling. The correlation between the design variables and the related model object is bound in a simulation to realize the rapid change of parameters, thereby facilitating the comparative analysis of various solutions.

The parameterization of behavior relations refers to the parametric expression of various behaviors and motion relations in the entire coupled system modeling, such as the position of the load, the position of the geometric constraint, etc. The input and boundary conditions of the simulation model are quickly configured by parameterizing this globally critical location point.

Expression parameterization refers to the definition of function expressions when defining the specific correlations among submodules or subobjects in a coupled system, without precuring the correlation between objects or modules, so that the entire modeling method can abstract the entire coupled system at a higher level and improve the versatility and application scope of the model.

In the coupled system dynamics simulation platform, these parametric modeling methods have been applied accordingly. Fig. 3.10 shows an example of the parametric modeling interface for the platform. Various parameter symbols are established and performed before the simulation modeling and direct assignment or assignment through an expression forms a list of parameters, which is shown in Fig. 3.10A. In the actual modeling process, the definition of the model can be completed by referencing this parameter as shown in

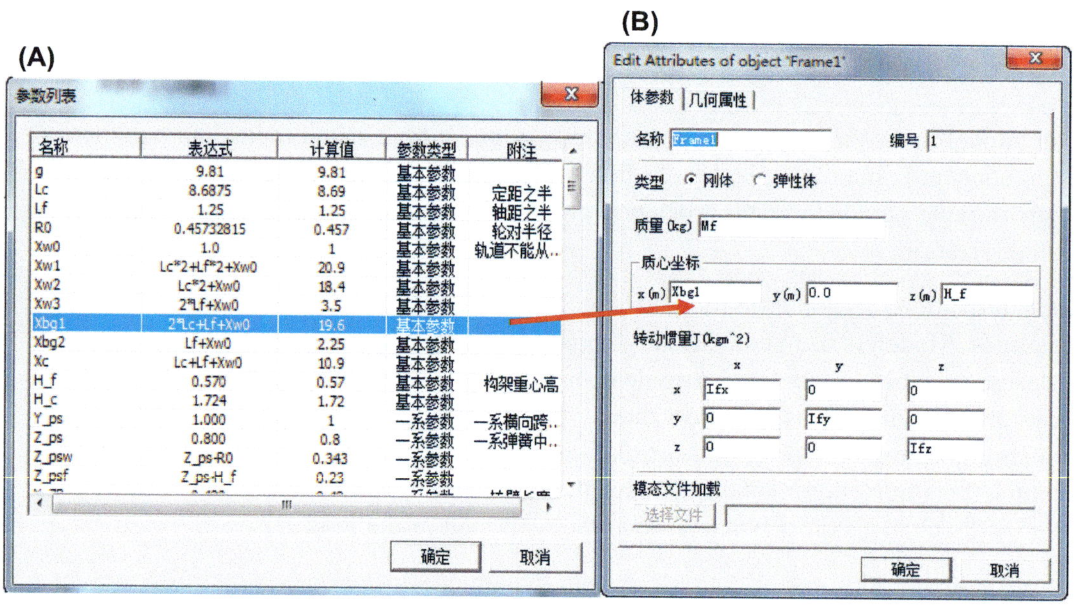

Parameters list

Parameters reference

Figure 3.10

Parametric expression and modeling. (A) Parameters list. (B) Parameters reference.

Fig. 3.10B. When the model needs to be modified, only the corresponding parameter value in the parameter list needs to be modified, because the system can automatically modify the part of the model that is associated with the parameter, and at the same time, the graphics display window of the model can also respond to the update of the parameter in real time and give feedback to the user in a timely manner.

In coupled system dynamics modeling, multidiscipline collaborative simulation is also involved. The parametric design concept can greatly increase the level of abstraction of the entire system model, make the model more adaptable, and greatly increase the efficiency of modeling and simulation analysis.

3.2.3 Graphical dynamic modeling of high-speed trains

With the continual enhancement of computer graphics and image processing capabilities, graphical modeling is a widely used modeling method in the field of system simulation. Most existing commercial simulation software systems use modeling techniques based on 2D graphics to assist in system modeling. In addition to providing 2D graphical modeling methods in a simulated platform for the dynamics of coupled systems in high-speed trains, it also very important to realize a more visual and intuitive 3D visualization of dynamics modeling.

The premise of graphical modeling is to divide the entire system into several suitable functional modules and then visualize each module in a 2D or 3D graphical manner. Finally, the interaction between the modules is set through the human—computer interaction operations such as selection and parameter setting among the graphical modules, and a simulation model is established.

Module packaging and interface definition are the key to modularization. The module packaging abstracts out the function set of the module. The interface definition abstracts the input and output parameter sets of the module. By defining the correlation of the entire simulation system by connecting these different module interfaces, the simulation flow and simulation logic correlation of the entire system are formed.

The abstract module functions are graphically encapsulated and implemented. During the simulation analysis of the coupled system dynamics, the user first uses drag, copy, etc., to interactively operate the corresponding graphics module to realize the topology modeling of the simulation model, and then inputs the corresponding parameters of each module and sets the conditions to complete the construction of a dynamics simulation model. The user-built simulation model can also be saved as a simulation template, which is convenient for reuse when a similar simulation analysis is performed next time. Fig. 3.11

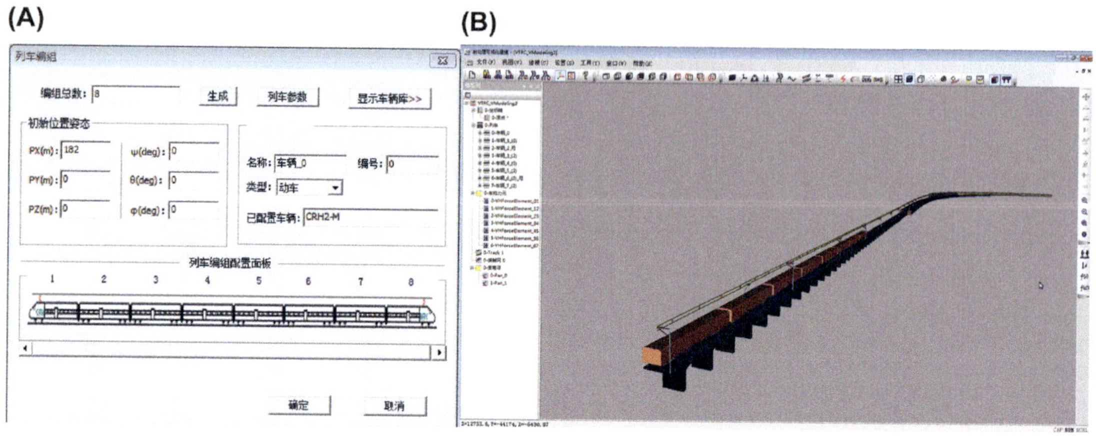

Figure 3.11

Graphical modeling of high-speed trains. (A) Two-dimensional graphical modeling of trains. (B) Three-dimensional modeling of trains.

gives an example of graphical modeling of trains. Fig. 3.11A shows the 2D graphical modeling of trains: first, set the number of vehicles, then set the position of each vehicle separately, then set each vehicle by dragging it from the "vehicle model library" on the right, and set the train formation mode and train characteristic parameters to complete the establishment of the train dynamics model. Fig. 3.11B is a 3D graphical modeling of a train: first, a vehicle model is established, then other vehicles are modeled by creating or copying operations, and finally the coupling device and train parameters are defined so as to realize 3D graphical modeling of trains.

In the 3D graphical modeling, the system provides a simple model by default to ensure that the model is clearly expressed. This makes it easy to understand and verify and it is also highly efficient. At the same time, the CAD model can also be imported through the interface provided by the system to replace the simple model of each module to enhance the intuitiveness and authenticity of the model.

3.3 The calculation method of the dynamics of coupled systems in high-speed trains

The dynamics model of coupled systems in high-speed trains consists of five subsystems; they are the train dynamics subsystem, track dynamics subsystem, pantograph–catenary dynamics subsystem, train aerodynamics subsystem, and tractive power supply and traction drive subsystem. The correlation and the transfer of mechanical behavior among these subsystems are described as shown in Fig. 3.12.

For realizing the coupling simulation described in Fig. 3.12 the following two key issues need to be solved.

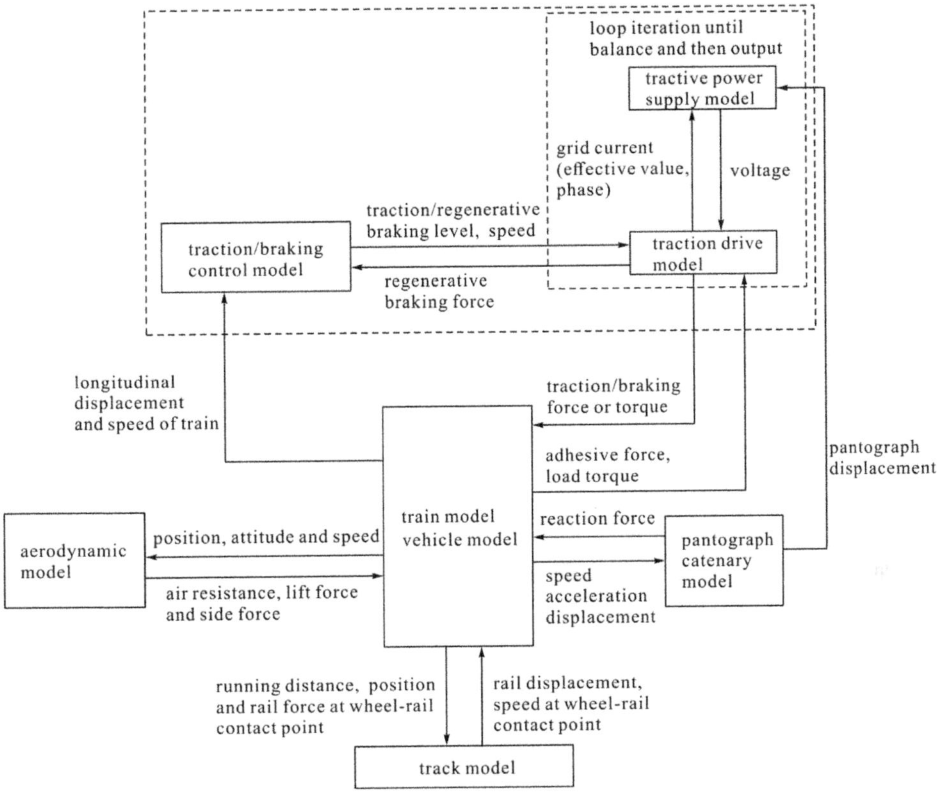

Figure 3.12
Coupling correlation model of a high-speed train system.

3.3.1 Modeling method for coupled subsystems with different study scales

The dynamics model of coupled systems in high-speed trains is the kernel for realizing coupling simulation. The study scales of the subsystems involved in the high-speed train coupled system are varied. For example, the study scale of high-speed trains is at the 100-m level, the study scale of wheel—rail contact spots is at the millimeter level, and the study scale of the tractive power supply systems is at the kilometer level. To find the coupling mechanism among these subsystems with different study scales, the modeling method for establishing a dynamics model of coupled systems in high-speed trains is the primary issue that needs to be solved.

3.3.2 Time—space synchronization control method for coupled subsystems with different integration steps

In addition to the different study scales, the calculation method, calculation step, and calculation time of each coupled subsystem are also different. For example, the calculation steps of each subsystem range from 10^{-5} to 10^{-1} s (space is out of synchronization), and the one-step calculation time varies from 10^{-2} to 10^{2} s (time is not synchronized). Therefore, another key issue that needs to be addressed is coordinating the simulation step for each subsystem to ensure that the simulation process of each subsystem keeps time—space synchronization in the same integral coordinate system.

3.3.2.1 *Integrated modeling technology for dynamics of coupled systems*

For constructing the coupled model for subsystems with different research scales, a modeling method based on substructure technology is proposed. First, the simulation model of each subsystem is established, then through the coupling correlation among various subsystems, the integrated model of the dynamics of coupled systems in high-speed trains is established. For each subsystem, the only change for the simulation model is increasing a boundary condition, which is defined as a coupling interface with other related subsystems. This multiple subsystem simulation modeling method not only effectively inherits the research methods and research results of various subsystems, but also expresses the interaction among subsystems and solves the problem of multiple subsystem coupled simulation modeling flexibly. In the software, a component-oriented computer simulation modeling technology is adopted. Each subsystem model is packaged as a component.

In the computer field, a component is a package of data or methods. Components have their own properties and methods. Its principle is the change of the component input set that causes the component output state to change. In the application of the component, users are normally more concerned about the input and output sets of the component, while weakening the study and its internal working mechanism. Utilizing component technology instead can fully encapsulate and apply existing research results of the subsystems, and greatly reduces the difficulty of coupling simulation modeling. The component-based modeling method can increase the flexibility of modeling. Therefore, various coupling simulation models that consist of parts of subsystems can be established quickly and easily.

Learning from the concept of object-oriented programming, the various subsystem models are abstracted as objects. When modeling, components are used to instantiate each subsystem model object. As shown in Fig. 3.13, system A—system E represents the subsystem components of the coupled systems in high-speed trains. The arrows represent the interface correlations between two components, and the arrows point to the flow of data; for example,② means the interface data flowing from system A to system B, and ① means the interface data flowing from system B to system A.

Taking the subsystem model object A as an example, it is instantiated as component A, where Ai represents the input sets of system A, including the coupling interface data flowing from other subsystems that have a coupling correlation with system A. And Ao represents the output sets of system A, including the coupling interface data flowing to

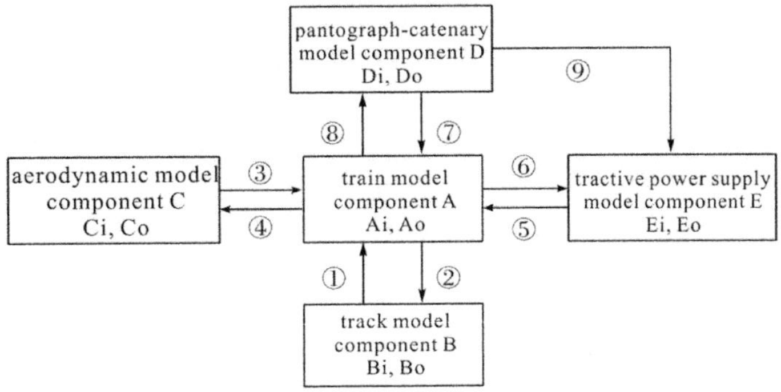

Figure 3.13
Component-oriented modeling of coupled systems in high-speed trains [6].

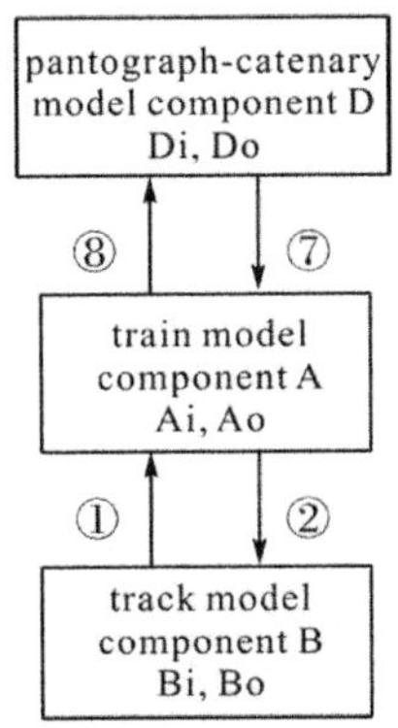

Figure 3.14
Component-based train–track–pantograph–catenary coupling simulation model.

other subsystems that have a coupling correlation with system A. In Fig. 3.13, the coupling interface data ①, ③, ⑤, and ⑦ are contained in Ai, and the coupled interface data ②, ④, ⑥, and ⑧ are contained in Ao.

Using component-based modeling methods, various coupled simulation models among parts of subsystems are established to study their coupling degree, such as train–track coupling simulation, train–track–pantograph–catenary coupling simulation, train–track–aero coupling simulation, train–track–tractive power supply coupling simulation, etc. Taking the train–track–pantograph–catenary coupling simulation as an example, the model of the coupled simulation component is shown in Fig. 3.14.

3.3.2.2 Coupling calculation method

Since the simulation step lengths of each subsystem in a coupled high-speed train system is inconsistent, it is necessary to adopt the synchronization control in the coupling calculation. Taking a coupling simulation condition as an example, the simulation step size of the train subsystem is 5×10^{-5} s, the simulation step length of the track subsystem is 5×10^{-5} s, the simulation step length of the pantograph—catenary subsystem is 5×10^{-5} s, the simulation step size of the aerodynamic subsystem is 1×10^{-3} s, and the simulation step length of the tractive power supply subsystem is 1×10^{-1} s. The simulation step size of each subsystem is distributed over 3 orders of magnitude, and for this purpose, a three-layer coupling control simulation method is adopted in the coupling simulation, which is shown in Fig. 3.15 with its description as follows:

(1) According to the simulation step size of each subsystem, all subsystems are divided into three layers: the first layer includes the train subsystem, the track subsystem, and the pantograph—catenary subsystem with simulation step size $S_{min} = 5 \times 10^{-5}$ s and defines the coupling simulation control step as $SI1 = 5 \times 10^{-5}$s. The second layer includes the aerodynamic subsystem with coupling simulation control step $SI2 = 1 \times 10^{-3}$ s and the third layer includes the tractive power supply subsystem with the coupling simulation control step $SI3 = 1 \times 10^{-1}$ s. After that, the parameters named t_e are set to terminate the coupling simulation.
(2) A parameter for recording the simulation step size of the coupling simulation system is defined as SA, and some other parameters are set for accumulating each level of the subsystem. The first layer accumulated simulation step length is defined as SA1, the second layer accumulated simulation step length as SA2, and the third layer accumulative simulation step length as SA3, and initialization $SA = 0$, $SA1 = 0$, $SA2 = 0$, and $SA3 = 0$.
(3) The interface data of the coupling correlation among the related subsystems are initialized.
(4) The simulation model of each subsystem is loaded and initialized.
(5) When the coupling simulation begins, the accumulated simulation step value of each layer adds S_{min} and makes the following discrimination.

If the first layer accumulated simulation step value is greater than or equal to the first layer coupling simulation control step value, that is, $SA1 \geq SI1$, then update the coupling correlation interface data among the systems in the first layer, and activate them to perform a single-step simulation. After that the value of the first layer accumulated simulation step SA1 minus SI1. The same work is done for the second and third layers.

(6) After that determine whether the simulation is finished by comparing the SA with t_e. If $SA < t_e$, then the coupling simulation continues, otherwise ends it.

When a subsystem is activated to perform a single-step simulation, the calculation flow is the same. Take the train subsystem as an example; the flow is as follows:

(1) preparing to start a simulation for the train subsystem;

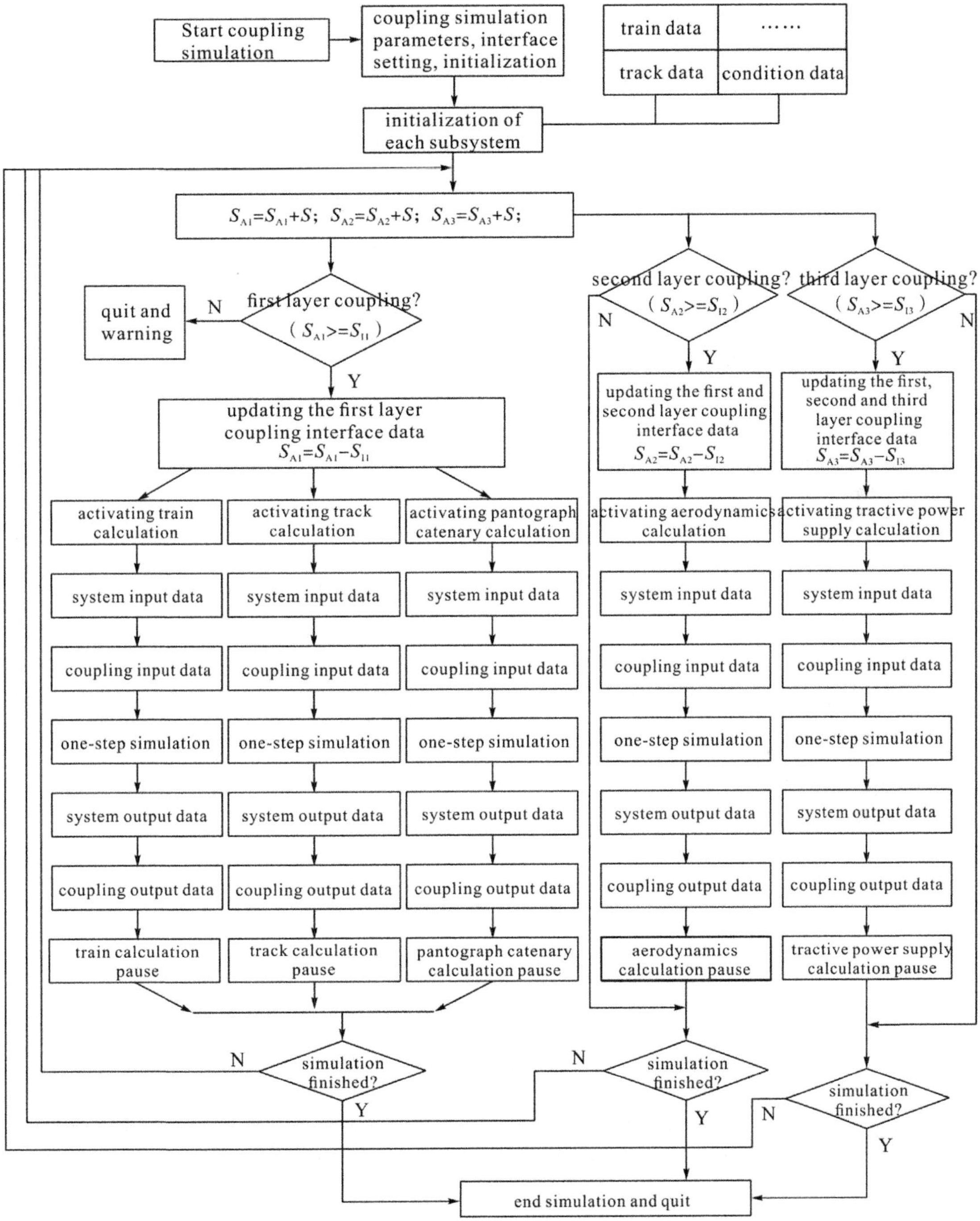

Figure 3.15

Three-layer coupling control method for high-speed train system dynamics.

(2) obtaining input data of the train subsystem, which is from the initialization data or last simulation results;
(3) obtaining the input coupling interface data from the related subsystems that have a coupling correlation with the train subsystem by the coupler;
(4) performing the single-step simulation of the train subsystem with the given simulation step size, input data, and coupling interface data;
(5) generating the single-step simulation result of the train subsystem, which contains two parts, one for the next step simulation and the other for analysis;
(6) generating the output coupling interface data and sending to the coupler to update the input coupling interface data of related subsystems;
(7) the train subsystem waits for the next step simulation.

The method to update the coupling input interface data of the subsystem uses an interpolation algorithm to process the last step simulation result of related subsystems. The appropriate interpolation algorithms can be used for different subsystems to improve the accuracy of the coupling input interface data so that ultimately the coupling calculation accuracy of the whole coupling simulation system is improved.

3.3.2.3 Coupled calculation implementation

In the distributed simulation of the dynamics of coupled systems in high-speed trains, the coupling calculation method mentioned earlier is implemented in a coupler, whose main function is solving the coupling correlation model of system dynamics in high-speed trains and updating the simulated calculation input interface data of each subsystem. All the subsystems are driven synchronously by the coupler to realize the coupled simulation among high-speed train, track, pantograph—catenary, tractive power supply, and train aerodynamics subsystem at each integration step.

The basic functions of the coupler include:

(1) controlling the work mode of coupling subsystems: it can send the start, pause, and stop commands to the subsystem of the coupled subsystems of a high-speed train to change its work mode;
(2) organizing the coupling calculation conditions: it can perform the coupled calculation with a self-suitable coupling correlation model according to the user's full coupling or partial coupling calculation requirements;
(3) coordinating the integration process of the coupling subsystem: owing to the different integration steps of the coupled subsystems of a high-speed train, a certain strategy (such as the multilevel simulated steps synchronization and coordination strategy) can be used to coordinate the simulation process of each subsystem to achieve coupling calculation;

(4) updating the coupling interface data of coupling subsystems: during the coupling calculation, the coupling interface data of all subsystems of a high-speed train can be processed, and the input data for each subsystem at this step can be provided. It can also provide the classified data interpolation algorithm to guarantee the accuracy of coupled calculation;
(5) sending the coupling interface data to a related subsystem: real-time communication with each coupled subsystem is accessible. It can receive the coupling output interface data from each subsystem and send the processed coupling input interface data to each subsystem in time.

The simulating procedures of the dynamics of coupled systems in high-speed trains under the control of the coupler are as follows:

(1) Each subsystem performs a one-step integration calculation independently.
(2) Each subsystem sends the coupling output interface data to the coupler.
(3) The coupler solves the coupling correlation model of all subsystem and processes coupling interface data via a suitable interpolation algorithm.
(4) The coupler sends coupling input interface data to each subsystem to update its computational boundaries.
(5) The coupler compares the total simulation step with the set termination condition.
(6) If the condition is satisfactory, the coupling simulation is stopped; otherwise, it returns to step 1.

The simulation process of the coupled subsystems in high-speed trains is divided into three phases: calculation waiting and data preparation, single-step integration calculation, and calculation output. Driven by the coupler, three phases are cycled, as shown in Fig. 3.16. In the coupling calculation, the input of each subsystem contains two parts: the first is the data obtained within itself, which is called the subsystem input; the second is the data obtained from other coupled subsystems, which is called coupling input. Only when the

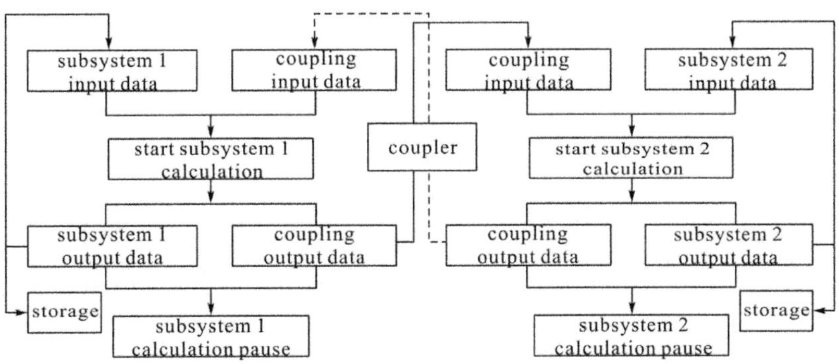

Figure 3.16
Simulation process of a coupled subsystem under the coupler [6].

subsystem input and the coupling input are both ready can the subsystem start the integral calculation at this step. The output also includes two parts, named the subsystem output and the coupling output. The subsystem output is the calculation results at this step. Part of the results are used for the next step calculation, while others are used for monitoring, analysis, or evaluation.

The coupling output provides the next coupling input data for the subsystem related to this subsystem. The output of the subsystem is the input of the coupler as well as the basis for solving the coupling correlation model. The calculation results of the coupling correlation model are the output of the coupler and also the input of each coupled subsystem. It is the coupler that coordinates and acquires the required data from the coupled subsystems, solves the coupling correlation model, processes the relevant data, and sends the required results to the corresponding subsystems, ultimately controlling and advancing the simulation of the coupled system.

3.4 The postprocessing display technology

The movement simulation of a high-speed train is an important postprocessing display approach to the simulation platform for the dynamics of coupled systems in high-speed trains. The independent development of 3D visualization software is employed to demonstrate the virtual operation of the simulated results of the high-speed train as a coupled system of dynamics.

3.4.1 The simulation display technology of the high-speed train movement with different granularity

The primary problem of the postprocessing system is the automated modeling for real conditions: a detailed postprocessing model or scenario should be quickly and automatically generated by the given data, such as initial points and lengths of tunnels and bridges and the curves and slopes of a line, and then the simulation operation displayed, which is driven by the simulation results or measured data. Fig. 3.17 is a part of a 3D view that is automatically generated by the user-defined data files. In addition to the bridges, tracks, and catenaries shown in the diagram, the roadbeds, tunnels, and catenaries in the tunnels are also included.

The dynamic simulation results of high-speed train, line, pantograph–catenary, traction power supply, and transmission system can be displayed by the postprocessing display system with space–time synchronous integration, including the combination display of 2D data and 3D entity.

Compared with the traditional 2D curve display, the 3D visual display in this system is more vivid. Also compared with other similar commercial softwares, this system is

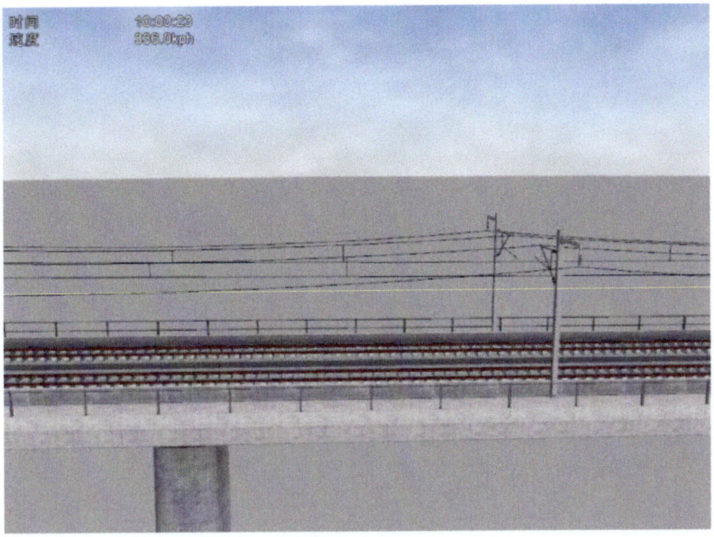

Figure 3.17
The three-dimensional scene automatically stitched on the data files of a line.

specifically designed for the movement simulation of high-speed trains; hence its specialized function is more comprehensive.

Fig. 3.18 is a screenshot of the postprocessing display system, showing the incline and vibration during the train's running. The vibration during running is usually small and difficult to observe, but it can be viewed interactively after amplification.

Figure 3.18
A screenshot of the postprocessing display system.

Figure 3.19
Driver's cab view.

In addition to the front view shown in Fig. 3.18, the system also predefines multiple views such as the driver's cab view, pantograph view, wheel/rail view, rear view, passenger view, pedestrian view, and free view. These views can be switched at one's discretion. At each view, the user can pan and rotate in all directions. All viewpoints have a motion-following function except the free view. Some views can switch from carriages back and forth.

Fig. 3.19 is a screenshot of the driver's cab view. It can reflect the incline and vibration of the driver's cab, and it also can pan and rotate in all directions. The train control, traction drive, and traction power supply are dynamically displayed on the control panel in the driver's cab. Users can use the mouse to click the buttons on the control panel for real-time interactive switch.

3.4.2 Dynamic state display techniques for different domains

Fig. 3.20 is a screenshot of the wheel–rail force display. To clearly observe the wheel–rail force, the body of the vehicle is displayed semitransparently. The wheel–rail force is displayed in the direction of the arrow (means the direction of the force) and the length (means the magnitude of the force) at the wheel–rail contact point, and the wheel–rail contact state can be locally enlarged to observe clearly. The train body, frame, and wheelset components can also be hidden or translucently displayed. What's more, the transparency can be modified.

Figure 3.20
Wheel—rail display and translucent body.

Based on the simulation results, the system can simulate arcing between the pantograph and the catenary. Fig. 3.21 shows the instant of arcing between the pantograph and the catenary in the tunnel. This phenomenon is simulated with a particle system in computer graphics. The particle system is a technique that simulates specific fuzzy phenomena in computer graphics, and it is difficult to implement the realism of these phenomena with other traditional rendering techniques.

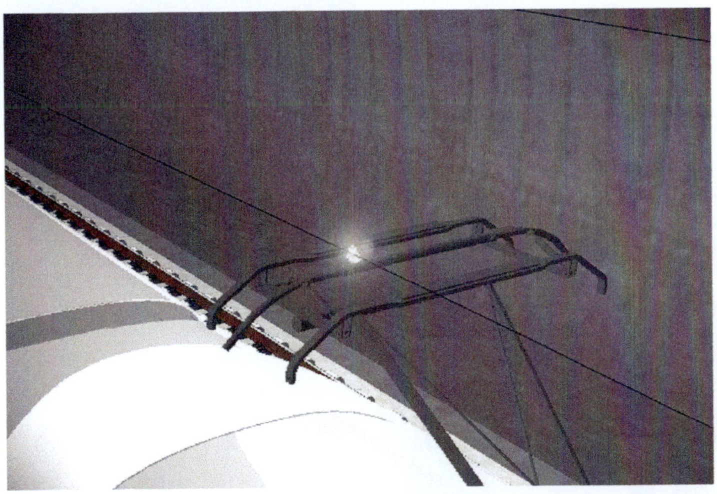

Figure 3.21
Simulation of arcing between pantograph and catenary.

3.4.3 Diversified display technology of Dynamic data

To display multiple views at the same time for the purpose of expanding the field of view and performing the distributed synchronization display with other subsystems, this system can be displayed collaboratively with other subsystems. In the cooperative display mode, several 3D display terminals, 2D curve display terminals, and other subsystem display terminals may be simultaneously enabled. The collaborative display server simultaneously sends synchronized updating instructions to each terminal; every terminal will synchronize and update after receiving the instructions. During this process, you can simultaneously pause, continue, or stop and exit the display. Each terminal can join or exit at any time during the display.

Fig. 3.22 is a diagram of the cooperative display; each 3D display terminal can present a different view at the same time. When needed, the output of each terminal can also be projected on a large screen with a video matrix. Compared with a single-window multi-viewport, the multiterminal mode does no harm to the frame rate, so it runs more smoothly. For time—histories data, the 2D curve display terminal in the system can display

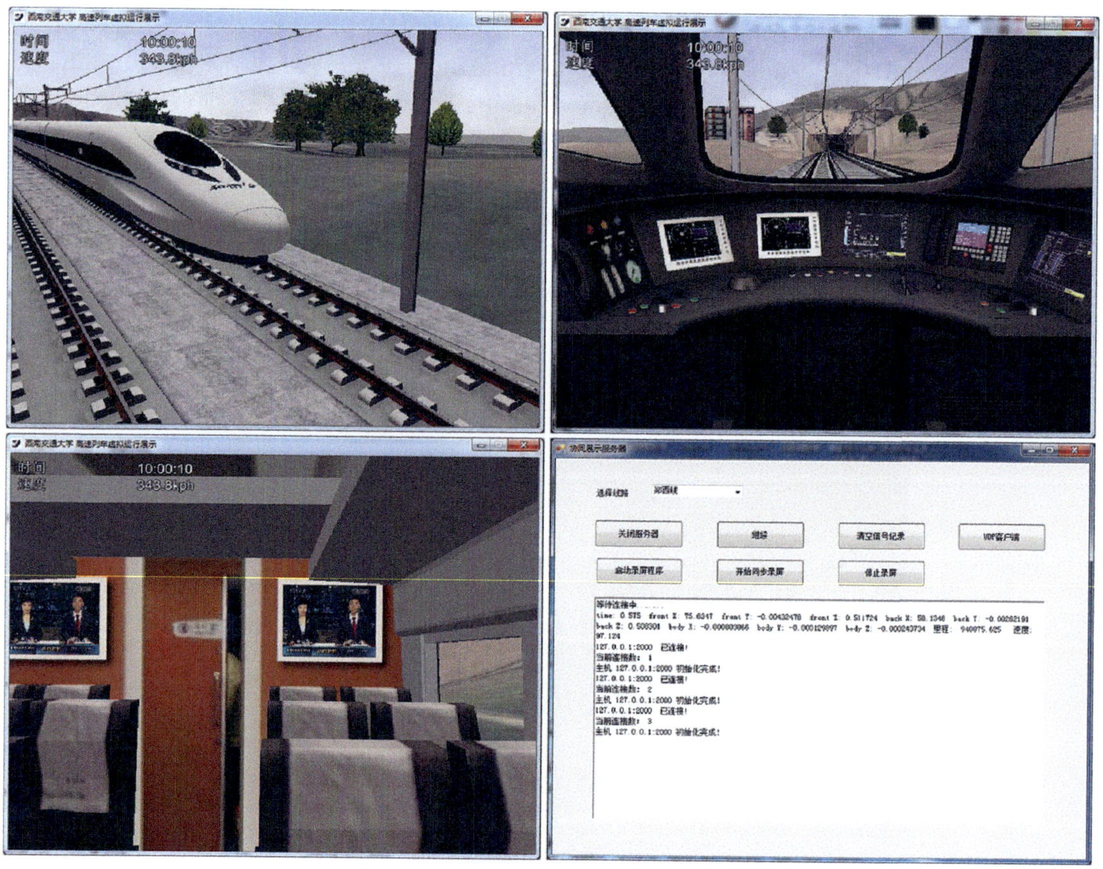

Figure 3.22
Cooperative display.

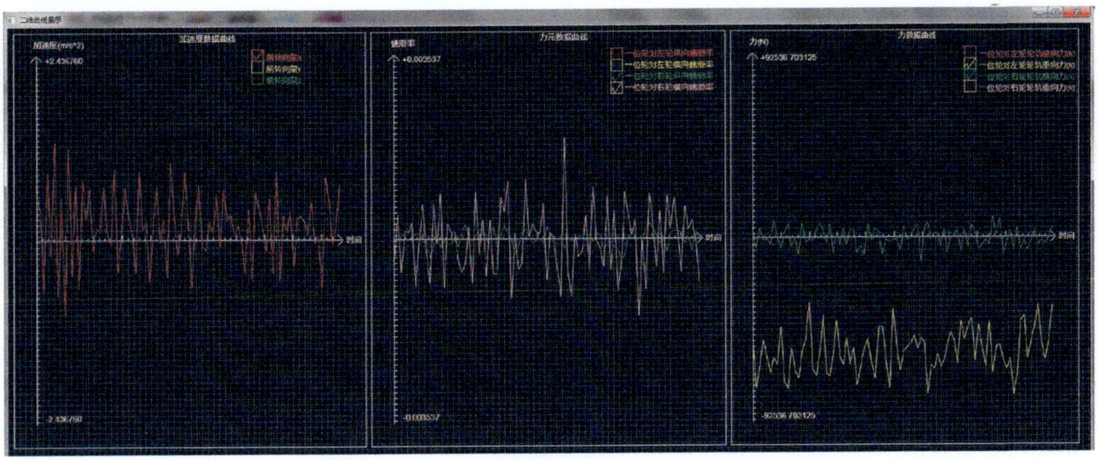

Figure 3.23
The 2D curve synchronization display.

these data synchronously. As shown in Fig. 3.23, the 2D and 3D synchronizations are combined to make it easier for users to understand and apply.

To present the simulated motion state of the high-speed train as a coupling system more vividly, the system can use the simulation data to drive the 6DOF (6 degrees of freedom) motion platform, as shown in Fig. 3.24. The motion platform can simulate the vibration during the train's operation. When passing through a curve, bend, or turn, the simulator, which is located above the motion platform, can reflect the incline of the vehicle body caused by the superelevation of the outer rail. At the same time, it is also possible for one to experience the impact of the disturbance caused by high-speed airflow when two trains pass each other.

Figure 3.24
A 6DOF motion platform driven by simulation data.

To further improve the presence, immersion, and reality, this system implements the sound display of the train movement simulation. The system can also switch between stereo and nonstereo modes. In the next step we are considering adopting naked eye stereo technology.

Apart from the visual display, the system also implements sound simulation. Different conditions have different sound effects. For example, different scenes, such as the driver's cab, passengers' compartments, pedestrians, tunnel entrance and exit, bridges, and passing trains, will have corresponding unique sound effects, and we are able to handle the reverberation of sounds. Different trains create a variety of different combinations of sounds; therefore users may utilize corresponding sound materials for different trains during simulation. The sound generated by the computer simulation is synchronized with the visual display and can vary with the speed of the train and with the operating section. The sound output is stereophonic, and it can also reflect the frequency, loudness, and sense of orientation. The system can further process some special sound effects, such as the reverberation, reflection, and Doppler effect of sound.

The Doppler effect on the sound is quite obvious when two trains pass each other or when pedestrians watch the train passing by. The Doppler effect is a phenomenon in which the frequency of the wave received by the observer is different from the frequency of the wave source if the wave source and the observer have relative motion. The Doppler effect can be observed when a train is rushing toward us and the whistle becomes sharp (i.e., the frequency becomes higher and the wavelength becomes shorter), and the whistle of a leaving train becomes deeper (i.e., the frequency becomes lower and the wavelength becomes longer). The frequency correlation between the observer and the source is as follows [5]:

$$f' = \left(\frac{v \pm v_O}{v \mp v_s}\right) f$$

In the equation, f' is the observed frequency and f is the frequency of the wave source; v is the wave velocity; v_o is the observer's velocity and v is the velocity of wave source. The upward and downward calculations of the numerator and denominator in the brackets indicate "near" and "away".

3.5 Case study and verification of the simulation platform for the dynamics of coupled systems in high-speed trains

3.5.1 Case study of the simulation platform for the dynamics of coupled systems in high-speed trains

The simulation platform for the dynamics of coupled systems in high-speed trains is a distributed simulated application platform with a C/S structure. It supports multiuser and multioperation simulations. In addition it makes allowance for flexible combinations of coupled calculations among train (vehicle), track, pantograph–catenary, and tractive power supply and transmission, as well as aerodynamic subsystems. The platform's client is the simulation condition management module with the functions of setting and submitting coupled computing conditions, managing various results and models of the condition, etc. The preprocessing module, postprocessing module, and system service module are integrated in it. And the server is the coupled computing module, which works in the background.

The simulation workflow of this platform is shown in Fig. 3.25.

(1) The simulation condition management module is the entry of the system.
(2) By starting the preprocessing module in the system interface, we can generate the calculation models of train (vehicle), track, pantograph–catenary, and tractive power

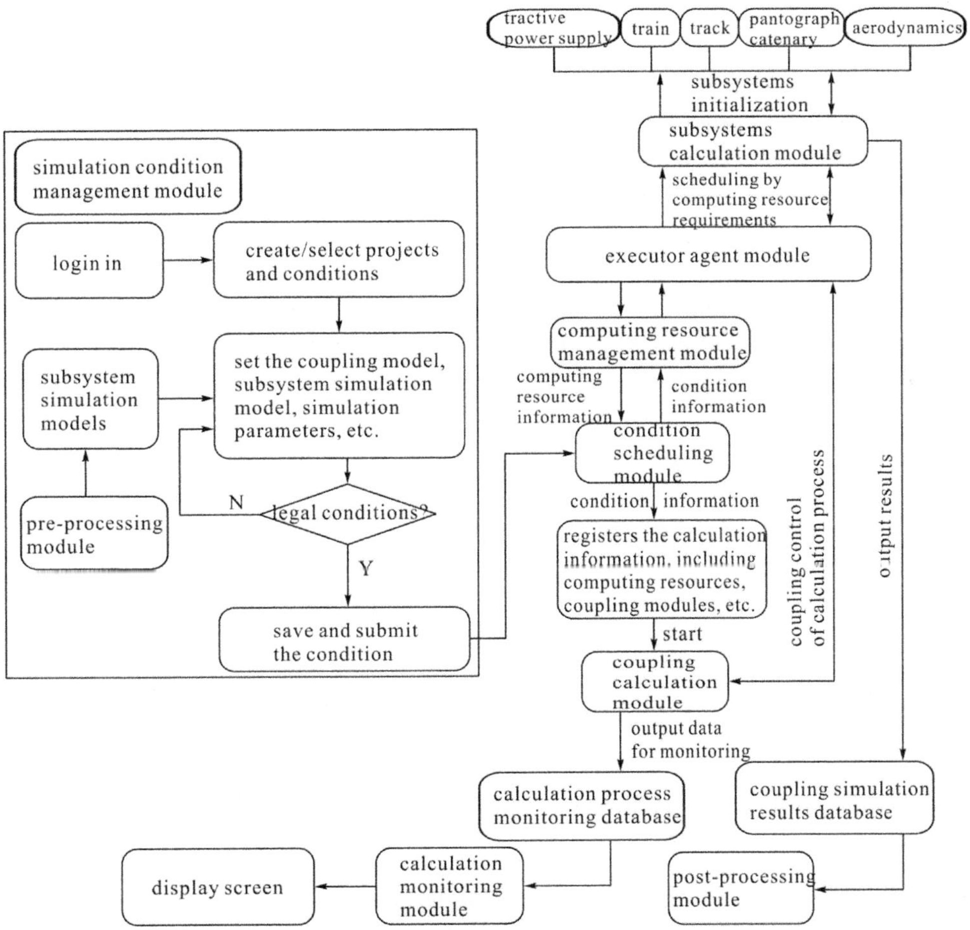

Figure 3.25
The workflow of the simulation platform for the dynamics of coupled systems of high-speed trains.

supply and transmission, as well as aerodynamic subsystems, and then store them in the calculation model database of each subsystem.

(3) Setting a coupled calculation condition and submitting it: When setting a coupled calculation condition, first we should extract the required calculation model from the module library of each subsystem and then set the simulation controlling parameters for each subsystem and coupled system to generate a coupled calculation model. After that we will store it in the database and submit it at the same time to the coupled calculation module.

(4) Starting the coupled calculation: The process is as follows:

 1) When the task-scheduling submodule in the coupled calculation module receives the coupled calculation request, it calls the computing resource management submodule to query the status information of the computing resource, and the computing resource management submodule obtains the related resource information by calling the executor agent submodule.

 2) The task-scheduling submodule applies a certain scheduling strategy, which sends the task to the appropriate computing resource via the executor agent submodule, and registers the scheduling information in the coupler. Then the subsystem calculation module is started to complete the initialization, and the status information and the initial result will be sent to the coupler.

 3) The next steps are starting the coupled calculation task under the control of the coupler, coordinating the calculation process of each subsystem, and applying the coupled control strategy and data-processing algorithm to realize the coupled calculation among related modules. The coupler and each subsystem executor will sort out the data in the calculation process and store them in the corresponding database.

 4) Next, the calculation process monitoring module starts to monitor the parameters of the subsystems in the simulation process, and raises the alarm in the event of anomalies.

(5) Then the postprocessing module starts to display and evaluate the calculation results. The platform provides general postprocessing submodules, train virtual operation display submodules, and subsystem-specific postprocessing submodules. Through an integration of 2D and 3D, the simulation results of each subsystem are fully displayed.

According to the mentioned workflow, the application case of the platform is described as follows:

(1) The simulation condition management module is started and the user interface of the simulation platform is entered, as shown in Fig. 3.26. The system adopts a three-level document management structure, namely project, conditions, and subsystems. That is,

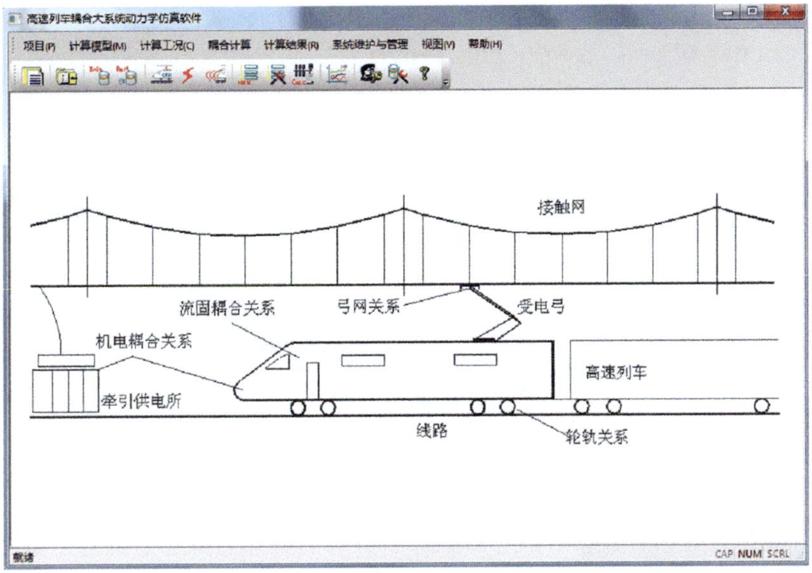

Figure 3.26

The interface of the simulation platform for the dynamics of coupled systems in high-speed trains.

the project needs to be established first, and then the conditions in the project are established, and finally the subsystems involved in the coupled simulation are organized, and the simulation inputs and outputs of the subsystems are saved under these conditions. The project is a description of a certain type of simulation task, while the condition is a description of one of the specific simulation tasks. The subsystem is the combination of the subsystem that participates in a specific coupled simulation and its input and output. For example, a series of simulations on a certain type of vehicle can be defined as the name of the project. Each specific simulation task is defined as the name of the condition. The vehicle–track coupled simulation condition is taken as an example. The condition includes vehicle subsystem and track subsystem.

(2) The preprocessing module is started and simulation models are built for each subsystem of vehicle/train, track, pantograph–catenary, tractive power supply, and aerodynamics, then the simulation models of each subsystem are stored in the database or in the user's local computer system, as shown in Fig. 3.27. In the preprocessing modeling of subsystems, the user is able to obtain the vehicle's CAD solid model through the interface provided by the modeling system and then set the dynamic parameters based on the solid model to enhance the model's authenticity. In the figure, the system's default model expression is adopted because this kind of modeling is easy to build with high efficiency. It can meet the requirements of preprocessing model structure expression and inspection of the model.

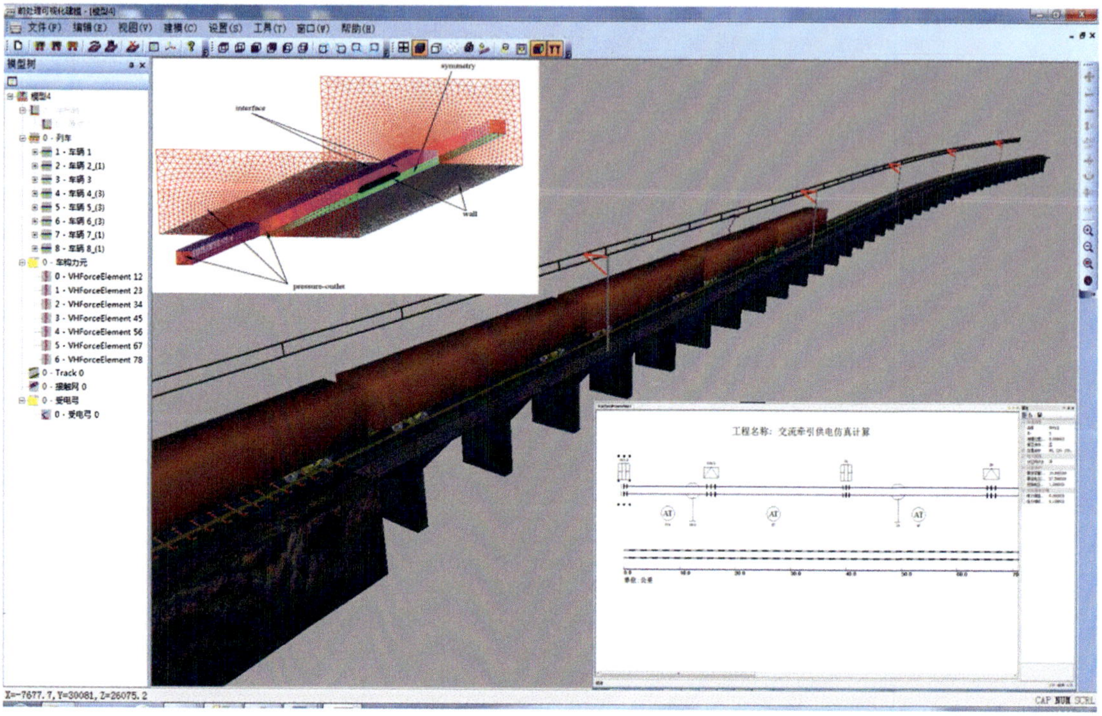

Figure 3.27
The preprocessing simulation modeling of train, track, pantograph–catenary, aerodynamics, and tractive power supply subsystem.

(3) Setting the coupled simulation condition: After creating or selecting a project, we should create or select a simulation condition in the selected project, and select the coupled correlation model. Then we can select the simulation model of the coupled subsystem from the database or from the user's local computer system, and set the simulation parameters of the subsystem as well as the coupled control parameters. Meanwhile, we shall store the condition in the simulation condition database, and then submit it to the server for coupled simulation calculation, as shown in Fig. 3.28.

(4) Starting the coupled simulation calculation: When the task-scheduling module receives the coupled calculation request, it sends the task to the appropriate computing resources and starts the coupling simulation. The monitoring data in the calculation process are written to the simulation process monitoring database and displayed in the monitor, as shown in Fig. 3.29. The anomalies in the process can also be alarmed. The results of the coupled simulation are stored into the simulation result database according to the coupled subsystem classification.

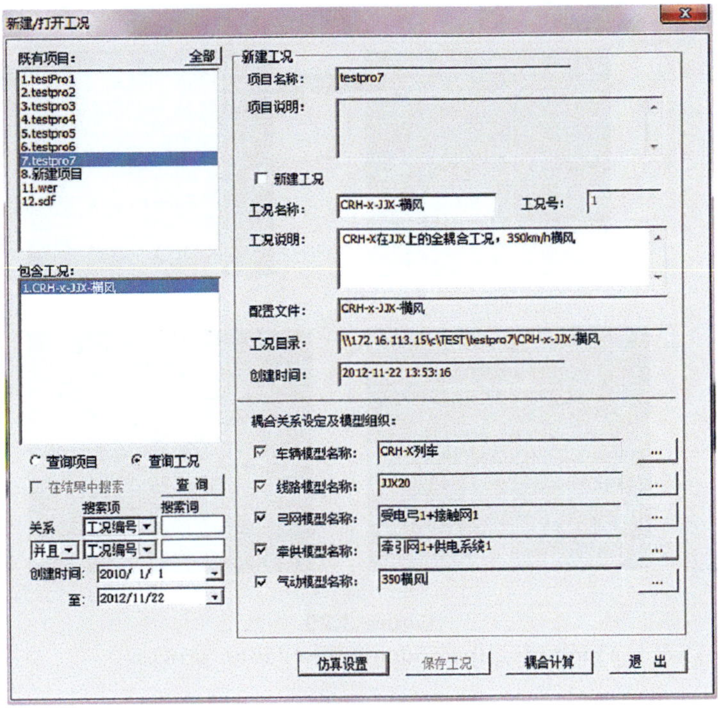

Figure 3.28

The settings of the coupling simulation conditions.

(5) Postprocessing of the coupled simulation results: After the data have been taken out from the simulation results database, we can carry out the postprocessing of the simulation results in a 2D or 3D manner according to the subsystem. The platform provides a general 2D postprocessing module that can automatically parse the format of the postprocessing results of each subsystem and display it in the form of a result tree. We can drag one or more sets of indicator data from the tree to create a chart. It is also possible to process the simulation results visually in a designer and provide evaluation methods commonly used in a dynamic index, and result reports are automatically generated according to the format, as shown in Fig. 3.30.

The platform also provides a 3D operation simulation module. This module analyzes the model data generated by the preprocessing and automatically generates vivid models of the track and the catenary. It also calls up a vivid train model and conducts the virtual display of train operation driven by simulation data. The module can perform 2D and 3D simultaneous display, multiwindow display, and multiple subsystem collaborative display, as shown in Fig. 3.31.

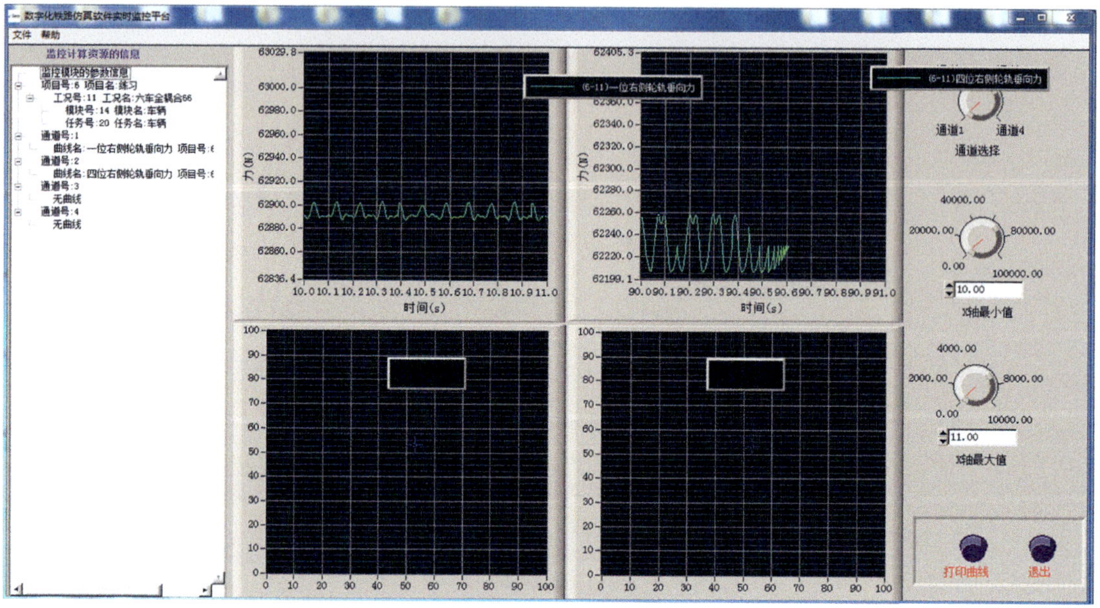

Figure 3.29
Monitoring the coupling simulation process.

3.5.2 The simulation calculations and verification of the dynamics of coupled systems in high-speed trains

3.5.2.1 The parameters of the simulation system model

3.5.2.1.1 Vehicle dynamics model

The vehicle system consists of 15 individual bodies, including one vehicle body, two frames, eight axle boxes, and four wheelsets. There are 6 independent degrees of freedom for the body and the frame, only 1 rotational degree of freedom for the axle box, 4 independent degrees of freedom for the wheelset, and a total of 42 degrees of freedom for the whole vehicle. The primary suspension of the vehicle adopts the arm-positioning method, installed with suspension components such as steel springs and vertical dampers. The second suspension is equipped with air springs, lateral stops, lateral dampers, antiyaw dampers, and traction rods. In the multibody dynamics model of the vehicle, a rigid body model is used for the body, frame, axle box, and wheelset. A corresponding force element model is established in compliance with the position and characteristics of the primary and secondary suspension components. The models are shown in Table 3.6.

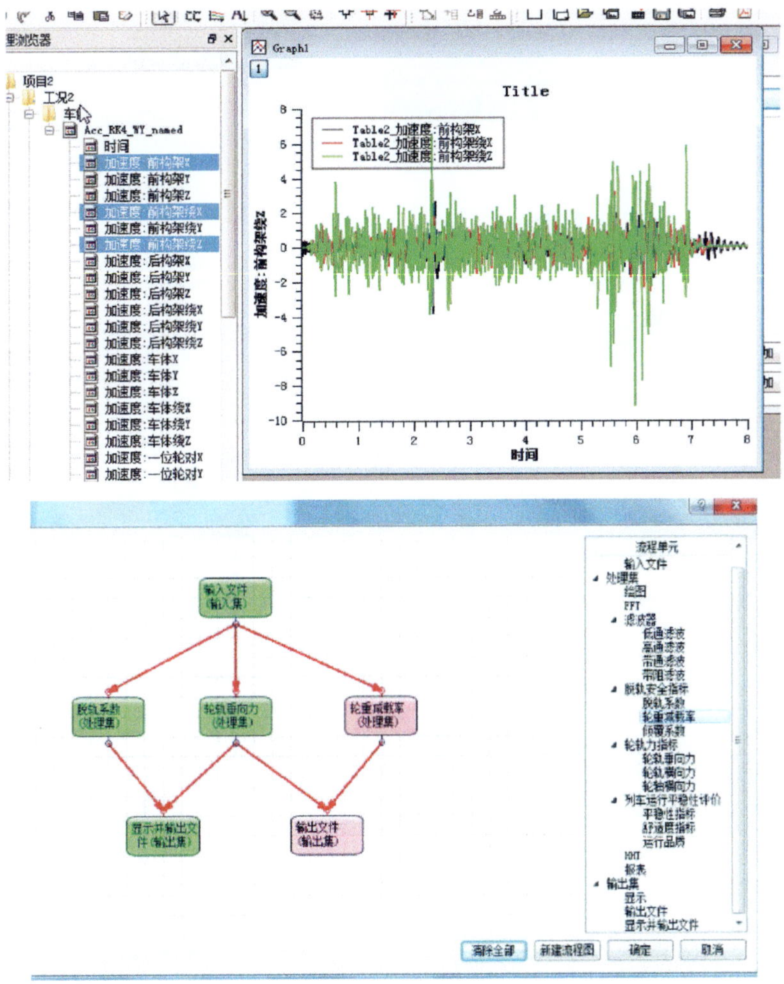

Figure 3.30
The general two dimensional postprocessing module.

3.5.2.1.2 Train dynamics model

The train system consists of six power vehicles and two trailers. The performance parameters are shown in Table 3.7. The traction/braking characteristics are shown in Fig. 3.32.

3.5.2.1.3 Track dynamics model

The track dynamics model adopts a plate-type ballastless track, and the type of support under the rail is roadbed and bridge. The main parameters of the track dynamics model are shown in Table 3.8.

Figure 3.31
The three-dimensional operation display module of a high-speed train.

Table 3.6: Parameters of the vehicle dynamics model.

Vehicle system	Degrees of freedom	Quantity	Total degrees of freedom
Vehicle body	6	1	6
Bogie	6	2	12
Axle box	6	8	8
Wheelset	4	4	16

Table 3.7: Parameters of the train simulation model.

Parameter name	Parameter value
Power supply system	Single-phase AC 25 kV, 50 Hz
Transfer method	Direct handed
Traction power	7200 kW
Maximum operating speed	350 km/h
Intended maximum speed	370 km/h
Grouping method	6M2T
Capacity	610 persons
Group length	201.4 m
Group weight	345 t
Bogie axle load	$\leq$14 t
Starting acceleration	1.46 km/h/s
Braking method	Regenerative braking and electropneumatic connective braking system
Basic resistance	$\omega_0(v) = 8.63 + 0.07295v + 0.00112v^2$ (N/t)

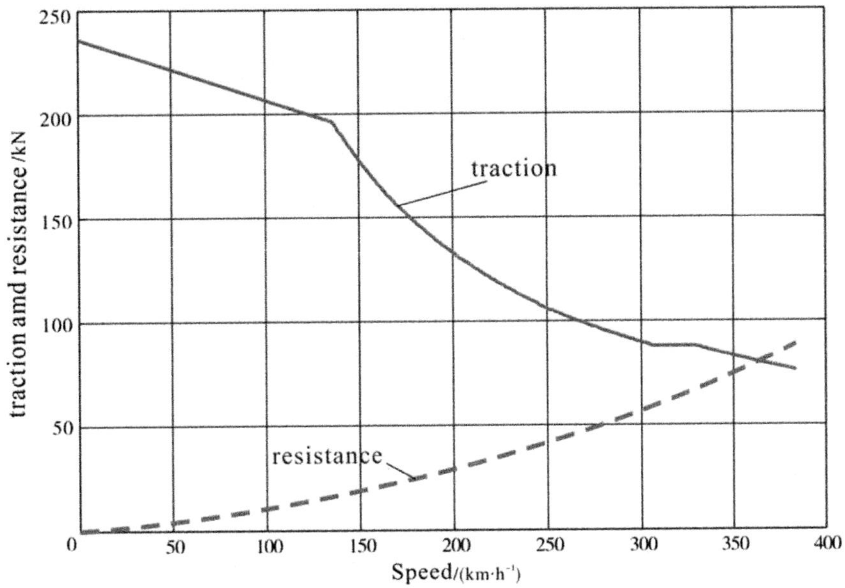

Figure 3.32
Traction characteristics of a certain electrical multiple unit.

Table 3.8: Main parameters of plate-type ballastless track.

Parameter name	Parameter values (roadbed/bridge)
Rail elastic modulus (N/m^2)	2.059×10^{11}
Rail vertical antibending inertial moment (m^4)	3217×10^{-8}
Rail horizontal antibending inertial moment (m^4)	524×10^{-8}
Rail antitor inertial moment (m^4)	215.1×10^{-8}
Rail mass of the unit length (kg/m)	60.64
Track plate size (m × m × m)	$5.0 \times 2.5 \times 0.2$
Cement Asphalt mortar elastic modulus (MPa)	100/7000
Concrete base size (m × m × m)	$4.93 \times 3.2 \times 0.3$
Concrete base elastic modulus (MPa)	3.25×10^4
Poisson's ratio of concrete base	0.2
Fastener spacing (m)	0.6
Dynamic stiffness of fastener joints (kN/mm)	60
Vertical stiffness of rubber cushion under rail (N/m)	50×10^6
Vertical damping under rail (N·s/m)	7.5×10^4
Horizontal stiffness of rubber cushion under rail (N/m)	30×10^6
Horizontal damping under rail (N·s/m)	7.5×10^4

3.5.2.1.4 Track irregularity

Since there is no standard track spectrum for high-speed railways in China, the power spectrum density is shown in Figs. 3.33 and 3.34 in accordance with the random data measured on the Beijing–Tianjin high-speed railway and the Wuhan–Guangzhou high-speed railway.

Regression processing of the aforementioned power spectrum density function can yield a random irregular regression curve of spectrum spectral density. In the actual simulation calculations, time samples of different wavelength ranges can be obtained in compliance with the random irregular power spectrum density with the frequency domain power spectrum equivalent algorithm, as shown in Fig. 3.35.

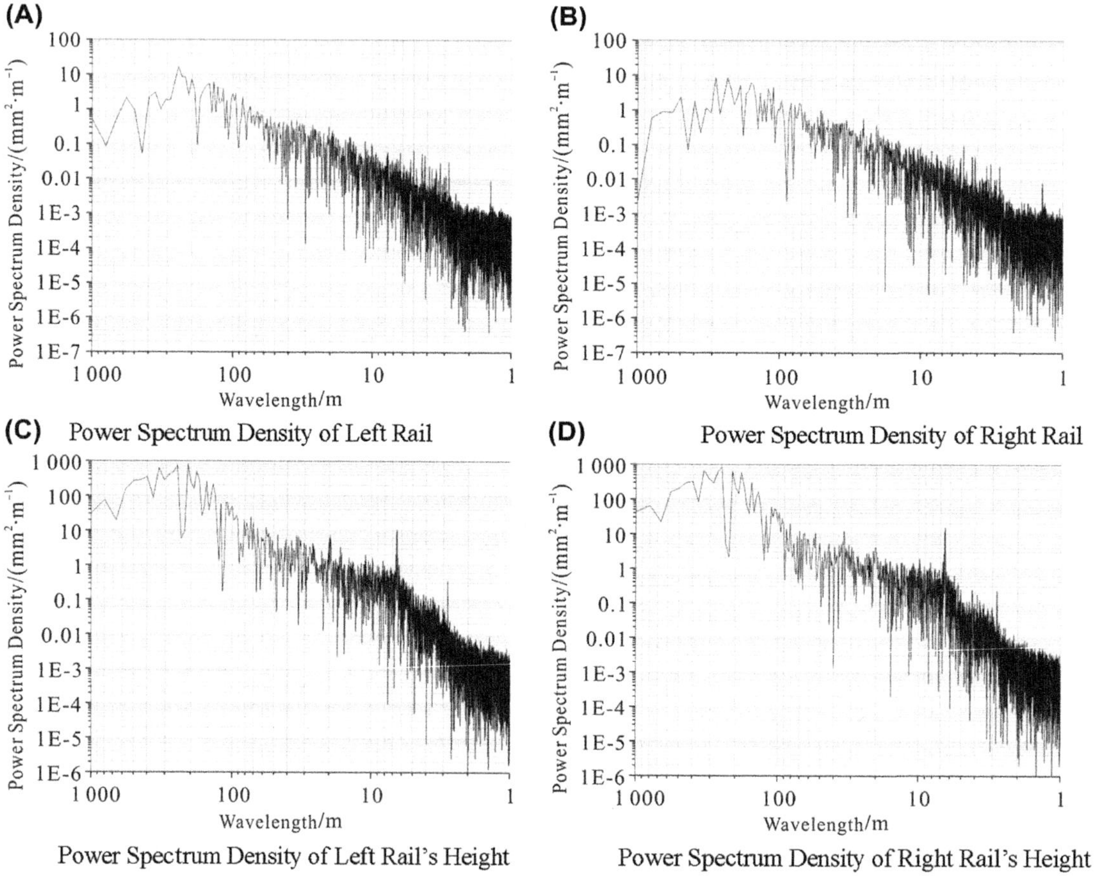

Figure 3.33

The random data measured on the Beijing–Tianjin high-speed passenger dedicated line. (A) Power spectrum density of left rail. (B) Power spectrum density of right rail. (C) Power spectrum density of left rail's height. (D) Power spectrum density of right rail's height.

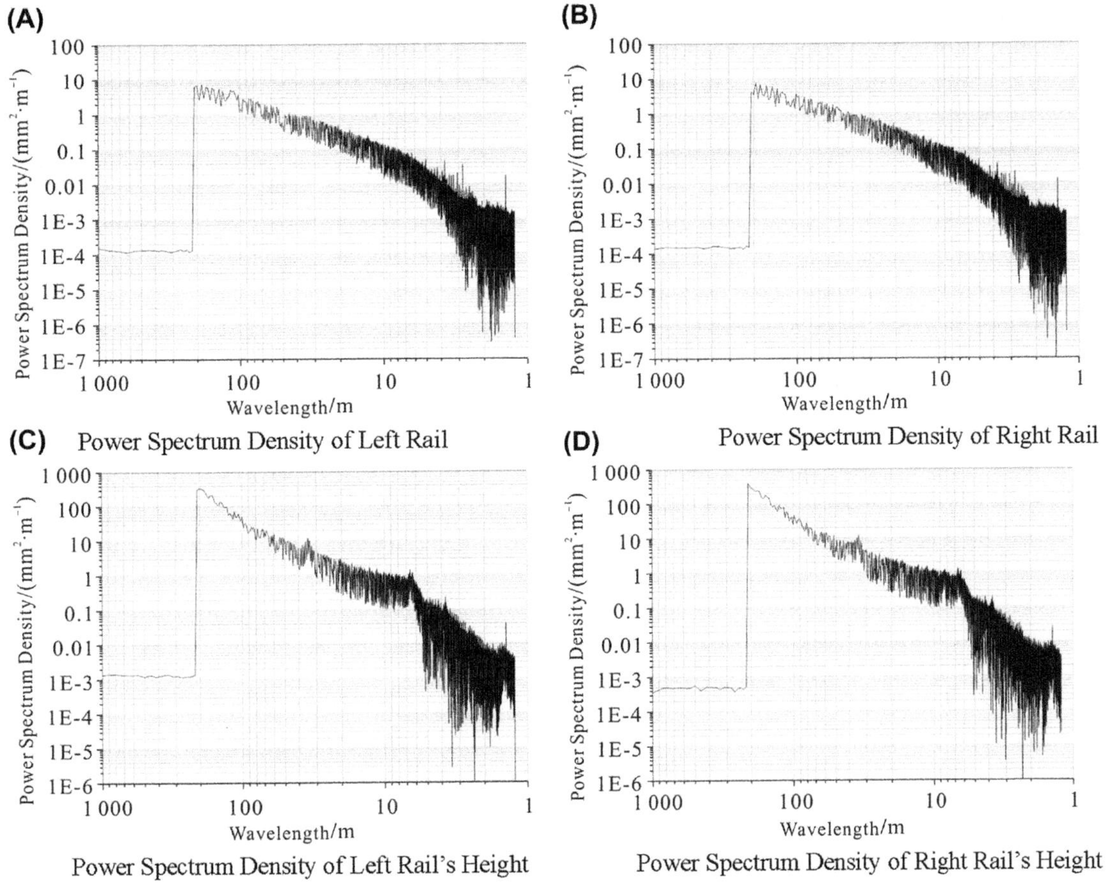

Figure 3.34

The random data measured on the Wuhan—Guangzhou high-speed -railway. (A) Power spectrum density of left rail. (B) Power spectrum density of right rail. (C) Power spectrum density of left rail's height. (D) Power spectrum density of right rail's height.

3.5.2.1.5 Pantograph—catenary dynamics model

In this case, a catenary dynamics model including messenger wires, assistant wires, contact wires, and droppers is established for a stitched suspension catenary system. The structure and parameters of the catenary are shown in Fig. 3.36. At the same time, an equivalent mathematical model for high-speed pantographs is established with the mechanical parameters shown in Table 3.9. The pantograph and catenary models are combined and the corresponding contact algorithm is defined to obtain the dynamics model of the pantograph and catenary system.

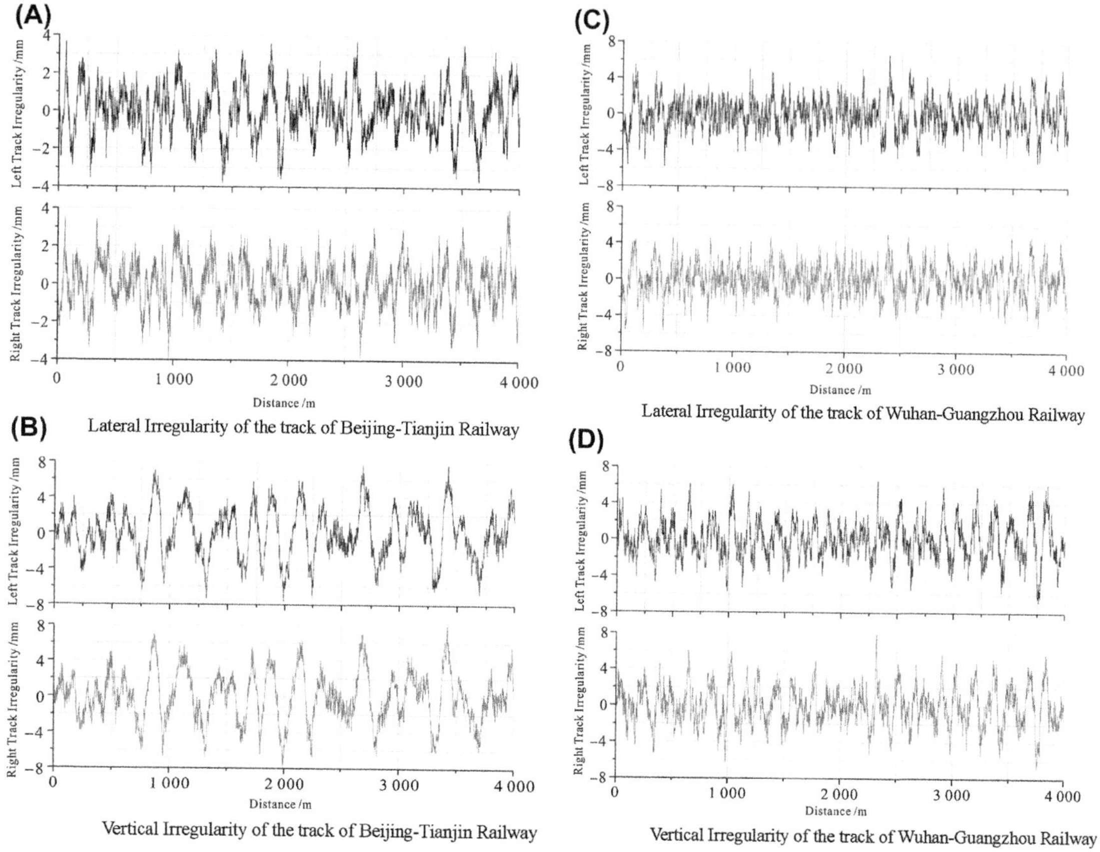

Figure 3.35

Samples of high-speed railway irregularity. (A) Lateral irregularity of the track of the Beijing–Tianjin railway. (B) Vertical irregularity of the track of the Beijing–Tianjin railway. (C) Lateral irregularity of the track of the Wuhan–Guangzhou railway. (D) Vertical irregularity of the track of the Wuhan–Guangzhou railway.

3.5.2.1.6 Train aerodynamics model

The aerodynamics model of two trains meeting is shown in Fig. 3.37. The line spacing is 5 m, and the longitudinal distance between the two trains is 50 m. The speed of both trains is 350 km/h. The aerodynamics model of a train passing through a tunnel is shown in Figs. 3.38–3.40. The distance from the nose of the train to the entrance of the tunnel is 80 m, the cross-sectional area of the tunnel is 100 m^2, the line spacing in the tunnel is 5 m, and the speed of the train is 350 km/h. The meshes in the computational domain are composed of structured meshes and unstructured meshes.

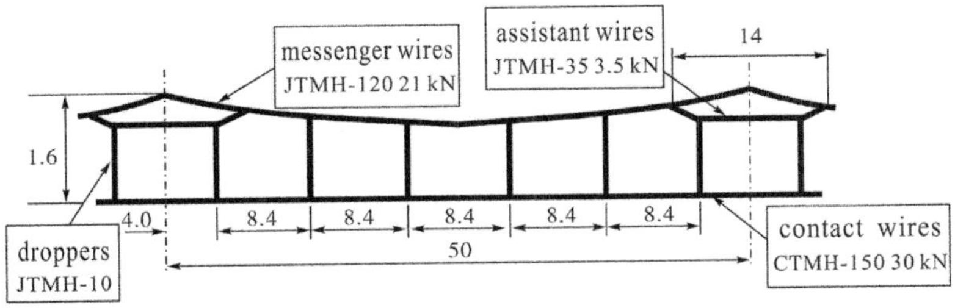

Figure 3.36
Catenary parameters.

Table 3.9: Pantograph parameters.

Parameter name	Parameter Values
Pantograph head mass m1 (kg)	6.1
Pantograph head stiffness k1 (N/m)	10,400
Pantograph head damping c1 (N·s/m)	10
Pantograph frame mass m2 (kg)	10.154
Pantograph frame stiffness k2 (N/m)	10,600
Pantograph frame damping c2 (N·s/m)	0
Pantograph lower-frame Mass m3 (kg)	10.3
Pantograph lower-frame stiffness k3 (N/m)	0
Pantograph lower-frame damping c3 (N·s/m)	120
Pantograph static lifting force F0 (N)	70
Pneumatic pantograph lifting coefficient (N·h2/km^2)	0.00097

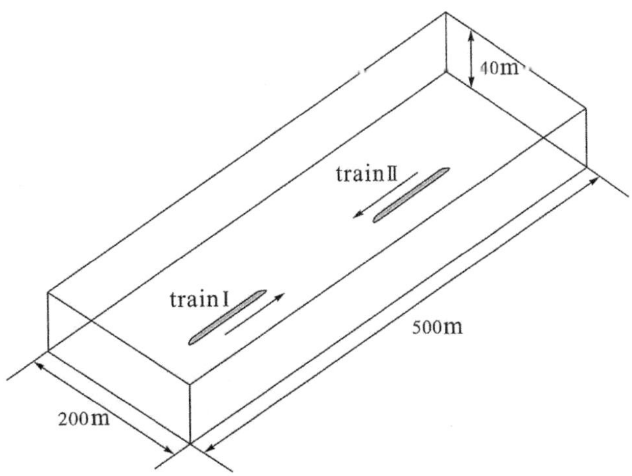

Figure 3.37
Parameters of a two-car intersection model.

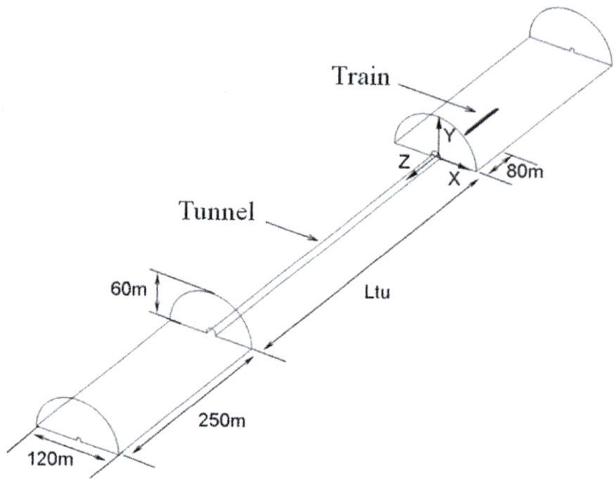

Figure 3.38
The computational domain of the tunnel.

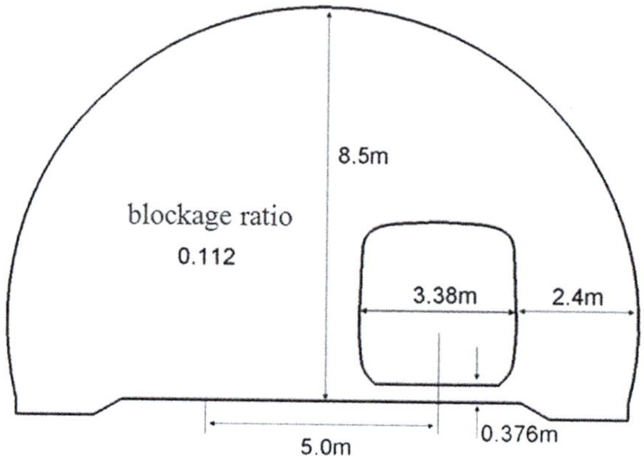

Figure 3.39
The cross-sectional area of the tunnel.

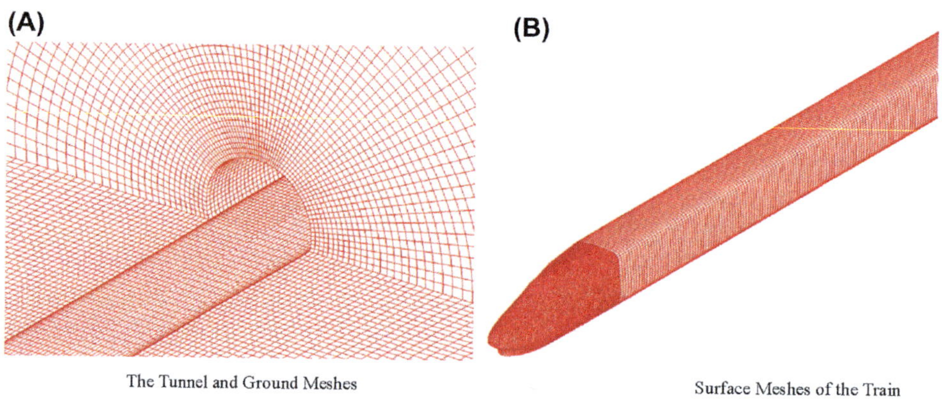

Figure 3.40
Meshing. (A) The tunnel and ground meshes. (B) The surface meshes of the train.

3.5.2.1.7 Settings of the coupled calculation parameters

Based on the simulated calculation model of the dynamics of coupled systems in high-speed trains, we can set the calculation parameters of each subsystem and the coupled control parameters.

The calculation parameters of each subsystem are as follows: the integration step of the vehicle subsystem, track subsystem, and pantograph—catenary subsystem is set to be 5×10^{-5} s, and the integration step of the aerodynamic subsystem is set to be 2×10^{-3} s. The integration step of the tractive power supply and the transmission subsystem is set to be 1×10^{-1} s.

The coupled parameters are set as follows: the coupled control level and the coupled steps are automatically and dynamically set according to the subsystems that are involved in the coupled simulation condition.

3.5.2.2 Comparison between the numerical and the test results of the high-speed train dynamic system

To verify the numerical model and method, the calculation results will be compared with the measured data, based on the example of vehicle dynamic performance under open-line operations, pantograph dynamic performance, aerodynamic performance of a train in the open air, and aerodynamic performance of a train passing through a tunnel

3.5.2.2.1 Comparison of vehicle dynamic performance under the open-line operation

Through the full-scale tests, vehicle dynamic performance data can be obtained, such as three-way acceleration of the vehicle body and the frame. Fig. 3.41 shows the comparison results of a vehicle dynamics simulation and test at a speed of 350 km/h. It can be seen

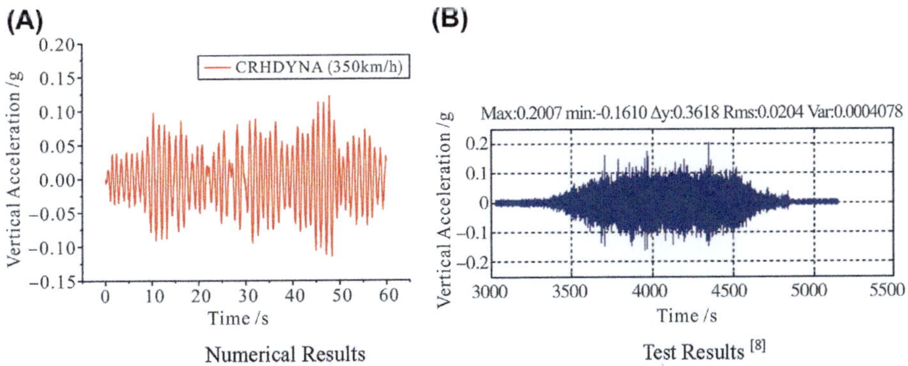

Figure 3.41
Comparison between (A) numerical results and (B) test results of the vertical vibration acceleration response of the vehicle body at a speed of 350 km/h [7].

from the figure that the trend of the numerical results is basically consistent with the test results. The overall amplitude is slightly smaller than the test result because the simplification of the vehicle model and the railway model leads to incomplete reflection of the contribution of some frequencies. The fluctuation of the test results is due to the fluctuation of the speed during operation.

3.5.2.2.2 Comparison of the dynamic performance of the pantograph—catenary under the open-line operation

The pantograph—catenary current collection is obtained by full-scale testing, which includes contact force, arcing, and height of contact wire measurement. Fig. 3.42 shows the test results of contact force when the pantograph is running in the closed direction at a speed of 350 km/h. The numerical results are shown in Fig. 3.43. It can be seen from the figure that the peak and the trend are basically the same.

Table 3.10 shows the aforementioned test results of contact pressure as well as the average, minimum, and maximum of the numerical calculations. From Table 3.10, it is plain to see that the simulation results of pantograph contact pressure for the high-speed train dynamic systems are close to the test results.

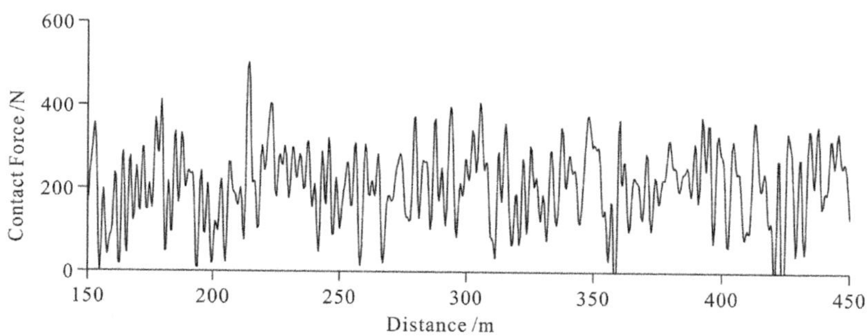

Figure 3.42
Test results of the contact force of the pantograph at a speed of 350 km/h [4].

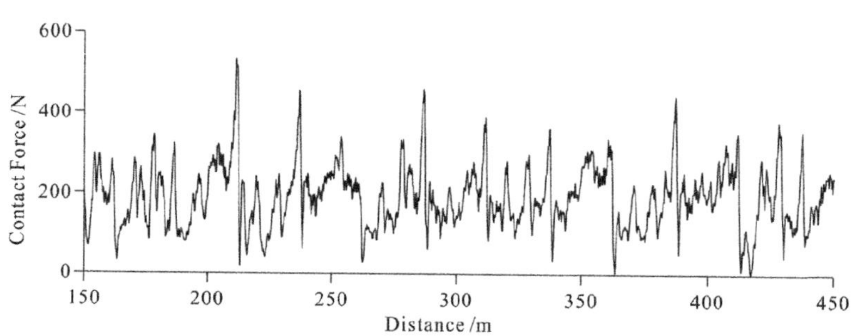

Figure 3.43
Numerical results of the contact force of the pantograph at a speed of 350 km/h.

Table 3.10: Comparison of contact pressure results at a speed of 350 km/h.

Contact pressure/N	Test results	Numerical results
Average	223.87	192.40
Minimum	0	0
Maximum	502.11	534.95

3.5.2.2.3 Comparison of aerodynamic performance of a train in the open air

Under conditions where two trains pass each other in the open air, the comparison between the pressure curve of a certain numerical simulation at a measurement point of a train model and that of a real vehicle is shown in Fig. 3.44. It can be seen from the figure that the numerical and practical test results are basically the same comparing peaks and trend.

3.5.2.2.4 Comparison of aerodynamic performance of a train passing through a tunnel

The measured data show the pressure curve of a measurement point on the window in the middle of a train when the train passes a double-track tunnel with a line space of 5.0 m and a length of 2989 m at a speed of 350 km/h. The pressure curves of the vehicle test results and numerical results are shown in Fig. 3.45. It can be seen that the pressure fluctuation is basically the same and the peak values are very close.

3.5.2.2.5 Comparison of running performance and energy consumption at the maximum operating conditions of trains

For when the train runs at its maximum operating conditions, Fig. 3.46 shows the comparison between the simulated performance curve and the testing data. As can be seen in the figure, the two lines are almost identical.

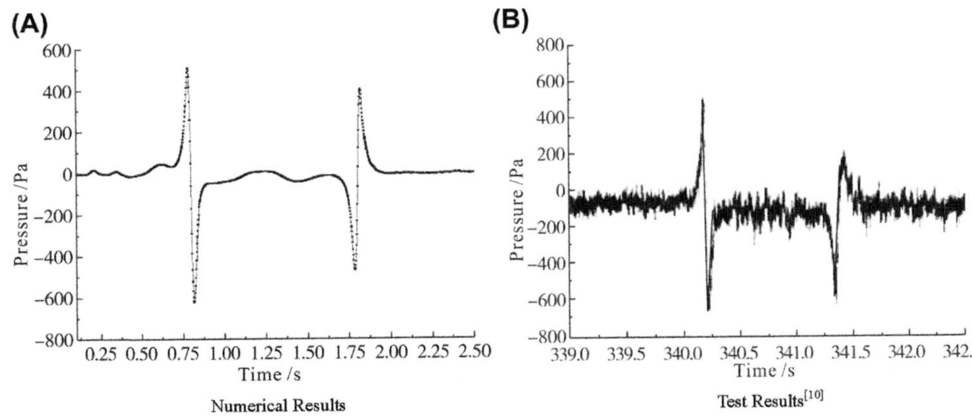

Figure 3.44
Comparison between (A) numerical and (B) test results of curves for two trains meeting [8].

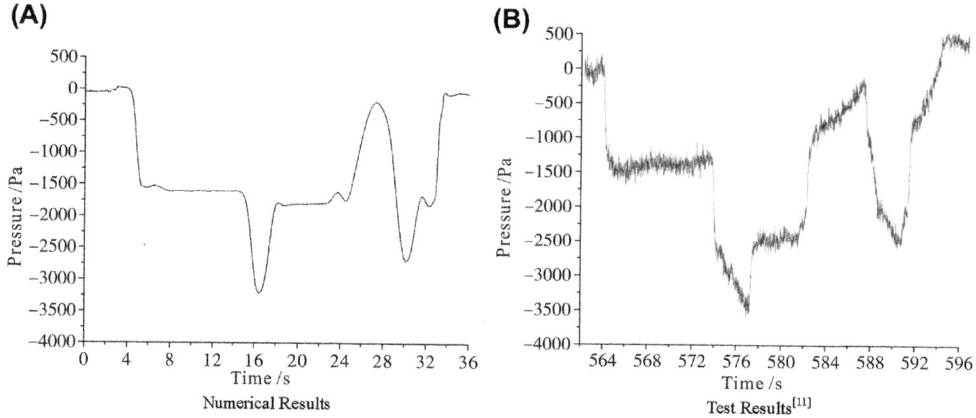

Figure 3.45
The pressure curve of a measurement point on a window in the middle of a train (vehicle speed: 350 km/h). (A) Numerical results. (B) Test results [9].

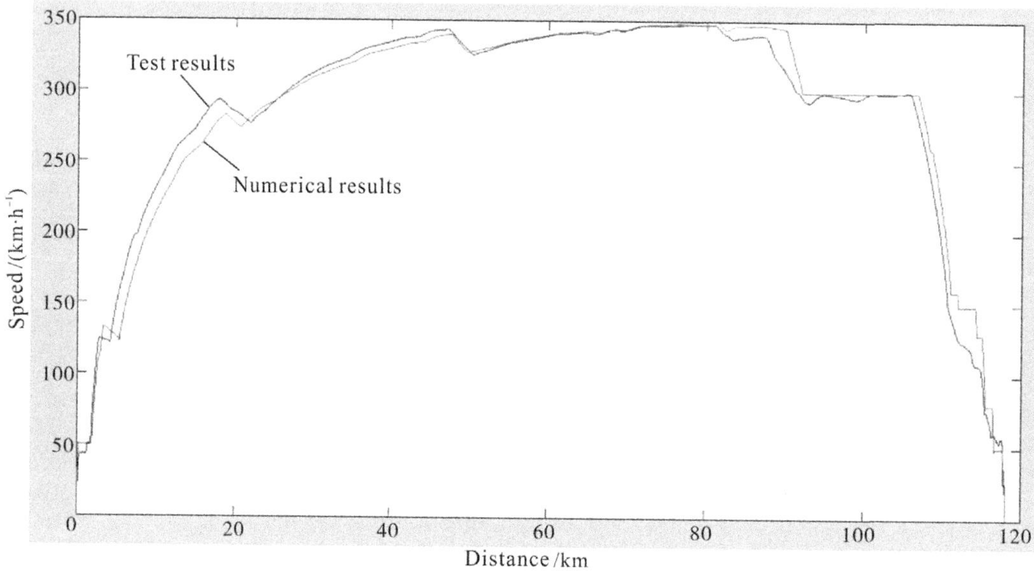

Figure 3.46
Speed—distance comparison.

The comparison of energy consumption is shown in Table 3.11. The total energy consumption deviation between numerical results and testing data is 0.9%, and the traction energy consumption deviation is 2.46%. The deviation in the regeneration energy is 25.28%. The large deviation of the calculation of regeneration energy is mainly due to the fact that the braking method during the operation depends on the maneuvering habits of

Table 3.11: Comparison of energy consumption.

	Testing data	Numerical results	Deviation
Total energy consumption (kW·h)	2667	2691	0.9%
Traction power consumption (kW·h)	2844	2914	2.46%
Renewable energy (kW·h)	−178	−223	25.28%

the train driver, which are difficult to simulate. Due to the small proportion of energy feedback during the operation of regeneration braking, there will be no large deviation in the calculation of the total energy consumption.

From the comparison and analysis of the numerical results and testing data under the aforementioned typical operating conditions, the numerical results and test results are basically the same, indicating that the modeling method and the coupled simulation method used by the platform are feasible. Therefore, the platform is well equipped to support the simulation analysis of the dynamics of high-speed train coupled systems.

References

[1] Ding G, Yan K, Zhang W, Yan W, Wang T. Research on extraction of product properties based on virtual prototyping design. Computer Integrated Manufacturing Systems 2006;12(1):14–20.

[2] Hong J. Calculation of multibody system dynamics. Beijing: Higher Education Press; 1999. p. 30–3.

[3] Zhang H, Rong L. Research report on virtual prototyping system technology of high-speed train. Chengdu: Laboratory for Rail Transportation of Southwest Jiaotong University; 2012.

[4] State Key Laboratory of Traction Power of Southwest Jiaotong University. Test report on system dynamics of pantograph and current collection of Wuhan-Guangzhou high-speed railway. Chengdu: State-Key Laboratory of Traction Power of Southwest Jiaotong University; 2009.

[5] Wikipedia. The Doppler effect, 12.19; 2012. http://zh.wikipedia.org/wiki/The Doppler effect.

[6] Zou Y, Zhang W. Research on the calculation platform for the dynamics of coupled systems of high-speed train. Chengdu: Post-doctoral Research Report of Southwest Jiaotong University; 2011.

[7] State Key Laboratory of Traction Power of Southwest Jiaotong University. Beijing-tianjin intercity high-speed railway research report (report on CRH2 vibration acceleration analysis and test). Chengdu: State-Key Laboratory of Traction Power of Southwest Jiaotong University; 2008.

[8] State Key Laboratory of Traction Power of Southwest Jiaotong University. Research report on Beijing-Tianjin intercity high-speed railway -aerodynamics. Chengdu: State-Key Laboratory of Traction Power of Southwest Jiaotong University; 2008.

[9] State Key Laboratory of Traction Power of Southwest Jiaotong University. Research report on aerodynamics of Wuhan-Guangzhou high-speed railway. Chengdu: State-Key Laboratory of Traction Power of Southwest Jiaotong University; 2009.

Further reading

[1] Yu H, Ding B. Schema integration and query decomposition in multidatabase environment. Comput Eng 2000;26(10):124–6.

[2] Song B, Zhang B, Wang G, Ge Y. Approach of maintaining integrated data consistency in superbase. Mini-Micro Syst. 2000;(08):858–61.

CHAPTER 4

Basic characteristics and evaluation of the dynamics of the coupling systems of the high-speed train

Chapter Outline

Dynamics of Coupled Systems in High-Speed Railways. https://doi.org/10.1016/B978-0-12-813375-0.00004-2

4.1 Dynamics and parameters of the high-speed train coupling system

4.1.1 Parameters of the high-speed train

The train is composed of multiple vehicles, and the dynamic calculation model of each vehicle is presented in Figure 4.1. When performing dynamic analysis, we need to obtain the mass parameters of every major component of the rolling stock, including parameters of the suspension components and geometrical parameters. For the train dynamics, the parameters of geometry and properties of the coupling devices between vehicles should also be obtained.

Table 4.1 lists the basic parameters required for the train dynamics simulation analysis. The vehicle type also needs to be refined in line with the actual situation. Therefore, this book will specify some models and parameters that need to be refined.

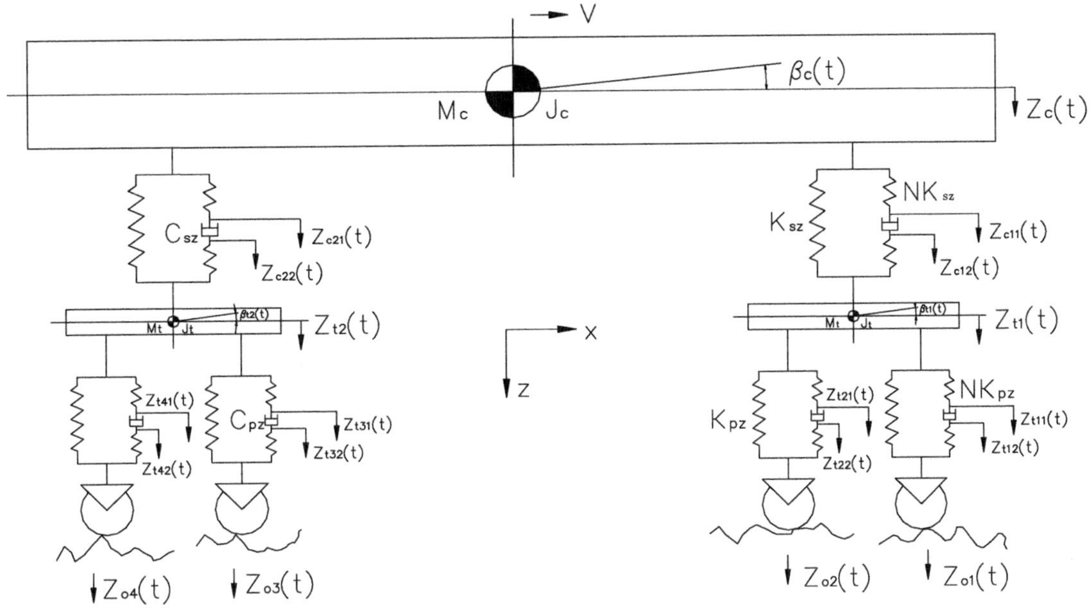

Figure 4.1
Dynamic calculation model of the rolling stock.

4.1.1.1 Description of air spring calculation parameters

In the calculation of train dynamics, there are primarily three kinds of air spring models [1]: simplified model, linear model, and nonlinear model. The calculation parameters required for the different air spring models are also different.

In the calculation of traditional vehicle dynamics, the simplified model of the air spring (Figure 4.2) is commonly used. It consists of a spring k and damper d in parallel, which is generally obtained using a test. This model is basically no longer used because it cannot truly reflect the variable frequency and phase characteristics of a real air spring.

Figure 4.3 shows the linear model of an air spring. This linearized model is based on an ideal gas polytropic equation, a linear equation of flux in the orifice, and linearizing by ignoring the infinitesimal of a higher order. The stiffness parameters of this model consist of three parts: First, the effective area, pressure and volume-related stiffness K1 of the air spring; second, stiffness K2 of the supplementary air chambers relating to the volume ratio between air spring and supplementary air chambers; and thirdly, stiffness K3 caused by the change rate of the effective area. Parameter d2 in Figure 4.3 is the orifice damping.

The nonlinear model of an air spring, consisting of the air spring mechanical model and the connecting pipe mechanical model, is presented in Figure 4.4. According to the Helmholtz equation, the gas dynamics behavior in the connecting pipe can be simplified to

Table 4.1: Train Dynamics simulation input parameters.

Geometric parameters	
Vehicle distance	A series of vertical shock absorbers
Bogie wheelbase	Secondary vertical damper action point position
Gauge	Second-order lateral shock absorber position
Wheel center distance	Anti-snake shock absorber action point
Wheel profile	Anti-rolling torsion bar point position
Rail Profile	Traction device position
Wheel rolling circle diameter	Lateral stop position
Frame center height	Horizontal stop free clearance
Height of vehicle center of gravity	Hook device position
A series of suspension point locations	Workshop shock absorber location
Secondary suspension point location	Windshield equivalent position
Quality parameters	
Wheelset quality	Empty body mass
Wheel-to-roll rolling moment of inertia	Empty Body Side Rolling Moment of Inertia
Wheel pair nod rotation inertia	Empty Body Nodding Moment of Inertia
Wheel pair shaking head inertia	Empty body shaking head inertia
Frame quality	Heavy body mass
Frame Rolling Moment of Inertia	Heavy body roll moment of inertia
Frame nodal moment of inertia	Heavy body nod rotation inertia
Swing head inertia	Heavy body shaking head inertia
Suspension parameters	
A series of longitudinal positioning stiffness	Two-way transverse damping
A series of lateral positioning stiffness	Secondary vertical damping
A series of vertical stiffness	Anti-roll-roller side roll angle rigidity
A series of vertical damping	Traction device equivalent stiffness
Second series longitudinal stiffness	Lateral stop characteristic curve
Second-order transverse stiffness	Hook device characteristics curve
Secondary vertical stiffness	Workshop damper damping
Anti-snake damper damping	Windshield equivalent stiffness and damping

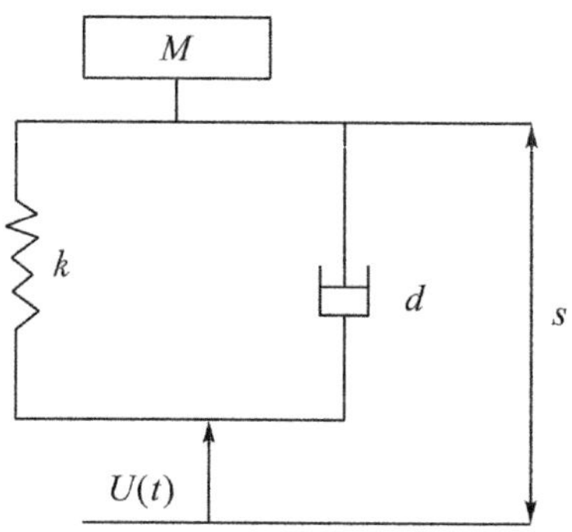

Figure 4.2
Simplified model of air spring.

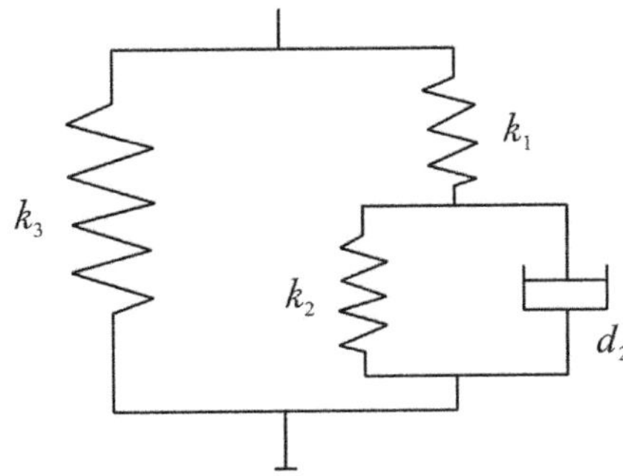

Figure 4.3
Linearized model of an air spring.

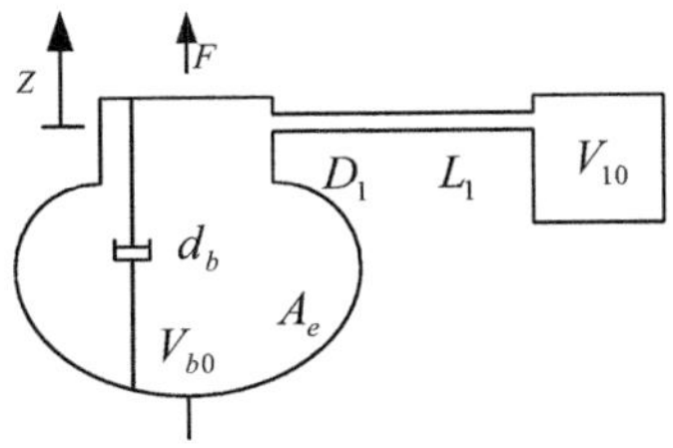

Figure 4.4
Nonlinear model of an air spring.

Table 4.2: Main calculation parameters required for the nonlinear model of the air spring.

Body volume	Additional chamber volume
Effective bearing area Body to additional chamber line diameter	Throttle Damping Body to additional air chamber line length

a spring-mass system. The mass is considered as the gaseous mass in the connecting pipe, and the spring effects are obtained by adding the equivalent of the gas in the air chamber. The calculation parameters required for the nonlinear model of the air spring are presented in Table 4.2.

4.1.1.2 Description of the calculation parameters for hydraulic buffers

There are rubber joints at both ends of the hydraulic buffers of locomotives and vehicles. When the equivalent damping of shock absorbers is large, the rubber nodes and the elasticity of the liquid itself cannot be ignored.

Domestic and foreign shock absorber manufacturers and scholars have conducted long-term research on the dynamic model of the hydraulic buffer, and have established many types of dynamic simulation models. Hydraulic buffers require different parameters

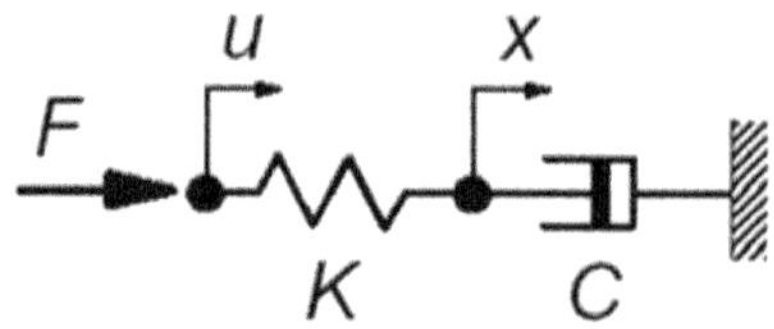

Figure 4.5
Maxwell Damper Series Model.

based on the different degrees of complexity of various hydraulic buffer models. Based on the DIN prEN 13802, the simplest Maxwell model (with dampers and springs in series) is suitable for application as a standardized hydraulic buffer model for vehicle dynamics simulations. A diagrammatic representation is shown in Figure 4.5. Both the damping component *C* and the equivalent stiffness *K* of the model can adopt nonlinear properties.

Hydraulic buffers can be described using static and dynamic properties. Only the dynamic properties of hydraulic buffers can accurately reflect the essential characteristics of buffers.

Assuming that the coefficient of the buffer's equivalent stiffness in series is k; damping coefficient is c; displacement of the shock absorber piston is x_0; end of the buffer subjected to vibration amplitude is A; and the sinusoidal displacement input is x with a frequency of w, i.e., $x = A \sin(wt)$.Then the differential equation of vibration of the shock absorber system can be expressed as:

$$k(x_0 - x) + c\dot{x}_0 = 0 \tag{4.1}$$

Figure 4.6 shows the relationship between the displacement $\dot{x}$, velocity $\dot{x}$, piston displacement x_0, piston velocity x_0, relative displacement of the spring in series $x_{sti} = x - x_0$, and damping force F.

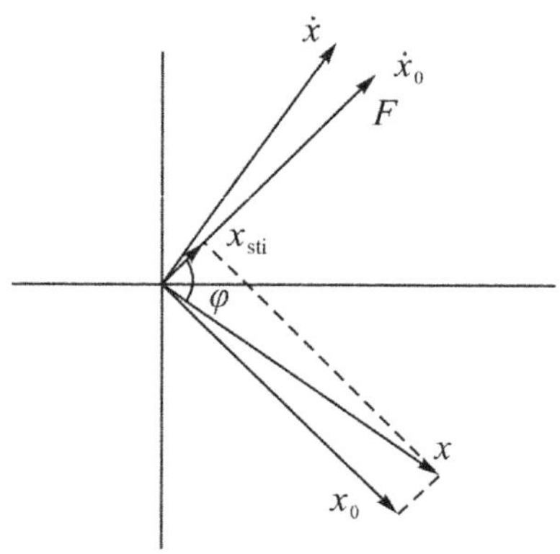

Figure 4.6
Relationship between damping force-velocity-displacement.

Figure 4.6 shows that when we consider the series stiffness of the shock absorber, the damping force F is no longer in phase with the excitation speed $\dot{x} = Aw\cos(wt)$, and a lag in phase $\pi/2 - \varphi$ occurs. The damping force amplitude then reduces commensurately, becoming a multiple, $k\Big/\sqrt{k^2 + c^2w^2}$, of the ideal damping force. According to the requirements of the standardized BS EN 13802:2004 *Railway applications suspension components hydraulic dampers*, the measuring dynamic damping coefficient in the experiment is:

$$c = \frac{k}{w\tan\varphi} \tag{4.2}$$

where w is *the* excitation frequency, φ is the phase angle between damping forces, and k is dynamic stiffness.

Figure 4.6 also shows that the dynamic stiffness of the buffer is the ratio of the amplitude of the axial force $F_{\max}$ of the shock absorber to the deformation $x_{\text{sti,max}}$ of the spring deformation.

$$k = \frac{F_{\max}}{x_{sti,\max}} = \frac{F_{\max}}{x_{\max}\cos\varphi} = \frac{F_{\max}}{x_{\max}}\sqrt{1 + \tan^2\varphi} \tag{4.3}$$

The dynamic stiffness is not to test the ratio between the measured axial force $F_{\max}$ and the displacement amplitude $x_{\max,}$ but to test the influence of dynamic effects. The dynamic stiffness and damping of the hydraulic buffers can be processed based on the test results described above.

4.1.2 Parameters of wheel-rail coupling

The parameters of wheel-rail coupling include mainly: (1) The geometry and profile parameters of the wheels and rails; (2) Geometry and dynamic properties of railways, and important components such as sleepers, ballast bed, and track slab;(3)Track irregularity.

4.1.2.1 Wheel

The wheels currently used on rolling stocks in China are basically rolled solid wheels, consisting primarily of the flange, tread, wheel rim, radial plate, and wheel hub. The tread is the contact surface between the wheel and the rail and is the main working area of the wheel. The flange is inside of the wheel and consists primarily of a protruding circular arc. It is an essential component that ensures the safe operation of the vehicle along the rail, thus preventing derailment. A wheel profile consisting of a wheel tread and a flange is an important parameter for wheel-rail coupling. A typical wheel profile [2], shown in Figure 4.7, includes the loss of metallic cross-sectional area (LMA) type tread and the S1002-type tread.

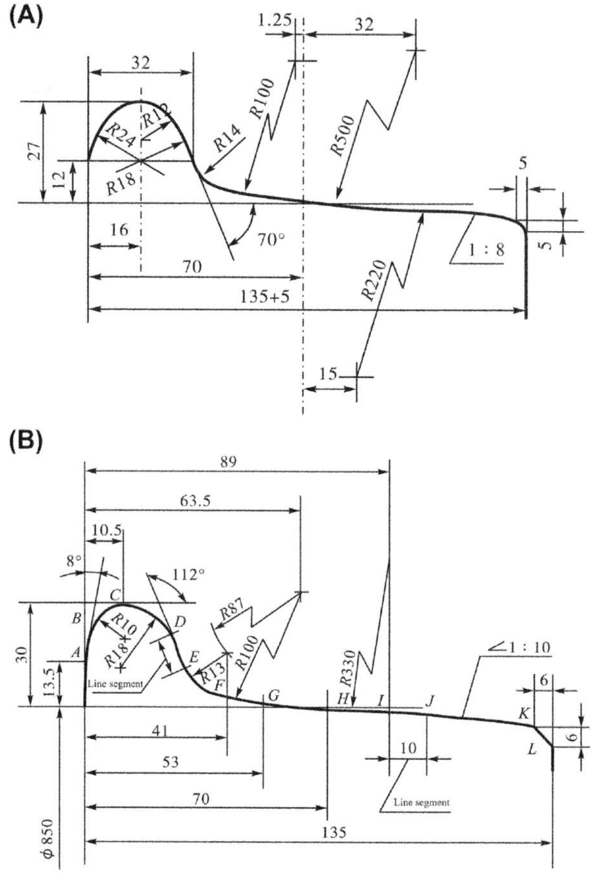

Figure 4.7
Typical wheel tread profile. (a) LMA-Type Tread (b) S1002-Type Tread.

4.1.2.2 Rail

High-speed railways around the world use 60 kg/m rails. The cross section of a typical rail is shown in Figure 4.8. The corresponding main parameters are shown in Table 4.3.

4.1.2.3 Sleeper

The concrete sleeper is widely used in the ballasted track of high-speed railways. The structural form and applicable speed range [3] of concrete sleepers for high-speed railways in foreign countries are presented in Table 4.4, while the main design parameters of China's high-speed railway sleepers [3] are listed in Table 4.5.

4.1.2.4 Fastener

The critical parameters for high-speed railway fasteners consist primarily of clamping force, elasticity, tensioning, and insulation. Adequate buckle pressure is a basic

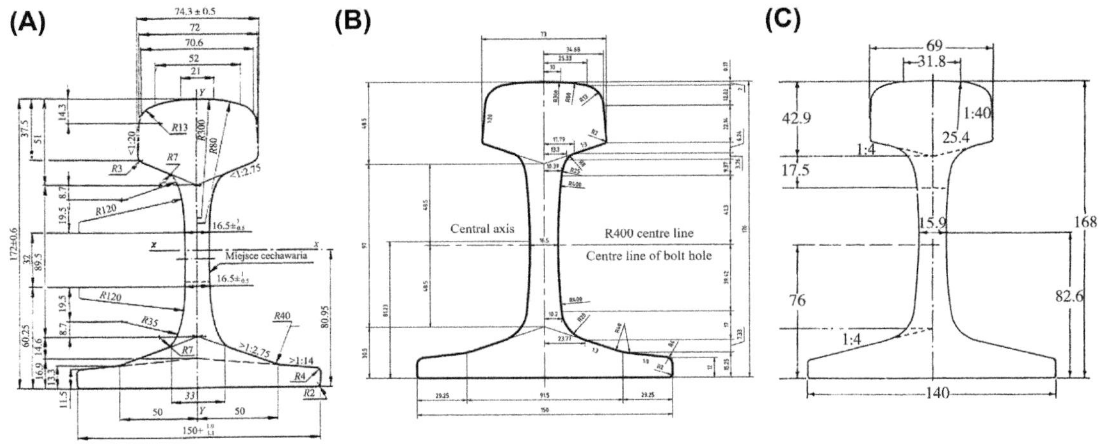

Figure 4.8
Typical rail tread profile. (A) Europe UIC 60 track (B) China CHN 60 track (C) Japan JIS 60 track.

Table 4.3: Typical rail section parameters.

Rail type	UIC 60	CHN 60	JIS 60
Quality | (kg/m)	60.34	60.64	60.80
Area (cm^2)	76.86	77.45	77.50
Horizontal axis moment of inertia (cm^4)	3055	3214	3090
Upper section modulus (cm^3)	335.5	339.4	321.2
Lower section modulus (cm^3)	377.4	396.0	397.2
Rail height (mm)	172	176	174
Rail bottom width (mm)	150	150	145
Railhead height mm	51.0	48.5	49.0
Rail width mm	73	73	65
Rail thickness (mm)	16.5	16.5	16.5

requirement for ensuring the longitudinal and lateral stability of the rail. Good elasticity can ensure that the railway system has good vibration reduction and noise reduction performance. A fastener with good tensioning performance is an important factor for reducing daily maintenance on the line. The insulating performance of a fastener is a key factor for improving the reliability of the circuit, lengthening the track circuit, and reducing investment in the track circuit. The parameters of the typical domestic and foreign high-speed rail fastener [3] are listed in Table 4.6.

4.1.2.5 Sleeper

The sleeper of a high-speed railway ballasted track is composed primarily of a crushed rock layer of about 300 mm and a gravel layer of about 200 mm. Cross sections of the foremost high-speed railway lines in the world [3] are presented in Figure 4.9.

Table 4.4: Main parameters of a typical sleeper for high-speed railways abroad.

Country	Sleeper type	Sleeper Model	Length (mm)	Rail under section size		Intermediate section size		Pillow area (cm^2)	Weight (kg)	Prestressed tendons		Maximum speed (km/h)
				Height (mm)	Bottom width (mm)	Height (mm)	Bottom width (mm)			Number	Weight (kg)	
Japan	Whole body formula	3T	2400	190	283	175	230	6430	260	16ϕ2.9 ×3	6.00	< 210
		3H	2400	220	310.5	195	250	7040	325	20ϕ2.9 ×3	7.50	210 ~ 270
		4H	2400	220	310.5	195	250	7040	325	4ϕ13	9.98	
Germany	Whole body formula	B70W	2600	210	300	175	220	5930	304	4ϕ9.7	6.42	250
		B90W	2600	210	320	180	240	6680	330	4ϕ9.7	6.42	
		B75	2800	240	330	200	290	7560	380	4ϕ9.7	6.42	
France	Double piece formula	U31	2245	220	290	-	680	3944	218	-	-	160
		U41	2415	220	290	-	840	4827	248	-	-	300

Table 4.5: Typical sleeper parameters of the China high-speed railway.

Sleeper type	I		II		III	
Sleeper length (mm)	2500		2500		2500	
section	Under the rail	In track	Under the rail	In track	Under the rail	In track
Height (mm)	201	175	201	165	230	185
Surface width (mm)	165	155	165	161	170	200
Underside width (mm)	275	250	275	250	300	280
Sleeper mass (kg)	250		251		320, 340	
Sleeper bed area (cm^2)	6588		6588		7720	
Head area (cm^2)	490		490		590	

Table 4.6: Typical fasteners at home and abroad.

Country	Fastener type	Buckle pressure (kN)	Pad thickness (mm)	Pad Static stiffness (kN/mm)	Stretch range (mm)
Japan	120 dual elasticity	6	10	60	-
France	Nabal	11	9	70	8.1–9.1
Germany	HM	11	-	70 ~ 80	14
The United Kingdom	Pandrol	11	10	30 ~ 50	12
China	I	9	10	55–80	8
	II (60-12-17)	10	12	40–60	10
	II (60-12-11)			90–115	
	III	11	10	55–80	13

The mechanical properties of the structure of a sleeper are related to the material that the sleeper structure uses. In general, the modulus of elasticity is 70 ~ 270 MPa, density is 1700 ~ 2200 kg/m^3, equivalent damping of the track bed is approximately 6.0×10^4 N·s/m, equivalent shear stiffness of the track bed is approximately 8.0×10^7 N/m, and the equivalent shear damping of the ballast bed is approximately 8.0×10^4 N·s/m.

4.1.2.6 Slab track

A typical slab track structure is shown in Figure 4.10. The main structural parameters include: fastener spacing, 0.625 m; dynamic stiffness of fastener joints, 60 kN/mm; dimensions of track slab, $5.0 \times 2.5 \times 0.2$ m; elastic modulus of track slab concrete, 3.6×10^4 MPa (C60 cement); Poisson ratio of track slab concrete, 0.2; cement and asphalt (CA) adjustment mortar layer under the track plate, 50 mm; elastic modulus of the CA mortar, 100 ~ 300 MPa; external dimensions of the concrete base, 3.2×0.3 m; elastic modulus and Poisson ratio of the base concrete are 3.25×10^4 MPa (C40 cement) and 0.2, respectively.

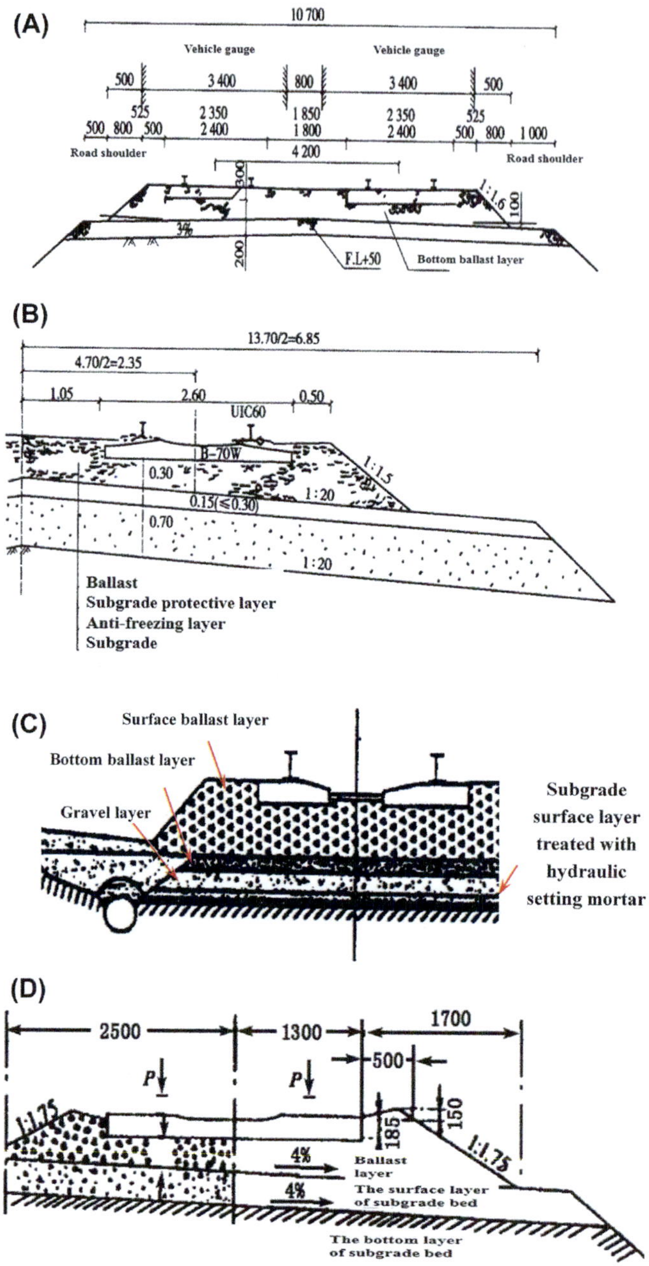

Figure 4.9

Sections of domestic and foreign high-speed railway lines. (A) Section of Japan's Tōkaidō Shinkansen line (units: mm).(B) Section of the new German InterCity Express (ICE) line (units: m). (C) Section of the French Train à Grande Vitesse (TGV) line (D) Section of China's High-Speed Railway Line (Units: mm).

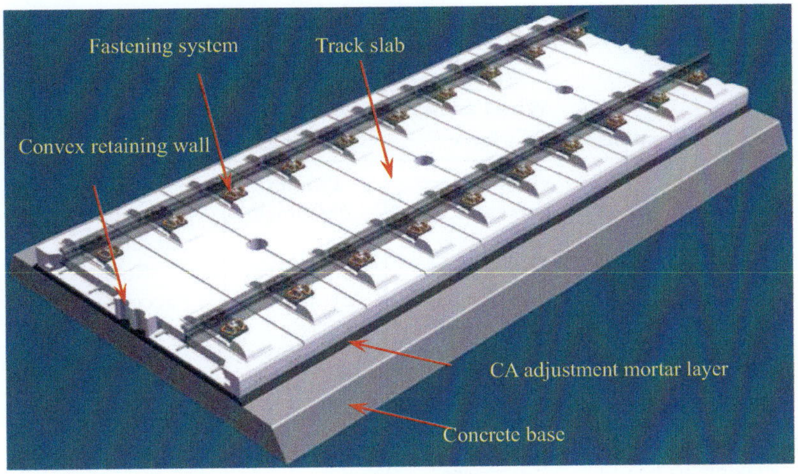

Figure 4.10
Typical slab track structure.

4.1.2.7 Track Irregularity

Track irregularity is the primary source of excitation in the wheel-rail system and is also increasingly the source of the dynamic response of high-speed train coupled large systems to the dynamic wheel-rail force. It has a significant influence on safety and stability during train operation and is also a major factor among components of the railway that can limit the safe running speed of the train. The major effects of different track irregularities on a large-scale coupling system [3] are presented in Table 4.7, and the reference values of the laying accuracy for track irregularities are presented in Table 4.8. For track stochastic irregularity, please refer to section 3.6.2.1 of this book.

4.1.3 Dynamic Pantograph-catenary interaction parameters

4.1.3.1 Catenary

1 Catenary structure parameter

The catenary system is one of the key technologies of electrified railways. Its structure and parameters have a great influence on the dynamic performance of the pantograph-catenary system. At the beginning of this century, China used the Beijing-Tianjin Intercity Railway as a starting point for putting into operation the 350 km/h maximum running speed for current collection with a single pantograph high-speed catenary system. The catenary system of the Beijing-Tianjin line primarily adopts auto-tensioned simple catenary technology. Its structure and parameter configuration are presented in Table 4.9.

Because the running speed is greater than 120 km/h, the main line, station track, and transition line contact suspension of the Beijing South and Tianjin railway stations both

Table 4.7: Impact of track irregularities.

Influences species	Vibration response	Wheel and rail force	Nature		
			Safety	Stability	Service Performance
High and low	Ups and downs, nodding	Vertical force increases or decreases	Promote derailment	Increased vertical acceleration	Structural life
Level	Roll	Vertical force increases or decreases	Promote derailment	Rolling acceleration increases	Structural life
Distortion	Roll	Vertical force increases or decreases	Trigger Suspension derailment	Rolling acceleration increases	Structural life
Track	Side swing, shaking his head	Lateral force increases	Trigger Crawl rail derailment	Lateral acceleration increases	Service status
Gauge	-	-	Trigger Derailed off the road	-	-
Rail level Compound Irregularity	Side roll, roll	Vertical force increases or decreases Lateral force increases	Crawl the track Suspension derailment	Vertical and horizontal Increased acceleration	Structural life Service status
Rail short wave	Wheel and rail high frequency	Increased vertical impact force	Promote broken rails Short axis	noise	Loss of Looseness
Rolling irregularities	-	Periodicity Increased wheel and rail forces	-	Increased vertical acceleration	Bed status

Table 4.8: Track irregularity control target value.

Species	Sweden (No fragments)	Germany (No fragments)	France (with fragments)	Spain (with fragments)	Japan (No fragments)	China	
						With fragments	No fragments
Level	2	2	3	4	2	3	2
Distortion	2	-	1‰/3 m	1.3‰	1.5/2.5 m	3	-
High and low	2	2/5 m	3	3	2	3	3
Track	2	2	2	3	2	3	3
Gauge	±2	-	-	±3	-	+3/−2	±2

Table 4.9: Catenary structure and parameter configuration of the Beijing-Tianjin line.

Technical Parameters			Units	Parameter values
Catenary system	Contact line	Contact line type		CuMg-120
		Tension	KN	27
	Messenger wire	Contact line type		BzII-120
		Tension	KN	21
	Dropper	Contact line type		JTMH-10
	Dropper distance		m	4,8,8,8,8,8,4
	Span		m	≤ 50
	Structural height		m	1.6
	Suspension mode			simple stitched

use BzII 70 + CuAg0.1 AC120 (15 kN+15 kN) contact wires. The other intermediate station tracks and transition line contact suspensions are also BzII 70 + CuAg0.1 AC120 (15 kN + 15 kN) conductors.

The opening of the Wuhan-Guangzhou high-speed railway, signaled that the high-speed pantograph system with a maximum running speed of 350 km/h and current collection with double pantographs had been put into use. The main line of the Wuhan-Guangzhou railway catenary system primarily adopts auto-tensioned elastic catenary technology. Its structure and parameter configuration are presented in Table 4.10.

The auto-tensioned simple catenary is used for the tie-line, the transition line of the main line, the station track, and the running line of multiple depots of the Wuhan-Guangzhou railway line catenary system.

In 2012, the Beijing-Shanghai high-speed railway was officially opened. Its catenary system is similar to that of the Wuhan-Guangzhou line, and its main line also adopts auto-tensioned elastic catenary technology primarily. The structure and parameter configuration are presented in Table 4.11.

Table 4.10: Catenary structure and parameter configuration of the Wuhan-Guangzhou line.

Technical parameters			Units	Parameter values
Catenary system	Contact line	Contact line type		CTMH-150
		Tension	KN	28.5-30
	Messenger wire	Contact line type		JTMH-120
		Tension	KN	21-23
	Elastic stitch wire	Contact line type		JTMH-35
		Tension	KN	3.5
	Dropper	Contact line type		JTMH-10
	Dropper distance		m	4,8.4,8.4,8.4,8.4,8.4,4
	Span		m	50-55
	Structural height		m	1.6
	Suspension mode			elasticity stitched

Table 4.11: Catenary structure and parameter configuration of the Beijing-Shanghai line.

Technical parameters			Units	Parameter values
Catenary system	Contact line	Contact line type		CTMH-150
		Tension	KN	31.5
	Messenger wire	Contact line type		JTMH-120
		Tension	KN	20
	Elasticity stitch wire	Contact line type		JTMH-35
		Tension	KN	3.5
	Dropper	Contact line type		JTMH-10
	Dropper distance		m	5,8,8,8,8,8,5
	Span		m	50-55
	Structural height		m	1.6
	Suspension mode			elasticity stitched

According to the current structure and parameter configuration of the high-speed catenary system, the developmental trends are as follows:

- Simplify the structure of the catenary as much as possible to improve its reliability.
- Use high-strength, highly conductive, lightweight materials for the catenary.
- Increase the tension of the contact line as much as possible to increase the wave propagation speed on overhead contact lines under certain conditions, and thus satisfy the requirements for high-speed operation.

2 Parameters of catenary irregularity

Due to manufacturing processes, assembly errors, and wear, the catenary is inevitably subject to irregularities. Catenary irregularity can be simply expressed by a sine formula that includes wavelength and amplitude. The exact catenary irregularities should be the combination of all kinds of harmonic waves: that means there is a network spectrum as well as track irregularity. The network spectrum needs a large number of actual measurement statistics to be obtained [4].

The acquisition method for the vertical irregularity of a contact line can be summarized as follows [5]: First, obtain the relationship curve between the spatial frequency k of the corresponding line and the power irregularity of the vertical irregularity of the contact line by performing a maximum entropy spectrum estimation of the vertical irregularity of the contact line. Second, select the smooth function to perform numerical fitting on the obtained power spectral density curve, to achieve the functional form of the power spectral density and facilitate the theoretical calculation. The selected smooth function is presented in equation (4.4).

$$S(k) = e^{\sum_{i=1}^{n} a_i (\ln(k))^i + a_0} \tag{4.4}$$

In the equation, $S(k)$ is the power spectral density of spatial frequency, k *is* spatial frequency, i is a constant ($i = 1,2,\ldots n$), and a_i is the fitting coefficient. The values of i *and* a_i depend on the type and status of the contact line.

Finally, a trigonometric series method is used to simulate and obtain the power irregularity of the contact line vertical irregularity from a random sample of the known power spectral density. The basic model of the trigonometric series method is presented as a formula in equation (4.5).

$$X(t) = \sum_{k=1}^{N} \sqrt{2S(\omega_k)\Delta\omega}\cos(\omega_k t + \Phi_k) \quad (4.5)$$

In equation (4.5), $X(t)$ is the random sample describing the vertical irregularity of the contact line; N is a positive integer; $S(\omega_k)$ is the power spectral density of the contact strip temporal frequency; $\Delta\omega = (\omega_n\text{-}\omega_1)/N$ *is* the frequency interval; ω_1 and ω_n, respectively, are the initial and cut-off frequencies of $S(\omega)$; Φ_k is an independent random variable that is uniformly distributed within $(0,2\pi)$.

$$S(\omega) = \frac{1}{V}S(k) = \frac{1}{V}S\left(\frac{\omega}{V}\right) \quad (4.6)$$

$$\omega_k = \omega_1 + (k - 0.5)\Delta\omega \quad (4.7)$$

In equation (4.6), V *is the* running speed of the train.

According to the above method of obtaining the vertical irregularity of the contact line, we can obtain samples of the vertical irregularity of the contact lines of Wuhan-Guangzhou high-speed railway, Beijing-Shijiazhuang-Wuhan high-speed railway, Zhengzhou-Xian high-speed railway, and Harbin-Dalian high-speed railway successively.

1) Sample of contact line vertical irregularity of the Wuhan-Guangzhou high-speed railway
Obtaining the relationship curve of the spatial frequency and power spectral density by performing the maximum entropy spectrum estimation on the vertical irregularity of the contact line of Wuhan-Guangzhou high-speed railway yields the relationship curve shown in Figure 4.11.

The fitting of the power spectral density curve using a smooth function is shown by the dotted line in Figure 4.11. The correlation coefficients of the curve fitting are presented in Table 4.12. The coefficient is substituted into equation (4.4), and we can obtain the specific form of the smooth function.

The trigonometric series method used is presented in equation (4.5), and the sample of calculated contact line vertical irregularity of the Wuhan-Guangzhou high-speed railway is shown in Figure 4.12.

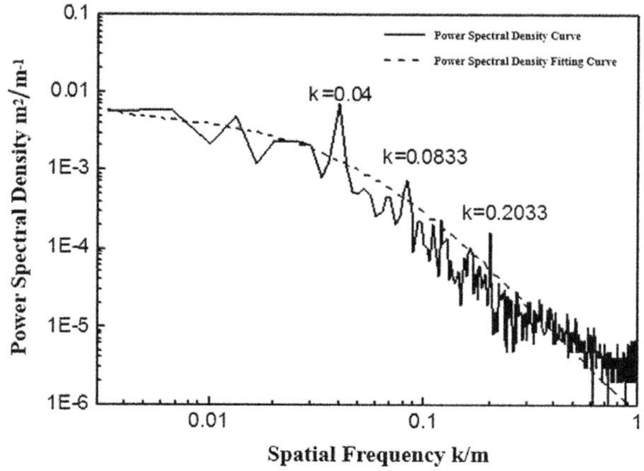

Figure 4.11
Contact line vertical irregularity spectrum of the Wuhan-Guangzhou high-speed railway.

Table 4.12: Fitting coefficients of the power spectral density of the catenary of the Wuhan-Guangzhou line.

a_0	a_1	a_2	a_3	a_4
−14.02096	−2.31114	0.5047	0.21689	0.01846

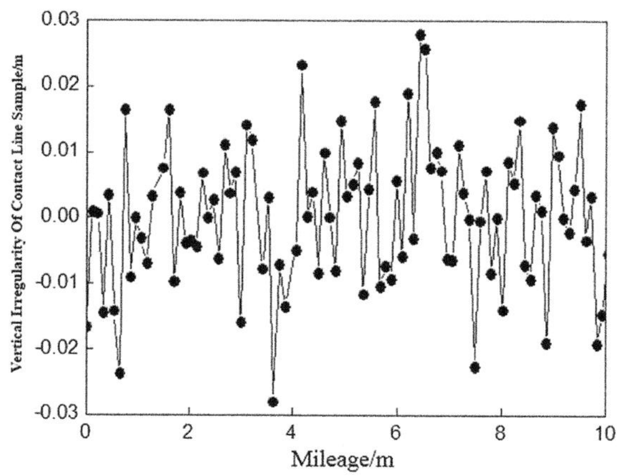

Figure 4.12
Contact line vertical irregularity of the Wuhan-Guangzhou high-speed railway.

2) Sample of contact line vertical irregularity of the Beijing-Shijiazhuang-Wuhan high-speed railway Performing maximum entropy spectrum estimation on contact line vertical irregularity data collected from the Beijing-Shijiazhuang-Wuhan high-speed railway to obtain the relationship curve of the spatial frequency and power spectral density yields the curve shown in Figure 4.13.

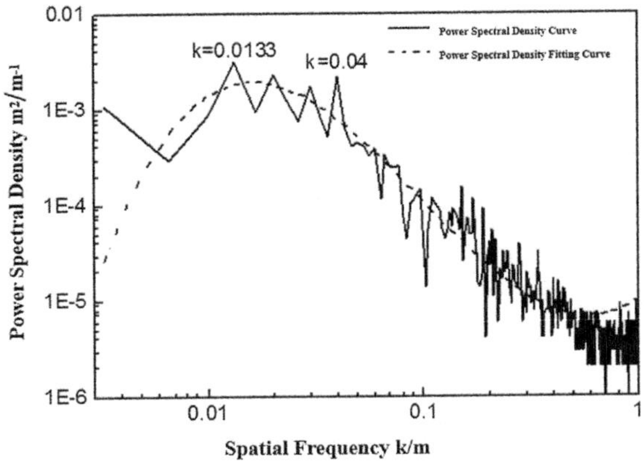

Figure 4.13
Contact line vertical irregularity spectrum of the Beijing-Shijiazhuang-Wuhan high-speed railway.

The smooth function is used to fit the power spectral density curve, shown by the dotted line in Figure 4.11. The correlation coefficients of the curve fitting are presented in Table 4.13.

The trigonometric series method is used to obtain the sample of contact line vertical irregularity of the Beijing-Shijiazhuang-Wuhan high-speed railway shown in Figure 4.14.

Table 4.13: Fitting coefficients of the power spectral density of the Beijing-Shijiazhuang-Wuhan high-speed railway.

a_0	a_1	a_2	a_3
−11.49002	1.62849	1.71538	0.24555

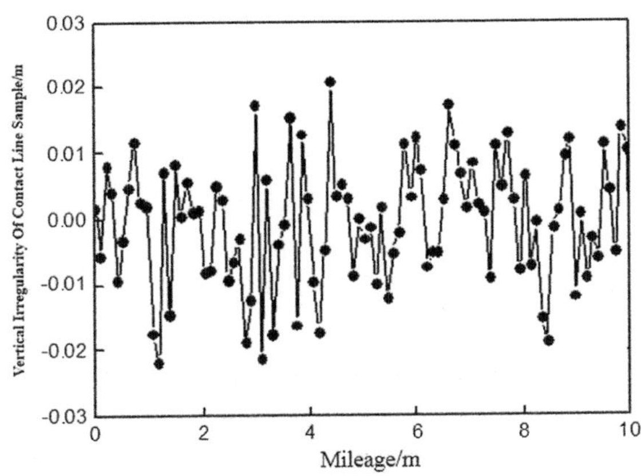

Figure 4.14
Contact line vertical irregularity of the Beijing-Shijiazhuang-Wuhan high-speed railway.

3) Sample of contact line vertical irregularity of the Zhengzhou-Xian high-speed railway Maximum entropy spectrum estimation is performed on the contact line vertical irregularity data collected from the Zhengzhou-Xian high-speed railway to obtain the relationship curve between spatial frequency and power spectral density. The curve is shown in Figure 4.15.

The smooth function is used to fit the power spectral density curve, as shown by the dotted line in Figure 4.15. The correlation coefficients of the curve fitting are presented in Table 4.14.

The trigonometric series method is used to obtain the sample of contact line vertical irregularity of the Zhengzhou-Xian high-speed railway shown in Figure 4.16.

4) Sample of contact line vertical irregularity of the Harbin-Dalian high-speed railway Maximum entropy spectrum estimation is performed on the contact line vertical irregularity data collected from the Harbin-Dalian high-speed railway to obtain the relationship curve between spatial frequency and power spectral density. The curve is shown in Figure 4.17.

The smooth function is used to fit the power spectral density curve, as shown by the dotted line in Figure 4.17. The correlation coefficients of the curve fitting are presented in Table 4.15.

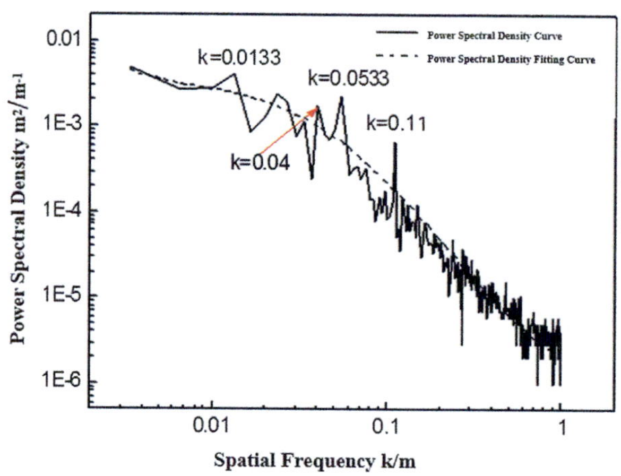

Figure 4.15
Contact line vertical irregularity spectrum of the Zhengzhou-Xian high-speed railway.

Table 4.14: Fitting coefficients of the power spectral density of the Zhengzhou-Xian high-speed railway.

a_0	a_1	a_2	a_3	a_4
−13.01765	−0.93061	1.07807	0.3251	0.02605

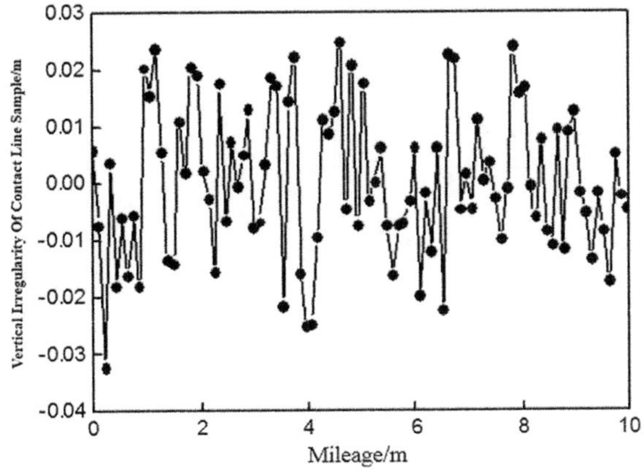

Figure 4.16
Contact line vertical irregularity of the Zhengzhou-Xi'an high-speed railway.

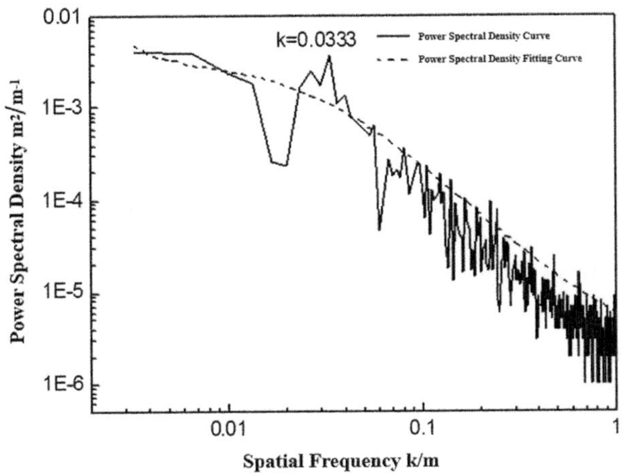

Figure 4.17
Contact line vertical irregularity spectrum of the Harbin-Dalian high-speed railway.

Table 4.15: Fitting coefficients of the power spectral density of the Harbin-Dalian high-speed railway.

a_0	a_1	a_2	a_3	a_4	a_5	a_6
−12.08488	−1.31444	0.04404	−0.13996	−0.05765	−0.00568	−4.13305E-5

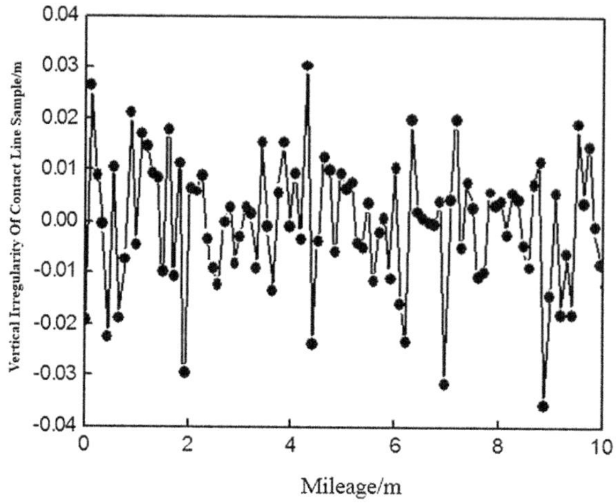

Figure 4.18
Contact line vertical irregularity sample of the Harbin-Dalian high-speed railway.

The trigonometric series method is used to obtain the sample of contact line vertical irregularity of the Zhengzhou-Xian high-speed railway shown in Figure 4.18.

4.1.3.2 Pantograph

The structure and parameters of the pantograph also influence the dynamic performance of the pantograph-catenary system. Presently, there are primarily three types of high-speed pantographs used in China: CRRC Zhuzhou Locomotive Co. Ltd., introduced the TSG19 pantograph (SSS400+), shown in Figure 4.19(A). It is produced by the German company Siemens. Beijing CED Railway Electric Technology Co, Ltd., introduced the DSA380 pantograph, shown in Figure 4.19 (B). It is produced by the German company Stemmann. CRRC Qingdao Sifang Rolling Stock Research Institute Co, Ltd., introduced the CX25 pantograph, shown in Figure 4.19(C). It is produced by the French company Faiveley.

The parameters of a high-speed pantograph are categorized into static and dynamic parameters. The static parameters primarily include pantograph geometry, static contact force, lifting pantograph time, housing force. The static parameters of the TSG19 pantograph are presented in Table 4.16.

The dynamic parameters of the pantograph primarily include mechanical parameters such as lumped mass, stiffness, and damping of the major components. The three lumped masses model of the TSG19 and DSA380 are shown in Figure 4.20. The lumped parameters are presented in Table 4.17 and Table 4.18, respectively. m_1, m_2, and m_3 are the lumped masses of the pantograph head, upper arm, and the lower arm's lumped mass. c_1, c_2, and c_3 are the lumped damp of the pantograph head, upper arm, and lower arm. k_1, k_2, and k_3 are the equivalent stiffness of the pantograph head, upper arm, and lower arm.

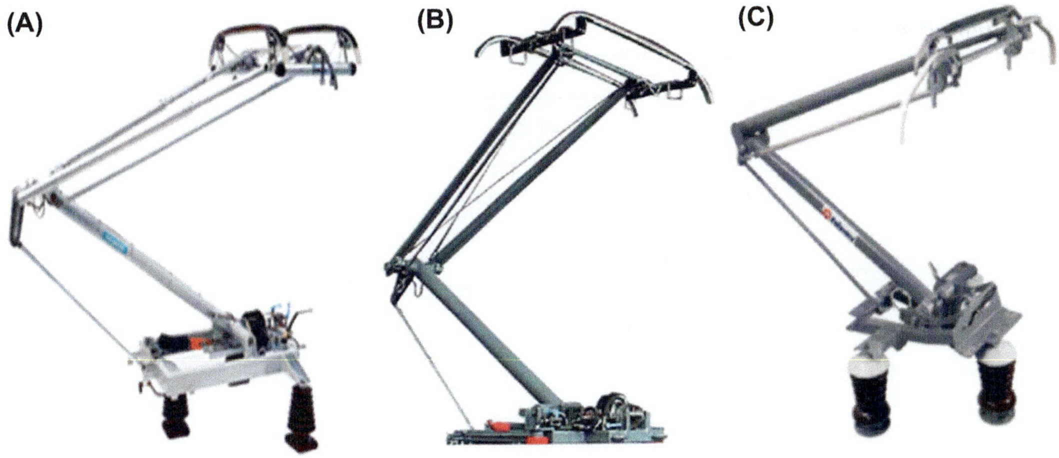

Figure 4.19
High-speed pantographs . (A) TSG 19 (B) DSA 380 (C) CX 25.

Table 4.16: Primary Static Parameters of the TSG19 Pantograph.

Structure type of the pantograph	Single-arm pantograph
contact strip material	integral carbon contact strip
driving method	compressed air
Pantograph raising mechanism	air spring
static contact force	70 N(adjustable)
maximum pantograph height	2800
height at "upper operating position"	2500
housed height	615
maximum collector head length	1950
contact strip length	1573
distance between two contact strips	597

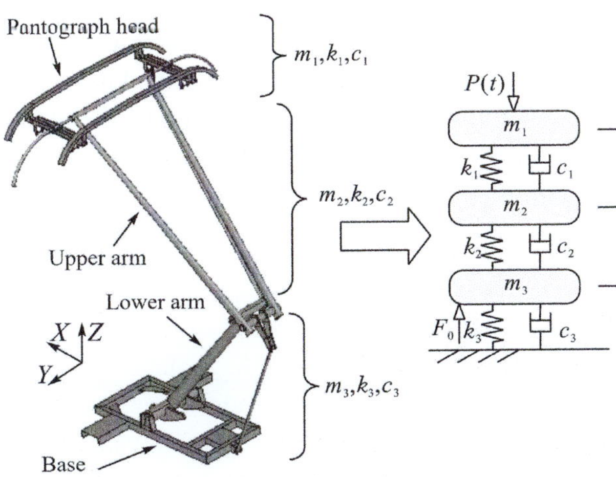

Figure 4.20
Three lumped masses model of the pantograph.

Table 4.17: Equivalent mechanical parameters of the TSG19 pantograph.

m_1 (kg)	6.10	c_3 ($N \cdot s \cdot m^{-1}$)	120
m_2 (kg)	10.15	k_1 ($N \cdot m^{-1}$)	10400
m_3 (kg)	10.30	k_2 ($N \cdot m^{-1}$)	10600
c_1 ($N \cdot s \cdot m^{-1}$)	10	k_3 ($N \cdot m^{-1}$)	0
c_2 ($N \cdot s \cdot m^{-1}$)	0	F_0 (N)	70

Table 4.18: Equivalent mechanical parameters of the DSA380 pantograph.

m_1 (kg)	7.12	c_3 ($N \cdot s \cdot m^{-1}$)	70
m_2 (kg)	6	k_1 ($N \cdot m^{-1}$)	9430
m_3 (kg)	5.8	k_2 ($N \cdot m^{-1}$)	14100
c_1 ($N \cdot s \cdot m^{-1}$)	0	k_3 ($N \cdot m^{-1}$)	0.1
c_2 ($N \cdot s \cdot m^{-1}$)	0	F_0 (N)	70

4.1.4 Parameters of fluid and structure interaction

The high-speed train is running in the atmospheric boundary layer. The train appearance, running ambience, and parameters of fluid flow have a significant influence on performance based on fluid and structure interaction. The fluid and structure interaction parameters primarily include the following:

(1) The geometry of train appearance and operation parameters include the length, width and height of the vehicle, the streamlined length of the head car, vehicle marshaling mode, and running speed. For example, Figure 4.21 shows the geometric model of a China Railway High-speed (CRH) high-speed electric multiple units (EMU), which is a train model with eight cars: including a head car, six middle cars, and a tail car. The total length of the train is 201.4 meters.
(2) The running ambience and parameters of geometric appearance include rail spacing, the height of the bridge from the ground, bridge width and thickness, embankment height, width and slope, shape of the tunnel's section and length of the tunnel. For example, Figure 4.22 shows the simplified geometric model of flat ground, embankment, and bridge under certain conditions.
(3) Parameters of fluid flow include air density, viscosity, ambient temperature, crosswind speed, and angle.

Figure 4.21

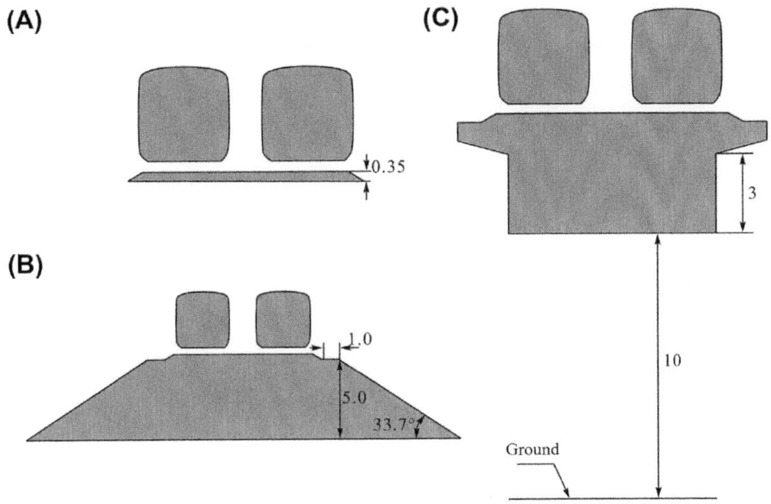

Figure 4.22

For unsteady wind, the wind speed is often decomposed into mean velocity and fluctuating velocity (gust wind speed). There are two main models of fluctuating wind. One is the deterministic model, and the other is the random wind model. In the deterministic model, the fluctuation between the mean wind speed and the extreme wind speed is called fluctuating wind velocity. The loading and unloading of a deterministic fluctuating wind are in accordance with predefined functions, as shown in Figure 4.23. These are mainly the step function (shown in the left column in Figure 4.23) and the pulse function (shown in the right column in Figure 4.23). The exponential function loading used for the fluctuating wind model can be seen in Technical Specification for Interoperability (TSI). The 1-cos form is commonly used in aeroacoustics, and the pulse function is also a commonly used function. In the random wind model, the wind histories are first converted from a fluctuating wind power spectral density function and then applied to the train as aerodynamic loads. For fluctuating wind, it can be a load to both fixed and moving points. The fluctuating wind load to the running train is always obtained based on Cooper theory and the harmony superposition method.

Based on the Von Karman spectrum, the power spectral density expression of the fluctuating wind at a point fixed to the moving train [6] is:

$$\frac{nS_w}{\sigma_w^2} = \left[\frac{4(nL'/\overline{u})}{\left(1+70.8\left((nL'/\overline{u})^2\right)^{5/6}\right)}\right] \times \left[\left(\frac{\overline{w}}{\overline{u}}\right)^2 + \left(1-\left(\frac{\overline{w}}{\overline{u}}\right)^2\right)\frac{0.5+94.44(nL'/\overline{u})^2}{1+70.8\left((nL'/\overline{u})^2\right)}\right] \tag{4.8}$$

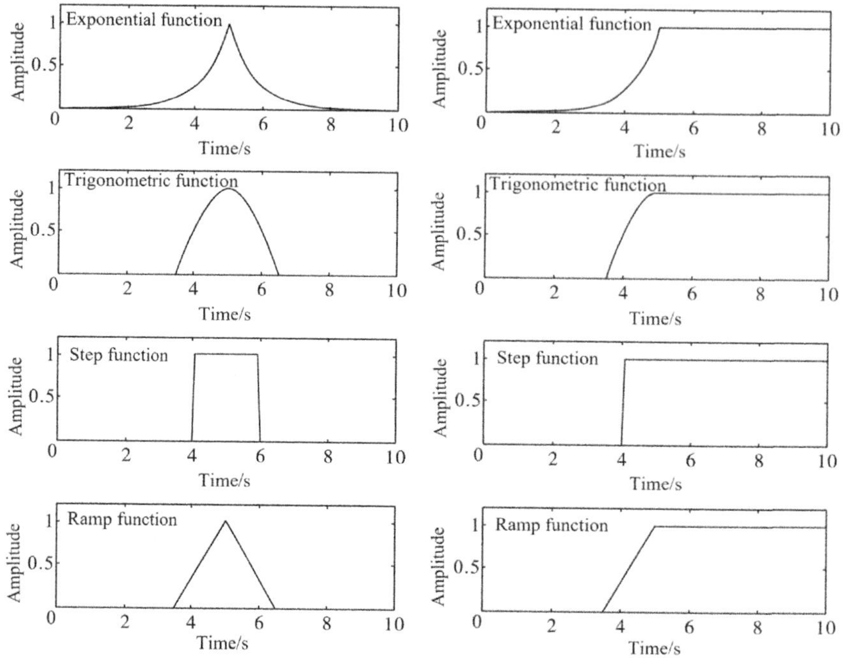

Figure 4.23

where:

$$L' = L_w^x\left((\overline{w}/\overline{u})^2 + 4\left(L_w^y/L_w^x\right)^2\left(1 - (\overline{w}/\overline{u})^2\right)\right)^{0.5} \tag{4.9}$$

In the expression, nS_w/σ_w^2 is the dimensionless power spectral density of the fluctuating wind.

n: frequency

S_w: power spectral density of the fluctuating wind speed

σ_w: the standard deviation of the fluctuating wind speed, $\sigma_w = I_z\overline{w}$

I_z: turbulence intensity

L_w^x: longitudinal turbulence integral scale

L_w^y: lateral turbulence integral scale

$\overline{u}$: mean synthetic wind speed, $\overline{u} = \sqrt{v_t^2 + \overline{w}^2}v_t$: train speed

Figure 4.24 shows the corresponding time history of fluctuating wind when the running speed is 200 km/h and 400 km/h, and the average wind speed is 20 m/s. Figure 4.25 shows

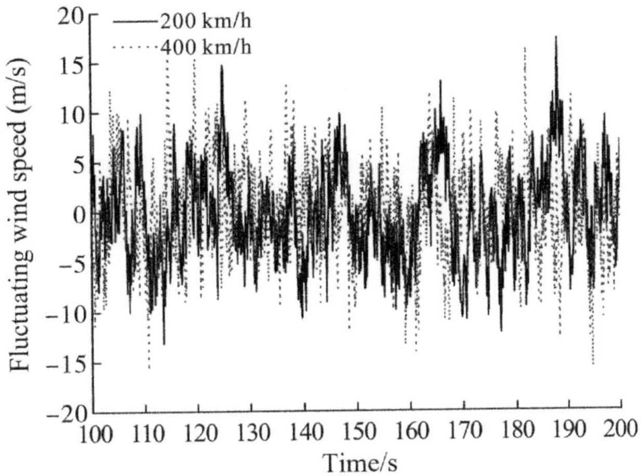

Figure 4.24
Fluctuating wind speed at different running speeds.

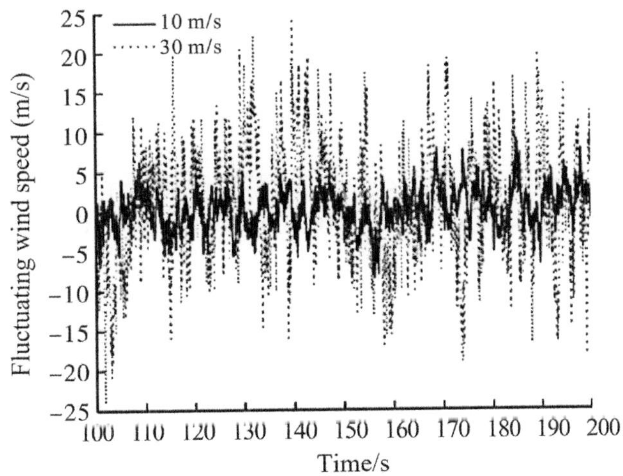

Figure 4.25
Fluctuating wind speed at different wind speeds.

the corresponding time history of fluctuating wind when the running speed is 350 km/h, and the average wind speed is 10 m/s and 20 m/s.

When the average wind speed is constant, the fluctuation degree of the fluctuating wind speed does not change significantly with an increase in running speed; and when the running speed of the train is constant, with an increase in the average wind speed the degree of fluctuation of the fluctuating wind speed increases significantly.

4.1.5 Electro-mechanical coupling parameters

The operation of the train is constrained by the operation control system and the traction power supply system. The performance parameters of train traction/braking, traction drives, and traction power supply have important effects on the electromechanical coupling dynamic performance of the train. The electro-mechanical coupling parameters primarily include the following parts.

4.1.5.1 Train traction/braking performance parameters

High-speed trains have important timeliness in railway passenger transport, of which travel time is an important measurement. Furthermore, travel time also largely depends on the traction/braking performance of the train at the operational control level of the system. Table 4.19 shows the train traction/braking performance parameters.

4.1.5.2 Traction drive system electrical parameters

The traction drive system of a high-speed train is composed of a main drive circuit and a control system. The main drive circuit includes a traction transformer, auxiliary equipment, a traction converter, and a traction motor. Taking the drive system of a CRH EMU as an example, the relevant parameters are shown below [7–9].

(1) Traction transformer and auxiliary equipment

A traction transformer converts a high voltage (25 kV) obtained on a catenary to a voltage suitable for the operation of traction converters and other motors and electric appliances.

Table 4.19: Train traction/braking performance parameters.

Category	Parameter name	Unit
Traction performance	Starting acceleration	(m/s^2)
	Average acceleration on a flat and straight road	(m/s^2)
	Gradeability	(‰)
	Residual acceleration	(m/s^2)
Braking performance	Emergency braking distance	(m)
	Service braking distance at all levels	(m)
	Braking deceleration - Speed characteristics	(m/s^2 - km/h)
Fault/Rescue mode	Balancing speed running on flat and straight roads under power loss	km/h
	Maximum launch ramp in rescue mode (condition train AW0 (empty train) trails AW3)	‰

Table 4.20: Rated performance parameter values of traction transformers.

	Parameter name	Parameter value
Primary winding	Capacity (kVA)	3060
	Voltage (V)	25000
	Current (A)	122
Traction winding	Capacity (kVA)	2570
	Voltage (V)	1500
	Current (A)	857
Auxiliary winding	Capacity (kVA)	490
	Voltage (V)	400
	Current (A)	1225

Its operating principle is the same as that of a common power transformer. The main parameters are presented in Table 4.20.

(2) Traction converter

A pulsed rectifier is a power-side converter of a traction drive system. When a train starts, it acts as a rectifier, and when the train brakes, it acts as an inverter. Therefore, a pulsed rectifier can realize rapid and smooth conversion between traction and braking. An Intermediate DC circuit is the connecting component between a grid-side converter and a motor side converter and functions as an accumulator to satisfy the need of the AC motor for reactive power. A traction inverter is a motor-side converter of a traction drive system. When a train starts, it acts as an inverter, and when the train brakes, it acts as a rectifier. Therefore, a traction inverter can also realize rapid and smooth conversion between traction and brake. Its parameters are presented in Table 4.21.

Table 4.21: Rated performance parameter values of traction converters.

	Parameter name	Parameter value
Input side	AC voltage (V)	1500
	AC current (A)	857
	AC frequency (Hz)	50
	AC power (kVA)	1285
Rectifier	switching frequency (Hz)	1250
	power factor	0.97
Intermediate DC side	DC voltage (V)	3000
	current (A)	432
Outlet side	AC voltage (V)	0 ~ 2300
	AC current (A)	432
	AC frequency (Hz)	0 ~ 220
	AC power (kVA)	1475
Inverter	switching frequency (Hz)	500 ~ 1000
Converter	overall efficiency	0.96

(3) Traction motor

A traction motor is the core component to realize the conversion between electric energy and mechanical energy. It operates as a motor during train traction to convert electric energy into mechanical energy and acts as a generator during braking to convert mechanical energy into electrical energy. Its parameters are presented in Table 4.22.

4.1.5.3 Traction power supply system electrical parameters

A traction substation provides power for a traction power supply system. In the modeling and simulation of a traction substation, the main parameters include the connection modes of the traction transformer, the rated capacity, nominal voltage, short-circuit voltage percentage, and the short-circuit loss. The compensation device parameters of the traction substation include the type and capacity of the compensation device, presented in Table 4.23.

The common traction power supply system includes a direct feeding system, a direct feeding system with return wire, and an AT power supply mode. In different power supply modes, the traction power grid wires are in different spatial distribution patterns. The spatial distribution of the traction power grid wires in the case of a double track AT power supply mode is shown in Figure 4.26. In Figure 4.26, MW, CW, R, PF, PW, and AW, respectively, represent the messenger wire, contact line, rail, positive feeder, guard line, and strengthened line. The impedance of the traction power grid is calculated based on the texture and spatial distribution of the traction power grid wires. The main parameters of the traction power grid are presented in Table 4.24.

In the AT traction power supply mode, the parameters of the AT substation include mainly the AT substation's position, capacity, and short-circuit voltage percentage. The simulation parameters of the section post include the start and end positions of the section post phase segregation, the AT transformer parameters in the section post, and whether supplying power over-zone or not. The main parameters of an AT substation and section post are shown in Table 4.25.

Table 4.22: Rated performance parameters of the traction motor.

Parameter name	Parameter value
Phase	3
Poles	4
Power (kW)	300
Current (A)	106
Rated frequency (Hz)	140
Slip ratio	1.4%
Rated Speed (r/min)	4140
Efficiency	0.94
Power factor	0.86

Table 4.23: Main parameters of traction substation.

Spatial location	km
Power supply mode	All-parallel at power supply
Connection mode of transformer	Three-phase vx connection
Rated capacity	$2\times(50+50)$MVA
Primary/secondary voltage level	220 kV/2 $\times$ 27.5 kV
Open circuit losses	35 kW
Load loss	124 kW
No-load loss	0.28%
Short-circuit impedance	10.5%
The start and end positions of the phase segregation	km
Primary-side phase sequence	ABC

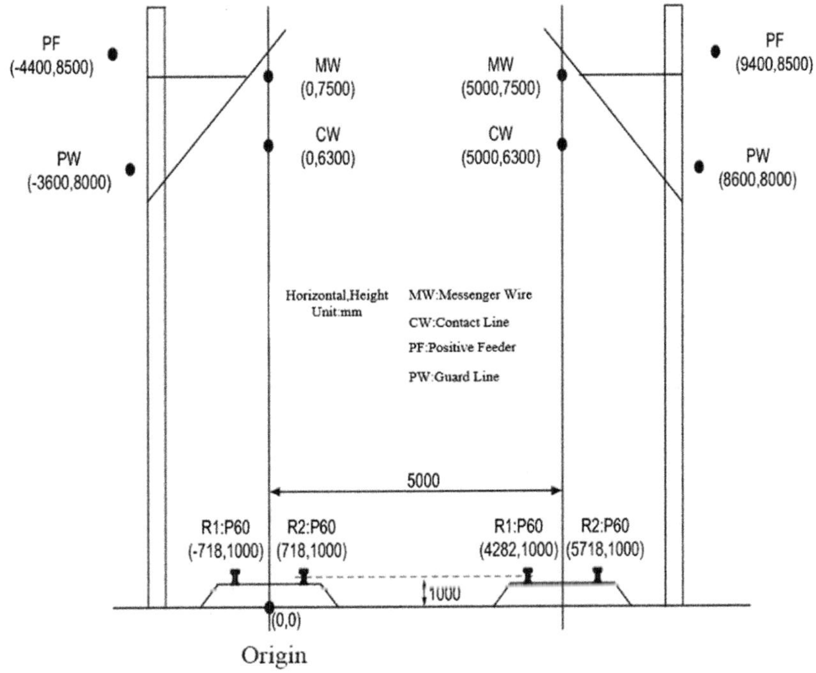

Figure 4.26
Spatial distribution of traction power grid wires in the case of a double track AT.

4.2 Dynamic performance evaluation index

4.2.1 Train vibration evaluation index

4.2.1.1 Motion stability evaluation

Influenced by the creep force between the wheels and the rail, an unstable self-excited vibration called hunting occurs when the vehicle reaches a certain critical speed. Hunting at high speeds manifests as intense transverse vibration between the wheel and the bogie,

Table 4.24: Main parameters of traction power grid wires.

Contact line	CTMH-150
Messenger wire	JTMH-120
Positive feeder	(JL/LB1A-200-26/7) × 2
Guard line	JL/LB1A-125-22/7
Integrated grounding wire	JT70
CPW line interval	1.5 km
Integrated grounding wire buried depth	0.7 m

*Note: The CPW line is a guard line and connecting auxillary track line.

Table 4.25: Main parameters of an AT substation and section post.

AT substation position	km
AT variable capacity	32 MVA
nominal voltage	55/27.5 kV
short-circuit impedance	1.72%
no-load loss	0.086%
open circuit losses	10964 W
load loss	56170 W
rated current	581.8/1163.6 A
section post position	km
the start and end positions of the section post and the phase segregation	km

which seriously threatens the vehicle operation safety. The motion stability of a vehicle is generally characterized by the critical speed, but for a strongly nonlinear rolling stock, the critical speed is defined differently. According to Zhang Weihua [10], the critical speed of the vehicle system includes the linear critical speed, the nonlinear unstable velocity, the nonlinear stable velocity, and track irregularity induced unstable velocity. Specific definitions, symbols, and explanations are presented in the hunting limit cycle diagram in Figure 4.27 and Table 4.26. In general, vC2 < vC < vC0 < vC1. For some vehicles, the nonlinear unstable velocity vC1 may not occur. If the running speed is less than vC2, the vehicle system could be in absolute stability. Therefore, if the conditions permits, the nonlinear stable velocity vC2 should be found and used as an evaluation index of the high-speed train's critical instability speed.

In both tests and practical operation, because the rolling stock always runs along a track with irregularity, it is infeasible to evaluate the stability by determining whether each rigid body of the vehicle system has converged to its balance position. Therefore, the lateral acceleration of a bogie frame is generally used to evaluate the running stability of the bogie.

In accordance with UIC518 and EN1436 standards, it is recommended that vehicle stability be determined based on the following method:

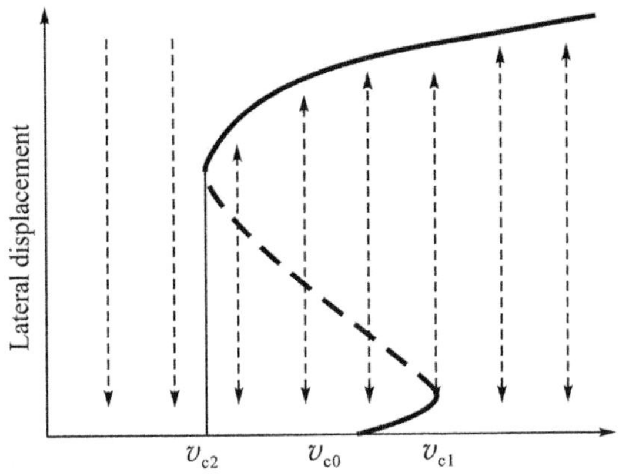

Figure 4.27
Hunting limit cycle.

Table 4.26: Different types of critical speed assessment.

Assessment	Symbol	Definition
Linear critical speed	v_{C0}	As the running speed increases, the critical speed at which the hunting of the wheel occurs under minor disturbances.
Nonlinear unstable velocity	v_{C1}	After a small amplitude instability occurs in a nonlinear vehicle system, and as the running speed increases, the speed at which the periodic motion of the wheel's instability suddenly diverges and jumping occurs.
Nonlinear stable velocity	v_{C2}	After a complete unstable movement of the vehicle system, and as the running speed decreases, the speed at which the hunting of the wheel disappears.
Track irregularity induced unstable velocity	v_C	Under track irregularity, and as the running speed increases, the speed at which there is noticeable hunting.

$$\ddot{y}^{+}_{\text{rms,lim}} = 6 - \frac{m^{+}}{10} \tag{4.10}$$

where m^{+} is the mass of the bogie (including the wheel) measured in tons, and $\ddot{y}^{+}_{\text{rms}}$ is the root mean square value of the lateral acceleration. When this indicator exceeds the limit value, the lateral instability of bogie can be determined.

TSI L84 recommends that stability be assessed using the following method: conduct continuous monitoring and sampling of the lateral acceleration of the bogie frame above the axle box, with 3 to 9 Hz band-pass filtering. If the peak acceleration has reached or exceeded the limit value of $8m/s^2$ more than 10 times in a row, it is determined that the bogie losses lateral stability.

In China, the *Test methods and evaluation criteria for dynamic performance test of 200 km/h EMUs* stipulates that when the lateral acceleration of the frame is filtered by a 10 Hz filter, and the peak value has reached or exceeded the limit of 8 ~ 10 m/s^2 (in accordance with the design of the bogie) more than 6 times in a row, it is determined that the bogie losses lateral stability.

4.2.1.2 Operation stability evaluation

At present, the smooth running of high-speed trains is mainly evaluated by the indexes of stability, comfort, and vibration acceleration.

(1) Ride Index

According to the requirements of GB5599-1985, ride index measurements for each speed level must be collected for at least 10 to 20 segments, and each segment lasts for 18 seconds. The vehicle body acceleration includes several frequency components, and the calculation formula for the ride index of a single frequency is:

$$W = 7.08 \sqrt[10]{\frac{A_i^3}{f_i} F(f_i)} \quad (4.11)$$

where W_i: ride index;

A_i: acceleration (unit: g);

f_i: frequency (unit: Hz);

$F(f_i)$: coefficient of correction (See Table 4.27).

The ride index of acceleration at different frequencies is synthesized according to equation (4.12), and the frequency range used to calculate the ride index is between 0.5 and 40 Hz.

$$W = \sqrt[10]{\sum_{i=1}^{n} W_i^{10}} \quad (4.12)$$

The levels of the ride index of rolling stock are presented in Table 4.28, and the same criteria are applied to both the vertical and the horizontal.

Table 4.27: Coefficient of correction.

Vertical vibration		Horizontal vibration	
0.5 ~ 5.9 Hz 5.9 ~ 20.0 Hz > 20.0 Hz	$F(f) = 0.325f^2$ $F(f) = 400/f^2$ $F(f) = 1$	0.5 ~ 5.4 Hz 5.4 ~ 26.0 Hz > 26.0 Hz	$F(f) = 0.8f^2$ $F(f) = 650/f^2$ $F(f) = 1$

Table 4.28: The levels of ride index.

Stability level	Measurement	Stability index
Level 1	Excellent	W < 2.5
Level 2	Good	2.5 < W < 2.75
Level 3	Qualified	2.75 < W < 3.0

(2) Degree of Comfort

The degree of comfort defined by UIC513 is most often used. It is recommended to divide the time history of the acceleration into segments, each of them is summed with weight, and then the degree of comfort is obtained by adopting a statistic treatment to the weighted summation.

The formula for simplifying calculation methods defined by UIC513 is:

$$N_{MV} = 6\sqrt{\left(a_{XP95}^{W_d}\right)^2 + \left(a_{YP95}^{W_d}\right)^2 + \left(a_{ZP95}^{W_b}\right)^2} \tag{4.13}$$

where N_{MV} is degree of comfort;

a is the root mean square value of acceleration;

$W_{d,}$ W_b are superscripts related to the frequency weighting value of the weighted curves d and b (see Figure 4.28);

iP95 are indices related to the location of the measuring point and the statistical probability, while $i = X, Y, Z$ represents the direction of sensitivity of the acceleration sensors in longitudinal, transverse, and vertical directions, respectively; P indicates that the measuring point is on the carriage floor; 95 indicates that the distribution probability quantile is at 95%.

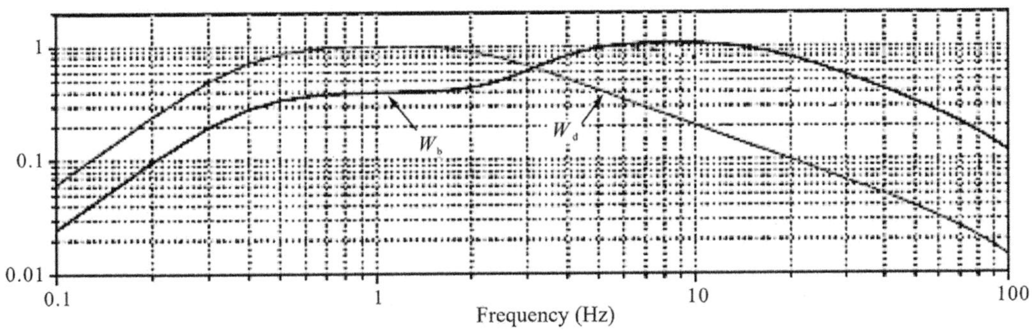

Figure 4.28
Comfort Weight Function.

In the degree of comfort measurement, the samplings of each speed level last five minutes. The effective values of longitudinal, transverse, and vertical accelerations are calculated according to Figure 4.28 for each 5 s interval, and the frequency range is 0.4 ~ 80Hz. Then, the 95% confidence limit of these 60 accelerations for every direction are collected to calculate the final degree of comfort. The levels of comfort are presented in Table 4.29.

(3) Acceleration measurements

Acceleration measurements A_y and A_z are objective measurement parameters of the vibration behavior of the rolling stock. For a high-speed EMU with speeds above 200 km/h, China has specifically made the following provisions for its maximum vibration acceleration:

$$A_{ymax} \leq 2.5\ m/s^2 \tag{4.14}$$

$$A_{zmax} \leq 2.5\ m/s^2 \tag{4.15}$$

In practical application, one or more of the above evaluation indexes can be selected in the evaluation of vibration behaviors of the EMU to make up for shortcomings, thus facilitating a comprehensive evaluation of the vibration behaviors of EMUs.

4.2.1.3 Vibration intensity evaluation

In order to test the vibration behavior of the train, the acceleration of the rolling stock's main components should be evaluated. However, current assessment of operation stability is primarily based on the vibration of the rolling stock, and there is no specific vibration evaluation index for the separate components yet.

China's existing temporary regulation, *Specifications for Strength and Dynamic Performance of Passenger Cars of High-Speed Test Trains* (95J01-M), gives requirements for the level of vibration intensity evaluated based on the acceleration of suspended components below the carriage floor and acceleration of bogie frame components. The components of the rolling stock are required to withstand a longitudinal acceleration of 3.0 g, lateral acceleration of 1.0 g, and vertical acceleration of 3.0 g at the end of the vehicle and 1.5 g at the central portion of the vehicle. Components of the bogie frames are

Table 4.29: Levels of degree of comfort.

Comfort Level	Comfort Index	Evaluation
Level 1	$N < 1$	Very comfortable.
Level 2	$1 \leq N < 2$	Comfortable.
Level 3	$2 \leq N < 4$	Ordinary
Level 4	$4 \leq N < 5$	Uncomfortable.
Level 5	$N \geq 5$	Very uncomfortable.

required to withstand a longitudinal acceleration of 5.0 g, lateral acceleration of 1.0 g, and a vertical acceleration of 3.0 g. These regulations are just guiding requirements and still lack practical application, but they are reasonable attempts in this field.

It has been suggested that a corresponding vibration limit standard be put forward based on the operational states of different components. On the one hand, such a standard value is necessary for the design of high-speed trains. On the other hand, while in operation, the vibration behavior of vehicle components must also be controlled within a reasonable and design allowable range to ensure that the components can safely operate.

4.2.1.4 Wheel-rail relationship evaluation index

On the existing railways and newly built high-speed railways, the dynamic interaction between wheel and rail becomes more intense due to improvements in train running speed. Excessive vertical and lateral contact forces between wheel and rail may not only damage rails, fasteners, sleepers, track slabs, and other components, but may also lead to a sharp increase in track irregularity, increasing the maintenance workload and costs of the line, and significantly impacting operational safety. Therefore, it is necessary to limit the dynamic interaction between wheel and rail.

(1) Vertical Contact Forces Between Wheel and Rail

Japan has clearly specified the maximum of wheel weight in test specifications for speeding up existing lines, and it is required to be smaller than the design load of the track components, including the PC sleeper and track slab. The design load takes into consideration the induced load caused by the wheel-rail dynamic impact due to the flat scar on the wheel tread. Based on research and analysis, the wheel load ratio between static and dynamic situations caused by the flat scar on the wheel tread is about 1.7. After taking into account a particular safety factor, it is 2.0. The flat scar induced dynamic wheel weight can be taken as three times the static weight and used as a design load. The Shinkansen uses P standard load with an axle weight of 170 kN and a design load of 270 kN while the existing line use a load with an axle weight of 160 kN and a design load of 255 kN.

As early as the 1970s, the British Rail (BR) noted that when rolling stocks passed low rail joints, there would be a severe wheel-rail dynamic impact in the rail joint region. The characteristics of the wheel-rail impact force were first discovered through experiments and P1, P2, and their limitation standards were thus defined as:

$$\begin{cases} P_1 \leq 400 \quad \text{kN} \\ P_2 \leq 250 \quad \text{kN} \\ P_1 + P_2 \leq 600 \quad \text{kN} \end{cases} \tag{4.16}$$

Deutsche Bahn in Germany stipulates that, in terms of the track load, the vertical contact force between wheel and rail must not exceed a limit of 170 kN. According to China's *Specifications for Strength and Dynamic Performance of Motor Cars of High-Speed Test Trains* (95J01-L), when motor vehicles pass straight lines, curves, turnouts, and bridges, the peak value of the vertical force acting on the track through each wheel of the guide wheelset is $P_{max} = 170$ kN. The *Vehicle Test Specifications for High-Speed EMU* also specifies $P_{max} = 170$ kN.

(2) Lateral contact forces between wheel and rail

The railway track structure has substantial strength reserve in the vertical direction, while in the lateral direction an exponential strength is proposed to basically ensure the track has a smooth alignment. However, when the line condition deteriorates, excessive wheel-rail lateral force may lead to fastener breakage, track irregularities, and even rail collapse, causing train derailment. The wheel-rail lateral force limit is determined primarily by the transverse force limit that a sleeper track spike can withstand, the track elasticity, and the transverse design loads of the fastener. Under the effect of excessive wheel-rail lateral force, the track fixed using the spike knocked-on sleeper has two failure modes: when the spike is extruded, the track gauge is enlarged; and when the spike is pulled up, it causes rail rotation (squeeze over rails). Japan made a detailed study of this and proposed a lateral force limit of spike extrusion and pulling up, shown in Table 4.30. The first limit is the spike limit yield stress, which is the limit that cannot be exceeded. The second limit is the elastic limit of the spike stress, which is the limit value at which countermeasures should be taken.

For tracks with elastic fasteners, the lateral force between wheel and rail should be less than the fastener lateral design load. The fastener lateral design load limit adopted for the Shinkansen in Japan is 0.4 times the axle weight. According to test results, European and American railways also generally adopt 0.4 times the axle weight as the required limit of the lateral force between the wheel and rail. That is:

$$Q \leq 0.4(P_{st1} + P_{st2}) \tag{4.17}$$

where P_{st1} and P_{st2} are the static load of the left and right wheel, respectively, measured in kN.

Table 4.30: Lateral force limit of Japan Rail (JR) spike extrusion and pulling up.

Type	Lateral force limit of spike extrusion (kN)		Lateral force limit of spike pulling up (kN)
	First limit (yield limit)	Second limit (elastic limit)	
Iron backing plate	$Q \leq 29 + 0.30P$	$Q \leq 19 + 0.30P$	$Q \leq 17.3 + 0.51P$
Iron-free backing plate	$Q \leq 18 + 0.30P$	$Q \leq 12 + 0.30P$	$Q \leq 20.9 + 0.28P$

For the standards of railway lines with sleepers, China's *Specifications for Dynamic Performance Evaluation and Test Evaluation of Railway Rolling Stock* (GB5599-85) also follows the Japanese standard with regards to iron backing plates:

a) spike pulling up spike stress is the elastic limit of the spike stress:

$$Q \le 19 + 0.3P_{st} \tag{4.18}$$

b) spike pulling up spike stress is the limit yield stress:

$$Q \le 29 + 0.3P_{st} \tag{4.19}$$

In equation (4.18) and (4.19), Q is the lateral contact force between wheel and rail (kN), and P_{st} is the static load on the wheel (kN). In recent years, elastic fasteners have been used instead of spikes in the design of railway line and switch in China, and recommendations require that the maximum lateral contact forces of spike pulling up be appropriately adjusted.

(2) Lateral force of wheel axle

The study of jointless track stability shows that excessive lateral force of wheel axle is the main reason for the rail sliding in the lateral direction and the dynamic instability of the jointless track, which results in the rail expansion phenomenon. Therefore, further to ensuring longitudinal and lateral resistance of the line, restricting the maximum lateral force on the wheelset acting on the line is also very important. Table 4.31 shows the limit value formula for the lateral force of wheel axle used in European countries, the United States, UIC, and Japan.

In China's *Specifications for Dynamic Performance Evaluation and Test Evaluation of Railway Rolling Stock* (GB5599-85), the limits of the lateral force of wheel axle that may cause severe deformation of the track are defined as follows:

For wooden sleeper:

$$Q \le 0.85[10 + (P_{st1} + P_{st2})/2] \tag{4.20}$$

Table 4.31: Limit standards for the lateral force of wheel axle used in European Countries, the United States, UIC, and Japan.

Standard used in European Countries and the United States(kN)		Standard used in Japan(kN)	
Limit value $Q \le 10 + P/3$	Recommended value $Q \le 0.85(10 + P/3)$	First limit $Q \le 10 + 0.35(P_{st1} + P_{st2})$	Second limit $Q \le 0.85[10 + 0.35(P_{st1} + P_{st2})]$

For concrete sleeper:

$$Q \leq 0.85[15 + (P_{st1} + P_{st2})/2] \tag{4.21}$$

It can be seen from Table 4.31 that the lateral force of wheelset limits used by European countries, the United States, UIC, and Japan are generally the same, and significantly different from China's standards in earlier years. At present, China's *Vehicle Test Specifications for High-Speed EMU* stipulates that the lateral force of wheel axle evaluation limit adopt the following formula:

$$\mathrm{H} \leq (10 + \mathrm{P_0}/3) \tag{4.22}$$

where $\mathrm{P_0}$ represents the static axle weight.

4.2.1.5 Derailment safety index

Operation safety of vehicles focuses primarily on derailment and overturning of vehicles. Depending on the process, vehicle derailment can be roughly divided into wheel flange climbing derailment, wheel jumping derailment, and sliding derailment. During wheel rotation, the flange gradually climbing up to the railhead and separating the wheel from the track is called wheel flange climbing derailment. This is a common form of derailment and is also the focus of research by scholars in various countries. Generally, the derailment coefficient, rate of wheel load reduction, and overturning coefficient are used to evaluate the safety of vehicle operation.

(1) Derailment coefficient

The derailment coefficient measures the derailment safety by using the ratio of lateral force and vertical force. This approach was first proposed by Nadal. Nadal assumed that there is a wheel that has already started the wheel climb and reached the critical point (i.e., the maximum incline angle of the wheel flange angle). To simplify the analysis, the role of the wheelset attack angle and the wheel-rail contact point-ahead are not considered.

The wheel-rail contact spot separated from the flange, shown in Figure 4.29, is taken as the unit of analysis. The wheel vertical force acting on the contact spot is defined as P, the lateral force is defined as Q, the force acting on the contact spot has a normal force N, and the friction force of the rail on the contact spot that prevents the wheel from sliding downwards is defined as μN. Assuming the flange angle is α, the contact spot is in balance under the action of the forces above it (i.e., the wheel is in a state where it has a downward trend but cannot slide), the wheel climbing condition can be obtained by decomposing forces acting on the contact spot A into components in the normal and tangent directions:

$$\begin{cases} P \sin \alpha - Q \cos \alpha = \mu N \\ N = P \cos \alpha + Q \sin \alpha \end{cases} \tag{4.23}$$

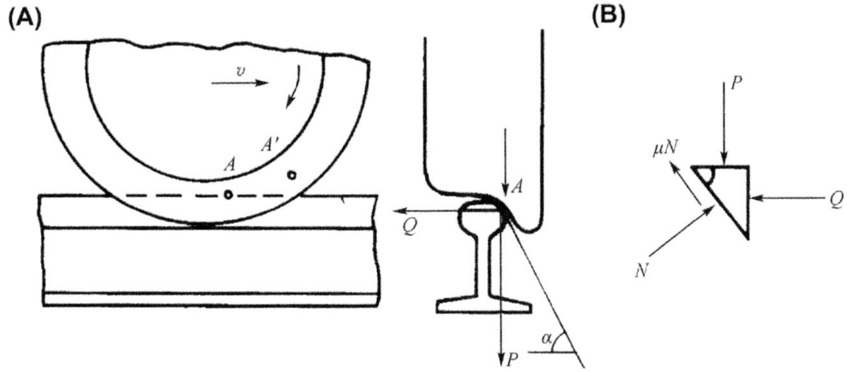

Figure 4.29
Wheel-rail contact and force.

where α is the maximum incline angle of the wheel flange angle, and μ is the friction coefficient of the force between the flange and the rail.

Through the solution of equation (4.23), equation (4.24) can be obtained:

$$\frac{Q}{P} = \frac{\tan\alpha - \mu}{1 + \mu\tan\alpha} \tag{4.24}$$

Equation (4-24) represents the balance of the wheelset at a critical point of wheel climbing. If $\frac{Q}{P}$ is greater than the right-hand term of equation (4.24), the wheel may climb on the rail, and if otherwise, it will slide downward. Therefore, the wheel climbing should meet the following equation:

$$\frac{Q}{P} \geq \frac{\tan\alpha - \mu}{1 + \mu\tan\alpha} \tag{4.25}$$

The ratio $\frac{Q}{P}$ is the wheel derailment coefficient. $\frac{\tan\alpha-\mu}{1+\mu\tan\alpha}$ is the critical value at which the wheel climbing starts, and is also referred to as the critical value of wheel derailment coefficient. The magnitude of the critical value is related to the flange angle α and friction coefficient μ between flange and rail. Figure 4.30 shows the critical values of wheel derailment with different friction coefficients $\mu$and different flange angles α. As can be seen from the figure, the smaller the flange angle α, the greater the friction coefficient μ, and the easier it is for wheel climbing to occur.

The International Union of Railways (UIC) stipulates $Q/P \leq 1.2$, test standards for German ICE high-speed trains stipulate $Q/P \leq 0.8$, the test standards for speeding up the existing railways in Japan also stipulate $Q/P \leq 0.8$, while the North American Railway Association standard stipulates $Q/P \leq 1.0$.

The derailment coefficients formulated by China are presented in Table 4.32. The first limit in Table 4.32 is the eligibility criteria, and the second limit is the criteria that increase the

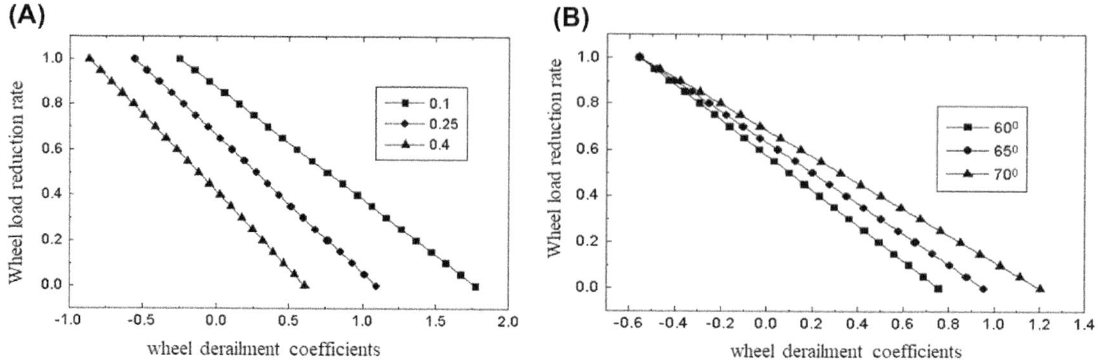

Figure 4.30
Influence of the friction coefficient μ and the flange angle δ_L on train derailment. (A) Influence of the friction coefficient μ (B) Influence of the flange angle δ_L.

Table 4.32: Derailment coefficient safety limit values formulated by China.

Index	GB5599-1985		TB/T2360-93		95J01-L (M)	Vehicle Test Specifications for High-Speed EMU
Derailment Coefficients	First limit $\leq$ 1.2	Second limit $\leq$ 1.0	Good 0.8	Qualified 0.9	$\leq$ 0.8	$\leq$ 0.8

safety margin. Furthermore, the derailment coefficient safety limit values of the climbing wheel in the *Test Methods and Evaluation Criteria for Dynamic Performance of Railway Locomotives* (TB/T2360-93), *Specifications for Strength and Dynamic Performance of Motor Cars of High-speed Test Trains* (95J01-L), *and Specifications for Strength and Dynamic Performance of Passenger Cars of High-Speed Test Trains* (95J01-M) are also presented in Table 4.32. It can be seen that with the development of high-speed trains in China, the Chinese requirements for derailment coefficients are becoming increasingly stringent and is gradually converging with foreign requirements.

The coefficients in Table 4.32 are deduced from the derailment coefficients based on the state of wheel climbing, referred to as wheel flange climbing derailment coefficients (duration of the lateral force is greater than 0.05 second). Furthermore, instantaneous impact between wheel and rail may also cause the wheel to jump on the rail; its coefficient is called the wheel jumping derailment coefficient (duration of lateral force is less than 0.05 seconds). As stipulated in Japan, when the duration of the lateral force between wheel and rail is less than 0.05 s, the wheel jumping derailment coefficient (safety factor of jumping derailment) can appropriately be represented as:

$$\frac{Q}{P} \leq \frac{0.04}{\mathrm{t}} \tag{4.26}$$

where t is the lateral force between wheel and rail, and the unit is s.

$$\begin{cases} \dfrac{Q}{P} \leq \dfrac{0.04}{t} & t < 0.05s \\ \dfrac{Q}{P} \leq 0.8 & t \geq 0.05s \end{cases} \tag{4.27}$$

It can be seen from equation (4.27) that when t = 0.05 seconds (the boundary time between wheel jumping and climbing), and the allowable value of Q/P is 0.8, which is consistent with the limit value of the Nadal wheel flange climbing derailment coefficients of high-speed trains. Therefore, the wheel jumping derailment coefficients and the wheel flange climbing derailment coefficients can be unified by the duration of the lateral force. The derailment coefficients related to the duration of the lateral force are shown in Figure 4.31.

The formula for the wheel jumping derailment coefficients shows that the lateral force of wheel-rail instantaneous impact may be large, but the duration is very short. The wheel does not climb the rail. Although the derailment coefficient at that point may be greater than 0.8, as long as the wheel jumping derailment coefficients limit (0.04/t) is not exceeded, the wheel will not derail. This corresponds with the actual situation.

There is presently no clear regulation on the wheel jumping derailment caused by instantaneous impact between wheel and rail. The main reason is that the mechanism of wheel jumping is still being researched. Consequently, China simply adopts a conservative limit value for judging wheel jumping derailment safety, without considering the duration time of the lateral force between wheel and rail. However, it is believed that with the development of high-speed trains and the gradually intensifying research on high-speed

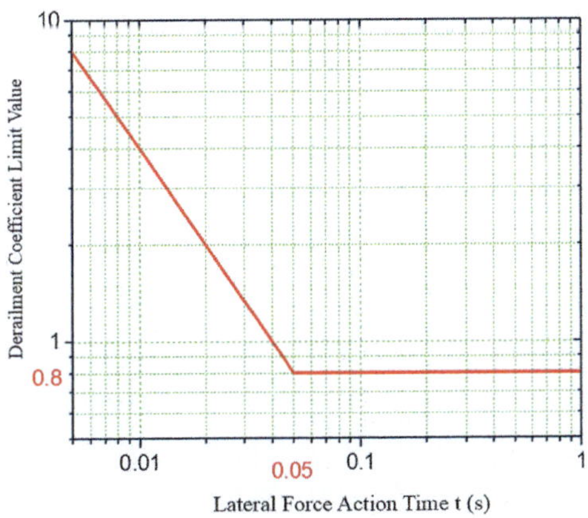

Figure 4.31
Derailment coefficients related to the duration of the lateral force.

trains, the assessment of derailment will be more scientific in the future, and consequently a definite stipulation will be made on the wheel jumping derailment coefficients.

(2) Wheel load reduction rate

The influence of the derailment coefficient on wheelset derailment is analyzed above. This derailment occurs because the lateral force Q_1 is large, but the vertical force P_1 is small. However, in practical operation, it is found that there is also a possibility of derailment when the lateral force is not large, but the load on one wheel is significantly reduced. The heavy wheel load reduction situation is analyzed as follows:

If the axial lateral force H is very small (let $H= 0$), P_2 is large and P_1 small, i.e., $P_2 >> P_1$. For some reason, the left wheel flange has already contacted the rail at the maximum angle of the wheel flange. Due to the large tread friction $\mu_2 N_2$ of the right wheel, the left wheel can still maintain a critical state of derailment.

The average wheel-rail vertical force P and wheel load reduction amount ΔP can be expressed as the following formula:

$$\begin{cases} P = \dfrac{1}{2}(P_1 + P_2) \\ \Delta P = \dfrac{1}{2}(P_2 - P_1) \end{cases} \tag{4.28}$$

where P is the average wheel-rail vertical force and ΔP is the wheel load reduction amount.

Substituting equation (4.28) into equation (4.25):

$$\frac{\Delta P}{P} \geq \frac{\dfrac{\tan\alpha_1 - \mu_1}{1 + \mu_1 \tan\alpha_1} - \dfrac{\tan\alpha_2 + \mu_2}{1 - \mu_2 \tan\alpha_2}}{\dfrac{\tan\alpha_1 - \mu_1}{1 + \mu_1 \tan\alpha_1} + \dfrac{\tan\alpha_2 + \mu_2}{1 - \mu_2 \tan\alpha_2}} \tag{4.29}$$

In equation (4.29), $\frac{\Delta P}{P}$ is the wheel load reduction rate, and the right value is called the critical value of the wheel load reduction amount.

When the wheel load reduction rate exceeds its critical value, there is a tendency for the wheelset to derail. The Japan Rail (JR) standard uses 0.8 as the standard value of the dynamic wheel load reduction rate, while Germany and the United States use a limit value of 0.9. China's *Railway Vehicle Dynamic Performance Evaluation and Test Appraisal Standards* (GB5599-1985), *Specifications for Strength and Dynamic Performance of Motor Cars of High-Speed Test Trains (95J01-L)* and *High-Speed EMU Vehicle Test Standard* (2008) stipulate that the wheel load reduction rate should meet the standard value, shown in Table 4.33.

Table 4.33: China's wheel load reduction rate safety limit.

Index	GB5599-1985		95J01-L (M)	High-Speed EMU Vehicle Test Standard
Load Reduction Rate	First limit $\leq$ 0.65	Second limit $\leq$ 0.60	$\leq$ 0.6	Quasi-static $\leq$ 0.65 Dynamic $\leq$ 0.8

It should be pointed out that the derailment coefficient and wheel load reduction rate are deduced from the equilibrium condition of the wheel vertical force and lateral force under the conditions of the wheelset lateral force, $H > 0$ and $H = 0$, respectively, which are indicators for evaluating wheel derailment in two different situations. It is impossible to ensure safety through the wheel load reduction rate alone. Only under the premise that the derailment coefficient is used together with the wheel load reduction rate can this value fully ensure the safety of the vehicle operation.

(3) Overturning coefficient

The overturning coefficient is used to evaluate whether the vehicle will overturn on one side in the most unfavorable combination of lateral wind, centrifugal force, and lateral inertial force. The wheel-rail pressure of the outside rail of the vehicle is set as P_2, the wheel-rail pressure of the inside rail of the vehicle as P_1, and the overturning coefficient D is defined as:

$$D = \frac{P_2 - P_1}{P_2 + P_1} \tag{4.30}$$

When the wheel load at one side of the vehicle is reduced to zero ($P_1 = 0$), the vehicle reaches the critical state of overturning. At this time, $D = 1$. Therefore, to prevent the vehicle from overturning, the condition $D < 1$ must be met. China's *Specifications for Dynamic Performance Evaluation and Test Evaluation of Railway Rolling Stock* (GB5599-85) and *Specifications for Strength and Dynamic Performance of Passenger Cars of High-Speed Test Trains* (95J01-M) stipulate that $D < 0.8$.

4.2.2 Evaluation Index of the interaction between pantograph and overhead contact line

The main goal of studying the pantograph-catenary relationship is to ensure that the pantograph has good current collection performance. However, how to evaluate the current collection performance of the pantograph-catenary system is a problem that must first be solved within pantograph-catenary relationship research. In Europe and Japan, standards related to current collection have been widely used as part of the railway series standards.

On the basis of the foreign evaluation systems of the pantograph-catenary relationship, combined with the actual operation of the pantograph-catenary system, China also gradually formed an evaluation system of the pantograph-catenary relationship to address the requirements of Chinese railway operation. The evaluation primarily includes the dynamic contact force, percentage loss of contact, the hard spot (vertical acceleration), and dynamic contact line height.

4.2.2.1 Dynamic contact force

The contact force of the pantograph-catenary system refers to the dynamic contact force between the contact strip and contact line. The contact line and the contact strip perform relative sliding frictional motions and tare a friction pair. The interaction force between pantograph and catenary can be decomposed into the longitudinal (in line direction) force, transverse force, and vertical force. The longitudinal force primarily includes the friction force between the contact strip and the contact line, and the longitudinal impact force caused by contact line defects from manufacture, construction, and operation. The transverse force is primarily the friction force formed by the relative movement of the contact line near the contact strip center due to catenary stagger. The vertical force is the contact force, which primarily affects the contact line uplift and the wear of contact lines and contact strips. The contact line uplift influences the safety of the pantograph-catenary system, and the wear of contact lines and contact strips affects the service life of the catenary and the contact strips. To achieve good contact between the pantograph and the catenary, the contact force must be kept within a limited range. Large contact force furthers the wear of the contact wire and causes wire fatigue. On the other hand, contact force that is too small gives rise to arcing. Therefore, the contact force describes the contact state between the pantograph and the contact line under high-speed sliding contact conditions of the pantograph and catenary and is an important indicator for evaluating the current collection quality.

The evaluation of the contact force is usually characterized by its statistical value. The maximum, minimum, and standard deviation of the contact force basically characterize the current collection quality. A maximum of the contact force that is too large will increase the wear, while a minimum contact force that is too small will cause contact loss. An excessive standard deviation of the contact force will cause abnormal changes in contact force, directly affecting the stability of the current collection. The mean value of the contact force is also an important indicator for evaluating current collection quality, which can characterize the integrated state of the contact force during the running of the train.

Therefore, the evaluation criteria for each indicator in the *Electric Traction Overhead Contact Lines* (EN50119:2001) are as follows:

1) Maximum (N): $F_{\max} = \begin{cases} 300 & (v \leq 200km/h) \\ 350 & (v > 200km/h) \end{cases}$;
2) Minimum (N): $F_{\min} > 0$;
3) Mean (N): $F_m \leq 0.00097v^2 + 70$ (v is running speed in *km/h*).

4.2.2.2 Contact loss

Contact loss is the mechanical disengagement of the pantograph from the contact line. With good interaction between the pantograph and the overhead contact line, contact loss spark does not occur or is at a minimum during pantograph and catenary operation. The contact loss spark is the ultimate performance indicator of current collection quality and state, which reflects the effects of various running states on the current transmission. There are many causes of continuous sparks during the operation of the pantograph. For example, when a hard bending or a wave bending occurs in the contact line, the current transmission at the contact point of the contact line and the contact strip is discontinuous. Furthermore, an uneven contact line, uneven static stiffness of the contact line, and the mismatch of the pantograph-catenary system dynamic parameters will also cause contact loss sparks. Because there are many causes of the contact loss sparks, the incidence of sparks is a comprehensive and important evaluation indicator of the pantograph-catenary relationship. The main parameters for evaluating contact loss sparks include maximum contact loss time, numbers of sparks per unit distance, and the percentage loss of contact.

The contact loss duration characterizes the duration of each mechanical disengagement. The number of sparks indicates the frequency of mechanical disengagement. Evidently, under the same conditions, the longer the contact loss duration and the more times the contact loss occurs, the weaker the current collection performance. In general, contact loss duration with a time of 0.1 to 60 ms is categorized as small contact loss, and the contact loss duration with a time of more than 100 ms is considered large contact loss.
The percentage loss of contact is a comprehensive indicator of the contact loss and can be expressed as:

$$s = \frac{\sum t_i}{T} \cdot 100\% \tag{4.31}$$

where t_i is the duration of mechanical disengagement of the pantograph from the catenary, and T is the total time of detection.

The percentage loss of contact indicates the contact state between the pantograph and the catenary, which is generally required to be less than 5%, with a maximum contact loss duration of less than 100 ms. In this way, the pantograph-catenary system will have good current collection quality. In summary, the evaluation criteria for contact loss sparks can be summarized as:

1) Maximum contact loss duration should be less than or equal to 100 ms;
2) The number of contact loss sparks should be less than or equal to one per 160 m;
3) The percentage loss of contact should be less than or equal to 5%.

4.2.2.3 Hard spot

The hard spot is a collective term for an inhomogeneous state of a catenary suspension. The inhomogeneity of the catenary can be caused by several factors, such as some parts of the catenary suspension and contact line may have hard bending, bottom surface distortion, the installation of a section insulator and or mid-point anchor do not meet design requirements, dropper clamp and electrical connection clamp tilt, and excessive contact line slope. During high-speed operation of the train, these parts are abnormally raised or lowered and may cause pantograph impact (i.e., sudden changes in force, displacement, velocity, or acceleration will occur at these parts). Hard spots accelerate abnormal wear and impact damage on contact lines and contact strips. Continuous sparks and abnormal changes in contact force peaks often appear in the hard spot position, affecting the normal contact and current collection of the pantograph-catenary system. Serious hard spots may impact train safety. Thus, the hard spot is also an important indicator for evaluating the pantograph-catenary relationship.

The hard spot index is mainly evaluated by impact acceleration of the pantograph during operation. This acceleration has two directions: horizontal direction and vertical direction. Generally, the vertical acceleration is used to evaluate hard spots on the pantograph. The criteria for evaluating hard spots primarily include:

1) when V (running speed) $<$ 200 km/h, the vertical acceleration should be less than or equal to 490 m/s^2 (50 g);
2) when 200 km/h $\leq$ V $<$ 300 km/h, the vertical acceleration should be less than or equal to 588 m/s^2 (60 g);
3) when 300 km/h $\leq$ V $\leq$ 350 km/h, the vertical acceleration should be less than or equal to 686 m/s^2 (70 g).

4.2.2.4 Dynamic contact line height

The dynamic contact line height of the contact line is primarily characterized by the vibration displacement of the contact strip. It describes the dynamic changes in contact line height when the pantograph slides at high speed. By detecting this index, it is possible to evaluate the irregularity and dynamic elasticity of changes in the contact line, and whether the contact line height meets the catenary design requirements, especially the slope requirements of the contact line. The main evaluation parameters include the maximum dynamic contact line height D_{max}, the minimum height D_{min}, and the maximum vibration amplitude of the contact line $D_{max}-D_{min}$ in one span. The corresponding evaluation criterion is as follows:

Maximum vertical amplitude of the contact line $D_{max}-D_{min} \leq 150$ mm.

4.2.2.5 Uplift displacement of the contact line

In the static state, the contact line is raised due to the static uplift force of the pantograph. The uplift displacement, in this case, is called static uplift displacement. During high-speed operation, the uplift displacement of the contact line caused by the pantograph-catenary interaction is referred to as vibration uplift displacement. Under the influence of the pantograph, the sum of static uplift displacement and vibration uplift displacement forms dynamic uplift displacement. The uplift displacement index of the contact line primarily refers to the dynamic uplift displacement. The evaluation method involves checking whether the value exceeds a certain limit. The specific evaluation is:

The uplift displacement of the contact line should be less than or equal to 120 mm.

4.2.3 Evaluation index of fluid-structure interaction

4.2.3.1 Requirements for pressure inside and outside of the carriage, and airtightness of the vehicle

The transient pressure in the carriage has an influence on passenger riding comfort. The evaluation criteria [11] are as follows:

For single track tunnel: pressure inside $<$ 0.80 kPa/3 s

For double track tunnel: pressure inside $<$ 1.25 kPa/3 s

The pressure difference between the inside and outside of the carriage has a significant influence on the strength of the vehicle. For a railway vehicle, the pressure difference between the inside and outside of the carriage should be less than 4000 Pa [12].

4.2.3.2 Evaluation criteria for micro pressure wave at tunnel exit

The maximum allowable peak values of the micro pressure wave at a tunnel exit [11] are presented in Table 4.34.

4.2.4 Electro-mechanical coupling evaluation

4.2.4.1 Energy conversion efficiency evaluation

When the train is running under traction conditions, the traction drive system, through the pantograph, converts the electric energy of the traction power supply system into the mechanical energy that drives the train's operation, and the mechanical energy is

Table 4.34: Criteria for micro pressure wave.

	Conditions of the entrance		Calculating datum points	Peak value of micro pressure wave
1	No building on the entrance (The building is more than 50 m away from the entrance)		20 m from the entrance	$< 50Pa$
2	There is a building on the entrance	Buildings have no special environmental requirements	The building	$< 20Pa$
		Buildings have special environmental requirements		As required

converted into electrical energy under braking conditions. Combined with the mechanical braking system, the traction or braking of the train is completed with the effect of the wheel-rail coupling facilitating the train operation on the track. Thus, the train operation is a process of interconversion between electric energy and mechanical energy. Each device inevitably has a specific energy loss in the process of energy transmission or conversion, which is described by the electromechanical efficiency of the train and the efficiency of each component. The electromechanical efficiency of an entire vehicle is an equivalent concept and is a product of the efficiency of transformers, rectifiers, inverters, motors, gear drives, etc. Taking the CRH EMU as an example, the efficiency indexes of the entire vehicle and various transmission components are as follows:

(1) The efficiency of the entire vehicle should not be lower than 0.85,
(2) The traction transformer efficiency should not be lower than 0.95,
(3) The traction converter efficiency should not be lower than 0.96,
(4) The efficiency of the traction motor at the continuation point should not be lower than 0.94.

4.2.4.2 Electrical characteristics evaluation index of the traction drive system [13]

1) Power supply mode: single-phase AC 25 kV, 50 Hz;
2) Power supply quality: The highest voltage is 29 kV, the lowest voltage is 22.5 kV, and others conform to the GB1402 *Mainline Traction Power AC Traction Voltage Standard*;
3) Harmonic values are in accordance with Table 4.35;

Table 4.35: Target value of the higher harmonics (unit: %).

Voltage	5 times	7 times	11 times	13 times	17 times	19 times	23 times	over 23 times
6.6kV	4.0	2.8	1.8	1.5	1.1	1.0	0.87	0.80
22 kV or more	6.7	4.8	3.1	2.6	1.9	1.8	1.5	1.4

4) Equivalent disturbing current (Jp) is less than 4.0 A (8 unit grouping);
5) When the EMU is at the rated tractive power, the total power factor (λ) at the grid-side is not less than 0.97 (except for the rated load and auxiliary circuit);
6) The total harmonic distortion (THD) of the traction transformer primary side is less than 10% (Condition: excellent power without the influence of other vehicles and equipment and the rated load).

4.3 Dynamic performance of the high-speed train's coupling system

Traditional vehicle dynamics primarily studies the influence of the vehicle's structure and suspension on the vehicle's dynamic performance. The interactions between vehicles, vehicle and rail, vehicles and the flow field, pantograph and catenary, are not considered. Increase in EMU running speed not only necessitates new dynamic performance design requirements at higher levels but also requires closer evaluation of the increased friction and wear from wheel-rail and pantograph-catenary relations that could cause failure or impact operation safety. High-speed airflow is no longer just the running resistance but also affects the operation performance of EMU. Therefore, technological innovation for the high-speed EMU requires the study and resolution of not only the dynamic behavior of the EMU but also the coupling between the EMU and other systems. Then system optimization and matching of the high-speed EMU and its coupled large systems will be realized. Thus, it is absolutely necessary to study the dynamics of coupled large-scale systems of the high-speed EMU.

4.3.1 Dynamic characteristics of interaction between vehicles

For the single vehicle model, the influence of the connecting device between vehicles (including draft gears, absorber, and windshield) cannot be considered. However, the connecting device between vehicles impacts the dynamic performance of the vehicle. The CRH EMU is taken as an example to analyze the influence of the connecting device between vehicles (including draft gears, absorber, and windshield) on the dynamic performance of the vehicle (aerodynamics effects are not considered).

The calculation cases for motion stability and operation stability are straight railways, and the calculation cases for safety are curved tracks (radius of the curve is R 7000 m, and superelevation is 180 mm). The track spectrum of the Beijing-Tianjing line is applied, and the calculation results are presented in Figures 4.32 to 4.39.

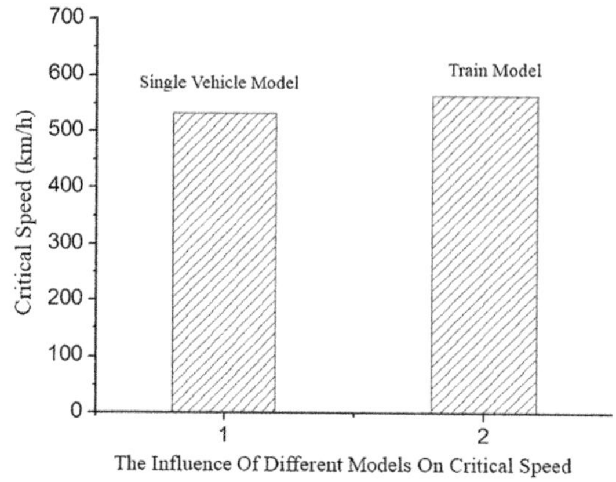

Figure 4.32
Comparison of critical speed.

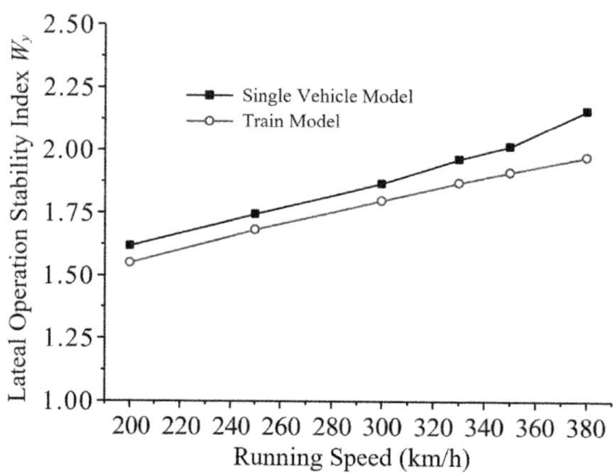

Figure 4.33
Comparison of lateral operation stability.

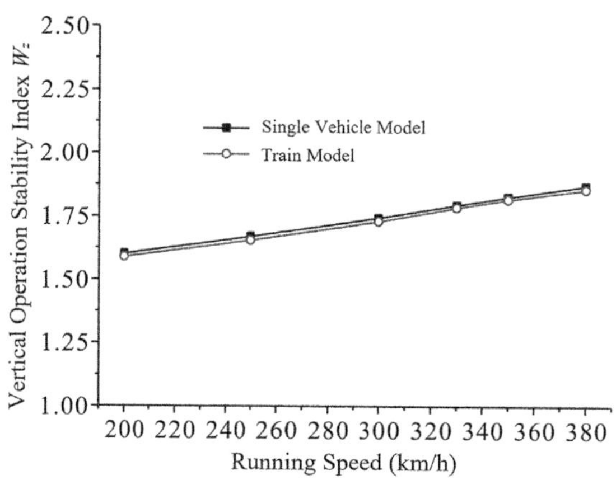

Figure 4.34
Comparison of vertical operation stability.

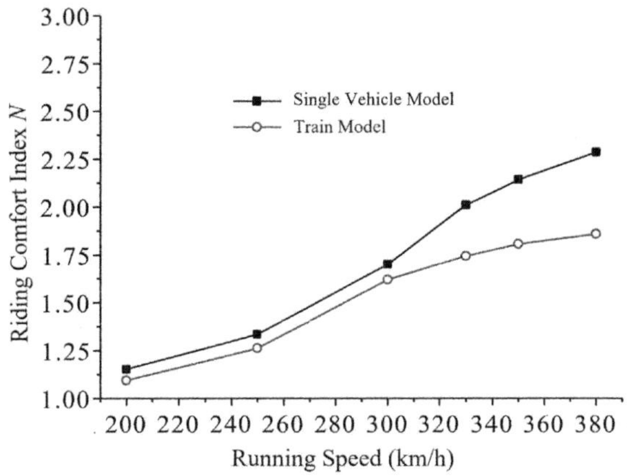

Figure 4.35
Comparison of riding comfort.

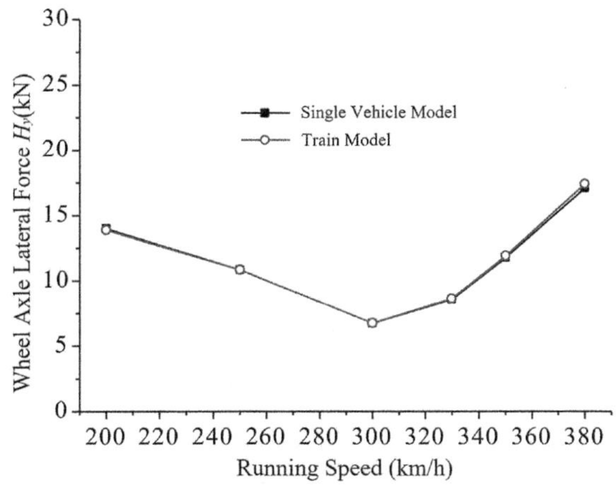

Figure 4.36
Comparison of wheel axle lateral force.

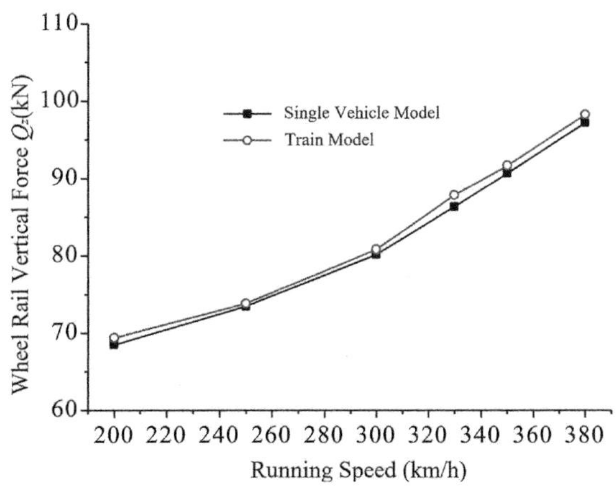

Figure 4.37
Comparison of wheel-rail vertical force.

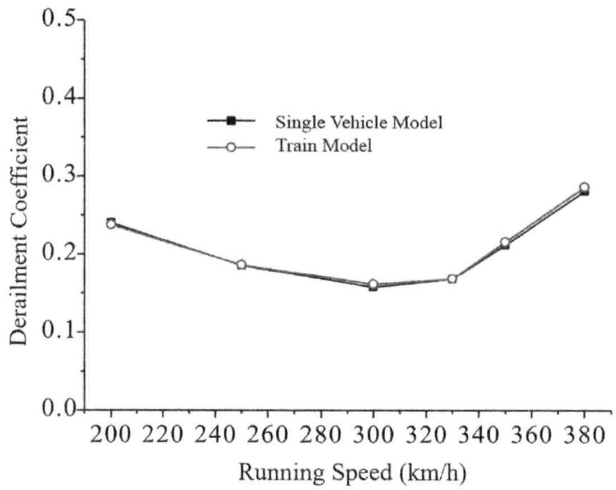

Figure 4.38
Comparison of derailment coefficient.

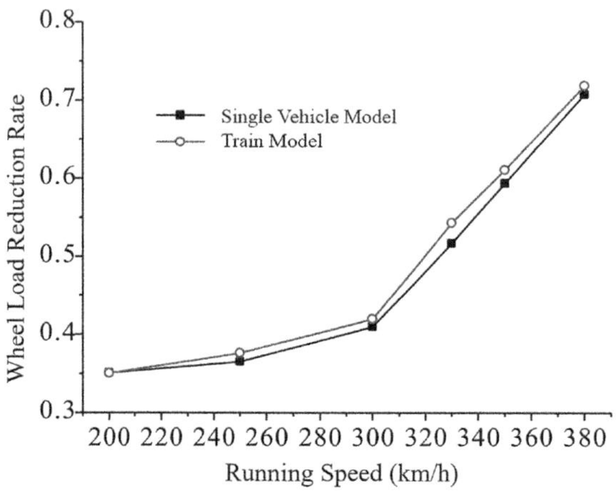

Figure 4.39
Comparison of wheel load reduction rate.

Based on these calculation results, it can be concluded that:

(1) The critical speed of the train model is higher than that of the single vehicle, which is increased by approximately 30 km/h (a 6% increase).
(2) The lateral operation stability of the train model is significantly better than that of the single vehicle model and can be increased by approximately 5% to10%.
(3) The riding comfort of the train model is much better than that of the single vehicle model. The higher the speed, the greater the improvement. The riding comfort can be increased by 5% at speeds below 300 km/h and by 10–20% at speeds above 300 km/h.

(4) The vertical operation stability of the train model and the single vehicle model have few differences.
(5) There is little difference between the safety indexes of the train model and single vehicle model (wheel axle lateral force, wheel-rail vertical force, derailment coefficient, and wheel load reduction rate).

Based on the calculations on the train model and the single vehicle model, it is found that the connecting device between vehicles (draft gears, absorber, and windshield) has little impact on vertical operation stability and safety, and has greater impact on lateral operation stability, riding comfort, and the critical speed of the system. To accurately analyze the dynamic performance of the train system, priority should be given to the train dynamic model for analysis.

4.3.2 Dynamic characteristics of interaction between wheel and rail

Taking China's CRH EMU as an example, the single vehicle model is adopted for calculation, with a primary focus on the influence of the rigid track, elastic ballast track, and elastic ballastless track on the wheel-rail coupling dynamics. The calculation and analysis results take the wheel-rail normal force as an example to analyze whether the vehicle is on a straight track and whether it is not affected by the input excitation of other track irregularities.

Fig. 4.40 shows the time histories of the wheel-rail normal forces for the rigid track, elastic ballast track, and elastic ballastless track. For the rigid track, the model takes only

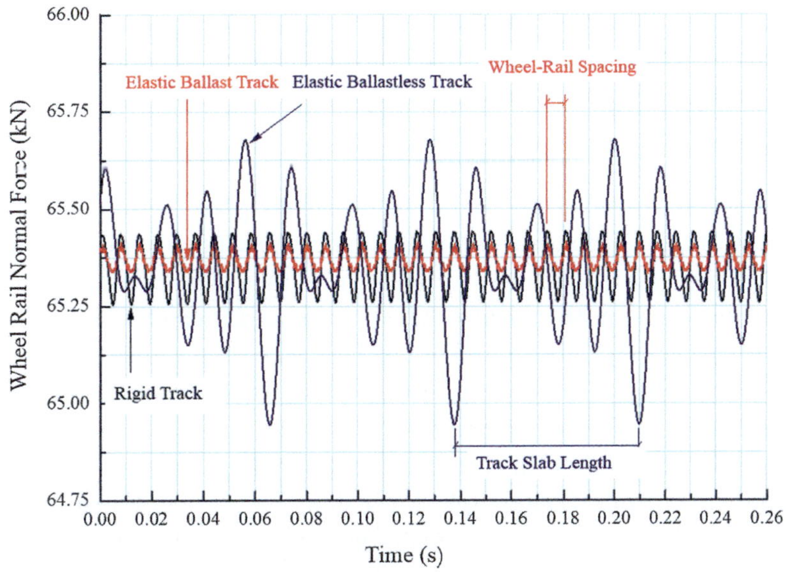

Figure 4.40
Positive pressure between wheel and rail under different track models.

the sleeper, ballast, or track slab as a rigid foundation. The elastic supporting effect of the rail elasticity and fastening system are considered.

It can be seen from Fig. 4.40 that the mean values of the positive pressure between wheel and rail for the three orbital models are close at approximately 65.35 kN. In the absence of track irregularity excitation, the fluctuation of the positive pressure between wheel and rail is very small, especially in terms of the elastic ballast track. It can be seen that there is significant track cross-frequency fluctuation from the positive pressure between wheel and rail under the rigid track. This is the practical process of the vehicle passing through a track reflected in the excitation interaction between the vehicle and track of the sliding model, and it also verifies the accuracy of the slip model from another perspective. The positive pressure between wheel and rail for the elastic ballastless track is characterized by a significant periodicity fluctuation. Based on the analysis of the parameters such as the period of fluctuation, running speed (300 km/h), and the length of the track slab, it is found that the fluctuation period of positive pressure between wheel and rail for the elastic ballastless track corresponds to the track slab passing frequency. This characteristic frequency is clearly reflected in track spectrum test data on the Beijing-Tianjin-Hebei-Guangzhou-Guangzhou passenger dedicated line.

4.3.3 Dynamic pantograph-catenary interaction characteristics

4.3.3.1 Dynamic performance of the pantograph-catenary system

When the catenary is under a steady condition, the pantograph slides along the catenary at a definite speed, and the parameters that characterize the dynamic performance of the pantograph-catenary system are obtained through analysis. Typical dynamic performance parameters include uplift displacement of the catenary, contact force, pantograph vibration acceleration, percentage loss of contact, and dynamic stress.

(1) Uplift Displacement

When the pantograph slides along the contact line at high speed, corresponding dynamic uplift displacement of the contact line will occur under the influence of the pantograph. Fig. 4.41(A) shows the dynamic lifting displacement of the catenary at the pantograph contact point when a typical pantograph passes through, and it can be seen that the uplift displacement of the pantograph contact point changes according to the span period. The maximum value of the uplift displacement of the contact line in each span occurs at the mid-span position, and the minimum value occurs near the positioner. Fig. 4.41(B) shows the dynamic uplift displacement of the catenary on the mid-span, which presents a certain degree of fluctuation. When the pantograph slides along the contact line to the mid-span position, the displacement of this node gradually increases until the pantograph slides to the mid-span position of the catenary and the uplift displacement reaches the maximum

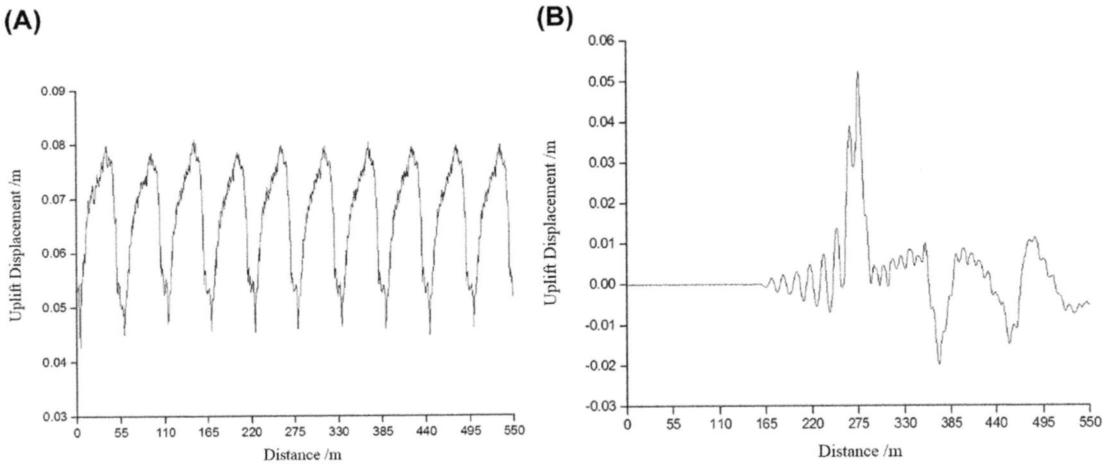

Figure 4.41
Dynamic uplift displacement of the catenary. (A) pantograph and catenary contact position (B) mid-span position.

value. As the pantograph gradually slides away from the mid-span position, the uplift displacement gradually decreases. It can be seen that the elasticity value of the catenary at the mid-span position is maximum, and at the positioner, it is the minimum.

(2) Contact force

In the process of sliding contact between pantograph and catenary, contact force must be maintained between the contact strip and the contact line. The dynamic contact force between the pantographs reflects the contact state between the pantograph and the contact line under high-speed operating conditions. It is an important indicator reflecting the current collection quality of the pantograph and catenary. Figure 4.42 shows the contact force curves at two different speeds. It can be seen from Figure 4.42 that the contact force also changes with the span within a period. The average value of the contact force is approximately the sum of the static uplift force and the aerodynamic force and fluctuates around the average value. With increase in speed, the maximum and average values of the contact force are basically increasing, but the minimum value is decreasing. Figure 4.42 also shows that the fluctuation of the contact force tends to increase with the increase in speed.

(3) Vibration acceleration

In the process of the interaction between pantograph and catenary, the lateral acceleration of the pantograph is slow, and the vertical and longitudinal accelerations are usually involved. Figure 4.43 and Figure 4.44, respectively, present the measured data of the

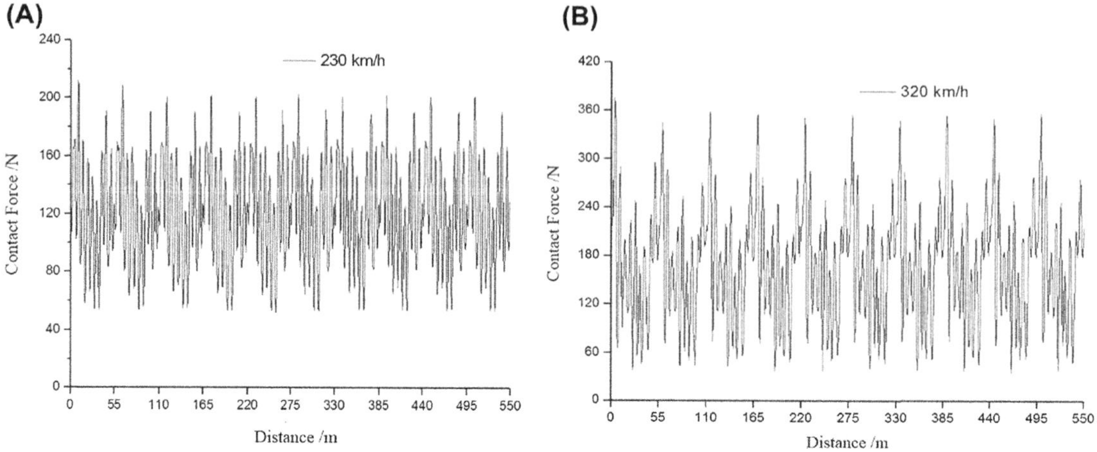

Figure 4.42
Contact Pressure of Catenary and Pantograph. (A) 230 km/h (B)320 km/h.

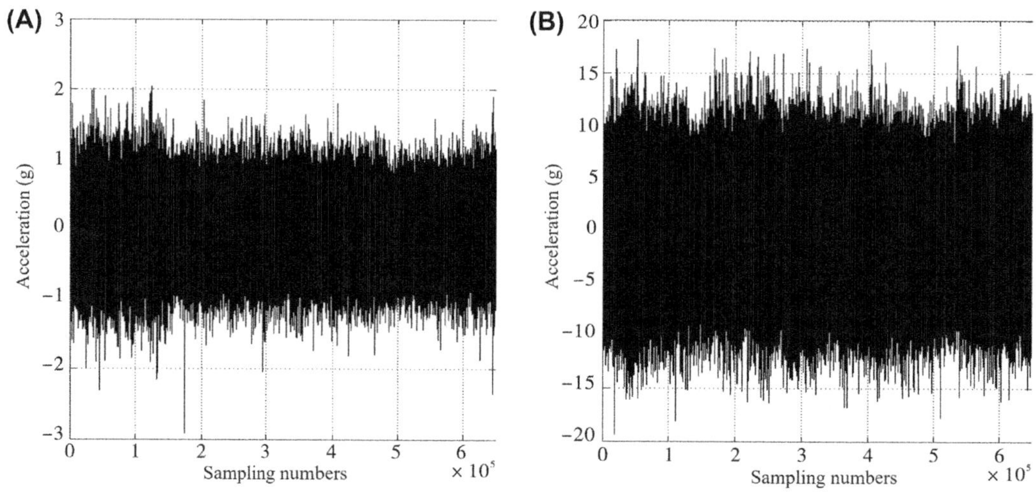

Figure 4.43
Vertical acceleration of pantograph. (A) Lower Arm of Pantograph (B) Pantograph Head Support.

vertical and longitudinal accelerations at the pantograph head support position and the lower arm position of a high-speed pantograph at a running speed of 330 km/h. As can be seen from the figures, the vertical and longitudinal accelerations of the lower arm position are much smaller than the acceleration at the pantograph head position. The acceleration value is about 1 g to 2 g, and the vertical and longitudinal accelerations at the pantograph

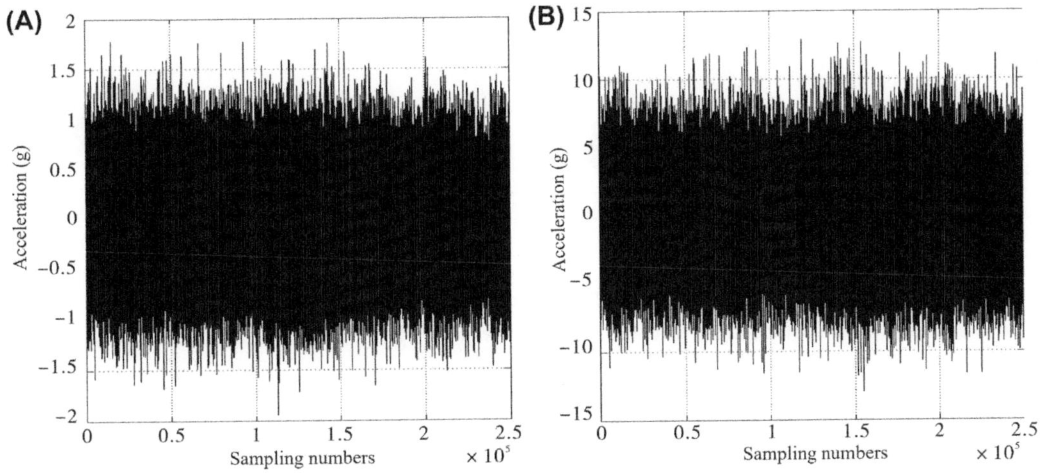

Figure 4.44
Longitudinal acceleration of pantograph. (A) Lower arm of pantograph (B) Pantograph head support.

head support are relatively large, about 10 g to15 g. Typically, the vertical and longitudinal acceleration value of the pantograph, from contact strip to base frame, shows a downward trend from top to bottom. For this type of pantograph, when the speed is between 300 km/h and 350 km/h, the maximum vertical and longitudinal acceleration at the contact strip can exceed 60 g, while the vibration acceleration at the base frame of the pantograph and the vibration acceleration at the roof are basically the same.

(4) Dynamic stress

To increase the propagation speed of the catenary, a method of increasing the tension of the contact line is generally used. However, the stress in the contact line is simultaneously increased, which affects the reliability of the catenary. Fatigue failure of the main components, such as messenger wires, clamps, and positioners occur from time to time during the use of the catenary. For pantographs, due to the strong vibration in the interaction process between the pantograph and catenary, the dynamic stress in the pantograph components is drastically increased, resulting in fatigue failure of the components. Fatigue failures are exacerbated by an increase in the running speed. Therefore, pantograph dynamic stress is also one of the important indicators for evaluating the dynamic performance of the pantograph-catenary system. Faults in the pantograph and catenary will easily occur if the dynamic stress is too high, which will directly affect the reliability of the current collection.

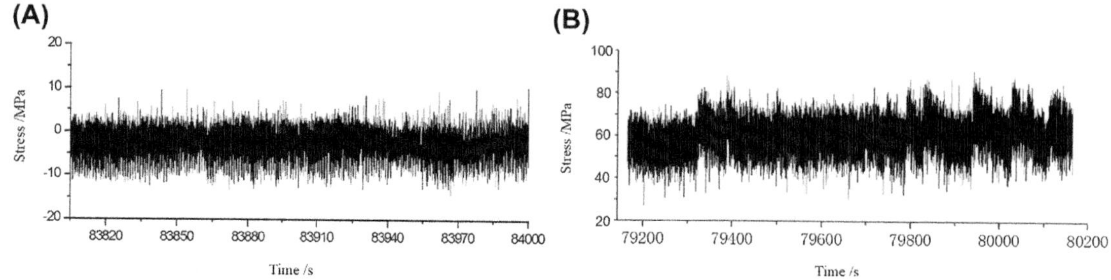

Figure 4.45
Dynamic stress in the pantograph. (A) Lower arm of pantograph (B) Pantograph head support.

Due to the complexity of the pantograph structure and service environment, the dynamic stress distribution at different locations varies. Figure 4.45 shows the measured data of the dynamic stress at the pantograph lower arm and the head support position at a running speed of 330 km/h. As can be seen from the Figure 4.45, the dynamic stress values are quite different, and the dynamic stress at the pantograph head support position is significantly higher than that at the lower arm position. In general, the dynamic stress distribution of the pantograph presents the following trends: the pantograph head that directly interacts with the contact line, especially some key components such as the pantograph head support, pantograph head crossbeams, rotating shafts, and pantograph head suspension systems, has a high dynamic stress and is prone to fatigue failure. Dynamic stress in the pantograph gradually reduces from top to bottom with the weak influence of contact force and aerodynamic force on the components below the pantograph head.

4.3.3.2 Effect of fluid-solid coupling on dynamic performance of the pantograph-catenary system

The pantograph has a large vertical motion during its operation, and the motion inevitably affects the surrounding flow field. Concurrently, change in the flow field will also cause a change in the aerodynamic force of the pantograph and the interaction between the two forms a complex fluid-solid coupling. To study the influence of fluid-solid coupling on the dynamic performance of the pantograph-catenary system, a simulation analysis of the simple stitched catenary and TSG19 pantograph used on the Beijing-Tianjin high-speed railway is conducted. Finite element modeling method is used for the catenary, and a catenary model with a nine-span tension section is established; the three lumped masses pantograph model is used to calculate the dynamic pantograph-catenary interaction; the three-dimensional solid model is used for the aerodynamic simulation of the pantograph; and the time step for the calculation of the aerodynamics and dynamic pantograph-catenary interaction is taken as 0.001s for the joint simulation of pantograph-catenary system and fluid-solid coupling. Considering the influence of the boundary conditions of the catenary on the calculation results, only the calculation results within a distance of 100 m to 350 m are given when the simulation results are analyzed.

In the joint simulation of the pantograph-catenary system and fluid-solid coupling, the unsteady computation is used for the pantograph aerodynamics, and the $k - \varepsilon$ model is adopted for the turbulence model. The calculated aerodynamic force on the pantograph is shown in Figure 4.46 for both the contact loss simulation and the fluid-solid coupling simulation. Figure 4.47 shows the calculated vibration velocity of the pantograph head in the fluid-solid coupling simulation. It can be seen from Figure 4.47 that the influence of the pantograph head vibration on the flow field around the pantograph is not considered in the contact loss simulation, and the aerodynamic force basically remains constant. In the fluid-structure coupling simulation, the flow field changes caused by the vibration of the pantograph head are taken into consideration, and the aerodynamic force fluctuates greatly. Because the catenary static stiffness varies according to the span periods, the aerodynamic force of the pantograph and the vibration speed of the pantograph head basically also change according to the span period.

Figure 4.48 and Fig. 4.49, respectively, show the uplift displacement of the contact line and the contact force of the pantograph-catenary system obtained from the contact loss

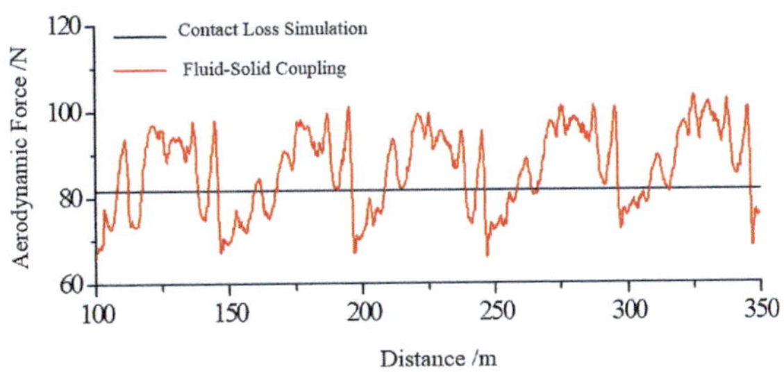

Figure 4.46
Aerodynamic Force on the Pantograph.

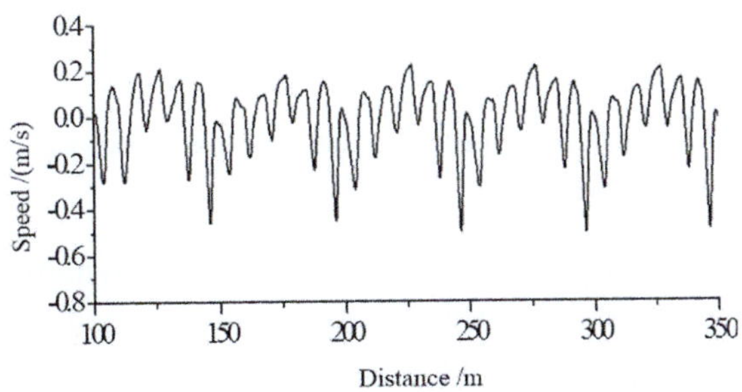

Figure 4.47
Vibration Speed of the Pantograph Head.

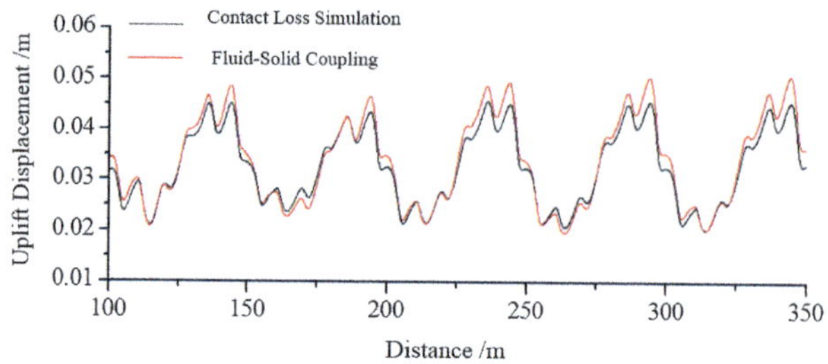

Figure 4.48
Uplift displacement of the contact line.

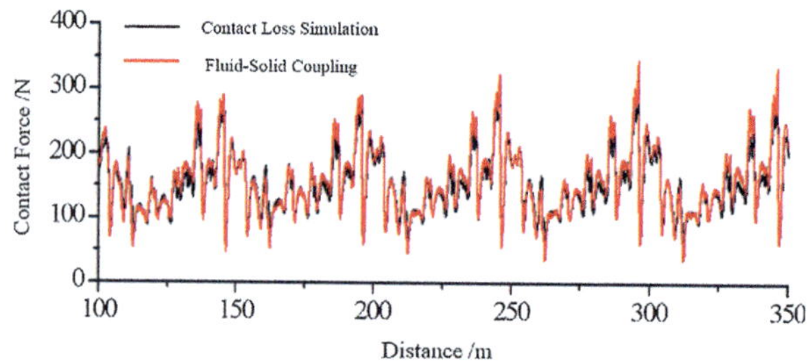

Figure 4.49
Contact force of the pantograph.

simulation and the fluid-solid coupling calculations. It can be seen that the maximum value of the uplift displacement of the contact line in the fluid-solid coupling calculation is larger than that in the contact loss simulation. The maximum value of the contact force becomes larger while the minimum value becomes smaller. According to the statistics on contact force, the average, minimum, maximum, and standard deviation of the contact force in the contact loss simulation are 152.10 N, 53.87 N, 329.44 N, and 45.68 N, respectively. The average, minimum, maximum, and standard deviation of the contact force in the fluid-solid coupling simulation are 157.38 N, 35.74 N, 346.26 N, and 52.23 N, respectively. According to the statistical results on the contact force, the fluid-solid coupling effect of the pantograph has a significantly large influence on the dynamic performance of the pantograph-catenary system. The fluctuation of the aerodynamic force makes the current collection quality of the pantograph weak.

4.3.3.3 Effect of vehicle-bridge interaction on the dynamic performance of the pantograph-catenary system

To examine the influence of vehicle-bridge interaction on the dynamic performance of the pantograph-catenary system, a vehicle-bridge model is established of a continuous box girder bridge of the high-speed railway in China and high-speed railway multiple units. A nine-span equidistant continuous beam is adopted for the bridge with a span of 30 m.

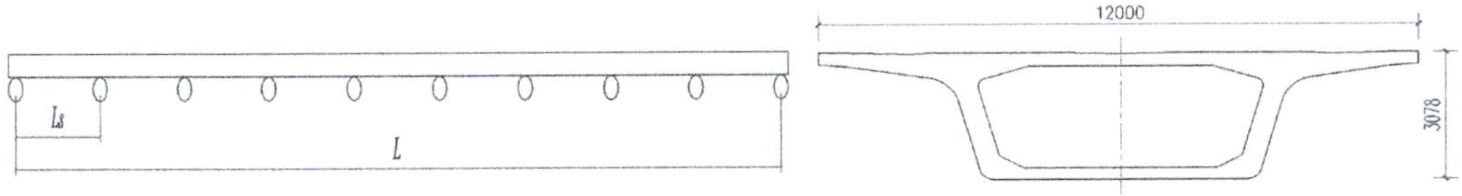

Figure 4.50
High-speed railway box beam.

As shown in Figure 4.50, the bridge surface is 12000 mm wide and 3078 mm high. The high-speed railway multiple units and center distance Lw are both 2.5 m, the length between truck centers Lc is 17.5 m, and the vehicle is about 60 t in weight.

Based on the vehicle-bridge model, the dynamic model for the pantograph-catenary system is established on the viaduct. Figure 4.51 and Figure 4.52 show the different dynamic uplift displacements of the contact force and the contact line with considering the vehicle-bridge vibration and without considering the vehicle-bridge vibration.

It can be seen that there is a big difference in the contact force of the pantograph when the vehicle-bridge vibration is taken into consideration and when it is not considered, while the difference in the dynamic uplift displacement of the contact line is small. Statistics show that the average, minimum, and maximum values of the contact force are 127.46 N, 25.99 N, and 317.16 N when the vehicle-bridge vibration is not considered. When the vehicle-bridge vibration is considered, the average, minimum, and maximum values of the contact force are 135.52 N, 22.01 N, and 350.01 N, where the difference in maximum values of the contact force is the largest and the change rate is 10.36%.

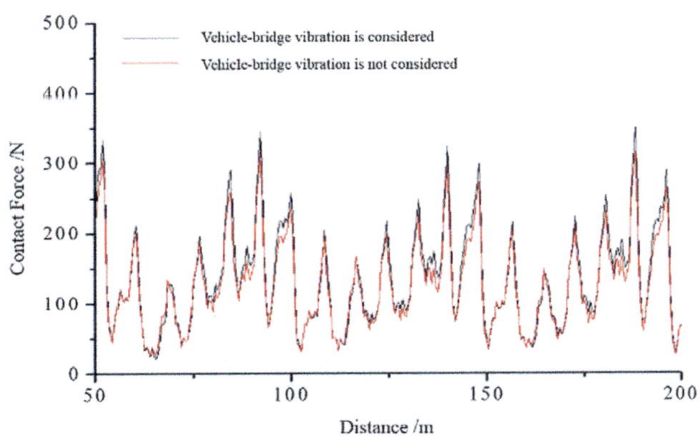

Figure 4.51
Comparison of contact force.

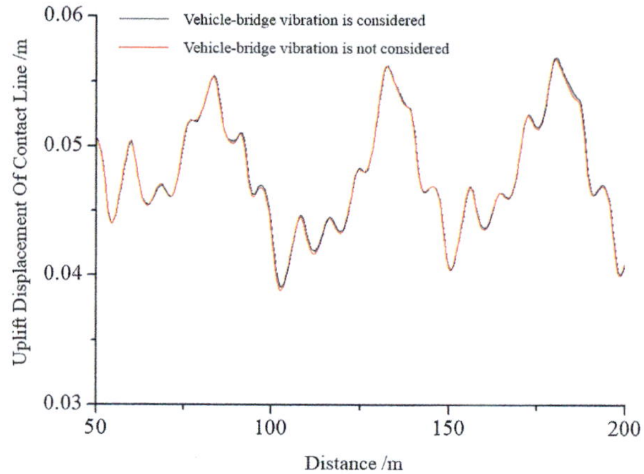

Figure 4.52
Comparison of dynamic uplift displacement of the contact line.

4.3.4 Dynamic characteristics of fluid-structure interaction

The fluid-solid coupling effect of high-speed trains under ambient wind is more distinct when compared with the fluid-structure interaction effects of vehicles passing by each other or passing through tunnels. Therefore, the interaction between the airflow and the train is analyzed by taking the dynamic characteristics of fluid-structure interaction under ambient wind disturbance as an example.

4.3.4.1 Fluid-structure interaction effects

To analyze the influence of altitude change on the aerodynamic performance of high-speed trains under ambient wind, the aerodynamic characteristics of the train is compared with consideration given to coupling and uncoupling. The comparison is as follows: when the speed of the high-speed train is 350 km/h, and the ambient wind speed is 13.8 m/s, the wind direction angle is 90°.

Figure 4.53 shows a side view of the pressure distribution on the vehicle surface of a high-speed train under ambient wind. It can be seen that regardless of whether the altitude

Figure 4.53
Pressure distribution on train surface. (A) Coupling (B) Uncoupling.

change is considered the pressure distribution on the car body surface is basically the same: the area with the largest pressure amplitude is at the tip of the head car; the area with the smallest amplitude is located near the cowcatcher; the surface of the windward side of the train has a positive pressure, while the leeward side primarily has negative pressure. When the altitude change is taken into consideration, the maximum pressure amplitude and the minimum pressure amplitude on the vehicle surface increases, and the surface dislocation in the connection zone between vehicles makes the area near the windward side become a negative pressure area, and positive pressure occurs near the leeward side of the stop junction.

Figure 4.54 is a comparison of aerodynamic force and vehicle altitude with and without altitude changes being taken into account. The aerodynamic force and altitude of the head

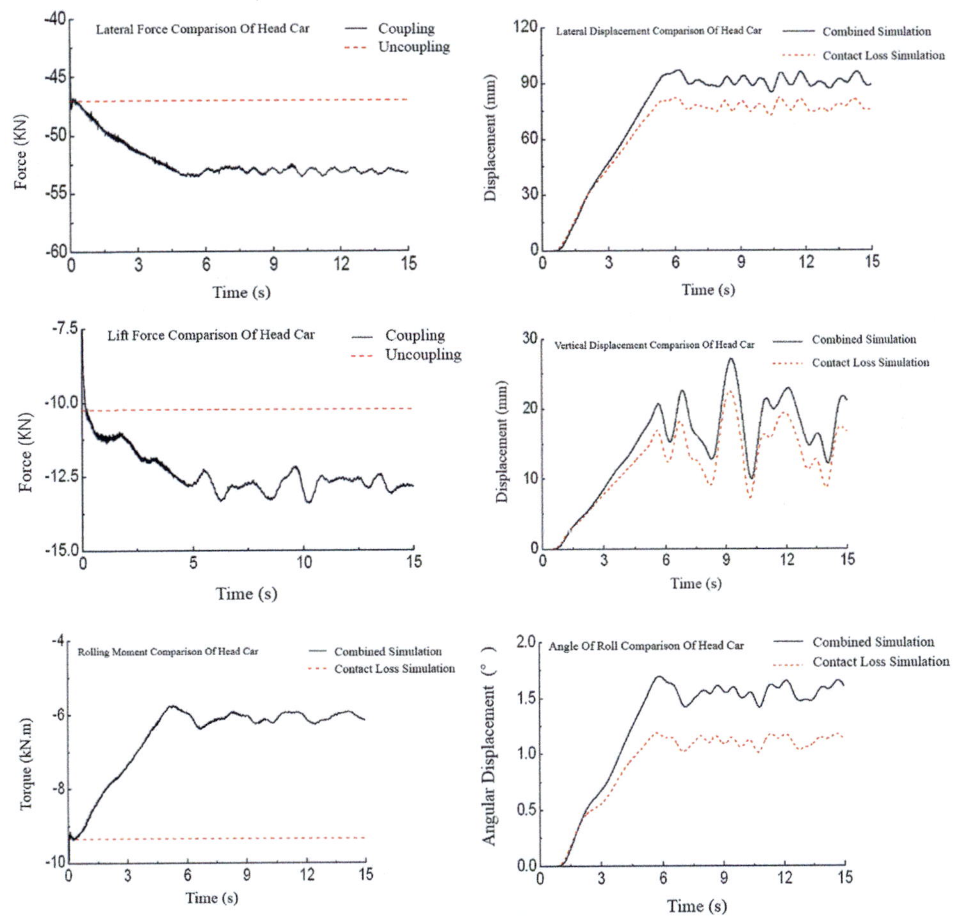

Figure 4.54
Comparison of aerodynamic force and vehicle altitude.

car change as follows when the interaction between the airflow and the train is taken into consideration [14,15]:

(1) The lateral force amplitude of the head car increases by about 6.5 kN (approximately 13.8%), and the transverse displacement of the head car increases by about 15 mm.
(2) The lift force amplitude of the head car is increased by approximately 2.4 kG (approximately 15.7%), and the magnitude of the heaving motion of the head car is increased by approximately 4.2 mm.
(3) If the centroid of the head car is used as the torque reference point, the rolling moment of the head car is negative, and the roll moment is reduced by about 3.5 kN.m. The angle of roll increases by about 0.6°, and the direction rotates around the centroid to the leeward side.

Figure 4.55 shows the pressure distribution on a cross-section of the head car with or without altitude change taken into account. The cross section in the left figure is located at the streamline position x = 5m, and the cross section in the right figure is at the streamline position x = 18 m, which is the non-streamlined position of the head car. Considering the interaction of the airflow and the train, the pressure on the surface of the windward side increases, and the pressure changes on the leeward side of the streamlined train surface is slightly different from that on the non-streamlined surface. Considering the interaction of the airflow and the train, the section pressure on the head car changes as follows [14]:

(1) The pressure on the body surface of the windward side increases.
(2) Because the vehicle turns to the leeward side, and the area at the junction of the windward side and the bottom of the car body becomes the windward side, the negative pressure amplitude at the junction at which the windward side meets the bottom of the vehicle body reduces.

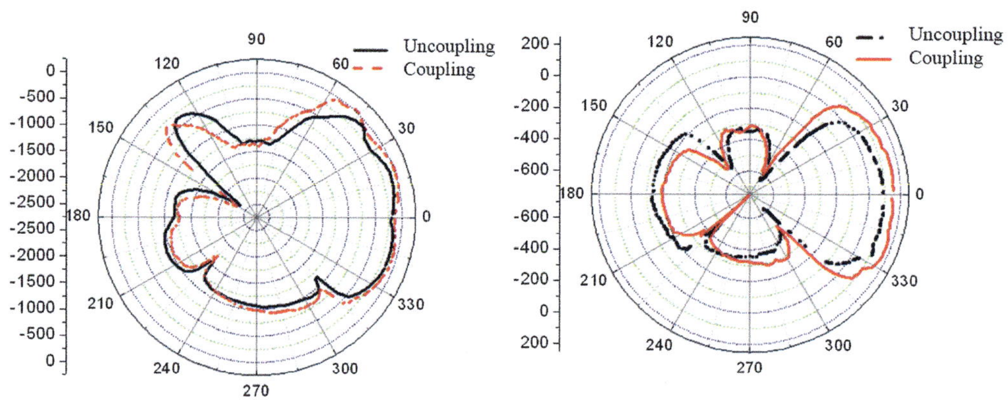

Figure 4.55
Polar chart of pressure distribution on cross-section of the head car (left: $x = 5$m, right: $x = 18$m; unit: Pa).

(3) Because the vehicle turns to the leeward side, the gap between the bottom of the vehicle and the ground becomes narrow, vortex shedding increases around the junction, and the negative pressure amplitude at the junction between the leeward side and the bottom of the vehicle increases.

4.3.4.2 Distribution of the pressure on the vehicle body surface and the flow field

Under ambient wind, when the airflow flows past the high-speed train vortex shed from the surface of the leeward side of the vehicle and the surface pressure of the vehicle body will change. This section primarily discusses the effect of the yaw angle on the characteristics of the pressure at the circumference of the vehicle cross section and flow field. Here are three cases: ① running speed of the train is 200 km/h, and ambient wind speed is 25 m/s; ② running speed of the train is 300 km/h, and ambient wind speed is 20 m/s; ③ running speed of the train is 350 km/h, ambient wind speed is 15 m/s, and wind attack angle is 90°. The composite yaw angle corresponding to the three cases is 24.23°, 13.50°, and 8.77°, respectively.

Figure 4.56 shows the cross-sectional pressure and flow field distribution of the train body under cases ①, ②, and ③. The pressure and flow field distribution for each section are as follows [15]:

Pressure distribution: For the section of non-streamlined parts, the pressure at the windward side of the vehicle body surface is almost positive. The pressures at the bottom of vehicle, roof, and leeward regions are almost all negative, and the areas where the negative pressure amplitude is relatively large are located primarily at the four corners of the vehicle body. In contrast, the pressure for the section of streamlined parts of the tail car are different as the negative pressure appears on the windward side of the vehicle body while the pressure on the leeward side is positive, which is caused by shedding of the vortex at the tail stream and is also an important reason for the inconsistency between the direction of the lateral force of the tail car and the other vehicles' side forces. As the speed of the ambient wind increases, the amplitude of the positive pressure on the windward side of the vehicle section in the non-streamlined part increases. Due to the interaction between the airflow and the train, surface dislocation in the connection zone between vehicles causes the flow field in this area to be disordered (the pressure changes violently).

Flow field distribution: The yaw angle is an important factor affecting the main vortex structure of high-speed trains in ambient winds. For the vortex shed from the leeward side of high-speed trains, a different yaw angle results in different separation points and turbulence intensity. For example, at the body section 10 m, there are three situations including the non-presence of the main vortex, the main vortex just separates from the leeward side of the vehicle body and, the main vortex of the leeward side is developing into the main vortex. The yaw angle is related to the maximum height dimension of the main vortex shed from the leeward side. As the yaw angle increases, the maximum height of the main vortex on the leeward side of the vehicle body also increases. For example, the

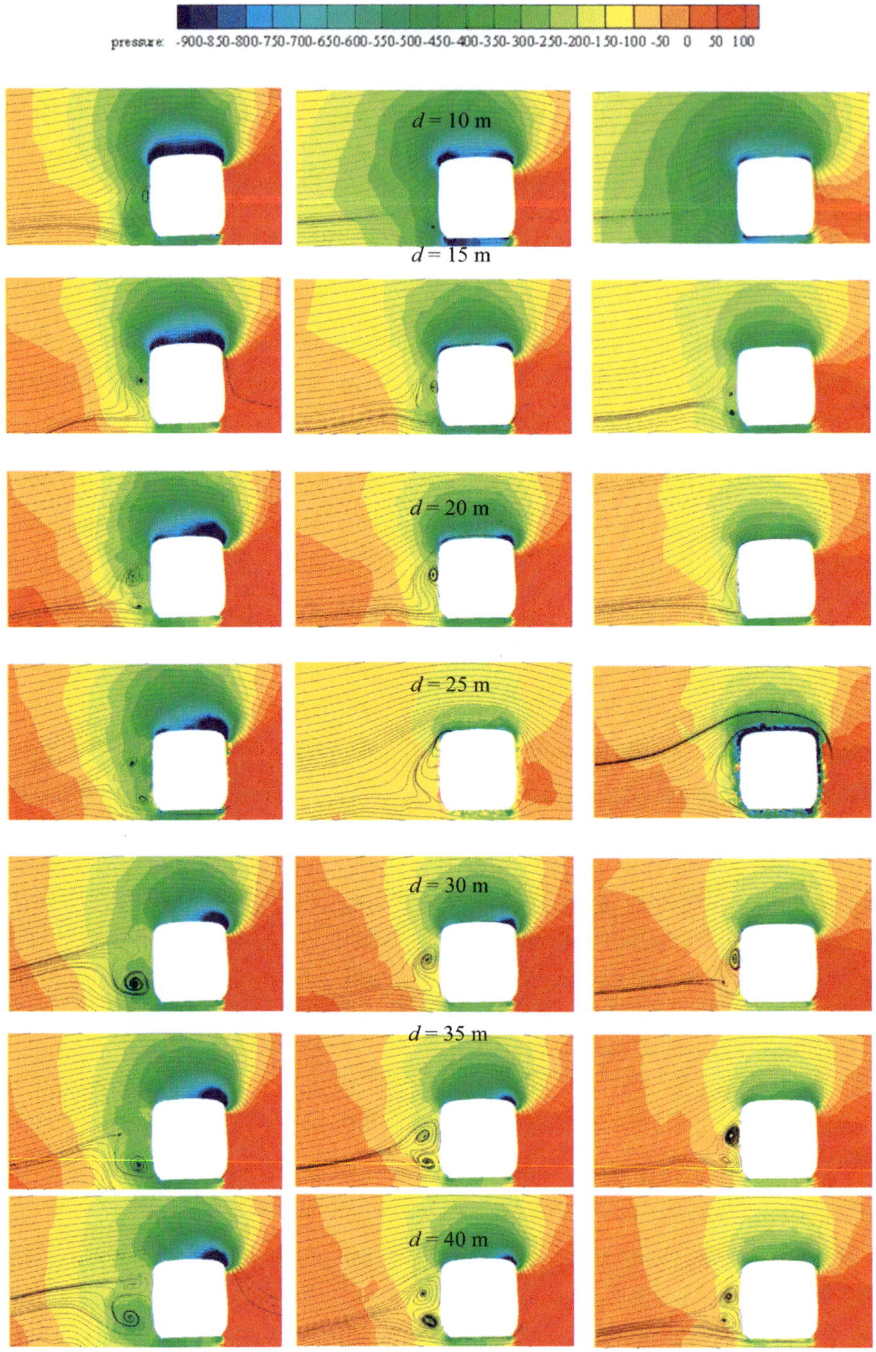

Figure 4.56
Cross-sectional pressures of train and flow field characteristics at different yaw angles (unit:Pa).

Figure 4.56
Cont'd.

maximum height dimension of the main vortex on the leeward side of the vehicle body is less than one-third of the train height in case ③, the maximum height is approximately one-third the height of the train in case ②, the maximum height exceeds half the train height in case ①.

4.3.4.3 Aerodynamic force and dynamic performance of the train

The speed of the train and the speed of the ambient wind are important factors affecting the aerodynamic force on the high-speed train. Due to the different appearances of the head car, tail car, and the middle car, the aerodynamic forces on the vehicles differ greatly. This section focuses on the relationship between aerodynamic force, altitude, running speed of the train, and the speed of the crosswind while considering the interaction of the air and vehicle altitude [16].

Figure 4.57 shows the trend of the lateral force of the high-speed trains in equilibrium with changes in train speed and wind speed. The lateral force amplitudes of the head and middle cars both increase with an increase in the train speed and crosswind speed. The change in the lateral force amplitude of the tail vehicle is relatively complex and may decrease with an increase in the running speed of the train. The lateral force direction of the head car and the middle car both point to the leeward side, and the lateral force direction of the tail vehicle is more complicated. Generally, the lateral force direction of the tail vehicle is opposite to that of the head car and the middle car; the magnitude of the head car is the largest, followed by that of the middle car, and that of the tail car is relatively small.

Figure 4.58 shows the trend of the lift force of the high-speed train in equilibrium with the change in train speed and wind speed. Generally, the lift force amplitudes of the middle car and tail car are larger than those of the head car, the lift force amplitude of the head car varies with increase in the running speed of the train, and the lift force amplitudes of the middle car and the tail car increase as the speed of the train increases. The lift force direction of the head car is related to the running speed of the train and the wind speed, and the change is relatively complex. The force direction of the middle car and the tail car is upward, i.e., a lifting force.

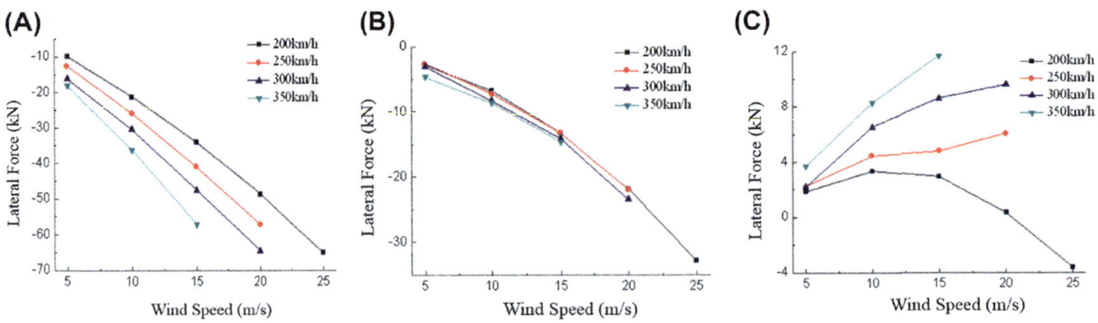

Figure 4.57
Lateral force of the vehicle body. (A) Head car (B) Middle car (C) Tail car.

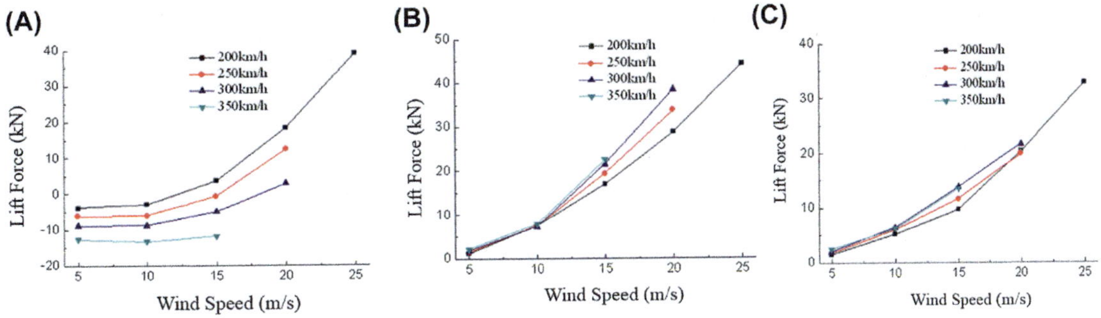

Figure 4.58
Lift force of vehicle body. (A) Head car (B) Middle car (C) Tail car.

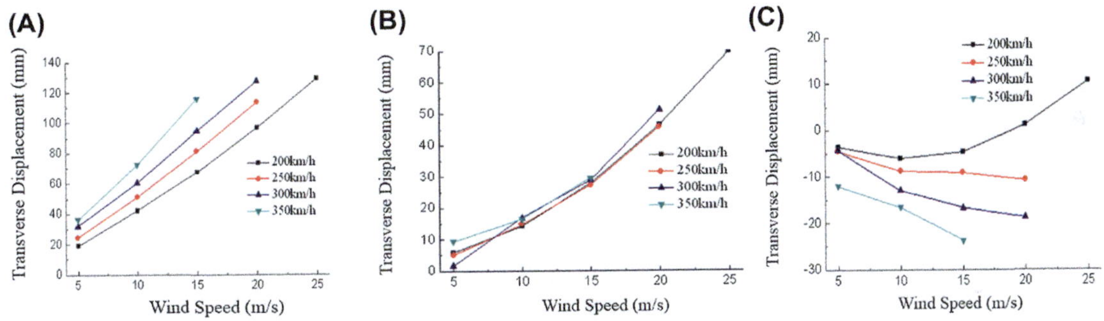

Figure 4.59
Transverse displacement of vehicle body. (A) Head car (B) Middle car (C) Tail car.

Figure 4.59 shows the trend in the variation of the vehicle's transverse displacement with change in train speed and wind speed in equilibrium. The head car has the largest transverse displacement amplitude, followed by that of the middle car, and that of the tail car is the smallest. The variation trend of the transverse displacement of each car is almost consistent with that of the lateral force. The direction of traverse of the head car and the intermediate vehicle is on the leeward side. The direction of the tail traverse is related to the running speed and wind speed of the train, and the change is complicated. The direction of traction of the head car and the middle car both point to the leeward side. The transverse displacement direction of the tail car is related to the running speed of the train and the wind speed, and the change is complicated.

Figure 4.60 shows the trend of the rolling angular displacement of the high-speed train in equilibrium with change in train speed and wind speed. The rolling angular displacement of the head car and the middle car are both positive, i.e., rotating from the centroid to the leeward side. The rolling angular displacement of the tail car is related to the running speed of the train and wind speed, and the change is complicated. The rolling angular

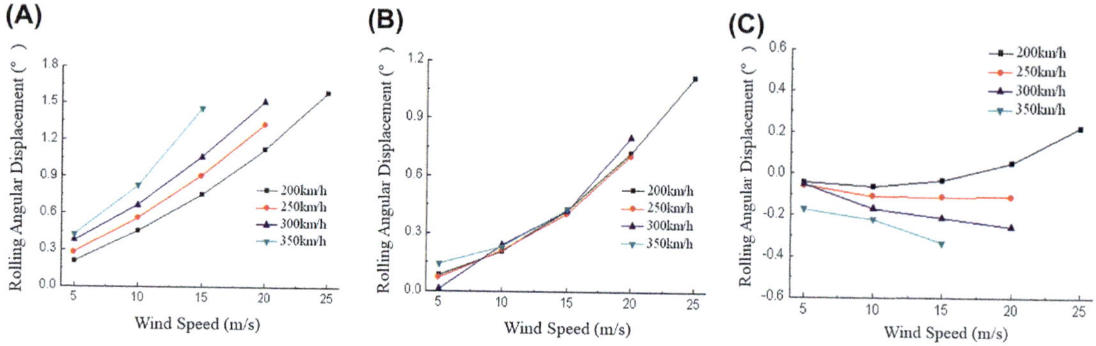

Figure 4.60
Rolling angular displacement of vehicle body. (A) Head car (B) Middle car (C) Tail car.

magnitude of the head car is the largest, followed by that of the middle car, and that of the tail car is the smallest. The rolling angular magnitude of the head car increases with increase in the running speed of the train; change in the train running speed has a small effect on the amplitude of the rolling angular magnitude of the middle car.

Because the influence of track irregularities in equilibrium is not taken into consideration, the displacement fluctuations of the wheelset gradually converge around a constant value. Under ambient wind, the horizontal displacement of the head carriage is the largest [15]. Figure 4.61 shows the variation in the transverse displacement of the wheelset of the head car in equilibrium with the change in running speed of the train and wind speed. Wheelset 1 of the head car has the largest transverse displacement amplitude; the transverse displacement direction of wheelset 1 and wheelset 2 points toward the leeward side, which is opposite to that of wheelset 3 and wheelset 4. The transverse displacement amplitude of wheelset 2, wheelset 3, and wheelset 4 increases with increase in running speed of the train and wind speed. The variation of the transverse displacement amplitude of wheelset 1 is complex, mainly due to the large transverse displacement amplitude of wheelset 2 affecting the transverse displacement of wheelset 1.

4.3.4.4 Margin of operation safety in ambient wind

The train safety indexes include wheel load reduction rate, derailment coefficient, the vertical force of wheel and rail, and lateral force of the wheel axle. The maximum safety index of the head car is larger than that of the middle car and the tail car in ambient wind. Figure 4.62 shows the variation of the maximum safety indexes for the head car in ambient wind with change in the running speed of the train and wind speed. Changes in the safety indexes are as follows: the magnitude of the transverse force of the wheel axle, vertical force between wheel and rail, derailment coefficient, and wheel load reduction rate increases with increase in the running speed of the train and in the speed of the crosswind.

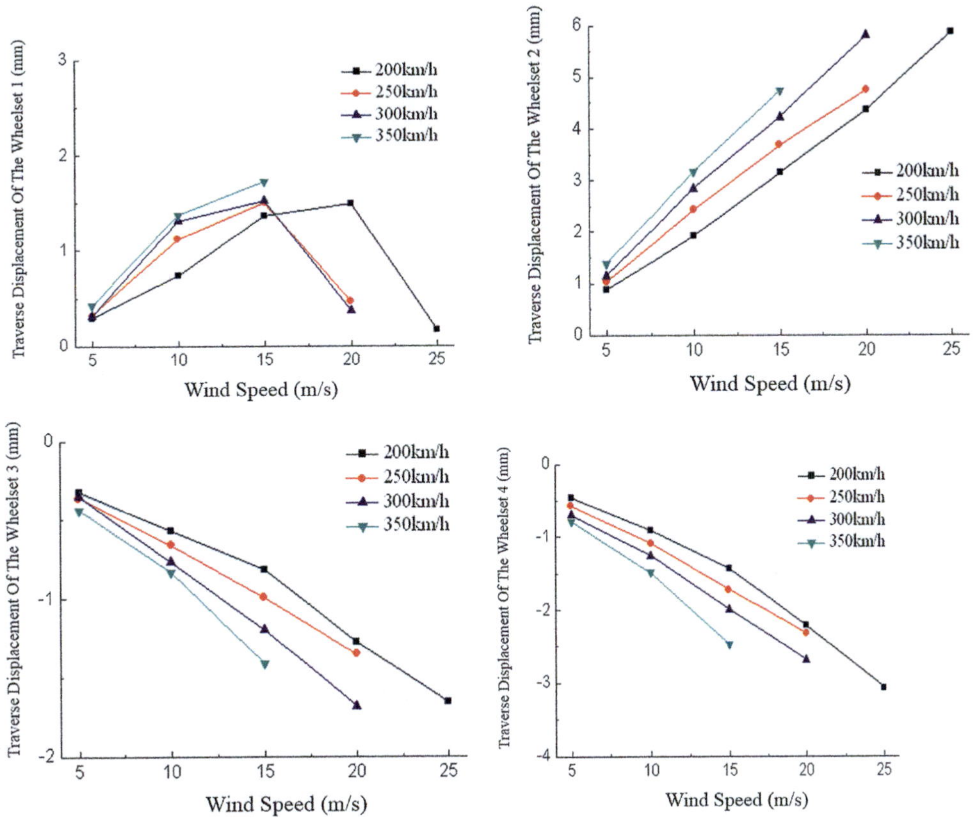

Figure 4.61
Transverse displacement of wheelset of the head car.

When both the lateral force of the wheel axle and the wheel load reduction rate have exceeded the standard, the magnitude of the derailment coefficient is close to the critical value. It is relatively difficult for the magnitude of the vertical force between wheel and rail to exceed the current standard.

When the running speed of the train is equal or more than 300 km/h, the wheel load reduction rate is more likely to exceed the relevant standards and is thus used as a safety assessment index. While in the condition that the running speed of the train is equal or less than 250 km/h, the lateral force of the wheel axle is more likely to exceed the relevant standard; thus, the lateral force of the wheel axle is used as a safety assessment index. According to the head car safety indexes in Figure 4.62 and relevant standards, the margin of operation safety of the high-speed train in ambient wind can be concluded [15] as shown in Figure 4.63.

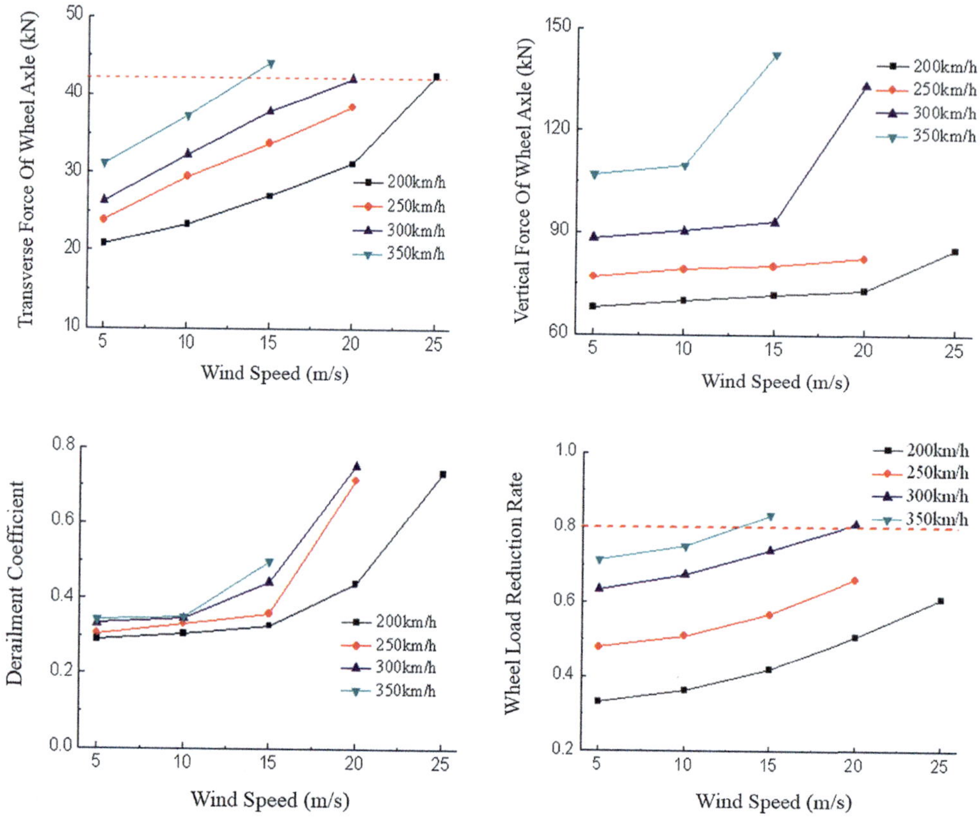

Figure 4.62
Safety indexes of the head car.

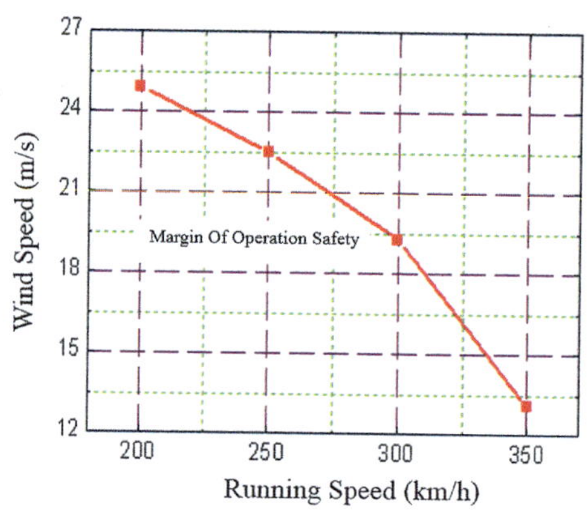

Figure 4.63
Margin of operation safety in ambient wind.

4.3.5 Electro-mechanical coupling dynamics characteristics

4.3.5.1 Electro-mechanical coupling system characteristics

Train speed-net voltage-power characteristics, as shown in Figure 4.64. When the net voltage fluctuates, the traction drive system guarantees that:

- The net voltage is rated power in the range of 22.5 to 29 kV;
- When the net voltage is 22.5 ~ 19 kV, the power is linearly reduced to 84% of the rated power;
- When the net voltage is 19 ~ 17.5kV, the power is linearly reduced to zero;
- When the net voltage is 29 ~ 31 kV, each device can work normally.

Figure 4.65 shows the curve of power demand and position for a single-train model. The AT power supply mode is adopted for the traction power supply system. The section post is set at the head of the section, and the traction substation is set at the end. The AT substation is located between the section post and the traction substation. Figure 4.66 shows the curve of the minimum net voltage of the traction power grid when a single train passes through the power supply section in the AT power supply mode. When the train is in the operating condition, far from the AT substation and the traction substation, the traction power grid has a large voltage loss.

Figure 4.67 shows the change in the minimum net voltage of the traction power grid at different departing time intervals. It can be seen that in the AT power supply mode, the change in the minimum net voltage of the traction power grid at different departing time

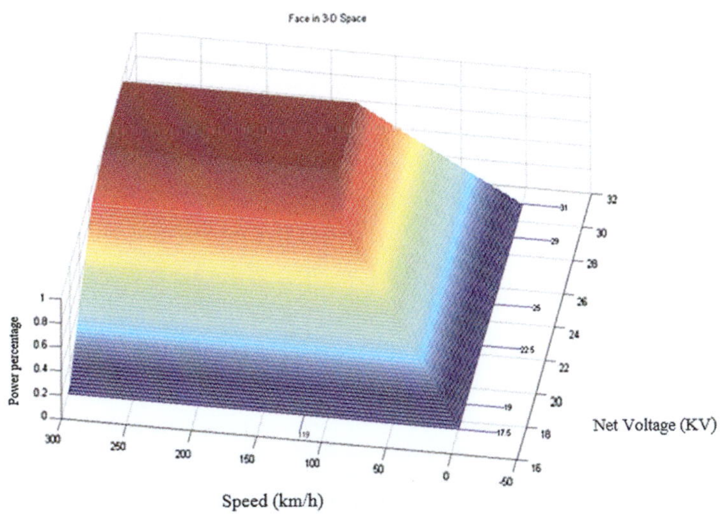

Figure 4.64
Train speed-net voltage-power characteristics.

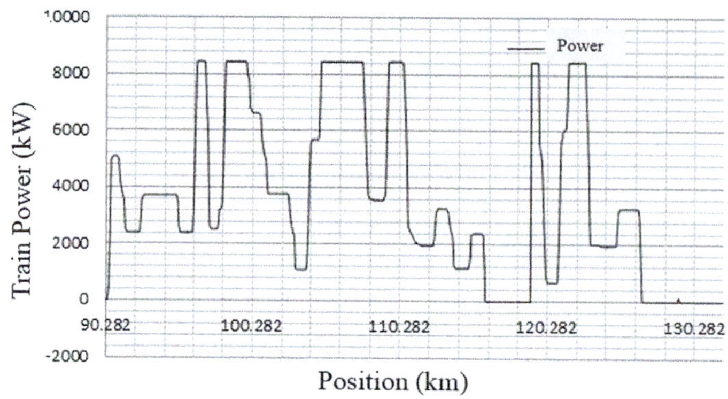

Figure 4.65
Curve of power demand and position for single-train operation.

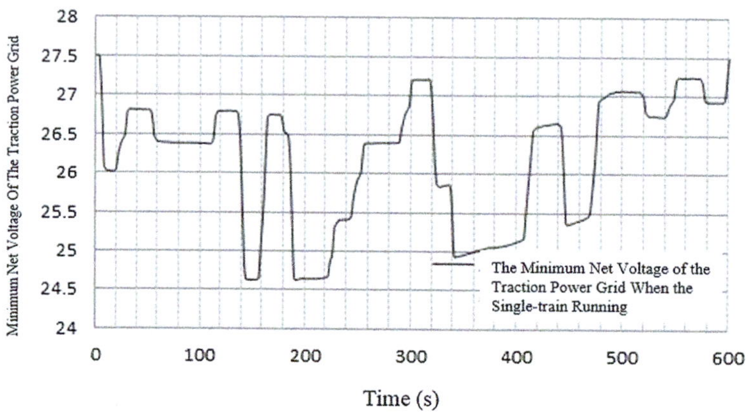

Figure 4.66
Variation of minimum net voltage of the traction power grid when a single train passes through the power supply section.

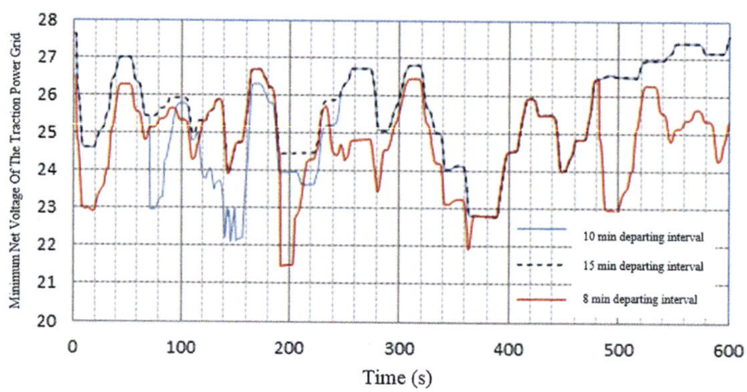

Figure 4.67
Variation of the minimum net voltage of the traction power grid.

intervals on the same power supply section of a traction substation has the following trends: The minimum net voltage of the traction power grid is closely related to the train number (train density) on the power supply section, the running state, and the power demand, etc. As the train density increases, the net voltage of the traction power grid decreases, and the output power of the traction drive system also decreases accordingly.

4.3.5.2 Effect of contact loss arcs on the traction power supply system

To analyze the effect of the contact loss arcs on the traction power supply system, a dynamic pantograph-catenary interaction simulation model is first established to calculate the vibration gap *d* between the contact strip and the contact line when a contact loss occurs at a certain moment in the pantograph. Secondly, a contact loss arc model is constructed based on parameters such as the vibration gap *d*, traction current, pantograph head voltage, and contact shape. Finally, the entire traction power supply system simulation model is established based on parameters such as the traction substation, catenary, rail, and high-speed multiple units, to research the effect of contact loss arcs on the traction power supply system, as shown in Fig. 4.68.

The parameters of the traction transformer, catenary, rail, and multiple units of a high-speed rail line in China are presented in Table 4.36 to Table 4.39.

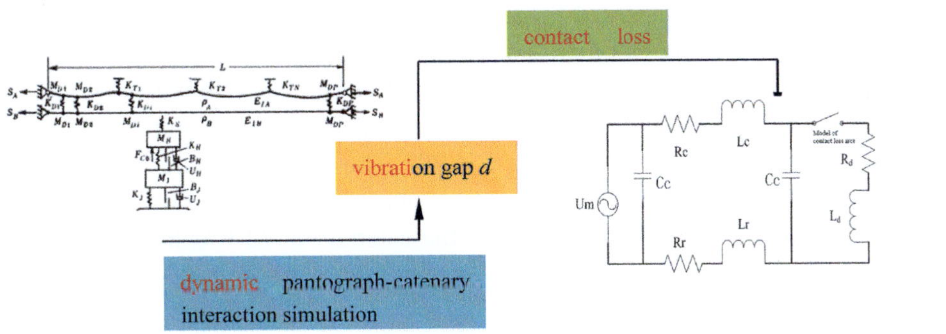

Figure 4.68
Simulation of the effect of the contact loss arcs on the traction power supply system.

Table 4.36: Parameters of the high-speed railway traction transformer.

Rated capacity	50000 kVA
Nominal voltage	220/2 × 27.5 kV
Rated current	227.27/909.08 A
Open circuit losses	43.566 kW
Load loss	146.488 kW
Short-circuit impedance	16.48%
No-load loss	0.23%
Equivalent loss resistance	0.177 Ω
Equivalent inductance	31.7 mH

Table 4.37: Parameters of the high-speed railway catenary.

Name	Contact line	Messenger wire
Type	CTMH-150	JTMH-120
Unit length resistance (Ω/ km)	0.191	1.45
Equivalent radius (mm)	8.92	12.79
Average height from catenary to the rail surface	5.3 m	

Table 4.38: Parameters of the high-speed railway rail.

Rail type	60 kg/m
Unit length resistance	0.135 Ω/km
Gauge	1435 mm
Equivalent radius	12.79 mm

Table 4.39: Parameters of the high-speed railway multiple units.

Power capacity of the entire vehicle	8800 kW
Power factor	0.89
Rated operational voltage	25 kV

The equivalent circuit parameters of the traction substation, catenary-rail circuit, and high-speed multiple units are calculated based on the parameters presented in Table. 4.36 to 4.39. Thus, the simulation model for the entire traction power supply system is established. Figure 4.68 shows the effect of the contact loss arcs on the traction power supply system when the initial arcing phase is 90°. The simulation results show that the overvoltage amplitude of the pantograph head is approximately 60.7 KV, the harmonic distortion is 26.7%, overvoltage amplitude of the catenary is 72 KV, harmonic distortion is 11.43%, overvoltage amplitude of the transformer outlet is 64.97 KV, and the harmonic distortion is 6.62%.

Figure 4.70 and 4.71 show the time domain characteristics and frequency domain characteristics of the overvoltage in the traction power supply system when the initial phase is 27° and 153°, and the arc is reignited. When the initial phase of arcing is over 90°, the overvoltage peak appears at the 270° phase. Based on the overvoltage at different initial arc phases and the corresponding harmonic distortion characteristics, the simulation data in Figure 4.72 and Figure 4.73 can be obtained. It can be seen from the results that the overvoltage amplitude and its harmonic distortion are related to the initial phase of the arcing. When the initial phase is around 90°, the overvoltage amplitude is large; the overvoltage amplitude on the contact line is 1.8 times the normal value; the harmonic distortion is the largest, with the harmonic distortion at the pantograph head as high as 27.5%.

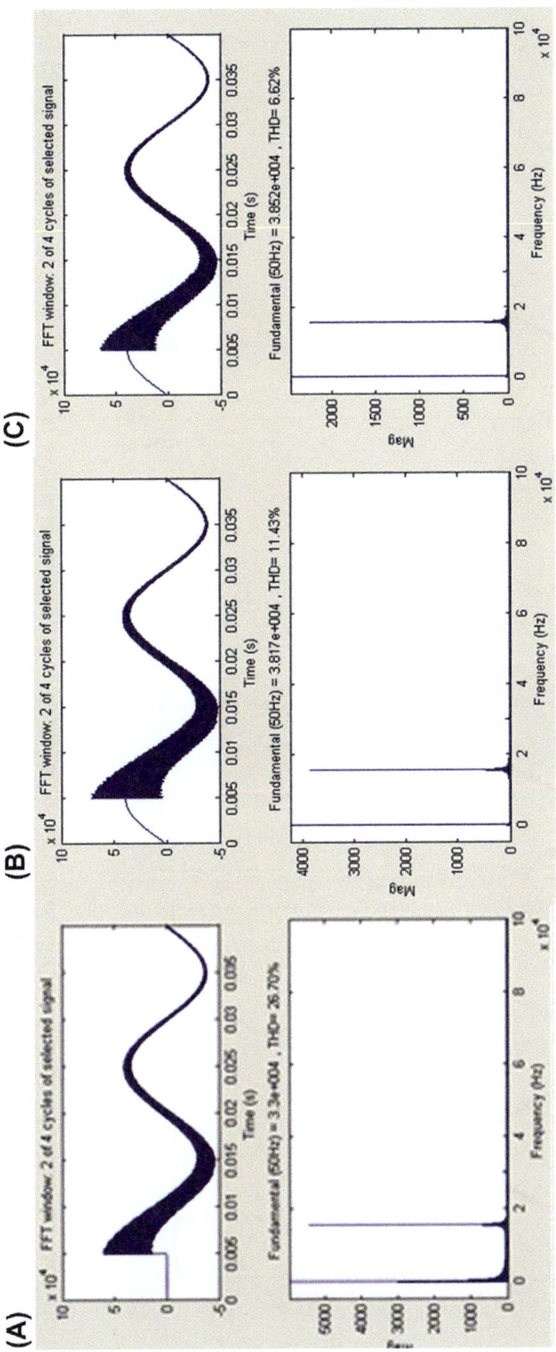

Figure 4.69
Traction power supply system overvoltage and harmonic characteristics (initial phase is 90°). (A) Pantograph head (B) Central catenary (C) Traction transformer outlet.

(A) (B) (C)

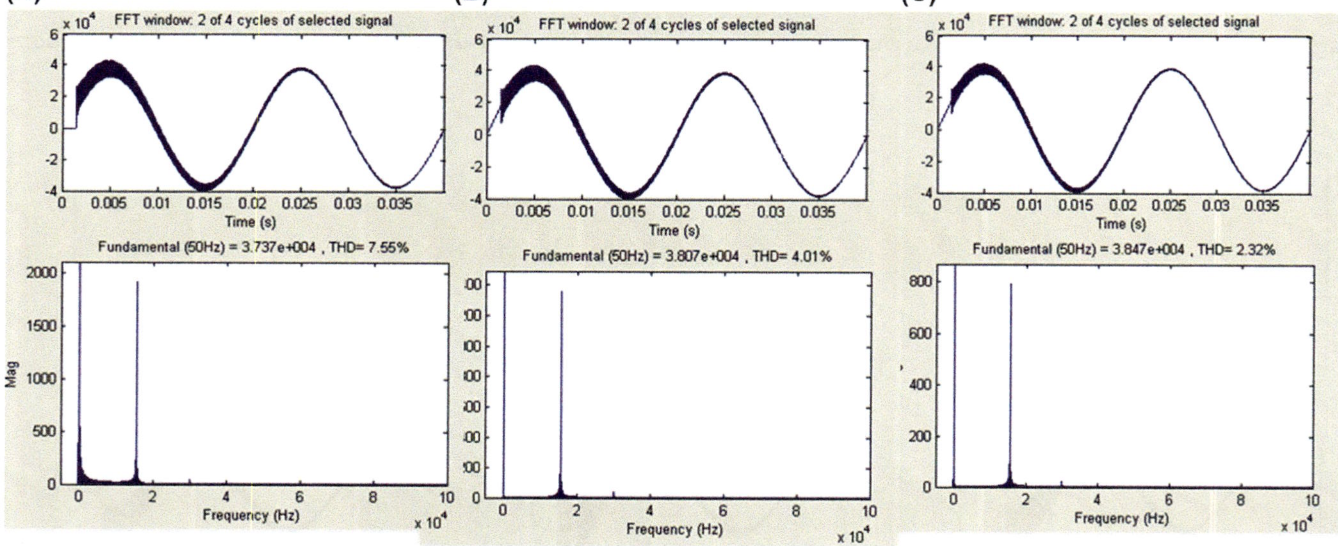

Figure 4.70

Traction power supply system overvoltage and harmonic characteristics (initial phase is 27°). (A) Pantograph head (B) Central catenary (C) Traction transformer outlet.

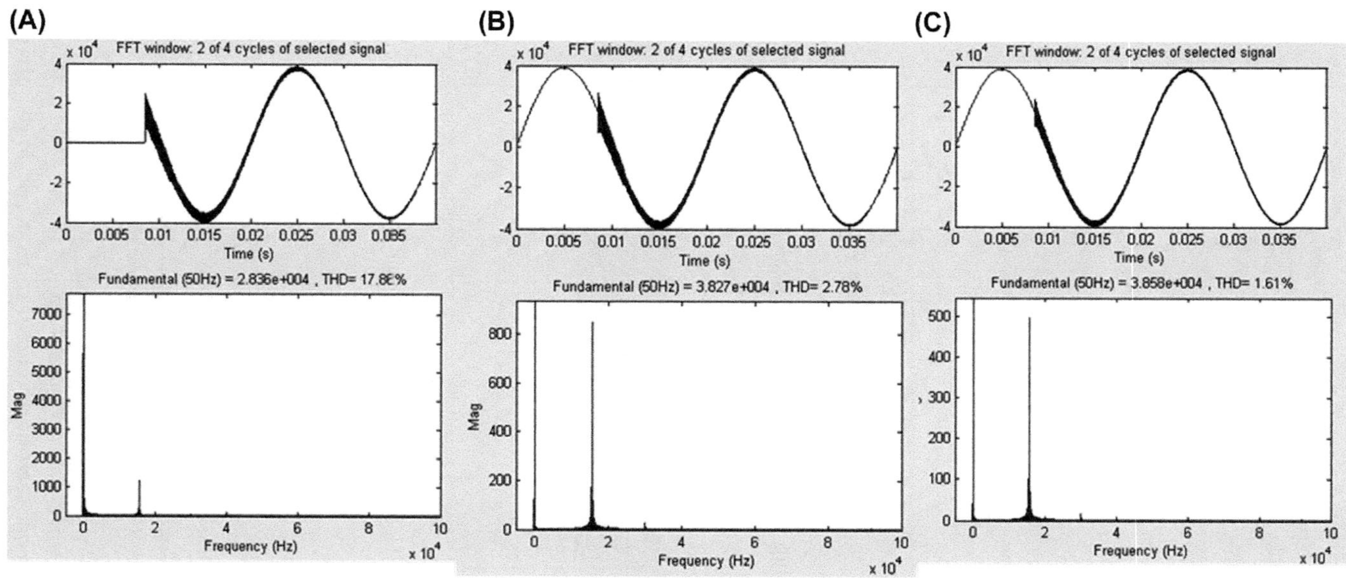

Figure 4.71
Traction power supply system overvoltage and harmonic characteristics (initial phase is 153°). (A) Pantograph head (B) Central catenary (C) Traction transformer outlet.

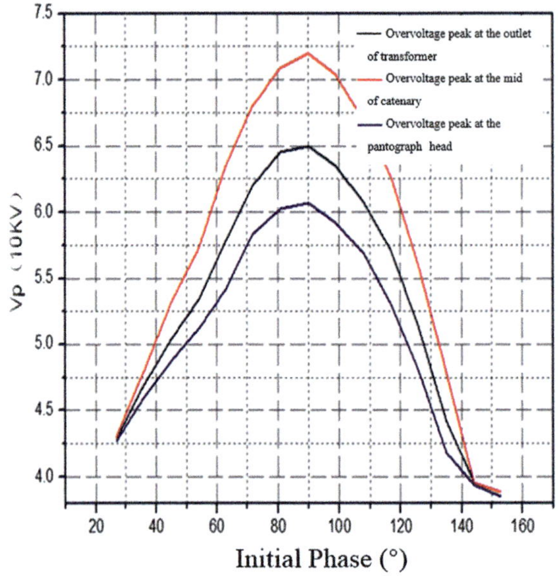

Figure 4.72
Relationship between overvoltage peak and initial phase.

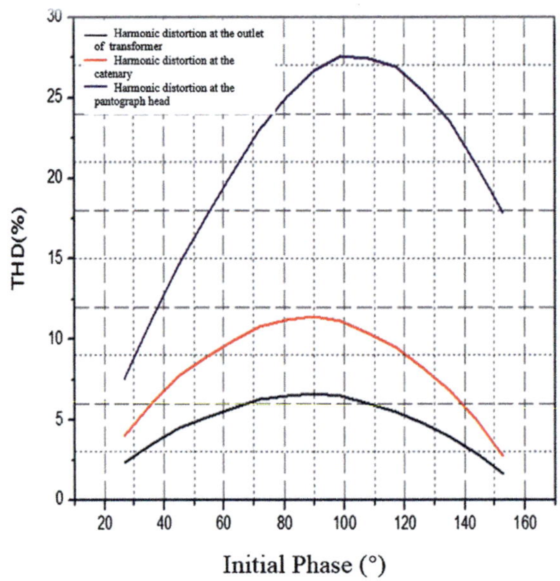

Figure 4.73
Relationship between harmonic distortion and initial phase.

It can be seen that the occurrence of contact loss arcs causes overvoltage, waveform distortion, and harmonic distortion of the traction power supply system. If the initial reignition phase of the pantograph arc is different, then its overvoltage amplitude and harmonic distortion are also different, and when the initial phase of the arcing is around 90°, its influence is most significant.

References

[1] Zhao Changlong, Maoru Chi, Wu Xingwen, et al. Study on Air Spring Model of Railway Vehicles. *Rolling* Stock 2012;50(9):1–3.
[2] Yu Jinfan, Miao Liwen, et al. The New High-speed Railway Design, Construction, and Line Speed-up and Transformation of New Technology Practice. Beijing: Contemporary China Audio and Video Publishing House; 2004.
[3] Practical Handbook Editorial Committee. The latest high-speed railway construction technology application and key quality control and construction safety management practical manual. Beijing: Contemporary China Audio and Video Publishing House; 2006.
[4] Zhang Weihua, Mei Guiming, Chen Liangqi. Effects of contact line slackness and surface irregularities on contact currents[J]. Journal of the China Railway Society 2000;22(6):50–4.
[5] Yu Mengge, Zhang Jiye, Zhang Weihua. Unsteady aerodynamic loads of high-speed trains under random winds. Journal of Mechanical Engineering 2012;48(20):113–20.
[6] Zhai Ronghua, Jiao Jinghai, Su Guanghui, et al. Coupling dynamics analysis of the pantograph considering the vertical irregularity of the contact line. Journal of the China Railway Society 2012;34(7):24–9.
[7] Deng Xueshou. CRH2 type 200 km/h traction drive system. Locomotive electric drive 2008;(4):1–7.
[8] Xu Yunwu, Deng Xiaojun, Yu Guizhen, et al. CRH2 200km/h EMU. Locomotive Electric Drive 2007;(3):39–45.
[9] Xiaoyun Feng. Electric Traction AC Drive and Its Control System. Higher Education Press; 2009.
[10] Zhang Weihua. *Dynamic* Simulation of Locomotives and Vehicles. Beijing: China Railway Publishing House; 2006.
[11] High-speed Rail Construction. No. 88. Supplementary provisions for relevant standards for the design and construction of railway tunnels. 2007.
[12] Science and Education Equipment. No. 21 Provisional Regulations for Strength Design and Test Appraisal of Speed Class Rail Vehicles at 200km/h and Above. 2001.
[13] Zhang Shuguang. CRH2 EMU. Railway Press; 2008.
[14] Tian Li, Zhang Jiye, Zhang Weihua. Coupling Dynamic Simulation of Fluid-Structure Interaction of High-Speed Train under Cross Wind. Journal of Vibration Engineering 2012;25(2):138–45.
[15] Tian Li. Fluid-Structure Coupling Computation Method and Dynamics Performance for High-Speed Trains. In: Ph.D. Thesis, Southwest Jiaotong University; 2012.
[16] Deng Yongquan, Tian Li, Zou Yisheng, Zhang Jiye, Zhang Weihua. Equilibrium Characteristics of High-speed Train in Crosswind. Applied Mechanics and Materials 2013;275-277:532–6.

CHAPTER 5

Optimization design method for the dynamic performance of high-speed trains

Chapter Outline

Dynamics of Coupled Systems in High-Speed Railways. https://doi.org/10.1016/B978-0-12-813375-0.00005-4

5.1 Design of Optimization targets and priority indexes of high-speed trains

5.1.1 Optimization targets of the dynamic performance of high-speed train

In high-speed train operation, a line forms the base or foundation for train operation, the airflow is the medium where-in it operates, and the catenary together with its power supply system are the power source. The dynamic performance and effects of boundary conditions directly impact safety, comfort, environmental protection, and the economics of vehicle operation. The optimization targets of dynamic performance will help solve such problems.

In line with the operating requirements of high-speed trains and the key factors restricting high-speed trains, the core optimization targets of high-speed train dynamic performance include:

1) Safety requirements related to high-speed kinetic stability
2) Safety requirements related to long-distance and long-term operational reliability
3) Requirements on ride quality and passenger comfort related to effective isolation with minimal external disturbance
4) Passenger comfort requirements under high-speed and long-distance travelling conditions
5) Environmental requirements related to low noise performance
6) Environmental requirements related to low vibration performance
7) How energy conservation and low energy consumption performance impacts economic requirements
8) Optimal low resistance performance of operational system

Fig. 5.1 shows the core optimization targets of dynamic performance listed above. As shown in the figure, environmental protection, stability, ride quality, safety, comfort, and

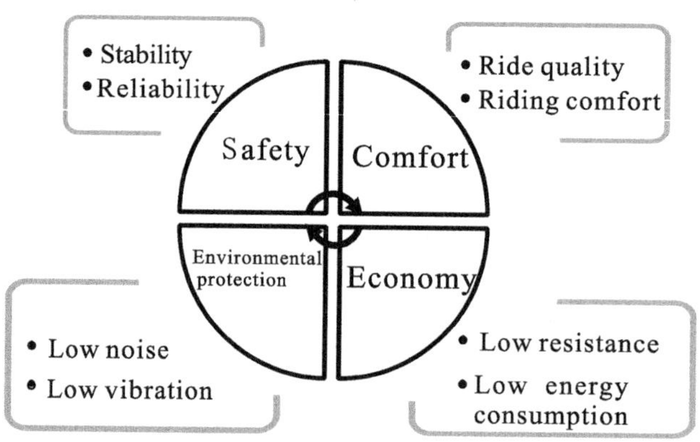

Figure 5.1
Core optimization targets of high-speed train dynamic performance.

economic factors are grouped together. Energy conservation, low energy consumption performance, and economic targets may reflect on environmental protection because these factors are related to some degree. Reliability, in most cases, will be seen as a total system problem; however, for this research the dynamics performance, design reliability focuses on the fatigue of the structure.

5.1.2 Priority design indexes of high-speed train

China is a vast territory with diverse regions, climates, and passenger demands. Questions on the type of high-speed trains and construction methods formed the basis of consideration for the development of high-speed rail systems. Priority design indexes [1] is the index system reflecting the general condition, core technology, and system interface of a high-speed train. Priority design indexes are also fundamental bases of high-speed train design.

Currently, there exists no research on the unified understanding or a fixed prescribed format for the priority design indexes of high-speed trains. The technical requirements for the supply of EMU train imported in China include [2]:

1) General provisions (type, speed)
2) Application conditions (geographical conditions, climatic conditions, line conditions, power supply conditions, communication signals, transportation organization, service facilities, etc.)
3) Basic requirements (basic principles of EMU train design, basic performance requirements, standards, and norms of EMU train)
4) Basic characteristics, compositions, and the main technical parameters of the EMU train (basic characteristics, main technical parameters, and traction braking characteristics of EMU train)
5) Signs, appearances, and decorations.

It was necessary to consider several aspects before setting priority indexes because of the technical conditions related to supply, the technical items being very specific, and reasons for setting priorities being unclear. A basis for design requirements needs more thought and understanding.

All designs are created to meet the requirements. Thus, setting design indexes is an inseparable task from demands. The transportation capacity of the passenger vehicle, passenger comfort, and technical aspects are all design factors to be considered; however, safety remains a key factor to be emphasized. The environment and environmental-friendly design become additional requirements in modern scientific design efforts. All the above factors form the basis for the successful development of high-speed trains in China. [3,4]

5.1.2.1 Transportation capacity indexes

Speed and passenger capacity are two factors that determine the transportation capacity of high-speed trains. Speed is the core index of high-speed train design, which is not only the direct factor improving transportation capacity but also the key basis for passengers to choose the transportation medium and operating schedule. Speed, however, is normally not confirmed or set by the designer, but it is rather set based on the high-speed railway system construction or larger transportation planning. In terms of design, speed is the target of high-speed train design or the input condition of design. In fact, the key technologies of high-speed train design are all conducted around speed.

In terms of transportation capacity, the speed index needs to be confirmed. It is necessary to first identify how to reach such a defined high speed, and what acceleration process is to be used reach that speed. The basic considerations are the traction power of the train and motor-trailer ratio of power allocation. Of course, in order to confirm the power, speed is only one of the input indexes, specific power determination also requires operating resistance at maximum operating speed, the inertial force and gradient resistance caused by residual acceleration, and so on. For realizing such speed, the ability to stop the train safely should also be considered. Under the premise of ensuring the braking distance, the braking method and braking power must both be determined as part of priority indexes. Although, the modes of traction and braking are not part of the direct numerical index, they are the core technology modes that reflect the technology level, and therefore, they are used as technical indexes to reflect the technical direction.

Another direct key factor of the transportation capacity design is passenger capacity, which depends on the length of train formation and the layout of the passenger compartment. As for the layout of the passenger compartment, what needs to be considered is not only the transportation capacity, but also the function and weight configuration of the entire train. The passenger capacity directly affects the weight of train or axle load. The train weight finally impacts the allocation of traction and braking power.

The index system of high-speed train for transportation capacity is shown in Fig. 5.2; the figure shows that the speed and passenger capacity are the priority indexes of demand, followed by technical priority indexed to meet demand.

5.1.2.2 Safety index

Safety is the basic requirement of high-speed train operation. As a priority index, safety problems consist of operational safety elements. Two factors affect the operating safety of high-speed trains: the safety factor affected by EMU trains itself, and the other is that affected by the operating environment of EMU trains.

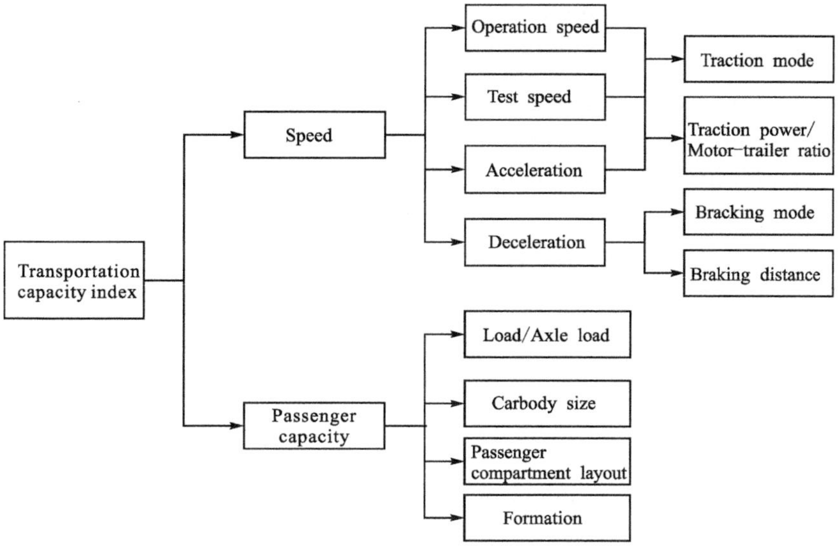

Figure 5.2
Transportation capacity index system of high-speed train.

The main safety performance factors affecting the EMU train are the kinetic stability and the reliability of the structure and system. The degree of kinetic stability is reflected by the critical speed of instability; if the critical speed is low, when instability appears, the vehicle will present a dramatic lateral periodic shake that will not only severely affects passenger's riding comfort, but also threatens operation safety. Thus, the degree of kinetic stability and its sensibility to parameters can reflect the safety quality of a high-speed train.

The system reliability problems affecting safety are related to structural reliability. It is difficult to provide a specific design index for structural reliability and therefore the service life of vehicle and the main structure data are generally used. The structural reliability can be represented by the life cycle index, which also includes failure rate in one million kilometers or MTBF as indication.

The increase in the operating speed of the high-speed train intensifies the external excitation suffered by the train. The airstream of relative motion not only greatly increases the air resistance, but also leads to aerodynamic noise, which is an important factor restricting the development of the high-speed train. The airstream is an important factor influencing the safe operation of high-speed trains because the high-speed airstream disturbs the motion of the high-speed train. The lateral force and lifting force within the aerodynamic force will lead to safety problems in operation. The key factor determining the value of aerodynamic force is the shape of train. Line conditions and operating environment correlated at the same time includes line spacing, line (embankment, elevated railway, bridge, and tunnel) conditions and environmental wind, etc. The operating environment index involving the safety of the high-speed train should also be considered as part of the priority index of the high-speed train design.

The increase in the train operating speed and the line condition not only affects ride quality but also the operating safety. The parameters of the line condition that affect operating safety and riding comfort include horizontal and vertical sections and line irregularity, among others. Indexes for line horizontal section are the curve radius at the maximum passing speed and the minimum radius allowed for passing. Indexes for line vertical section are the gradient and the change rate between different gradients. Traction power is the characteristic most influenced by gradient while line irregularity affects vehicle vibration and riding comfort, thereby significantly affecting the wheel-rail force and derailment safety index.

Climate, weather conditions, and the environment directly affect the operating safety of trains. Wind, rain, and sand may influence visibility, mechanical malfunctions of moving parts, passenger inhalation problems through the ventilation system, and even derailment or overturning. Snow accumulation on the bogie may lead to frozen conditions resulting in damper failure, congestion of motion space, or mechanical failure, especially in Alpine or similar regions.

An accident history such as that of the "723" EMU ensured the recognition of crashworthiness indexes. According to collision test standards of the vehicle body, deformation evaluation under standard speed conditions or maximum collision speed conditions are essential for safety.

The indexes system around the safety of the high-speed train is shown in Fig. 5.3. Indeed, stability and reliability are priority indexes for the performance quality of the high-speed train. However, the line excitation, airflow turbulence, and environment are basic requirements for high-speed train operation, which are also the comfort indexes considering safety.

5.1.2.3 Comfort index

Comfort is a basic requirement when choosing any form of transportation. The traditional comfort index was a one-dimensional index considering decibels of noise, quality index of vibration, etc. The proposed higher standards of passenger comfort require a multi-dimensional index considering noise, atmospheric pressure (and its rate of change), vibration, temperature, humidity, illumination, etc. as part of the list of priority indexes.

A specific index is proposed to determine riding comfort that includes ergonomics, multimedia applications, and the riding environment. It further considers the effect of these factors on passenger activity, space height, corridor width, seat spacing, luggage rack height, etc.

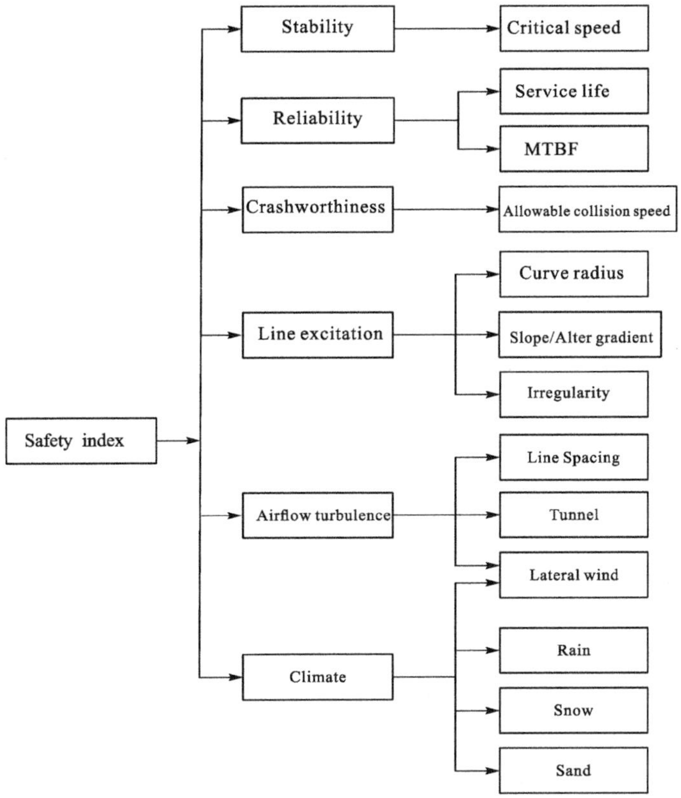

Figure 5.3
Safety indexes system of high-speed train.

A second level of indexes may look at seating comfort including aspects such as seat shape, material, or softness that may support passenger tiredness during long journeys. Other level indexes include color schemes, sleep, and read lighting, class status of vehicle compartments as well as aspects that may be connected to traditions, geographic conditions, education, or society. Multimedia systems such as audio or video can also increase travel quality. All of the above factors influence passenger mood and add to the satisfaction of the travel experience.

External factors such as easy pick-ups or drop-offs around stations, access to facilities or clear direction indications supports the passenger experience before and after travel. Air quality and volume control inside trains supports the physical and mental health conditions. These factors form part of the health and safety indexes which also look at electromagnetic radiation levels as part of the larger framework.

The index system for high-speed train comfort is shown in Fig. 5.4. It is not only a people-orientated design concept but also an intelligence requirement as part of an

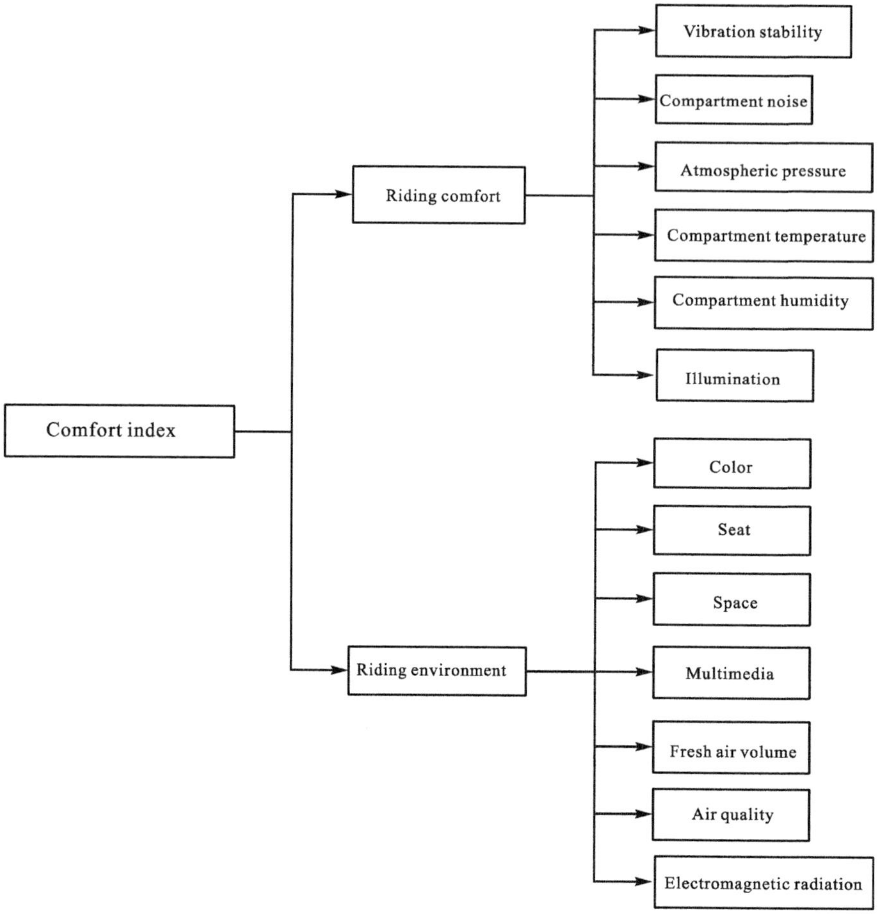

Figure 5.4
Comfort indexes system of high-speed train.

information orientated society and an indication of the advanced technical level of high-speed trains.

5.1.2.4 Friendly environment index

The friendly environment index considers the adaptability of high-speed trains to its operating environment and its effects on the environment. The operating environment refers to the adaptability of the high-speed train to the natural environment in which it operates, and the external environment refers to the adaptability to other operating systems within the same operating environment. These are considered in the design as part of the interface correlation or interference.

The adaptability of EMU to the natural environment, altitude, environmental temperature, and other environmental indexes will be considered during the design capability and correlation processes of the EMU. Wind, sand, rain, fog, and other environmental factors are index requirements that need to be considered; however, these indexes form part of operational safety, and therefore, they are considered in the safety index.

The friendly environment index also considers the effect of high-speed train operation on the environment. An external factors index will consider the influence of the train on the outside environment in which it operates. The influence of high-speed train operation on surroundings includes noise, vibration, discharge, and electromagnetic radiation; the design will also consider mechanisms to minimize these effects.

Energy conservation is one measure of sustainable development; energy conservation and the capability of the high-speed train to conserve energy or utilize energy efficiently form a key part of the design and evaluation capabilities of the high-speed train. As a measure of energy conservation, reducing the operating resistance and heat transfer coefficient through the design, and adopting an energy saving regenerative braking technology is essential.

The system environment of the high-speed train operation includes platform, line (gauge and line spacing), contact line, power supply system, signal system, etc. The main interface correlations need to be clearly defined in the design of the priority index.

The indexes system around the friendly environment is shown in Fig. 5.5; it includes natural environment adaptation, system environment adaptation, and environmental influence, in addition to energy conservation, as four aspects.

5.1.2.5 Priority design index

The priority design indexes of high-speed trains are proposed according to the analysis of priority design indexes and the index system established above. The requirements and overall parameters for high-speed train operation and maintenance are considered and they include the index attributes, index names, and index units, etc., as shown in Table 5.1.

These indexes roughly correspond to the index names in Figs. 5.2–5.5, very few of which will be classified further. Some indexes appear in the index system for the first time, such as comprehensive comfort index or track irregularity index because there are no existing uniform expressions in the industry; annotations are provided to better understand these indexes. It should be noted that these specific indexes will be adjusted according to different design requirements.

5.2 Design methods of high-speed train kinetic stability

In the development of the high-speed EMU, the focus is on improving the three main indexes of high-speed train dynamic performance: kinetic stability, ride quality, and

Table 5.1: Priority design indexes of high-speed train.

Index attribute	Index name	Unit	Note
Transportation capacity indexes	Operating speed	km/h	
	Design speed	km/h	
	Test speed	km/h	It is 1.1 times the design speed
	Starting acceleration	m/s^2	
	Residual acceleration	m/s^2	
	(Maximum) Braking deceleration	m/s^2	
	Traction mode		
	Traction power	kw	
	Motor-trailer ratio		
	Braking mode		
	Braking distance	m	
	Marshaling		
	Length×Width×Height of train	m×m×m	The width of the train takes the design of the platform for the line into account
	Axle load	t	
	Capacity	people	
Safety indexes	Critical hunting speed	km/h	
	Failure rate	time/million km	
	Curve radius	m	
	Minimum curve radius	m	
	Slope and slope change rate	‰	
	Horizontal irregularity	mm	Referring to the maximum amplitude in specific irregularity spectrum
	Vertical irregularity	mm	As above
	Gauge irregularity	mm	As above
	Direction irregularity	mm	As above
	Twist irregularity	mm	generally expressed as a twist depth under specific chord length
	Line spacing	m	
	Tunnel area	m^2	
	Crosswind speed	m/s	Cross wind limit under different operating speed
	Surface water height	cm	(to rail surface)
	Surface snow height	cm	(to rail surface)
	Surface sand height	cm	(to rail surface)
Comfort indexes	Comprehensive comfort index		Comprehensive comfort index considering the combined action of vibration, noise, pressure, temperature, humidity and illumination

Table 5.1: Priority design indexes of high-speed train.—cont'd

Index attribute	Index name	Unit	Note
Friendly environment and system comfort indexes	Vertical vibration acceleration	m/s^2	
	Lateral vibration acceleration	m/s^2	
	Longitudinal impact	m/s^3	
	Vibration stability index or vibration comfort index		
	Compartment noise index	dB(A)	
	Compartment pressure change	Pa/s	
	Compartment temperature range	°C	
	Compartment humidity range	%RH	
	Compartment illumination	Lux	
	Fresh air volume	l/min	
	Air quality		Content of dust, CO_2, poison gas, etc.
	Seat		
	Channel width	Mm	
	Color		User requirement
	Multimedia		Audio and video entertainment system
	Communication inside vehicle		Internet and mobile communication system
	Electromagnetic inside vehicle	V/m	
	Altitude	M	
	Environment temperature range	°C	
	Noise outside vehicle	dB(A)	
	Vibration of earth	dB	
	Discharge		
	Electromagnetic	V/m	
	Operating resistance	kN	
	Heat transfer coefficient		
	Regenerative braking rate	kW	
	Platform height	M	
	Distance between platform and line centerline	M	
	Gauge	M	
	Line spacing	M	
	Voltage of contact line	kV	Minimum and maximum allowable voltage should be marked
	Conductor height of contact line	M	
	Signal mode		

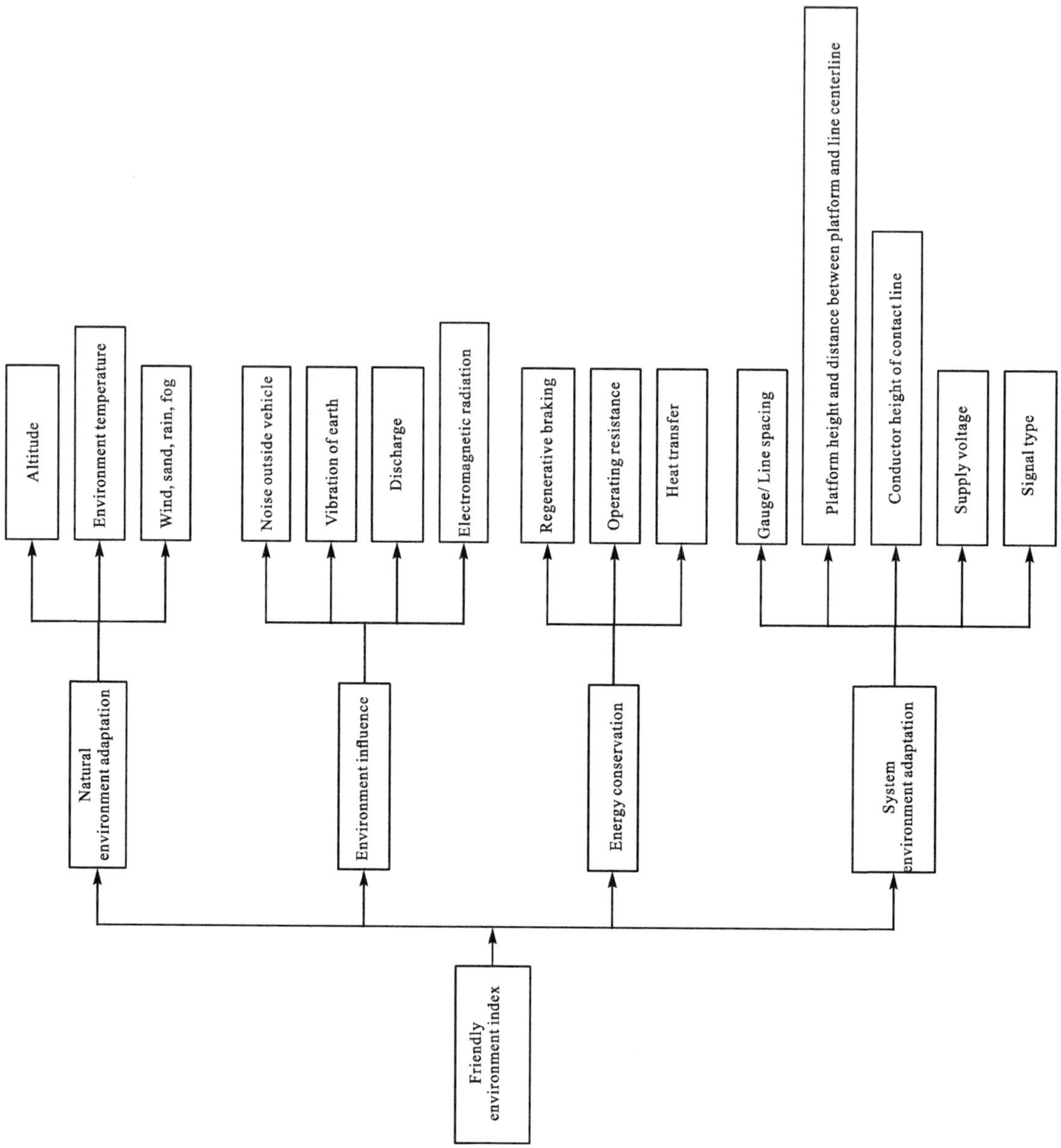

Figure 5.5
Friendly environment indexes system of high-speed train.

curve negotiation. Improving kinetic stability remains the fundamental task because it not only reflects the instability of behavior and motion but also the ride quality; more importantly, it may affect operational safety.

Stability as concept was formulated as early as the 17th century during the Torricelli period; however, in 1892, Lyapunov [5] provided a definition. The wider concept of stability may be broken down to static stability and dynamic (kinetic) stability. In terms of railway vehicles, the hunting stability is the inherent attribute of the locomotive system itself that forms part of the dynamic (kinetic) stability [6].

In the early research of locomotive kinetic stability, linear or linearized models were adopted [6–9] to study the kinetic stability because of the limitations of calculation methods. In these studies, experiments were performed under the assumption of small displacement. Linear wheel-rail interaction and linear suspension parameters were considered and linear differential equations with constant coefficients eigenvalues assessing stability methods were used to determine system stability [10].

In reality, many nonlinear characteristics are active within the operating environment of the locomotive. Wheel-rail correlation, suspension characteristics, and braking characteristics are some examples of the above. Lately, several new methods have been created to solve problems in the study of kinetic stability through direct numerical integration [11–12]. The author once used the numerical integration method [13] to study the stability of the fixed point in the periodic solution of the locomotive vehicle system and to analyze its nonlinear motion stability; the author found the limit cycle of motion and then discussed the stability of the limit cycle.

The rapid improvement in computation speed enables big data capability processes to be used for comparison and selection in the optimization of the locomotive dynamic parameter design. It is not always easy to provide scientific explanations as to why parameter matching gained from optimization and to achieve greater stability performance for high-speed trains. Thus, the author proposed control strategies for kinetic stability related to the railway vehicle's structure and parameter characteristics [14] and then aimed to provide more rational research methods and scientific explanations.

5.2.1 Kinetic stability control strategy

The kinetic stability performance of the vehicle system depends on the characteristics of vehicle system, and it is influenced by external excitation simultaneously.

External excitation includes train turbulence caused by line irregularity and airflow. In the wheel-rail system of the railway vehicle, because of the taper effect of the wheelset tread, periodic lateral motion characteristics of the wheels occur on two rails, which is called the hunting instable motion. As for the vehicle system, the hunting motion is the self-induced turbulence of the wheelset. The caused frequency f_w is given by

$$f_w = \frac{v}{2\pi}\sqrt{\frac{\lambda}{AR_0}} \tag{5.1}$$

where λ is the equivalent taper, R_0 is the wheel rolling radius, A is the half spacing between the left and right contact points, and v is the rolling speed of the wheel. The values of R_0 and A are commonly fixed or are in a certain range, and the equivalent taper of the wheel-rail contact varies and is changeable. The equivalent taper can be calculated as

$$\lambda = \frac{R_R - R_L}{2y_w} \tag{5.2}$$

where R_L and R_R are the rolling radii of the left and right contact points, and y_ω is the transverse shift of the wheelset. The taper of the wheelset under the condition of the 3 mm transverse shift of the wheelset is always set as the equivalent taper [15,16]. R_L and R_R in formula (5.2) depend on the wheel-rail profile, wheelset back distance, and transverse shift.

When different wheelset treads match with the CN60 rail, the equivalent tapers under the wheelset back distance of 1353 mm and 1360 mm are shown in Fig. 5.6 [17]. When the wheelset back distance is 1353 mm for the China high-speed train tread LMA, the difference between the Japan high-speed train tread JAPA and Europe high-speed train tread S1002 is tiny. However, when the distance comes to 1360 mm, the equivalent tapers of the tread profile significantly improve in the same time. In particular, the peak value appears in the S1002 tread when the transverse shift is 0.5 mm, which reaches above 0.5.

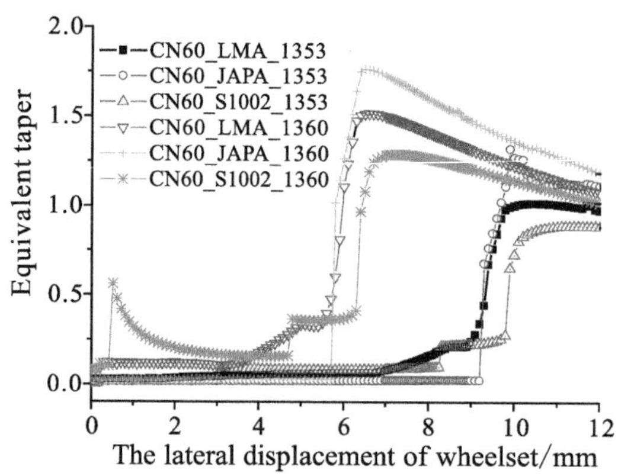

Figure 5.6
Equivalent tapers corresponding with different profile.

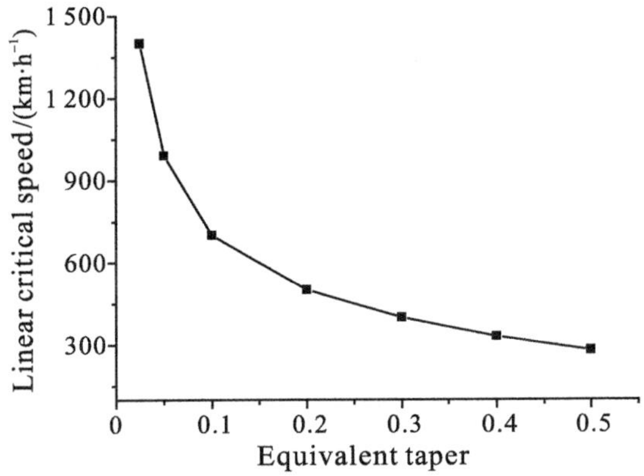

Figure 5.7
Effect of equivalent taper on linear critical speed.

The linear critical speed corresponding with different equivalent tapers processed by equivalent linearization is shown in Fig. 5.7. It is based on the parameters of the CRH2-300 EMU; the wheelset back distance is 1353 mm, and LMA tread is adopted. The maximum critical speed of vehicle can exceed 1000 km/h, and it decreases with an increase in the equivalent taper. The linear critical speed is on the high side under the condition of linear assumption. When the nonlinear factors are considered, the nonlinear critical instable speed is only 551.3 km/h.

In the research on high-speed train stability, the wheel tread profile and the wheel-rail correlation need to be determined first, i.e., an appropriate wheel-rail match equivalent taper needs to set.

The hunting motion of the wheelset shown in formula (5.1) is performed without constrains. Considering the wheelset constrains, the hunting motion characteristics of the wheelset will be changed, and the hunting motion is limited. Once the wheel profile is determined, the limitation to the wheelset hunting motion comes from the longitudinal constraint of the wheelset, which is the longitudinal stiffness setting of the axle box. Under the wheel self-excitation of the hunting motion, the bogie frame can be in the hunting motion with the wheelset, and the hunting motion of the bogie cannot be limited by only adding the longitudinal positioning of the wheelset relative to the bogie frame. Thus, kinetic stability depends on the car body of bogie. The kinetic stability of the bogie is ensured through the constraint from the car body to the bogie, and the yaw damper is mainly used to achieve this. As for freight train, the hunting motion of the bogie can be limited through the rotation friction torque provided by the center plate or the side bearing. However, this is not enough because the car body is not fixed, which can be in the hunting motion with the bogie. To ensure enough of the "calm down" effect of the car body and to limit the hunting motion of the bogie, certain mass and inertia are usually

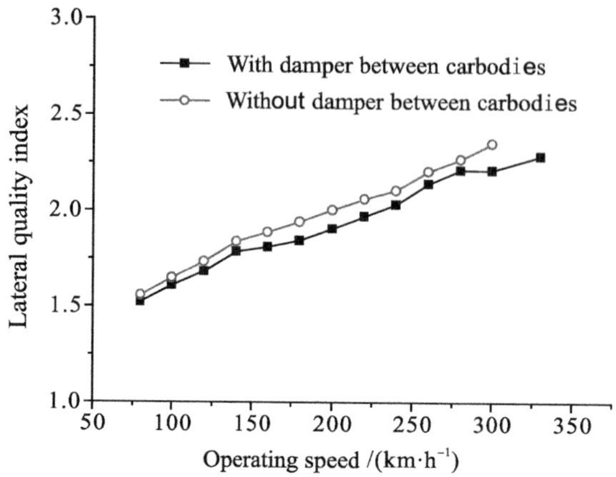

Figure 5.8
Effect of coupling damper between car bodies on lateral ride quality of vehicle under different speeds (bench test).

used to achieve the "calm down" effect. When the mass of the car body is too small to achieve the "calm down" effect to resist the hunting motion, the anti-hunting capacity of the car body can be improved by using the coupling between the car bodies.

In CRH2-300, which has a lightweight car body, the coupling dampers are set between the car bodies. This not only improves the stability of the vehicle but also improves the ride quality of the vehicle. The bench test result of the CRH2-300 EMU dynamics performance is shown in Fig. 5.8. The lateral ride quality index significantly decreases after installing the coupling dampers between the car bodies. The line test result of the CRH2-300 EMU dynamics performance is shown in Fig. 5.9 [17].

Through the analysis above, the control strategy of the vehicle kinetic stability can be concluded, as shown in Fig. 5.10. In the design of the high-speed train kinetic stability, a sufficient "calm down" capacity should be first ensured for the car body. When the anti-hunting capacity of the car body is not enough, and we need to consider using coupling dampers between the car bodies to improve stability. By limiting the car body to the bogie motion, the anti-hunting capacity of the bogie frame can be realized. Finally, the axle box positioning is considered to limit the hunting motion of the wheelset.

To illustrate the control strategy for the vehicle kinetic stability, CRH2 EMU is considered as an example. The sensitivity of the effect of the mass parameters of the car body (sprung), bogie frame (between primary suspension and secondary suspension), and the wheelset (unsrpung) on kinetic stability are shown in Fig. 5.11, which indicates that the sensitivity is very high when car body weight is light, which is up to 6.0. With an increase

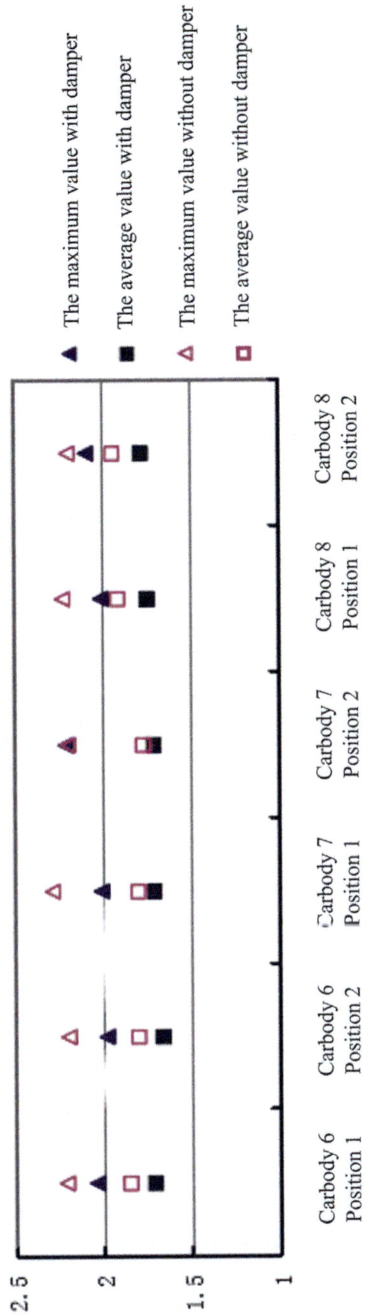

Figure 5.9

Effect of coupling damper between car bodies on lateral ride quality of vehicle under different speeds (line test).

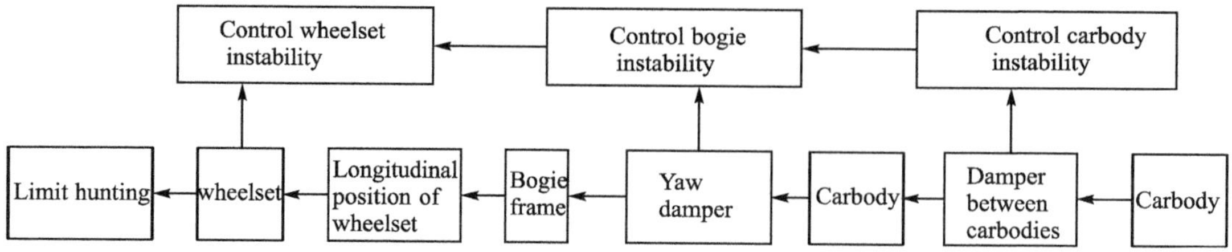

Figure 5.10
Control strategy of vehicle kinetic stability.

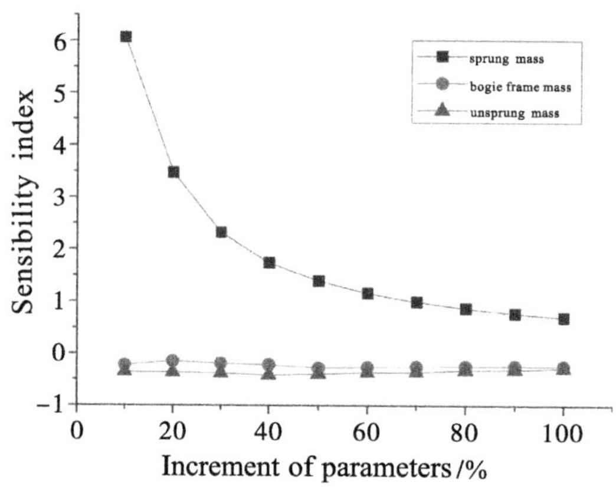

Figure 5.11
Sensitivity of the kinetic stability to vehicle mass.

in the mass of the car body, its sensitivity decreases, and it can be seen that the anti-hunting capacity of car body is enhanced. The sensitivity of the effect of mass between the primary and the secondary suspension and the sprung mass on the kinetic stability of vehicle is relatively small, which is almost unchangeable. The sensitivity of the effect of mass between the primary and the secondary suspension, such as the bogie frame, on the kinetic stability of the vehicle is the smallest (<0.5).

After analyzing the mass, the sensitivity of the effect of the secondary lateral damper, anti-hunting damper, and primary longitudinal positional stiffness on the kinetic stability of vehicle are shown in Fig. 5.12. The sensitivity of the anti-hunting damper is higher than that of the primary positioning longitudinal stiffness; the maximum coefficient of the anti-hunting damper is higher than 3.0, while it is about 1 for the primary longitudinal positioning stiffness. From a sensitivity of 6.0 for the effect of the car body mass on the kinetic stability of vehicle, to 3.0 for anti-hunting damper, and about 1.0 for the primary longitudinal positioning, it can be proved that the control strategy of the vehicle kinetic stability shown in Fig. 5.10 is appropriate.

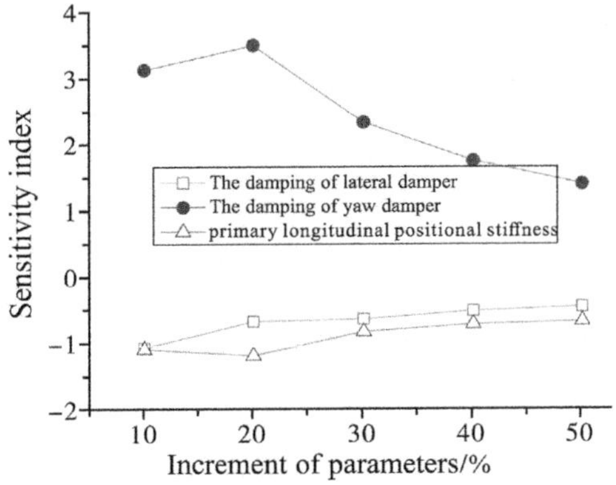

Figure 5.12
Sensitivity of the kinetic stability to different suspension parameters.

5.2.2 Method of parameter optimization design

The previous section suggested that the sensitivity of the vehicle kinetic stability to the parameters reflects the importance of parameter optimization. The sensitivity analysis is in fact part of the parameter optimization design content. Sensitivity analysis is widely adopted in engineering optimization design, which is also studied in the design of the railway vehicle design. For general system, the state equation can be written as [18,19]

$$\begin{cases} \dot{\mathbf{X}} = \mathbf{F}(\mathbf{X}, \mathbf{P}, t, \mathbf{u}) \\ x(t_0) = x_0 \end{cases} \tag{5.3}$$

where X is the state vector, F is nonlinear vector function, P is parameter vector, $P = [P_1, P_2, ...P_r]^T$, U are input vectors in the system, t is the time variable, and x_0 is the state value at initial time t_0.

When the performance function $\mathbf{Y} = [y_1, y_2, ..., y_n]^T$ express the system performance, if performance function y_i exists, the k-order partial derivative for parameter P is given by

$$\frac{\partial^k y_i(t, \mathbf{P})}{\partial P_1^{k_1} \partial P_2^{k_2} ... \partial P_r^{k_r}} \quad k_1 + k_2 + ... + k_r = k \tag{5.4}$$

The formula above is called the kth-order sensitivity function of parameter P for the performance function y_i. In general, only the first-order sensitivity function is considered.

Because of the nonlinear characteristics of the structure, parameter, and the performance effect in the dynamic performance of railway vehicles, the correlation between sensitivity and parameters is difficult to express using any function, even in linear conditions.

With the linear model, the sensitivity condition of the suspension stiffness and the damping parameters of high-speed train under different speeds to the kinetic stability of vehicle (the real part of maximum eigenvalue) is given by reference [20], as shown in Fig. 5.13, where an obvious nonlinear characteristic appears in the figure. It can be seen from the calculation results that the primary stiffness coefficient K_{pt} is the most sensitive to stability under all stiffness values, and the secondary lateral stiffness takes the second place. The damping of the anti hunting damper is the main sensitivity of damper to the stability, consistent with the previous control strategy analysis. Thus, the sensitivity of the vehicle kinetic stability to its parameters is usually solved numerically.

Although sensitivity research is an important method to optimize and determine dynamic parameters appropriately, the optimal selection of parameters can be conducted according to the control strategy of the vehicle kinetic stability illustrated above. However, the influence of the parameters on the vehicle system kinetic stability is always nonlinear, and it does not always have a peak but a monotonic correlation. The calculation results of the effect of the high-speed train's parameter variety on nonlinear hunting instability are shown in Fig. 5.14. Only the primary (axle box) positioning longitudinal stiffness in the figure has an extreme value. The damping of the anti-hunting damper, the damping (secondary) lateral damper, and the stiffnesses of primary longitudinal and lateral positioning have a considerable effect on critical speed.

Because of the monotonic characteristic, extremes cannot be sought during parameter optimization design, and the existing optimization of extremes is no longer appropriate. Moreover, the monotonic effect does not mean that a parameter can be increased or decreased infinitely to seek for stable extremes. Thus, the optimization of the high-speed train kinetic stability performance (or optimization of parameters) needs to be based on conditions or a regular method, as mentioned below.

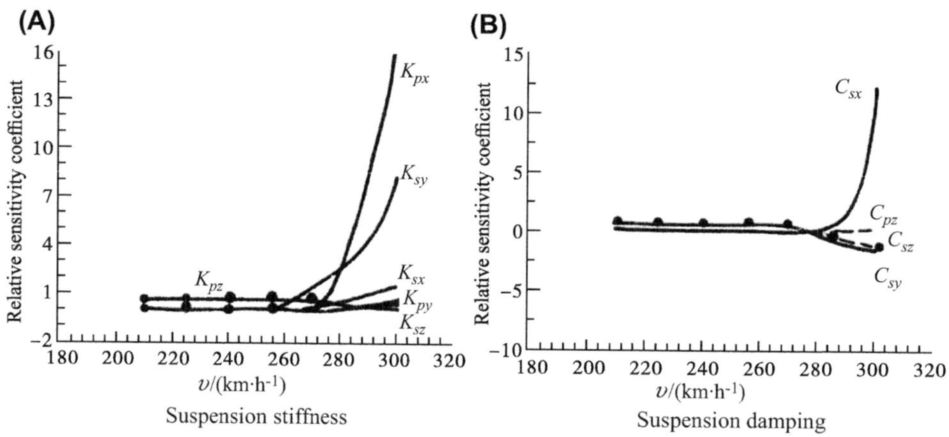

Figure 5.13
Sensitivity of suspension parameters to vehicle kinetic stability. (A) Suspension stiffness (B) Suspension damping.

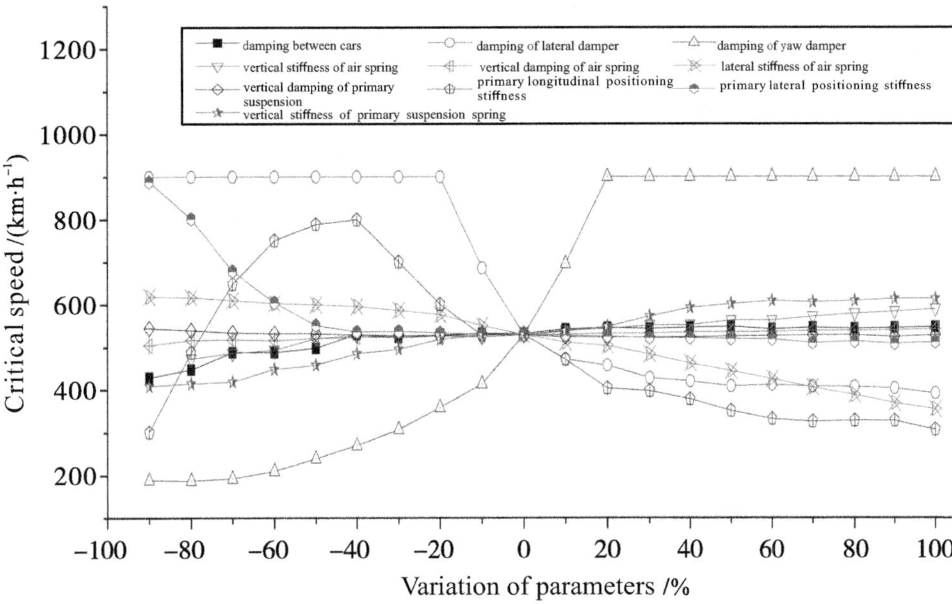

Figure 5.14
Effect of parameter variety on critical instable speed.

5.2.2.1 Determination principle of target value of critical instable speed

The design target is the critical design speed the vehicle should achieve; therefore, the critical instable speed parameter is used to evaluate the kinetic stability performance of the high-speed train. After the speed of the high-speed train is determined in design, there is no recognized method to identify the design target value of the critical speed and only a consensus that the critical instability speed of the kinetic stability index of the high-speed train's critical instable speed should be higher than an appropriate value of its running speed to ensure that the train has a certain margin of safety.

At present, there are two methods to determine critical speed.

(1) Professor Shen Zhiyun, Academician of Chinese Academy of Sciences and academician of Chinese Academy of engineering, proposes that in order to ensure the stability and safety of the high-speed train, the line test speed should be 100 km/h higher than the design speed without instability; the bench test speed should add another 100 km/h to ensure stability in the same time. That is, the margin of the critical instable speed is 200 km/h [21]. For the design of the new generation high-speed train, CRH380A, the continuous operating speed designed is 350 km/h, and the target value of the nonlinear instability critical speed is 550 km/h [17], which happens to be coincident with the viewpoint of Academician Shen Zhiyun.

(2) State Key Laboratory of the Traction Power of Southwest Jiaotong University is one of China's most important bases studying railway vehicle dynamics. Almost all railway vehicle dynamics research and parameter optimization in China are carried out in this laboratory. With the reference of stability design experience in mechanical systems, the critical speed is required to be more than 25% higher than the operation speed [22]. According to the actual condition of the high-speed train, it is considered that critical instable speed target should add 30% of speed based on line test speed. As for CRH380A, because the maximum designed operating speed is 380 km/h, the speed determined by line test is 420 km/h. The target value of CRH380A critical instable speed is 550 km/h, which is higher than 420 km/h $\times 1.3 = 546$ km/h.

The effectiveness of the two methods above can be proved through engineering application. However, in the strict sense, it is not scientifically defined according to the requirements of the railway vehicle kinetic stability.

from the author's point of view, the key to determine the critical speed target of the high-speed train kinetic stability index is full consideration to the declining extent resulted from service process. According to the application and maintenance principles of the EMU, both the parameters that may change during train operation and the resulting decreased value of critical speed should be considered first. Moreover, the ride quality and safety of vehicle operation should be ensured by considering a certain margin of critical instable speed. The requirement of the system motion to damping should be ensured to ensure the kinetic stability of the vehicle. Thus, the design target value for the critical speed is based on the designed operating speed by adding it to the declining extent resulting from the service process, and by adding to critical speed margin requested by system damping to ensure the ride quality of the vehicle. The details of the idea are shown in Fig. 5.15. It is noted that the target value of the critical speed should at least exceed the line test speed added to the critical speed margin requested by system motion damping to

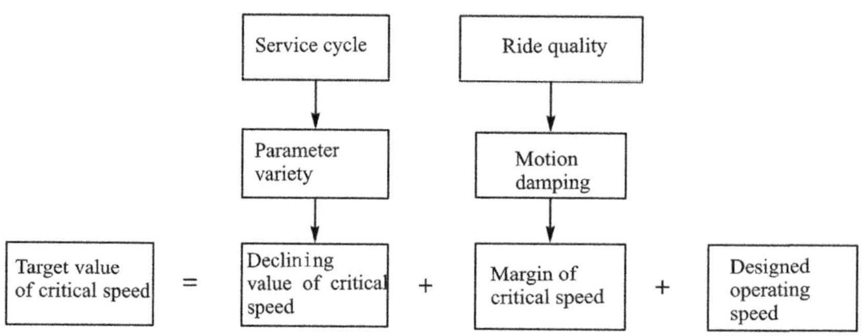

Figure 5.15
Determination principle of critical speed target value.

ensure the ride quality of the vehicle. Indeed, because the line test speed is generally only 10% higher than the design running speed, this "greater than" condition is generally satisfied.

According to the repair class and repair system of the high-speed train, the repair regulation of the high-speed train is divided into 5 levels [23]. The wheel tread and the exterior of the suspension components are inspected in the first level repair by checking if there is a definite failure. Besides the exterior, the wheel tread profile is measured in second level repair, and if it exceeds certain limits, lathing should be conducted. The bogie is disassembled and overhauled in the thrid level repair, wherein the contents are exterior inspection and tread measurement. The bogie is disassembled and overhauled in the fourth level repair as well; however, the parameter measurement to suspension components are added, and once the dynamic parameter deviates outside the allowable range, repair or renewal to the components will be conducted to recover the suspension parameter. In the fifth level repair, bogies and wheelsets are renewed. From the dynamics point of view, the vehicle is brand-new.

In the operation process, the dynamics parameters are not checked actively, especially the values of the suspension parameters. However, the operating quality is used to judge whether it is abnormal when an EMU is in operation, and a maintenance program will be triggered once abnormality occurs. There is still no complete information on the suspension parameter variety in service for the EMU in China. In the third-level maintenance of the CRH2 EMU, the results of the parameter measurement indicate that the damping of the damper decreases gradually during the operation. When the EMU's achieved 600,000-kilometer running distance, its anti-hunting damping is about 90% of the new car.

The most significant effect on vehicle stability is the effect of the tread wear on the equivalent taper. In the operation of the high-speed train, the wheelset wear cannot be avoided, and it exceeds the estimate. The tread conditions of the left and right wheel of the same wheelset of CRH2 EMU gained by the tracking measurement (measured every 50,000 km) are shown in Fig. 5.16. The train operates on the existing lines at speeds of

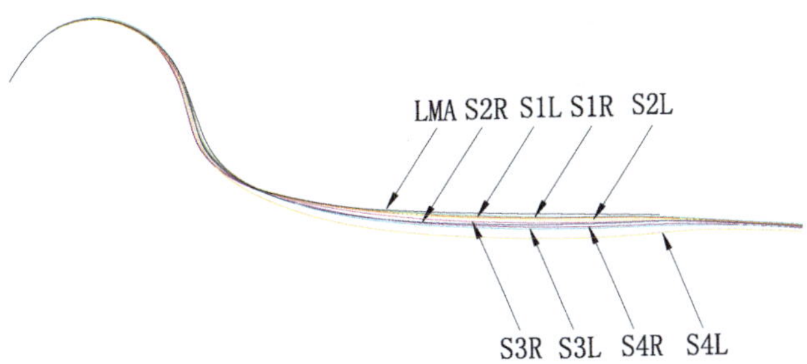

Figure 5.16
Tracking measurement results of tread wear.

200–250 km/h. In Fig. 5.16, LMA represents the initial tread profile, S is the measured tread profile, the numbers 1–4 are the measured numbers; L and R denote the left and right wheel, respectively. It can be seen that the tread wear concentrates around the rolling circle, and it presents a concave profile. The wears on left and right wheels are different while the EMU is under operation.

The wheel wear characteristics of different EMUs are different, and the same wear characteristics will not appear even when the same EMUs operate on different lines. The comparisons of the average value of the flange wear and the average value of the tread wear of CRH2 and CRH3 are shown in Fig. 5.17 [17]. The figure indicates that

1) The tread wear value is lower than the average wear value of 0.1 mm/10,000 km, especially for CRH3.
2) The flange wear of CRH3 is higher than that of CRH2-300.
3) The tread wear of CRH3 is lower than that of CRH2-300.

Whether the wheel tread wears to the concave profile or the wear difference exists between the left and right wheel, the equivalent taper of the wheelset will change, and then, the critical speed will be changed. Reference [24] compares the variation of the equivalent taper corresponding with the measured profile to the initial profile. It can be found that the equivalent taper greatly increases after wheel wear; the equivalent taper (3 mm) measured in the fourth measurement is higher than 0.4. According to long-term operation and analysis, the equivalent taper of 200–250 km/h EMU and the equivalent taper of a 300–350 km/h EMU are limited to 0.4 and 0.3 in their operation, respectively.

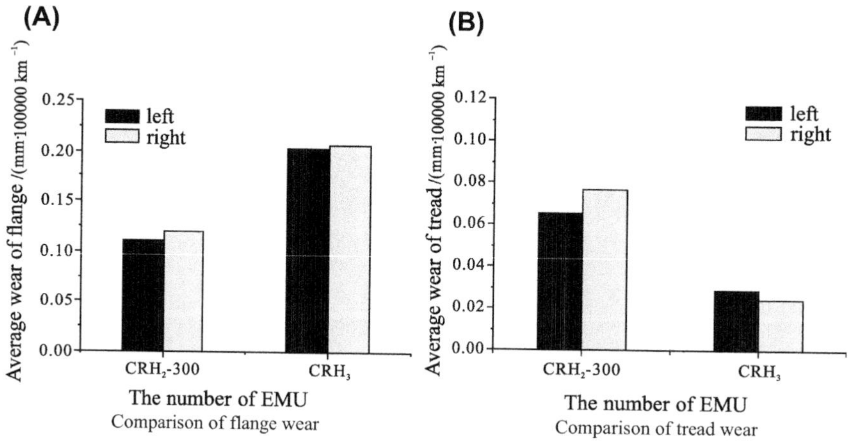

Figure. 5.17
Comparison of wheel wear of CRH2 and CRH3. (A) Comparison of flange wear (B) Comparison of tread wear.

Table 5.2: Effect of tread wear on critical speed.

Working condition number	Profile of left wheel tread	Profile of right wheel tread	Diameter of left wheel(mm)	Diameter of right wheel(mm)	Radius difference(mm)	Critical speed (km/h)
0	LMA	LMA	860.0	860.0	0	543
1	S1L	S1R	858.99	859.73	0.74	480
2	S2L	S2R	858.3	856.71	1.59	475
3	S3L	S3R	855.73	857.82	2.09	472
4	S4L	S4R	853.68	856.68	3	344

Critical speeds [24] obtained through calculation for the measured worn treads are shown in Table 5.2. The table indicates that the decline in the value of the critical instable speed of measured CRH2 EMU is close to 200 km/h in the fourth measurement, which is much more than expected. Because the maximum operating speed of this EMU is only 250 km/h, there should be a 100 km/h margin. What needs to be discussed is the critical instable speed margin to ensure the ride quality. Because the parameters of each vehicle are different, its vibration modes and modal damping ratio of the hunting motion are different.

Thus, the margin value needed for the critical speed should be calculated under the condition that the modal damping ratio of the hunting instable motion for ensuring the ride quality is achieved. According to the linear vibration theory, a 5.00% margin is usually required for damping the ratio of system vibration. It is suggested in reference [25] that a 5.00% damping ratio is applied for examining the calculation of the linear critical speed of the railway vehicle, as shown in Fig. 5.18. It is suggested that the margin of damping ratio should be set to be greater than 0.05 according to the years of calculation and experimental experience.

5.2.2.2 Engineering range conditions of the parameter

For high kinetic stability, the parameters may improve as it decrease or increases under the condition that the parameters have a monotonic effect on the hunting instable speed.

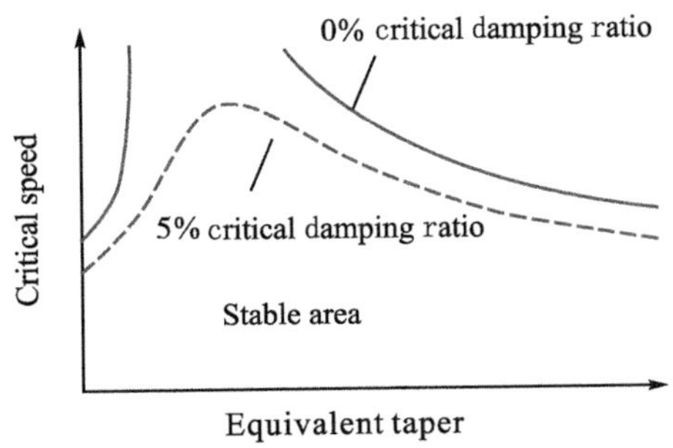

Figure 5.18
Stable areas.

However, as an engineering product, because of the limitation of product structure, material, structure and size, some parameters cannot be set arbitrarily. Some discussion around key parameters are provided below.

(1) Mass of car body

Stability has the largest sensitivity of the stability to car body mass, and an appropriate increase in car body mass is advantageous for the kinetic stability of the vehicle, but it is not economically desirable. In fact, under the limitations of materials, structure, and size, the car body mass has an economic goal, and thus, making it lightweight is a priority. When the car body mass cannot achieve the "calm down" effect for the kinetic stability of high-speed train, coupling dampers between the car bodies need to be considered to improve the anti-hunting capacity of the car body.

(2) Secondary air spring

Because the air spring is required to support the weight of car body, the action area needs to be ensured under a certain air pressure. Considering the probability of suspension structure size, the suspension stiffness provided by the air spring is always in a certain range under the conditions of certain air spring area and air pressure.

(3) Secondary damper

The secondary damper includes the lateral damper and the anti-hunting damper. According to the results in Fig. 5.14, the effects of the lateral damper and the anti-hunting damper on the critical speed are all monotonic with considerable effect. The required damping value can be easily found based on the critical speed target of hunting instability. In addition, the damper can be chosen or customized from the view of engineering.

(4) Primary lateral positioning stiffness

As for the bogie that depends on the spring axle box position, the primary positioning stiffness is restricted by the vertical load capacity of the spring and the required vertical stiffness. The spring that has a large capacity in the vertical load cannot have a small vertical stiffness because of space limitations, in which the lateral positioning stiffness is large. Under the condition of certain space and vertical load capacities, the correlation between the vertical stiffness K and the lateral positioning stiffness K_r is shown as formula (5.5) according to reference [26].

$$K_r = \frac{\sqrt{1 + 4.29\left(\frac{D}{H}\right)^2} - 1 - \frac{f}{1.3H}}{\left[0.295\left(\frac{H}{D}\right)^2 + 0.384\right]\left[\sqrt{1 + 4.29\left(\frac{D}{H}\right)^2} - 1\right]} K \tag{5.5}$$

where D is the pitch diameter of the spring, H is the effective working height of the spring, and $H = H_0\text{-}f\text{-}d$, where f is the vertical displacement, H_0 is the free height (fixed value) and d is the bar diameter of the spring.

It can be indicated that the lateral positioning stiffness is not independent under the condition of the vertical stiffness.

For the positioning method of the rotary arm, the lateral positioning stiffness is not only correlated to the lateral stiffness of the axle box spring, but it is also correlated to the longitudinal stiffness k_y, torsional stiffness k_φ, and the rotary length l of the rotary rubber bush. According to calculation model for the additional axle box lateral positioning of the rotary arm rubber bush in Fig. 5.19, the additional lateral positioning stiffness K_{py} of the rotary arm rubber bush is

$$K_{py} = \frac{k_y k_\varphi}{k_y l^2 + k_\varphi} \tag{5.6}$$

The lateral positioning stiffness of the rotary arm rubber bush is correlated to the axial stiffness of rotary arm rubber bush and the torsional stiffness of the bush. The additional lateral positioning stiffness of the rotary arm rubber bush is connected in parallel with the

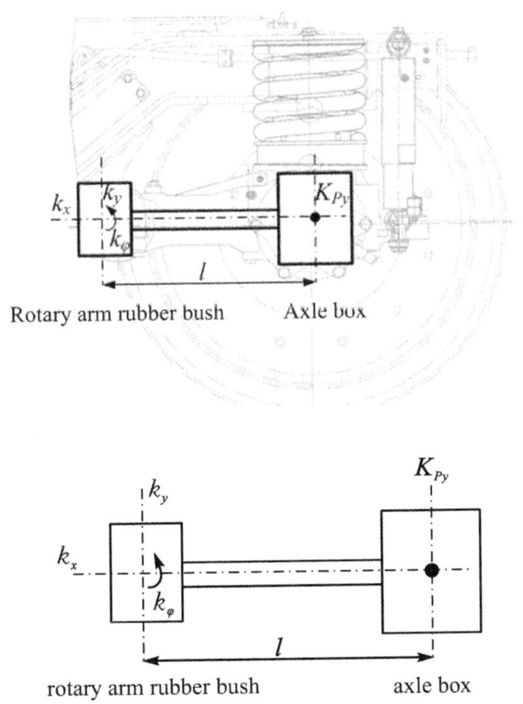

Figure 5.19
Calculation model for additional axle box lateral positioning of rotary arm rubber bush.

lateral stiffness of the axle box spring, which obtains a comprehensive lateral positioning stiffness of the axle box. Owing to the characteristics of the rubber bush and the process characteristics, the large longitudinal positioning stiffness of the rotary arm rubber bush controls the axial stiffness of the rotary arm rubber bush and the torsional stiffness of the bush, which indicates that it controls the lateral positioning stiffness of the axle box.

(5) Primary longitudinal positioning stiffness

Rotary arm positioning is widely used for the high-speed train. However, the space of the rotary arm positioning bush is limited, and the primary longitudinal positioning stiffness is mainly controlled by the hardness of a single rubber and thickness of rubber, which can provide a large range for longitudinal positioning stiffness. For example, the primary longitudinal stiffness of the axle box of CRH2 is only 13.7 MN/m, while the primary longitudinal stiffness of the axle box of CRH3 can reach 120 MN/m.

5.2.2.3 Optimization principle based on sensitivity

According to the service characteristics of the high-speed train and the sensitivity analysis results, the following aspects should be considered during parameter design:

(1) The sensitivity of the chosen parameter should be minimized to ensure good stability for the railway vehicle system.
(2) Considering the performance and structure characteristics of components, if the components has great stability in service or has characteristics of self-stabilization, the sensitivity can be a little high to ensure lower sensitivity to other parameters.
(3) Since the damper is prone to failure, the high-speed train operation should be ensured to be as safe as possible under the failure of the damping parameters, i.e., the kinetic stability should be ensured.
(4) From Fig. 5.14, it can be seen that some parameters have a slight influence on stability, and some have a dominating effect. It is clear that the primary position stiffness, anti-hunting damper, and secondary lateral damper are control parameters and take priority. Other parameters have little influence on the instable critical speed, and therefore, in keeping with engineering standards, the priority control parameters will be adjusted.
(5) From Fig. 5.7, it can be seen that the effect of the equivalent taper of the wheel tread on the kinetic performance of vehicle is significant. This can easily be changed during operation. Therefore, the tread profile is set as a priority; however, the effect of the variance of the tread equivalent taper on the kinetic stability should also be considered. The component performance and structure also influence the parameters; however, if the components are stable or have characteristics of self-stabilization, the sensitivity may be set higher to ensure lower sensitivity to other parameters.

5.2.2.4 Equilibrium principle of dynamic performance

To achieve excellent dynamic performance of the high-speed train requires excellent levels of kinetic stability, ride quality, and curve negotiation when correlated to safety. Even though the three major dynamic performance parameters may have the same levels of importance, the effect of parameters on the performance differ substantially. The process of selecting vehicle parameters to realize the equilibrium for dynamic performance follows a philosophical method of compromise. The kinetic stability of the EMU should be ensured first, curve negotiating performance improved second, and ride quality should be optimized last.

The main parameters that have contradictory effects on the dynamic performance of the railway vehicles are:

(1) Primary longitudinal stiffness; contradiction between stability and curve negotiation.
(2) Anti-hunting damper; contradiction between stability and curve negotiation.
(3) Axle spacing; contradiction between stability and curve negotiation.
(4) (Side bearing) rotational friction torque; contradiction between stability and curve negotiation.
(5) Primary vertical stiffness; (might appear) contradiction between comfort and wheel unloading rate
(6) Stiffness of traction rod; contradiction between stick-slip vibration and stability or quality.

There are other methods to solve the contradiction of parameter selection, such as adopting a radial bogie or independent wheelset structure, which can realize improvements in the critical speed and curve negotiation performance. Thus, using the change in the structure to solve the dilemma of the contradiction of parameter selection is also an alternative.

5.2.2.5 Reliability design of kinetic stability

The kinetic stability of the high-speed train is decided based on the structure parameters and suspension parameters of the vehicle. These parameters are allowed to have tolerance in manufacture, which will be changed in application. Therefore, the distribution law of its parameters can be obtained through the measurement of the actual product. For a parameter with a nominal value e_0, the parameter range under the allowable tolerance is (e_0- e_i, e_0+ e_s). The parameter in the allowable range will be a probability distribution, the designed nominal value v_{cr0} of instable critical speed can be obtained through the corresponding parameter nominal value e_0. The value range (v_{cr0}- v_{ei}, v_{cr0}+ v_{es}) of the instable critical speed can be obtained through the corresponding selection range (e_0- e_i, e_0+ e_s) of the parameter. The distribution law P_e corresponds with the parameters, and the distribution law P_v corresponds with the critical speed. Indeed, because of nonlinearity,

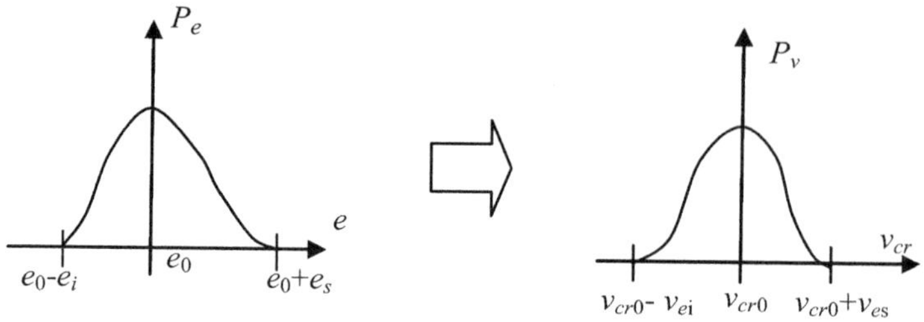

Figure 5.20
Distribution of parameters and its corresponding critical speed.

the mapped correlation between them is not linear, which reflects in the differences in the symmetry of the range of the corresponding nominal value and the distribution law, as shown in Fig. 5.20. The distribution law of the critical speed and the reliability based on probability should be considered in the design.

5.3 Optimal design of high-speed train ride quality performance

The ride quality of the high-speed train is indeed reflected by vibration intensity. For a high-speed train, the ride quality depends not only on the damping properties of the train, but also on the disturbances from the railway line and airflow. It is necessary to optimize the vibration reduction and vibration isolation performance of the train to improve the ride quality of the high-speed train. Since the excitation is often not controlled by vehicle engineering but by the irregularity, requirements proposed for the track profession are based on the ride quality requirement of the train. To improve the smooth running of high-speed trains, it is necessary to optimize the train's vibration reduction and isolation performance.

5.3.1 Vibration quality control

The vehicle suspension system can be classified into primary and secondary suspensions. For a mass spring damper system, improving the train's ride quality is the system's own vibration isolation performance, in particular, to prevent the resonance of the system. Such resonance is called the suspension resonance (rigid body resonance). However, the car body and the bogie of the high-speed train is actually a flexible body, and therefore, under certain vibration conditions, the structure will generate structural vibration, which also affects the ride comfort.

The airflow disturbance is not considered in the study of the general vehicle system dynamics, because it is generally low frequency disturbance that has little effect on the ride quality of the vehicle. However, the airflow tends to cause disturbance to the lateral movement of the tail car of the train, which should be considered in the study of the ride quality of the high-speed train. In general, the disturbance of the line and airflow is correlated to the speed of the train, and in particular, its frequency. In the excitation, the tendency of the hunting motion of the wheelset should also be considered. This excitation frequency is not only correlated to the running speed of the train, but also to the increase in the equivalent taper caused by the wear of the wheel tread profile (Eq. 5.1).

Fig. 5.21 shows the frequency isolation control strategy for the high-speed train. As the operating speed increases, the excitation frequency of the line and airflow increases, and simultaneously, the hunting frequency of the wheelset increases with the running speed and tread wear. In order to ensure effective vibration isolation, the car body and bogie structure of the EMU and the natural frequency of the suspension (rigid body) should be effectively isolated from the line excitation frequency and the hunting frequency of the wheelset. Even if it is impossible to isolate, it is necessary to ensure that the damping ratio of each vibration mode in the damping design to assure the attenuation of the vibration. For line disturbances with a fixed cycle, the resonance frequency should be avoided within the operating speed range during the design of the suspension parameters to effectively isolate the vibration caused by the line disturbance.

The ride quality of the high-speed train can be optimized based on the frequency isolation control strategy for the high-speed train. The first step is to obtain the excitation frequency, i.e., the excitation frequency of the line. Track irregularity is composed of the irregularities caused by the track slab, sleeper, rail joint, other irregularities, and random rail irregularities. Hence, the track irregularity includes random excitation and fixed frequency excitation formed by the structure, as shown by the track spectrum of the Beijing-Tianjin railway shown in Fig. 5.22. With the excitation spectrum signature, the suspension parameter design of the vehicle can be carried out, i.e., the transfer function of

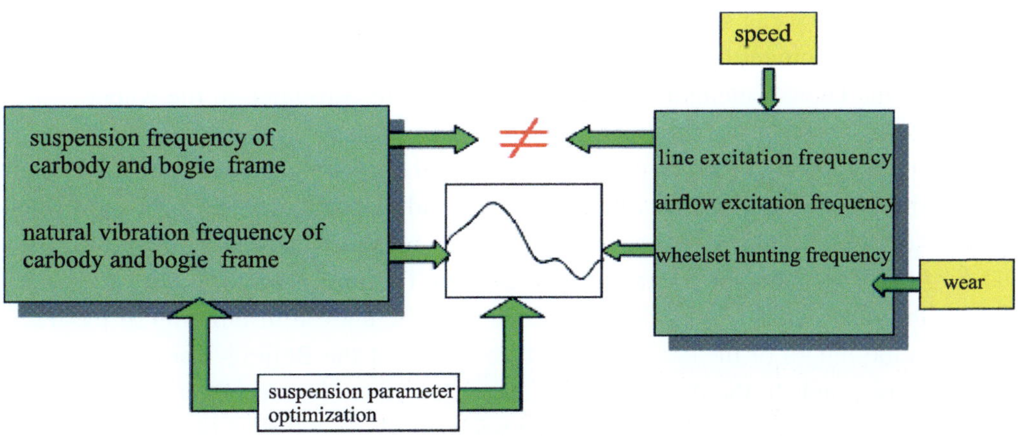

Figure 5.21
Frequency isolation control for EMU.

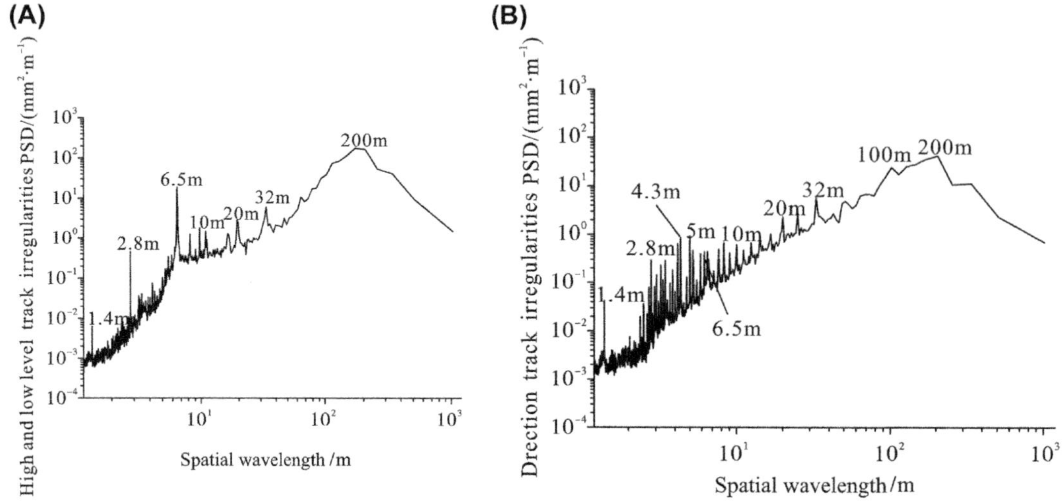

Figure 5.22
Track Spectrum of Beijing Tianjin Railway. (A) High and low level track irregularities (B) direction track irregularities.

the mechanical vibration is optimized because one, it avoids resonance, and two, it ensures reasonable damping of the system and effectively attenuates vibration.

5.3.2 Optimal design of parameter

5.3.2.1 Resonance control

The method to control resonance is avoiding resonance as far as possible. The priority is to identify the excitation frequency caused by the vehicle or carriage structure. Fig 5.22 (a) shows the track irregularities spectrum of the Beijing-Tianjin railway line on the high and low level where we can see that the main frequency of this line is very obvious. The wavelength of the dominant frequency is distributed at 1.4 m, 2.8 m, 6.5 m, 10 m, 20 m, 32 m, and 200 m. The wavelengths at 1.4 m and 2.8 m are correlated to the wheel perimeter or semi-perimeter of the track inspection car and are caused by the "out of roundness" or eccentricity of the wheelset. Theoretically, it should be removed from the line irregularity spectrum. However, there is also a kind of excitation input for the EMUs because of the same roundness and eccentricity of the wheels; 6.5 m is the length of the track slab of the Bögl unballasted track in the Beijing-Tianjin railway line. It forces vibration with the length of the track plate because of the uneven installation of the track plate; 32 m is the length of the elevated structure bridge of the Beijing-Tianjin railway line, which corresponds to the dominant frequency due to the forced vibration caused by the bridge. The length of the rail is 100 m, the wavelength of 200 m is twice the length of the rail, and the forced vibration is caused by the rail section [17]. Fig. 5.22 (b) shows the direction of the irregular track spectrum with the peak frequency of the excitation similar to that of the irregular track spectrum at high and low levels.

The line irregularity spectrum as shown in Fig. 5.22 adopts the channel spectrum, expressed in terms of the wavelength. The EMU spectrum is characterized by frequency (Hz). The excitation frequency of the line irregularity can be calculated based on the wavelength.

$$f_j = \frac{v}{L_j} \tag{5.7}$$

In equation (5.7), v is the running speed of the vehicle and L_j is the excitation wavelength. Indeed, as the speed increases, the excitation frequency of the relevant wavelength of the track irregularity is comprehensively increased, exacerbating the excitation of the bogie and leading to a decrease in the ride quality of the train. Simultaneously, the increase in the excitation frequency also increases the dynamic interaction between the wheel and the rail, which aggravates the vibration and damage to the track system.

After determining the characteristics of the track spectrum, it is followed by the natural vibration characteristics of the EMU. For a full-degree-of-freedom vehicle model, the modal shape of the bogie and car body include three translations and three rotary motions. The vibration mode that affects the ride quality of the vehicle includes the traverse, bounce motion (vertical displacement), rolling, pitch, and yaw; the longitudinal motion is generally not considered.

In order to avoid resonance, a simple method is to provide the line excitation frequency at different speeds, the vehicle suspension frequency, and the structure natural frequency forming the vibration spectrum of the vehicle system (as shown in Fig. 5.23).
The correlation between the natural frequency of the vehicle system and the rail line exciting frequency can be seen from the vibration spectrum below. The high-speed train running speed shown in Fig. 5.23 is in the range of 350 km/h–400 km/h, the forced vibration caused by the length of the track slab and the length of the bridge will not resonate with the vertical natural frequency of the vehicle. The hunting frequency of the bogie does not resonate with the lateral natural frequency of the vehicle system. It should be noted that the suspension and structural vibration of the car body and the frame have various vibration modalities, and the dominant frequency of the disturbance is more than one. Fig. 5.23 is only an example of the low frequency range. In fact, in order to avoid certain disturbing frequency points, it can be achieved by the design of the suspension stiffness.

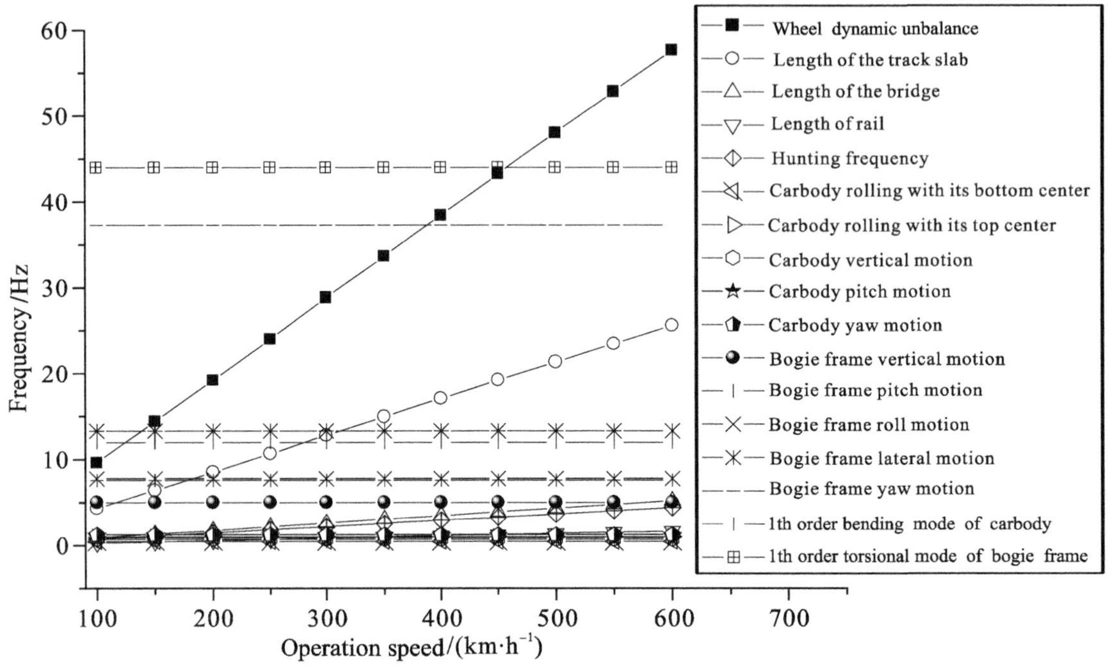

Figure 5.23
Low frequency vibration spectrum of vehicle system.

5.3.2.2 Optimal design of the transfer function

In order to make passenger trains better in terms of ride quality and to provide comfort for the passengers, the doubled suspension system is generally adopted. The primary suspensions from the axle box to the frame is generally hard, and the secondary suspensions from the frame to the car body is relatively soft generally because it has air springs with little stiffness. The main frequency of the car body is a low frequency, which is only correlated to the total static deflection of the suspensions and has nothing to do with the allocation of static deflection in the two suspension systems. The displacement of the car body and bogie during low frequency vibration is into the same direction. The main frequency of the bogie frame is high frequency, in addition to the total static deflection, and it is also correlated to the distribution of static deflection in the two suspension systems and their mass ratio. The displacement of the car body and bogie during high frequency vibration is in the reverse direction [17].

Rail excitation is the main source of vibration for the high-speed train. In addition to the fixed structure frequency, there are stochastic irregularities. Hence, to optimize the ride quality performance of the high-speed train requires effective separation and containment of excitation from the line in the full frequency domain, which is to carry the optimal design of transfer function as shown in Fig. 5.17. After the vehicle system is determined, the system vibration transfer law (transfer function) is constant regardless of the change in the parameters during operation. The response of the vehicle to the excitation is analyzed by the transfer function, which can fully reflect the vibration transmission of each frequency component. Assuming that the transfer function of the system is $\boldsymbol{H}(\omega)$, the response spectrum of the system under the action of the stochastic excitation spectrum $\boldsymbol{S}_x(\omega)$ is [17]:

$$\boldsymbol{S}_y(\omega) = |\boldsymbol{H}(\omega)|^2 \boldsymbol{S}_x(\omega) \tag{5.8}$$

Fig. 5.24 compares the primary and secondary vibration acceleration transmission rates of a CRH high-speed train under the condition of the vehicle linear model. The transmission rate characteristics of the primary and secondary suspensions are different. The secondary suspension has a lower transmission rate to the high frequency; it can isolate the high

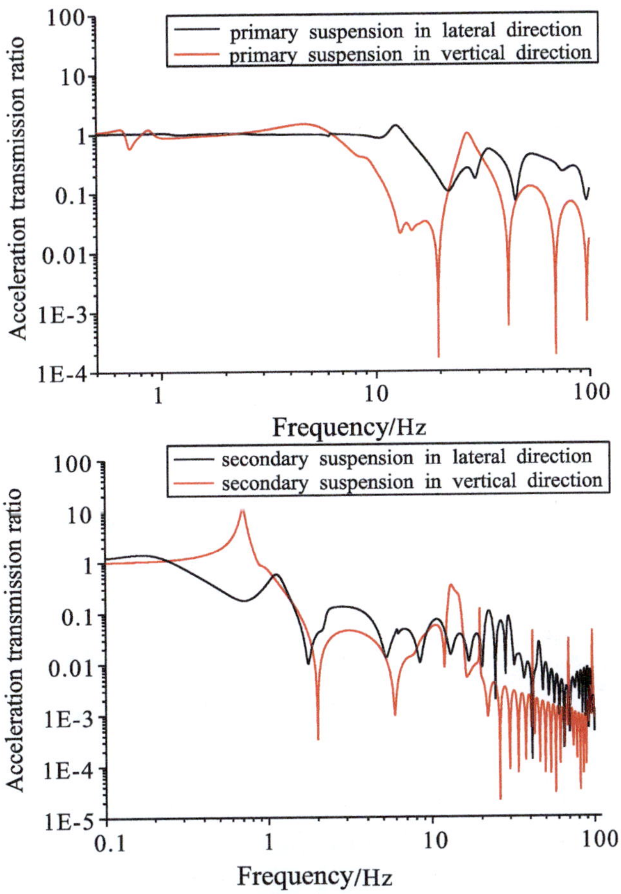

Figure 5.24
Acceleration transfer rate of primary and secondary suspension.

frequency excitation, but the resonance of the natural frequency of the vehicle body is strengthened, such as the resonance peak value about 1 Hz.

The purpose of controlling the vehicle vibration is not only to reduce the ride quality and the comfort index of the vibration state over the entire frequency range, but also to reduce the abnormal vibration in a certain frequency range, or to attenuate intrinsic excitation with a certain frequency. In general, different attenuation ratios at different frequencies can be obtained by changing the damping ratio of the system. To change the amplitude of the frequency response implies changing the damping. Fig. 5.25 reflects the vibration reduction effect of the vertical damping of the secondary suspension on the vertical vibration of the car body. Indeed, as the vertical damping of the second system increases, the amplitude of the low frequency response below 1 Hz decreases; however, the amplitude of the high frequency response increases. Therefore, the vehicle response of the different excitation frequency segments can be controlled by the damping design.

All optimal designs of the ride quality performance must be specific to the spectrum characteristics because the track irregularities and the response to vehicle disturbance are different under different line conditions. Fig. 5.26 shows the power spectral density of the vehicle acceleration under the action of Germany low interface track spectrum, Wuhan-Guangzhou track spectrum, Beijing-Tianjin track spectrum, and 300 km/h level temporary repair track spectrum (track spectrum are generated by the requirements of *Rules of Maintenance for Unballasted Track of High-speed Train (trial)*). It is clear that the same transfer function differs in the response spectrum under different track spectrum

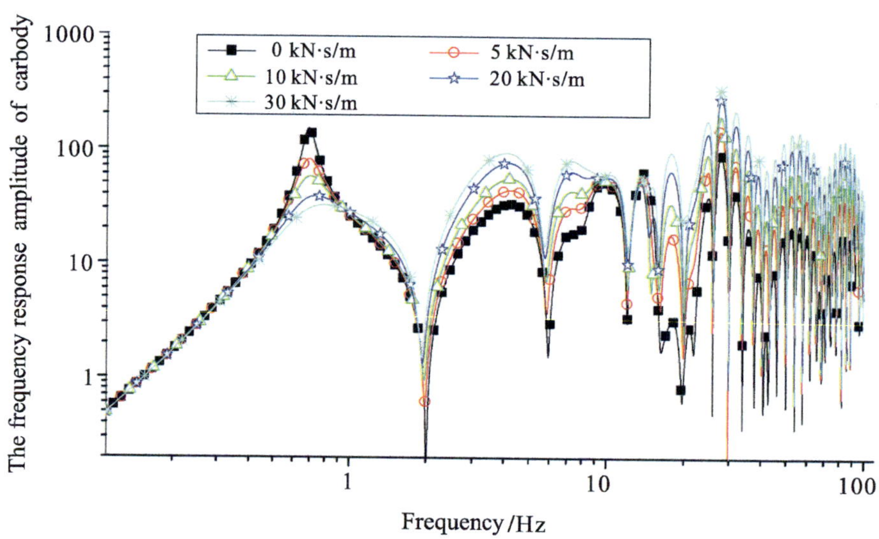

Figure 5.25
Influence of the vertical damping of the second system on the vertical acceleration of the car body.

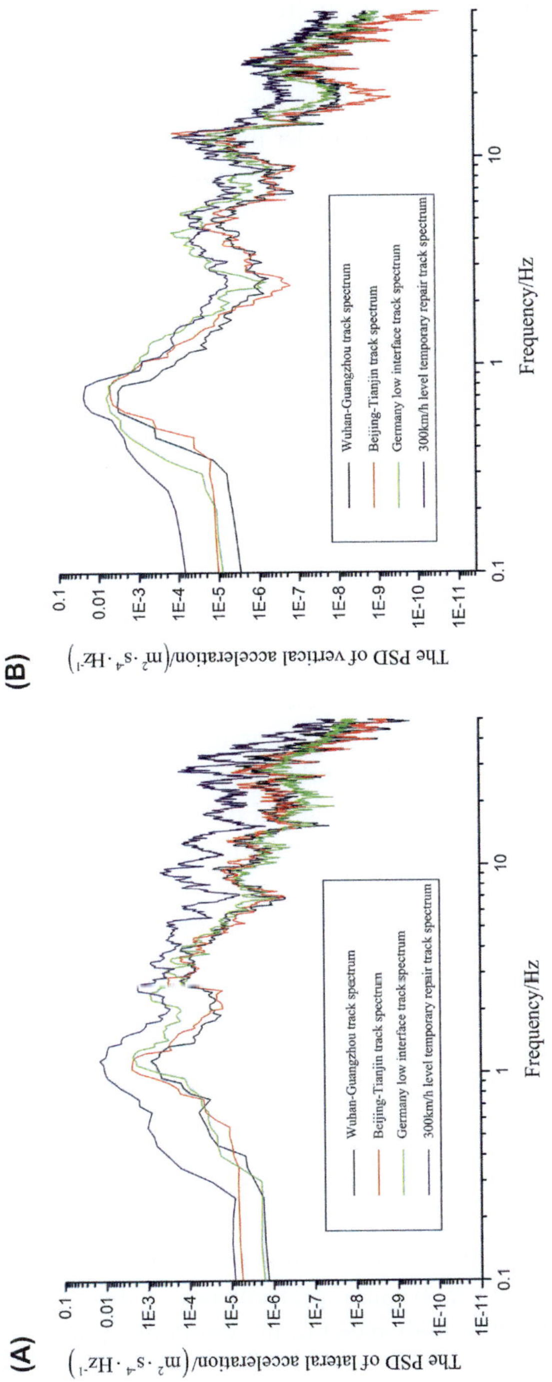

Figure 5.26
Vehicle Acceleration response under different excitation spectrum. (A) lateral (B) vertical.

excitations. For a specific vehicle model, a targeted vehicle ride quality design should be carried out in conjunction with the new repair status of the railway line and the track characteristics of the service state.

For the optimal design for the transfer function of a high-speed train ride quality performance, a more active control method can be adopted, i.e., under the same transfer function design. Under the action of the stochastic excitation spectrum $S_x(\omega)$, the desired response spectrum of car body is $S_y(\omega)$. In this way, it is theoretically possible to derive the desired transfer function as $\boldsymbol{H}(\omega)$, and then design the primary and secondary suspension parameters of the EMU according to the transfer function. It is not possible to arbitrarily design the desired transfer function as $\boldsymbol{H}(\omega)$, since the vehicle structure of EMU is restricted by the constraints of primary and secondary suspensions and parameter engineering applications.

5.4 Safety design of running high-speed train

Operating safety is always priority in the development of high-speed trains. The speeds of 300–350 km/h are similar to the take-off and landing speeds of aircraft, and therefore, derailment or collision will be catastrophic. The worst incident to date happened on June 3 1998 between Munich and Hamburg in Germany. The derailment caused by a wheel fracture while crossing the bridge at 200 km/h. The train and bridge collapsed leaving 101 people dead, 88 injured, and economic losses of more than 100 million mark. On July 23 2011, a Chinese EMU had a collision incident on the Yong Wen railway line. Although the EMU only had a rear-end collision with the 16 km/h train at a speed of 99 km/h, 40 people were killed, and 200 people were injured. The Japanese railway, known for its safety, on April 27 2005 suffered a major train derailment incident in Hyogo, killing 106 people and injuring more than 400 people. These incidents are reasons why the safety of high-speed trains receive considerable attention.

Ensuring the operational safety of high-speed trains is not only the basic task of researchers, designers, and manufacturers, but also the top priority for operators. Assuming responsibility for the basic performance guarantee of the high-speed train must also be priority from within the field of engineering dynamics.

Before discussing the basic control strategy for the operation safety of the high-speed train, the first thing is to understand the basic elements and characteristics that affect it. The most serious accidents in high-speed train are collision and derailment. Collision is often caused by operational management, and not by the dynamic performance of the high-speed train, while derailment may be correlated to high speed train system dynamics.

In order to understand the derailment, we first consider the basic characteristics:

- Inherency: The wheel rail correlation is an open constraint. On the one hand, the train has a space for upward movement on the track; on the other hand, although the left and right motions have a rim effect, the lateral movement of the train wheelset is not completely constrained because of the wheel rail wear. Thus, from the perspective of restraint, the train has an inherency of derailment.
- Complexity: Derailment is not isolated but results from a number of factors. In terms of derailment due to failure, the degree of impact on the train operation safety varies depending on the location and the form of failure. In addition to the impact of natural disasters, the analysis of derailment becomes very complicated.
- Controllability: Although derailment is inherent in railways, it can be controlled. Once we have mastered the mechanism and influencing factors, and we can influence the laws of derailment, we can completely avoid any derailment accidents. It is the contradictory role of the inherency and controllability of derailment that makes the derailment happen by chance.

Derailment can be divided into four categories [17], the first category is the derailment caused by manmade disasters, which includes derailment caused by abnormal conditions such as damage to railway facilities and train operation management; this type is not part of dynamics. The second category is the derailment by disasters caused by natural disasters such as earthquakes, sandstorms, landslides, and mudslides. For this type of derailment, it is necessary to study the impact of disaster types and disaster intensity on derailment, and finally propose measures to actively reduce the damage of natural disasters. This is another important task in the study of the dynamics of coupled system of high-speed train, which is an extension of the application of dynamics. The third category is the derailment by failure, which is due to the failure of the vehicle structure or the contact surface. Failure is a process from accumulation to suddenness and is also a multidisciplinary study of dynamics research, structural strength research, friction and wear research, and reliability research. The fourth category is the derailment by the intrinsic property, which is caused by its inherent properties such as wheel-rail correlation, catenary-pantograph correlation and vehicle structure, or often under normal operating conditions, due to the characteristics of the railway system, especially due to changes in characteristics (parameter failure) caused by long-term operation. Since this type of derailment is a non-human derailment, it is necessary to guarantee the running safety of the train in terms of mechanism, theory, and features, which needs to be solved at the beginning of the train design, i.e., from the source.

Although there are four types of derailments, their degrees of influence are different. Among them, failure causes the most accidents, including the German ICE and France's TGV derailment accidents due to structural failure. The study of high-speed train safety service should pay considerable attention to the impact of structure and material failure on the running safety the high-speed train.

Failure is a ubiquitous problem, and the failure of the running safety of the high-speed train has its own importance. Failure is basically caused by vibration or relative motion; for high-speed trains, the form of the relative motion is very special and it presents diversity. The basic relative motion has four forms that are high-speed rolling (wheel-rail), high-speed sliding (bow net and brake), tight fit with fretting (axe wheelset axel and threaded thread connection), and structural vibration (steel structure) [17]. Under vibration conditions, different relative motion modes have corresponding failure modes, such as wear and contact fatigue in high-speed wheel and rail rolling, friction and wear in high-speed pantograph sliding and arc ablation, fretting fatigue and surface bonding in an interference fit of axles and fatigue damage of structural parts, etc. In the same relative motion mode, there are often different forms of failure such as friction and wear and contact fatigue in the wheel-rail system. Different failure modes have different failure mechanisms, and there are also different failure modes of coexistence and mutual transformation problems. Therefore, in the study of the dynamics of coupled system in a high-speed train, the similar research is classified into the generalized kinetic research category.

The safety control of high-speed train is divided into two levels. The first is the design level, which guarantees the safety of the basic performance of the dynamics, i.e., theoretically not to derailment. The second is the application level, which is to study the safety of the following vehicles in service failure, parametric transformation, and natural disasters. The most basic research content of basic vehicle system dynamics is kinetic stability, ride quality, and curing performance. First, at the design level, the priority is to ensure the kinetic stability of the train so that the running safety of the train can be guaranteed, but only the EMU is sufficiently stable (including in the service process). The curving performance is a test of the steering capability of the bogie. Although the curve radius of the high-speed railway during operation is relatively large and the curve passing problem is no longer prominent, the tendency of derailment caused by wheel-rail lateral force always needs attention. In particular, for high-speed EMUs, small curves such as station yards should be considered. China's high-speed EMUs have experienced derailment accidents when it passes small curves. Although traditional research on the curve passing ability is no longer relevant, the high-speed switch still needs more attention. This is not only the verification of the derailment coefficient such as the traditional derailment coefficient and the wheel unloading rate, but more attention needs to be paid from the perspective of high-speed motion and impact.

Although the ride quality of the train will be affected by the load reduction of the wheel weight, the running safety is still the first consideration. The study of the influence of vibration on structural failure has a higher priority than the influence of vibration on ride quality. Hence, it is a problem at the application level. Thus far, the research on the vibration of parts and components of the EMU has not received considerable attention. In fact, the vibration of the components of the EMU causes accidents such as the axle box spring breakage, transmission gearbox housing cracking, cowcatcher fracture, and so on. The reason is that the vibration severity of the components is not yet clear; the design and calculation basis are insufficient. The ride quality performance of the EMU in the application is reduced and the vibration is aggravated. Realizing a high ride quality for the EMU as well as reducing the vibration intensity of components are fundamental measures to minimize the failure of EMU parts and components. At the application level, the second problem of train running safety is the non-interference performance. This is for strong external excitation, including strong crosswind and strong earthquakes. The dynamics of coupled systems in high-speed trains couples the effects of the railway line and airflow into the dynamics study. On the one hand, through the train's configuration design and suspension design, it can reduce the impact of external excitation so as to achieve higher anti-interference performance. On the other hand, taking the train derailment as the goal, the dynamics of the coupled systems in high-speed trains is used to study the threshold to ensure the range of running safety. For example, under different excitation levels, such as the magnitude of winds and earthquakes, it is determined that the train's speed range of running safety is guaranteed to provide a basis for the operation of the train. The safety control strategy for high-speed trains is shown in Fig. 5.27.

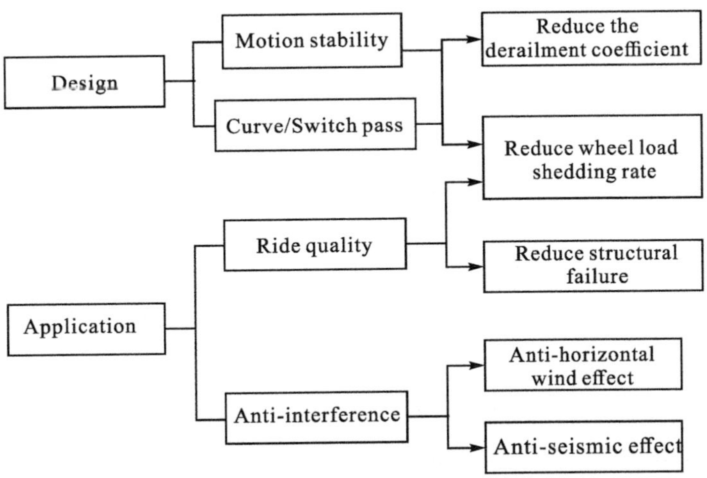

Figure 5.27
Safety Control strategy for high-speed train.

5.5 Comprehensive design of high-speed train parameters

The high-speed train consists of power cars and trailers. Since the power car needs to arrange a driving system, there are more traction motors, gear couplings, and gearboxes than trailers. In addition, the foundation brake cannot be used for disk braking, and only the disc brakes on both sides of the wheel are used. Other structures are similar to those of the trailer. For the power and trailer bogies, other structural parameters and suspension parameters associated with dynamic performance generally follow the same design principles to manage the exchange of components.

5.5.1 The influence of parameters of high-speed train on dynamic performance

The basic performance requirements of high-speed trains are kinetic stability, ride quality, and curving performance involving safety; the correlated control strategies have been discussed above. Although the same parameters may be equally important for vehicle system dynamics, the effects on different performance may be reversed. Table 5.3 shows the effect of suspension parameters on vehicle dynamic performance [27].

It can be seen from Table 5.3 that both the primary suspension and secondary rotary damping moment (the damping of yaw damper) have strong influence on the kinetic stability and curing performance. Parameter changes have an opposite effect on the kinetic stability and the curing performance. As for how to choose the parameters, a compromised treatment has been proposed in Section 5.2.2.4 to balance the dynamic performance. In addition, structural optimization can be used to resolve conflicts such as the innovative structural design of the freight car bogie. The famous Z8A bogie has poor stability and curing performance. In order to improve its stability, the bogie side frame is added with a

Table. 5.3: The effect of suspension parameters on vehicle dynamic performance.

Suspension parameters	Ride quality	Curve pass performance	Lateral ride quality	Vertical ride quality
Primary longitudinal stiffness	●	●	○	—
Primary transverse stiffness	●	○	○	—
Primary vertical stiffness	—	—	—	○
Primary vertical damping	—	—	—	○
Secondary transverse stiffness	○	—	●	—
Secondary vertical stiffness	—	—	—	●
Secondary shaking angle stiffness	○	○	—	—
Secondary transverse stiffness	○	—	●	—
Secondary vertical stiffness	—	—	—	●
Secondary rotary damping moment/ Damping of yaw damper	●	●	○	—

Note: ● Strong; ○ Medium; - Weak

cross bar to improve the anti-warp performance of the bogie, self-guided, self-steering, and force-steering radial bogie technology are used. As a result, the motion stability and curing performance are both improved. Fig. 5.28 shows the critical velocity and wheel-rail lateral force and derailment for balancing the curing performance of the original Z8A bogie (model 1), force-steering radial bogie (model 2), self-steering radial bogie (model 3), and the cross-bar support (model 4) [28]. It is clear that the improved three structural schemes achieve a double increase in critical velocity and curving performance. Therefore, it is the best choice to use structural changes to obtain parameters under permitted conditions.

As can be seen from the previous examples, using a little philosophy and methodology, the vehicle system dynamic performance can be optimized.

5.5.2 Parameter optimal design of high-speed train

There are many factors that affect high-speed train dynamics, but the core focus is the bogie. The bogie is the running gear of the train—similar to human legs—that bears the

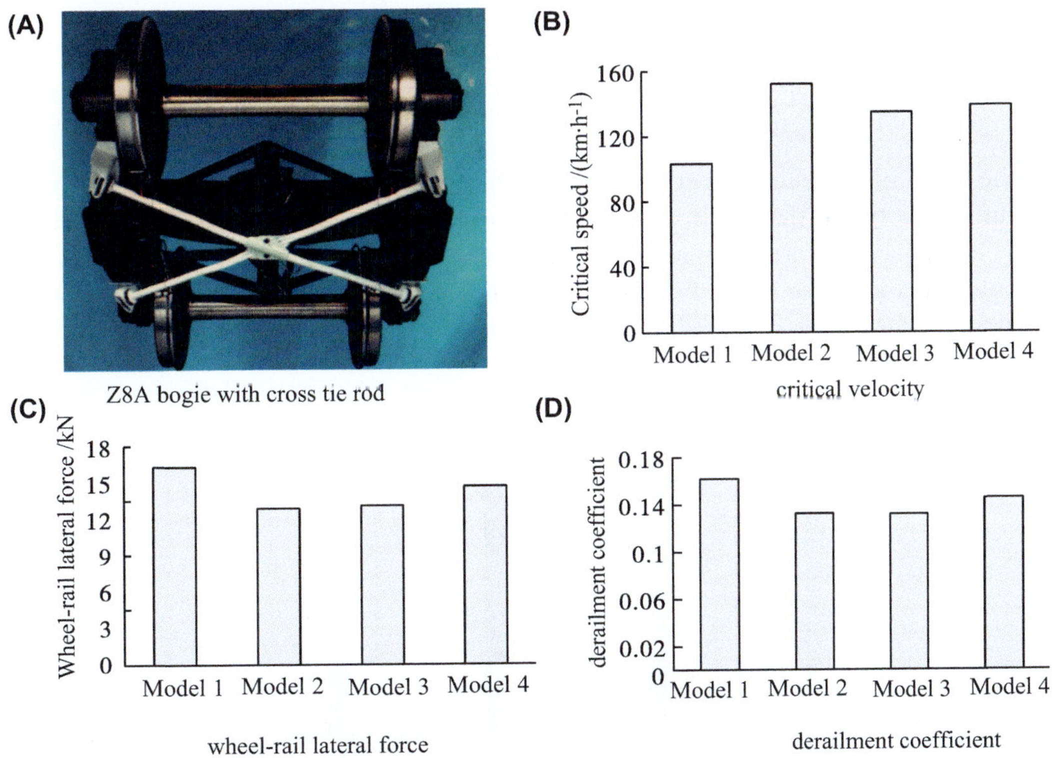

Figure 5.28
Performance comparison of Z8A bogies under different improvement schemes. (A) Z8A bogie with cross tie rod (B) critical velocity (C) wheel-rail lateral force (D) derailment coefficient.

bearing, the guidance, and the vibration isolation function of the train. It ultimately implements the drive trains and the braking force and further determines the kinetic stability of the vehicle. If the bogie is not designed well, as the running speed increases, the train may experience hunting instability, resulting in large vibration or even derailment. This may endanger the operation safety and deteriorate the riding quality of the train.

There are many parameters affecting the dynamic performance of the rolling stock, such as the tread shape of the wheel-rail, the suspension stiffness, and damping. These parameters depend on and support each other, and therefore, the high-performance of the high-speed train is realized through the system optimization of parameters.

5.5.2.1 Design of wheel tread

As seen from Fig. 5.29, as the equivalent taper increases, the critical velocity of the vehicle system decreases. From Fig. 5.30, we can see that in the curve radius of the R7000m with a super-elevation of 100 mm, when the vehicle running speed is over 250 km/h (deficient super-elevation), the derailment becomes smaller as the equivalent taper increases. It proves that the equivalent taper has the opposite effect on motion stability and curve performance. Thus, in order to increase the critical velocity of the vehicle, a smaller equivalent taper is required; for the curve performance of the vehicle, a larger equivalent taper is required.

Since the effect of equivalent taper on kinetic stability and curve performance is opposite, the traditional notion is that the tread design cannot be used to balance the kinetic stability with the curve pass performance; however, this is not so. First, the wheel tread we are using now is not a simple tapered tread but a worn wheel tread (the equivalent taper of different positions can be designed to different values). Second, the wheel-rail contact

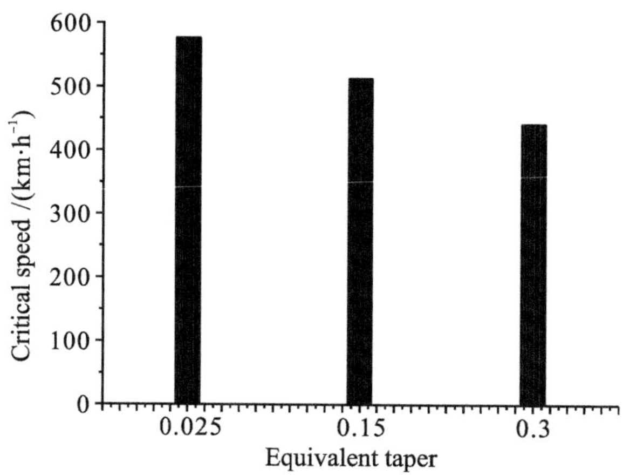

Figure 5.29
Effect of equivalent taper on critical velocity.

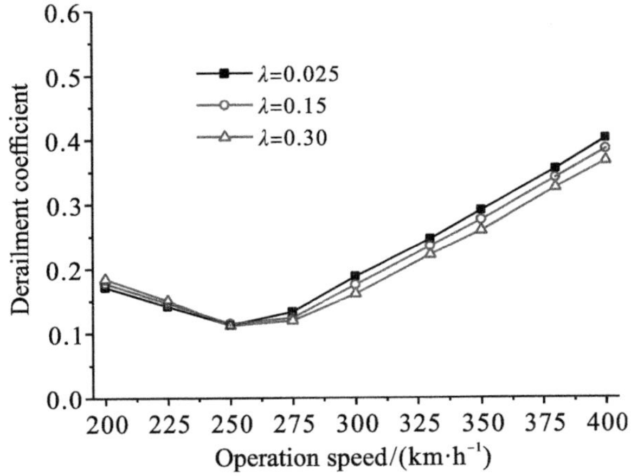

Figure 5.30
Effect of equivalent taper on derailment coefficient.

position when the vehicle is running on a straight line and on a curve is different (as shown in Fig. 5.31). We can design the wheel tread profile according to the equivalent taper target curve in Fig. 5.32 (the equivalent taper on the straight line is small, the equivalent taper on the large radius curve is larger, and the equivalent taper on the small radius curve is larger). In this way, the contradiction between the kinetic stability and the curve performance of rail vehicle can be solved by the special design of the wheel tread shape.

Through the traditional tread design method (first designing the tread and then calculating its equivalent taper), it is really difficult to design the target equivalent taper for the wheelset shown in Fig. 5.32. However, according to Shevtsov's RRD method of inverse curve (release the tread profile according to the required roll radii difference) [29], it is easier to solve this problem.

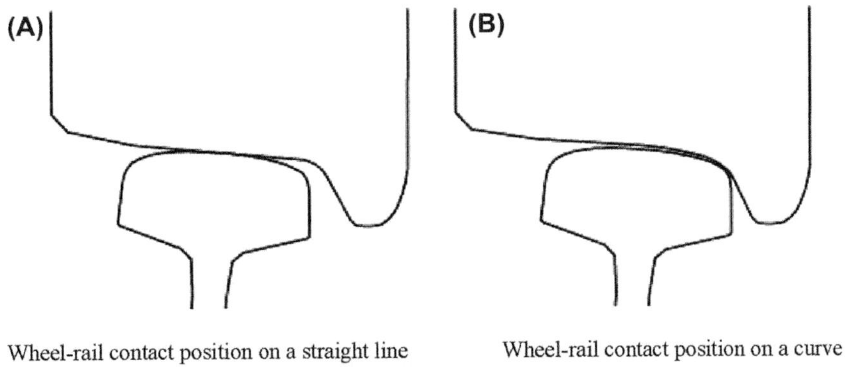

Figure 5.31
Wheel and rail contact position. (A) Wheel-rail contact position on a straight line (B) Wheel-rail contact position on a curve.

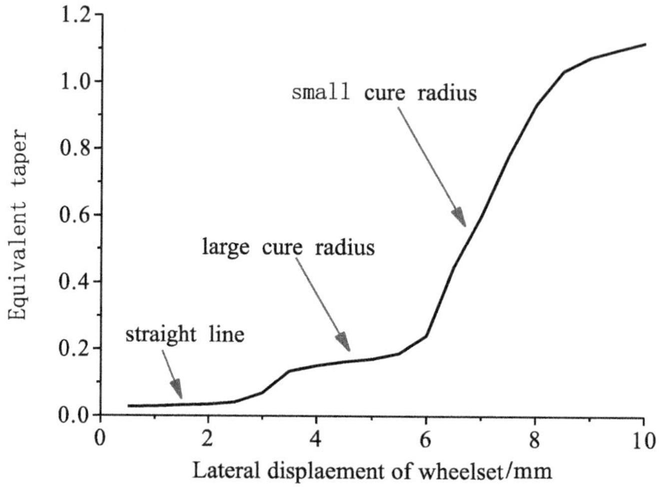

Figure 5.32
Wheelset equivalent taper design.

5.5.2.2 Selection of wheel diameter

The wheel diameter of the CRH EMU is available in three diameters: 860 mm, 915 mm, and 920 mm owing to the introduction of different technology platforms. The selection of the wheel diameter is limited on the one hand by the speed limit of the axle box bearing and the gearbox bearing, and also by the layout of the bogie construction space. From the perspective of dynamics, by increasing the wheel diameter, the bearing speed and the wheel-rail stress can be reduced. In addition, it is helpful to improve the critical speed of the vehicle system (as shown in Fig. 5.33). However, while increasing the diameter of the

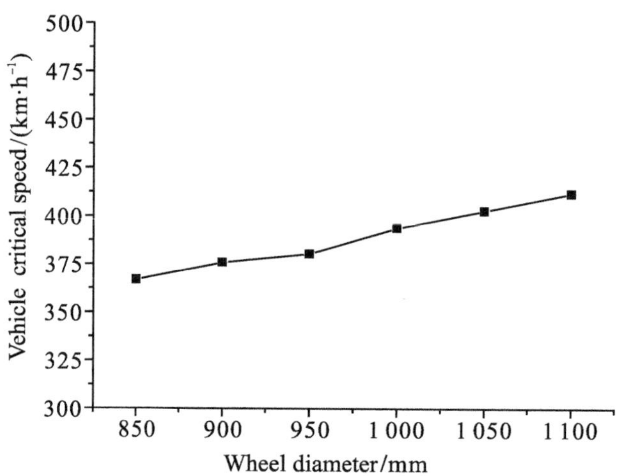

Figure 5.33
Effect of wheel diameter on critical velocity.

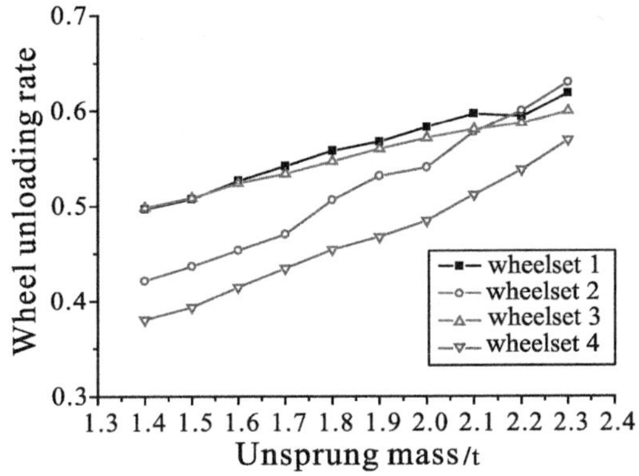

Figure 5.34
Effect of unsprung mass on wheel load reduction ratio.

wheel, attention should be paid to the change in the unsprung mass. As the unsprung mass increases, the wheel-rail dynamic force and the wheel unloading rate (wheel unloading rate) will also increase [27] (as shown in Fig. 5.34). Under such a contradiction, it is recommended to increase the wheel diameter appropriately if the structural space allows it, while controlling the unsprung mass by the careful design of the wheel web and the hollow axle. The CRH EMU axles generally use hollow axles with central bore diameters of 65 mm, 60 mm, and 35 mm. The central hole is arranged to reduce the unsprung mass, and more importantly, to facilitate the detection of internal cracks of the wheelset axle.

5.5.2.3 Selection of wheelbase

The wheelbase in most cases is determined by the bogie structural size, such as for the arrangement of motors and gearboxes. In fact, the bogie wheelbase has a considerable effect on the stability of the vehicle system: as the wheelbase increases, the critical speed of the vehicle system increases (as shown in Fig. 5.35); however, the curve performance is

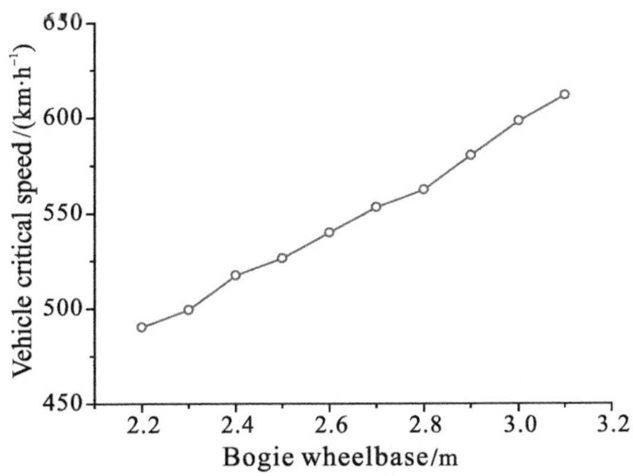

Figure 5.35
Effect of wheelbase on critical speed.

reduced, and the wheel-rail lateral force and wheel wear are increased. If the kinetic stability of the EMU is prioritized, the wheelbase of the high-speed EMU can be designed to be larger. However, as the wheelbase increases, the inertial radius of head shaking of the bogie frame usually increases. As the inertial radius of head shaking increases, the critical speed of the vehicle system decreases (as shown in Fig. 5.36). Therefore, while increasing the wheelbase, the change in the inertia of head shaking of the frame needs to be considered, and the moving moment of inertia of the frame can be controlled by a careful layout design of the bogie structure and accessory components [27]. In CRH EMUs, most bogies use a 2.5 m wheelbase form, which is intended to reduce the longitudinal dimension of the bogies in the structure. This is conducive to the arrangement of the undercarriage equipment and to reduce manufacturing cost.

5.5.2.4 Design of primary positioning stiffness

In general, appropriately increasing the longitudinal and transverse positioning stiffness of primary suspension can increase the critical speed of the vehicle system. However, the positioning stiffness cannot be as large as possible because when it increases beyond a certain extent, the critical speed increases more slowly and then decreases. This trend can be seen from Fig. 5.37. It can be seen from Fig. 5.38 that the primary longitudinal positioning stiffness has a greater influence on the curve performance, while the influence of primary lateral positioning stiffness is relatively small. Primary positioning stiffness has little effect on the vertical ride quality, and the primary lateral positioning stiffness has little effect on the lateral ride quality. There is an optimization range for the influence of the primary longitudinal positioning stiffness on ride quality. As shown in Figs. 5.39 and 5.40, too large or small longitudinal positioning stiffness are not conducive for improving the lateral ride quality.

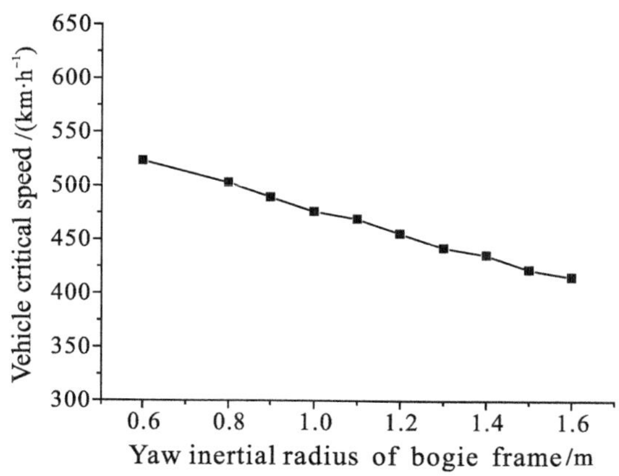

Figure 5.36
Effect of frame inertia radius on critical speed.

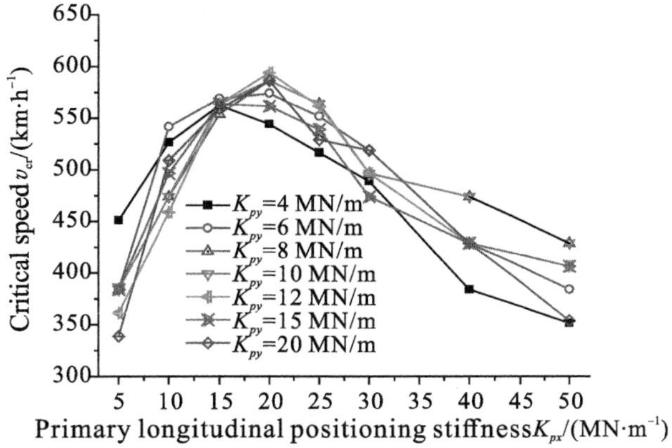

Figure 5.37

Effect of primary longitudinal positioning stiffness on critical speed.

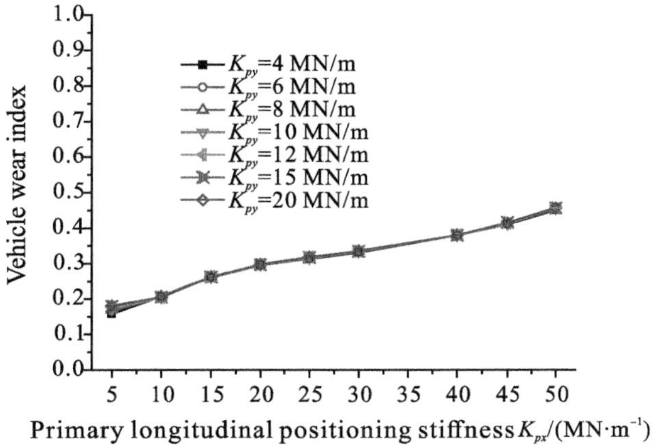

Figure 5.38

Effect of primary longitudinal positioning stiffness on wear index.

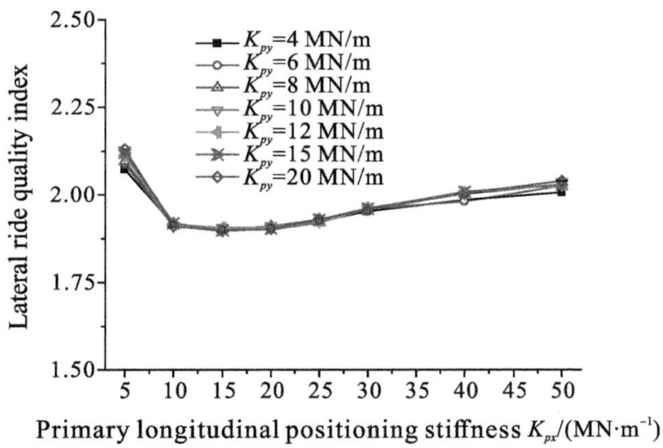

Figure 5.39

Effect of primary longitudinal positioning stiffness on lateral ride quality.

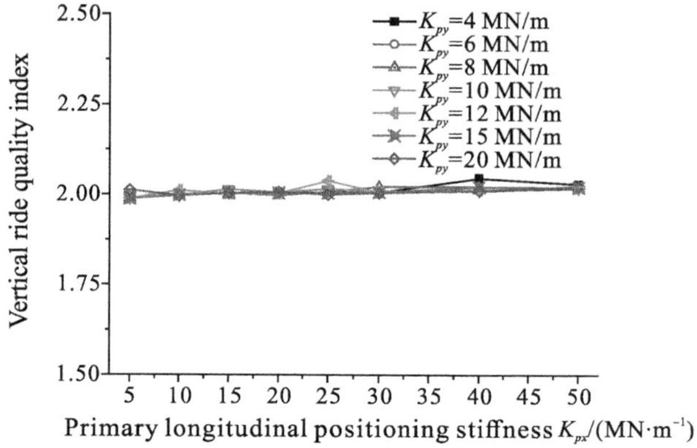

Figure 5.40
Effect of primary longitudinal positioning stiffness on vertical ride quality.

The curve radius of a high-speed train is generally large, and therefore, the curve performance is not too bad. The running speed of the high-speed train is relatively high, leaving the train stability as the main problem that needs to be solved. Therefore, the selection of the bogie suspension parameter should be based on train stability. To synthesize the influence rules of the primary longitudinal and transverse positioning stiffness on vehicle dynamics, i.e., if the high-speed bogie is to be operated at a speed of 350 km/h, the primary longitudinal positioning stiffness is in the range of 10 to 20 MN/m, while the primary transverse positioning stiffness is suitable in the range of 5–10 MN/m.

5.5.2.5 Parameter design of air spring

The secondary suspension of a high-speed EMU evolved from a swing bolster to a no swing bolster, from a bolster to bolster-free, and finally, it evolved into a bolster-free air spring suspension system.

Air springs have many advantages: the extreme equivalent spring deflection, vibration frequency, and working height that does not vary with loading, the internal set orifice for damping, etc. With the development of the air spring technology, the type of the air spring and the control mode have changed considerably. The most striking difference of air springs is the large airbag air spring and small airbag one [27], as shown in Fig. 5.41.

The stiffness of the air spring is directly correlated to the size of the airbag, and it depends on the size of the air spring attachment chamber and the orifice. As can be seen in Fig. 5.42, increasing the volume of the airbag and the additional chamber can reduce the stiffness of the air spring. In addition, the response of the large airbag air spring and the small airbag one is different for different frequencies. In the low frequency phase, the stiffness of the large and small airbag of the air spring is equal; however, as the frequency increases, the difference in the stiffness of different airbags gradually increases (Fig. 5.43).

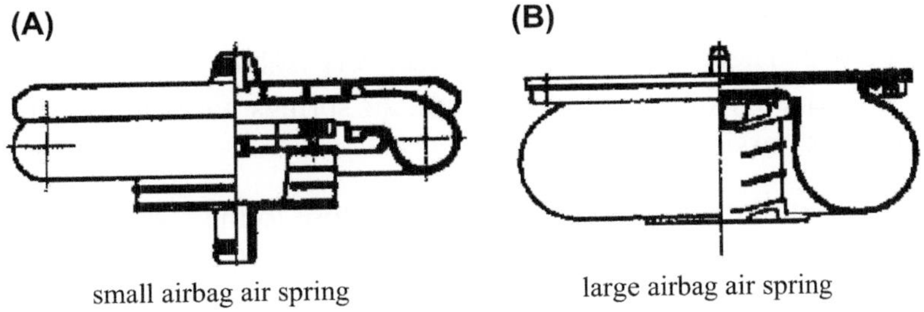

Figure 5.41
Structure of air spring. (A) small airbag air spring (B) large airbag air spring.

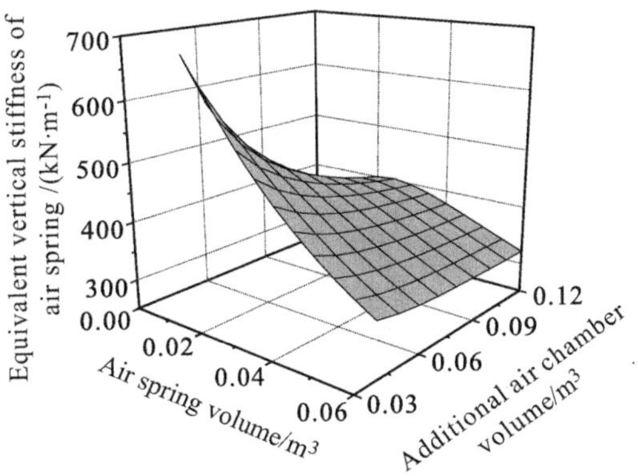

Figure 5.42
Effect of airbag and additional air chamber volume on vertical stiffness.

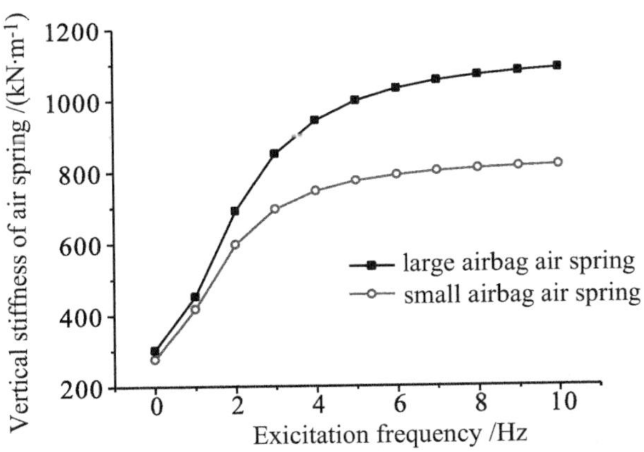

Figure 5.43
Variation of different airbag stiffness with frequency.

The effect of the high-frequency vibration of the large airbag air spring is better than that of the small airbag one. Therefore, in order to effectively isolate high-frequency vibration and improve the ride quality of the vehicle system, air springs with larger airbags should be selected as much as possible [27].

The dynamic stiffness of the air spring is particularly close to the size of the air spring orifice. As can be seen from Fig. 5.44, the smaller the orifice diameter, the greater is the difference between the static and dynamic stiffness of the air spring. When the diameter of the air spring orifice increases to 30 mm, the difference between the static and dynamic stiffness is small. However, when the diameter of the air spring orifice is increased to 25 mm, substantially no equivalent damping is provided. Increasing the air spring orifice is useful for reducing the dynamic stiffness of the air spring; however, its damping is greatly reduced, and the attenuation of the vertical vibration is reduced [27].

Taking one CRH type of the EMU as an example, the diameter of the air spring orifice is changed to adjust its equivalent damping and thereby adjust the vertical damping of the secondary suspension to achieve the best damping effect. Fig. 5.45 shows the effect of the air spring orifice diameter on the vertical ride quality index and the comfort index of the car body [27]. It can be seen that the orifice should not be too large or too small, and instead, it should be reasonably selected in conjunction with the actual situation of the vehicle.

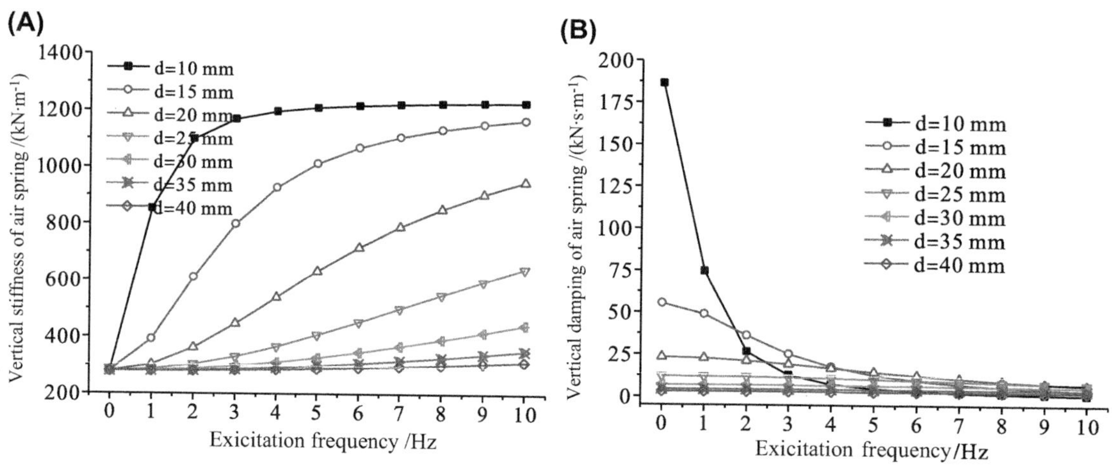

Effect of air spring orifice on dynamic stiffness damping Effect of air spring orifice on equivalent

Figure 5.44

Effect of the air spring orifice on the air spring stiffness and equivalent damping (d means the orifice diameter). (A) Effect of air spring orifice on dynamic stiffness (B) Effect of air spring orifice on equivalent damping.

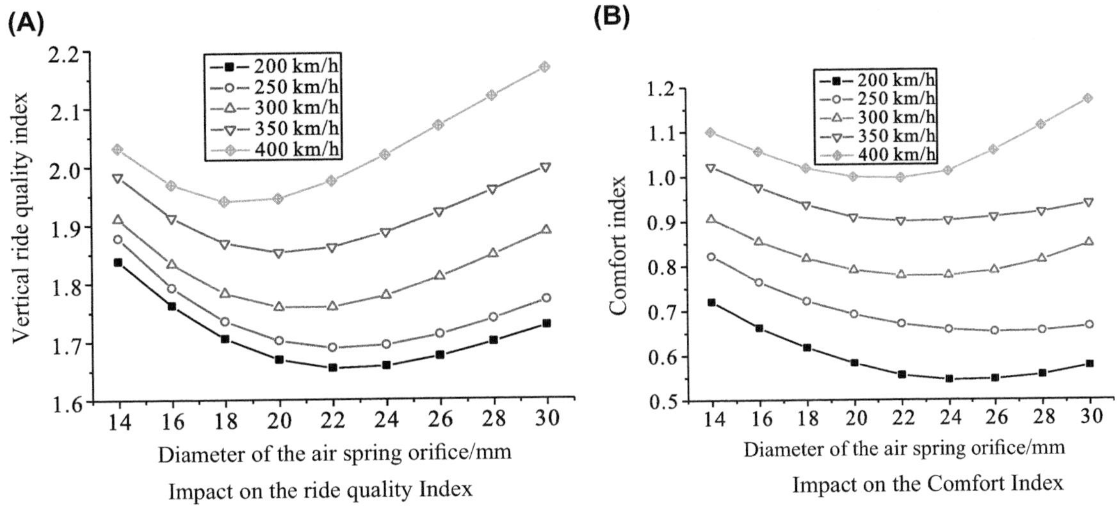

Figure 5.45
Effect of air spring diameter on vertical ride quality and comfort index. (A) Impact on the ride quality Index (B) Impact on the Comfort Index.

5.5.2.6 Parameter design of damper

In order to reduce vibration and improve comfort, a variety of dampers are provided on the high-speed EMU bogie, mainly the primary vertical damper, secondary lateral damper, secondary vertical damper (it can be replaced by air spring orifice) and yaw damper, as shown in Fig. 5.46.

The bogie of the EMU damper generally uses a hydraulic shock absorber; the resistance of a hydraulic shock absorber is correlated to the piston tensile or compressive velocity. The higher the speed, the greater is the resistance. In a mechanical vibration study, it is

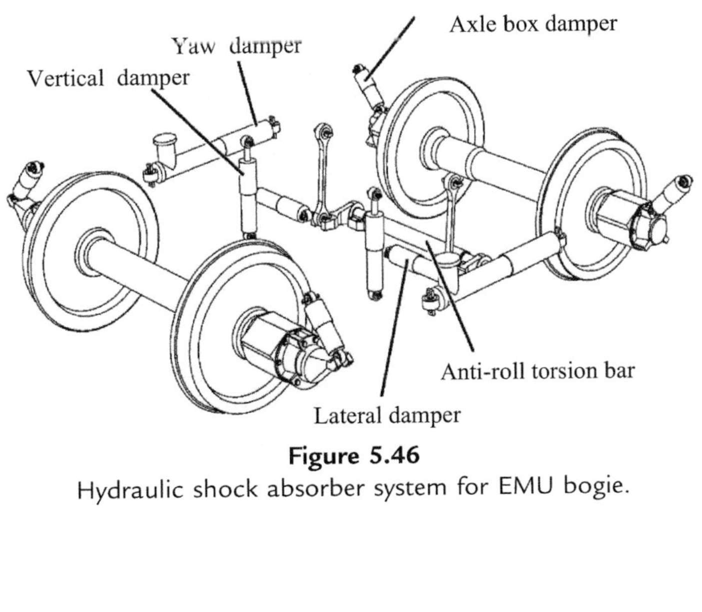

Figure 5.46
Hydraulic shock absorber system for EMU bogie.

generally considered that the viscous damping is proportional to the first power of the vibration velocity, and that, it is so-called linear damping. In order to prevent the damper from the high oil pressure caused by the vibration speed, the damper is required to have a pressure limiting function. The typical damper unloading features can be designed as shown in Fig. 5.47. The *OA* segment of the curve represents the normal working area and the *AB* segment represents the unloading zone. Within this range, after the unloading hole is opened, the area of the orifice increases and the resistance rises slowly. Typically, the damper operates before the unloading zone, so this so-called "unloading" is actually the limitation of the safety value.

(1) Vertical damper

In order to attenuate the vertical vibration transmitted upward from the wheel-rail interface, a hydraulic shock absorber is vertically placed between the axle box and the frame, and thus, partial attenuation and suppression are achieved such that the larger wheelset vibrations are transmitted to the bogie frame. In modern high-speed passenger trains, the secondary suspension mostly uses air springs, and an orifice can be provided in the air spring structure to provide a vertical damping force (such as in CRH2). A secondary vertical damper can also be designed to provide vertical damping force (such as in CRH1 and CRH5), and the CRH3 reserves the position of the secondary damper (but currently no vertical damper is installed).

The primary and secondary vertical damper is used to improve the vertical ride quality of vehicle; however, increasing vertical damping does not always produce a better output [27]. As can be seen from Figs. 5.48 and 5.49, the higher the running speed of the vehicle, the smaller is required vertical damping. For primary vertical damping, the optimal vertical damping at 200 km/h is 20 kN.s/m, the optimal vertical damping at 300 km/h is 15 kN.s/m, and the optimal vertical damping at 350 km/h is 10 kN.s/m. For the secondary

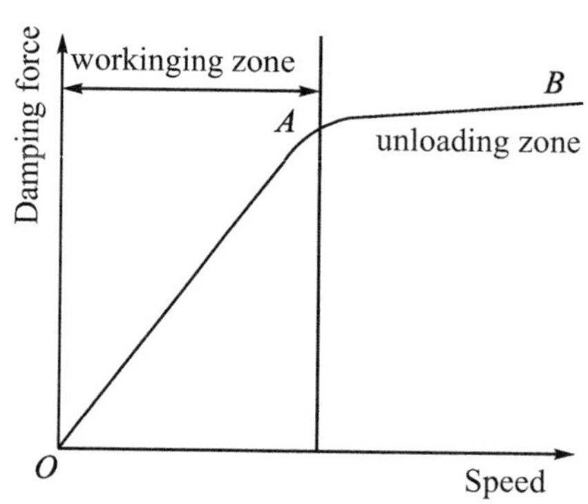

Figure 5.47
Unloading features of hydraulic shock absorber.

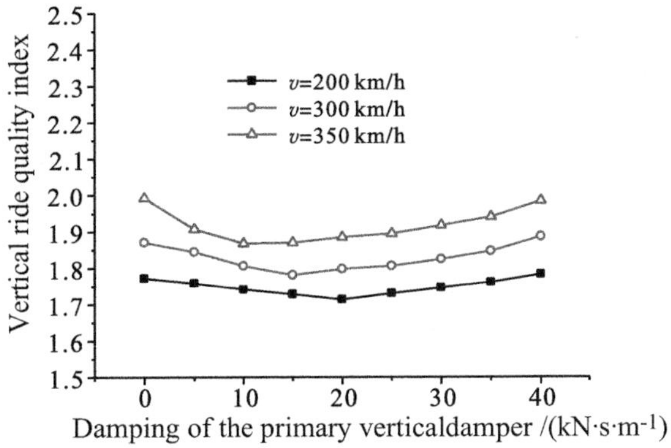

Figure 5.48

Effect of the primary vertical damping on vertical ride quality.

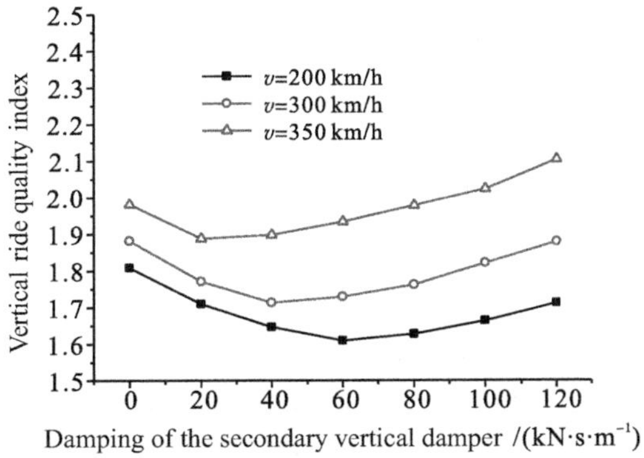

Figure 5.49

Effect of the secondary vertical damping on vertical ride quality.

vertical damping, the optimal vertical damping at 200 km/h is 60 kN.s/m, the optimal vertical damping at 300 km/h is 40 kN.s/m, and the optimal vertical damping at 350 km/h is 20 kN.s/m.

In general, damping suppresses low frequency vibration better; however, it does not favor high frequency vibration isolation. Therefore, the higher the running speed of the vehicle (the higher the excitation frequency), the higher should be the reduction in the vertical damping . Moreover, the vertical damping of primary suspension should be smaller than the vertical damping of the secondary suspension because the track excitation transmits more high frequency vibration components to the primary suspension. After the primary suspension vibration isolation, the high-frequency components transmitted to the secondary suspension are attenuated considerably. It can also be seen from Figs. 5.48 and 5.49 that the effect of the vertical damper on the ride quality is optimal, which simplifies the design in terms of the damping value of the damper. It should be noted that the vertical damping of the primary or secondary suspension system has little influence on the kinetic stability and the running safety of the EMU, and only the impact on the ride quality needs to be considered.

(2) Secondary lateral damper

In general, the lateral dampers of the primary suspension are not provided, and mostly, a primary vertical damper is used to tilt the damper at an angle to achieve a certain lateral damper performance [27], as shown in Fig. 5.50. Therefore, the lateral damper on the bogie refers to the secondary lateral damper between the car body and the bogie.

As can be seen from Fig. 5.50, the secondary lateral damper is indispensable for attenuating the lateral vibration of the CRH2 vehicle. But not the lateral damper damping is bigger, the better [27]. As can be seen from Fig. 5.50 (b), the lateral damping of the high-speed bogie shall not be too large. For the lateral damping is too large, the critical speed and ride comfort of the vehicle system decrease. Through the comparison of the simulation and test results in Fig. 5.51, it is found that the damping of the lateral damper should not be too large. If two lateral dampers are installed per bogie, the 20–30 kN·s/m damping of each lateral damper can achieve better results. In order to avoid transmitting too much high frequency vibration to the vehicle body, the joint stiffness of the secondary lateral damper should not be too large, and it should be generally about 10 MN/m.

(3) Yaw damper

We mentioned in the kinetic stability design method that the yaw damper has a strong influence on the critical speed of the EMU. At the same time, we know from the

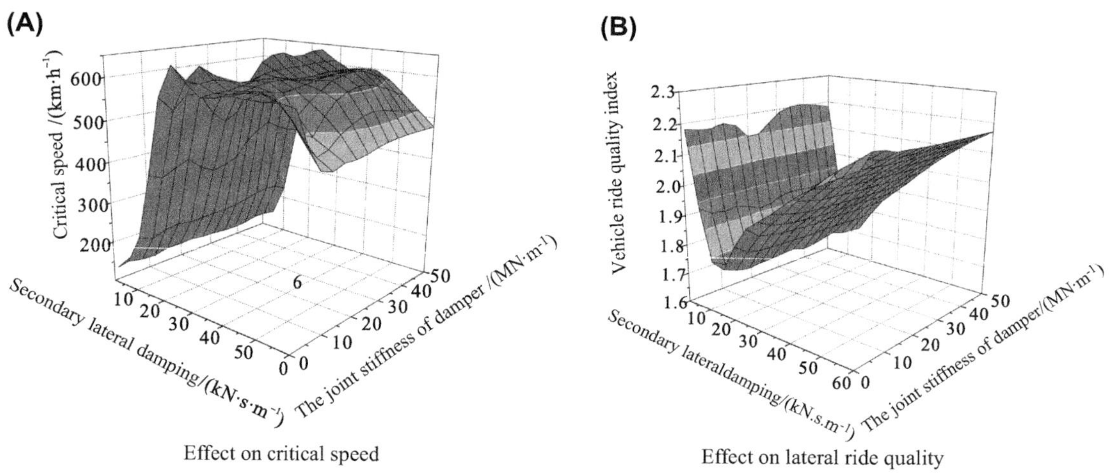

Figure 5.50
Effect of damping and joint stiffness of the secondary lateral damper on the critical speed and lateral ride quality. (A) Effect on critical speed (B) Effect on lateral ride quality.

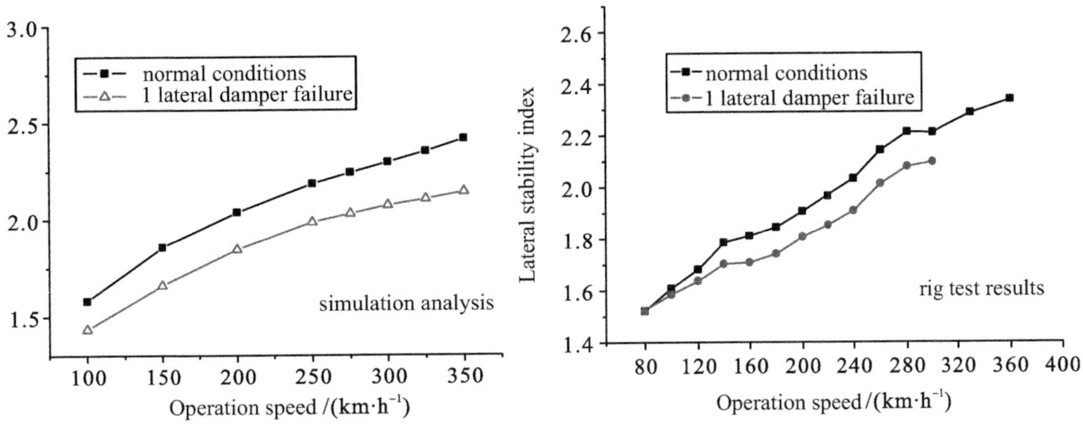

Figure 5.51
Comparison of simulation results and test results.

contradiction of the influence of parameters on the dynamic performance index that the damping force of the yaw damper hinders the steering of the bogie, resulting in a decrease in the curve performance of the EMU.

The yaw damper has the characteristics of a hydraulic shock absorber, except that the throttling structure and the throttling characteristic are different from other dampers. The resistance force curve of the yaw damper is generally shown in Fig. 5.52. The abscissa V is the relative speed of the end points of the damper, and the ordinate F represents the shock absorber force. The resistance force curve can be expressed by the damper saturation resistance F_{max} and the damper unloading velocity V_0. Therefore, the selection of the damping parameter of the damper F_{max} can be attributed to the selection of F_{max} and V_0. Under the condition that the saturation resistance F_{max} is constant, the reduction of the unloading speed V_0 is equivalent to increasing its damping value.

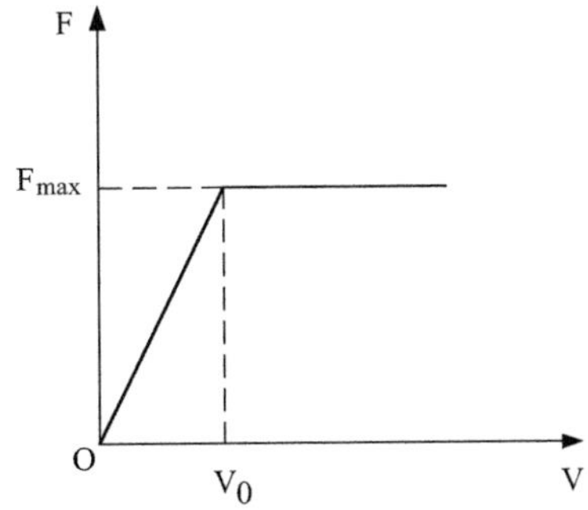

Figure 5.52
Yaw damper damping characteristic curve.

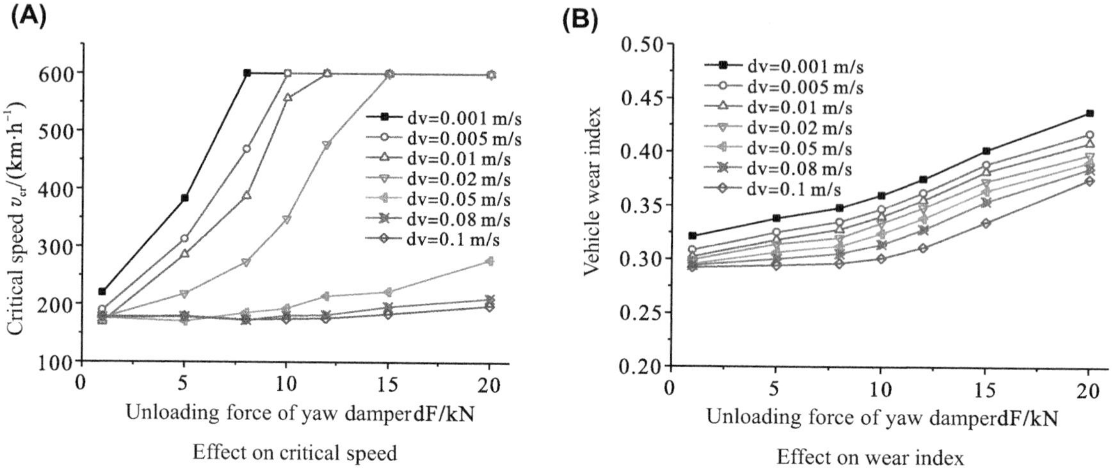

Figure 5.53
Effect of yaw damper on critical speed and wear index. (A) Effect on critical speed (B) Effect on wear index.

Fig. 5.53 (a) is the calculation result of vehicle critical speed [27]. It can be seen from this figure that when the equivalent damping of the yaw damper is insufficient, the critical speed of the vehicle system will drop drastically, which indicates the necessity of installing the yaw damper on the high-speed EMU bogie. Fig. 5.53 (b) shows the calculation result of the wear index of a vehicle running along a R2200m curve at a speed of 200 km/h. It can be seen that as the equivalent damping of the yaw damper increases, the wheel-rail wear of the vehicle increase, which indicates that the yaw damper damping of the bogie cannot be increased indefinitely. From the results shown in Fig. 5.53, the unloading force of the EMU yaw damper is selected to be around 10 kN with the unloading velocity about 0.005 m/s. This will ensure that the EMU has a higher critical speed without excessive wheel-rail wear.

As can be seen from Fig. 5.54, both the simulation and the rig test results show that when the yaw damper of the bogie fails, the lateral ride quality at the high-speed stage will be seriously deteriorated, resulting in hunting instability. Therefore, it is necessary to monitor the working condition of the yaw damper from time to time during the operation. If necessary, a redundancy design can be applied to the yaw damper of the bogie (Fig. 5.55), i.e., four yaw dampers are installed per bogie.

As can be seen from Fig. 5.56, if the yaw damper adopts a redundant design, even if the yaw damper of each bogie accounts for half of the failures (one failure on each side of the bogie), the ride quality of high-speed train can still be guaranteed.

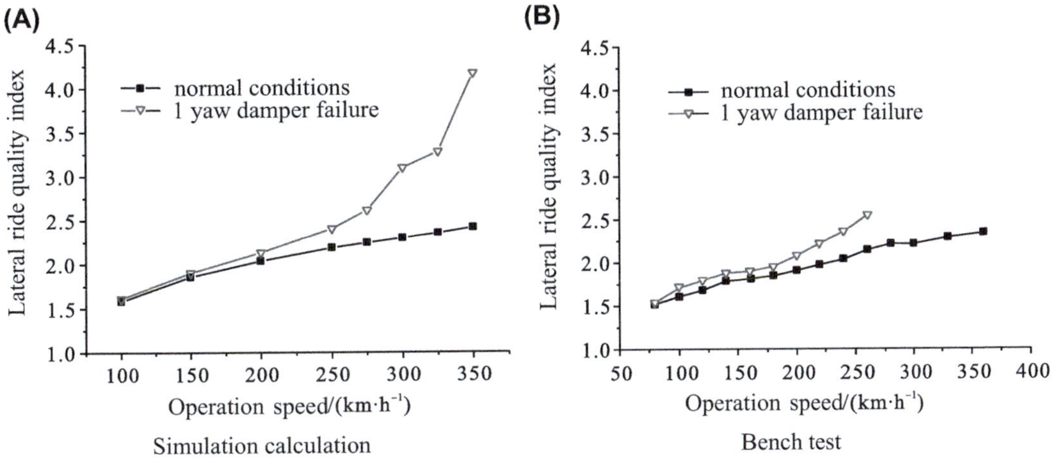

Figure 5.54
Effect of yaw damper failure on lateral ride quality. (A) Simulation calculation (B) Bench test.

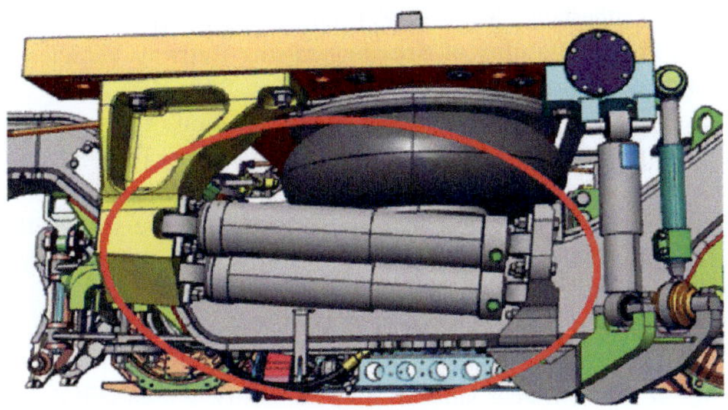

Figure 5.55
Schematic of redundancy design for yaw damper.

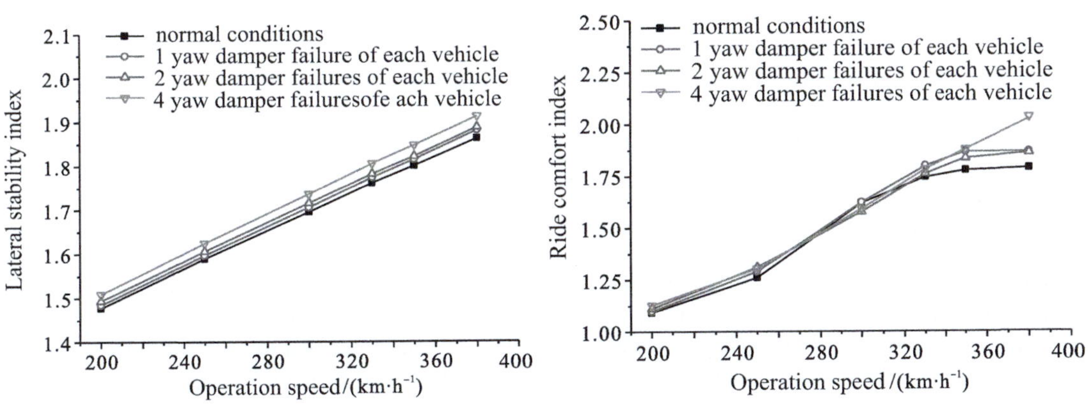

Figure 5.56
Redundancy design for yaw damper.

References

[1] Zhang Weihua. Study on Top-level Design Specifications of High-speed Trains. Journal of the China Railway Society 2012;34(9):1–5.

[2] CRH3 EMU Project Contract Documents - Annex 7 Technical Conditions for EMU Supply. Beijing: Ministry of Railways; 2005.

[3] CRH2 Type Overall Technical Plan for 300-350 km EMU. Beijing: Ministry of Railways; 2007.

[4] CRH3 Type Overall Technical Plan for 300-350 km EMU. Beijing: Ministry of Railways; 2007.

[5] Shu Zhongzhou, Zhang Jiye, Cao Dengqing. Running stability. Beijing: China Railway Publishing House; 2001.

[6] Absel SG, Cooperrider NK. An equivalent linearization algorithm for nonlinear system limit cycle analysis. Journal of Dynamics Systems, Measurement, and Control 1985;107(2):117–22.

[7] Alfred J. On the application of similarity laws to a scaled railway bogie mode. Netherlands:TU Delft; 1990.

[8] Pater AD. The exact theory of the motion of a single wheelset moving on a perfectly straight track. In: Proceedings of 6th IAVSD Symposium on Vehicle System Dynamics. Netherlands: Swets & Zeitlunger; 1979. p. 397–407.

[9] Yokose Keiji. An analysis of running stability of high speed railway trucks connected in series with elastic and frictional forces working against truck turning. In: Proceedings of 10th IAVSD Symposium on Vehicle System Dynamics. Netherlands:Swets & Zeitlunger; 1987. p. 543–54.

[10] Meng Hong. Research on Lateral Stability of Speeding-raising Railway Vehicles and Its Applications. Chengdu: Southwest Jiaotong University; 2009.

[11] Zeng Jing, Wu Pinbo. Algorithm Study for the Convergent Solution of Steady State Equations of Vehicle Systems. Journal of the China Railway Society 1994;16(Supplement):125–30.

[12] True H, CHR KP. A bifurcation analysis of nonlinear oscillations in railway vehicles. In: Proceedings of 8th IAVSD Symposium on Vehicle System Dynamics. Netherlands:Swets & Zeitlunger; 1984. p. 655–65.

[13] Zhang Weihua, Shen Zhiyun. Nonlinear Stability Analysis of Railway Vehicle System. Journal of the China Railway Society 1996;18(1):29–34.

[14] Zhang Weihua, Yan Li, Song Dongli. Design Method for Motion Stability of High-Speed Trains. Journal of Southwest Jiaotong University 2013;48(1):1–9.

[15] Polach O. Comparability of the nonlinear and linearized stability assessment during railway vehicle design. Vehicle System Dynamics 2006;44(Sup):129–38.

[16] UIC518. Testing and approval of railway vehicles from the point of view of their dynamic behavior safety track fatigue ride quality [S. l.]. 3rd Edition. Union Internationale des Chemins de Fer Publications; 2005.

[17] Zhang Shuguang. Design Method of High-Speed Train. Beijing: China Railway Publishing House; 2001.

[18] Luo Jian. Introduction to System Sensitivity Theory. Xi'an: Northwestern Polytechnical University Press; 1990.

[19] NALECZ AG. Application of sensibility methods to analysis and synthesis of vehicle dynamic systems. Vehicle System Dynamics 1989;18(1-3):1–44.

[20] Zeng Jing, Wu Pingbo. Parameter Sensitivity Analysis of Dynamic Performance of High-Speed Passenger Cars. Journal of Southwest Jiaotong University 1996;31(Supplement):186–98.

[21] Shen Zhiyun. Since July, the National High-speed Rail Has a Speed Limit of 300 Kilometers [N]. West China City Daily, 20110521 (4).

[22] Qiu Xuanhuai. Mechanical Design. 4th edition. Beijing: Higher Education Press; 2004. p. 325.

[23] Wu Di, Zhong Yan. Research on the Maintenance System of Railway Passenger Cars in China. Railway Transport and Economy 2006;28(1):50–1.

[24] Yan Li, Zhang Weihua, Maoru Chi. etc. Influence of Wheel Tread Profile and Rolling Diameter Difference on Dynamic Performance of Vehicles. Journal of the China Railway Society 2010;32(1):104–7.

[25] Polach O, Vetter A. Methods for running stability prediction and their sensibility to wheel/rail contact geometry. In: 6th International Conference on Railway Bogie and Running Gears. Budapest: [S.l. n.]; 2004. p. 13–6.
[26] Guo Rongsheng. Lateral Elastic Properties of Helical Springs. Railway and Vehicles 1976;(4):7–17.
[27] Zhang Weihua. EMU and Its Bogie. Beijing: China Railway Publishing House; 2011.
[28] Wang Yong. Study on the Coupling System Dynamics for Three-piece Bogie Tank Cars Taking Into Account the Liquid Sloshing. Chengdu: Southwest Jiaotong University; 2005.
[29] Shevtsov IY, Markine VL, Esveld C. Optimal Design of wheel profile for railway vehicles. Wear 2005;258(728):1002–30.

CHAPTER 6

Optimal design for coupled systems parameter of high-speed train

Chapter Outline

Dynamics of Coupled Systems in High-Speed Railways. https://doi.org/10.1016/B978-0-12-813375-0.00006-6

6.1 Optimal design of high-speed railway line parameters

A detailed study on the optimal design of high-speed railway line parameters was carried out in the National 973 Program "Research on Basic Science Problems of High-Speed Train Service Safety" and the National 863 Program Sub-project "Research on the Overall Technical Conditions and Integration Technology of 400 km/hr High-Speed Train Detection." A basic optimization method for high-speed railway line parameters was proposed based on the analysis of the influences of line parameters on the safety and comfort of the high-speed train.

6.1.1 Plane and vertical section design

The plane and vertical section design parameters include curve radius, transition curve length, type of the knick point of the vertical section, etc. These parameters directly affect the vertical dynamic performance of a running high-speed train.

6.1.1.1 Effect of the curve radius

Fig. 6.1 presents the wheel-rail lateral force, derailment coefficient, overturning coefficient, lateral vibration acceleration of the train and safety and comfort performance indexes of a CRH train passing the curved trajectories of 8000 m, 7000 m, 5500 m and 4500 m at 250 km/h, respectively; it indicates that the wheel-rail lateral force, derailment coefficient, overturning coefficient, and lateral vibration acceleration of the train reduce with an increase in the curve radius. It is worth noting that, when the curve radius is less than 5500 m, the lateral acceleration and overturning coefficient of the train are greater than their respective qualification limits. This is insufficient to meet the safety and comfort requirements of high-speed trains running at a speed of 250 km/h, and 5500 m is the minimum curve radius of the train running at 250 km/h.

6.1.1.2 Effect of transition curve length

Fig. 6.2 shows the derailment coefficient and the lateral stability index value of the high-speed train passing the curve (V = 250 km/h, R = 7000 m) under different transition curve lengths; further, Fig. 6.2 shows that longer the transition curve length, the better the dynamic performance of the wheel and rail, which is more conducive to the safe and smooth transition of the high-speed train from a linear track to a circular curved track. For example, when the transition curve length is 150 m, the derailment coefficient and the lateral stability index value are 0.33 and 2.68 respectively, and those for a curve

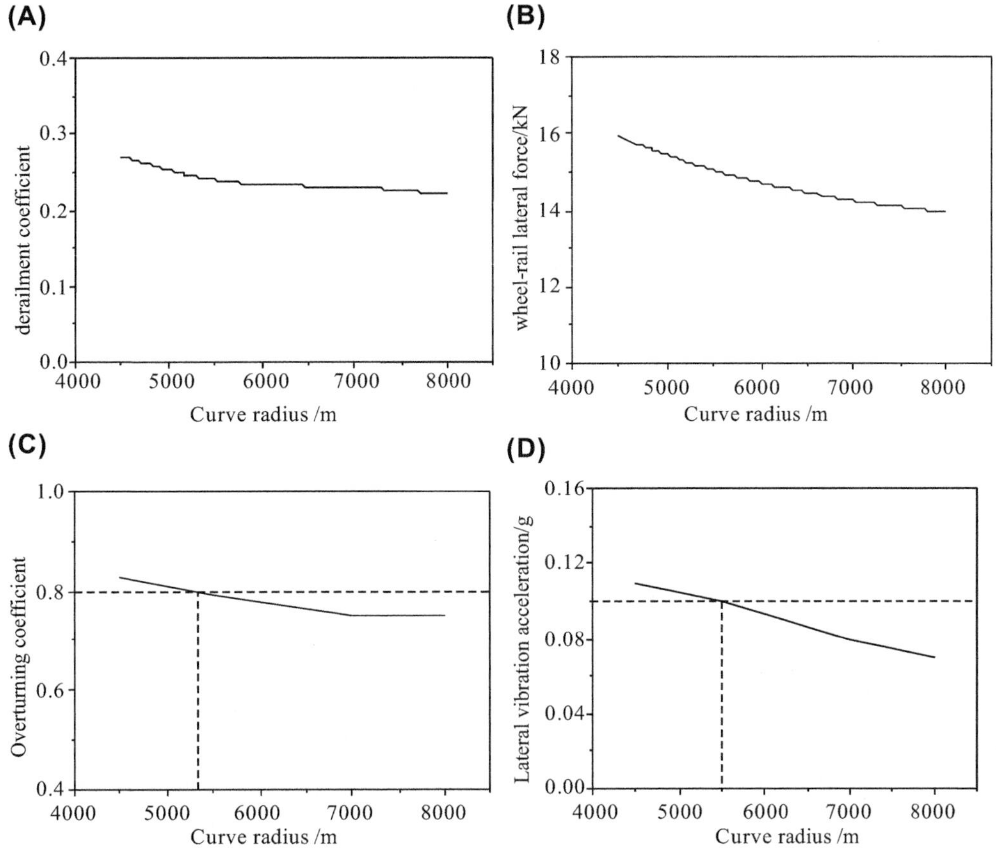

Figure 6.1
Correlation between the curve pass performance index and curve radius of a CRH high-speed train. (A) Wheel-rail lateral force (B) Derailment coefficient (C) Overturning coefficient (D) Lateral vibration acceleration of the train.

length of 450 m are 0.18 and 2.42, respectively. Thus, the running safety improves by about 40%, and the ride comfort improves from a good level to a superior level.

6.1.1.3 Effect of changing slope point

The cost of line construction of the railway depends on the terrain; therefore, the transition between the up and down slopes should be considered in the lateral and vertical plane. An inappropriate transition section would seriously affect the operation stability of the train. Fig. 6.3 shows the vertical vibration acceleration waveforms (10 Hz filtering) of CRH-A and

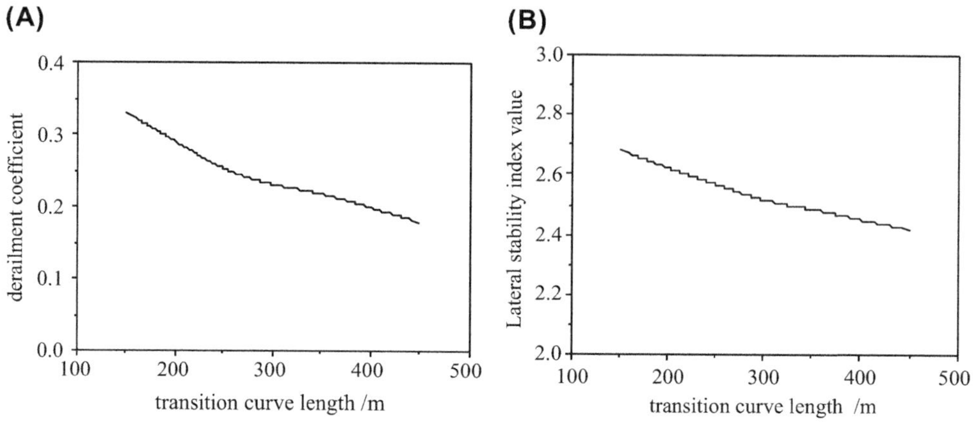

Figure 6.2
Correlation between the curve passing performance index of a CRH EMU and the transition curve length. (A) Derailment coefficient (B) Lateral stability index value of the train.

CRH-B trains when passing the same line [2]. Both these trains exhibit shock vibrations while passing through the slope changing point; however, the vibration waveforms of the two trains are very different owing to the different suspension parameters. The CRH-A train only produces one obvious vibration wave; however, the CRH-B produces 2–5 obvious vibration waves. The main reason for the slower vibration attenuation of the CRH-B train through the slope point is that the two-stage vertical damping is too small. In order to improve the matching between the vehicle and the line and to reduce the effect of the slope change point on the vehicle vibration response, the follow-up series of CRH-B trains have optimized the secondary suspension parameters and achieved good performances.

Optimal design of line stiffness [1]

The plane and vertical sections and the irregularity of the line are the external disturbances for train operation. Line stiffness is part of the high-speed train coupling large system model and it contributes to the vibration response of the system. The line stiffness non-uniformity directly affects the static deformation of the line when the train passes and acts as an external disturbance to the train; therefore, it is equivalent to the internal vibration source of the system.

(1) Reasonable stiffness of the track Excessive track stiffness will lead to an increase in wheel-rail contact force, increased vibration of the track structure, and accelerated deformation of the overall structure of the track and its components. However, if the track stiffness is too small, the deformation will be abnormal under the train load. The line state is not easy to maintain, and the stability is deteriorated, which affects the normal

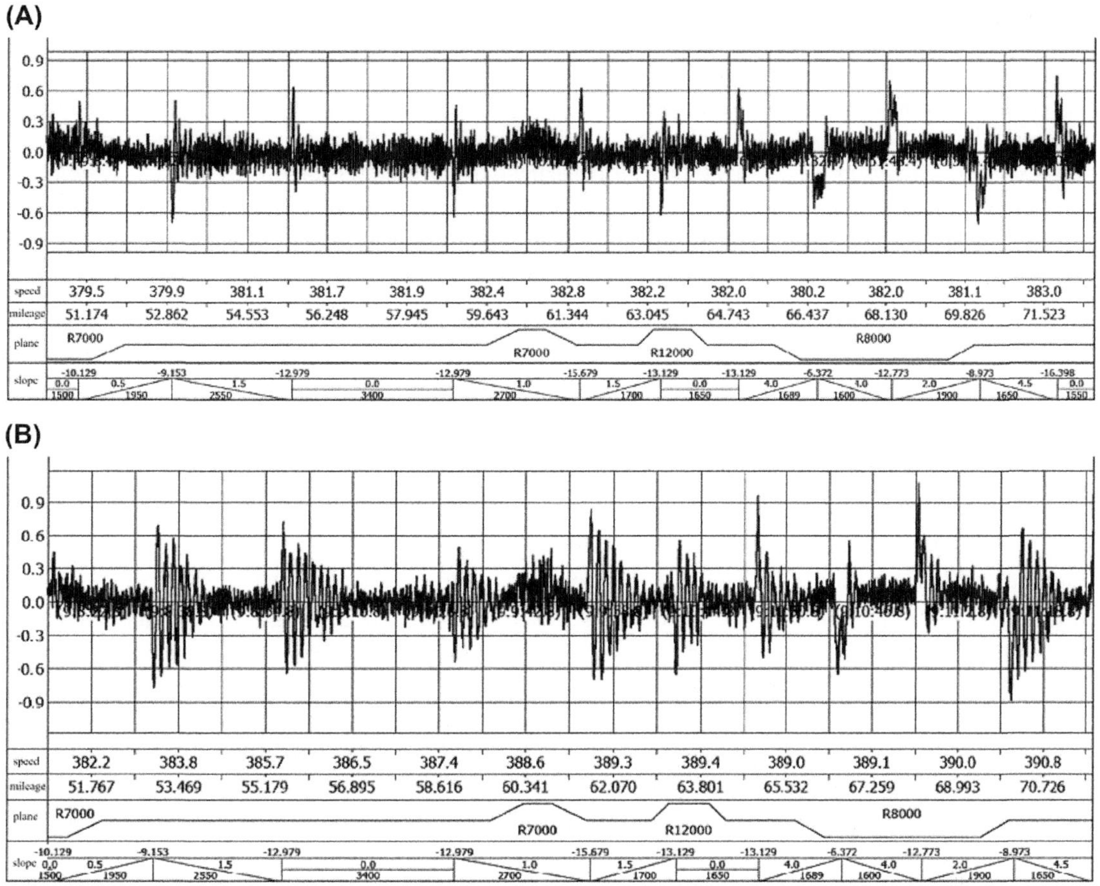

Figure 6.3

Effect of changing slope point on vertical vibration acceleration of the train body (10 Hz filtering). (A) CRH-A EMU (B) CRH-B EMU.

operation of the line. Therefore, the rational design of a ballastless track stiffness for a passenger dedicated line is of great significance.

Table 6.1 lists the simulation results of the wheel-rail system dynamics of a high-speed train running at 350 km/h under the excitation of high-interference orbit spectrum of Germany. The results show that with the increase in the foundation stiffness under the track, the wheel-rail vertical force, pillow pressure, wheel unloading ratio, wheel rail wear index increase, and the dynamic bending stress and track deformation decrease. To reduce the wheel-rail force, ensure the track deformation requirements, reduce the wheel-rail wear, and improve the safety of the train, the track stiffness should be kept in a reasonable range from the perspective of dynamics. In other words, the track stiffness should be kept within a reasonable range.

Table 6.1: Effect of foundation stiffness on the dynamic performance of wheel-rail system.

Under-rail foundation stiffness/ ($MN\cdot m^{-1}$)	Wheel and rail wear index/ ($N.m\cdot m^{-1}$)	Vertical force of wheel and rail /(kN)	Rail vertical displacement/(mm)	Pillow pressure/(kN)	Load shedding rate	Rail bending stress/ (MPa)
10	10.66	102.42	2.71	29.09	0.3	98.3
20	12.55	108.79	1.68	34.56	0.38	90.57
30	13.77	114.36	1.3	39.53	0.45	88.12
40	14.24	118.37	1.09	44.23	0.51	87.15
50	14.45	122.29	0.96	49.2	0.56	86.43
60	14.58	126.27	0.86	52.98	0.61	85.89
70	14.66	129.68	0.8	55.16	0.65	85.31
80	14.74	132.51	0.74	56.68	0.69	85.08

In order to analyze the influence of the foundation stiffness of the ballastless track on the dynamics of the wheel and rail, the calculation result in Table 6.1 are non-dimensionalized, as shown in Fig. 6.4. It shows that the reasonable stiffness of the foundation under the track of the ballastless track structure of the high-speed passenger trains should be 20–30 MN/m, with the range of deformation of the rail between 1.3–1.7 mm.

(2) Reasonable stiffness of the subgrade surface of the transition section Since the structure of the abutment and the subgrade in the connection between the road and the bridge are very different, a transition section is necessary for the stiffness and deformation of the line. The elasticity of the ballastless track structure is provided by fasteners, and the supporting stiffness of the subgrade surface directly affects the force, deformation, and stability of the ballastless track structure. The static stiffness of the ballastless track fastener is 20–30 MN/m. Considering the most unfavorable situation in the calculation, the stiffness of the fastener on the bridge is set as 30 kN/mm, the stiffness of the fastener on the roadbed is set as 20 MN/m, and the stiffness ratio is 1.5. Table 6.2 lists the influence of the support stiffness of the roadbed on the vibration response of the wheel-rail system when the train passes the junction of the road and the bridge at a speed of 350 km/h. The calculation results are nondimensionalized, as shown in Fig. 6.5.

Fig. 6.5 shows that when the bearing stiffness of the subgrade surface exceeds 1000 MPa/m there is no influence on the wheel-rail system. Judging from the calculation results of the wheel-rail force, pillow pressure, and rail displacement, the appropriate range of the subgrade surface stiffness of the ballastless bridge (tunnel) transition section should be 500–1000 MPa/m.

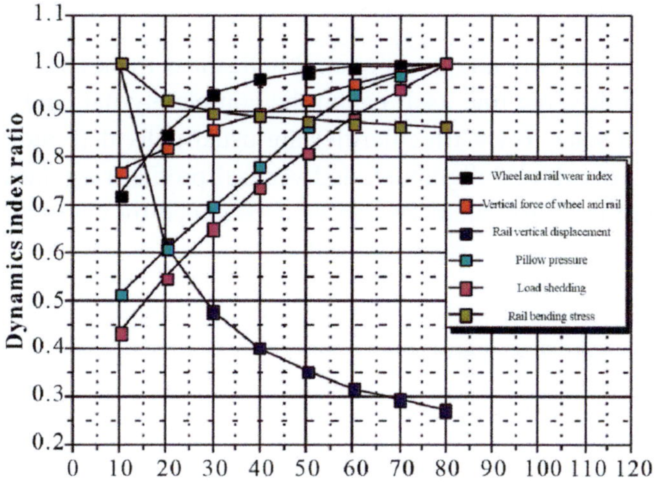

Figure 6.4
Reasonable stiffness range of the ballastless track for the passenger dedicated line ga.

Table 6.2: Influence of subgrade bearing stiffness on vibration response of wheel-rail system.

Subgrade bearing stiffness/ $(MN \cdot m^{-1})$	Vertical force of wheel and rail /(kN)	Pillow pressure/ (kN)	Rail deflection/ (mm)
50	81.89	30.68	1.23
100	80.94	30.30	1.17
200	79.94	27.80	1.09
500	79.12	26.81	0.98
1000	78.90	26.74	0.94
5000	78.52	26.71	0.91

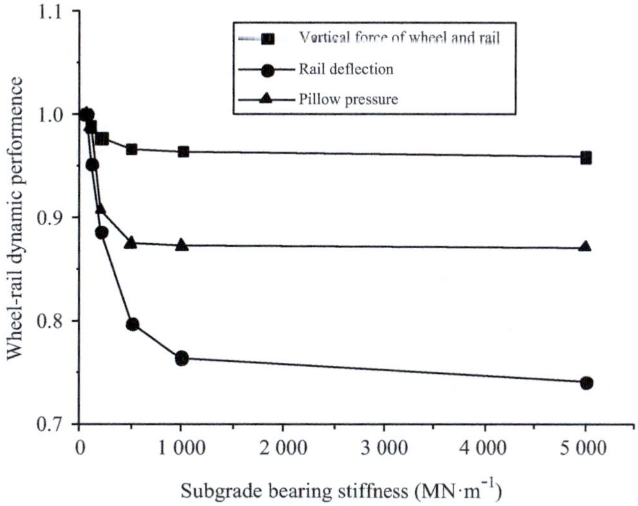

Figure 6.5
Effect of subgrade bearing stiffness on wheel-rail dynamic performance.

Track irregularity control

Although the high-speed railway adopts an overall ballastless track structure of high-regularity and high-stability, the line itself cannot be completely flat, and it always has a deviation with the ideal state. Track irregularities are usually divided into several categories [3]: gauge, cross level, alignment, and vertical profile irregularities, as shown in Fig. 6.6. These kinds of track irregularities can be determined by the vertical and lateral errors (z_L, y_L, z_R, y_R) of the left and right rails [4]:

Gauge irregularity $y_g = y_L - y_R$

Level irregularity $y_a = (y_L - y_R)/2$

Alignment irregularity $z_c = z_L - z_R$,or $\theta_c = (z_L - z_R)/2l$, l is the half of the horizontal spacing of the wheel

Vertical profile irregularity $z_p = (z_L + z_R)$

The Beijing–Tianjin track shown in Fig. 5.22 indicates that, in addition to the fixed frequency excitation caused by the line structure, the track irregularity is a broad spectrum of random irregularities. In addition, inevitable service problems such as line settlement, deformation of the rail welded joint, and sinking of elevated line piers will cause twist irregularity and harmonic irregularity. Here, we study the effects of different types of track irregularities on the safety of train operation.

Effects of track random irregularity [5]

(1) Sensitive track irregularities at a speed of 250 km/h In the 6th raising of speed in large scale, comprehensive safety tests showed that high-speed trains, especially trains with

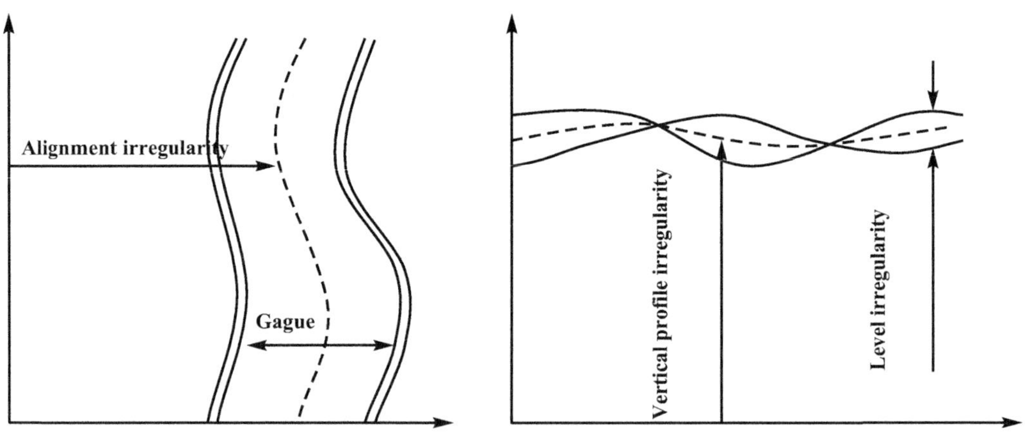

Figure 6.6
Definition of gauge irregularity.

small axle load, were sensitive to line conditions. In addition, the wheel load shedding rate was found to exceed 0.8 on the existing lines. In order to analyze the adaptability of high-speed trains to the irregularity of the line, specific calculations of the wheel load shedding index of the trains under the excitation of one-way random irregularity (high and low or track direction) are carried out when the train is running at the speed of 250 km/h on the passenger dedicated line. The aim of this calculation is to find out the extent of the impact of high and low or track direction irregularity on the wheel load reduction rate. The wavelength range of the irregularity is 1 to 45 m, and the track spectrum is set as the German high interference spectrum.

Fig. 6.7 shows the calculation results of the wheel load shedding rate under the excitations of track high and low, track direction, and comprehensive irregularities. It shows that under the single effect of track high and low irregularity, the phenomenon that the wheel load reduction rate exceeds the dynamic acceptance limit (0.8) is very common. The wheel load shedding rate is sometimes close to 1.0, indicating a suspected instantaneous detachment between the wheel and the rail. However, the time for wheel load reduction rate or vacating is shorter than 0.015 s, which is required by the interim scheme for derailment safety standards proposed by Japan [6]. The results in Fig. 6.7(B) show that the track direction irregularity has little effect on the wheel load reduction rate. The results of the wheel load reduction rate in Fig. 6.7(C) (combined effect of high and low and track direction) show that the wheel load reduction rate under the excitation of single track high and low irregularity are very close to the value under that of comprehensive irregularity. There is no significant difference between these values, which indicates that the unloading phenomenon of the high-speed train is caused by the track high and low irregularity, and the tack direction irregularity contributes little to the load shedding rate.

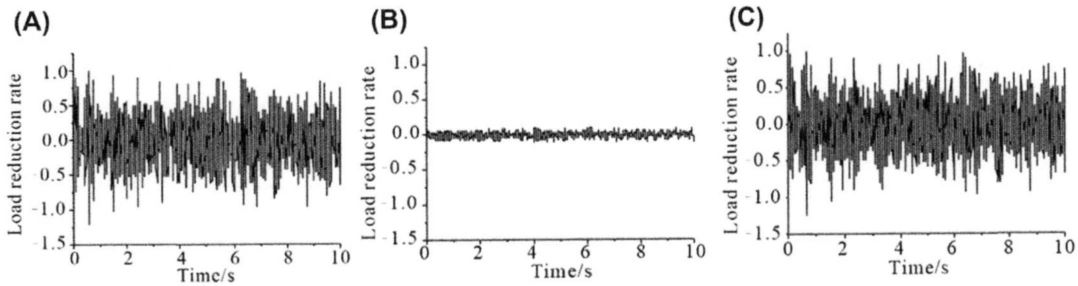

Figure 6.7

Calculation results of time domain response of load shedding rate under excitation of irregularity. (A) Excitation of the longitudinal irregularity (B) Excitation of track direction irregularity (C) Combined excitation of track high and low and direction irregularity.

Table 6.3: Main frequency and sensitive wavelength of the lower wheel load shedding rate at different speeds.

Speed/(km/h)	160	180	200	210	220	230	240	250	260	270
Main frequency/Hz	32.2	34.0	36.5	34.8	35.8	36.5	37.3	37.3	37.8	35.0
Wavelength/m	1.38	1.47	1.52	1.68	1.71	1.75	1.79	1.86	1.91	2.14

Under the excitations of the irregularity of wavelength 1−45 m, the indexes of the wheel load shedding rate are calculated when the trailer (empty) passes the vertical curve line at 160, 180, 200, 210, 220, 230, 240, 250, 260, and 270 km/h. As listed in Table 6.3, the main frequency value of the wheel load shedding rate in the speed range of 160−270 km/h can be obtained by the spectrum analysis. The calculation results show that the main frequency of the wheel load reduction rate at the speed of 160−270 km/h is concentrated at 30−40 Hz. According to the correlation among wavelength, speed, and frequency, the sensitive wavelength of the wheel load reduction rate at 160−270 km/h is 1−2.5 m.

The above analysis shows that the 2.5 m wavelength is a limit value; therefore, the irregularities of three different wavelength ranges—1 to 2.5 m, 2.5 to 45 m, and 1 to 45 m—are drawn to discuss the impact of irregularity of different wavelength ranges on the safety of the high-speed train. The calculation result of the time domain and frequency domain responses of wheel rate shedding rate of the trailer at the speed of 250 km/h are shown in Figs. 6.8−6.10.

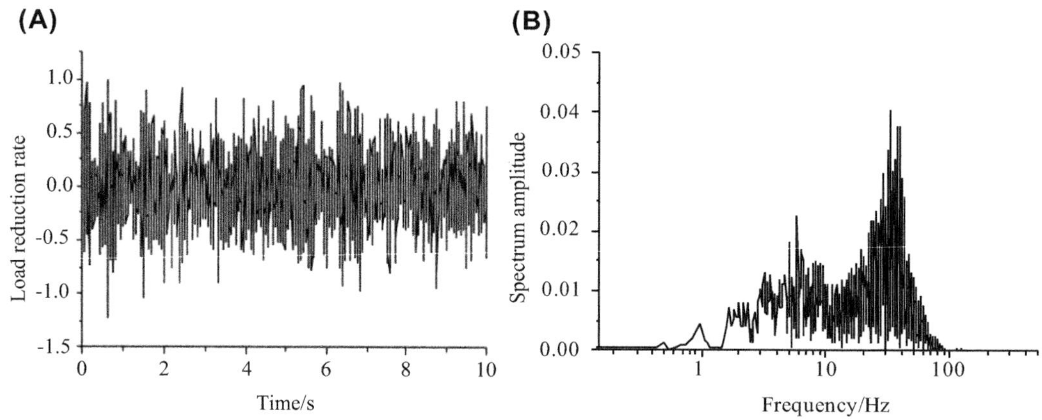

Figure 6.8
Calculation results of the wheel load reduction rate under the excitation of irregularity in wavelength range of 1 to 45 m. (A) Time domain response of load shedding rate (B) Frequency domain response of load shedding rate.

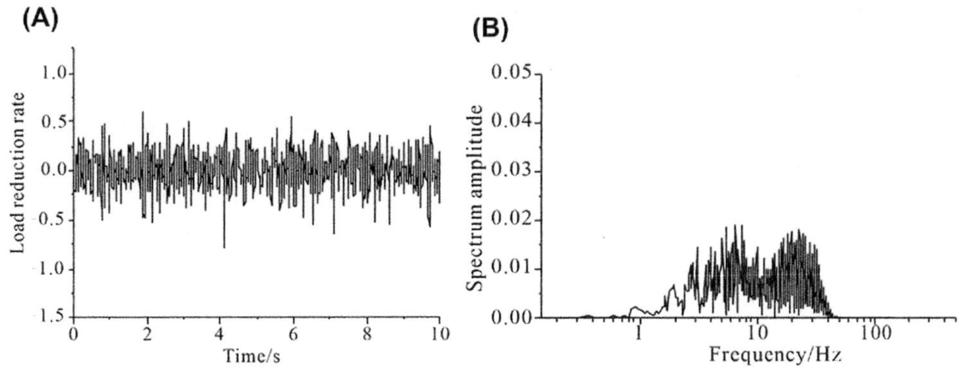

Figure 6.9
Calculation result of the wheel load reduction rate under the excitation of irregularity in wavelength range of 2.5 to 45 m. (A) Time domain response of wheel load shedding rate (B) Frequency domain response of wheel load shedding rate.

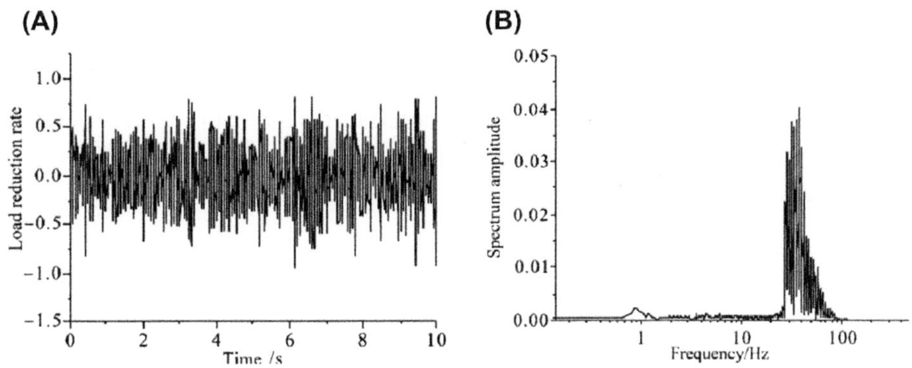

Figure 6.10
Calculation results of the wheel load reduction rate under the excitation of irregularity in wavelength range of 1 to 2.5 m. (A) Time domain response of wheel load reduction rate (B) Frequency domain response of load shedding rate.

Comparing the calculation results of Fig. 6.8 with those of Fig. 6.10, it can be found that under the excitations of irregularity in the wavelength range of 1–45 m, the wheel load reduction rate is more serious at a speed of 250 km/h. There are more instances when the wheel load reduction rate reaches 1.0. The load shedding rate greater than the dynamic load shedding rate limit of 0.8 is very common; however, it is unfavorable in terms of safe operation. The amplitude is larger in the main frequency range. For example, under the harsh excitation of irregularity in the 2.5–45 m wavelength range, the wheel load reduction rate below 0.65—less than the steady-state limit of the wheel load reduction rate—meets the safe driving requirements. The amplitude at the main frequency range is much smaller than the value at wavelengths between 1−45 m. In the case of the excitation of irregularity in the wavelength range of 1−2.5 m, the wheel load reduction rate reaches the dynamic wheel load reduction rate limit of 0.8 several times. The amplitude at the dominant frequency range is very close to the value at the wavelength of 1−45 m. Under the condition of a speed of 250 km/h, the track irregularity in the wavelength range of 1−2.5 m is the reason for the heavy load shedding of the running EMU.

(2) Sensitive track irregularity at speed of 350 km/h Owing to the high operation speed, the wavelength is initially set as in the range of 1−120 m. Taking the high-speed train with a design speed over 300 km/h as the calculation object, Fig. 6.11 presents the results of the wheel load reduction rate by using the high interference spectrum of Germany as the track spectrum.

Fig. 6.11 shows that under the line conditions of the high interference spectrum of Germany, the maximum peak of the wheel load reduction rate reaches 0.85 and exceeds the dynamic acceptance limit of 0.8. Therefore, under this excitation of irregularity, the operation safety of high-speed train cannot be satisfied.

As shown in Fig. 6.12, the frequency domain responses of the wheel load reduction rate can be obtained by spectrum analysis; further, Fig. 6.12 shows that the main frequency of the wheel load reduction rate is 37 Hz.

According to the correlation among wavelength, speed, and frequency, the wheel load reduction rate at a speed of 350 km/h is about 3 m.

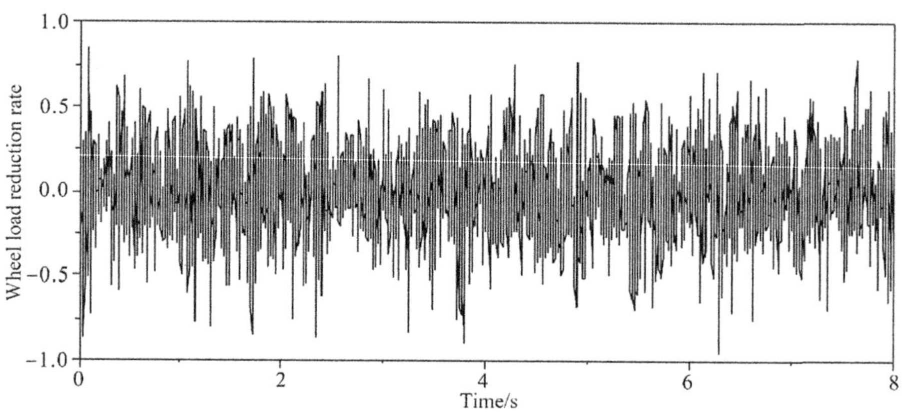

Figure 6.11
Time response of the high-speed EMU load shedding rate index.

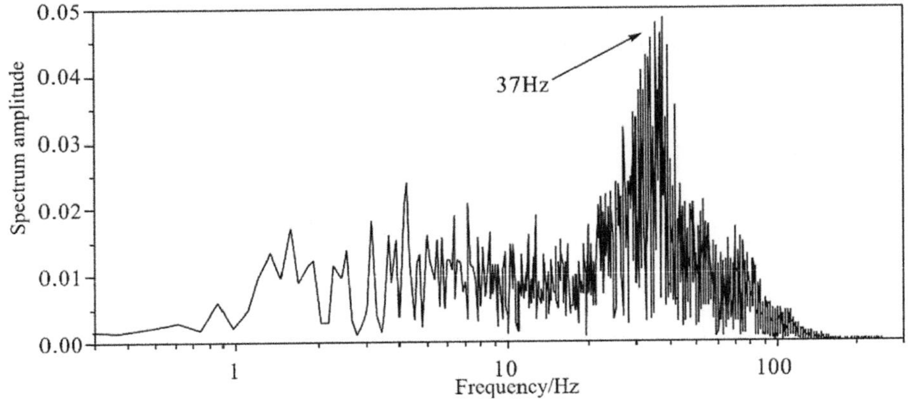

Figure 6.12
Frequency domain response of the high-speed train wheel load reduction rate indexes.

Effects of Line Harmonics Irregularity [5]

The effect of line harmonics irregularity are analyzed at a speed of 250 km/h.

(1) The effect of the irregularity of the twist of track on the wheel load reduction rate Fig. 6.13 depicts the irregular geometries of a common twist of the track on a line.

The irregular waveform function is

$$Z = \frac{A}{2}\left(1 - \cos\frac{2\pi x}{L}\right) \tag{6.1}$$

where L and A are the wavelength and amplitude of the irregularity, respectively; L is set as 12.5 m for the twist of track.

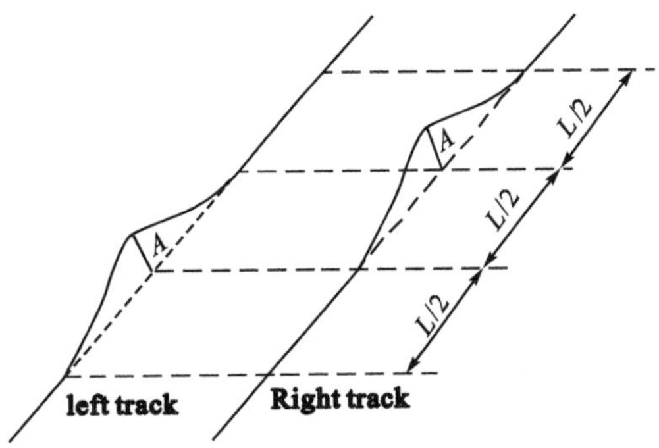

Figure 6.13
Schematic of the Twist of Track.

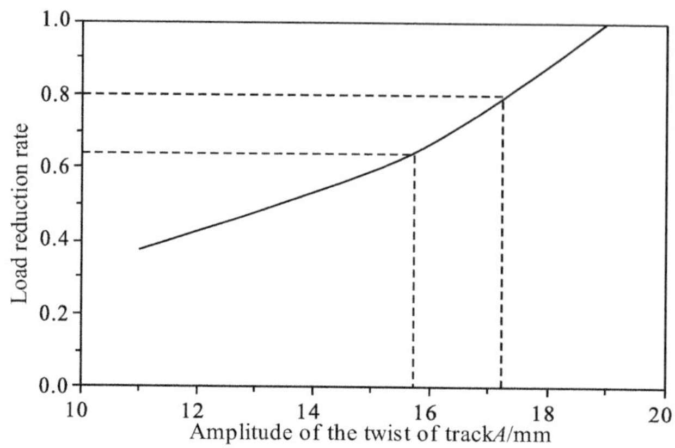

Figure 6.14
Effect of the amplitude of the twist of track on the load shedding rate.

The effect of the amplitude of the twist of track

Without considering the effects of random irregularity, Fig. 6. 14 presents the variation of the wheel load reduction rate of a trailer (empty) with the amplitude of the twist of track under an operation speed of 250 km/h. It can be seen from Fig. 6.14 that if amplitude A is less than 15.7 mm, the wheel load reduction rate is less than 0.65 (quasi-static qualification limit); further, if amplitude A is less than 17.2 mm, the wheel load reduction rate is less than 0.8 (dynamic qualification limit).

In order to meet the quasi-static standard requirement of the load reduction rate for high-speed trains, the irregularity amplitude of the twist of track on the line must be controlled within 15.7 mm.

(2) Influence of the wavelength of the twist of track Similar to random irregularities, the twist of the track wavelength affects the excitation frequency of the wheel-rail system. This section only studies the influence of the twist of track with an amplitude of 15 mm on the wheel load reduction rate of the high-speed train. The random irregularity of the track is not considered in calculation.

Fig. 6.15 shows the wavelength effect of the wheel load reduction rate when the trailer (empty) of the high-speed train is running at 250 km/h; further, Fig. 6.15 shows that, when the amplitude of the twist of track is fixed, the shorter the wavelength, the larger is the value of the wheel load reduction rate, and the greater is the possibility that the wheel will be off the rail. When the wavelength is greater than 10.8 m, the value of the wheel load reduction rate is less than the quasi-static qualification limit (0.65), and when the wavelength is less than 7.6 m, the wheel load reduction rate will exceed the dynamic acceptance limit and be unable to meet the high-speed running requirements.

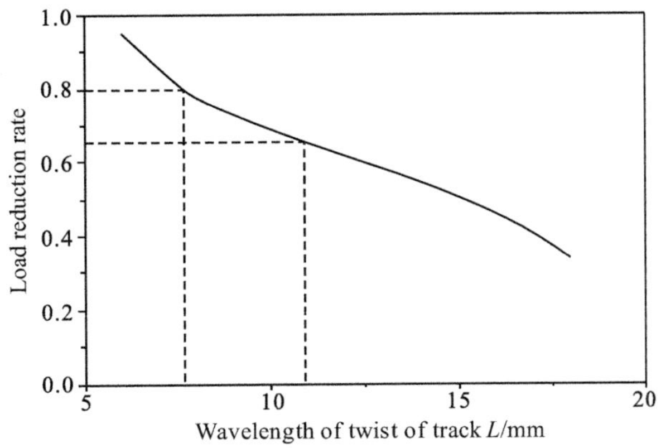

Figure 6.15
Effect of the wavelength of twist of track on load shedding rate.

The above calculation and analysis is obtained considering only one example with a 15 mm wavelength.

(2) Effect of the shape of the twist of track From the perspective of a single-sided rail, the cosine shape of the above-mentioned twist of track is convex. This section studies the influence of the twist of track with a cosine shape on the wheel load reduction rate of the high-speed train.

Fig. 6.16 shows the responses of the trailer (empty) of a high-speed train at a speed of 250 km/h passing through the convex and concave shaped twist of track ($A = 14$ mm, $L = 12.5$ m).

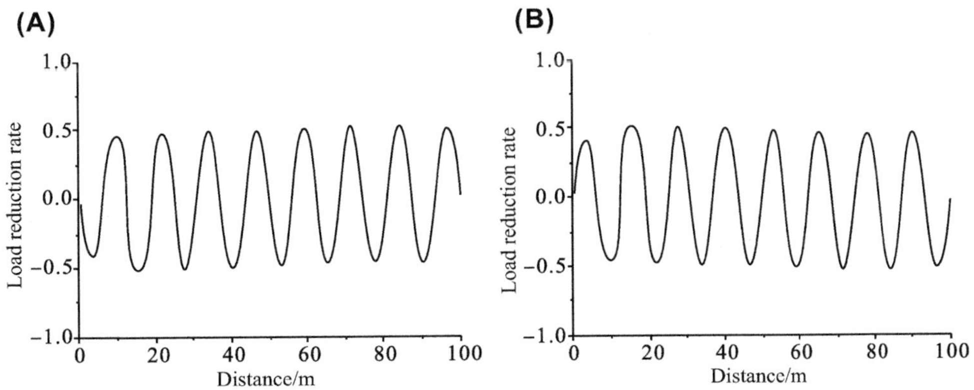

Figure 6.16
Calculation result of wheel load shedding rate under different shape twist of track conditions. (A) Convex shape twist of track (B) Concave shape twist of track.

Fig. 6.16 shows that the convex and concave twists of track have little effect on the wheel load shedding rate; the maximum wheel load reduction rate under both conditions is 0.53.

(2) Effect of high and low harmonics irregularity on wheel load reduction rate The common high and low harmonic irregularity on the line is shown in Fig. 6.17. The irregularity waveform function Z can be expressed as follows:

$$Z = A\left(1-\cos\frac{2\pi x}{L}\right) \quad \text{or} \quad Z = 2A\sin\frac{2\pi x}{L} \tag{6.2}$$

where L and A stand for irregular wavelength and half peak, respectively; L is set as 10 m.

(1) Effect of amplitude of high and low harmonics irregularity on amplitude Fig. 6.18 shows the wheel load reduction rate of the trailer (empty) of a high-speed train at the speed of 250 km/h under different amplitudes of high and low harmonics irregularity. It shows that the load shedding rate increases almost linearly with the amplitude. When the

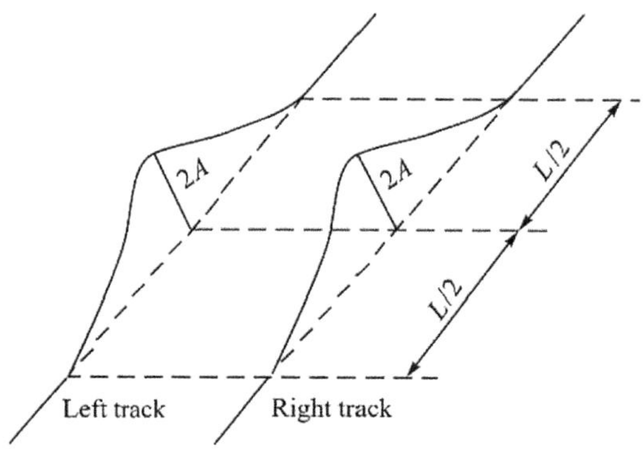

Figure 6.17
Schematic of High and Low Irregularity.

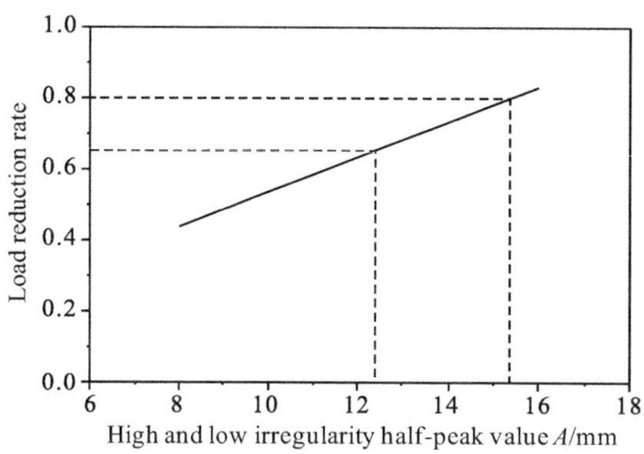

Figure 6.18
Effect of high and low irregularity half-peak value on the load shedding rate.

amplitude is less than 12.2 mm, the wheel load reduction rate is less than the quasi-static qualification limit of 0.65; when the half-peak is between 12.2 mm and 15.3 mm, the wheel load reduction rate varies from 0.65 to 0.8; when the half-peak value is greater than 15.3 mm, the wheel load reduction rate exceeds the dynamic limit of 0.8, failing to meet the requirements of running safety.

Fig. 6.18 shows that, under the action of a single high and low harmonics irregularity ($L = 10$ m), the amplitude within 12.2 mm can meet the quasi-static standard of wheel load reduction rate at 250 km/h.

(1) Effect of wavelength of high and low harmonics irregularity Considering the high and low harmonics irregularity with an amplitude of 8 mm as an example, this section studies and analyzes the correlation between the wavelength L and the wheel load reduction rate of the high-speed train. The random irregularity of the track is not considered in this section. Fig. 6.19 presents the wheel load reduction rate indexes of the trailer of a high-speed train at 250 km/h under different wavelengths.

Fig. 6.19 shows that when the amplitude of the high and low harmonics is fixed, the wheel load reduction rate increases as the wavelength becomes shorter. When the wavelength is less than 7 m, the wheel load reduction rate exceeding 0.8 is unqualified; when the wavelength is 7–7.7 m, the wheel load reduction rate is between 0.65–0.8; when the wavelength is greater than 7.7 m, the load wheel load reduction rate less than the quasi-static qualification limit of 0.65 meets the requirements for a high-speed train.

Similarly, the results of the wavelength effect analysis of the wheel load reduction rate are correlated to the amplitude of the irregularity. In general, the larger the amplitude, the larger is the wavelength limit.

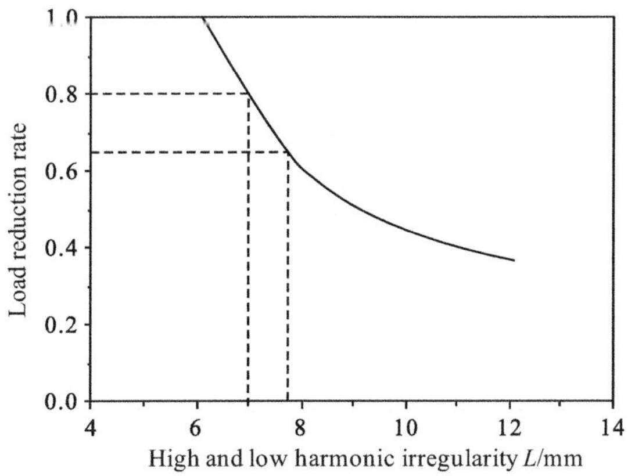

Figure 6.19
Effect of high and low harmonic irregularity on wheel load reduction rate.

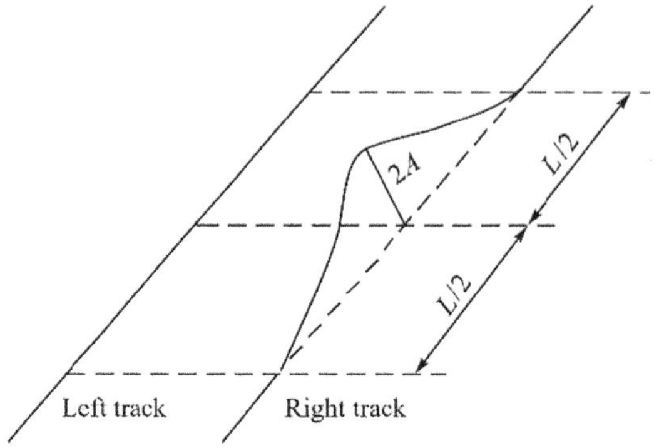

Figure 6.20
Horizontal irregularity.

(3) Effect of horizontal harmonics irregularity on load shedding rate The geometry of the high and low harmonic irregularity is shown in Fig. 6.20, and the corresponding waveform (L = 12.5 m) is similar to Equation (6.2).

(3) Effect of amplitude of horizontal harmonic irregularity Fig. 6. 21 presents the wheel load reduction rate indexes of the trailer (empty) of a high-speed train passing the horizontal harmonics irregularities with different amplitudes at 250 km/h. It shows that the wheel load reduction rate is linear with the growth trend of the irregularity amplitude. When the amplitude is less than 11.7 mm, the wheel load reduction rate is less than the quasi-static qualification limit of 0.65; when the amplitude is between 11.7 mm and 15 mm, the wheel load reduction rate varies from 0.65 to 0.8; when the amplitude is greater than 15 mm, the wheel load reduction rate exceeding the dynamic acceptance limit of 0.8 fails to meet the running safety requirements.

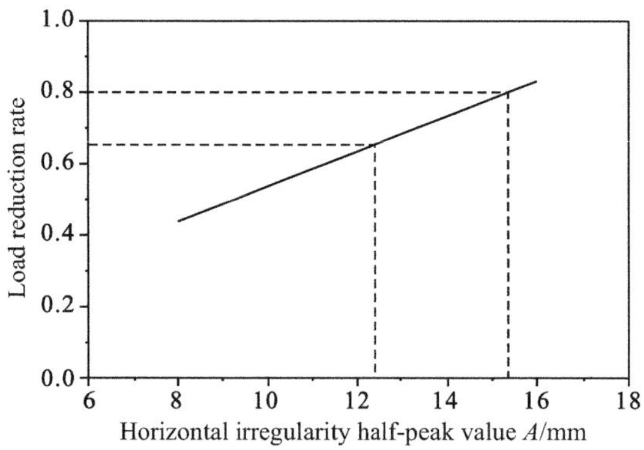

Figure 6.21
Effect of horizontal irregularity half-peak value on wheel load reduction rate.

Fig. 6.21 shows that under the condition of the single horizontal harmonic irregularity ($L = 12.5$ m), only the half-peak value is controlled within 11.7 mm; the wheel load reduction rate can meet the requirement of 250 km per hour.

(2) Effect of wavelength of horizontal harmonics irregularity This section studies the correlation between the horizontal harmonics wavelength L and the wheel load reduction rate index of the high-speed train. In the analysis, the amplitude is set as 10 mm, and the random irregularity of the track is not considered. Fig 6.22 presents the wheel load reduction rate indexes of the trailer (empty) of high-speed train under different wavelengths at 250 km/h.

Fig. 6.22 shows that as the wavelength becomes shorter, the wheel load shedding rate increases. When the wavelength is less than 7 m, the wheel load reduction rate reaches a dangerous level, exceeding 0.8; when the wavelength is 7−10.5 m, the wheel load reduction rate is between 0.65 and 0.8; when the wavelength is greater than 10.5 m, the wheel load reduction rate less than the quasi-static qualification limit of 0.65 meets the running safety requirements.

Similarly, the results of this analysis are correlated to the amplitude of irregularity. The larger the amplitude, the larger is the wavelength limit.

(4) Study on the influence of random track irregularity on the comfort performance of running EMU For a certain type of 250 km/h EMU, the random vibration characteristics of the running train comfort index values are studied in depth from the frequency domain. The commonly used running train comfort index include the lateral and vertical vibration accelerations of the vehicle body and its sperling index.

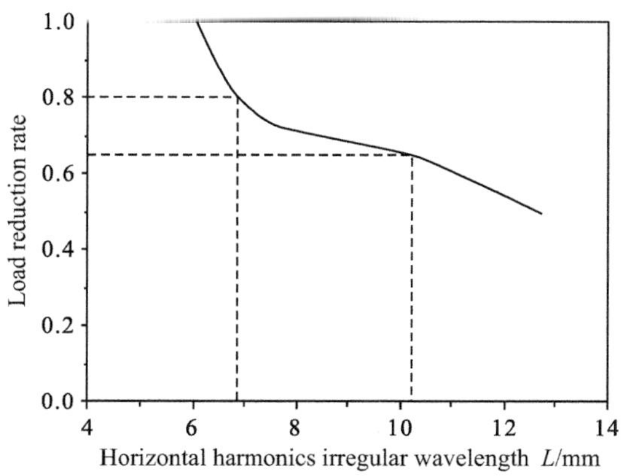

Figure 6.22

Effect of horizontal harmonics irregular wavelength on wheel load reduction rate.

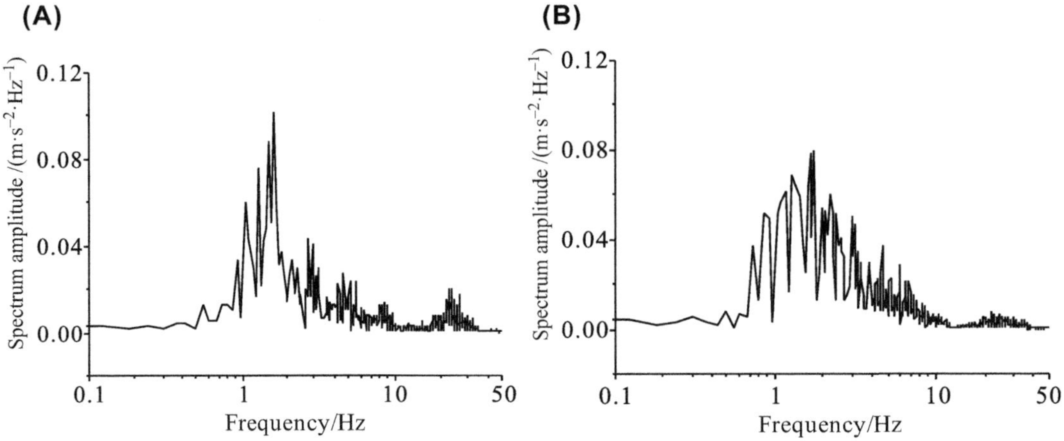

Figure 6.23
Frequency domain response of vehicle body vibration acceleration (V = 250 km/h). (A) Lateral acceleration (B) Vertical acceleration.

(1) Sensitive wavelength analysis of comfort index Fig. 6.23 shows the frequency domain response of the lateral and vertical vibration acceleration of one type of EMU running at 250 km/h under the excitation of the passenger railway irregularity in China. According to Fig. 6.23, the lateral and vertical vibrational energy of the vehicle body is concentrated in the low frequency region, distributed in the range of 1–10 Hz.

Table 6.4 lists the main vibration frequencies and the sensitive wavelengths of the lateral vibration acceleration of the vehicle body when the high-speed train at 160–280 km/h (According to the correlation among speed, frequency, and wavelength, it is easy to calculate the irregular wavelength corresponding to the main frequency at each speed, i.e., the sensitive wavelength). The calculation results of the vertical indexes are shown in Table 6.5.

Tables 6.4 and 6.5 show that in the calculation speed range, the main frequency of the lateral vibration of the vehicle body is mainly concentrated in the range of 1 to 2 Hz, and the main frequency increases slightly as the speed increases. In the speed range of 160 to 280 km/h, the sensitive wavelength of the irregularity is from 40–50 m. In this case, the lateral vibration energy of the body is the largest.

Table 6.4: Main lateral vibration frequency and sensitive wavelength of the vehicle body.

Speed /(km/h)	160	180	200	220	250	280
Main Frequency /Hz	1.0	1.2	1.2	1.4	1.5	1.6
Wavelength /m	44.44	41.67	46.30	43.65	46.30	48.61

Table 6.5: Main vertical vibration frequency and sensitive wavelength of the vehicle body.

Speed /(km/h)	160	180	200	220		250	280
Main Frequency /Hz	1.1	1.2	1.2	1.3		1.5	1.5
Wavelength /m	40.40	41.67	46.30	47.00		46.30	51.85

Table 6.5 shows that, for the vertical vibration of the vehicle body, the main frequency at each speed is very close to the main frequency in the horizontal direction. The sensitive wavelength range is also close to the horizontal range.

In addition, it is worth noting that the vibration of the high-speed train body is centered on the low frequency range (1–2 Hz). The human sensitive frequency ranges in lateral and vertical are 0–2 Hz and 4–8 Hz, respectively. Therefore, it is necessary to pay enough attention to the lateral vibration acceleration in the study and analysis of the ride comfort of high-speed trains.

(2) Comparative analysis of comfort performance at different wavelengths This section takes the lateral vibration of the body as the research object and compares the stability indexes under the random irregularity of different wavelength ranges. According to the previous analysis, 40–50 m is the sensitive wavelength range. Three different irregularity wavelength ranges in a passenger line are randomly set in simulation calculations. The first kind of wavelength range is from 1–40 m, and the sensitive wavelength is not considered; the second type is the combination of the first type and the sensitive wavelength, ranging from 1–50 m; the third type is the combination of the second type and irregularity of the longer wavelength, with the maximum wavelength reaching 100 m.

Table 6.6 lists the calculation results of the vehicle's lateral stability index at 250 km/h. Without the consideration of sensitive wavelength and long wavelength, the lateral stability index of the vehicle body is 2.45 in the irregularity wavelength range of 1–40 m; it is superior in dynamic performance. Owing to the imposed sensitive wavelength irregularity, the lateral stability of the vehicle body drops sharply, and the index value increases to 2.56. This is good in the dynamic performance evaluation standard under the excitation of 1–50 m wavelength irregularity. If the wavelength irregularity is further increased to 100 m based on the wavelength range of 1 to 50 m, the lateral stability of the body slightly increases to 2.59. It shows that the sensitive wavelength irregularity plays a major role in the lateral stability of the body, and the long-wave irregularity higher than the sensitive wavelength has less influence on the lateral stability of the body.

Table 6.6: Calculation result of lateral stationarity index.

Wavelength range /m	1~40	1~50	1~100
Lateral stability index	2.45	2.56	2.59

6.2 Optimal design of high-speed pantograph and catenary parameters

6.2.1 Optimized design of pantograph parameters

The parameters of the pantograph are one of the important factors affecting the dynamic characteristic of the pantograph-catenary system. After determining the structure and parameters of the catenary, if the parameters of the pantograph are inappropriate, not only will the pantograph-catenary vibration and the offline (offline state or offline time) increase, the current collection will deteriorate, and this will lead to abnormal abrasion and a series of problems such as severe bowing and drilling. Therefore, to obtain the best pantograph design parameters, it is necessary to analyze and qualitatively understand the influence of parameters such as mass, stiffness, and damping of the pantograph head and frame on the dynamic performances of the pantograph-catenary system. Based on the parameters of the TSG19 pantograph, the optimal design of the pantograph parameters is briefly introduced.

6.2.1.1 Suspension stiffness of pantograph head

Varying the suspension stiffness of pantograph head from 8000 to 15000 N/m, Fig. 6.24 presents the statistical results of contact forces at 350 km/h; it shows that, as the suspension stiffness of pantograph head increases from 8000 N/m to 15000 N/m, the minimum contact force increases first and then decreases, and the maximum and standard deviations decrease. Therefore, for this type pantograph, appropriately increasing the suspension stiffness is advantageous for suppressing the excessive vibration displacement of the pantograph suspension system and for avoiding large contact force oscillation. However, owing to the increase in the suspension stiffness of the pantograph head, the minimum contact force is generally reduced to result in an offline state. Thus, the pantograph head suspension stiffness cannot be too large.

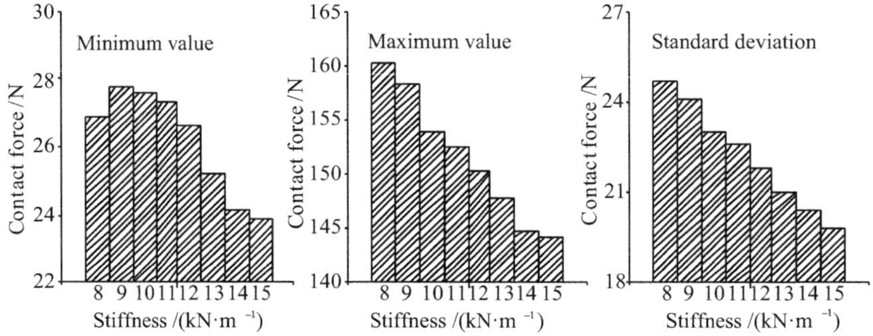

Figure 6.24
Effect of bow head suspension stiffness on contact force.

6.2.1.2 Suspension damping of pantograph head

The suspension damping of pantograph head is advantageous for reducing vibration and suppressing the oscillation of the pantograph head suspension system; however, excessive damping will hinder the movement of the pantograph head and affect its performance. Therefore, suspension damping should be reasonably selected to achieve a good dynamic performance of the pantograph-catenary system. Varying the suspension damping from 0 to 90 N.s/m, Fig. 6.25 presents the statistical results of contact force at 350 km/h. As the damping of the suspension system increases, the amplitude of the oscillation of the contact force decreases, and the quality of the current collection is greatly improved. When the damping value is increased to 50 N·s/m, the influence of damping on the contact force is gradually reduced. Therefore, for this type pantograph, it is reasonable to control the suspension damping of the pantograph head to about 50 N·s/m owing to the unobvious effect of the larger suspension damping on the contact force.

6.2.1.3 Pantograph head quality, frame quality

Varying the pantograph head quality from 4 to 11 kg, Fig. 6.26 presents the statistical results of the contact force at 350 km/h. It shows that as the quality of the pantograph head gradually increases, the minimum values contact force decrease; the maximum and standard deviation values increase, the performance of pantograph-catenary gradually deteriorates. Therefore, in terms of improving the dynamic performance of the pantograph head, the quality of the pantograph head suspension should be reduced as much as possible to ensure the followability of the pantograph head so that the pantograph can keep up with the vibration of the catenary. This is similar requirement for the optimal design of the frame quality.

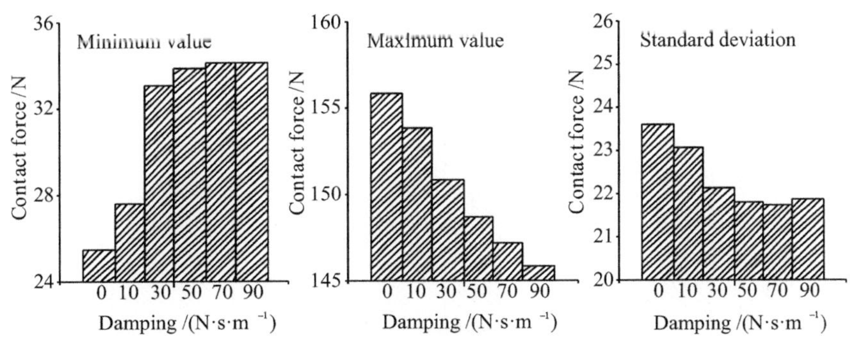

Figure 6.25

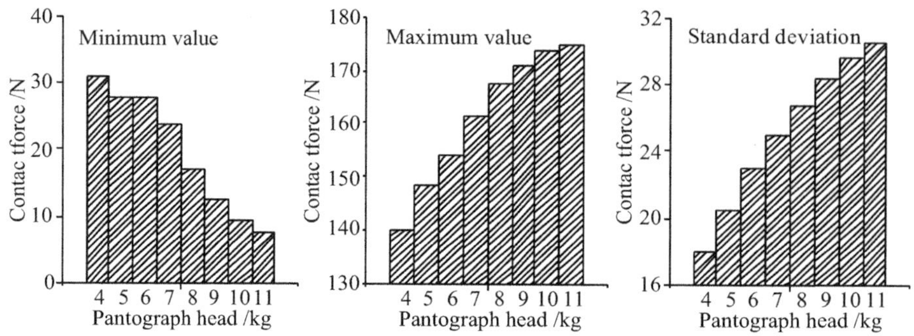

Figure 6.26
Effect of bow head suspension mass on contact force.

6.2.1.4 Frame stiffness

Varying the stiffness of the pantograph frame from 7 to 13 kg, Fig. 6.27 presents the statistical results of contact force at 350 km/h. It shows that as the stiffness of the frame increases, the minimum value of the contact force generally increases; the maximum value decreases, but the standard deviation does not change considerably. Therefore, it is advantageous to appropriately increase the rigidity of the frame for the pantograph to improve the vibration of the pantograph, reduce the fluctuation of the contact force, and improve the current collection quality.

6.2.1.5 Frame damping

Varying the stiffness of the pantograph frame from 0 to 80 N.s/m, Fig. 6.28 presents the statistical results of contact force at 350 km/h. It shows that the influence of frame damping on the contact force is not as obvious as the pantograph head damping. As the frame damping increases, the minimum value of contact force increases first and then

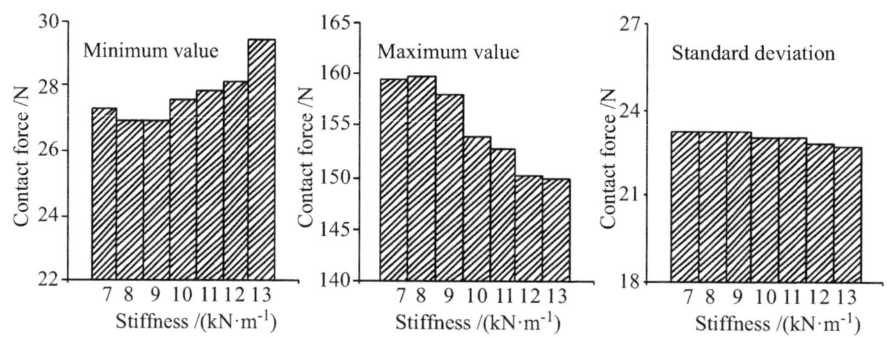

Figure 6.27
Effect of frame stiffness on contact force.

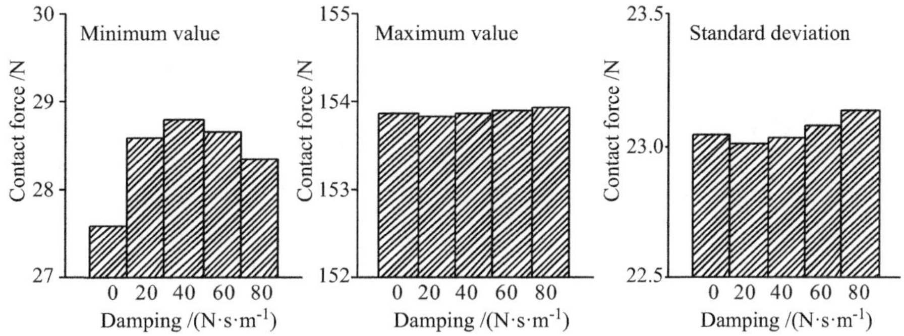

Figure 6.28
Effect of frame damping on contact force.

decreases; the maximum and standard deviation values do not change much. This indicates that too small or too large damping values are detrimental to the dynamic performance of the pantograph. For this type of pantograph, the damping value is about 40 N·s/m, which is the most reasonable.

6.2.1.6 Raising force

Varying the raising force from 70 to 140 N, Fig. 6.29 presents the statistical results of contact force at 350 km/h. It shows that as the raising force increases, the minimum value of contact force increases first and then decreases. The maximum value, average value, and standard deviation are constantly increase, which indicates it is beneficial to improve the dynamic performance of the pantograph by appropriately increasing the raising force of the pantograph under the condition of other constant parameters. However, when the raising force increases to a certain extent, since the average value, standard deviation, and

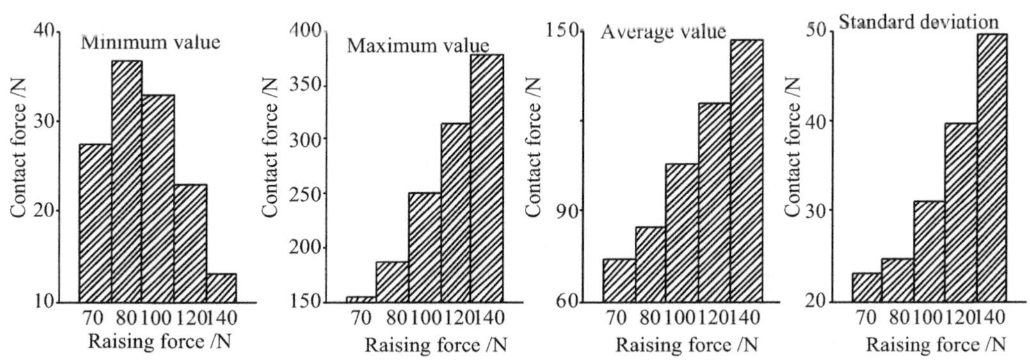

Figure 6.29
Effect of raising force on contact force.

maximum value of the contact force of the pantograph increase, the oscillation of the contact force increases, and thus, the quality of the pantograph-catenary degrades further.

6.2.2 Optimized design of catenary parameters

In order to meet the requirements of high-speed current collection, countries in the world have adopted different catenary systems based on the actual scenarios for their national high-speed railway. For different catenary systems, the structure and parameters are considerably different, and this has an important effect on the quality of the pantograph-catenary. Therefore, to provide a reference for the optimal design of catenary parameters, the following discussion is based on the effect of parameters of the Wuhan-Guangzhou high-speed railway catenary system, the parameters of the different suspension forms, tension, span and dropper spacing of the catenary on dynamic performances.

6.2.2.1 Suspension form

The suspension form of the high-speed catenary structure is composed of three types: a simple stitched catenary, an elastic stitched catenary, and a double chain suspension. The operation practice has proven that the above three kinds of catenary suspension forms can meet all requirements of running a high-speed train; however, each type has its own advantages and disadvantages. Because of the simple structure, convenient construction, low engineering cost, and low operation and maintenance cost, the simple chain suspension catenary has been recognized by all countries in the world and has become a common type of high-speed railway catenary. The more representative high-speed catenary systems are those in France, Italy, and China, as shown in Fig. 6.30.

Compared with the simple chain suspension catenary, the elastic chain suspension catenary has a better elastic uniform coefficient owing to the addition of an elastic suspender to achieve better current collection quality. Therefore, this type of high-speed catenary has also been adopted by many countries, including Germany, Spain, and China. The German high-speed railway catenary has always adopted elastic chain suspension. Based on the

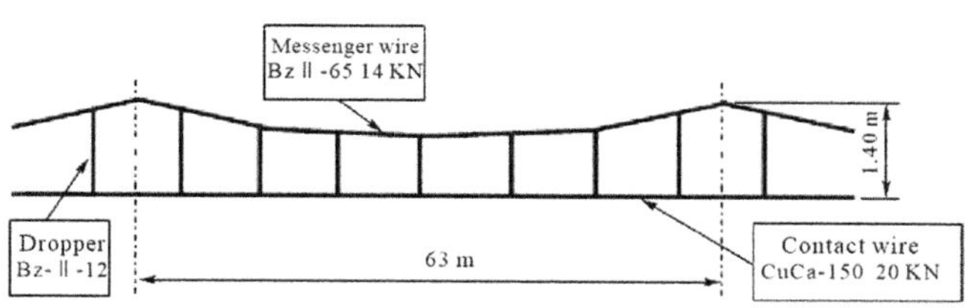

Figure 6.30
French TGV standard simple chain suspension catenary.

three standards of Re75, Re100 and Re160, the Re200, Re250, and Re330 standard series have been established. Re denotes a standard catenary, and the number after it denotes the maximum speed at which the trains can run in the standard catenary form.

The double-chain suspension catenary is used in Japan owing to multiple typhoons and many earthquakes. It basically follows its traditional catenary suspension method, multi-chain, since the first high-speed railway catenary system adopts the complex chain suspension as shown in Fig. 6.32. The elastic uniformity coefficient of this type of catenary is quite good; however, it is followed by high cost, inconvenient construction, and high maintenance. There is also a tendency to transit to a simple chain catenary on a line with a speed of less than 300 km/h from the new line.

In combination with the advantages and disadvantages of different catenary suspension forms, China now mainly adopts the type of catenary structure suspension with simple and elastic chain types. In order to compare the effect of the two suspension forms on the dynamic performance of the pantograph-catenary, the elastic sling is eliminated, and a simple chain catenary system is obtained on the basis of the prototype catenary system

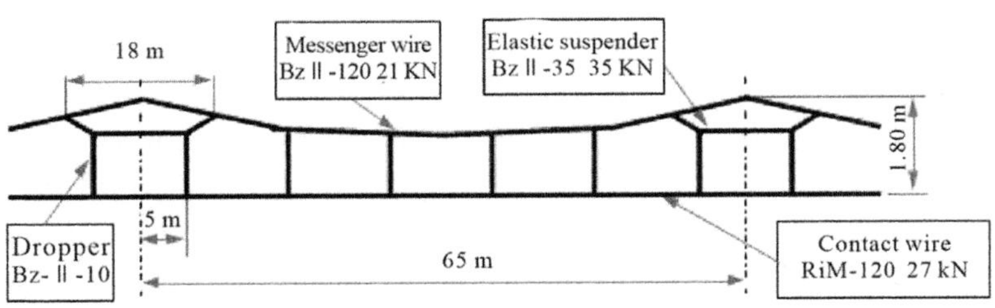

Figure 6.31
Germany Re 330 elastic chain suspension catenary.

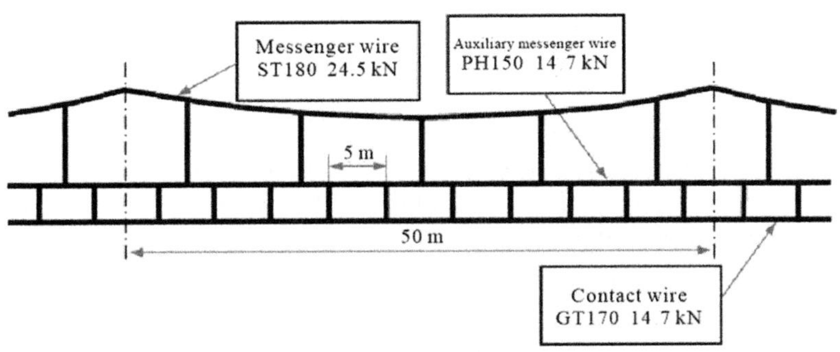

Figure 6.32
Japanese double chain suspension net.

Table 6.7: Contact force of different catenary suspension forms.

Speed/km·h^{-1}	Types of Catenary	Contact Force/N		
		Minimum Value	Maximum Value	Standard Deviation
350	Elastic Chain	31.53	159.83	23.86
	Simple Chain	16.35	160.41	26.95

of the Wuhan-Guangzhou high-speed railway. Table 6.7 shows the results of the contact force calculation result for the two types of catenary suspensions when the train is running at 350 km/h.

Table 6.7 shows that when the elastic chain suspension is used under other constant parameters, the minimum contact force of the pantograph is larger; however, the maximum and standard deviation are slightly smaller, and the fluctuation of the contact force of the catenary is smaller. This implies that if we do not consider other factors such as economical aspects, the elastic chain-type suspension pantograph is a better choice owing to its improved intra-elastic uniformity and better flow under the same conditions.

6.2.2.2 Span

Catenary elasticity and elastic uniformity are important performance indexes in the catenary optimal design with a direct impact on the dynamic performance of the pantograph-catenary system. In general, the better the elastic uniformity of the catenary, the better is the dynamics of the catenary and the quality of current collection. There are many parameters that affect the elasticity and elastic uniformity of the catenary, which include the span, the hanging string spacing, and the tension, in addition to the above-mentioned catenary suspension form. To analyze the effects of different spans of 58 m and 42 m on the elastic properties of the catenary and the dynamic performance of the pantograph-catenary, we keep the same hanging string spacing and increase and decrease one hanging string, respectively, based on the 50 m span of the catenary system of the Wuhan-Guangzhou line.

The static stiffness values of the catenary of three different spans (the reciprocal of the catenary elasticity) are shown in Fig. 6.33, and the elastic difference coefficient is calculated by the Equation (6.3); the results are shown in Table 6.8. It shows that there are obvious differences in the static stiffness and elastic difference coefficient of the catenary for different spans. The smaller the span, the greater is the static stiffness; the smaller the elasticity, the smaller is the elastic difference coefficient. Although the span is different, the static stiffness of the span has a similar distribution law, that is, the stiffness near the prop is larger, and the stiffness near the mid-span is smaller and the elasticity is larger.

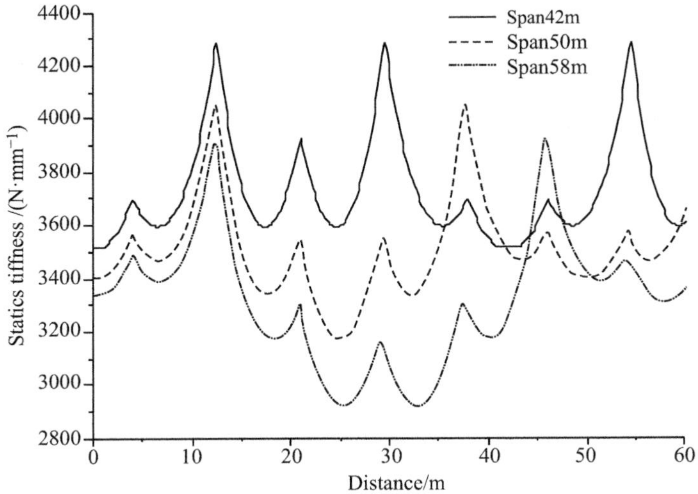

Figure 6.33
Static stiffness curve of the catenary span.

Table 6.8: Elastic difference coefficients of different spans.

	Span 42 m	Span 50 m	Span 58 m
K_{max}Maximum stiffness N/mm	4289.08	4059.11	3922.46
K_{min}Minimum stiffness N/mm	3512.98	3172.40	2925.27
εElastic coefficient of variation / %	9.95	12.26	14.56

$$\varepsilon = \frac{K_{max} - K_{min}}{K_{max} + K_{min}} \times 100\% \tag{6.3}$$

Table 6.9 lists the results of the contact force of three different catenaries when the train is running at 350 km/h and the distance between the hanging strings is constant.

It can be seen from Table 6.9 that as the span decreases, the coefficient of elasticity difference across the catenary decreases, and therefore, the minimum value of the contact force increases, and the maximum and standard deviation decrease. The change in contact

Table 6.9: Contact force of different catenary spans.

Speed/km·h^{-1}	Catenary span/m	Contact force/N		
		Minimum value	Maximum value	Standard deviation
350	42	34.93261	143.59668	21.20449
	50	31.52539	159.83049	23.85548
	58	30.69958	164.87238	26.44999

force tends to be stable with improved current collection quality. This proves that the elastic difference coefficient of the catenary has an important effect on the dynamic performance of the pantograph-catenary, and the short span is beneficial to improve the static stiffness near the middle of the span to ensure more uniform elasticity of the catenary and better flow quality. However, the span should not be too small; otherwise, the overall static stiffness of the catenary will be greater, and the overall elasticity lower. This is also detrimental to its dynamic performance and flow quality.

6.2.2.3 Hanging string spacing

The hanging string spacing is an important parameter that affects the elasticity, elastic uniformity of the catenary, and dynamic performance of the pantograph-catenary. In order to study the influence of different hanging string spacings on the dynamic performance of the pantograph-catenary, we keep the span constant, and we design two different hanging string arrangement schemes, as shown in Fig. 6.34.

It can be seen from Table 6.10 that compared with the first scheme of the hanging string arrangement, the second scheme shows a decreasing maximum and standard deviation of the contact force and an increasing minimum value; the change in the contact force decreases. Under the same constant span, the second scheme increases the spacing of the hanging strings at the pillars and reduces the spacing of the mid-span positions. The coefficient of elasticity difference of the inter-pantograph is reduced, which is beneficial to the improvement of the dynamic performance of the pantograph-catenary system. Therefore, under the constant span, the non-equal spacing hanging strings arrangement similar to that of the second scheme is adopted by appropriately increasing the spacing of

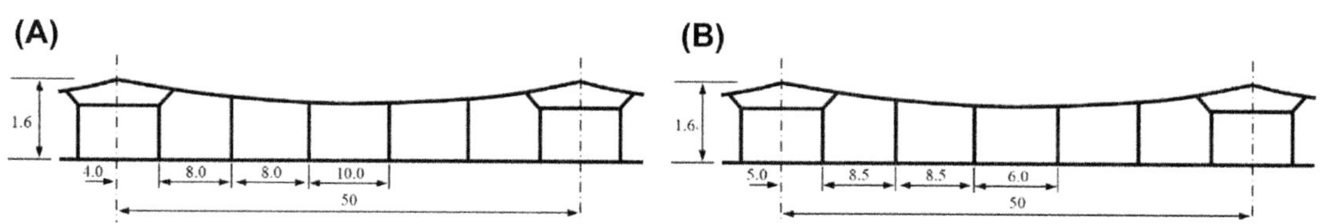

Figure 6.34
Catenary suspension arrangement. (A) First scheme (B) Second scheme.

Table 6.10: Contact force of different catenary hanging strings arrangement.

Speed/km·h^{-1}	Hanging string arrangement	Contact force/N Minimum value	Maximum value	Standard deviation
350	1	27.12363	161.00239	26.14716
	2	32.12524	155.12256	23.25614

the hanging strings near the pillars and reducing the spacing of the hanging strings near the middle. This is beneficial to improve the coefficient of elasticity difference across the span and improve the dynamic performance and current collection quality of the pantograph-catenary.

6.2.2.4 *Catenary tension*

Catenary wave velocity refers to the velocity of catenary wave propagation along the catenary. The propagation direction includes two directions: forward and backward. Catenary wave velocity is one of the important and effective parameters to describe the performance index of the catenary. The catenary wave velocity is correlated to the contact line tension and line density, and it can usually be determined by Equation (6.4). As the running speed of the train continues to increase to approach the wave velocity of the pantograph, the correlation between the pantographs significantly deteriorates; further, the fluctuation of its contact force increase greatly, and the current collection quality decrease significantly.

$$C_p = \sqrt{\frac{T}{\rho}} \tag{6.4}$$

where C_p is the speed of the catenary fluctuation (m/s), T is the contact line tension (N), and ρ is the contact line density (kg/m).

An increase in the tension of the catenary is always limited, and the increase in the tension causes other problems such as a decrease in the safety and reliability of the catenary. Therefore, currently, the tension of the catenary is designed to be 0.7–0.8 times the wave speed utilization rate, i.e., the ratio of the maximum running speed of the train to the wave speed of the catenary. At the same time, through the optimization of the structure and parameters of the pantograph-catenary, the wave speed utilization rate is improved as much as possible under the condition of the determined tension.

Fig. 6.35 shows the correlation between tension and speed of different contact lines based on the catenary system of the Wuhan-Guangzhou line. It shows that if the tension of the catenary is designed according to the wave velocity utilization ratio of 0.7–0.8 times, the speed of the catenary should be at least about 40 KN to reach a speed of 500 km/h. For this reason, based on the parameters of the prototype catenary system, we increase the tension of the contact line to 40 KN when other parameters do not change, and then, we verify the dynamics calculation of the pantograph-catenary system.

Fig. 6.36(A) presents the calculation result of the contact force of the pantograph-catenary system when the train runs at a speed of 500 km/h. It shows that under the condition that other parameters remain unchanged, even if the contact line tension is increased to 40 KN, the pantograph and the catenary are still offline. To this end, based on the optimization of

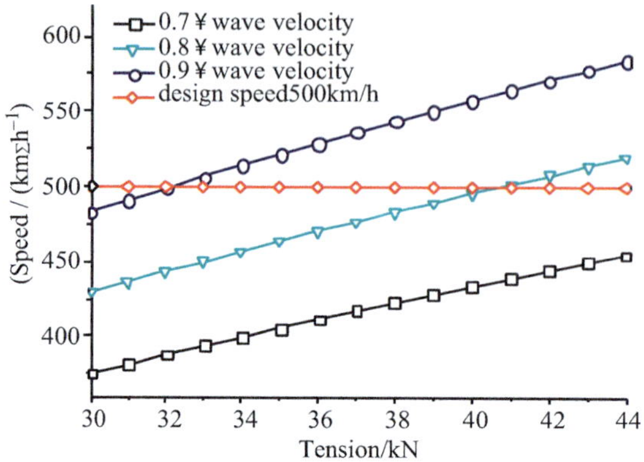

Figure 6.35
The correlation between contact line tension and speed.

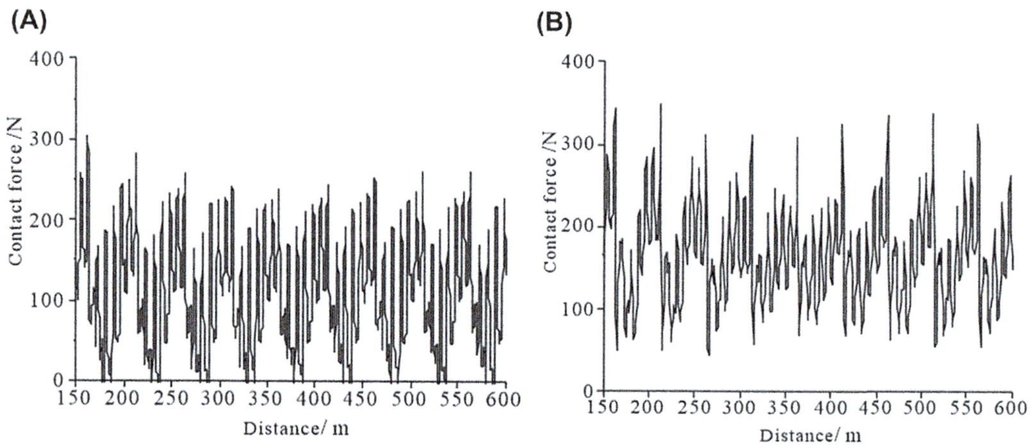

Figure 6.36
Pantograph-catenary contact pressure at the speed of 500 km/h. (A) Before Optimization (B) After optimization.

other parameters of the pantograph-catenary system, the optimized contact force is obtained, as shown in Fig. 6.36(B). It shows that offline does not occur at this time. This indicates that the contact line tension is designed according to be 0.7–0.8 times of the utilization ratio of wave velocity. In addition, based on the parameter optimization of the pantograph-catenary system, it is feasible to further improve the wave speed utilization rate to meet the requirement of the design speed.

6.2.2.5 Catenary material

Currently, high-strength, high-conductivity, lightweight materials are used for high-speed railway catenary. The tension of the contact line is increased as much as possible to ensure the reliability of the wire and increase the fluctuation speed of the contact wire. At present, China has been able to produce copper-silver alloy contact wire (CTA/CTAH), copper-magnesium alloy contact wire (CTM/CTMH), and copper-tin alloy contact wire (CTS) under independent property rights. Currently 150 mm^2 and 120 mm^2 are used for the high-speed contact wire cross-section, and the linear density of the former is between 1.33–1.35 kg/m, while that for the latter is between 1.07–1.08 kg/m. If the abrasion and ampacity contact wire meet the requirement, the lightweight 120 mm^2 contact wire is beneficial to improve the wave speed and improve current collection. Table 6.11 shows the contact forces for the two contact wire materials of 150 mm^2 and 120 mm^2 when the train is running at 350 km/h.

Table 6.11 indicates that when other parameters do not change, if the cross section of the contact wire is reduced from the original 150 mm^2 to 120 mm^2, the minimum value of the contact force increases, and the maximum value and the standard deviation both decrease. This indicates that the fluctuation of the contact force is reduced with improved dynamic performance of the pantograph-catenary system. Therefore, it is confirmed that the light contact wire is beneficial to improve the contact wire wave velocity and improve the dynamic performance and current collection quality under the requirements of the pantograph-catenary current and strength.

6.2.3 Optimal design of pantograph spacing

In order to improve the transportation capacity of the train, 16 vehicles in marshalling form are often used, which is equivalent to the two high-speed trains. Due to the limited conductivity of the pantograph, high-speed trains need be operated with double pantographs under this condition. When the train is running, the sliding of the pantograph along the catenary will cause vibrations in the contact wire, which will propagate in the form of waves along the contact wire. When the double pantograph operation occurs, due to the fluctuation of the contact wire, the dynamic performance of the rear pantograph of the train is often disturbed by the operation of the front pantograph, adversely affecting the current collection. If the position of the rear pantograph is different, this effect may be aggravated, resulting in unfavorable effects on the rear pantograph. Therefore, it is very necessary to explore how to properly choose

Table 6.11: Contact force of different contact wire materials.

Speed/km·h^{-1}	Cross-sectional area /mm^2	Contact force/N		
		Minimum value	Maximum value	Standard deviation
350	150	31.52539	159.83049	23.85548
	120	35.39313	155.35917	21.377

the double pantograph spacing to ensure good and stable flow, and a long service life of the contact line.

When the pantograph slides at high speed along the contact wire, the inconsistent static stiffness of the catenary will lead to the coupled vibration of the catenary and the pantograph that propagates the form of the wave on the contact wire. When the vibration wave reaches the two fixed ends, reflection occurs, and reflections cancel each other out; therefore, the main dominant effect on the vibration of the contact wire is the fundamental wave. The solid wire in Fig. 6.37 depicts the line vibration waveform caused at the contact point of the catenary. It shows that the change in the vibration waveform at the contact point of the catenary has a certain periodicity with the span as a period. Within each span, the maximum value of the vibration wave appears near the mid-span position, and the minimum value is near the pillar position [7]-[8].

When the two pantographs slide at high speed along the contact wire, due to the coupling between the pantograph and the catenary, the waveform of the pantograph vibrating in parallel with the vibration direction is bound to be caused. If the dotted line in Fig. 6.37 is regarded as the pantograph vibration waveform at the L_g behind the front pantograph, the solid line is regarded as the vibration waveform of the contact line at the contact point of the catenary. It indicates that the two vibration waves of the contact wire and the pantograph have a phase difference of 180°, which is in opposite phase. Due to the interference and superposition of waves, the vibration of the contact point of the catenary is weakened, and the inconsistency of the stiffness of the catenary is improved, thereby improving the current collection quality. It can be seen that when the double pantograph

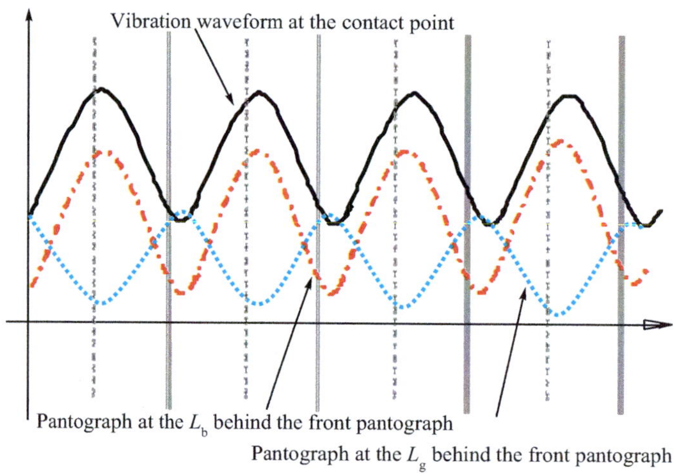

Figure 6.37
Vibration waveforms at different locations on the contact line.

spacing is optimal, the vibration waveform of the contact line at the contact point will be in an inverted position with the vibration waveform of the pantograph. Accordingly, this can be the case if the time required for the rear pantograph to travel to the L_g behind the front pantograph is equal to half the odd multiple of the contact line vibration wave period. Then, there is

$$\frac{L_g}{u} = \left(k + \frac{1}{2}\right)\frac{\lambda}{v} \quad (k = 1, 2, 3...) \tag{6.5}$$

where L_g is the optimal double pantograph spacing, u is the pantograph running speed, λ is the vibration wave wavelength, and v is the vibration wave velocity. On this basis, by determining the wavelength λ of the vibration wave and the wave velocity v, the formula for calculating the optimal double pantograph spacing can be derived as

$$L_g = (2k+1)\frac{Lu}{\alpha\sqrt{T/\rho}} \quad (k = 1, 2, 3...) \tag{6.6}$$

where L is the catenary span, T is the contact line tension, ρ is the contact line density, α is the correction factor, and 0.8 is taken for the elastic chain suspension catenary structure.

On the other hand, if the dotted line in Fig. 6.37 is regarded as the pantograph vibration waveform at the rear L_b of the front pantograph, the two vibration waves caused by the contact line and the pantograph are out of phase by 0°, which is in the same phase. The vibration of the contact point of the catenary is strengthened, the influence of the inconsistency of the stiffness of the catenary is aggravated, and the current collection quality is deteriorated. Therefore, the formula for calculating the worst double pantograph spacing is

$$L_b = 2k\frac{Lu}{\alpha\sqrt{T/\rho}} \quad (k = 1, 2, 3...) \tag{6.7}$$

For the elastic chain suspension pantograph system of the Wuhan-Guangzhou line, the maximum speed of the pantograph is 270 km/h, 300 km/h, 330 km/h, and 350 km/h, according to Equation (6.7). The calculation of the optimal double pantograph spacing is shown in Table 6.12.

It can be seen from Table 6.12 that the optimal double pantograph spacing can be obtained by Equation (6.7). When the train is running at around 330–350 km/h, the double pantograph spacing of 20 m is just near the optimal spacing with K = 2. It also shows that in the case of the pantograph system of the Wuhan-Guangzhou line, the double pantograph spacing of 200 m is reasonable for the current collection of 350 km/h double pantograph.

On the other hand, the dynamic simulation calculation of the catenary under the action of double pantograph is to study the influence of different double pantograph spacing on the

Table 6.12: Calculation results of optimal double pantograph spacing for the pantograph system of Wuhan-Guangzhou line.

Speed ($km \cdot h^{-1}$)	The optimal double pantograph spacing/m		
	K = 1	K = 2	K = 3
270	94.33	157.22	220.11
300	104.82	174.70	244.57
330	115.30	192.16	269.03
350	122.29	203.81	285.33

dynamic performance. Besides, the unfavorable and favorable spacing of the double pantograph can also be determined.

In order to determine the unfavorable and favorable spacing of the pantograph at different speeds, the dynamic performance of the catenary under different double pantograph spacing is calculated when the train is running at 270 km/h, 300 km/h, 330 km/h, and 350 km/h. Fig. 6.38 shows the results of how standard deviation of the rear pantograph contact force at four speeds varies with the distance of the double pantograph.

Fig. 6.38 shows that the double pantograph spacing has a great influence on the change of the contact force; when the double pantograph spacing is different, the contact force

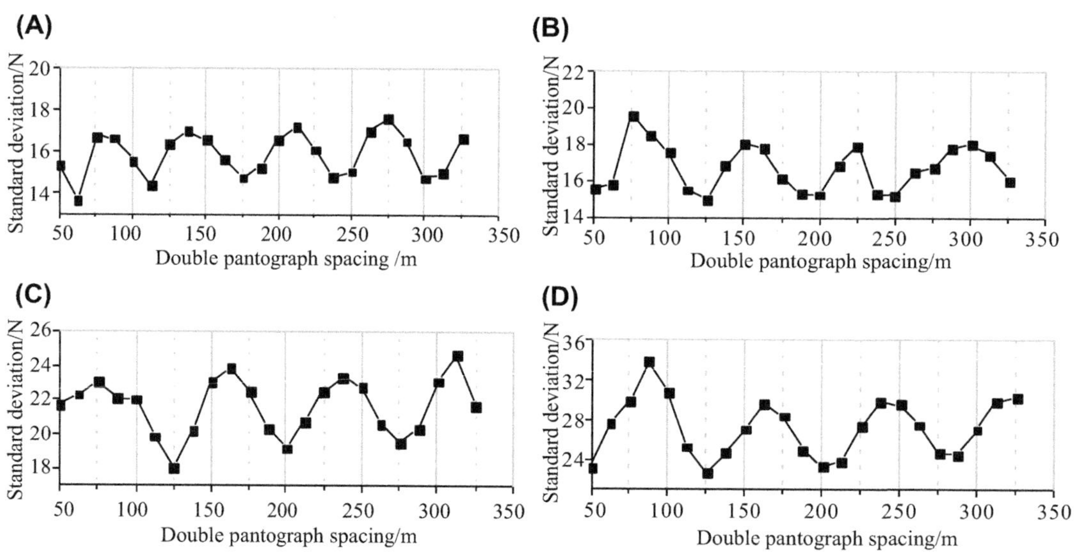

Figure 6.38
Standard deviation of the rear pantograph contact force for different double pantograph spacing at different speeds. (A) 270 km/h (B) 300 km/h (C) 330 km/h (D) 350 km/h.

change has an obvious difference. In a certain range of spacing, the double pantograph spacing is not as large as possible. At a certain speed, the influence of the double pantograph spacing on the change of the contact force has certain regularity, and the unfavorable and favorable spacing have a certain periodic distribution. When the rear pantograph is at an unfavorable distance, the contact force changes, the standard deviation increases, and the rear pantograph off-line rate increases.

Combined with the calculation result of the theoretical formula, it can be seen that in the favorable spacing given in Table 6.12, the standard deviation of the contact pressure of the rear catenary is at the minimum position. This shows that the fluctuation of contact pressure is small at this time, and the rear pantograph has better current collection quality.

It can be seen that the pantograph spacing of the double pantograph reconnection has an important influence on the current collection of the rear pantograph, and therefore, it is necessary to select the double pantograph spacing reasonably. When designing the parameters of the pantograph system, the optimal pantograph spacing can be obtained by theoretical calculation or the pantograph simulation under double pantographs to ensure the stable current collection quality of the rear pantograph.

6.3 Optimal design of the aerodynamic characteristics of high-speed train

6.3.1 Optimal design of high-speed train shape

The shape of high-speed trains is an important factor affecting the aerodynamic performance and running safety of high-speed trains. The head shape of the train affects the aerodynamic drag, aerodynamic lift, lateral force, aerodynamic performance of train passing, aerodynamic performance of pass wind, aerodynamic performance of tunnel dynamics and aerodynamic noise [9]. Optimal train streamlined design can reduce the negative impact of aerodynamic performance on a high-speed train.

6.3.1.1 Head profile control line

The longitudinal profile control line and the horizontal profile control line are key parameters for the design of the train head. In order to study the influence of the longitudinal profile control line and the horizontal profile control line on the aerodynamic performance of the train, three longitudinal profile control lines and three horizontal profile control lines as shown in Fig. 6.39 were established.

6.3.1.2 Head profile control line and steady-state basic aerodynamic characteristics

In order to study the correlation between the head profile control line and the basic aerodynamic performance of the steady-state operation of the train, the steady-state

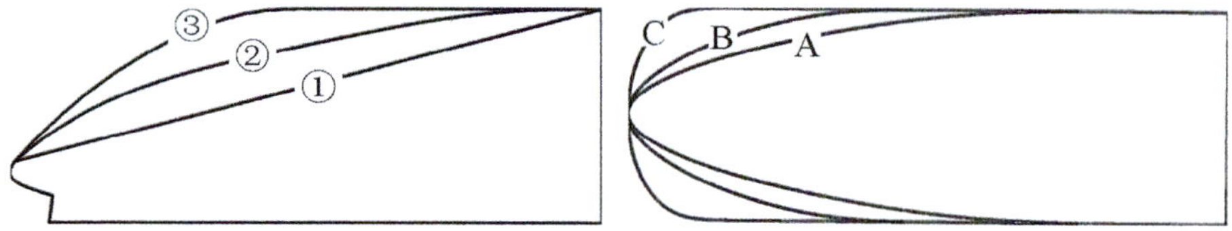

Figure 6.39
Head profile longitudinal control line and horizontal profile control line.

operation calculation model is established with the train head shape as the research object, as shown in Fig. 6.40. Figs. 6.41 and 6.42 show the aerodynamic drag and aerodynamic lift of trains of different shapes when it is running at 350 km/h.

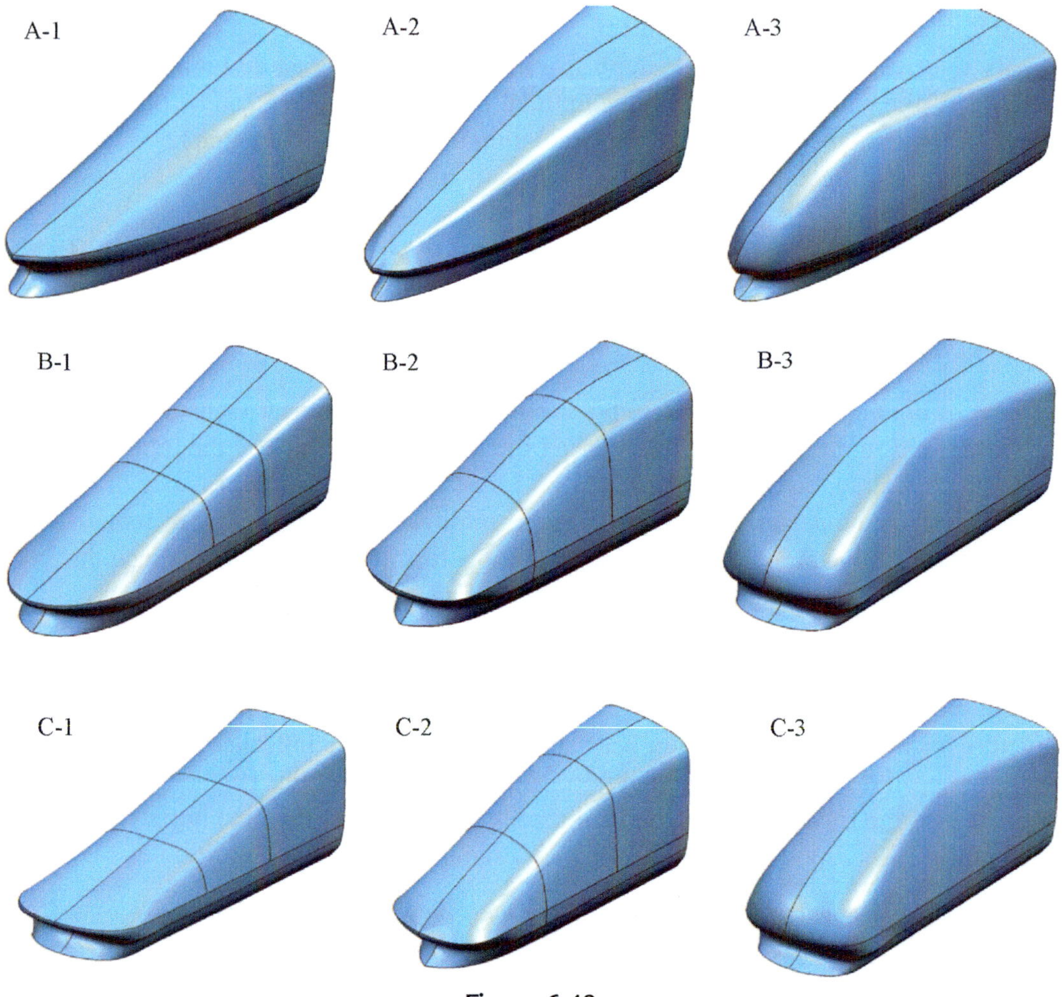

Figure 6.40
High-speed train head shape based on different longitudinal and horizontal control line shapes.

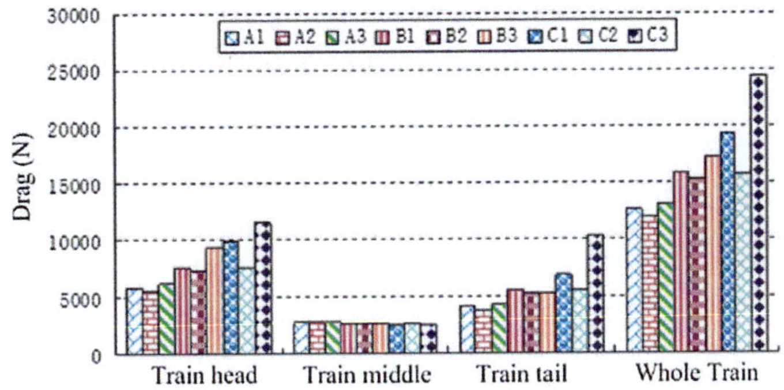

Figure 6.41
Train resistance.

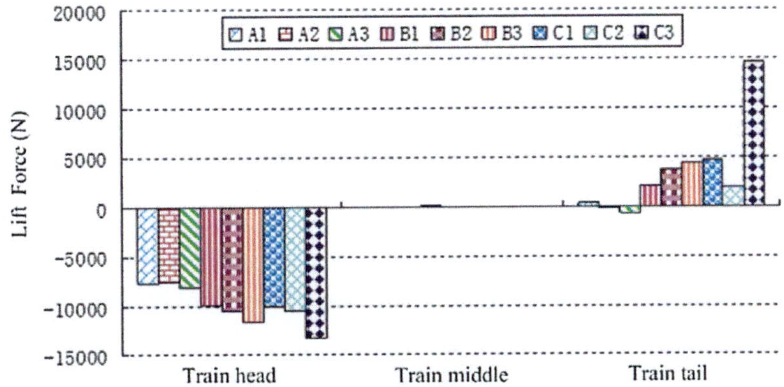

Figure 6.42
Train lift.

The first step is to analyze the influence of the head profile control line on the aerodynamic drag of the train. If the shape of the horizontal profile control line of the train head is the same, the train with the curved longitudinal profile control line ② has the smallest aerodynamic drag; the aerodynamic drag of the train with the straight longitudinal profile control line ① ranks second, and the train with the drum longitudinal profile control line ③ has the largest aerodynamic drag. If the shape of the longitudinal profile control line of the train head is uniform, the aerodynamic drag of the train with the tapered horizontal profile control line A is the smallest, and the aerodynamic drag of the train with the square horizontal profile control line C is the largest. By comparison, it is found that the influence of the horizontal profile control line on the aerodynamic drag is greater than that of the longitudinal profile control line. Based on the above analysis, the train with the combination of the tapered horizontal profile control line A and the curved longitudinal profile control line ② is shown to have the smallest aerodynamic drag.

The second step is to analyze the influence of the head profile line on the aerodynamic lift of the train. If the horizontal profile control line of the train head has the same shape, the aerodynamic lift amplitude of the head with the curved longitudinal profile control line ② is the smallest, and the aerodynamic lift amplitude of the head with the drum longitudinal profile control line ③ is the largest. If the longitudinal profile control line shape of the train head is uniform, the aerodynamic lift amplitude of the head with the tapered horizontal profile control line A is the smallest, and the aerodynamic lift amplitude of the head with the square horizontal profile control line C is the largest. For the lift of the tail, the aerodynamic lift of the tail with the tapered horizontal profile control line A may be negative, that is, the aerodynamic lift is downward. The aerodynamic lift performance of the tail with the tapered horizontal profile control line A is relatively good. Based on the above analysis, it can be seen that the combination of the tapered horizontal profile control line A and the curved longitudinal profile control line ② has a relatively good aerodynamic lift performance of the head and tail of the train.

Considering the correlation between the head profile control line and the aerodynamic drag and aerodynamic lift during steady-state operation of the train, the main conclusions are as follows:

(1) The influence of the horizontal profile control line on the aerodynamic drag is greater than that of the longitudinal profile control line. The train with the combination of the tapered horizontal profile control line A and the curved longitudinal profile control line ② has the lowest aerodynamic drag.
(2) The combination of the tapered horizontal profile control line A and the appropriate longitudinal profile control line enables the aerodynamic lift of the tail to be close to zero. The combination of the tapered horizontal profile control line A and the curved longitudinal profile control line ② has a relatively good aerodynamic lift performance on the head and tail of the train.

6.3.1.3 Head profile control line and pass-wind aerodynamic performance

In order to study the relationship between the head profile control line and the pass-wind aerodynamic performance of the train, the pass wind calculation model has been established with the train head shape in Fig. 6.42 as the research object. The train operation speed and the pass wind speed in this model are set as 350 km/h and 17.1 km/h, respectively.

Fig. 6.43 shows the variation trend of the side force of trains in different shapes under the action of pass wind. The amplitude of the side force of the head is the largest, and that of the middle is second, and that of the tail is relatively small. Under the action of pass wind, the longitudinal profile control line has a greater influence on the train side force, while the horizontal profile control line has less influence on the train side force. If the

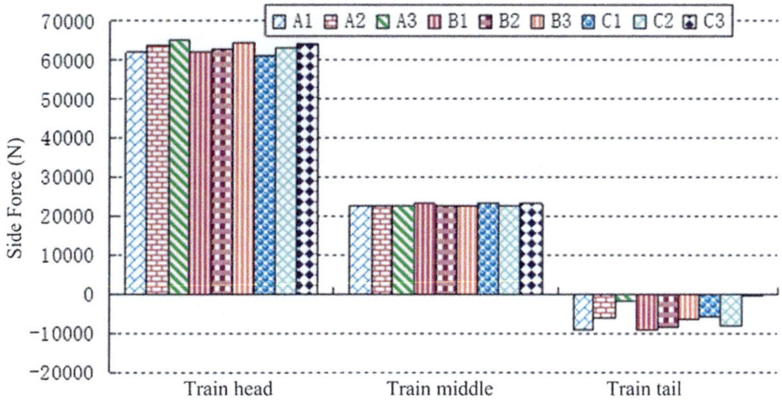

Figure 6.43
Side force of each train under strong pass wind.

horizontal profile control line shape of the train head is uniform, the head side force amplitude with straight longitudinal profile control line ① is the smallest, and the head side force amplitude with drum longitudinal section control line ③ is the largest.

Fig. 6.44 shows the change in the trend of the lift of trains in different shapes under the action of pass wind. Under the action of pass wind, the amplitude of the head lift is close to zero, and the lift amplitude of the middle and tail is relatively large. If the longitudinal profile control line shape of the train head is the same, the aerodynamic lift amplitude of the head and tail with the tapered horizontal profile control line A is the smallest, and the aerodynamic lift amplitude of the head and tail with the square horizontal profile control line C is the largest.

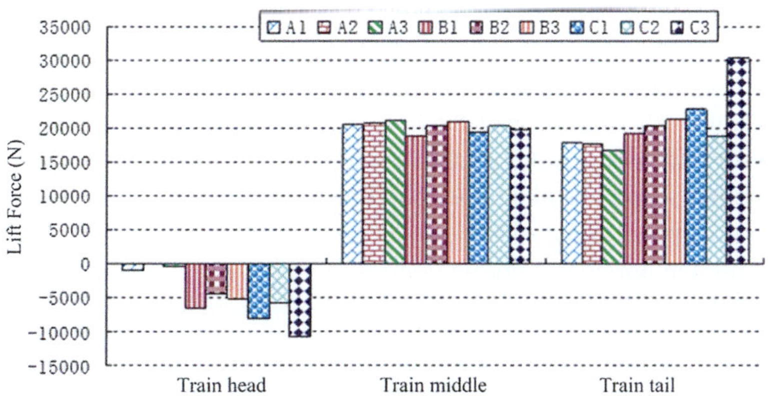

Figure 6.44
Lift of each train under strong pass wind.

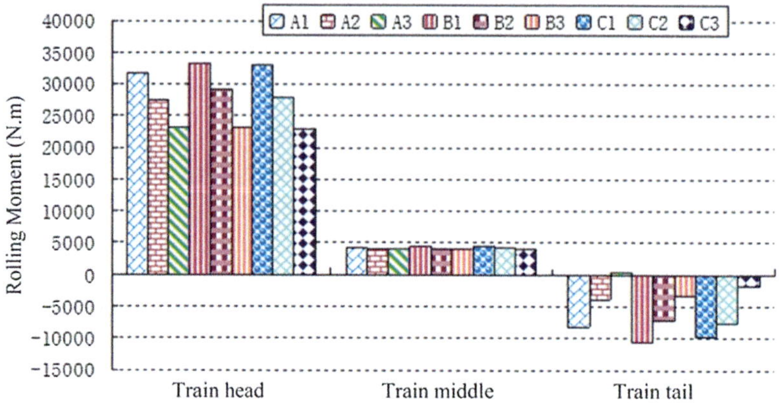

Figure 6.45
Rolling moment of each train under strong pass wind.

Fig. 6.45 shows the change trend of the rolling moment of different shape trains under the action of pass wind. The amplitude of the rolling moment of the head is the largest, and that of the middle is second, and that of the tail is relatively small. Under pass wind, the longitudinal profile control line has a greater influence on the train rolling moment, while the horizontal profile control line has less influence on the train rolling moment. If the horizontal profile control line of the train head has the same shape, the head and tail wheel rolling moments with the straight longitudinal profile control line ① have the largest amplitude, and the head and tail wheel rolling moments with the drum-shaped longitudinal profile control line ③ have the smallest amplitude.

Considering the correlation between the head profile control line and the pass wind aerodynamic performance of the train, the main conclusions are as follows:

(1) The longitudinal profile control line has a greater influence on the train side force, while the horizontal profile control line has less influence on the train side force. The head side force amplitude with the straight longitudinal profile control line ① is relatively small.
(2) The aerodynamic lift amplitude of the head and the tail with the tapered horizontal profile control line A is relatively small, and the aerodynamic lift amplitude of the head and the tail with the square horizontal profile control line C is relatively large.
(3) The longitudinal profile control line has a considerable influence on the train rolling moment, while the horizontal profile control line has less influence on the train rolling moment. The head and tail side force amplitudes with the drum-shaped longitudinal profile control line ③ are the smallest.

6.3.1.4 Head control line and train passing performance

In order to study the correlation between the head profile control line and the aerodynamic performance of the train, the train passing calculation model is established with the train head shape as the subject, as shown in Fig. 6.40. The pressure wave monitoring point distribution is shown in Fig. 6.46. The working conditions considered are: two trains will pass at the same speed on the flat at a speed of 350 km/h.

Fig. 6.47 shows the trend of pressure peak-to-peak value during the different shape trains passing at the constant speed. The longitudinal profile control line has a greater influence on the peak value of the pressure wave, while the horizontal profile control line has less influence on the peak value of the pressure wave. If the horizontal profile control line of the train head has the same shape, the peak of the passing pressure peak with the straight longitudinal profile control line ① and the curved longitudinal profile control line ② are relatively small. The peak of the passing pressure peak with the drum-shaped longitudinal profile control line ③ is relatively large.

Fig. 6.48 shows the variation trend of lateral force peak value of different shape trains passing at the same speed. It can be seen that the influence law of the head profile control line on the lateral force peak value is consistent with the influence of the head profile control line on the peak pressure value.

By analyzing the performance of the train heads of different head control line shapes, it can be known that,

(1) The longitudinal profile control line has a considerable influence on the peak value of the pressure wave, while the horizontal profile control line has little influence on the peak value of the pressure wave. The peak pressure value with the drum-shaped longitudinal profile control line ③ is relatively large.
(2) The influence law of the head profile control line on the lateral force peak value is consistent with the influence of the head profile control line on the peak pressure value.

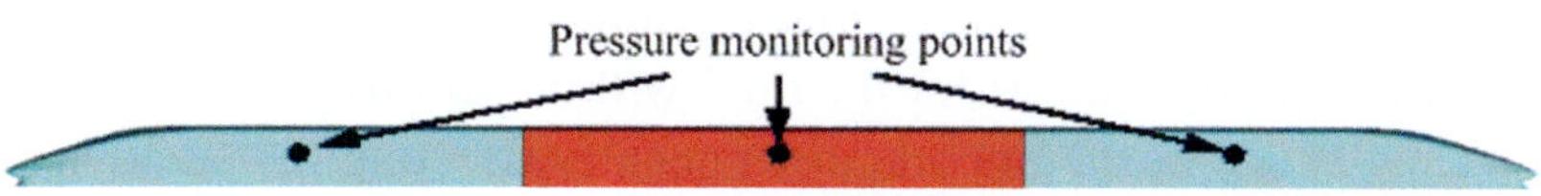

Figure 6.46

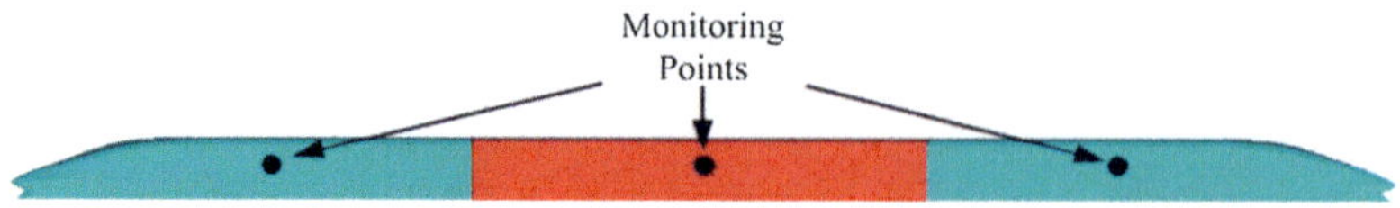

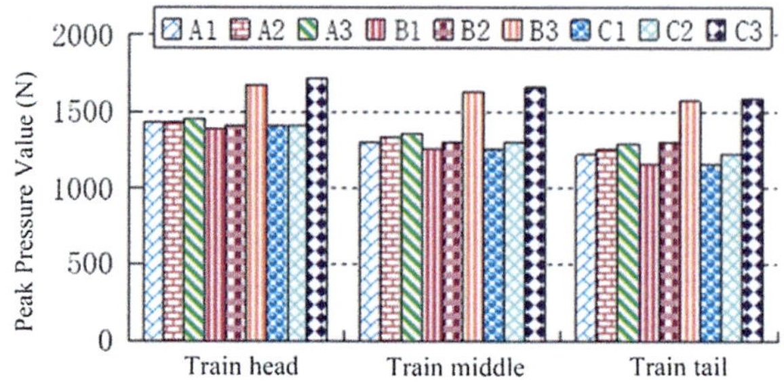

Figure 6.47
Peak pressure value of the train passing.

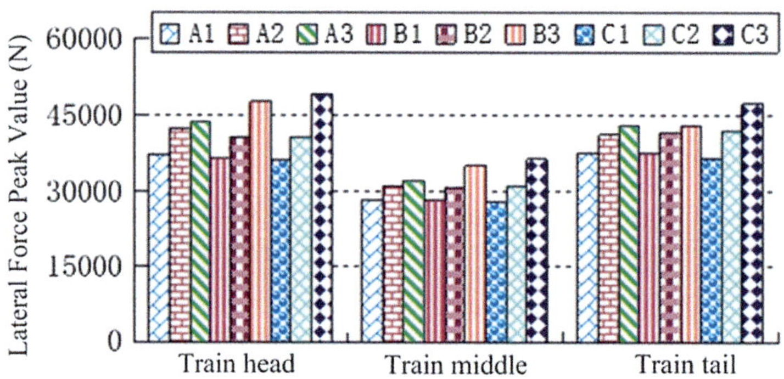

Figure 6.48
Lateral force peak value of the train passing.

6.3.2 Optimal design of line spacing

When two high-speed trains pass each other, it will cause fluctuations in the air around the train. At the moment when the head and tail of the two trains pass, the pressure on the side surface of the train and the side of the train will have a strong transient change. The running safety and passenger comfort for the train will be seriously affected by this change. Line spacing is one of the important parameters affecting the aerodynamic performance of the train passing. Therefore, it is necessary to study the influence of the line spacing on the aerodynamic performance of the train passing [10].

6.3.2.1 Aerodynamic Force and Pressure Waves of Train Passing

(1) Line spacing and train passing aerodynamic force

The first step is to study the passing aerodynamic performance of train at the same speed at different line spacings. The working conditions considered are: the train's running speed is 350 km/h, and the line spacing is 4.2 m, 4.5 m, 5.0 m, 5.5 m, and 6.0 m. Fig. 6.49 shows the trend of the lateral force and the rolling moment of the train head at different line spacings.

When the train is running at 350 km/h, the line spacing is 4.2 m, 4.5 m, 5.0 m, 5.5 m, and 6.0 m. The correlation between the train passing pressure wave and the line spacing of the measuring points at the head and tail windows of the train at different line spacings is shown in Fig. 6.50.

The correlation between the pressure peak value and the line spacing at the window of the train head and tail is as follows:

Measuring point at the train head window: $y = 2501.5e^{-0.1889x}$, $R^2 = 0.9969$

Measuring point at the train tail window: $y = 2462.3e^{-0.1831x}$, $R^2 = 0.9973$

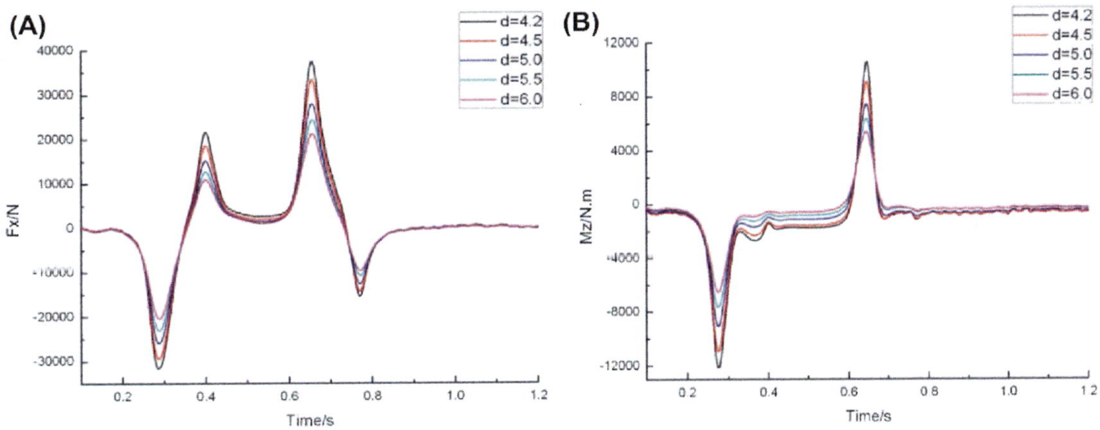

Figure 6.49
Line spacing and train passing aerodynamic force. (A) Lateral Force of the Train Head (B) Rolling moment of the train head.

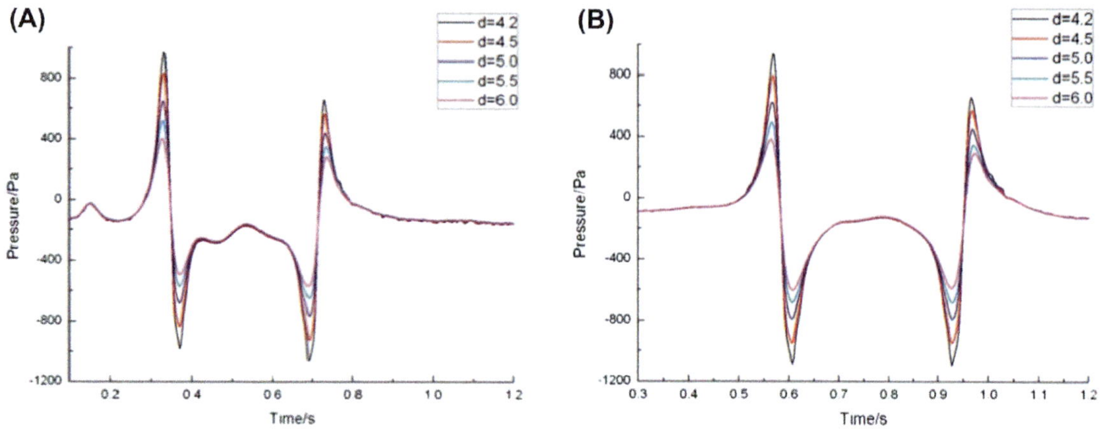

Figure 6.50
Line spacing and train passing pressure wave. (A) At the Train Head Window (B) At the Train Tail Window.

According to the formula, the pressure wave of the train passing has a negative exponential correlation with the line spacing.

(2) Train passing speed and aerodynamic force

The second step is to study the aerodynamic performance of the train at different speeds at fixed line spacing. The conditions are: line spacing is 5 m, train's running speed is 350 km/h, 380 km/h, 400 km/h, 420 km/h, and 450 km/h, respectively. Fig. 6.51 shows the trend of the lateral force and rolling moment of the train head at different running speeds.

When the trains pass at speeds of 350 km/h, 380 km/h, 400 km/h, 420 km/h, and 450 km/h, the correlation between the pressure wave and the train passing speed at different speeds is measured at the train head and tail windows, as shown in Fig. 6.52.

The correlation between the pressure peak value and the train passing speed at the window of the train head and tail is as follows:

Measuring point at the train head window: $y = 1249.6e^{0.1241x}$, $R^2 = 0.9885$

Measuring point at the train tail window: $y = 1249.6e^{0.1241x}$, $R^2 = 0.9885$

According to the formula, the pressure wave of the train passing is exponential with the train passing speed.

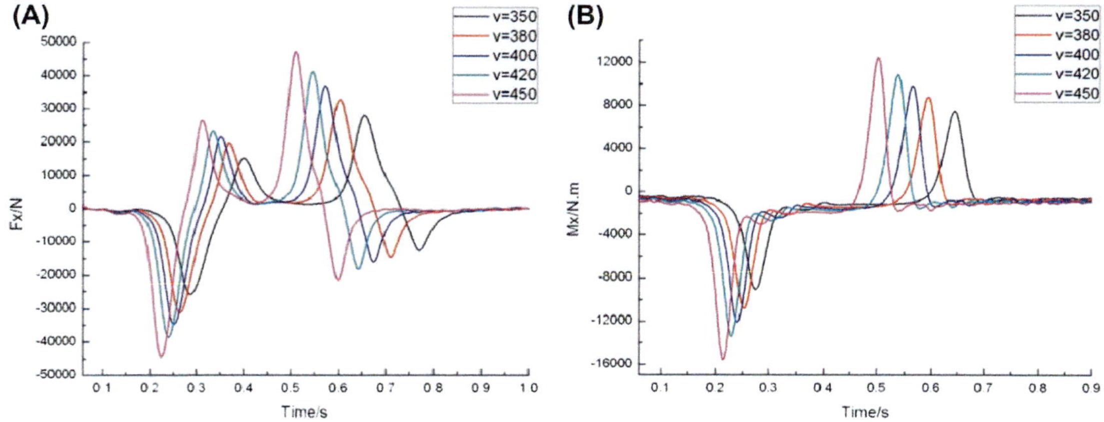

Figure 6.51
Train passing speed and aerodynamic force. (A) Lateral Force of the Train Head (B) Rolling Moment of the Train Head.

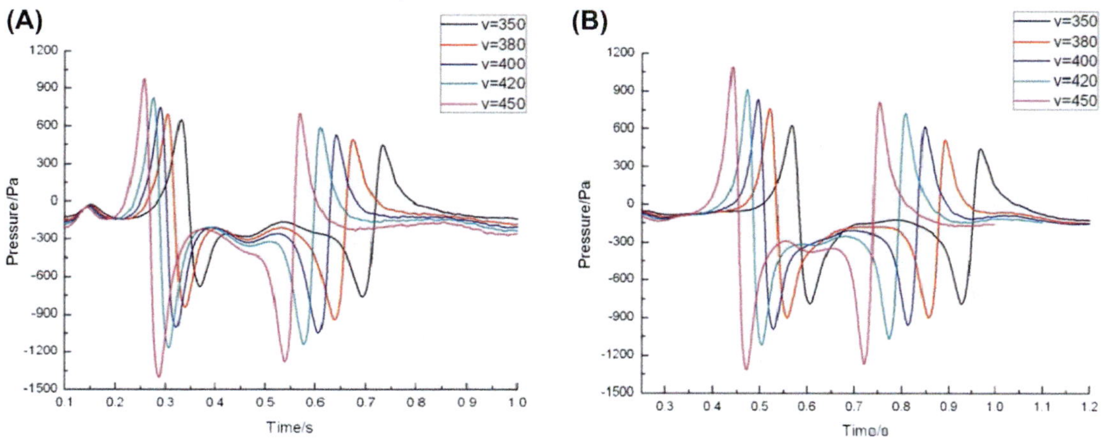

Figure 6.52
Train passing speed and pressure waves. (A) At the Train Head Window (B) At the Train Tail Window.

6.3.2.2 5 m line spacing recommended running speed

(1) Safety of train passing When the line spacing is 5 m and the train's constant velocity of passing is 350 km/h, the wheelset lateral force of the train is shown in Fig. 6.53. The lateral force of the wheel shaft fluctuates greatly during the car.

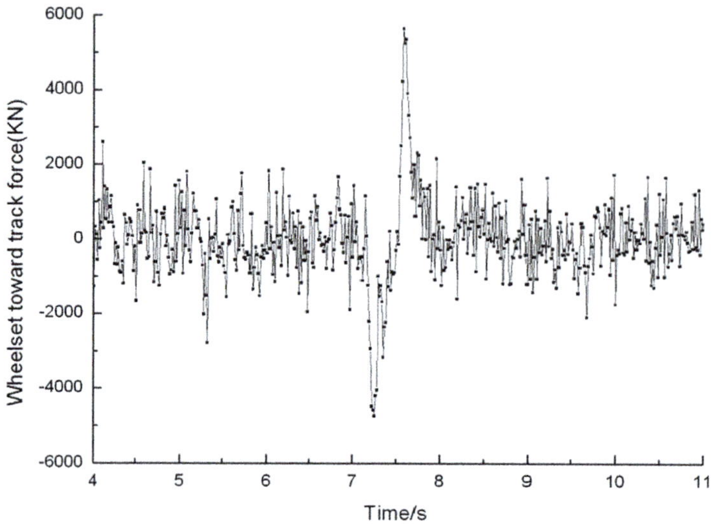

Figure 6.53
Lateral force of one wheel shaft of the train head in train passing.

When the train is running at 350 km/h, the correlation between the amplitude of the maximum lateral force of wheel shaft of the train head and tail and the line spacing is shown in Fig. 6.54; this figure shows that the lateral force amplitude of the wheel shaft of the train head, middle, and tail is far smaller than the standard limit value.

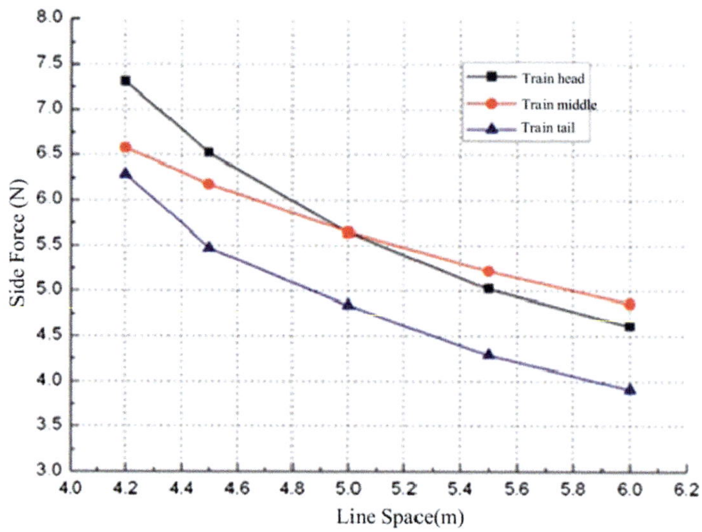

Figure 6.54
Correlation between the maximum lateral force of the wheel shaft and the line spacing (the passing speed is 350 km/h).

Therefore, when the trains pass at a constant speed of 350 km/h, the trains' running on the line between 4.2 m and 6.0 m are in a safe area.

In addition, when the train is running at 380 km/h and the line spacing is 5.0 m, the maximum wheel shaft lateral force amplitudes of the train head, middle, and tail are all within the limit value.

(2) Analysis of the stability of the train When the line spacing is 5 m and the train's constant velocity of passing is 350 km/h, the variation in the lateral and vertical acceleration of the right measuring point after tailing is shown in Fig. 6.55.

The correlation between the lateral and vertical acceleration of the train head, middle, and tail, and the line spacing, is shown in Fig. 6.56 when the train is running at 350 km/h. It can be seen that when the line spacing is 4.5 m, the maximum amplitude of the lateral acceleration of the tail is 2.52 m/s^2, which exceeds the limit. When the line spacing is 5 m, the maximum amplitude of the lateral acceleration of the train head, middle, and tail are within the limits.

In addition, when the train is running at 380 km/h and the line spacing is 5.0 m, the maximum lateral accelerations of the train head, middle, and tail car are 1.96 m/s^2, 1.63 m/s^2, and 2.38 m/s^2, respectively, all within the limits. The vertical acceleration values are 0.67 m/s^2, 0.51 m/s^2, and 0.88 m/s^2, respectively, all within the limits.

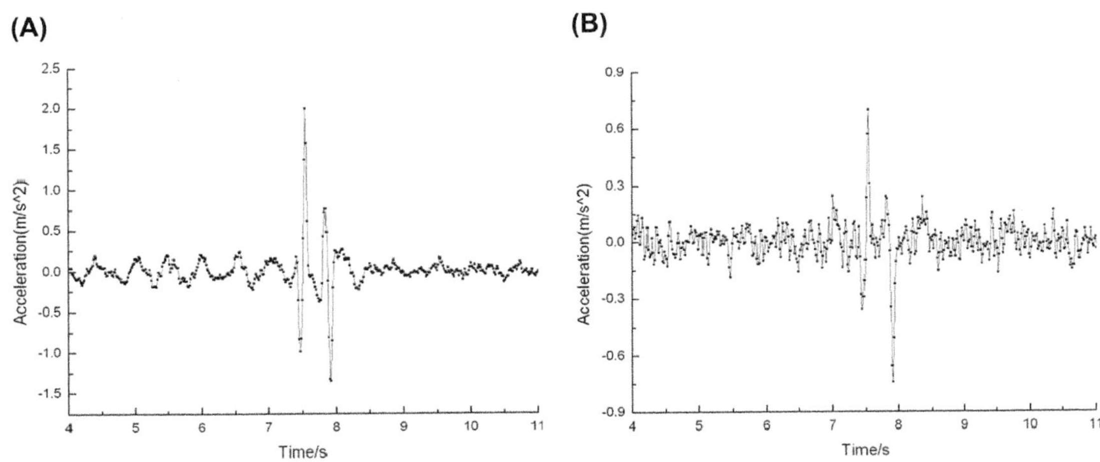

Figure 6.55
Lateral acceleration of the right measuring point after the tail at the passing speed 350 km/h. (A) Lateral Acceleration of the Right Measuring Point after the Tail (B) Vertical Acceleration of the Right Measuring Point after the Tail.

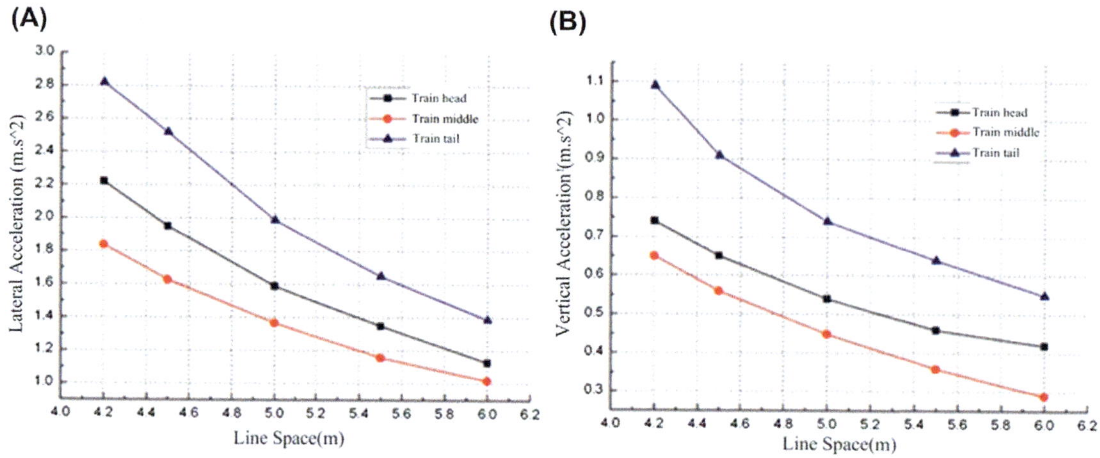

Figure 6.56
Correlation between train vibration acceleration and line spacing at the passing speed 350 km/h. (A) Lateral acceleration (B) Vertical acceleration.

To sum up: when the train is running at 350 km/h or 380 km/h, the line spacing should be 5.0 m to meet the safety and stability requirements of trains passing each other.

6.3.2.3 Recommendations for line spacing when the train's speed is over 380 km/h

(1) Safety analysis When trains pass each other at a speed of 400 km/h, the correlation between the maximum amplitude of the wheel shaft lateral force of the train head, middle, and tail, and the line spacing is shown in Table 6.13. It can be seen from Table 6.13 that the maximum amplitude of the wheel shaft lateral force of the train head, middle, and tail is far smaller than the standard limit value. Therefore, when the train is running at 400 km/h, the safety of the train passing line spacing should be 5.0 m, 5.2 m, and 5.5 m.

When trains pass each other at a speed of 450 km/h, the correlation between the maximum amplitude of the wheel shaft lateral force of the train head, middle, and tail, and the line spacing is shown in Table 6.14. It can be seen from the table that the maximum amplitude of the wheel shaft lateral force of the train head, middle, and tail is far smaller than the standard limit value. Therefore, when the train is running at 450 km/h, the safety of the train passing line spacing should be 5.5 m or 6.0 m.

Table 6.13: Lateral force of the wheel shaft at the passing speed 400 km/h (unit: kN).

Line spacing(m)	Train head	Middle	Tail
5.0	7.09	6.81	6.21
5.2	6.82	6.73	5.75
5.5	6.40	6.46	5.47

Table 6.14: Lateral force of the wheel shaft at the passing speed 450 km/h (unit: kN).

Line spacing(m)	Train head	Middle	Tail
5.5	6.49	6.02	6.85
6.0	5.89	5.72	6.07

(2) Stability analysis When trains pass each other at a speed of 400 km/h, the values of the vibration acceleration of the train head, middle, and tail at different line spacings are shown in Table 6.15. We can see that when the line spacing is 5.0 m, the lateral acceleration of the tail 2.64 m/s^2 exceeds the lateral vibration acceleration limit. When the line spacing is 5.2 m or 5.5 m, the lateral acceleration values of the train head, middle, and tail are all within the limits. When the line spacing is 5.0, 5.2, and 5.5 m, the vertical acceleration values of the train head, middle, and tail are all within the limits.

When trains pass each other at a speed of 450 km/h, the values of the vibration acceleration of the train head, middle, and tail at different line spacings are shown in Table 6.16. We can see that when the line spacing is 5.5 m, the lateral acceleration of the tail 2.85 m/s^2 exceeds the limit. When the line spacing is 6.0 m, the lateral acceleration values of the train head, middle, and tail are all within the limits. When the line spacing is 5.5 m and 6.0 m, the vertical acceleration values of the train head, middle, and tail are all within the limits.

To sum up: when the train is running at 400 km/h, the recommended line spacing is 5.2 m; when the train is running at 450 km/h, the recommended line spacing is 6.0 m.

Table 6.15: Vibration acceleration of the wheel shaft at the passing speed 400 km/h (unit: m/s^2).

	Train head		Middle		Tail	
Line spacing(m)	Lateral	Vertical	Lateral	Vertical	Lateral	Vertical
5.0	2.10	0.64	1.81	0.56	2.64	0.96
5.2	1.95	0.58	1.69	0.50	2.43	0.86
5.5	1.75	0.51	1.56	0.45	2.20	0.72

Table 6.16: Vibration acceleration of the wheel shaft at the passing speed 450 km/h (unit: m/s^2).

	Train head		Middle		Tail	
Line spacing(m)	Lateral	Vertical	Lateral	Vertical	Lateral	Vertical
5.5	2.47	0.74	1.85	0.76	2.85	1.01
6.0	2.08	0.58	1.64	0.63	2.41	0.80

6.3.3 Optimal design of windbreak

In the windy area along the railway, the construction of the windbreak is necessary to protect the safety of high-speed trains.

6.3.3.1 Calculation model

In order to analyze the influence of the height of the windbreak on the aerodynamic force of the train, a calculation model of the high-speed train under the pass wind has been established. The train consists of eight sections of vehicles, including the train head, six middles, and tail, with a total length of 201.4 m. The calculation model and grid map are shown in 6.57 and 6.58.

The windbreak used for the calculation is a reinforced pull-on windbreak [11], and the pass-sectional width of it is 1.5 m. Figs. 6.59 and 6.60 show a simplified geometric model and a meshed partial view of the windbreak, respectively.

In order to study the influence of the height of the windbreak on the aerodynamic force of the train, the height of the windbreak should be based on the lateral force, lift, drag, and rolling moment of the train. The running speed of the train is 350 km/h; the pass wind speed is 32.6 m/s; the wind direction angle is 90° with the most dangerous wind direction angle; the distance between the windbreak and the railway center line (Fig. 6.61) is $L = 3.5$ m; the calculation range of the height of the windbreak is $H = 2-4$ m.

6.3.3.2 Body pressure distribution law

Under the windbreaks of different heights, the pressure on the train surface are different under the railway condition of a windbreak. In order to have a more intuitive analysis of

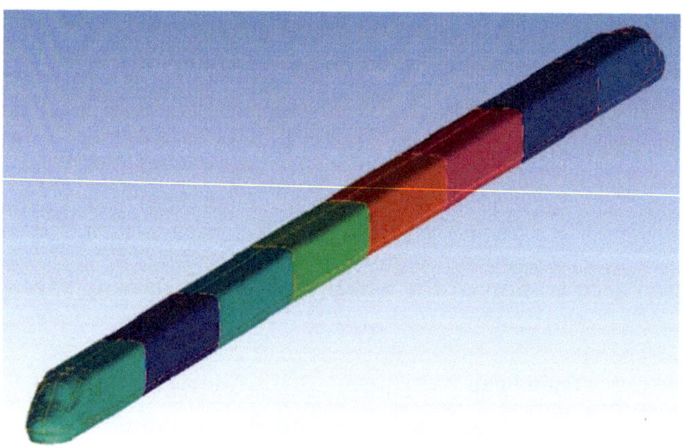

Figure 6.57
A type of CRH EMU model.

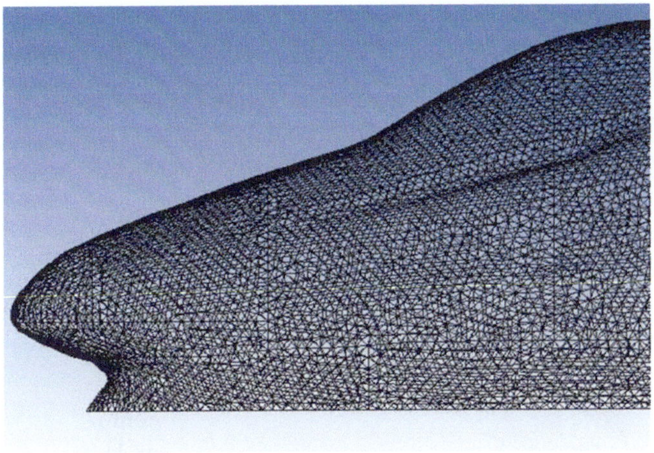

Figure 6.58
The front part grid diagram of the head end.

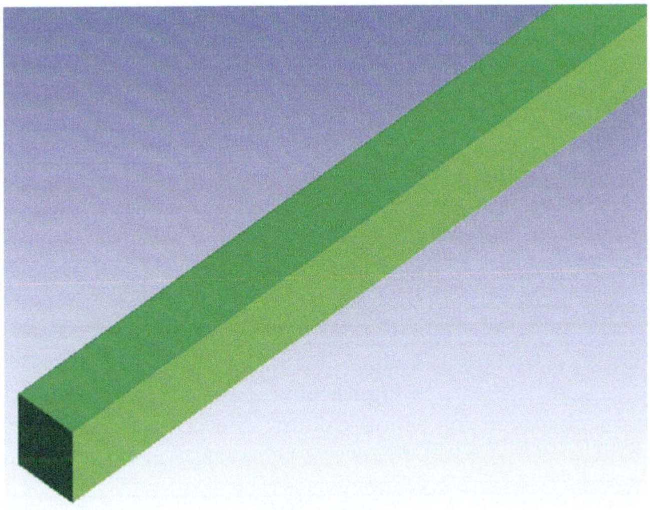

Figure 6.59
Windbreak model.

the pressure distribution on the train surface, the pressure cloud diagram at different pass sections of the trains is extracted. As shown in Fig. 6.62, pass-section 1 is at 5.7 m from the streamline to the nose of the train head; pass-section 2 is at 13.2 m from the non-streamline to the nose of the train head; pass section 3 is at the center of the middle train; pass section 4 is at the center of the seventh train body; pass-section 5 is at 13.2 m from the non-streamline to the nose of the train tail; and pass-section 6 is at 5.7 m from the streamline to the nose of the train tail. Since the middle train pressure cloud diagram is the same, only the second or the seventh train body are extracted.

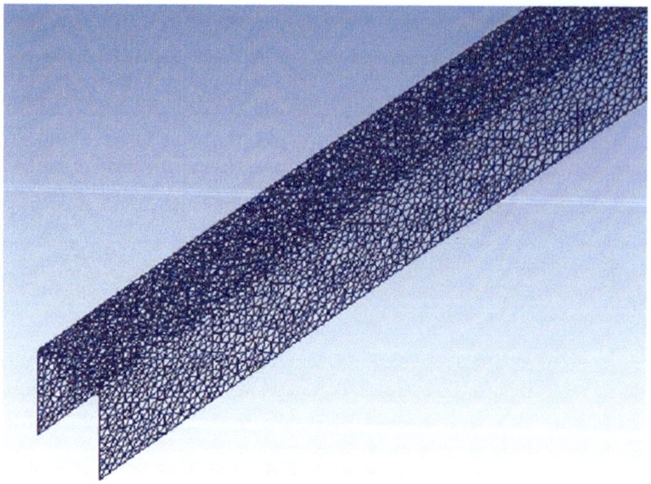

Figure 6.60
Windbreak meshing.

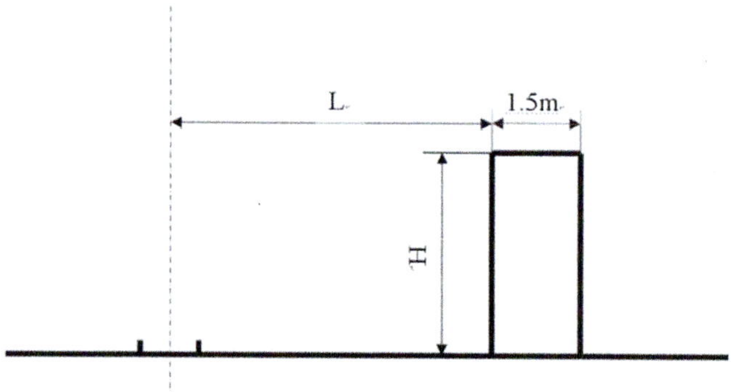

Figure 6.61
Distance between the windbreak and the center of the track.

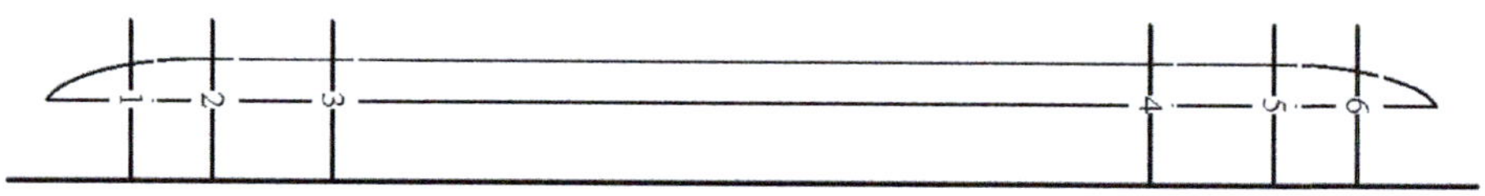

Figure 6.62
Pass section of the train.

In order to compare the pressure distribution on different pass sections, the height H of the fixed windbreak is 3.2 m. Fig. 6.63 shows the air pressure cloud diagram at six pass sections.

It can be seen from Fig. 6.63 that the pressure distribution of the streamlined part of the train head and tail differs considerably from the pressure distribution at other pass sections. The pressure distribution of the non-streamlined part of the train head and tail

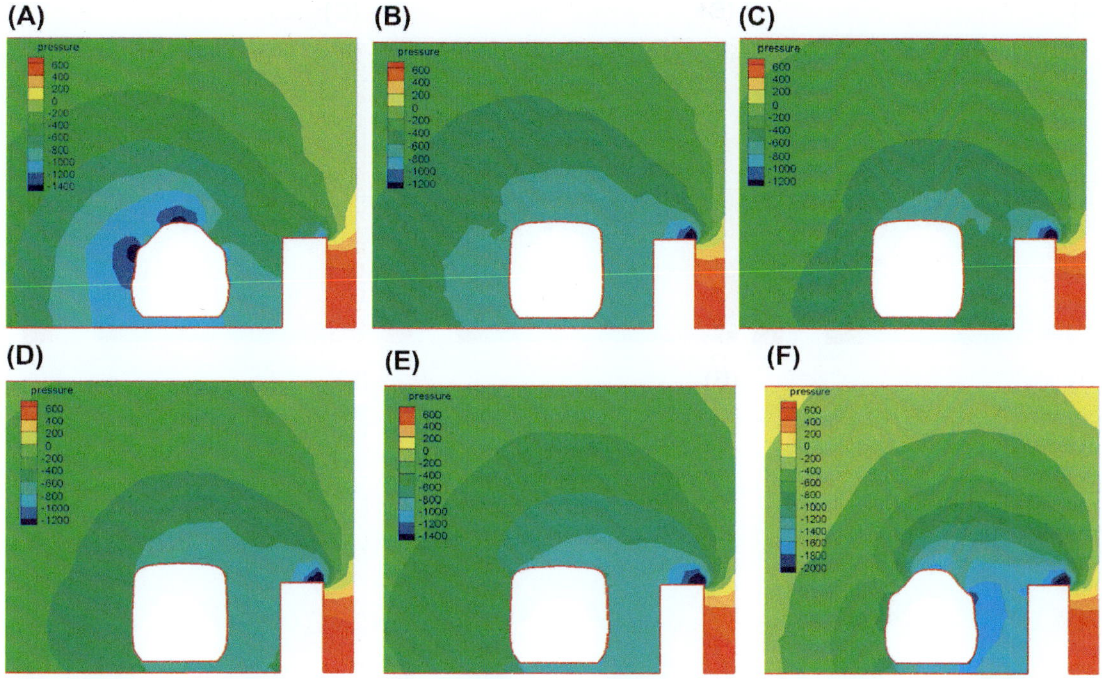

Figure 6.63

Pressure distribution of different train pass sections. (A) Pass Section 1 (B) Pass Section 2 (C) Pass Section 3 (D) Pass Section 4 (E) Pass Section 5 (F) Pass Section 6.

and that of the middle part are relatively close. The distribution of the streamlined part of the train head is negative pressure, and the negative pressure value on the side and top of the train on the leeward side reaches the maximum. The pressure distributed in the streamlined part of the train tail is also negative pressure; moreover, the negative pressure on the windward side has a tendency to become larger, and the maximum value of the negative pressure is distributed near the nose of the windward head. The surface distribution of the non-streamlined part of the train head and tail and the middle is negative pressure, and the negative pressure value changes little.

In order to compare the pressure distribution of the following train surfaces on different heights of the windbreak, taking pass-section 1 (i.e., the streamlined part of the train head) as an example, Fig. 6.64 shows the pressure distribution at the pass section 1 of the train at different heights of the windbreak. The height of the windbreak has a considerable influence on the pressure distribution on the train surface. When the height of the windbreak is small (not more than 2.8 m), the pressure distribution cloud diagram on the train surface is similar. As the height of the windbreak increases, the maximum negative pressure becomes smaller. When the height of the windbreak increases to 3.2 m,

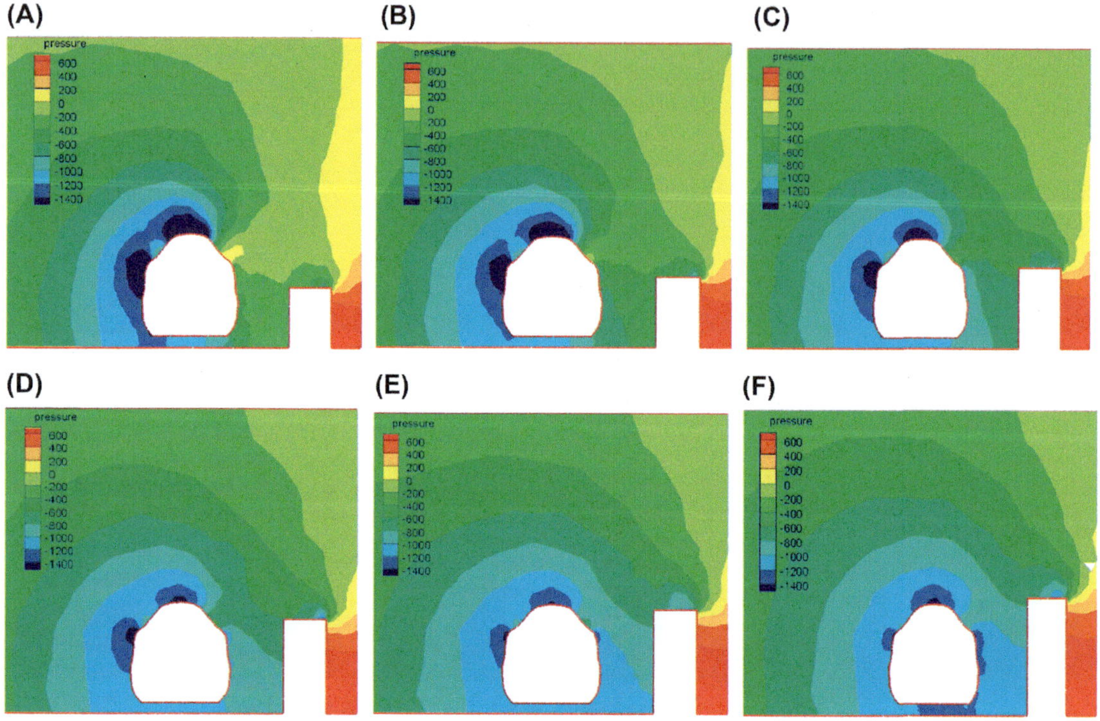

Figure 6.64
Pass-sectional pressure distribution of the trains under different heights of the windbreak. (A) H = 2.0 m (B) H = 2.4 m (C) H = 2.8 m (D) H = 3.2 m (E) H = 3.6 m (F) H = 4.0 m.

a strong negative pressure appears on the windward side of the train. As the height of the windbreak increases further, the negative pressure of the leeward side of the train becomes smaller, while the negative pressure of the windward side increases. This trend became more obvious when its height reaches 4.0 m.

6.3.3.3 Aerodynamic characteristics of train

Fig. 6.65 shows the variation trend of the aerodynamics of the train head as the height of the windbreak. When the height of the windbreak is less than or equal to 3 m, the side force and lift of the train head become positive values; when the height of the windbreak is more than 3 m, the side force and lift become negative. As the height of the windbreak increases, the side force, lift, and rolling moment decrease first and then increase in the opposite direction. When the windbreak is 3.3 m high, the side force of the train head is minimized; when the height of the windbreak is 3.2 m, the lift value of the train head is the smallest.

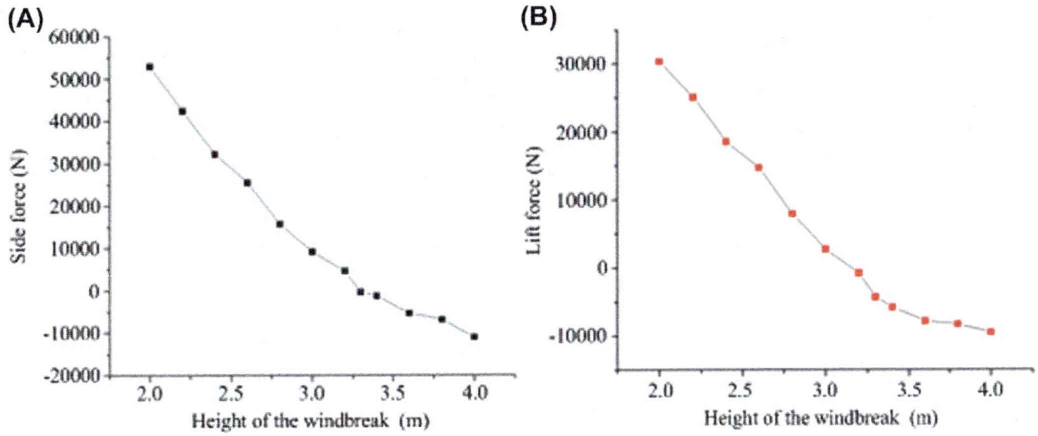

Figure 6.65
Change of the train head aerodynamic force with the height of the windbreak. (A) Side Force (B) Lift Force.

This section extracts the side force and lift received by the second train body and third train body of the train under the different heights of the windbreak.

Fig. 6.66 shows the variation of aerodynamic forces with the height of the windbreak in the second body and seventh train body. As the height of the windbreak increases, the aerodynamic trends of the second train body and seventh train body are the same, and the curve trends are not considerably different. When the windbreak is below 2.6 m, the side force direction of the second train body and the seventh train body is along the pass-wind direction. When the windshield is higher than 2.6 m, the side force changes direction. When the windshield is below 3.2 m, the lift of the second train body and seventh train body is positive and the direction is upward, and as the height of the windbreak increases, the lift becomes negative. As the height of the windbreak increases, the side and lift of the second train body and the seventh train body are first reduced and then increased in the opposite direction. When the windbreak is 3 m, the rolling moment acting on the two train bodies is the smallest; when the windbreak is 2.6 m, the side force acting on them is the smallest; when the windbreak is about 3.3 m, the lift value acting on them is the smallest.

Fig. 6.67 shows the variation in aerodynamic forces with the height of the windbreak in the train tail. It can be seen that when the windbreak is less than 3 m, the lift direction of the train tail is upward. As the height of the windshield increases, the lift of the train tail becomes downward. As the height of the windbreak increases, the side force of the train tail first increases and then decreases, and the lift of the train tail first decreases and then

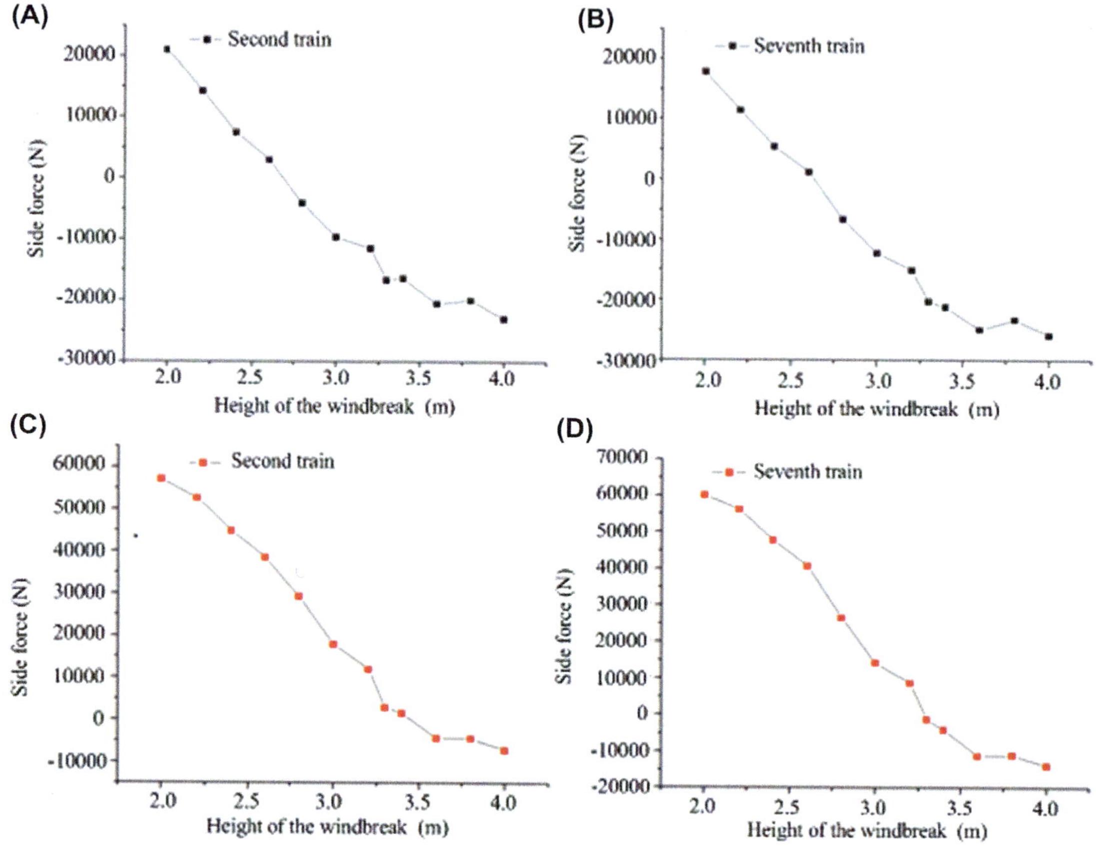

Figure 6.66

Change of the middle of the train aerodynamic force with the height of the windbreak. (A) The change of the lateral force of the second train body with the height of the windbreak (B) The change of the lateral force of the seventh train body with the height of the windbreak (C) The change of the lift of the second train body with the height of the windbreak (D) The change of the lift of the seventh train body with the height of the windbreak

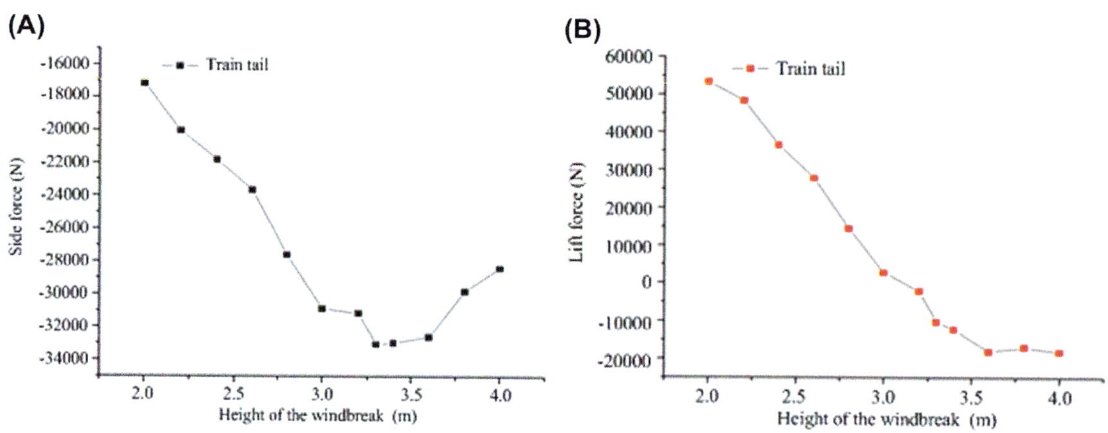

Figure 6.67

Change in the train tail aerodynamic force with the height of the windbreak.

increases in the reverse direction. When the windbreak is 3.2 m, the train tail receives the lowest lift value.

Fig. 6.68 shows the variation in aerodynamic forces with the height of the windbreak in the train. The curve of each middle part of the train is basically the same, and the aerodynamic force on the middle part of the train is the same as well. When the windbreak is less than 3.2 m, the side force on the train head is opposite to the side force on the train head, and the side force acting on the train head follows the pass-wind direction, and the side force acting on the train tail is opposite to the pass-wind direction. When the height of the windbreak is less than 3.2 m, the rolling moment on the train

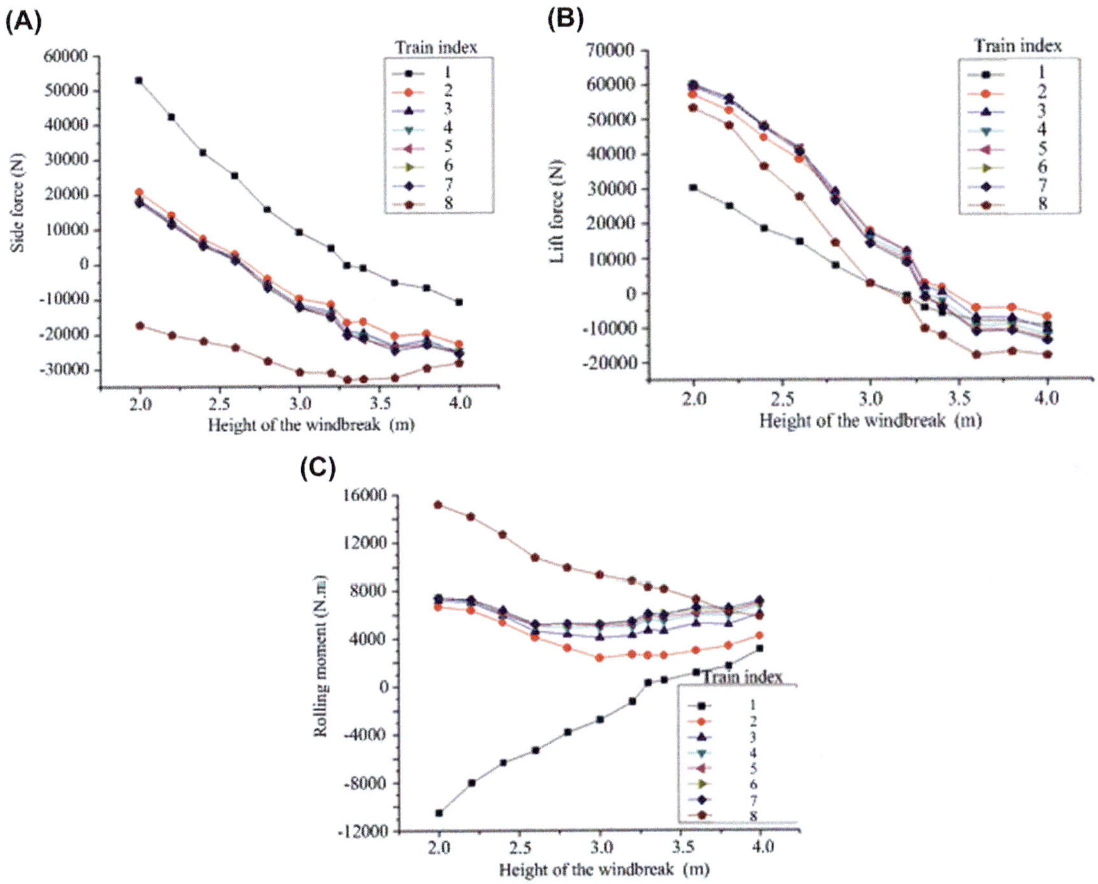

Figure 6.68

Aerodynamic force of the trains under different heights of the windbreak. (A) The change of the side force of each train body with the height of the windbreak (B) The change of the lift of each train body with the height of the windbreak (C) The change of the rolling moment of each train body with the height of the windbreak.

head is opposite to the direction of that of the train tail. The rolling moment of the middle part of the train and the train tail are always in the same direction, and the rolling moment of the middle part of train changes very little. As the height of the windbreak increases, the variation of the rolling moment of the train head and the tail is far greater than the change range of the rolling moment of the middle part of the train. The trend of the side force and lift of each train body is consistent. The side force, lift, and rolling moment values of each train body are comprehensively analyzed; when the windbreak is 3.3 m, the side force, lift and rolling moment of each train body are generally the lowest.

6.3.3.4 Recommendations for the height of the windbreak

The results of the analysis and comparison of the aerodynamic force of train under the windbreak of different heights indicate that when the height of the windbreak is 3.3 m, the aerodynamic force on the train is relatively small. It is recommended that the height of the windbreak be 3.2 m–3.4 m.

References

[1] National Key Basic Research and Development Program (Research on Key Issues of Safety Service for High-Speed Trains in the 973 Project, Project No.: 2007CB714701, Moderator: Zhang Weihua) Annual Progress Report 2008. Southwest University, Traction Dynamics National Key Laboratory; 2008.
[2] Zhang Shuguang. Research on Design Method of High Speed Train. Beijing: China Railway Publishing House; 2009.
[3] Zhang Weihua. Dynamic Simulation of Locomotives and Vehicles. Beijing: China Railway Publishing House; 2007.
[4] Wang Futian. Vehicle System Dynamics. Beijing: China Railway Publishing House; 1994.
[5] CRH EMU Line Adaptability Research Report (TPL2007-15). Southwest Jiaotong University, Traction Dynamics Laboratory; 2007.
[6] Zhang Hong, Yang Guozhen. On Derailment and Wheel Weight Reduction of Train Bogies. Railway Vehicles 2005;43(6).
[7] Zhou Ning, Zhang Weihua. Pantograph-catenary Dynamics under Double Pantograph. Journal of Southwest Jiaotong University 2009;44(4):552–7.
[8] Zhang Weihua, Zhou Ning, Li Ruiping. etc. Research and Application of Double Pantograph Flow Technology for High-speed Trains at the speed of 350 km/h and Above. In: Proceedings of the 7th World High Speed Railway Conference, volume 1; 2010. 12.7-9, China Railway Publishing House, Beijing: 91–95.
[9] Feng Zhipeng. Aerodynamic Performance and Shape Design of High-speed Trains. Master's Thesis of Southwest Jiaotong University; 2011.
[10] Zhao Yinghui. Research on Aerodynamic Performance and Vehicle Dynamics of High-speed Train Crossing. Master's Thesis of Southwest Jiaotong University; 2012.
[11] Li Tian, Zhang Jiye, Zhang Weihua. Dynamics of High-speed Train Passing Through the Windshield Under Crosswind. Journal of the China Railway Society 2012;34(7):30–6.

CHAPTER 7

Experimental technologies of high-speed trains for dynamic performance

Chapter Outline

Dynamics of Coupled Systems in High-Speed Railways. https://doi.org/10.1016/B978-0-12-813375-0.00007-8

7.1 Research platform construction of high-speed trains in National Laboratory of Rail Transit (NLRT) [1]

The innovative development of China's high-speed trains requires both innovative theories and research platforms. Southwest Jiaotong University has planned and constructed the NLRT under the proposal of China's Science and Technology Ministry, Ministry of Education, and Ministry of Railway. The construction aims at carrying out basic theory and application-foundation theory studies on the dynamics of high-speed trains systems such as wheel-rail interaction, pantograph-catenary interaction, operation safety, and other research with high-speed trains as the core topic. The NLRT was designed in accordance with systematic, experimental, and practical evaluation requirements for high-speed trains innovation and research.

(1) Systematic

High-speed trains form the core of high-speed railway systems, which are a means of transport directly serving passengers. In order to achieve stable and safe operation quality at high speeds, the train system combines high-speed wheel-rail technology, high-efficiency traction technology, and high-reliability brake technology. The operation of high-speed trains is influenced by the track line, catenary, and other fixed facilities while it is controlled and restricted by the traction power supply and train running control system; finally, it is limited by air flow and noise. Thus, a complex coupling system called dynamics of coupled systems in high-speed trains (DCSHT) was implemented.

With full consideration of the systematic requirement, the experimental research platform construction of DCSHT not only studies the response of the high-speed trains itself but also studies the responses of the relative coupling systems affecting the operation of the high-speed trains; this includes wheel-rail interaction, pantograph-catenary interaction, fluid-solid coupling relationship, and electromechanical coupling relationship.

(2) Experimental

Experiments are not only used to verify theoretic models, research, or innovative achievements but are also essential to disclose and prove scientific rules. The experimental research is focused on fundamental aspects including dynamic performance, safety, and the reliability of high-speed trains. In addition, the wheel-rail interaction, pantograph-catenary interaction, and the fluid-solid and electromechanical coupling relationship associated with other systems are considered. NLRT was established to carry out these studies. The fundamental research experimental platform uses high-speed trains as its research core for researching its principles, mechanisms, and regularity. Meanwhile, the experimental research platform of high-speed trains service performance has been established for carrying out systematic fundamental research to study the coupling relationship and optimize system performance.

(3) Practical

It is essential to apply the experimental method to explore and research the service behavior and service performance of high-speed trains, the performance law of track lines, bridges, tunnels, and other fixed equipment. Through the dynamic response of a long-term operation real-time monitoring system of high-speed railway system and comparative analysis of historical data in the time, spatial, and time and frequency domains, the performance evolvement rule of the high-speed trains and its coupling system has been mastered and disclosed.

The dynamic performance of the high-speed trains is determined based on the structure and parameters of the high-speed trains itself while also being influenced by the structure and parameters of track lines, bridges, tunnels, catenary, power supply, and other coupling systems. Therefore, it is necessary to study and optimize the structure and parameter of high-speed trains and its coupling system, and to further verify this via experiment and application. Based on the systematic, experimental, and practical guidelines above, the construction plan of NLRT has been implemented according to the structure described in Fig. 7.1. With three research methods for the fundamental experimental research, digital simulation, and service performance experiments, experimental studies on high-speed trains and its coupling system including dynamics in high-speed trains wheel-rail interaction, pantograph-catenary interaction, and fluid-solid and electromechanical coupling relationship can be conducted. Model verification can be applied to all these three research methods.

As mentioned above, the three platforms in NLRT include:

- Fundamental experimental research platform
- Digital simulation platform
- Service performance experimental research platform

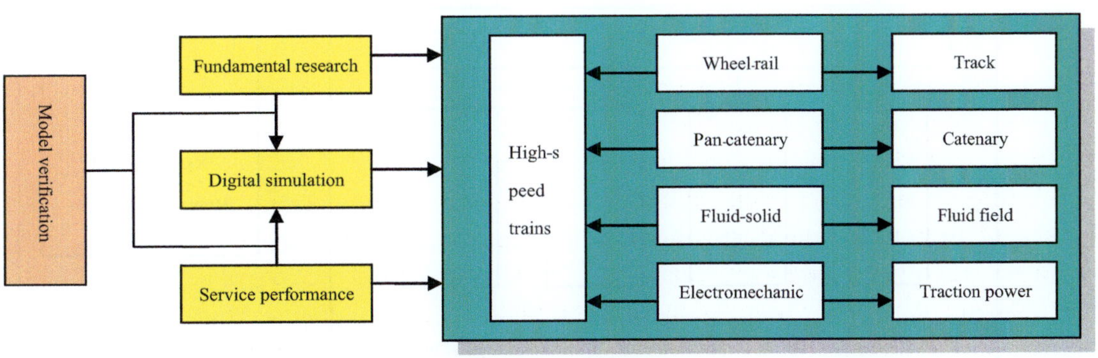

Figure 7.1
Research framework in NLRT.

According to the research framework and the three platforms as shown in Fig. 7.1, the construction content plan (Fig. 7.2) is designed based on the basic requirement of high-speed trains operation (safety and comfort).

Through the service performance experimental research platform, the dynamic response of high-speed trains is acquired by the experimental vehicle; the dynamic responses of track

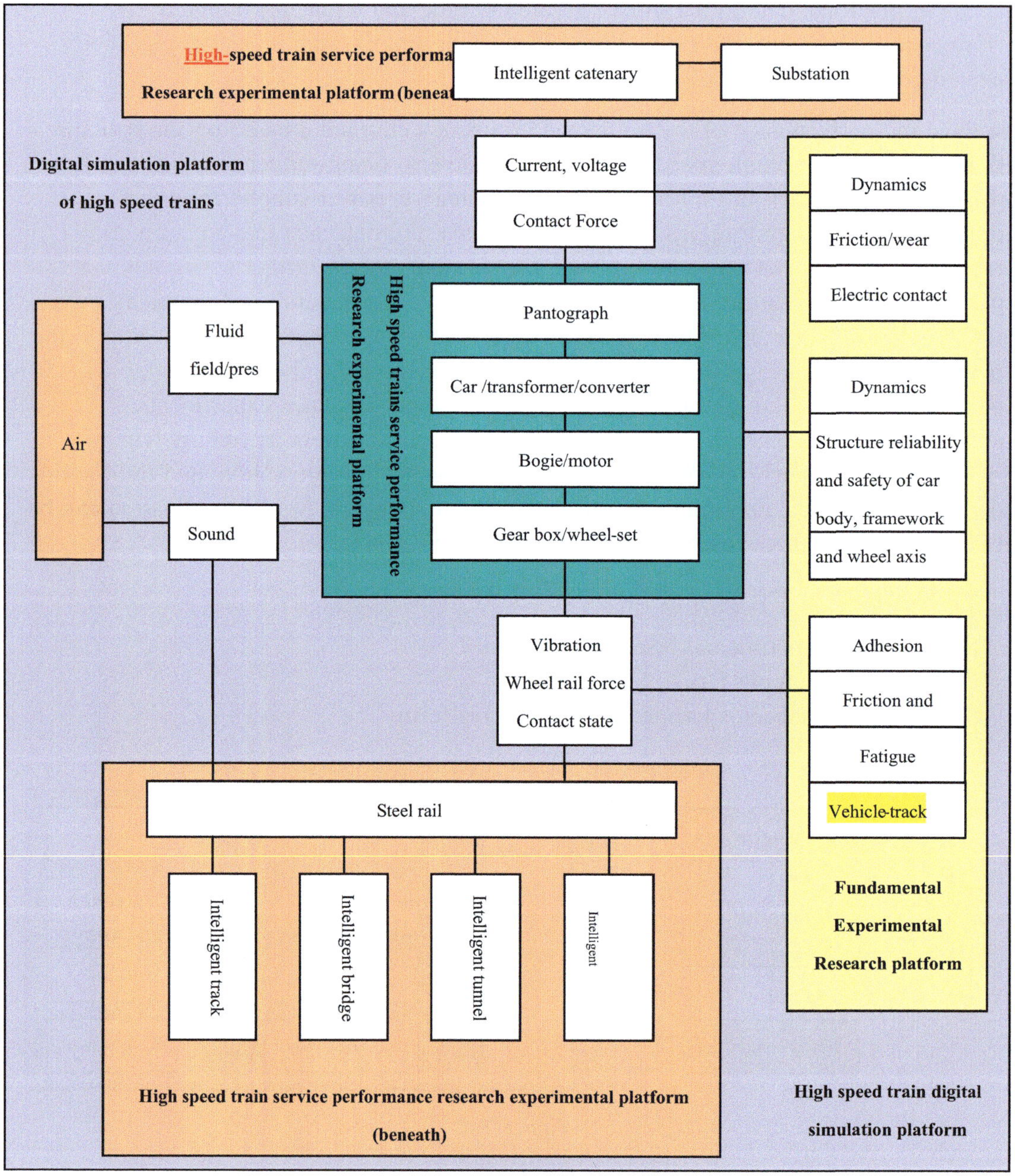

Figure 7.2
Construction frameworks of the three platforms.

(track lines, bridges, and tunnels) and catenary (catenary and traction power systems) are acquired by the experimental line. Response mapping is used to track the relationship between the track lines and the testing train, the wheel-rail interaction, pantograph-catenary interaction, fluid-solid and electromechanical coupling relationship.

The fundamental research experimental platform is utilized to study the stability, stationarity, safety, and reliability of high-speed trains; to find the friction and wear, fatigue, adhesion, and vehicle-track line couplings of wheel-rail contact; to explore the dynamics in pantograph-catenary contact, electric contact, and current carrying friction and wear; to research the aerodynamic drag, pressure wave and wind-induced safety performance in fluid-solid coupling; and to figure out the influences of snow and others on safe train operation. Besides, the fundamental research experimental platform is further used to identify the interaction mechanism of the high-speed trains system.

The dynamics models of high-speed trains and its coupling system are established based on acquired wheel-rail, pantograph-catenary, fluid-solid, electromechanical and other coupling models, and the digital simulation platform of high-speed trains is designed; moreover, the research of dynamics in high-speed trains system, in particular to the structure and parameter of high-speed trains and its coupling system, can be realized.

The experimental verification of the dynamics model and the simulation platform is performed through the service performance experimental research platform. The performance change in the high-speed trains under the service process and the evolvement rule of its coupling system in the service process can be acquired from the service performance experiment research platform through long-term operation detection. The service model is established and loaded in the simulation platform so that the service performance of high-speed trains can be simulated.

With the use of research processes mentioned above, the theoretical analysis and research of the high-speed trains and its coupling system are implemented practically, and thus, the system optimization and coupling relationship matching the research of the high-speed trains and its coupling system are conducted.

7.2 Bench test technologies of high-speed trains for fundamental research

7.2.1 Test technology of the whole vehicle dynamics performance

7.2.1.1 Real vehicle test [2–5]

The high-speed train bench test for the whole train is generally performed with a single vehicle. Through gradual research on the railway vehicles, many countries have designed various bench test devices to perform dynamics and power tests. The test bench adopts the test technology of a roller test rig and applies the rolling track wheel (referred to as roller)

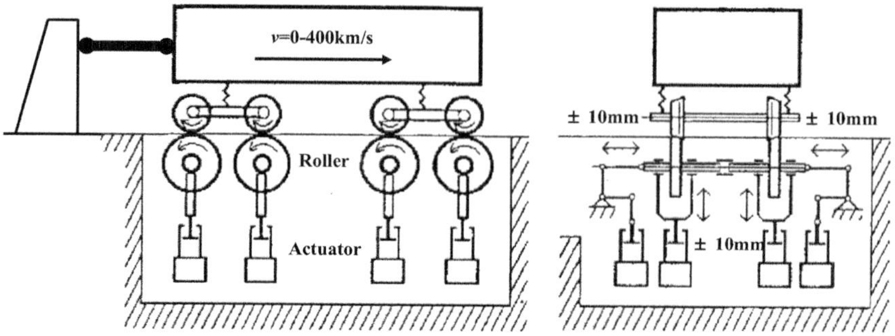

Figure 7.3
Test diagram of roller vibration test rig.

to simulate the steel rail; the wheel is rolled on the roller to simulate the advancing motion of the vehicle. For the research on the dynamics, the vibration of vehicle is simulated by the excitation of tracks. The roller test rig using the excitation device enables the roll wheel to achieve excitation motion, and it is closer to the operation state of railway vehicles, as shown in Fig. 7.3.

Table 7.1 describes how to use rollers to simulate the track. An ideal test rig can not only simulate the irregularity of the track, but also simulate the state of the curve; unfortunately, such an ideal test rig does not exist because of the complexity of the mechanical structure.

The roller test rig of the railway vehicle in the State Key Laboratory of Traction Power located at Southwest Jiaotong University played an important role in the innovative research on accelerated railway vehicle and high-speed trains. When the research platform was designed, the degree of the motion freedom of the roller test rig was confirmed based on a comprehensive consideration of design possibility, manufacturing difficulty, investment, and bench comprehensive performance requirements, as shown in Fig. 7.4.

a. left and right rollers are moved along the y-axis direction to simulate gauge change and track alignment irregularity.
b. left and right rollers are respectively moved along the z-axis to simulate the horizontal and vertical irregularities.
c. left and right rollers are rotated around the x-axis at the same time to simulate the curve super-elevation of the outer track.
d. left and right rollers are rotated around the y-axis to simulate the advancing motion of the vehicle. When the straight track is simulated, the left and right rollers work in sync; when the curve track is simulated, the left and right rollers are rotated at differential speeds.
e. left and right rollers are rotated around the z-axis to simulate the track curve direction.

Table 7.1: Relationship of track state and roller state.[2]

No.	Irregularity type	Track state	Roller state	Note
1	Horizontal irregularity	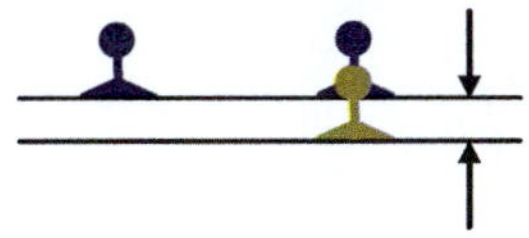	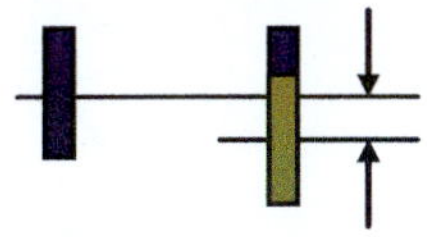	Left and right rollers are relatively moved in vertical direction
2	Vertical irregularity		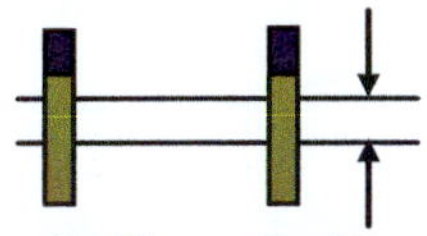	Left and right rollers are relatively moved in vertical direction
3	Direction irregularity	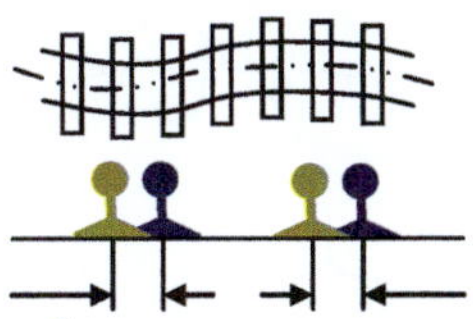	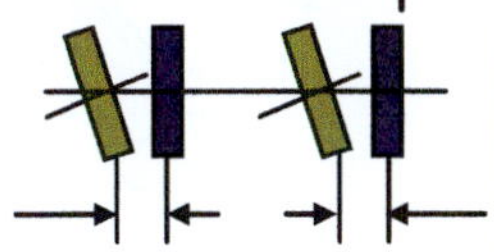	Left and right rollers are transversely moved and shook at the same time.
4	Gauge irregularity	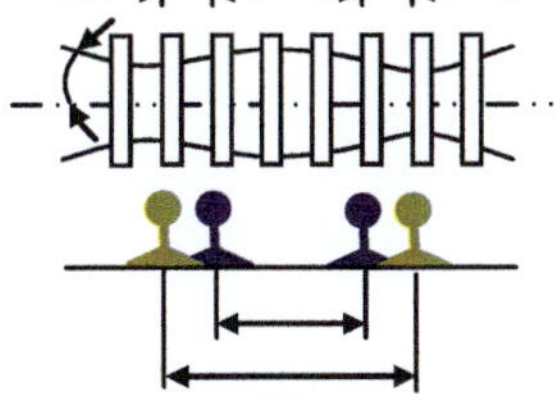	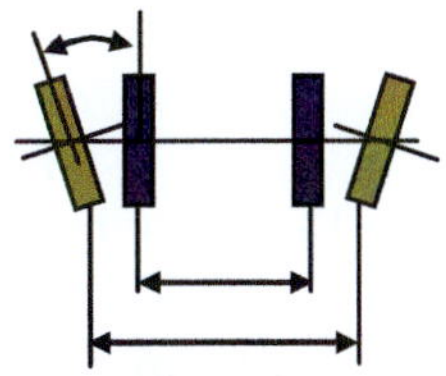	Left and right rollers are transversely moved in relative opposite direction and shook in relative opposite direction
5	Curve	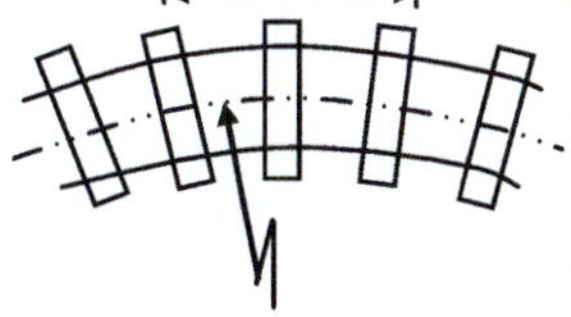	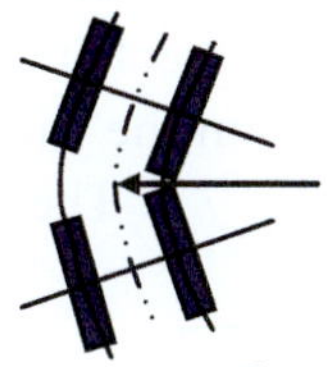	Left and right rollers are arranged on the curve to perform differential rolling.
6	Curve ultra-elevation	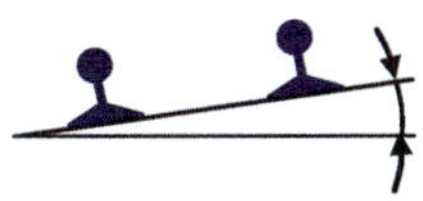	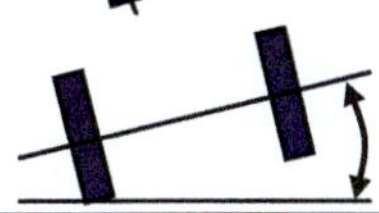	Left and right rollers are inclined at the same time.

For the roller degree of motion freedom mentioned above, the movement along the y and z axes and the rotation around the y-axis can be controlled in real time; the rotations around the x and z axes are preset before the test. The movement of every roller along the y and z axes directions is dynamically controlled through two electro-hydraulic servo exciters; the rotation in the y-axis direction is driven by direct current (DC) or alternating current (AC) servo motors; and dynamic rotation around the z-axis is achieved using the electro-hydraulic servo exciter. The advanced simulation method of roll-vibration combination is realized based on the above mentioned mechanism of movement and control. For meeting the test requirements of railway vehicles with different wheelbases and lengths between truck pivot centers, each bench unit can be moved and fixed in the x-axis direction.

The entire vehicle roller test rig of a railway vehicle is composed of roller wheelset units, mechanical drive units, hydraulic excitation systems, electric drive systems, data

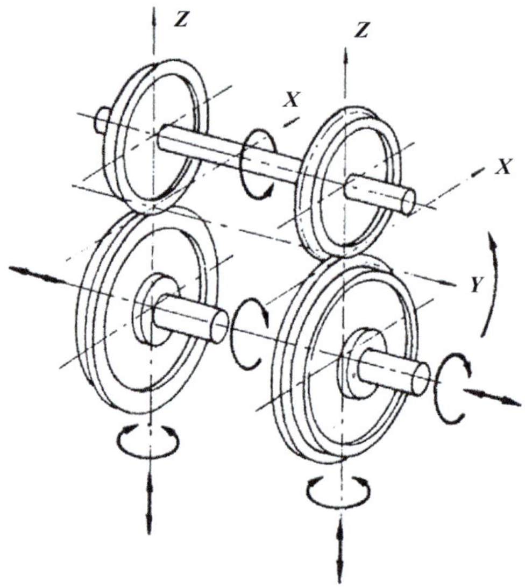

Figure 7.4
Degree of motion freedom of roller.

acquisition and processing systems, monitor systems, and power systems. This this test rig, the roller wheelset unit and the mechanical drive unit are formed to be a complete test unit, and each wheelset (shaft) in the trial run corresponds to one test unit. The system block diagram is shown as Fig. 7.5. The special electricity for railways (52 kV, 50 Hz) is

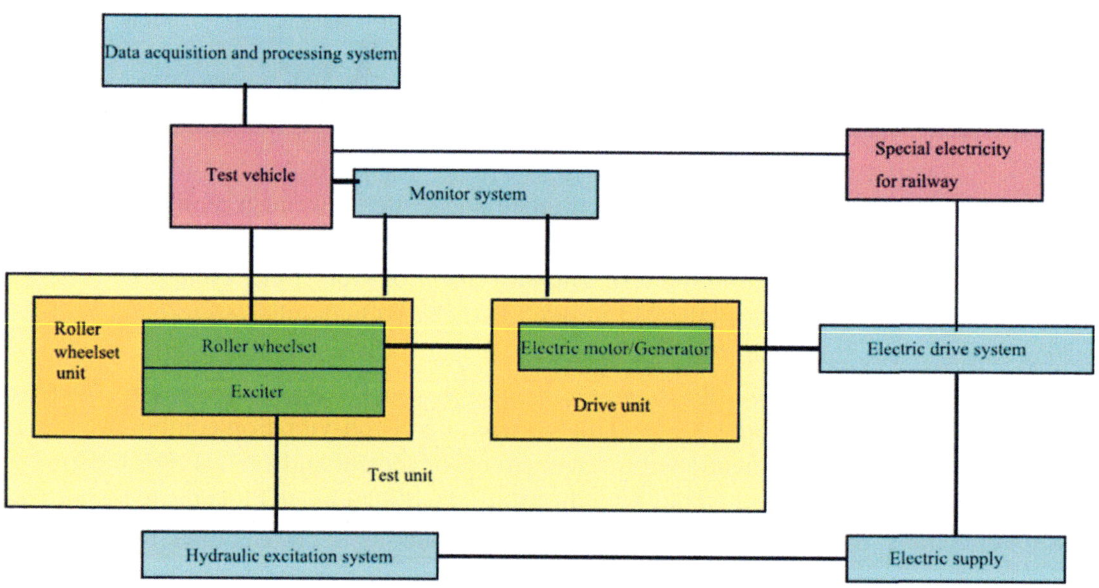

Figure 7.5
System block diagram of roller vibration test rig.

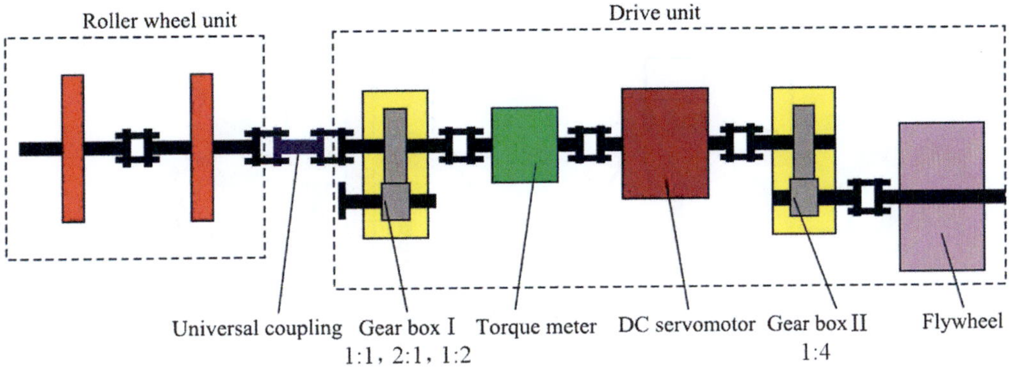

Figure 7.6
Test unit.

used to supply power to the locomotive or other motor vehicles, which is also applied to the power supply of the test rig drive system. The mains supply (380 V, 50 Hz) is applied to supply power to the excitation system, drive system, and other auxiliary systems.

The whole vehicle test rig is composed of six independent test units for easy adjustment and positioning during trial runs. The test units ensure adaptability according to the requirements of different wheelbases and lengths between the unit pivot centers. The test unit is composed of a roller unit and a drive unit, as shown in Fig. 7.6. According to different test tasks, the drive unit can provide different speeds and torques for the rollers through the universal joint; the drive unit is composed of a set of DC/AC servomotor, two gear boxes, a flywheel, and a torque meter, which are all fixed on a welded platform. The motor can be used as an electric motor or a generator upon the demand of drive or brake; the flywheel not only makes the drive run smoothly, but also simulates the weight of the train. Gear box II in Fig. 7.6 is used for improving the rotation speed of the flywheel and gear box I is used for providing different speeds and torques for the rollers. The drive ratio of the drive unit can be set to be 1: 1, 2: 1(for high torque) or 1: 2 (for high speed).

The roller unit is the key part of the roller test rig; all motions of the rollers are implemented through complex mechanical structures. The structure chart of the roller unit of the roller test rig is shown in Fig. 7.7; the roller shaft is supported by a U-shaped framework and the roller can rotate in the U-shaped framework. The U-shaped framework can create only vertical motion; under the drive of the hydraulic excitation system, the rollers can implement vertical motion, and the rollers are rotated with the roller shaft, which also implement cross motion with the roller shaft through a crank arm of the bearing base under the effect of the hydraulic exciter. The rollers can rotate around the z-axis in the roller body through an annular bearing to simulate tangent line directions of curves with different radii. The excitation forces in the vertical and cross directions are tested through force sensors at the top part of an exciter piston rod. The movement of the

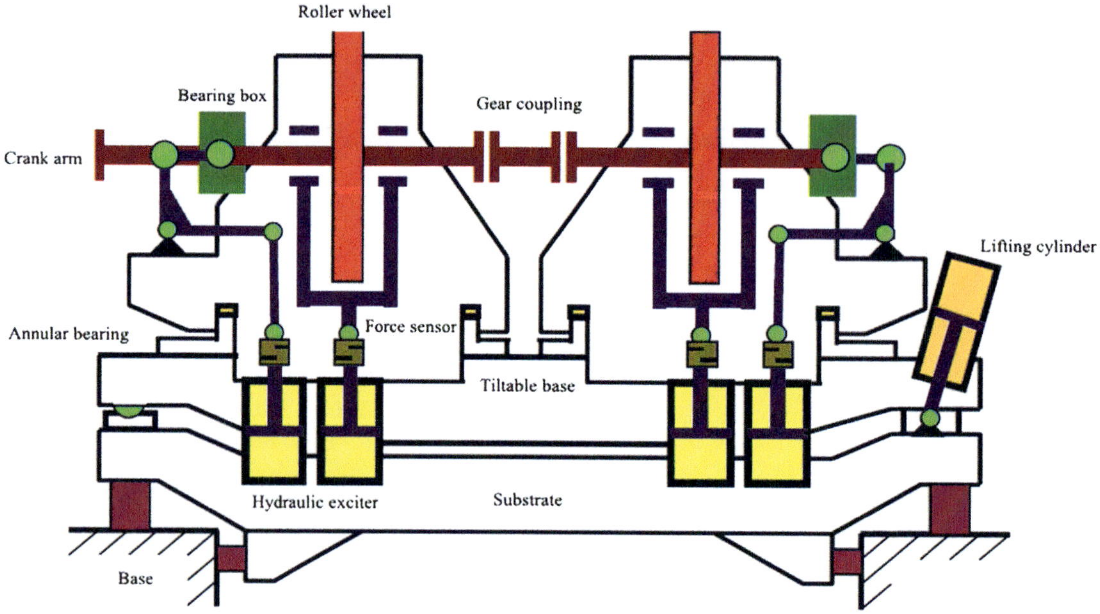

Figure 7.7
Structure chart of roller unit.

left and right rollers are independent, and a gear coupling is arranged in the middle to transfer rotation and torque; the left and right rollers are allowed to perform relative motion in the vertical and axial (cross) directions through the gear coupling. The left and right rollers are mounted on a base that can be titled, and this base can be inclined under the pushing and jacking of the hydraulic exciter, to simulate a curve super elevation. All devices are mounted on the platform, and the rollers can adapt to the requirements of the wheelbase of different railway vehicles by moving the platform.

The entire vehicle roller test rig of the railway vehicle was completed in 1993. It was a four-axis linkage with a design speed of 450 km/h. Owing to the demand for locomotives, the test rig was expanded in 2002, and the number of axes was increased to six-axis linkages; the axis power can reach 1200 kW in a short time. Besides, the differential function of the left and right rollers is added, and thus, the test rig became the first full vehicle test rig capable of performing roller (track wheel) rolling, vertical and transversal excitations, and differential of left and right tracks. It can form the four-dimensional motion coupling test rig for the whole vehicle, and thus, the curve negotiation test function is realized.

The schematic of the left and right rolling differential mechanisms is shown as Fig. 7.8; through an adjustable-speed motor, the rotation speeds of the input and the output of the gear box are different for the differential gear box, and thus, the rolling motion speed difference of the left and right rollers of the roller test rig is realized.

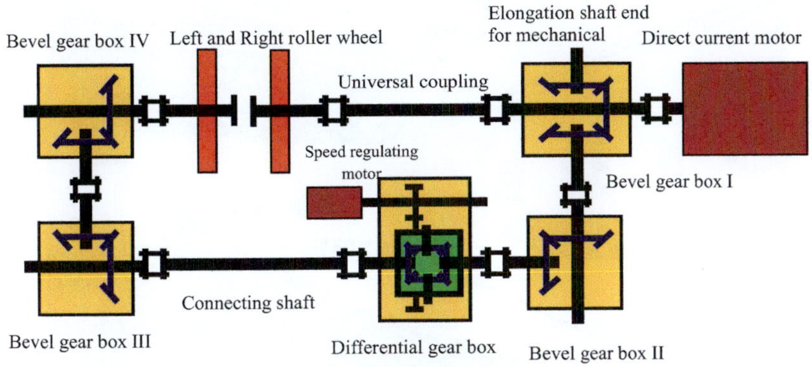

Figure 7.8
Block diagram of differential drive system.

With the rapid development of China high-speed railway in 2006, the original test rig could not meet the demand of high speed and high power. The test speed is increased to 600 km/h, and power is increased to 1800 kW in the NLRT; the technical indices of the test are summarized in Table 7.2.

In general, the dynamics performance test of one entire vehicle (motor train or trailer) is performed for high-speed trains; however, the test of coupling two vehicles is also performed for studying the influence between the vehicles on vehicle dynamics performance; one of the vehicles is placed on the ground, as shown in Fig. 7.9.

7.2.1.2 Proportional movement model (PMM) experiment

As the motion is simulated by the relative operation of the wheel and the simulated track, the vehicle does not advance realistically, and the influence of air flow on the dynamics

Table 7.2: Main technical indices of the full vehicle test rig of railway vehicle.

Vertical excitation	Maximum frequency f_{vmax}	30 Hz	Transversal excitation	Maximum frequency f_{vmax}	30 Hz
	Maximum amplitude A_{vmax}	±10 mm		Maximum amplitude A_{hmax}	±10 mm
	Maximum acceleration *avmax*	±4 g		Maximum acceleration *ahmax*	±5 g
Maximum longitudinal traction force F_e		10 t	Maximum axis weight *Mw*		25 t
Maximum lasting/short-time motor power		1800 kW	Maximum operation *V*		600 km/h
Maximum superelevation angleϕ_{max}		7°	Distance of railway vehicle *L*		4-22 m
Bogie wheelbase *l*		1680-2800 mm	Track gauge A_0		1000-1676 mm
Minimum curve radius *R*		200 m	Maximum axis number *N*		6

Figure 7.9
Bench test of high-speed trains.

performance cannot be simulated in the previous bench test rig. In order to simulate the influence of air flow on high-speed trains, the PMM experiment technology called ejection proportional model is employed.

The PMM test technology of ejection is applied to study the air dynamics of the train and to simulate the relative motion aerodynamic effects of a real vehicle and ground (also called ground effect). The Central South University (a part of former Changsha Railway University) built up 1: 16 movement wind tunnel test rig [6] driven by slingshot devices. The Mechanical Institute of Chinese Academy of Sciences planned to build a 1:10 movement model wind tunnel test rig, which can achieve launch using a high pressure air source during the study of high-speed trains' aerodynamic, and its speed is up to 500 km/h. A movement model test rig for dynamics performance of high-speed trains in NLRT is also under construction. The proposed test rig has the following characteristics:

- High-speed: more than 400 km/h;
- High ratio: 1:10;
- Tracks: double tracks, tunnel;
- Environment simulation: crosswind, raining, and snowing;
- Earthquake simulation: level 10 intensity earthquake simulation of 10-span bridge;
- Lauch: Controllable electromagnetic launch;
- Air pressure: 0.05-10 bar;

The total length of the launch test rig is 142.6 m. The main launch section is 50 m, test section is 43.6 m, brake section is 40 m, secondary launch section is 40 m, and the composite section for realizing crossing is 36.6 m. For the layout, see Fig. 7.10.

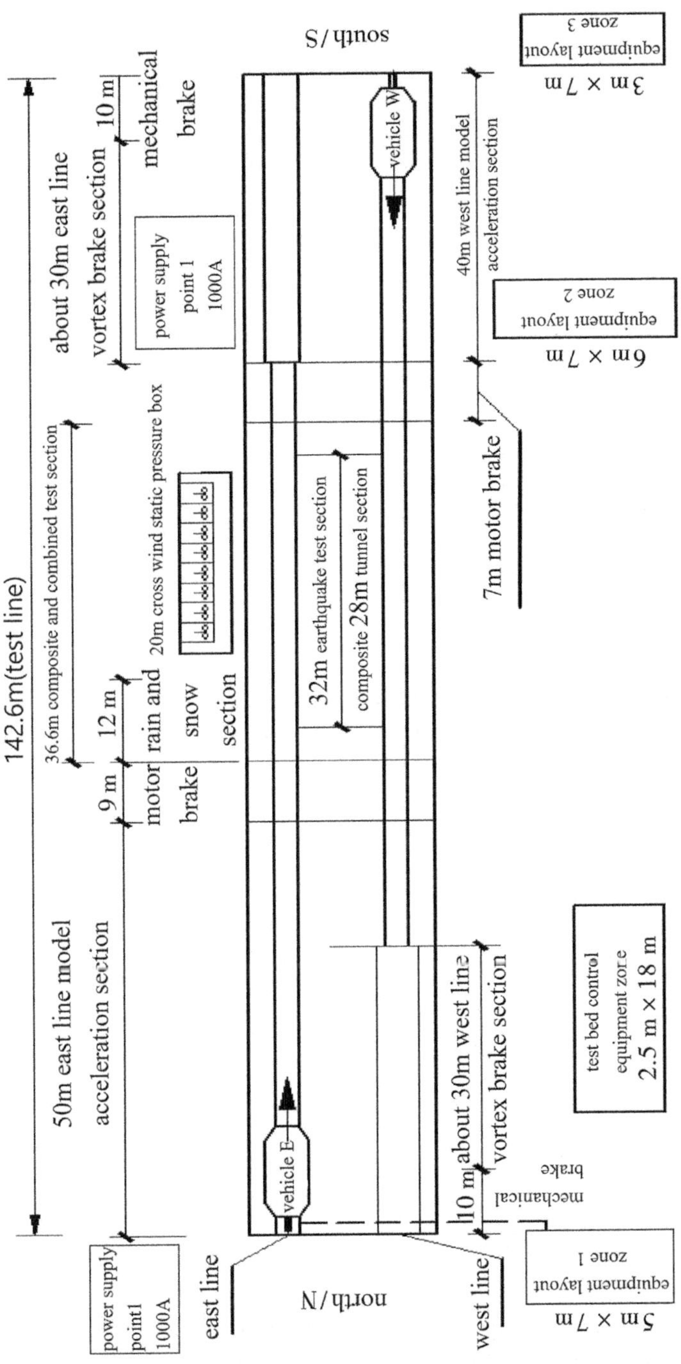

Figure 7.10
Operation simulation test rig plan of multifunctional high-speed trains.

Figure 7.11
Operation simulation test rig cross-section diagram of multifunctional high-speed trains.

The composite test rig comprises a 12 m rain & snow simulation section, 20 m crosswind simulation section, and 32 m (10 spans) earthquake simulation section.

Launching and braking technologies with large acceleration are required, as the test rig is the short in length. The drive acceleration needs to reach 8–15 g, and the brake deceleration needs to reach 20 g. For realizing such constantly high acceleration, the launch drive adopts an electromagnetic launch method, and the brake adopts the electric eddy current brake method under the permanent magnet condition. The whole test system is placed in a simulated vacuum and high-pressure pipeline, and the section layout is shown in Fig. 7.11. The pipeline diameter is 4.2 m, and dynamics analysis research under different air pressures can be carried out. In particular, the Reynolds number required by the proportional test can be realized under high pressure. The vacuum condition test is prepared to research on a vacuum pipeline train in the future. The model train for the experiment is three-vehicle model marshaling at the ratio of 1: 10, and the train is about 200 kg in weight. The model train is provided with internal sensors for detecting vibration, air pressure, and other related information; the data are collected during the test from these sensors.

With the use of this multifunctional test rig, the test research projects to be implemented include:

- Research of aerodynamics and fluid-structure interaction vibration response of high-speed trains in open line;
- Research of aerodynamics and fluid-structure interaction vibration response of high-speed trains running in tunnel;
- Research of aerodynamics and fluid-structure interaction vibration response of high-speed trains during the trains passing each other;
- Research of operation response of high-speed trains under rain, snow, and wind multi-phase medium;

- Research of safety of high-speed trains under crosswind;
- Dynamic response and safety research of high-speed trains under earthquake;
- Research of aerodynamic noise of high-speed trains;
- Aerodynamics research of high-speed trains under different altitudes;
- Research of high-speed trains operation response and aerodynamics in vacuum pipeline.

7.2.2 Test technology of wheel-rail interaction

The wheel-rail interaction is the basic characteristic as well as the most important interaction in high-speed railways. The wheel-rail interaction not only has the function of load bearing and guidance but also transfers the traction and braking forces; therefore, the research on wheel-rail interactions is always the most important field in the research of rail transit. The generalized wheel-rail interaction comprises wheel-rail geometrical contact relationship, wheel-rail force interaction, hard point, and match relation of the wheel-rail material. The fundamental research in wheel-rail interaction to be studied comprises the adhesion creep of the wheel-rail system, the contact fatigue of wheel-rail system, friction and wear, and the derailment mechanism.

(1) Test technology of wheel-rail creep force [2]

The wheel-rail creep force test aims at identifying the interaction of the wheel-rail contact force under the matching condition of a specific wheel-rail profile, wheel-rail appearance, material, and hardness; the most basic requirement is to test the wheel-rail creep force for one-wheelset. The State Key Laboratory of Traction Power in NLRT built an 1: 1 wheel-rail creep force testing device (see Fig. 7.12) using the entire vehicle roller test rig; the schematic is provided in Fig. 7.13. The testing device is

Figure 7.12
Picture of one-wheelset testing device.

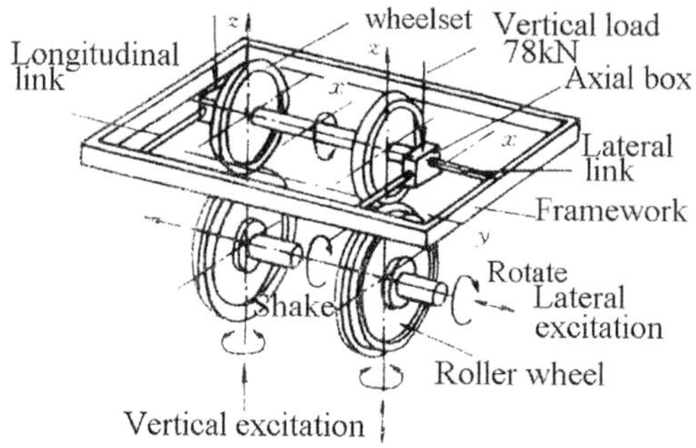

Figure 7.13
Schematic of one-wheelset wheel-rail creep testing device.

composed of ① a mechanical system, ②data acquisition and testing system, and ③hydraulic excitation and power control system. The mechanical system comprises a roller test rig, an instrumented wheelset, and a location frame of the instrumented wheelset; the data acquisition and testing system comprises a computer, gratings, a collecting ring, an elastic dynamometry link, a displacement gauge, an accelerometer, and a related secondary instrument. In test, the rollers (rail wheel) is driven by the motor, and friction between the wheel and the rail drives the wheelset rotation to simulate the advancing motion of the wheelset. Under the effect of the hydraulic exciter, the roller can perform independent vertical and lateral motions so as to simulate the vertical and lateral vibrations of the steel rail. The test wheel is transversely moved relative to the roller by the translational movement, the yaw angle of the wheelset relative to the roller can be arranged through the yaw motion of the location framework and the upper part of the wheelset bearing adapter; the hydraulic exciter at the top of the test rig can perform simulated loading with different axle loads, and the stiffness of the elastic instrumented bar between the wheelset and the location frame is changeable and can be used for simulating the longitudinal and lateral suspension stiffness between the wheelset and the bogie frame.

Due to the existence of the lateral movement and yaw angle of the wheelset, the contact geometric asymmetry of the left and right wheel-rails is formed. Thus, the upper part of the contact patch of the wheel and rollers or all mediums is slightly glided, and then, the creep force mechanism between the wheelset and the roller can be studied.

For testing the longitudinal and lateral creep force between the wheelset and the rail wheel, the instrumented wheelset was used. The instrumented wheelset can exactly measure the vertical and lateral creep forces between wheel and rail; however, the measurement precision of the longitudinal creep force cannot meet the requirements, and therefore, the strain of the longitudinal and lateral location link are collected to modify the creep force.

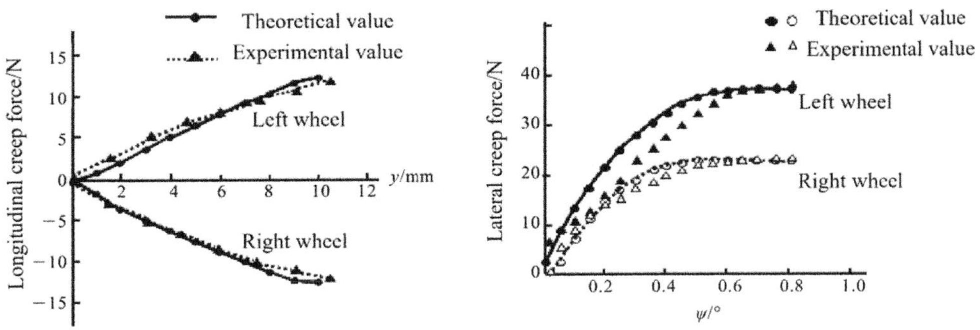

Figure 7.14
Kalker non-Hertzian rolling contact model verification.

The definition and calculation formula of the creep force is provided in Section 2.3.1.4. The diameters of the left and right wheel rollers can be altered via the wheelset lateral movement, and relationship between the wheel-rail contact longitudinal creep force and the wheelset lateral movement can be acquired. The relationship between the lateral creep force and the wheelset yaw angle is tested through the yaw angle (incidence angle) of the wheelset. When the test result is compared with the Kalker non-Hertzian rolling contact model, the wheel-rail force is acquired and compared with the creep force. Fig. 7.14(a) [7] indicates the change in the left wheel longitudinal creep force along with the lateral movement of the y-axis of the wheelset. Fig. 7.14(b) [7] indicates the change in the lateral creep force between the left and the right wheels along with the yaw angle ψ. The test result of the 1: 1 wheel-rail interaction is consistent with the calculation result of the Kalker non-Hertzian rolling contact model; thus, the Kalker non-Hertzian rolling contact model is verified.

(2) Test technology of wheel-rail adhesion

Wheel-rail adhesion refers to the traction force transfer ability between the wheel and the rail under the effect of traction power; the relationship between the wheel-rail adhesion factor and creepage of contact point between the wheel and the rail is required to measure the adhesion. The adhesion force is consistent with the creep force under the linear condition; however, the calculation result of Kalker's creep theory is different from that of the actual adhesion factor under the nonlinear area, and it is more obvious in the sliding area (Fig. 7.15). The main task of wheel-rail adhesion test is to confirm the adhesion status of the nonlinear area and the sliding area, in particular, to the adhesion factor of the wheel-rail contact from the adhesion to the sliding turning point.

The author introduced the adhesion test work in the State Key Laboratory of Traction Power in his monograph [2]. In addition, the experiment testbed with a wheel-rail adhesion is built in NLRT.

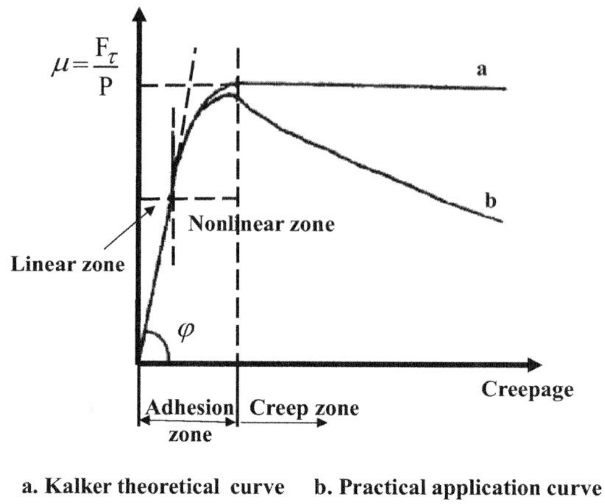

Figure 7.15
Difference of theory and actual adhesion model (a) Kalker theoretical curve (b) Practical application curve.

The test bed is a two-axle test bed (see Fig. 7.16) that realizes the rolling test of one bogie including the adhesion and fatigue tests, and the test speed reaches 600 km/h. In order to truly simulate the wheel-rail contact state, the test rig adopts a 1950 mm roller and the wheelbase of the tested bogie can be adjusted within 2000–3000 mm. The top of the bogie is provided with a load frame that can perform loading and dynamic excitation in the vertical, lateral, and longitudinal directions (Fig. 7.17). The flywheel with the same

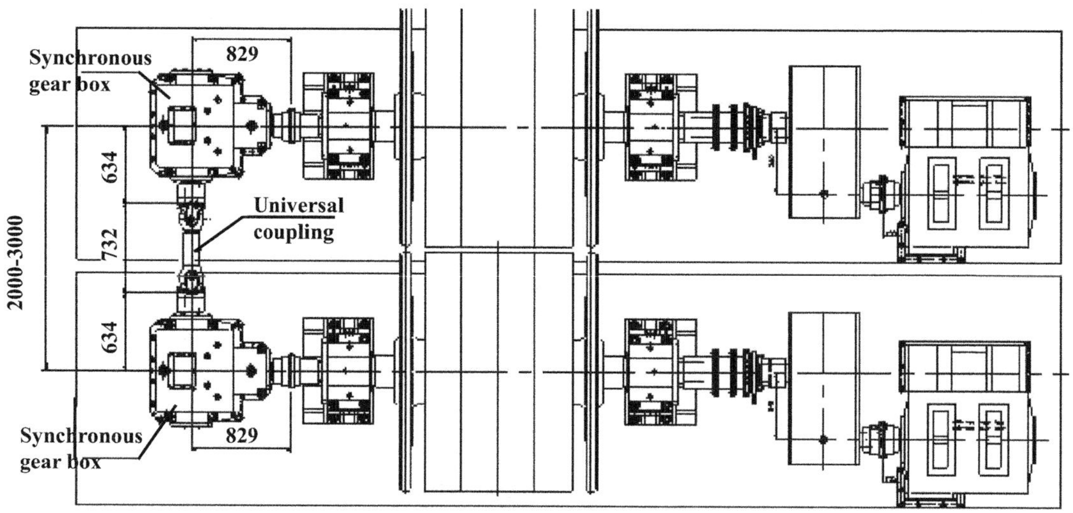

Figure 7.16
Plan layout of wheel-rail adhesion and wheelset fatigue test bed.

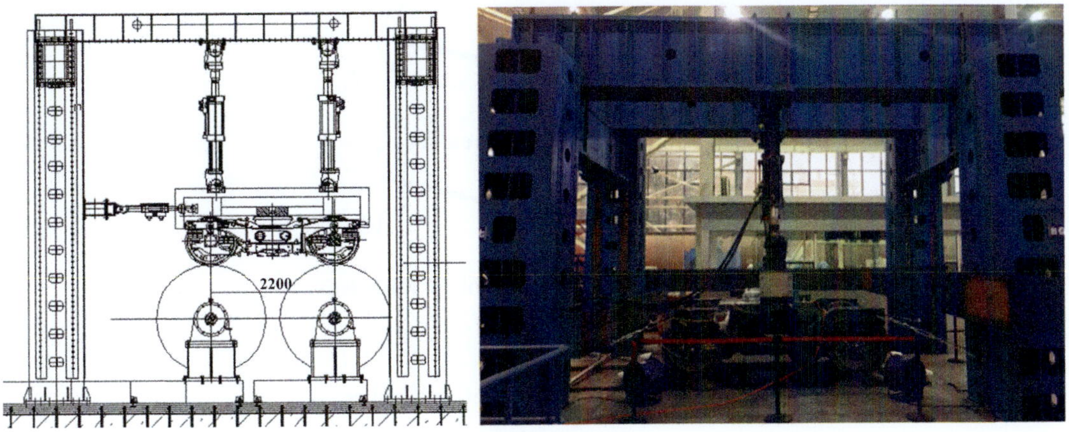

Figure 7.17
Test scene design chart and picture of wheel-rail adhesion and wheelset fatigue test bed.

diameter is set in the middle of the main roller to simulate the vehicle inertia and to meet the adhesion test requirement. The rotational inertia of the flywheel is not less than 18000 $\mathrm{kg\cdot m^2}$. When the adhesion test is carried out, the tested bogie wheelset must be equipped with a drive system with large power and torque to realize the test of the wheel-rail adhesion saturation point and the adhesion curve test after the slide occurs. The drive power of the test itself is 750 kW/axle, and it is suitable for a wheelset fatigue test. In the adhesion test, the rotational roller is kept stationary using the inertia of the flywheels; the relative sliding between the wheel and the rail is achieved by applying acceleration and deceleration on the bogie, and then, the adhesion factor under different creepage is obtained.

The creepage and torque of the roller and the wheelset are measured, and then, the relationship between the adhesion factor and the creepage is obtained. The adhesion factor is calaculated by

$$\mu = \frac{F}{N} = \frac{M_{Roller}}{R_{Roller}\cdot N} = \frac{M_{Wheelset}}{R_{Wheelset}\cdot N} \tag{7.1}$$

where F is traction force of contact point, N is the positive pressure of the contact point, M is the torque of a roller or wheel, and R is the radius of the roller or wheel.

Based on Formula (7.1), the adhesion factor can be calculated based on the torque on the roller and on the wheelset. It is very difficult to accurately control the sliding difference of the roller and the wheelset during the testing process, especially when the wheel-roller is sliding. Fig. 7.18 shows the torque of the wheelset and the roller in the adhesion test. The adhesion traction force of the wheel-rail contact point is F, and motion of inertia of roller

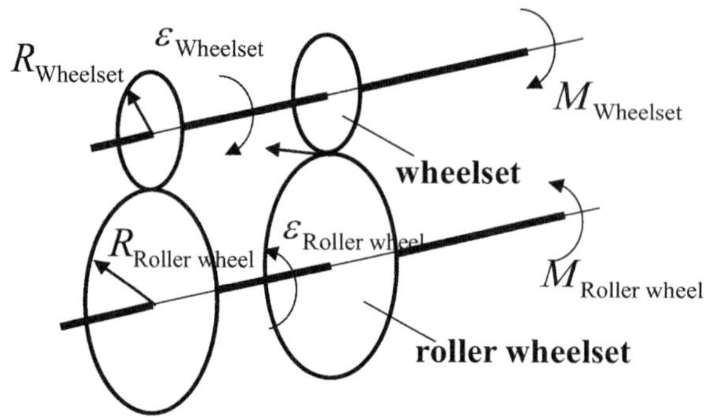

Figure 7.18
Torque stress of wheelset and roller.

pair or wheelset is denoted by J; ε represents rotating acceleration. The adhesion torque has the following relationships:

$$\begin{aligned} M_{Wheelset} &= FR_{Wheelset} + \varepsilon_{Wheelset} J_{Wheelset} \\ M_{Roller} &= FR_{Roller} + \varepsilon_{Roller} J_{Roller} \end{aligned} \tag{7.2}$$

The inertia force in the measured torque is removed when calculating the adhesion factor, and then, the traction force is used.

(3) Test technology of friction wear and contact fatigue

During the service process of high-speed trains, the most important parameter that is changed is the equivalent taper of the tread because of serious wheel-rail friction and wear. Therefore, the test of wheel-rail friction and wear is an important research aspect under wheel-rail interaction. There were three testbeds for wheel-rail friction and wear and contact fatigue in NLRT, and all these test rigs applied the proportion model.

The common method for this test is to rotate the rail or rail wheel to achieve the advancing motion of wheel relative to the steel rail; the wheel can simulate the dynamic action of the wheel-rail by applying the motion and the force in the vertical and lateral directions; meanwhile, the wheel also can apply the load torque to simulate the wheel-rail friction and wear in the traction or brake process, and thus, the traction force or brake force can be acquired.

Fig. 7.19 shows the test bed plan of the wheel and rail-roller; it is composed of incidence angle driving devices, an axle pin, a simulated rail-disk, a base platform, a vertical support, a DC driving motor, a diaphragm coupling, an acceleration speed change gear box, simulation wheels, load and force test part, excitation actor and excitation system,

Figure 7.19
Test bed of opposite rolling type wheel-rail friction and wear.

electric cylinder and driving system, a balance counterweight, braking devices, and a rolling supporting base. The test bed is characterized by small volume, high efficiency, and low cost; it can simulate the contact stress of the wheel-rail, and the real steel rail is supported by a sleeper dispersion.

However, the steel rail experiences heavy bending stress while wheel passing, and it is essential to consider the bending stress simulation to study the wheel-rail fatigue using the test bed. Fig. 7.20 shows the wheel-rail test bed considering the steel rail bending stress; compared with the test bed plan in Fig. 7.19, the main change in the test bed is the real simulation of the steel rail. The steel rail adopts the real cross section, which is paved on a round disk with a simulation sleeper support, and it is moved relative to the wheel along with the rotation of the round disk. The round disk diameter is 3 m, and the highest speed of the simulation reaches 380 km/h. The wheel with a reduced scale (1: 3) can apply vertical load and traction/brake torque while setting up the incidence angle ($\pm 4°$), and then, the lateral friction and wear resulting from the impact can be studied. The test bed in Fig. 7.20 considers the influence of the wheel-rail rolling contact stress and the steel rail bending stress on the service performance of the wheel-rail material. It can be used to

Figure 7.20
Test bed of wheel-rail rolling behavior simulation.

simulate the friction and wear and the fatigue of the high-speed wheel-rail damage, and to select the high-speed wheel-rail material. The main functions include

- Influence of rolling contact stress and steel rail bending stress on service performance of wheel-rail material;
- Simulation of wheel-rail contact fatigue, including collapse, crack, exfoliation, and oblique crack, under different wheel-rail profiles, materials, and hardness;
- Simulation of wheel-rail friction and wear, including uniform abrasion and corrugation, under different wheel-rail profiles, materials, and hardness.

(1) Test technology of derailment

Derailment is the one of the most serious accidents of high-speed trains in operation, and this is a concerns for all countries. The derailment mechanism based on the balance of friction force presented by NADAL is widely applied. Owing to different vehicle behavior and friction behavior at the moment of derailment, it is difficult to accurately judge the derailment behavior with the derailment rule; thus, research on derailment is necessary.

The author introduced research on derailment in his monograph [2], and the test rig applied is the one-wheelset creep test bed, as introduced above (Fig. 7.12 and Fig. 7.13). The difference is that the location framework is moveable for the derailment test. The framework is laterally loaded by the hydraulic exciter, and the vertical and offset loads can be applied through the vertical hydraulic exciter. In the test, the yaw angle of the rail wheel of the roller test rig relative to the wheelset can be set up. The derailment process is shown in Fig. 7.21 [2].

The derailment test technology in literature [2] is employed on the roller test rig. A derailment test device on a real track has been researched in NLRT, as shown in Fig. 7.22, through which, the research of full-scale derailment mechanism operated on the real track can be realized. The test parameters of the vehicle and the track can be adjusted, changed, and measured freely, and the research on the extreme working conditions of the vehicle derailment and road falling can be carried out in this test rig. As the dimensions of the bogie and the track and the shape and dimension of contact patch in wheel-rail contact state are the same as the contact state of the actual wheel-rail, the actual wheel-rail

Figure 7.21
Derailment in test process.

Figure 7.22
Derailment test bed.

interaction can be easily reflected. Therefore, the derailment test is important to provide real and effective test data for research on the derailment mechanism, and to control and ensure the safe operation of high-speed and heavy-load railways.

The track of the derailment test rig adopts the ballastless track, which is commonly used in high-speed railways. The gauge is 1435 mm, the rail is 60 kg, and the instrumented rail is set in the test section to acquire the wheel-rail interaction force. A guide rail of the traction vehicle is provided outside of the track, and the forward and backward motions of the traction vehicle are guaranteed. Through a steel rope, the traction vehicle is connected to the driving devices in its front and back ends, and the maximum running speed of traction vehicle reaches 35 km/h.

The bogie is placed in the traction trolley. The cylinder bars of the front and back lateral electric cylinders at the left side of the traction trolley are connected with front and back axle boxes of bogie through respective lateral steel ropes; the cylinder bars of the left and right longitudinal electric cylinders at the front of the traction trolley are connected with the bogie frame through the respective longitudinal steel ropes. The upper part of the bogie is provided with a balance weight box through which the vertical load and vertical offset load can be realized. The wheelset of the bogie is the instrumented wheelset.

The incidence angle of wheelset is adjusted through the left and right independent longitudinal acting forces using a relevant longitudinal electric cylinder. The independent lateral excitation forces are provided by relevant lateral electric cylinders. As the derailment mechanism at one side is symmetrical to the other side, the lateral excitation force is only applied to the left side on this test rig. The derailment of the wheelset may be occur when changing the offset load, incidence angle, and lateral force. The test parameters, including the lateral excitation force, longitudinal acting force, wheelset incidence angle, wheel-rail lateral force, and wheel-rail vertical force are tested and

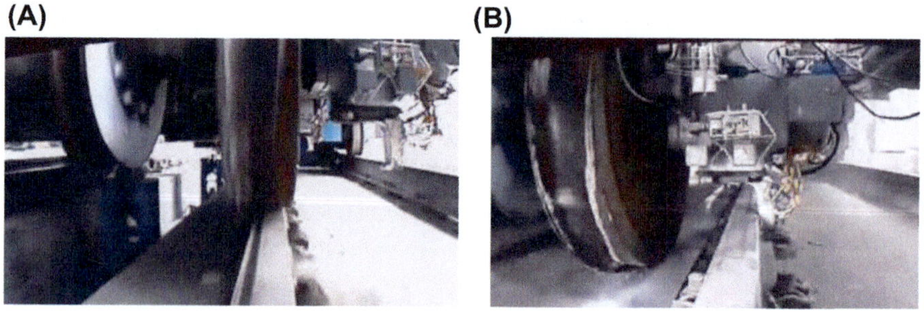

Figure 7.23
Derailment process. (A) before derailment (B) after derailment.

recorded in the testing process, which provides more real and effective test data for the research of the train derailment mechanism and control method.

The main functions of the test bed include:

- Simulation of low speed derailment process of wheelset;
- Experimental study of wheel derailment mechanism;
- Experimental study of match relationship of wheel-rail profile;
- Experimental study of wheel-rail interaction mechanism when derailment occurs;
- Experimental study of wheel-rail geometric contact state when derailment occurs;
- Experimental study of impact of incidence angle on derailment;
- Experimental study of impact of rail cant on derailment;
- Experimental study of static geometric contact relationship with different wheel-rail profiles;
- Experimental study of influence rule of track geometry (gauge, level, rail cant) on wheel-rail interaction.

The test bed was used to carry further derailment prevention research. During the test, a climbing derailer is arranged on the steel rail to derail the wheel, and then, the motion behavior after derailment and the interaction of the bogie and the tracks (steel rail, sleeper, buckler, and track plate) are studied, and the effectiveness of the derailment prevention device is inspected. Fig. 7.23 is the image screenshot of derailment process; it is obvious that further lateral movement of the wheelset is prevented because of the derailment prevention device beneath the axle box.

7.2.3 Test technology of fluid-solid coupling relationship

In the traditional aerodynamics model experiment, the train model was static, and therefore, it was difficuclt to move the model; this model experiment was called a static model experiment. In order to realize the relative motion between the train model and air, and to manually create a uniform wind environment, the wind tunnel emerged. In the wind tunnel, the uniform steady flow field and change of air flow speed can be easily changed, which meet the test requirements of higher speed. However, the wind tunnel experiment is carried out in the manual train operation environment, and the open atmospheric environment around the actual operation train does not always entirely conform to the enclosed seal environment of the wind tunnel body in the test section; the impact of the ground effect cannot be simulated if the train model is static. The operation simulation test bed of the multifunctional high-speed trains shown in Fig. 7.10 is built to realize the movement model aerodynamics experiment, called the movement model experiment (MME).

Although the MME solved the problem of the ground effect, it still belongs to the model experiment. The construction of the test bed and the setting of the test parameter allows meeting the requirements of the similarity criteria and considering the influence of the boundary layer in the wind tunnel.

(1) Similarity criteria

The similarity criteria implies that two phenomenon are similar to each other if the single valued conditions of the two phenomenon are the same and if these similar criterion numbers consisting of these single valued conditions are equal. To meet the similarity condition, the similarity of the geometric profiles of the model train and the actual train must be guaranteed. The geometric similarity only requires to reduce the model according to actual dimensions. The similar criterion numbers of the aerodynamics experiment include Prandtl number Pr, Froude number Fr, Mach number Ma, and Reynolds number Re. Among these, Prandtl number P represents the ratio of the fluid momentum transfer and heat transfer; the Froude number Fr represents the ratio of the fluid inertia and gravity; the Mach number Ma represents the ratio of fluid inertia and elastic force; and the Reynolds number Re represents ratio of fluid inertia and viscous force. The definition of each criterion number is given below [8]:

$$\mathrm{Pr} = \frac{\nu}{\alpha} \tag{7.3}$$

$$Fr = \frac{u}{\sqrt{gl}} \tag{7.4}$$

$$Ma = \frac{u}{a} \tag{7.5}$$

$$\mathrm{Re} = \frac{\rho u l}{\eta} \tag{7.6}$$

In the above formulae, ν is air kinematic viscosity (unit: m^2/s), α is the fluid thermal diffusivity (unit: m^2/s), u is the wind tunnel flow speed or operation speed of model train (unit: m/s), ρ is the air density (unit: kg/m^3), g is the gravity acceleration (unit: m/s^2), l is feature size of the train (unit: m), η is air dynamic viscosity (unit: $Pa \cdot s$), and a is the local sonic velocity (unit: m/s).

The mediums encircled by the train in the real train test and the wind tunnel test are air, and the Prandtl number Pr and Froude number Fr are the same when the model test and real train test are carried out under the condition of a small temperature difference. The operation speed of the model train in Fig. 7.10 reaches 400 km/h, which is the same as the Mach number Ma of the real train test.

The Reynolds number Re can be learned according to Formula (7.4), when the scale ratio of the model train is 1: 10, namely, $\frac{l_{Model}}{l_{Real \quad vehicle}} = \frac{1}{10}$; under the same temperature and pressure test conditions, $\frac{\mathrm{Re}_{Model}}{\mathrm{Re}_{Real \quad vehicle}} = \frac{1}{10}$.

(2) Structural parameters design of wind tunnel

Both the static model wind tunnel and the movement model wind tunnel create the uniform ambient wind conditions in the environment enclosed by the wind tunnel body, and the effect of wind tunnel sidewalls on the test results need to be considered. Fig. 7.24 (a), (b), and (c) describe the flow field distribution conditions when the real vehicle, the static model train, and the movement model train operate in the wind tunnel, respectively.

For the train operating in the open atmosphere, the air around the train moves along with the extrusion of the head vehicle and the friction of the vehicle body, resulting in the train induced air flow. Observing the train induced air flow on the moving train, the flow field around the train can be considered to be formed by air running in the reverse direction at the same speed as the train when the train is static; thus it can also be called as the air flow around a train, as shown in Fig. 7.24 (A) [9]. The figure shows that when the air flows around the train, the air close to the surface of the train remains static due to the viscosity of the air, and the air velocity increases from zero to undisturbed flow speed in the thin boundary layer between the vehicle and the air. As the air continues to flow around the train, the thickness of the boundary layer gradually increases. At the end of the last train, the air forms the strong turbulent wake. Owing to the low pressure at the end, there is a pair of strong vortices that are inwardly rolled. It shows that the aerodynamic effect of train running in the open atmosphere depends on the geometry and operating conditions of the train.

In the static model wind tunnel, the flow field around the train is the same as the train flow observed on the running train in the open atmosphere. However, the air flow near the wind tunnel sidewalls is different from that under real vehicle operation. When the train is running in open atmosphere, the far field air of the train is not disturbed and this does not affect the aerodynamic characteristics of the train. However, in the static model wind tunnel, when air flows through the sidewalls, a layer of laminar flow to turbulent flow is formed on the sidewalls, because of the effect of air viscosity, as shown in Fig. 7.24 (B).

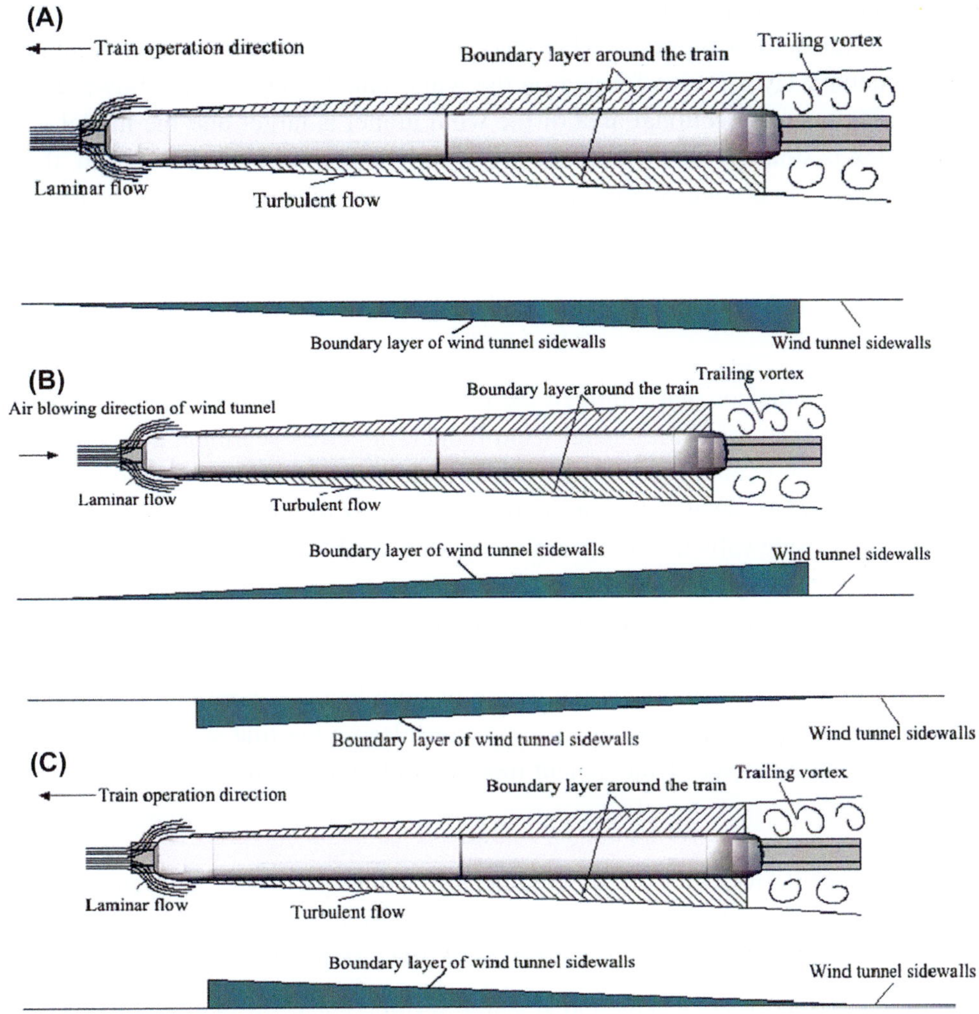

Figure 7.24
Distribution diagram of flow field around the Train. (A) Operation of real vehicle (B) Static model wind tunnel (C) Movement model wind tunnel.

The thicker the boundary layer near the wind tunnel sidewalls, the easier it is to interfere with the flow around the train. In addition, the more obvious the influence of the sidewall effect, the greater is the impact on the test results. Therefore, in the wind tunnel test, the congestion index between the model and the wind tunnel needs to be less than 5% and the corresponding measures should be taken to absorb the boundary layer of the sidewalls.

When the train is running in the movement model wind tunnel, the air flow around the train is the same as the train flow observed on a running train in the open atmosphere.

However, the situation is different when the ejection testing bed is placed in the pipeline in NLRT. In the far field of the train, when the train induced air flow passes through the sidewall, a boundary layer is also formed on the wind tunnel sidewalls, as shown in Fig. 7.24 (C). If the cross-sectional area of the wind tunnel is larger and the model's congestion index is smaller, the air near the sidewall is less disturbed by the train, the air velocity is low, and the boundary layer will be thinner. However, if the cross-sectional area of the wind tunnel is smaller, the model's congestion index is larger, the air near the sidewall is more disturbed by the train, the air velocity is high, and the boundary layer is thicker, and these may interfere with the boundary layer of air flows around the train and affect the test results.

The above analysis indicates that although the movement model wind tunnel is closer to the real vehicle test than the static model wind tunnel, the cross-sectional area of the wind tunnel must also be large enough to overcome the sidewall effect. Because the formation direction of the boundary layer between the boundary layer of air flows around the train and the wind tunnel sidewalls in the movement model wind tunnel is different compared with the static model wind tunnel test, the cross-sectional area of the movement model wind tunnel should be lower than that of the static model wind tunnel. Now, it is assumed that the model train is a certain type of CRH train with a maximum width of about 3.38 m, a maximum height of about 3.7 m, the length of train head is about 8.6 m, and the maximum cross-sectional area is about 11.5 m^2; considering the line spacing and the congestion ratio, the diameter of movement model wind tunnel is determined to be 4.2 m.

(3) Test condition

According to the similarity criterion, the number of similarity criteria between the model test and the real vehicle test must be equal. As mentioned above, the Pr, Fr, and Ma numbers are all automatically equal while Re is 10 times smaller.

According to the low-speed experimental research results, when $Re = 3.6 \times 10^5$, the Reynolds number Reenters the self-mode area. The train speed of the experimental platform reaches 400 km/h and the Re number far exceeds the critical Reynolds number in the self-mode area (critical). However, the results of the current vehicle test in China and abroad have shown that when the Ma number is greater than 0.3, the effect of air compressibility cannot be ignored. Due to the impact of compressibility, the error between the wind tunnel test and the current vehicle test is much larger than that at the low speed. To increase the accuracy of the model test, the Re number should be equal as far as possible.

According to Formula (7.6), when the fixed size of the train model is 1/10 of that of the real vehicle, the methods for improving the model test Re include increasing the operation speed of the model train, increasing the air density ρ, and reducing the aerodynamic viscosity η. Since the aerodynamic viscosity is a temperature function, the effect of changes in the pressure is less. Therefore, in the case where it is difficult to increase the speed of the model train, in addition to reducing η and increasing ρ to increase Re, the air temperature in the wind tunnel can be decreased. The other solution is to increase the pressure in the wind tunnel.

According to the equation for an ideal gaseous state

$$p = \rho RT \tag{7.7}$$

Gas density is proportional to pressure under constant temperature. If the air density in the wind tunnel is increased by 10 times, the pressure in the wind tunnel is required to be increased by 10 times, i.e., the absolute pressure in the wind tunnel reaches 10 atm so that $\mathrm{Re}_{Model} = \mathrm{Re}_{real\ vehicle}$ can be achieved and the similarity of the model experiment and the real vehicle test is reached.

If the researched speed of real train in operation is less than 400 km/h and the speed of model train is 400 km/h, the pipeline pressure required should be less than 10 atm. The actual pressure required in the pipeline is calculated as the following formula:

$$p_2 = 10\frac{p_1 U_{T1}}{U_{T2}} \tag{7.8}$$

In the formula, p_1 is the operation environment pressure of real vehicle, Pa; U_{T1} is the operation speed of real vehicle, km/h; U_{T2} is the operation speed of model train, km/h; p_2 is the absolute pressure required in the movement model wind tunnel, the unit is Pa.

It should be noted that Formula (7.8) gives the pressure conditions in the wind tunnel that are required for the train aerodynamic model experiments under normal pressure. If the experimental bench is used for vacuum aerodynamics experiments, the pressure in the pipeline needs to be set according to the degree of vacuum required to be simulated.

7.2.4 Test technique of pantograph-catenary interaction

The traditional research of pantograph-catenary interaction focuses on the dynamics performance. However, with the advancement of railways and the deepening of research, the pantograph-catenary interaction also includes the electric contact properties and the friction-wear relationship between the contact strips and the contact line. Both the electric contact properties and friction-wear of the material in the pantograph-catenary system are related to the contact state (arc) and conducting state (magnitude of current). The most important parameter describing the contact state is the dynamic contact force, and the pantograph-catenary dynamics is the most basic research field. The test technique of the pantograph-catenary dynamics is introduced.

The author has already introduced the simulation test techniques of pantograph-catenary dynamic interaction in the related works [2] and the papers [10–12]. Because the scale of the catenary system is several hundred meters or even thousands of meters, physical simulation cannot be fully conducted in the laboratory and the virtual simulation cannot achieve the purpose of engineering research on the pantograph performance. Thus, the author adopted the hardware-in-the-loop technique to establish a hybrid simulation test bed for the pantograph-catenary system, in which the pantograph is tested through the physical simulation and the catenary through a virtual mathematical model. The hybrid simulation pantograph-catenary test bed consists of three parts: the lower part is the excitation system that simulates vehicle vibration, the center part is the pantograph to be tested, and the upper part is the system that simulates the dynamic behavior of the catenary. The mathematical model of catenary is stored in a computer connected to the hydraulic excitation and control system. Fig. 7.25 shows the schematic of the hybrid simulation pantograph-catenary test bed. In the author's relevant references [2,10], only the pantograph-catenary hybrid simulation test bed that simulates the vertical movement of the pantograph-catenary is introduced. In fact, there are lateral movements (including tilt) in the operation of the pantograph. In addition, the catenary also has lateral vibration, including a special zigzag arrangement of the catenary. The simulation operation test bed considering the spatial movement of the pantograph is shown in Fig. 7.26 in NLRT.

Compared with the mass of the vehicle, the mass of the pantograph is negligibly small, which means that the vibration of the pantograph-catenary does not affect the movement of the vehicle. The vibration of the high-speed trains is reproduced through the motion platform under the pantograph of the test bed to simulate the vibration of the high-speed trains. Considering that the movement of the pantograph is reflected in the vertical, horizontal, and side roll motions, the motion platform below the pantograph is a

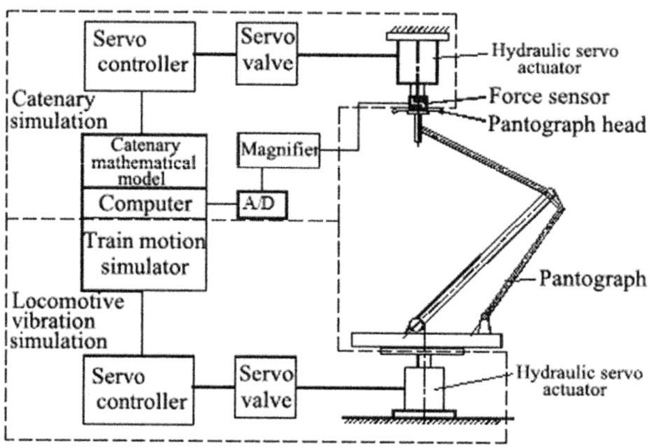

Figure 7.25
Schematic of the hybrid simulation pantograph-catenary test bed.

Figure 7.26
Hybrid simulation test bed of pantograph-catenary operation.

three-degree freedom motion platform and it is determined by the hydraulic exciter. The vertical displacement of the motion platform is ±50 mm, the side roll is ±3°, the lateral displacement is ±25 mm, and the highest excitation frequency is 30 Hz.

When the pantograph operates under the catenary, the interaction between the pantograph head and the contact line (contact force) causes coupling vibration between the catenary and the pantograph. However, the real factor of the pantograph-catenary vibration is the change in the contact force. The dynamic contact force between the pantograph-catenary is measured in the hybrid simulation pantograph-catenary test bed, and it is sent to the computer as the pantograph's uplift force to the catenary. According to the contact force and the position of the contact point, an integral method can be used to calculate the motion of the contact line. The motion of the contact line is reproduced by the piston movement of the vibrating cylinder of the hydraulic servo system. The piston movement of the vibrating cylinder is applied to the pantograph head through the force sensors. At this point, the piston movement causes the movement of pantograph and changes the contact force. A new contact force is detected and treated as a new input to the catenary motion integral, and then, the integral gets a new contact line movement displacement. When it is continued step-by-step, the hardware-in-the-loop hybrid simulation of pantograph-catenary is achieved. The key to the hybrid simulation is the real-time integral. The integral time of the catenary movement must be less than the real time so that the hybrid simulation can be achieved. The vertical motion simulation range of contact line is ± 50 mm and the maximum frequency of dynamic motion is 30 Hz. The maximum working load is 10 kN.

In the test bed, the lateral stagger of the catenary is considered to characterize the zig-zag arrangement of the catenary. The test bed is equipped with a simulating exciter of the stagger of the contact line. The maximum stagger of the contact line is 300 mm and the maximum motion frequency is 3 Hz.

In addition to conducting the hybrid simulation of the pantograph-catenary and researching the dynamic performance of the pantograph-catenary coupling system, the test bed has other basic detection functions, including:

➢ Hybrid simulation of pantograph-catenary: Simulate the operation quality of pantograph and catenary at different speeds, and find the optimal catenary parameters under the existing pantograph such as catenary height; catenary span; catenary structural height; material of contact line, tension force; bearing wire, tension; dropper wire and layout; sag of contact line; surface irregularity of contact line and the optimal pantograph model under existing catenary;
➢ Dynamic characteristics test: Test the equivalent parameters, catenary following behavior and frequency response characteristics of pantographs at different working height; longitudinal, lateral and vertical impact resistance of pantograph and lateral vibration of pantograph.
➢ Static characteristic test: Test the pantograph uplift and drop time, lateral stiffness, drop holding force, static contact force and working height relationship curve.
➢ Pantograph durability test: Conduct the typical test (such as uplift/drop operation, pantograph suspension device, lateral vibration test) and investigative test (such as vertical vibration), etc.

Friction and wear test bed with current performance ability have been established in NLRT as shown in Fig. 7.27. The basic structure is to simulate the relative motion of the

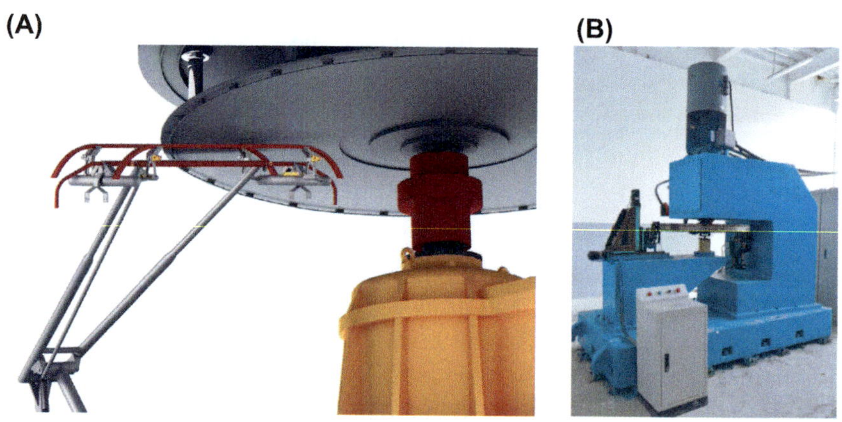

Figure 7.27
Test bed of pantograph current-carrying friction and wear in catenary motion method. (A) Real pantograph (B) Sliding plate material.

pantograph-catenary by means of a high-speed rotation of a disc equipped with contact line, and the relative speed can reach 300–350 km/h. There are two basic types of pantographs, one is the real pantograph, as shown in Fig. 7.27 (A), and the other is only the contact strip, as shown in Fig. 7.27(B). The contact strip moves laterally relative to the contact line to simulate the stagger of the contact line.

There is a pantograph-catenary test bed where the contact line does not move but the pantograph contact strip moves in NLRT, as shown in Fig. 7.28. It shows the ring-type contact lines, which are mounted on eight columns and they are in octagonal shaped. The contact line is equipped with an exciter actuator at the column to simulate the vertical

(A)

(B)

Figure 7.28

Test bed of pantograph-catenary electric contact of pantograph motion method. (A) Annular catenary method (B) Straight line catenary method.

motion of the contact line. The pantograph contact strip is mounted on the rotating arm and is driven by a motor to realize high-speed motion that simulates pantograph movement. Fig. 7.28(b) shows the straight-type contact line. Catenary is fixed in its front and rear end and the tension force can be adjusted. The physical pantograph, which is mounted on a movable trolley, is adopted. The pantograph operates back and forth within a certain distance, the designed operation speed is only 15 km/h. The test bed is used to research the electrical properties and the contact temperature rise characteristics of pantograph-catenary during high-current (max. 3000 A) receiving when train starting. It includes the steady-state electric contact test, the transient thermal effect in the process of sliding electrical contact, and the cause and effect of the arcing of the pantograph catenary system.

The current-carrying simulation is considered in the above-mentioned pantograph-catenary interaction test beds. There are two types of power supply: one is the railway power supply with power level of 25 kV, which considers the influence of voltage on the electrical contact characteristics of the pantograph-catenary and the other is the current. A low-voltage method is used to simulate the effect on the arc and the friction and wear of the pantograph-catenary when passing current. The voltage level is 100 −3000 V and the current level can reach 1000 A.

7.3 Line test technology of high-speed trains for dynamic service performance

7.3.1 Research platform of service performance

Research of the high-speed trains service performance is an extension of dynamics research; the service research includes:

1) Dynamics performance of DCSHT: under the condition of space-time synchronization, master the dynamic performance, vibration state, and transfer relationship of the high-speed trains and its coupling system through detection and simulation study.
2) Evolvement rule of high-speed trains service performance: study performance change under long-term service of high-speed trains through field tracking detection, and master the evolvement rule of performance in different operation rules.

In order to study the service performance of high-speed trains, the NLRT has established an experimental research platform for the high-speed trains service performance, see Fig. 7.29. For realizing the dynamics states of high-speed trains and its coupling system from the upper part to the lower part, from the front to the back, from the inside to the outside, the experimental research platform for the high-speed trains service performance is partitioned into an on-train measuring system and an off-train measuring system according to the operation features of the high-speed trains.

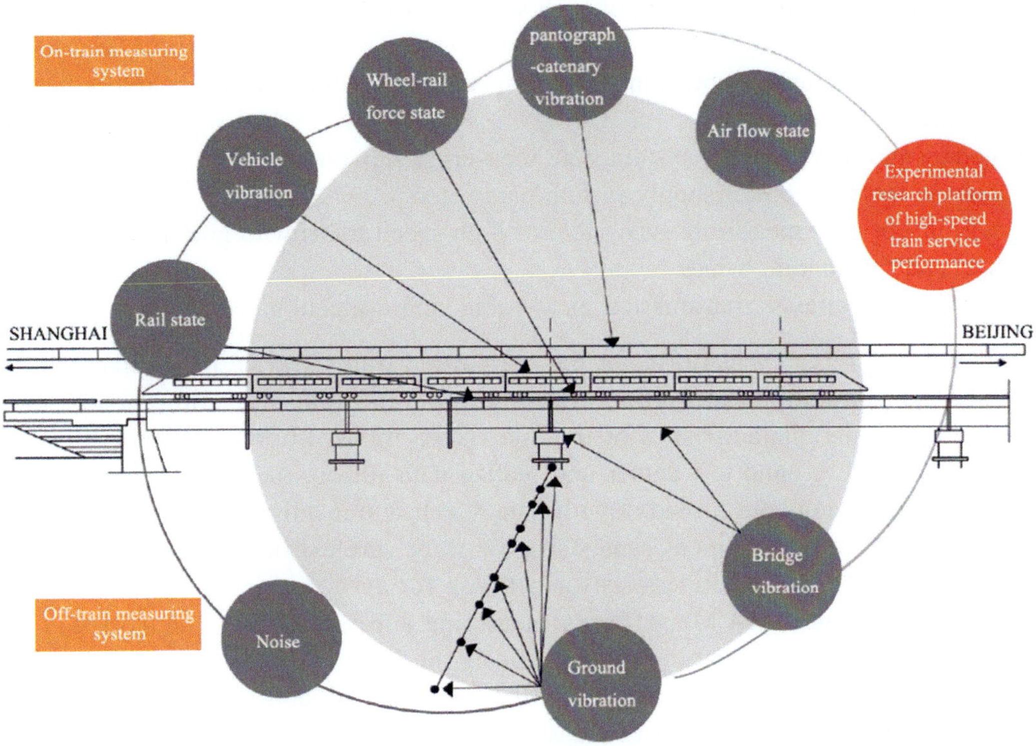

Figure 7.29
Experimental research platform of high-speed trains service performance.

The on-train measuring system comprises:

- Track condition measuring subsystem with irregularity, twist irregularity, and other special defects.
- Wheel-rail contact force measuring subsystem with wheel-rail contact point and contact force
- Airflow state measuring subsystem with side air pressure, space air flow state, train surface air pressure
- Catenary state measuring subsystem with aerodynamic force, pantograph-catenary contact state and arc
- Train vibration state and service performance measuring subsystem with vibration, mode, comprehensive comfort (noise, pressure, temperature, etc.)
- On-train data wireless transmission, processing, and assessment center

The off-train measuring system includes:

- High-speed trains running state (ground) measuring subsystem, measure interaction of wheel-rail

- Track state measuring subsystem of high-speed trains with vibration, stress, and deformation
- Bridge state measuring subsystem of high-speed trains-vibration, stress, and deformation
- Catenary and power supply system state measuring subsystem of high-speed trains with catenary vibration, power supply system parameters
- Environment state measuring subsystem of high-speed trains with ground vibration, noise, and surrounding wind.
- Off-train data wireless transmission, processing, and evaluation center

The experimental research platform for the high-speed trains service performance is partitioned into the on-train measuring system and the off-train measuring system based on the moving characteristics of the high-speed trains. In order to summarize all detected data to the ground test center and realize data fusion, the on-train and off-train measuring systems are provided with data wireless transmission center, data processing center, and assessment center. The on-train wireless data processing and assessment center adopts a wireless sensor network for every carriage, and its data transmission rate is up to 54 Mbps. Every measuring subsystem in the network can share data. WLAN network applied in the train is the information highway of the measuring subsystem of every carriage, and the data transmission rate is up to 54 Mbps. The train measuring system is accessed by public mobile communication network through GPRS MODEM, so as to realize the data exchange of remote survey and monitor the center of the on-train and off-train systems. The measuring subsystems of the head, middle, and last carriages out of carriages comprise an embedded computer, a sensor signal measuring systems, and a WSN communication module. The embedded computer is a server, which is used for applying a real-time operating system to receive the command from the monitor center, which controls the synchronous detection of every measuring subsystem on site, and preprocesses the measured data. Every measuring subsystem realizes the local share of measured data through WSN and shares train-grade internal measuring data. All computers of the high-speed trains measuring system form a local area network (LAN) through the information highway; the data processing terminal computer collects the measured data of every measuring subsystem of the carriage through the on-train wireless network, and it records, processes, displays, and reports an alarm. The on-train wireless network computer data processing system is shown as Fig. 7.30. The off-train data wireless transmission, processing, and assessment center also has a similar function. Since the ground measuring point may be far, wireless or wired transmission can be applied between the measuring point and the data wireless transmission, processing, and assessment center.

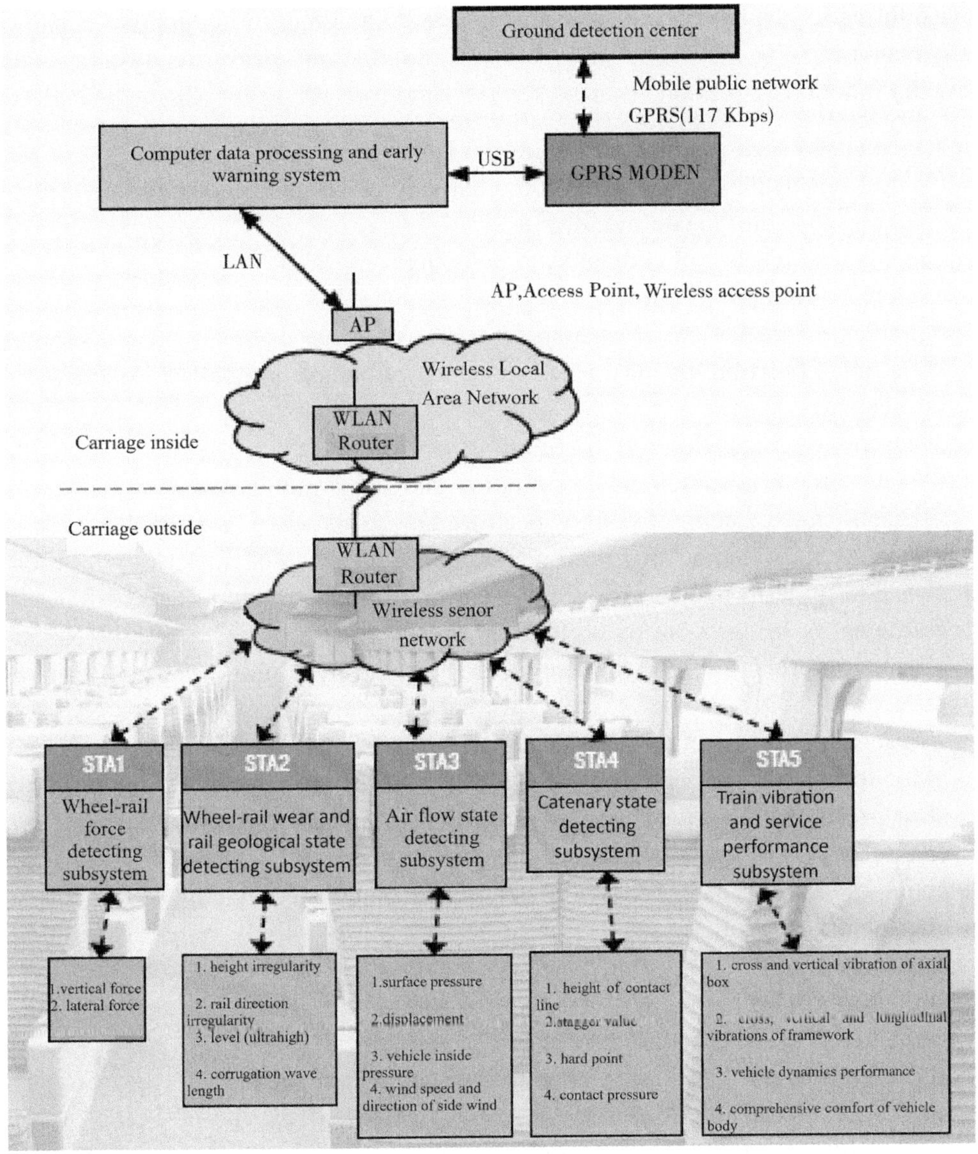

Figure 7.30
On-train measuring subsystem.

As the real train cannot be expected to take the signal as the experimental train, the tracking experiment is implemented on operating train so as to obtain the evolvement rule of the high-speed trains service performance. Therefore, the laboratory team developed a Zigebee-based wireless measuring system to achieve control of the measuring system (including power management) and data wireless transmission.

For more information on the on-train measuring system, the measuring content is shown in the block diagram in Fig. 7.30. The subsystems of the off-train measuring system include the following functions:

(1) Measurement of high-speed trains operation state (ground)
  - ➢ Wheel-rail dynamic interaction force detection of high-speed trains
  - ➢ Hunting instability state monitoring
  - ➢ High-speed wheel-rail contact state detection
  - ➢ Safety of high-speed trains

(2) Measurement of track state of high-speed trains
  - ➢ Dynamics function test of tracks (3-4 kinds of ballast and ballastless track structures)
  - ➢ Detection of track structure force and stress effect
  - ➢ Tracks structural settlement deformation and stress monitoring
  - ➢ Turnout state detection by high-speed trains
  - ➢ Geometrical state and structural performance detection of tracks
  - ➢ Long-term performance evolution detection of track structure

(3) Measurement of bridge state by high-speed trains
  - ➢ Detection of natural vibration characteristics and dynamic response of bridge structure
  - ➢ Static response detection of bridge structure
  - ➢ Monitoring and assessment of bridge structure damage
  - ➢ Long-term creep performance monitoring of railway pre-stressed concrete bridge
  - ➢ Detection of bridge temperature field

(4) Detection of states of catenary and power supply system by high-speed trains
  - ➢ Operation function state detection of traction power supply system by high-speed trains
  - ➢ Positive positioning point vibration waveform and negative positioning point vibration waveform measurement in specific zone
  - ➢ Operation overvoltage and resonant overvoltage detection
  - ➢ Vibration detection of different structural catenaries

(5) Measurement of environment state by high-speed trains
  - ➢ Real-time monitoring of wheel-rail noise, aerodynamic noise, collector system noise, and structure noise
  - ➢ Outside noise distribution and sound source identification of high-speed railway
  - ➢ Durability monitoring of vibration and noise reduction of various track structures (in particular to the damping type track)
  - ➢ Ground vibration test under effect of high-speed trains
  - ➢ Monitoring of sounding wind influence by high-speed trains

7.3.2 Tracking experiment of high-speed trains

The tracking experiment of the operating train aims at learning about the high-speed trains operating state and the change rule of the train in the long-term service process. In the completed tracking experiments of the Wuhan-Guangzhou, Shanghai-Ningbo, and Beijing-Shanghai high-speed trains, the tracking state is not only related to dynamics but also involved in the tracking measurement of the comprehensive comfort degree, traction and transmission, and other system performances. The tracking experiment is innovation research for mastering the performance and change rule and for providing supporting conditions for the design and system optimization at the early period of the development of high-speed trains. The tracking experiment of the operating train is introduced by considering the CRH high-speed trains tracking experiment of Wuhan-Guangzhou passenger special line. Tracking of one CRH high-speed train aims to research the performance state and performance evolvement rule of one CRH high-speed train operated on the Wuhai-Guangzhou railway line. Except for dynamics performance, the experimental research content further investigates the vibration characteristic of the off-train hanging devices, inside vibration characteristics, internal air pressure and temperature variation, variation of the off-train equipment cabin pressure difference, variation of bogie temperature, and electric parameters.

The entire tracking experiment system is consistent with the experimental research platform of the high-speed trains service performance in NLRT, and it is composed of a wireless data collecting system, a train-level measuring system, a ground data processing system, and a remote monitor center. Fig. 7.31 is the formation structure of the tracking experiment system.

The train-level measuring system is composed of a collector, switch board, vehicle-mounted computer, 3G router, and GPS (velocity measuring module); the vehicle-mounted computer can control the entire collecting system using the LAN in the train. One CRH high-speed train measures 1-4 vehicles, where each vehicle has an independent train-level measuring system.

Each vehicle forms an independent collection system, and the system is controlled by the vehicle-mounted computer data collecting software to collect and store data; the 3G router is used for realizing remote monitoring, and the structure of the 1-4 vehicle collecting systems is shown in Fig. 7.32.

The sensors used in the test include acceleration sensors, absolute pressure sensors, and temperature sensors. 2-axis (vertical, horizontal) acceleration sensors with a range of 100 g

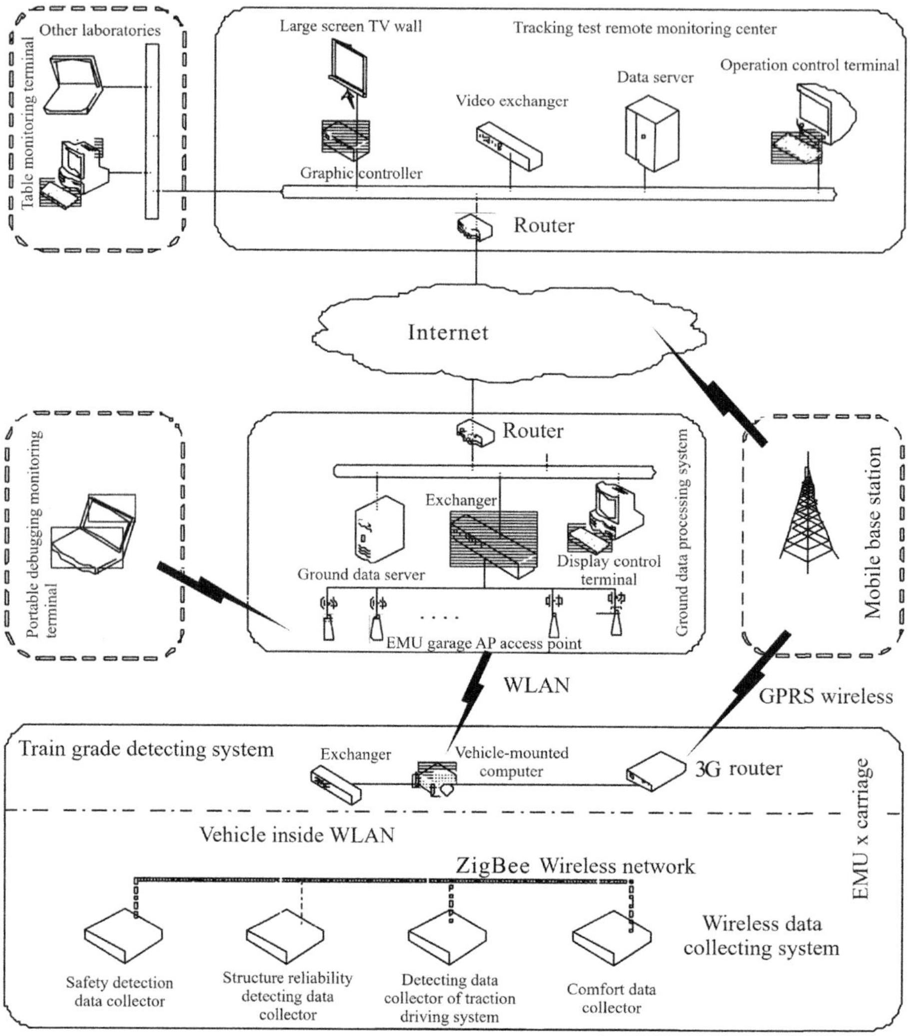

Figure 7.31
Tracking experiment system structure.

are used for axle boxes and frames. 3-axis acceleration sensors with a range of 18 g are used for air spring seats. 3-axis acceleration sensors with a range of 2 g are used for the suspension devices in and under the vehicle. The range of absolute pressure sensor is 0−0.11 MPa and the range of temperature sensor is −50°C−250 °C. The sampling frequency of the axle box acceleration is 5k, the sampling frequency of other acceleration is 2k, the sampling frequency of pressure is 2k, and the sampling frequency of temperature is 1 Hz.

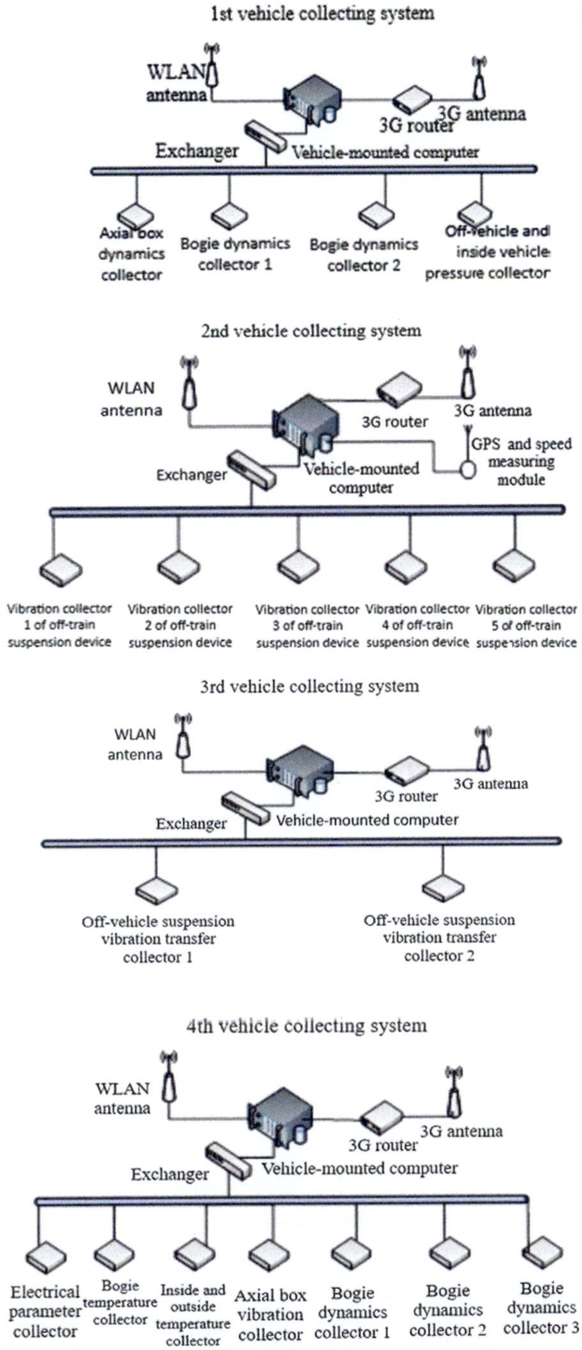

Figure 7.32
Structure chart of each vehicle collecting system.

Table 7.3: Dynamic measuring acceleration distribution point.

No	Measuring point position	Sensor type	Sensor direction	Number	
				Trailer	High-Speed Trains
1	Air spring base	Acceleration	XYZ	2	2
2	Axle box	Acceleration	XYZ	4	4
3	Motor	Acceleration	XYZ	0	1
4	Motor installing base	Acceleration	XYZ	0	1
5	Gear box	Acceleration	XYZ	0	1
6	Hanging base of gear box	Acceleration	XYZ	0	1
7	Framework spring base	Acceleration	YZ	2	2
8	Brake clamp base	Acceleration	XYZ	1	1
9	Inside floor	Acceleration	XYZ	1	1
10	Lower sleeper beam of vehicle body	Acceleration	XYZ	2	2

In the CRH high-speed trains tracking test, relevant dynamic state detection was performed on vehicle 1 (T) and vehicle 4 (M). Among them, there are six 3-axis vibration acceleration sensors and six 2-axis vibration acceleration sensors for the vehicle 1 bogie, and nine 3-axis vibration acceleration sensors and six 2-axis vibration acceleration sensors for the vehicle 4 bogie, as shown in Table 7.3.

7.4 Experiment research results of high-speed trains

7.4.1 Experimental research introduction of high-speed trains for the Beijing-Tianjin Intercity Railway

Opened in 2008 prior to the Beijing Olympic Games, the Beijing-Tianjin intercity high-speed railway is the first high-speed railway with a design speed of 350 km/h in China. As the technical director, the Author led the scientific research experiment on the Beijing-Tianjin inter-city high-speed railway and carried out systematic experimental research on high-speed trains. In 2008, domestic high-speed railway construction had just started, and the complete set of high-speed railway was not formed; in particular, the interaction of high-speed trains and related systems was under exploration. Since the Beijing-Tianjin intercity high-speed railway was put into use, unprecedented conditions were provided for carrying out the systematic experiment and research of high-speed railway technology, and the DCSHT was considered to be research object; on the basis of comprehensive and in-depth theory research, we confirmed the dynamic characteristics of high-speed trains under high-speed operating conditions, mastered the influence rule on fixed equipment and surrounding environment during the operation of high-speed trains, and explored the wheel-rail interaction, pantograph-catenary interaction, fluid-structure interaction, electromechanical coupling, and environment coupling in the DCSHT.

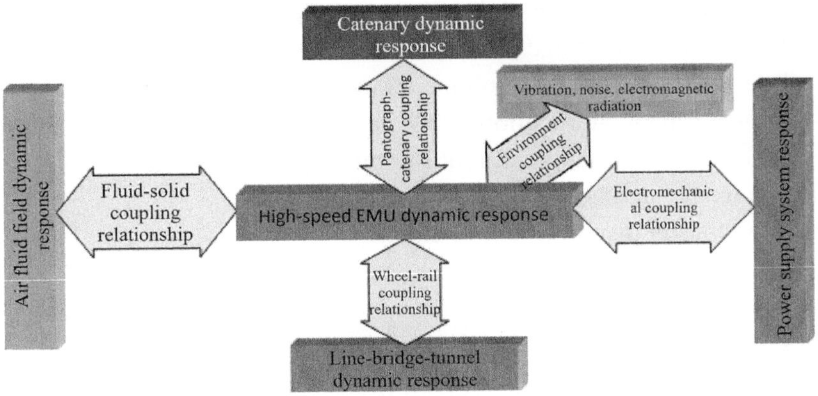

Figure 7.33
High-speed trains coupling system.

The scientific research experiment of the Beijing-Tianjin intercity high-speed railway took the dynamics experiment research of DCSHT as the core, and studied noise issues related to vibration and power supply issues related to traction at the same time; the system block diagram is shown in Fig. 7.33. The detailed research content includes:

(1) High-speed trains system dynamics
- ➢ Wheel-rail interaction force and location of action point
- ➢ Vibration acceleration of each vehicle body, bogie and axle box of HIGH-SPEED TRAINS
- ➢ Displacements of primary suspension and secondary suspension of HIGH-SPEED
- ➢ Rigid (suspension) mode and structure mode of HIGH-SPEED TRAINS

(2) Vibration of tracks and base
- ➢ Wheel-rail force (base measurement)
- ➢ Vibration displacement and acceleration of track structure
- ➢ Lateral movement test of wheelset in turnout
- ➢ Long-term service accumulated settlement of roadbed and transition section
- ➢ Vibration displacement and acceleration of bridge structure
- ➢ Ground vibration acceleration

(3) Traction power supply
- ➢ Traction and brake characteristics of high-speed trains
- ➢ Distributed ground resistance of steel rail, combined ground line distributed ground resistor, mutual resistance of steel rail, combined ground line, return path
- ➢ Voltage, current, power factor and electrical energy quality at outlet bus of traction substation
- ➢ Voltage, current and power energy quality of AT substation
- ➢ Voltage, current and power energy quality of terminal subzone of supply arm

- ➢ Catenary voltage and current distribution of catenary parallel point
- ➢ Neutral section overvoltage of catenary
- ➢ Distribution of catenary overvoltage along tracks
- ➢ Voltage and voltage attenuation coefficient of steel rail and combined ground line
- ➢ Distributed ground leakage current of combined ground line

(4) Pantograph-catenary interaction
- ➢ Vertical vibration displacement of catenary
- ➢ Pantograph-catenary arc
- ➢ Pantograph-catenary contact current-receiving amount
- ➢ Pantograph-catenary contact force
- ➢ Pantograph vibration displacement or acceleration
- ➢ Hard impact

(5) Fluid-structure interaction
- ➢ Air resistance
- ➢ Train surface pressure
- ➢ Vehicle body deformation
- ➢ Train wind field
- ➢ Crosswind

(6) Noise
- ➢ Sound field inside vehicle body
- ➢ Sound field when train passing through both sides
- ➢ Sound field of track environment
- ➢ High-speed wheel-rail noise and wheel-rail surface roughness

Since the time and space synchronization detecting method are used in test for the first time, the related testing result is introduced by taking system dynamics as the core.

7.4.2 *Vibration behaviors of high-speed trains from the leading vehicle to the last vehicle*

In the type test of railway vehicle, the representative vehicle of the train or the trailer of the railway vehicle or high-speed trains will be tested to obtain the dynamic performance of the locomotive or vehicle and to evaluate them according to the relevant standard. The traditional type test only grasps the dynamic performance of the tested vehicle. For high-speed trains disturbed by high-speed air flow during operation, the dynamics of the vehicle may be different at different positions in the train. The full train inspection program was adopted when conducting the scientific research test of the Beijing-Tianjin Intercity High-speed Railway to understand the performance of the entire train and grasp the performance characteristics of the vehicle in different locations. The measurement arrangement is as shown in Fig. 7.34 [13].

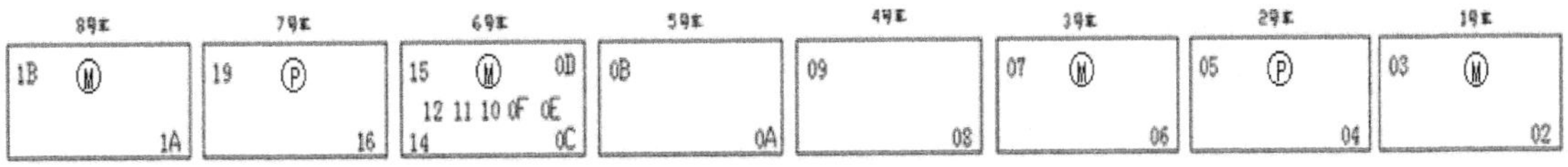

Figure 7.34
Test Layout Diagram of the HIGH-SPEED TRAINS of CRH. Ⓜ Indicates the motor vehicle and Ⓟ indicates the vehicle where the pantograph is located.

Fig. 7.35 shows the acceleration variation of the body at different parts of the CRH high-speed trains. It can be seen from the figure that there are similar statistical laws for both the vertical and lateral accelerations of the body at different positions of the CRH high-speed trains:

(1) Regularity of vertical vibration
- ➢ The vibration behaviors are similar when the trains are running in opposite directions;
- ➢ Due to the inconsistent vibration response of each vibration, it is preferable to use vehicles with the worst dynamics performance as the test vehicle rather than use one vehicle only.
- ➢ The vehicle body equipped with pantograph has greater vibration acceleration than that of its adjacent body. For the tested CRH3 high-speed trains, the vertical vibration of the body equipped with pantograph is the maximum in the train. It shows that the pantograph aerodynamic disturbance has a considerable effect on the dynamics of the vehicle.
- ➢ The vibration acceleration of the vehicle body in the middle of the train is obviously small, which means that the comfort of the central train is better.
- ➢ Vibration of the head vehicle is larger than that of other vehicle (except for the vehicles equipped with pantograph), which indicates that the aerodynamic

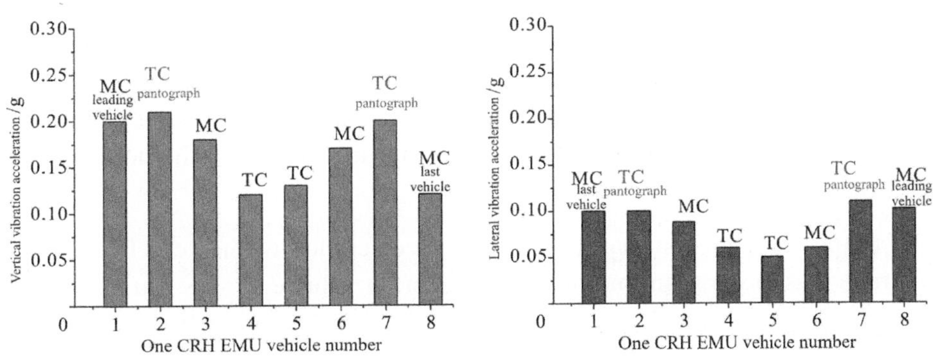

Figure 7.35
Acceleration variation of the body of different parts of the CRH high-speed trains.

disturbance of the head vehicle has a greater impact on vibration than that of the other vehicles.

- ➢ The vertical vibration response of the last vehicle is smaller than that of the head vehicle. It shows that the head type of this CRH high-speed trains is still more suitable for the last vehicle. However, this feature is different compared to other high-speed trains.

(2) Regularity of lateral vibration

- ➢ As the vertical vibration, the lateral vibration behaviors of the train operating in the forward and reverse directions are basically the same and the lateral vibration response of each vehicle is not the same.
- ➢ Lateral vibration acceleration is smaller than that of vertical acceleration.
- ➢ The vehicle body equipped with pantograph has greater lateral vibration acceleration than that of the adjacent vehicle.
- ➢ The lateral vibration acceleration of the vehicle body in the middle of the train is obviously small.
- ➢ Lateral vibration of the leading vehicle is also larger than that of other vehicle (except for the vehicles equipped with pantograph).
- ➢ With respect to the vertical direction, lateral vibration of the last vehicle is not significantly reduced, and it is almost equal to the lateral vibration of the leading vehicle. This shows that the lateral excitation of the tail vortex is still relatively large.

According to the test results, high-speed trains have the phenomenon that the lateral vibration acceleration of the last vehicle is greater than that of the mid-vehicle. To a certain extent, this can explain why people sitting on the leading and last vehicles of the high-speed trains obviously feel more shaking than those sitting in the mid-vehicle. For the reasons, there are two main factors:

1) Only one end of the head and end vehicle have an end constraint, while both ends of the mid-vehicle have end constraints.
2) The aerodynamic forces applied on the head and the last vehicles are obviously larger, and those on the front and rear ends of the train are different. However, the air flow disturbance at the front and rear ends of the mid-vehicle is smaller and the air disturbance force is particularly evident in the attitude motion of the head and last vehicles.

With regard to the phenomenon that the vibration acceleration of the vehicle equipped with pantograph was larger than that of the neighboring vehicle, it was caused by air flow disturbance of the pantograph. Thus, the design of the aerodynamic performance of the pantograph is very important.

7.4.3 Vibration transfer regulation from the catenary to the ground

In the previous type tests, different professional research institutes undertook relevant test content. The China Academy of Railway Sciences was responsible for the test work. The related performance of the high-speed trains was tested by the Research Institute of Vehicles; the related response of the tracks, by Railway Construction Institute; noise, by the Environmental Protection Station; and communication signal performance, by Communication and Signal Institute. In general, the tested results are not consolidated for analysis and a separate report is not issued any more. However, to research the dynamics of the coupling large-scale system of high-speed trains and fully grasp the vibration of high-speed trains and their interaction mechanism, the vibration response of the catenary, vehicle body, bogies, wheelset, rails, track slab, and bridge was tested in the scientific research test of Beijing-Tianjin Intercity high-speed railway (see Table 7.4 for the results). The following conclusions can be drawn through the test:

(1) The vibration of wheels and rails was most violent, which was caused by the track irregularity and surface irregularities of wheels. The vibration of the entire high-speed trains system originated from the dynamic interaction between the wheels and the rails.
(2) For the rails, the vibration was getting smaller when going down: As the vibration isolation effect of the track board: a support structure between the rail and the track slab, the vibration acceleration of the track slab dropped sharply. Because the damping layer was also provided between the track slab and the bridge and the large mass of

Table 7.4: Testing results.

No	Testing items	Testing value
1	Vertical vibration acceleration of catenary limiter/g	0.5~1.5
2	Vertical vibration acceleration of vehicle body/g	0.2-0.3
3	Vertical vibration acceleration of framework/g	2-3
4	Vertical vibration acceleration of axle box/g	20-30
5	Vertical acceleration of steel rail/g	150~380
6	Vertical acceleration of track slab/g	1.5~2.5
7	Vertical acceleration of bridge body/g	0.05~0.09
8	Vibration grade at ground (15 m) z VLz,max/dB	70~75
9	Vibration grade at ground (30 m) z VLz,max/dB	68~72
10	Steel rail vertical force/kN	75~96
11	Steel rail lateral force/kN	~10
12	Steel rail vertical displacement/mm	0.7~0.8
13	Steel rail lateral displacement/mm	0.6~0.9
14	Vertical displacement of track slab/mm	0.25~0.44
15	Vertical displacement of base/mm	0.03~0.08

the bridge (generally, about 500 tons), the vibration was very weak when it was transmitted to the bridge (more than three orders of magnitude were reduced compared to the rail).

(3) For the vehicles, the vibration was smaller going up: Owing to the vibration isolation of the primary suspension between the wheelset and the frame, the vibration acceleration of the frame was reduced by one order of magnitude compared to the axle box. For being subjected to the vibration isolation of the secondary suspension between the frame and the vehicle body, the vibration acceleration of the vehicle body was reduced by more than two orders of magnitude compared to the axle box.
(4) The vibration of the vehicle was not directly transmitted to the catenary through the pantograph. In addition, the vibration of the catenary originated from the coupled vibration of the pantograph-catenary system. From the test results, the vibration of the catenary was around 1 g.
(5) Ground vibrations near the bridge piers of the elevated tracks were relatively smaller and they were characterized by the dB number. Their vibration acceleration were within 0.05 g. Indeed, the vibration at 30 m was significantly lower than that at 15 m.
(6) The vibrations of the various parts of the tracks were within 1 mm.
(7) The wheel-rail interaction force detected from the rails was concentrated in the vertical direction, which reached 70–100 kN; however, the lateral wheel-rail force was smaller (about 10 kN). The results were obtained from tests on the straight line.

Through the test results of Table 7.4, we basically grasped the system response state when a high-speed train is operating on a ballastless rail. This provides a basis for the relevant theoretical research and system design.

The response of high-speed trains can be influenced under different operating speeds; by comparing the vehicle body vibration acceleration at the same speed with the vehicle body vertical and lateral acceleration responses of one CRH high-speed train under the speeds of 250, 300, and 350 km/h, as shown in Fig. 7.36, it can be seen that the vehicle body

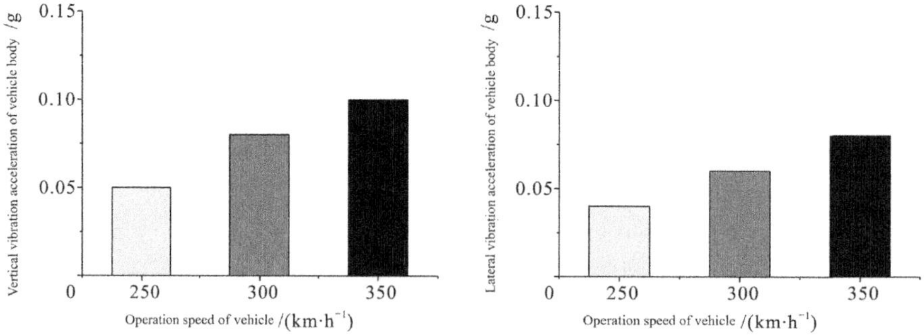

Figure 7.36
Comparison of vehicle body vibration acceleration under the same speed.

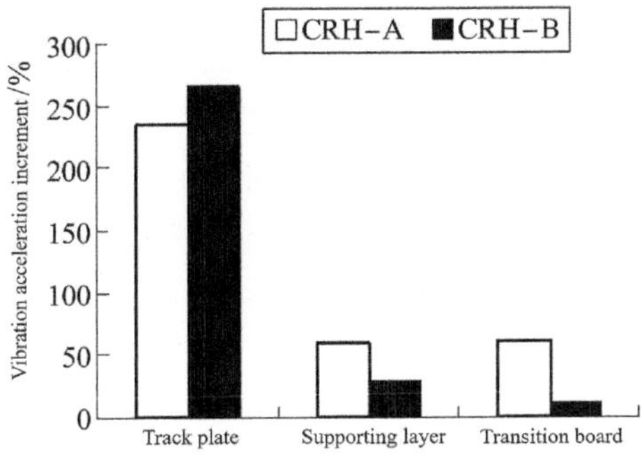

Figure 7.37
Vibration acceleration increment of track structure.

vibration is obviously increased along with the increase in the train operating speed even though the vibration acceleration value is small.

Different high-speed trains have different vibration influences on tracks. In the experiment of Beijing-Tianjin intercity high-speed railway, the driving speed was increased from 160 km/h to 350 km/h. By observing the vibration acceleration increment of its track structure (Fig. 7.37), it is obvious that the speed was significantly influenced by the vibration acceleration of the track plate, and the influence was obviously greater than that on the vibration acceleration of supporting layer and transition board. The effect of speed on the vibration acceleration of the supporting layer and the transition board is greater than that of one CRH-B high-speed trains, which indicates that the track structure beneath the steel rail contributes more to the vibration for one CRH-A high-speed trains.

7.4.4 Vibration behaviors of high-speed trains when passing each other

The train passing event is a typical working condition during which passengers feel strong pneumatic sound as well as transient vibration. By taking the high-speed trains passing on a pilot section of Beijing-Shanghai high-speed railway, the high-speed trains vehicle body vibration behavior characteristics under the conditions of train passing event at an open line and in the tunnel are provided, in particular, the lateral vibration characteristics of the vehicle body. The maximum value in train passing at the open line and in the tunnel is acquired. To acquire the average value of the vehicle's lateral vibration, the half of peak to peak value is defined, and the value is effective during the train passing. The sensors are arranged on each vehicle's first-end second-side inside floor in the CRH train.

7.4.4.1 Vibration behaviors of high-speed trains in the passing events on open line

Table 7.5 shows maximum acceleration value and effective value of lateral vibration inside the floor of each vehicle when one long marshaling high-speed train and one CRH train

Table 7.5: Lateral vibration feature value.

Measuring point	Maximum value(g)	Effective value(g)
1st vehicle floor	0.179	0.0395
2nd vehicle floor	0.132	0.0148
3rd vehicle floor	0.135	0.0175
4th vehicle floor	0.157	0.0177
5th vehicle floor	0.156	0.0135
6th vehicle floor	0.171	0.0195
7th vehicle floor	0.16	0.0172
8th vehicle floor	0.174	0.0167
9th vehicle floor	0.165	0.0166
10th vehicle floor	0.16	0.0165
11th vehicle floor	0.163	0.0161
12th vehicle floor	0.17	0.0169
13th vehicle floor	0.161	0.0170
14th vehicle floor	0.163	0.0154
15th vehicle floor	0.151	0.0151
16th vehicle floor	0.164	0.0147

pass on an open line (not in tunnel) at a speed of 380 km/h; the average maximum value of 16 vehicles is 0.16, and the effective value indicates the vibration energy.

Fig. 7.38 and Fig. 7.39 are the histograms of the maximum and effective values of the lateral vibration acceleration inside the floors from the 1st vehicle to the 16th vehicle at the speed of 380 km/h during the passing process. The 1st vehicle is the head vehicle in the

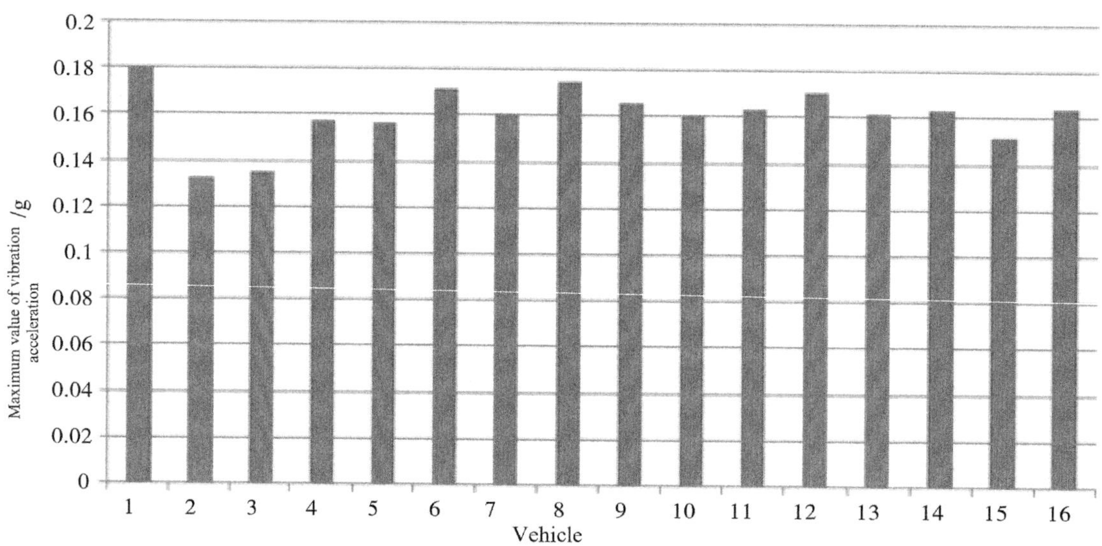

Figure 7.38
Comparison of lateral vibration acceleration maximum value of every vehicle in train passing.

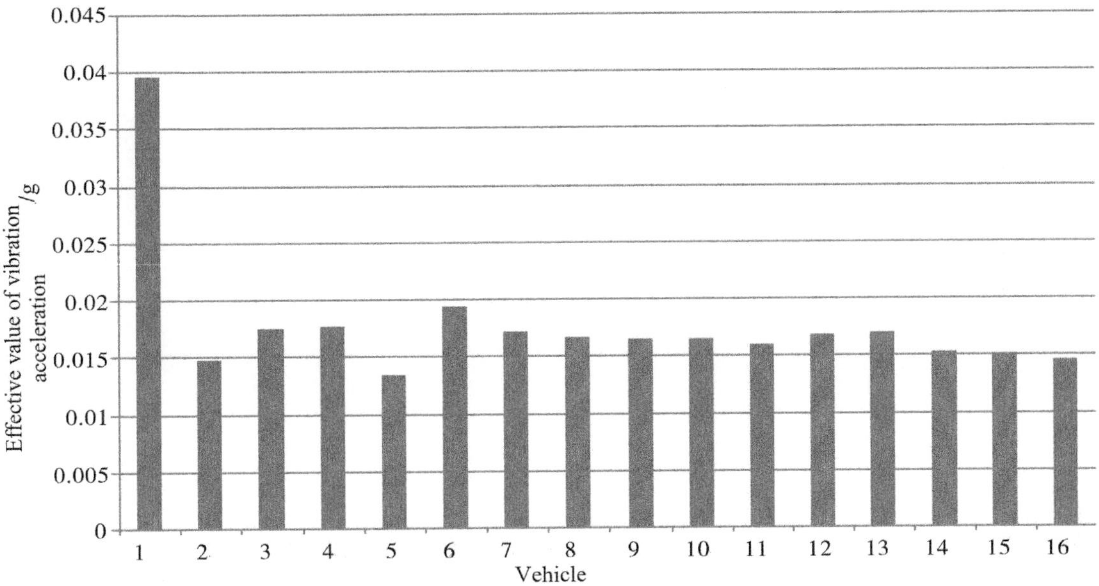

Figure 7.39
Comparison of lateral vibration acceleration of every vehicle in train passing.

passing event. The vibration of the first vehicle is obviously more than that of other vehicles, and the vibration effective value is almost doubled to other vehicles, as the head vehicle is considerably impacted by aerodynamics.

Fig. 7.40 is a time–space figure that shows the time domain lateral acceleration of the 1st vehicle floor when trains at 380 km/h are passing on an open line. It shows that the lateral vibration inside the floor will be impacted at the beginning and the end of train passing, the vibration amplitude of the measuring point is obviously increased. When the passing event is started, the maximum change in the vibration amplitude at the measuring point is about 0.179 g for the 1th vehicle; when the passing event is finished, the maximum change in the vibration amplitude of the measuring point is about 0.17 g.

7.4.4.2 Vibration behaviors of high-speed trains in the passing events in tunnel

Table 7.6 indicates the maximum value of the lateral vibration acceleration for the inside floor of each vehicle when one CRH long marshaling high-speed train and one CRH train at the speed of 380 km/h pass each other in a tunnel; the average maximum value of 16 vehicles is 0.236 g.

Fig. 7.41 and 7.42 show the histograms of the maximum and effective values of lateral vibration acceleration for the inside floor when trains pass across each other in the tunnel at a speed of 380 km/h. It can be seen from the histogram that the vibration for the inside

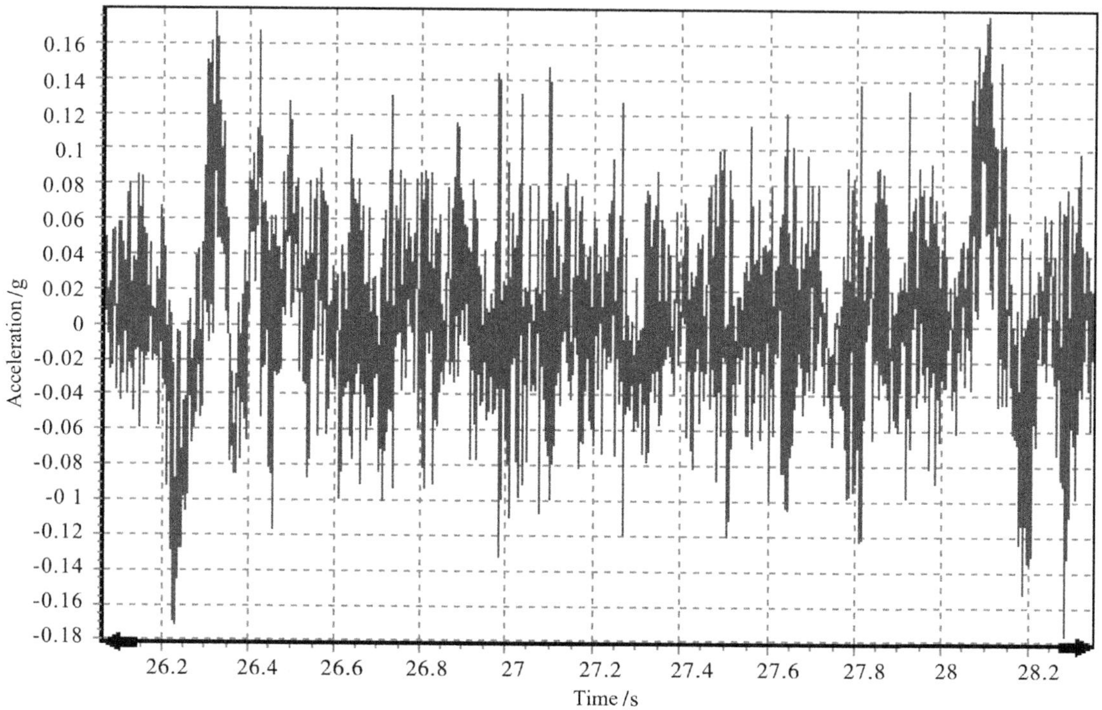

Figure 7.40
Lateral vibration time-domain figure of one-vehicle inside floor.

Table 7.6: Lateral characteristics of train passing in tunnel.

Measuring point	Maximum value (g)	Effective value (g)
1st vehicle floor	0.289	0.0260
2nd vehicle floor	0.188	0.0225
3rd vehicle floor	0.226	0.0222
4th vehicle floor	0.234	0.0208
5th vehicle floor	0.253	0.0216
6th vehicle floor	0.227	0.0249
7th vehicle floor	0.234	0.0179
8th vehicle floor	0.242	0.0245
9th vehicle floor	0.274	0.0211
10th vehicle floor	0.24	0.0149
11th vehicle floor	0.253	0.0179
12th vehicle floor	0.248	0.0167
13th vehicle floor	0.255	0.0210
14th vehicle floor	0.221	0.0178
15th vehicle floor	0.202	0.0190
16th vehicle floor	0.227	0.0185

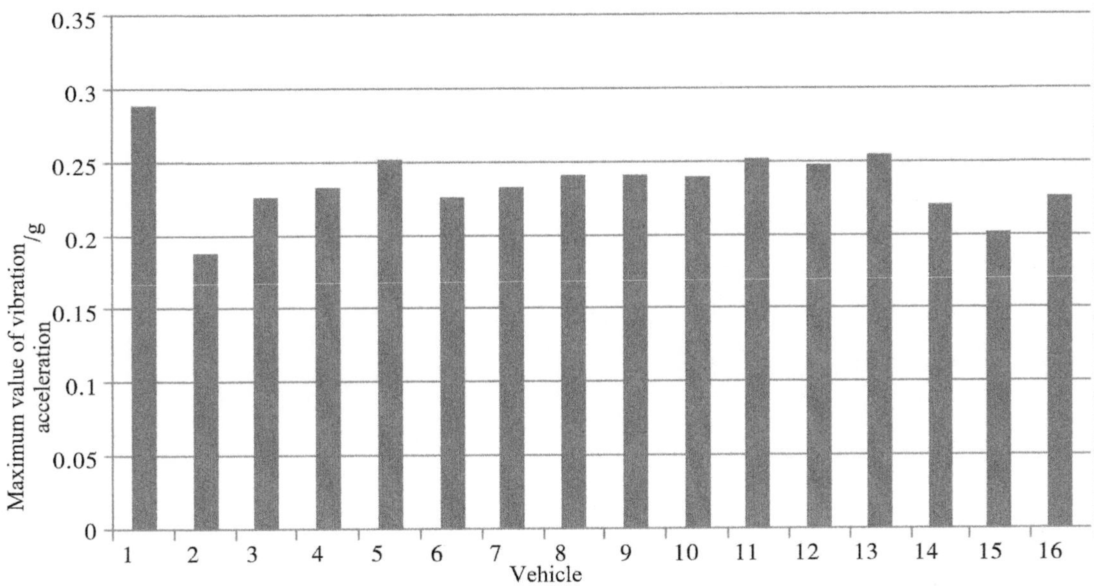

Figure 7.41
Comparison of every vehicle lateral vibration acceleration maximum value when passing cross in the tunnel.

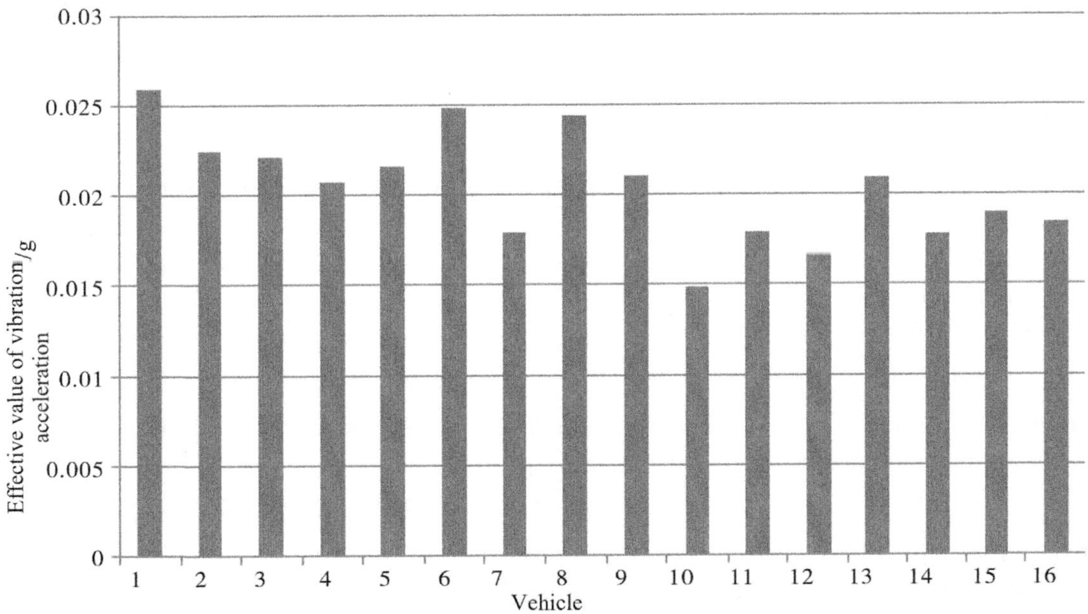

Figure 7.42
Comparison of every vehicle lateral vibration acceleration effective values when trains are passing cross in the tunnel.

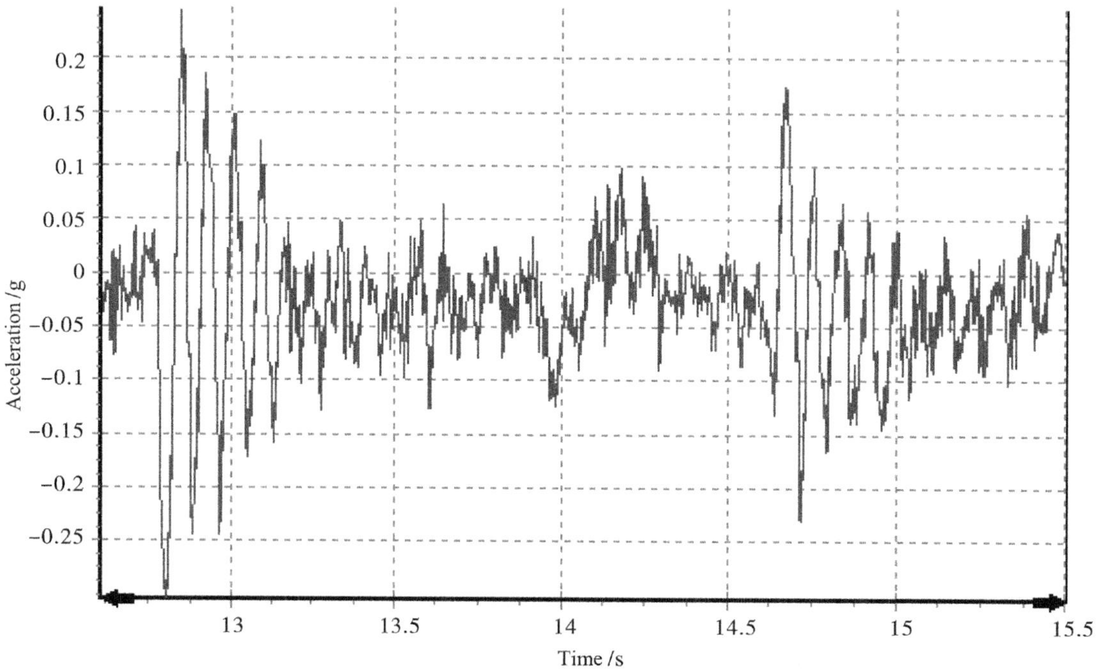

Figure 7.43
Lateral vibration time-domain figure of one-vehicle inside floor.

floor of the 1st vehicle is obviously larger than that of other vehicles, and the effective vibration value is almost double of that for other vehicles when the leading vehicle is more affected by aerodynamics.

Fig. 7.43 is the time-domain figure of the lateral vibration acceleration for the inside floor of the 1st vehicle running at the speed of 380 km/h when meeting at an open line. It can be seen that the lateral vibration of vehicle inside floor is certainly impacted at the beginning and the end of train passing, and the vibration amplitude of the measuring point is obviously increased. The maximum vibration amplitude of the measuring point is 0.179 g when the passing is started and it is about 0.17 g when the passing is finished.

Through comparison of the open line passing and the tunnel passing shown in Fig. 7.39-7.43, the following conclusion can be obtained:

For the passing, the pressure wave is influenced by the reflection of tunnel wall and is unable to effectively disperse as it does when passing on an open line. The lateral vibration amplitude of each vehicle is more than that in passing on the open line during the tunnel passing. Further, the passing waveform of the pressure wave acts repeatedly on the vehicle body because of the tunnel reflection during the tunnel passing.

7.4.5 Noises distribution regulation of high-speed trains inside and outside the vehicle

Noise is a main factor that limits the increase in train speed. Therefore, advanced detectors specific to noise research inside and outside the vehicle body have been adopted. Moreover, several fundamental detection and research works have been implemented with the debugging operation of Beijing-Tianjin, Wuhan-Guangzhou, Beijing-Shanghai, Harbin-Dalian, and other high-speed railways. Noise originates from vibration and is particularly caused by wheel-rail interaction and air flow around the vehicle. The inside and outside noises of high-speed trains are tested in the Beijing-Tianjin intercity high-speed railway system, and the sound source nephogram and the noise spectrum are described.

(1) Noises outside

Acoustic sensors are installed at different measurement locations as per traditional noise measurement and evaluation. However, only noise intensity of one point is provided, and the noise source cannot be identified. A multichannel array noise testing and analysis system is used in the test to identify the source of the noise. The external sound source identification system is shown in Fig. 7.44, which includes 78-channel spoke shape array sound source identification system with a diameter of 4 m; 60-channel wheel array frame with diameter of 1 m, which can perform regional refinement of sound source identification; 12 sets of acoustic intensity kits to study decreasing of frequency resolution

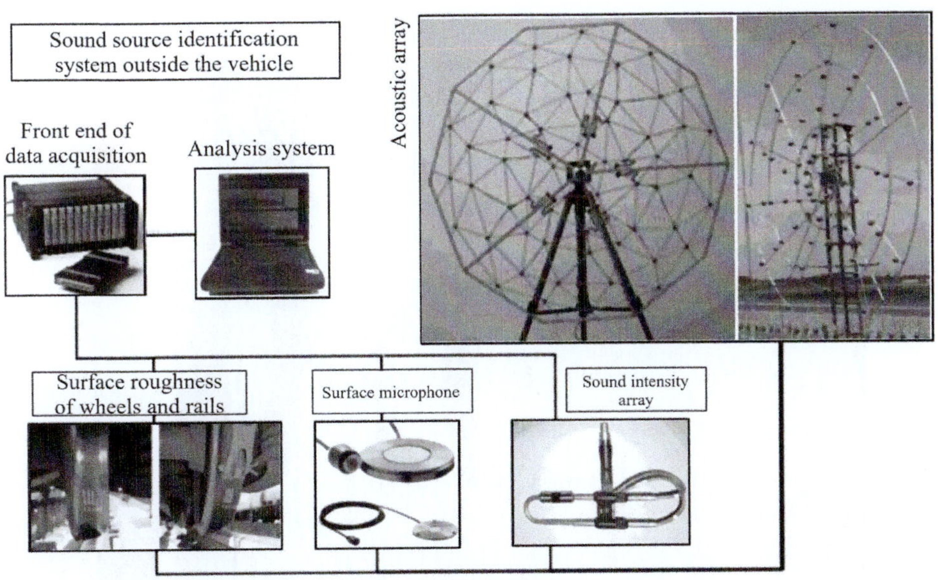

Figure 7.44
Multi- channel array noise detection photos.

and spatial resolution of acoustic pressure array in the low frequency band; and a surface microphone, which can be attached on the surface of high-speed trains to measure the aerodynamic noise, and to obtain outside vehicle noise with high sampling frequency.

Sound waves propagate exactly vibration energy. Sound energy, which passes through the unit area in a unit time, while being perpendicular to the sound wave propagation direction, is called sound intensity. Sound intensity is a vector with a strength value and a direction. In general, the noise cloud nephogram represented by the sound intensity is used as the preferred results of the phased acoustic array sound source identification system, which has clear and intuitive results. The corresponding sound power can be directly calculated based on the area of the sound source area. According to the acoustic array test data, the sound intensity cloud nephogram of the outside radiant noise of a high-speed CRH train with a speed of 394 km/h was measured based on the beamforming (SBA) sound source identification principle, as shown in Fig. 7.45. The transverse axis represents the train length direction and the vertical axis represents the vertical distance from the rail top surface. The corresponding value of the cloud nephogram is the sound intensity level. X represents the longitudinal distance from the point to the train nose, Y represents the vertical distance from the point to top surface of rail, and Level represents the sound intensity level of the point in decibels. This shows that the main sources of radiated noise outside the vehicle come from the bogies, the wheel-rail contact position, the pantograph and its base frame, and the vehicle connection. Compared to the aerodynamic noise at the bogies at other locations, the aerodynamic noise of the head vehicle's bogie is greater.

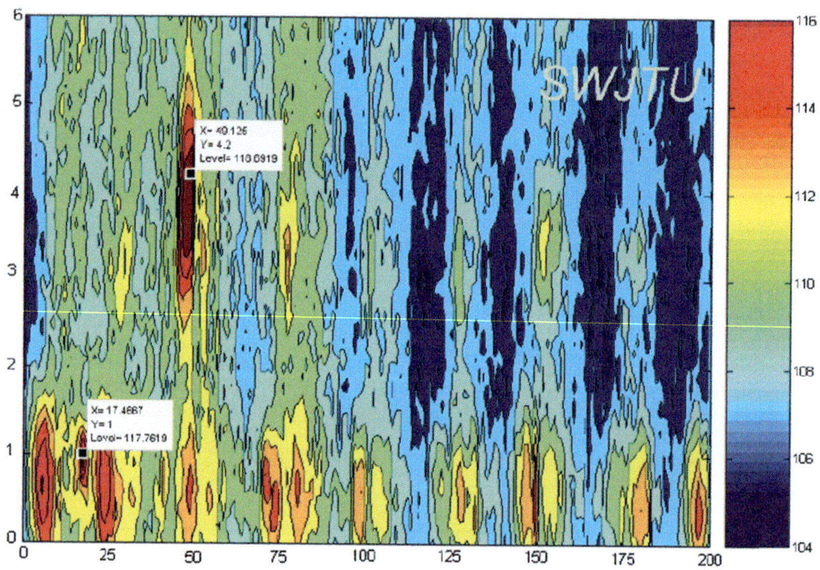

Figure 7.45
Could chart of outside noise of high-speed trains.

The sound intensity nephogram is divided into regions according to the structure of the high-speed trains and the range of the sound source. From the area size and the corresponding sound intensity value, the energy contribution of each part to the radiated noise outside the vehicle can be obtained. This is one of the core data sources for building the noise spectrum outside the high-speed trains, and it is the basis of establishing the sound source model corresponding to the noise spectrum outside the high-speed trains.

As the number of received sound sources, location, and audio frequency resonance influence the train passing noise, the most significant radiated noise point in Fig. 7.45 is not always the main source of the train passing noise, which means that the maximum value of the passing noise does not necessarily correspond to the most significant noise source. The vertical distribution of acoustic exposure levels at the plane of the sound intensity cloud nephogram in the corresponding high-speed trains is shown in Fig. 7.46. Numbers 1 to 8 indicate the 1st to 8th vehicles of the high-speed trains in Fig. 7.46. The maximum sound exposure level measured is located at the wheel-rail contact position of the head vehicle's, which reaches the maximum at the center of the wheel; meanwhile, the position corresponds to the maximum sound source point in the sound intensity cloud nephogram. The secondary loud sound exposure level is located in the second vehicle's pantograph position and the pantograph is in a raised state. The followings are located in the second vehicle's wheel-rail contact position, the upper positions of the head vehicle and the second vehicle. Fig. 7.47 shows the vertical distribution of the total sound exposure level of the high-speed trains at different operation speeds in the elevated tracks section. It is obvious that the distribution is consistent, which increase with speed. Fig. 7.48 shows the comparison of the total sound exposure levels between the elevated

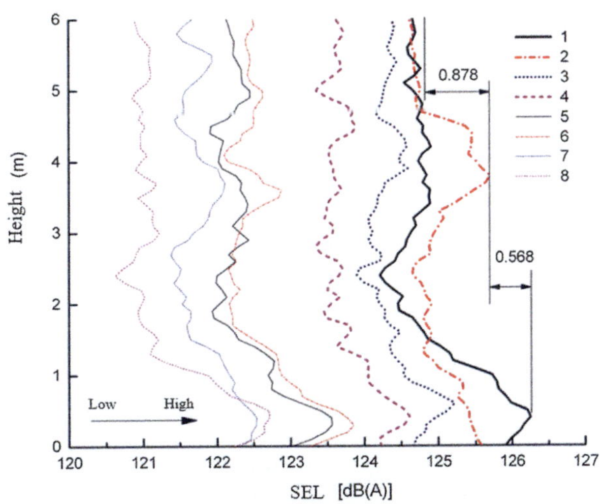

Figure 7.46
Sound exposure level of every vehicle of one CRH high-speed train at 394 km/h.

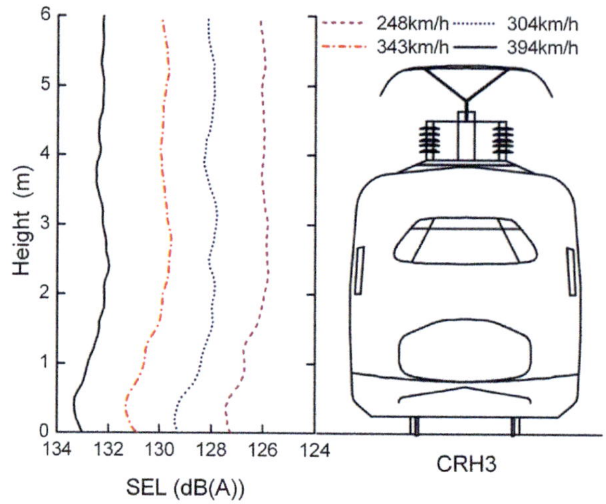

Figure 7.47
Vertical distribution chart of one CRH sound exposure level of overhead section under different speeds.

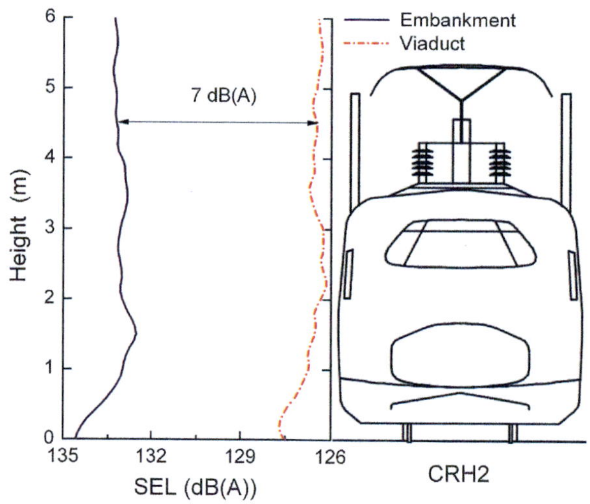

Figure 7.48
Sound exposure level comparison of overhead section and embankment section at 345 km/h.

tracks section and the embankment section. It shows that the noise outside the embankment is higher than that at the elevated tracks section.

Noise spectrum can be processed after the noise information from the tracks tests is acquired. Noise spectrum is the noise power spectrum at different frequencies. The noise spectrum outside the vehicle processed according to the test of Beijing-Tianjin intercity

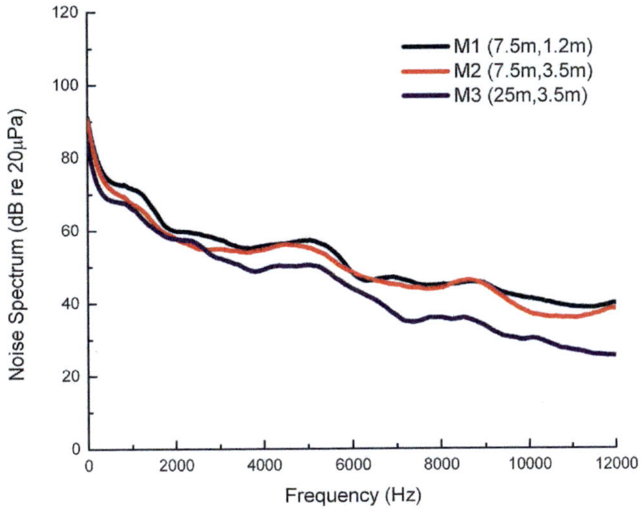

Figure 7.49
Outside noise spectrum with speed grade of 350 km/h.

high-speed railway and Wuhan-Guangzhou special line for the passenger train is shown in Fig. 7.49. It can be seen that the outside noise of the vehicle at the speed of 350 km/h has significant low frequency characteristics. Its most significant frequency occurs in the low frequency range, which originates from the aerodynamic noise contribution of the high-speed trains. In the 1000 Hz frequency range, there were more prominent local peaks, which comes from the noise contribution of the wheel and the rail. In the 5000 Hz and 8000 Hz high frequency sections, the noise comes from the aerodynamic noise of high-speed trains. In general, energy in test point 1 (7.5 m from the centerline of the track, with height of 1.2 m) was higher than that in test point 2 (7.5 m from the centerline of the track, with height of 3.5 m), indicating that noise at the lower part of the high-speed trains is more significant than that at the upper part. From the distance of the test point to the centerline of the track and the law of noise change, it shows that low frequency noise attenuation was smaller, and high frequency noise attenuation was larger. In the sensitive frequency range affected by the noise environment from 1000 Hz to 2000 Hz, the noise did not change significantly with distance. As seen in Fig. 7.50, the higher the speed, the higher is the noise spectrum. The situation of noise outside the vehicle in the case of the reconnection was more significant than that in the case of short-grouping, especially, in the 200–6000 Hz noise sensitive frequency range.

(2) Noises inside

Sound source identification is further detected and researched from the view of disciplinary study except for the point location sensor detected as the standard. As shown

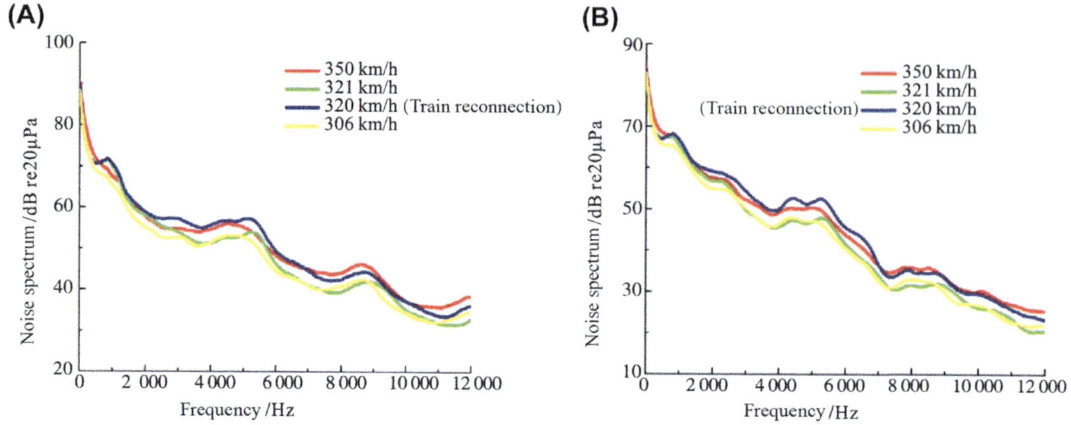

Figure 7.50
Outside noise spectrum of high-speed trains at speed of 350 km/h. (A) Distance to center line is 7.5 m, height is 3.5 m (B) Distance to center line is 25 m, and height is 3.5 m.

in Fig. 7.51, the inside sound source identifying system comprises spherical array sound source identifying systems of 50-ways microphones and 12-ways cameras; a double-faced 6 x 6 x 2 SONAH handheld array was configured to overcome the shortcoming caused by definition of frequency resolution and spatial resolution of the spherical array at the low frequency section. The refined test is conducted on local sound source position according

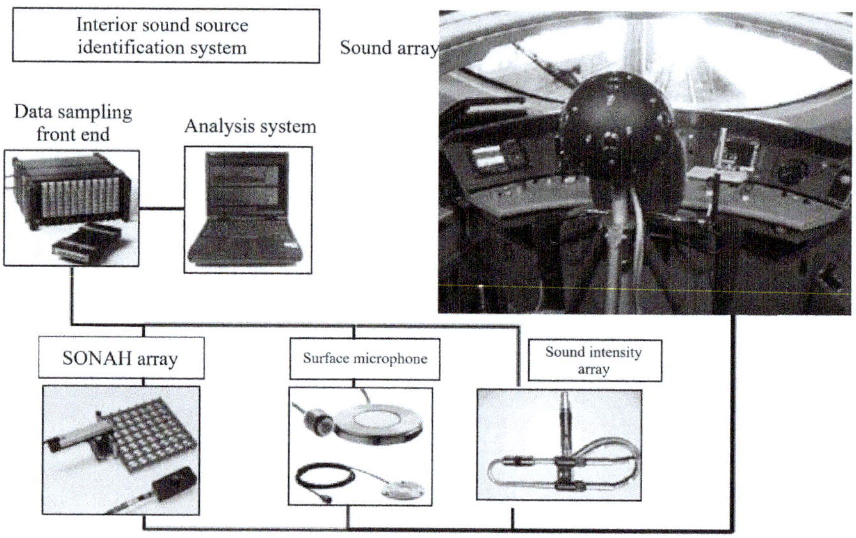

Figure 7.51
Inside noise detection and identification system.

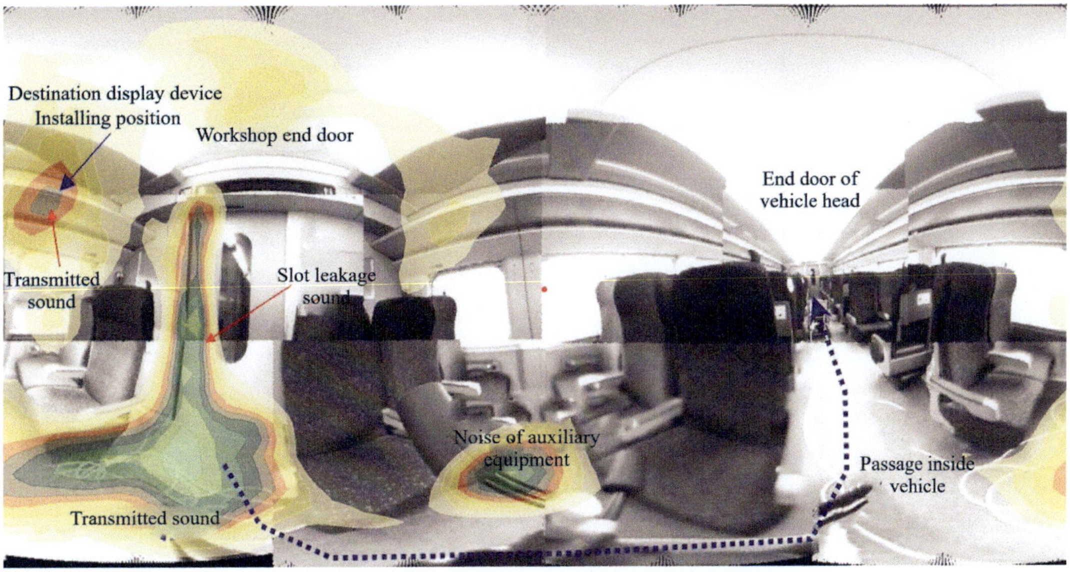

Figure 7.52
Sound source identifying result of high-speed trains.

to the sound source distribution zone of the spherical array sound source identifying the result.

The interior sound source identification of the high-speed trains is as shown in Fig. 7.52. It should be noted that the noise cloud nephogram result was the sound intensity distribution result in the interior surface of the high-speed trains. There were a series of surface field point sound intensity values in the cloud nephogram, which not only includes the value (source strength) and direction (directionality) information but also includes the spectral characteristics. The test results showed that the sound intensity was superimposed in the 360° interior view. Obviously, there was noise in the installation position of the display device on the upper left side of the carriage. The reason was that the outside of the double-decker vehicle body was drilled in order to install the display device, the double layer became a single layer, and the noise outside the vehicle was transmitted. At the end of the passenger compartment, there was also obvious leakage sound, which was related to the cable passing through the section during the test and the door was not well closed. Further, there was indeed noise near the bogie in the vehicle, which was transmitted through the bottom structure of the vehicle. The suspension auxiliary equipment under the vehicle also transmitted noise into the vehicle through the structure. The sound source can be found through the identification of the noise source distribution, which provided a strong basis for targeted vibration and noise reduction.

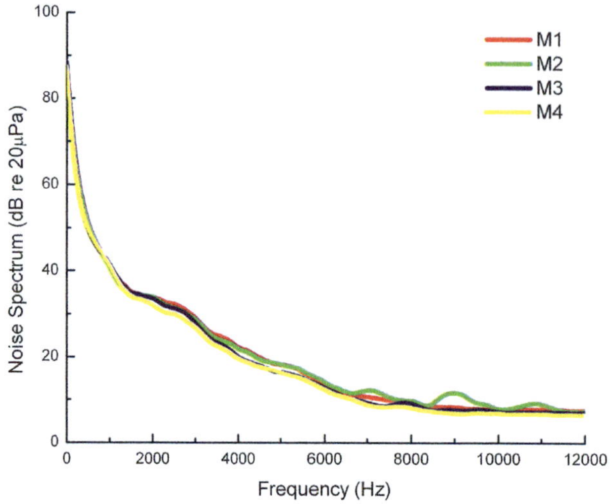

Figure 7.53
Inside noise spectrum with speed grade of 350 km/h.

Four measuring points (M1-M4) are detected at the upper part of the bogie and the middle of the guest room with heights of 1.6 m and 1.2 m. Fig. 7.53 is the noise spectrogram of four measuring points of one high-speed train under the speed of 350 km/h; obviously, the inside noise of the high-speed trains has very significant low frequency characteristics, and the difference of noise spectrum energy reflection is very small, and it shows that the inside reverberant field characteristics are significant. The test further detects the other speed working conditions, the change status of noise spectrum from 330 km/h to 300 km/h is basically similar to that of 350 km/h, as shown in Fig. 7.54. Compared with the other speeds that are less than 350 km/h, the inside noise over 2000 Hz high frequency is significantly increased; however, the total noise is controlled by low frequency zone and increased speed.

When the train passes through the tunnel, passengers can easily feel the increase in noise in the tunnel and this may become an important factor that restricts the speed of high-speed trains passing the tunnel. It is difficult for passengers to distinguish the direction of the noise. In addition, the general point pick-up sensors can only obtain the sound intensity after reverberation. The spherical array sound source identification system with a 12-channel camera can identify noise sources in the tunnel. The comparison result of the passenger compartment noise is shown in Fig. 7.55 when the high-speed trains pass through the tunnel, which is processed at 50 Hz. As the right side in Fig. 5.55 is close to the tunnel wall of the two-way tunnel, it shows that the noise near the right side of the tunnel wall is significantly increased from 75 dB to 83 dB, which passed in the vehicle through the window and the noise coms from the wheel track noise.

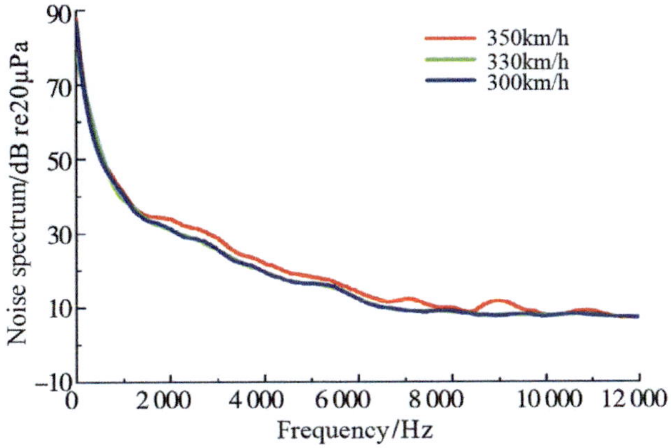

Figure 7.54
Inside noise spectrum of different speed grades.

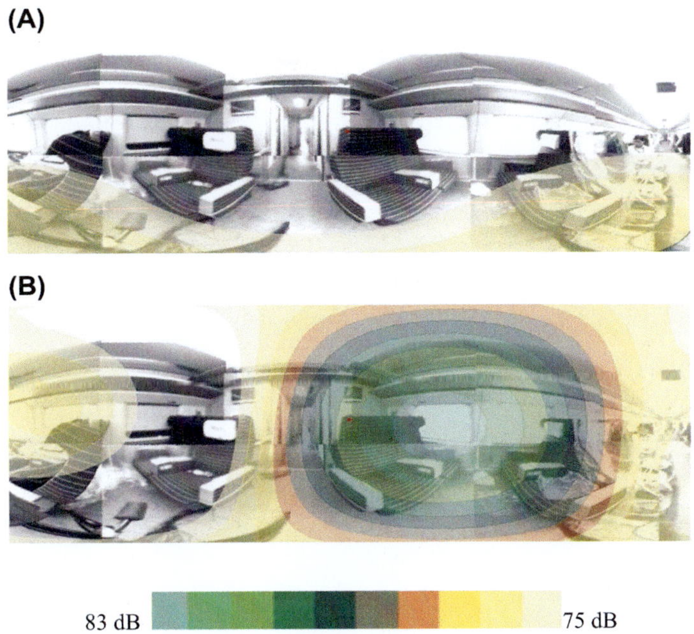

Figure 7.55
Comparison of guest room noises when high-speed trains passes through the tunnel. (A) Before passing through tunnel (B) When passing through the tunnel.

References

[1] Feasibility Study Report of National Laboratory of Rail Transit. Southwest Jiaotong University; 2009.
[2] Zhang Weihua. Dynamic Simulation of Locomotive. Beijing: China Railway Publishing House; 2006.
[3] Ma Shijun, Zhang Weihua, Chen Guangxiong, Zeng Jing. Full scale roller rig simulation for railway vehicles. Vehicle System Dynamics 1993;23(Suppl):346−57.
[4] Chen Liangqi, Zhang Weihua, Chen Jianzheng. A Roller Testing Rig of Railway Vehicles and Its Testing Procedures. Journal of Southwest Jiaotong University 1997;32(No.2):208−13.
[5] Zhang Weihua, Dai Huanyun, Shen Zhiyun, Zeng Jing. Handbook of railway vehicle dynamics, 14 Roller rigs. Taylor and Francis Group, LLC publication; 2006. p. 458−504.
[6] Tian Hongqi. Train aerodynamics. Beijing: China Railway Publishing House; 2007.
[7] Liu Jinsong, Liu Qiyue. Wheel-rail friction. Beijing: China Railway Publishing House; 2004.
[8] Liu Zhengchong. High-low speed wind tunnel pneumatic and structure design. Beijing: National Defence Industry Press; 2003.
[9] Ahmed SR, Fowthorpe RG, Markrodt P-A. Aerodynamics of Road and Rail Vehicles. Vehicle System Dynamics 1985;14:319−92.
[10] Zhang Weihua, Mei Guiming, Wu Xuejie, Shen Zhiyun. Hybrid Simulation of Dynamics for the Pantograph-Catenary System. Vehicle System Dynamics 2002;38(6):393−414.
[11] Mei Guiming, Zhang Weihua. Study on Influence of Catenary Parameters on Contact Force by Method of Hybrid Simulation. Journal of Railway 2004;26(1):23−39.
[12] Wu Xuejie, Zhang Weihua, Mei Guiming. Research on mixed dynamic simulation for the pantograph/catenary system. Journal of Railway 2002;24(1):14−8.
[13] National Key Laboratory of Traction Power of Southwest Jiaotong University. Test Report of CRH3 Beijing-Tianjin Railway Line.

CHAPTER 8

Service performance and safety control for the high-speed train

Chapter Outline

Dynamics of Coupled Systems in High-Speed Railways. https://doi.org/10.1016/B978-0-12-813375-0.00008-X

8.1 Development outcomes of service performance for the high-speed train

Based on a systematic analysis of test data gathered on the service conditions of electric multiple units (EMUs), the parameters related to safety, structural reliability, comfort, etc., are continuously tracked, monitored, and tested. The variation trends of every monitored parameter of trains are analyzed according to changes in service time under actual passenger-carrying conditions. Furthermore, the performance, reliability, and safety of key systems are studied to provide a basis for optimizing design, scheduling a reasonable maintenance period, and creating a database on key components along with a database on change of service time.

Considering actual running speed and running distances of high-speed railway networks, development rules for service performance are analyzed under three conditions: constant speed, variable speed, and different mileages.

8.1.1 Basic outcomes under constant speed

The basic outcomes of vibration acceleration measurement of the of axle box, frame, and car body at a constant speed of 300 km/h taken on a certain type of EMU (i.e., electric multiple units on the Wuhan-Guangzhou railway). The basic vibration outcomes are analyzed at 60 second time length measurements at a constant speed of 303 km/h on a horizontal straight line.

Table 8.1 presents an analysis of the maximum values and effective values of vibration acceleration of the axle box, frame, and car body within a 60 s vibration period at a constant speed of 303 km/h. The table shows that the vertical vibration acceleration of the axle box, frame, car body, etc., is greater than the corresponding lateral acceleration and longitudinal acceleration. The vibration energy transmitted from the axle box to the frame

Table 8.1: Vibration feature value sheet of axle box, frame, and car body under constant speed.

Objects		Maximum value (g)	Effective value (g)
Axle box	Vertical	11.35	1.25
	Lateral	9.86	0.67
Frame	Vertical	3.08	0.46
	Lateral	2.01	0.25
	Longitudinal	1.40	0.19
Car body	Vertical	0.24	0.04
	Lateral	0.16	0.03
	Longitudinal	0.10	0.02

is attenuated to one-fifth its initial strength, and the vibration from the frame to the car body is attenuated to one-tenth—one-twentieth. This shows that primary and secondary systems provide significant damping effects.

Time domain and frequency domain of the vertical and lateral accelerations of the axle box are presented in Fig. 8.1 and 8.2.

The main components of the dominant frequency of the vertical vibration acceleration of the axle box are 29.2 Hz, 58.9 Hz, 82.3 Hz, 215.5 Hz, and 353.93—618.9 Hz. The main components of the dominant frequency of the lateral vibration acceleration of the axle box are 2.15 Hz, 29.2 Hz, 58.9 Hz, 82.3 Hz, 117.7 Hz, 215.5 Hz, and 353.93—618.9 Hz.

In the vertical and lateral vibrations of the axle box, the vertical vibration energy is concentrated mostly in the low frequency (20 Hz—50 Hz) and high frequency range (350 Hz—650 Hz). The 29.2 Hz frequency is related to speed and is a dynamic unbalance component of the train (the dynamic unbalance frequency component corresponding to a speed of 303 km/h and a wheel diameter of 918 mm is just 29.18 Hz), and it is consistent with the forced frequency caused by wheelset unbalance.

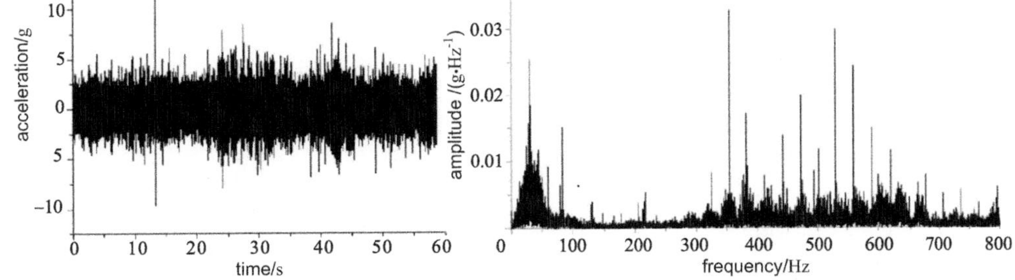

Figure 8.1
Time domain and frequency domain chart of vertical vibration acceleration of the axle box.

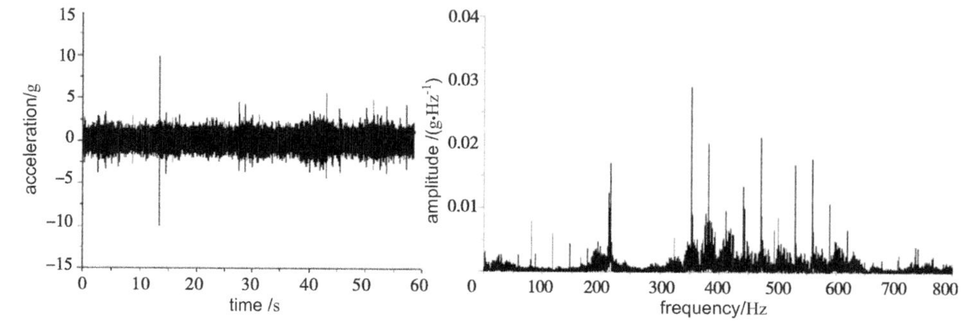

Figure 8.2
Time domain and frequency domain diagram of lateral vibration acceleration of the axle box.

The 58.9 Hz and 353.53−618.9 Hz frequencies are frequency components of 29.2 Hz at two times and 15−21 times its value. This shows that the wheelset is excited by 2nd and 15th to 21st order of magnitude wheel out-of-roundness. 82.3 Hz is the vibration frequency of the motor, and this frequency component is transferred to the axle box through the gearbox.

Time domain and frequency domain of the vertical, lateral, and longitudinal vibration acceleration of the frame are shown in Fig. 8.3−8.5.

The main components of the dominant frequency of the vertical vibration acceleration of the frame are 29.2 Hz, 82.3 Hz, 215.5 Hz, and 353.93−618.9 Hz. The main components of the dominant frequency of the lateral vibration acceleration of the frame are 2.15 Hz, 8.6 Hz, 29.2 Hz, 82.3 Hz, 131.9 Hz, 215.5 Hz, and 353.93−560.4 Hz. The main components of the dominant frequency of the longitudinal vibration acceleration of the frame are 29.2 Hz, 82.3 Hz, 117.7 Hz, 215.5 Hz, and 353.93−530.2 Hz.

In the vertical and lateral vibrations of the frame, the vertical vibration energy of the frame is concentrated primarily at the low frequency range (20 Hz − 50 Hz) and high

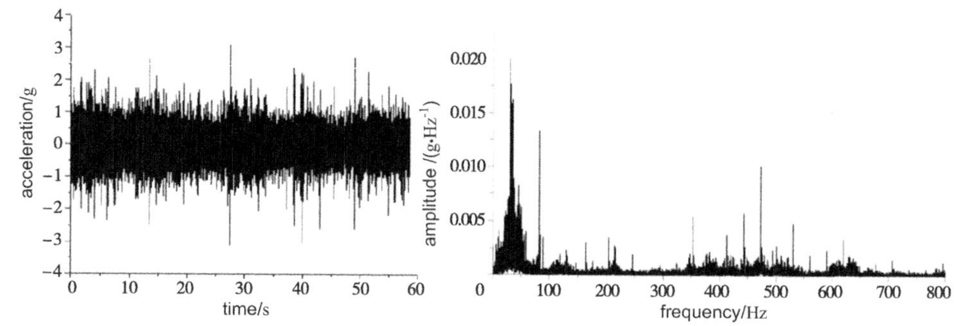

Figure 8.3
Time domain and frequency domain diagram of the vertical vibration acceleration of the frame.

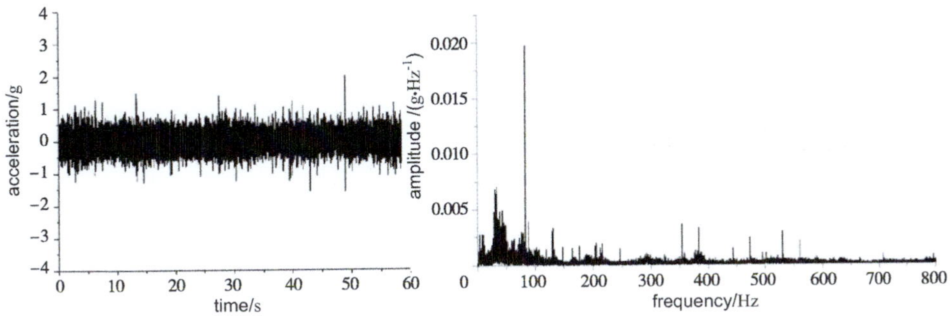

Figure 8.4
Time domain and frequency domain diagrams of the lateral vibration acceleration of the frame.

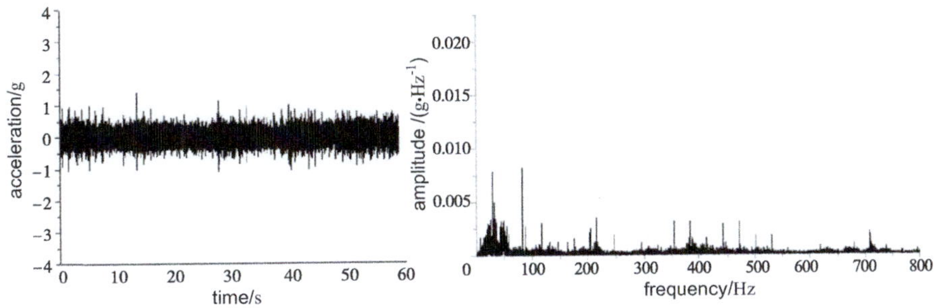

Figure 8.5
Time domain and frequency domain diagrams of the longitudinal vibration acceleration of the frame.

frequency range (350 Hz–620 Hz). The 29.2 Hz frequency is related to speed and consistent with the forced vibration frequency caused by wheelset unbalance. The low frequency-component decreases relative to the axle box, and the high-frequency component is one-sixth the high-frequency component of the axle box. This indicates that the primary system has an apparent damping effect on the high-frequency component, and the weaker damping effect is on a low frequency within 50Hz. The 82.3 Hz frequency component on the lateral plane is larger than that of the axle box because the motor is installed on the frame.

Time domain and frequency domain of vertical, lateral, and longitudinal vibration accelerations of the car body are shown in Fig. 8.6–8.8.

The main components of the dominant frequency of the vertical vibration acceleration of the car body are 1.6Hz, 10–20 Hz, and 215.5 Hz. The main components of the dominant frequency of the lateral vibration acceleration of the frame are 1.7 Hz and 82.3 Hz. The main components of the dominant frequency of the longitudinal vibration acceleration of the frame are 1.5 Hz, 7.8 Hz, and 29.2 Hz.

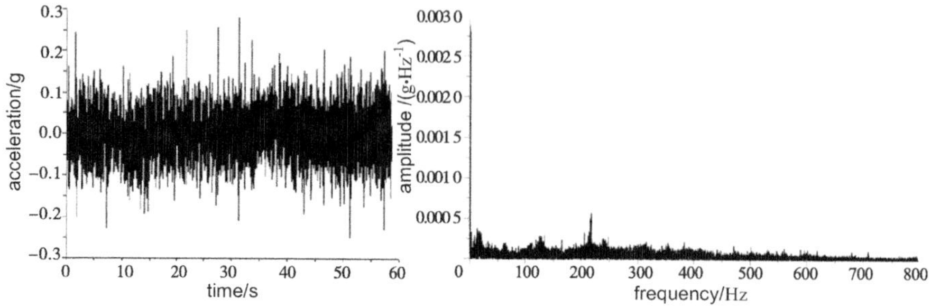

Figure 8.6
Time domain and frequency domain diagrams of the vertical vibration acceleration of the car body.

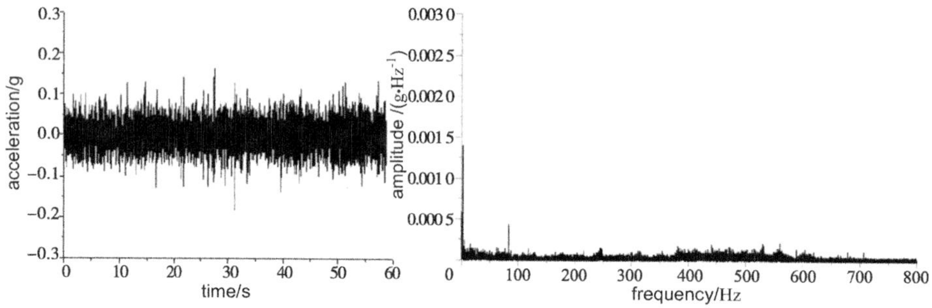

Figure 8.7
Time domain and frequency domain diagrams of the lateral vibration acceleration of the car body.

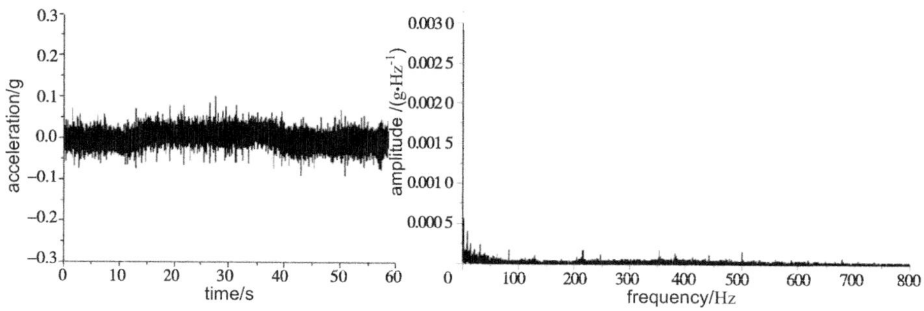

Figure 8.8
Time domain and frequency domain diagrams of the longitudinal vibration acceleration of the car body.

Among the vertical, lateral, and longitudinal vibration accelerations of all the parts of car body, the vibration energy of the frame is primarily concentrated within a low frequency of 2 Hz, which should be the natural frequency of the vertical suspension of the car body and the natural frequency of suspension of sway and head shaking of the car body.

There is a 215.5 Hz frequency component in the vertical direction, but its energy is only one-sixth of the low frequency, and thus has little effect on the car body. There is an 82.3 Hz frequency component in the lateral direction, which should be the vibration frequency of the motor transferred to the car body, and its vibration energy is one-fortieth that in the lateral direction of the frame. The energy of the frequency component in the vertical and longitudinal directions are smaller, and this indicates that the damping effect of the secondary system is apparent.

8.1.2 Change outcomes at different speeds

To analyze the vibrations of the axle body, frame, and car body at different speeds at an intersection, we analyze the changing outcomes of the maximum value and effective value of the measured vibration accelerations and speeds over the full rail line course from Guangzhou to Wuhan. The results are presented in Fig. 8.9–8.14.

From Figures 8.9 to 8.14, the following conclusions can be drawn:

(1) Vibrations of the axle box, frame, and car body increase as speed increases. Vibrations are also closely related to the rail line itself, especially at the entry and exit of a station where vibration impact exists because of turnout, rail gap, etc.

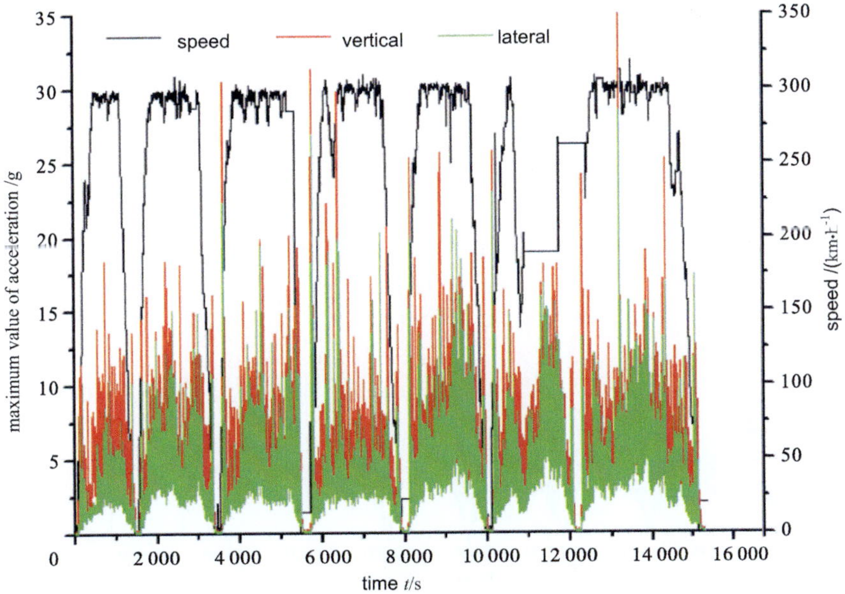

Figure 8.9
Full map of the maximum value of the vertical and lateral vibration acceleration of the axle box along with speed change.

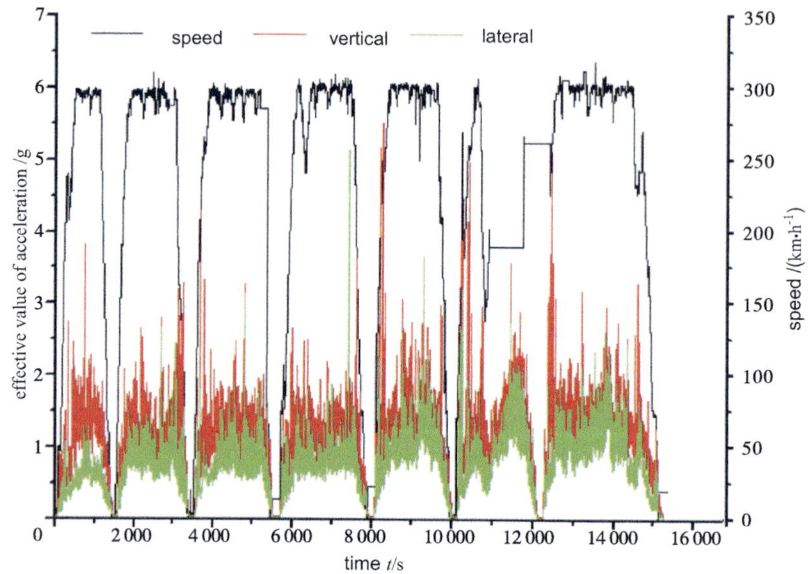

Figure 8.10
Full map of the effective value of the vertical and lateral vibration acceleration of the axle box along with speed change.

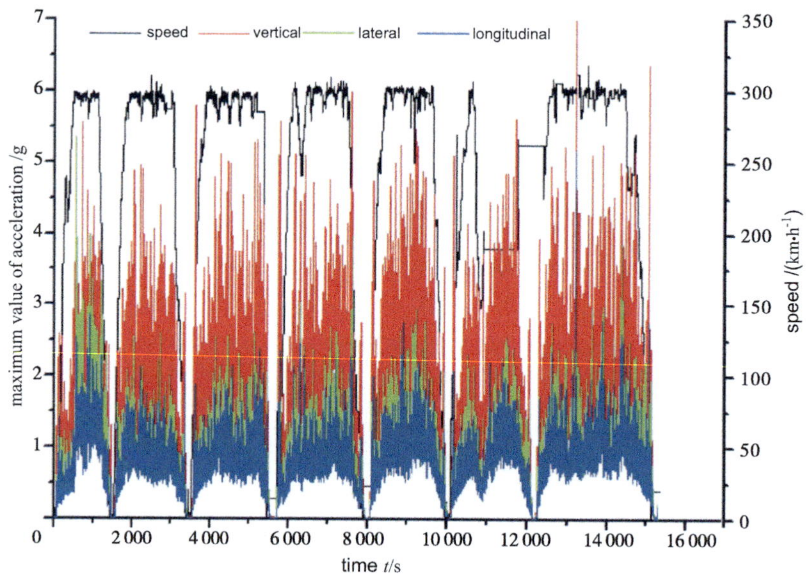

Figure 8.11
Full map of the maximum value of the vertical, lateral, and longitudinal vibration acceleration of the frame along with speed change.

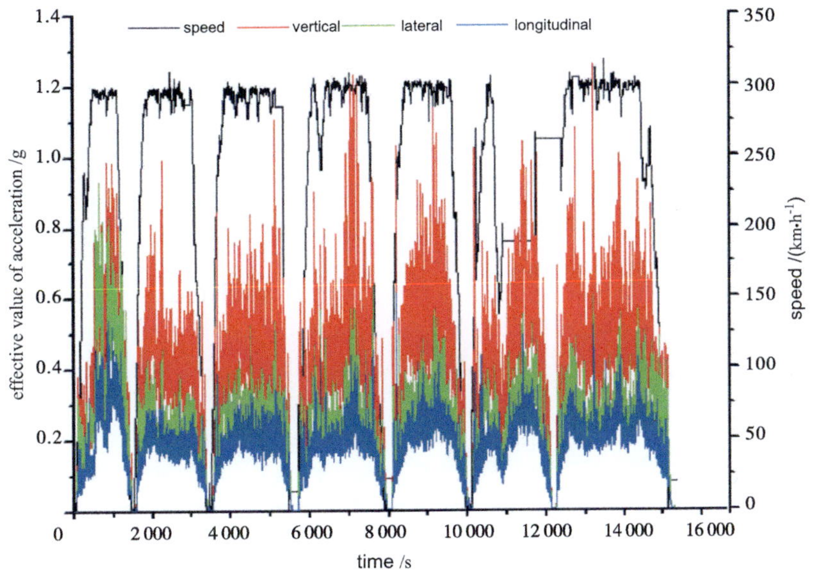

Figure 8.12
Full map of the effective value of the vertical, lateral, and longitudinal vibration acceleration of the frame along with speed change.

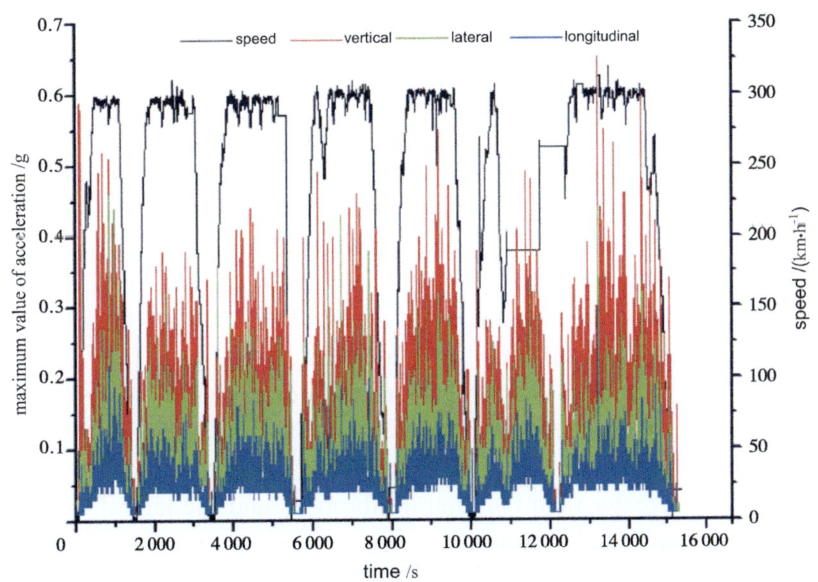

Figure 8.13
Full map of the maximum value of the vertical, lateral, and longitudinal vibration acceleration of the car body along with speed change.

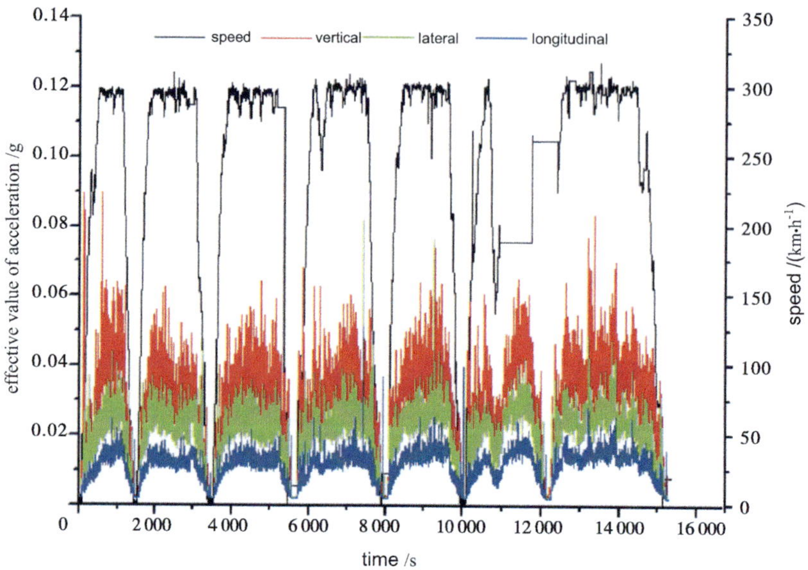

Figure 8.14
Full map of the effective value of the vertical, lateral, and longitudinal vibration acceleration of the car body along with speed change.

(2) At the 300 km/h speed grade, the maximum value of the vertical vibration of the axle box is 20−35 g and the effective value is predominantly 2−3 g; the maximum value of the lateral vibration is 15−25 g and the effective value is predominantly 1.5-2 g; the vertical vibration is greater than the lateral vibration. The maximum value of the vertical vibration of the frame is 5−7 g and the effective value is predominantly 0.8−1 g; the maximum value of the lateral vibration of the frame is 2−3 g, and the effective value is predominantly 0.3−0.4 g; the maximum value of the longitudinal vibration of the frame is 2−3 g, and the effective value is predominantly 0.2−0.3 g; the relationship between the three parts is such that the vertical vibration is greater than the lateral vibration, and the lateral vibration is greater than the longitudinal vibration. The maximum value of the vertical vibration of the car body is 0.3−0.4 g and the effective value is predominantly 0.05−0.06 g; the maximum value of the lateral vibration of the car body is 0.2−0.3 g and the effective value is predominantly 0.03−0.04 g; maximum value of the longitudinal vibration of the car body is 0.1−0.2 g and the effective value is predominantly 0.01−0.02 g; the relationship between the three parts is such that the vertical vibration is greater than the lateral vibration, and the lateral vibration is greater than the longitudinal vibration.

(3) Vibrations of the axle box, frame, and car body gradually decrease, with the vibration transmitted from the axle box to the frame decreasing to about one-fifth its original value and the vibration from the frame to the car body decreasing by one-twentieth.

This indicates that the vibration isolation effect of the primary and secondary systems is apparent and that the vibration isolation effect of the air spring of the secondary system is more effective than that of the rigid spring of the primary.

8.1.3 Evolution laws under different mileage

In order to analyze the influence of running mileage and lathing on the axle box, the frame, and car body, we analyzed five complete cycles of the wheelset of CRH EMUs of the Wuhan-Guangzhou Line from the installation of a new wheel up to the completion of 1.12 million km. Wheel lathing occurred three times during this period at 408,000 km, 746,000 km, and 974,000 km. The maximum values and effective values of the three-direction vibration acceleration of the axle box, frame, and car body were analyzed every 50,000 km for one section at a constant speed of 300 km/h on the Beijing-Shanghai line. The results for the vibration maximum value under different mileages are presented in Table 8.2, and the results for the vibration effective value are presented in Table 8.3.

By combining mileage with lathing, the change curves of the maximum value and effective value of the vibration acceleration of the axle box, frame, and car body, varying with mileage, are plotted and shown in Fig. 8.15–8.20.

From Table 8.2, Table 8.3, and Fig. 8.15–8.20, the following conclusions can be drawn:

(1) Within 1.2 million kilometers, mileage has an insignificant impact on the vibration of the axle box, frame, and car body because of the train's mechanical stabilization. As the wear increases, the vibration of every component increases gradually.

Table. 8.2: Vibration maximum values for the axle box, frame, and car body at different mileages.

Mileage (10,000 km)	Axle box		Frame			Car body		
	vertical	lateral	vertical	lateral	longitudinal	vertical	lateral	longitudinal
39.9	10.61	7.53	3.68	1.57	1.25	0.15	0.08	0.06
45.1	9.23	5.11	3.22	1.34	1.12	0.11	0.05	0.05
49.1	9.25	5.14	3.21	1.2	1.02	0.12	0.05	0.05
58.9	9.26	5.16	3.23	1.21	1.04	0.12	0.06	0.05
63.2	9.38	5.21	3.26	1.23	1.05	0.12	0.07	0.05
67.4	9.45	5.45	3.31	1.31	1.12	0.12	0.08	0.06
71	9.88	6.67	3.56	1.49	1.24	0.13	0.08	0.06
81.1	9.24	5.66	3.25	1.24	0.97	0.1	0.05	0.05
90.9	9.45	5.86	3.14	1.55	1.22	0.12	0.06	0.05
96.6	9.94	6.12	3.35	1.54	1.26	0.14	0.08	0.06
101.3	8.32	5.48	3.03	1.36	1.02	0.13	0.06	0.05
107.7	8.41	5.84	3.41	1.14	1.04	0.12	0.08	0.06
112	9.31	5.92	3.51	1.47	1.35	0.14	0.08	0.08

Table 8.3: Vibration effective values for the axle box, frame, and car body at different mileages.

Mileage (10,000 km)	Axle box		Frame			Car body		
	vertical	lateral	vertical	lateral	longitudinal	vertical	lateral	longitudinal
39.9	1.528	1.243	0.638	0.301	0.223	0.037	0.022	0.016
45.1	1.32	1.053	0.598	0.272	0.197	0.031	0.016	0.014
49.1	1.379	1.104	0.603	0.278	0.202	0.033	0.017	0.015
58.9	1.436	1.125	0.608	0.283	0.206	0.034	0.018	0.016
63.1	1.458	1.169	0.609	0.285	0.212	0.035	0.019	0.017
67.4	1.496	1.219	0.612	0.284	0.205	0.035	0.02	0.018
71	1.563	1.321	0.628	0.293	0.191	0.037	0.021	0.019
81.1	1.498	1.13	0.618	0.267	0.188	0.03	0.018	0.015
90.9	1.578	1.182	0.648	0.292	0.211	0.031	0.019	0.016
96.6	1.567	1.398	0.676	0.304	0.222	0.034	0.02	0.017
101.3	1.451	1.039	0.616	0.265	0.201	0.031	0.016	0.013
107.7	1.488	1.054	0.625	0.262	0.201	0.032	0.017	0.014
112	1.563	1.076	0.667	0.279	0.216	0.034	0.018	0.016

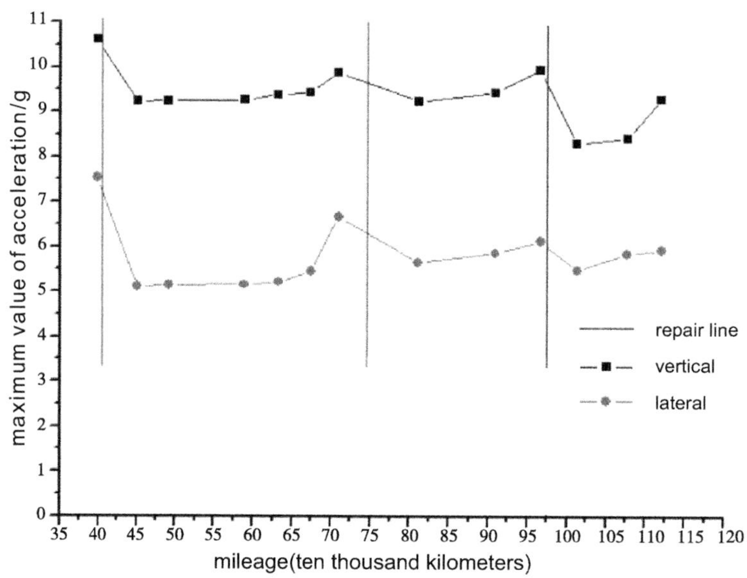

Figure 8.15
Maximum values of the vertical and lateral vibrations of the axle box in relation to mileage.

(2) In one lathing cycle, with an increase in wheelset wear, the vibration of the axle box, frame, and car body increased. However, the vibration of these parts decreased after lathing, which indicates that wheelset wear has a significant influence on the vibration characteristics of the train.

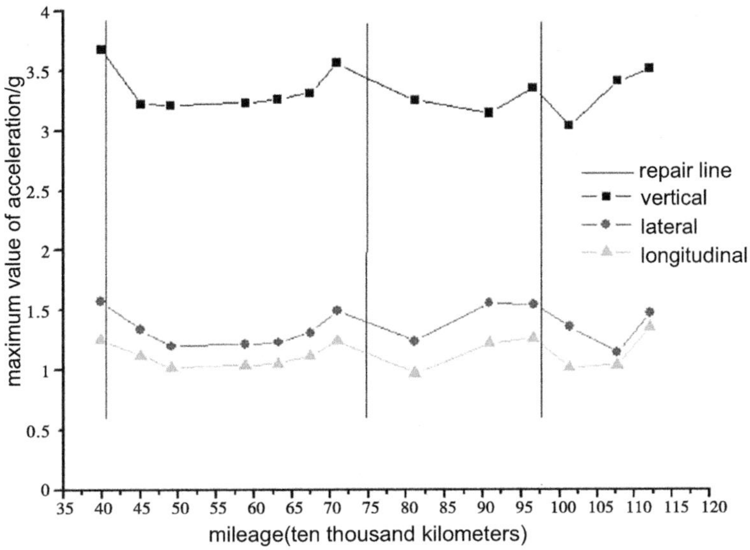

Figure 8.16

Maximum values of the vertical, lateral, and longitudinal vibrations of the frame in relation to mileage.

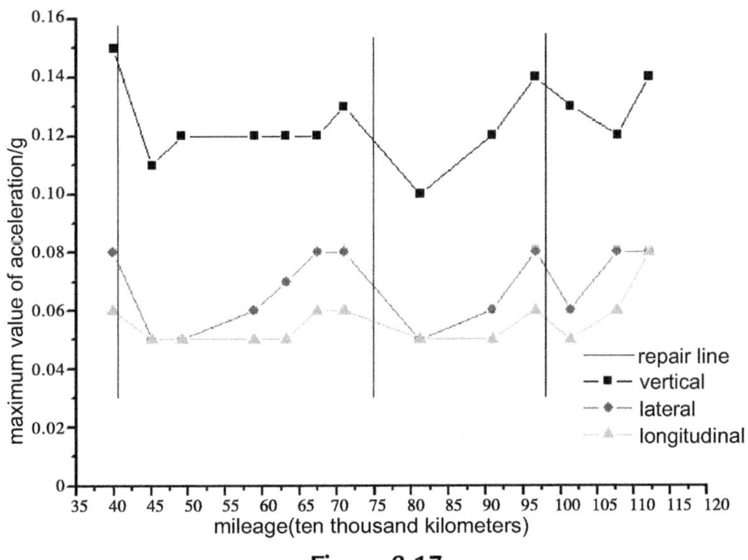

Figure 8.17

Maximum values of the vertical, lateral, and longitudinal vibration of the car body in relation to mileage.

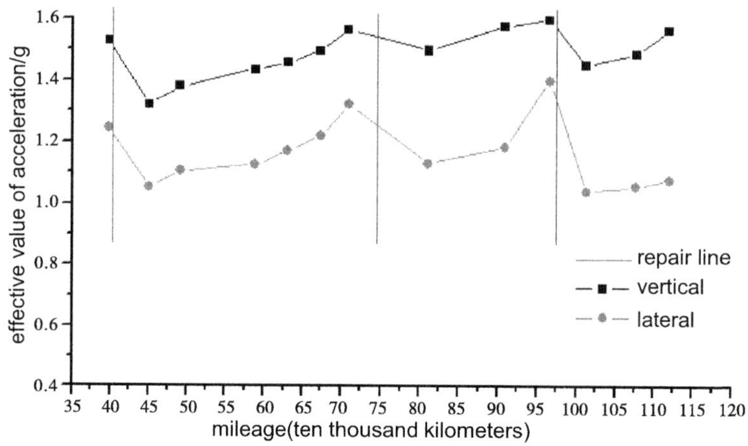

Figure 8.18
Effective values of the vertical and lateral vibration of the axle box in relation to mileage.

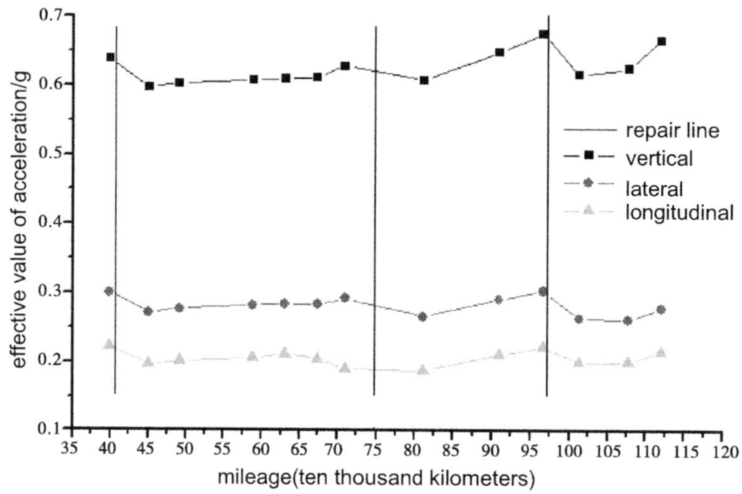

Figure 8.19
Effective values of the vertical, lateral, and longitudinal vibrations of the frame in relation to mileage.

(3) Considering the results of different lathing periods, wheelset lathing has an influence on the axle box, frame, and car body: after lathing, there is a decrease in the vibrations of these three components. Therefore, we can infer that the wheelset lathing has a significant influence on the vibration of the axle box, frame, and car body.

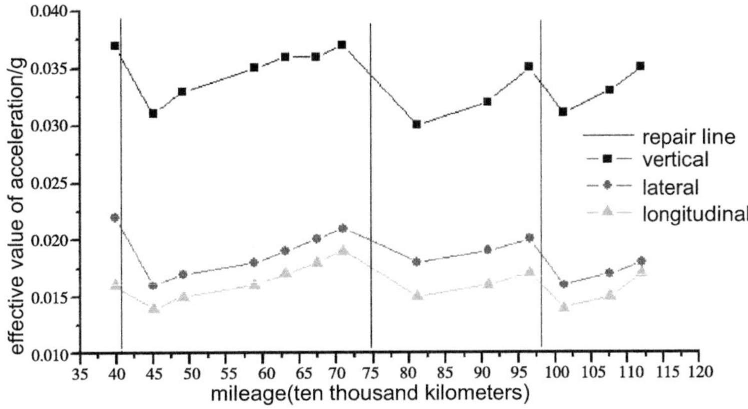

Figure 8.20

Effective value of the vertical, lateral, and longitudinal vibrations of the car body in relation to mileage.

8.2 Safety monitoring technology of the running gear of the high-speed train

8.2.1 Framework of the safety monitoring platform

Dynamic performance is the guarantee of the safety and ride quality of the high-speed train. Therefore, it is essential to conduct condition monitoring on the running gear of the train. However, condition monitoring of the high-speed train is difficult due to the complexity of its dynamic performance, the coupled response of the parts, the impact of the high-speed operation, violent vibration, and complex environment.

8.2.1.1 Main functions of the safety monitoring platform

The safety monitoring platform for the running gear performs the following functions:

(1) Comprehensive access to the state parameters and system response of key components of the running gear: Acquire the dynamic response of the system in real time through onboard monitoring technology; Build up a database on the condition of the running gear through a ground depot measurement system; Concurrently, systematically master the entire service process for the running gear using the digital maintenance archive.
(2) Analysis of the evolution rule of service performance of the high-speed train running gear: Develop the evolution rule of service performance of the high-speed train based on data from the long-term tracking and monitoring of the key components of the running gear, combined with the system dynamics of the high-speed train.
(3) State identification and estimation of the vehicle and track.: Identify and estimate the performance and condition of the running gear and track using the train-track coupled system dynamics model and the onboard state response system.

(4) Safety assessment and prediction of the running gear: Assess the safety and structure reliability of the running gear of the high-speed train based on real-time conditions of the vehicle and track, combined with the theory and simulation of coupled system dynamics; predict the development trend of the running gear safety by referring to historical information and the evolution rule of service performance.
(5) Safety control of the running gear: Achieve on-road security early warnings for the high-speed train based on safety assessment and prediction results. Provide assistive decision-making suggestions, including braking, stopping, and maintenance to further guarantee the safety of the high-speed train, the attainment of preferential highly-efficient maintenance, and reliable operation.

8.2.1.2 Framework of the safety monitoring platform

The safety monitoring platform of the high-speed train running gear is aimed at establishing an integrated monitoring and control system, covering omnidirectional detection, comprehensive analysis, early warning, prognostic and assistive decision-making for the running gear. A diagrammatic representation of the framework is presented in Fig. 8.21, including three major parts: condition monitoring, diagnosis and assessment, and safety control.

(1) Condition monitoring

Comprehensive data acquisition on the operating condition and dynamic response of the running gear is essential for safety monitoring of the high-speed train. Onboard monitoring (shown in Fig. 8.22) is a basic approach towards this objective. The monitoring scheme for the running gear should be designed based on comprehensively mastering the structure compositions, main service parameters, component interaction, running properties, operational conditions, etc. The objective of the monitoring is to acquire omnidirectional performance data on the running gear. The principal goal is to acquire as much data as possible, utilizing as few sensors as possible. Furthermore, it is aimed at ensuring easy installation, high efficiency, and good sensitivity when selecting key parameters and system responses closely related to safety during the service of the running gear. Ultimately, establishing a comprehensive, scientific, and optimally structured detection network for acquiring the status condition of the running gear.

(2) Diagnosis and assessment

The running gear of the high-speed train is a time-variant coupled system involving all its parts, and its response is apparently non-stationary. When a part is damaged, the response will become nonlinear and non-gaussian. Due to the violent vibration, complex environment, and long-term time-variant service of the high-speed train in the time and space domains, the running gear is more than a typical *three-non* (non-stationary,

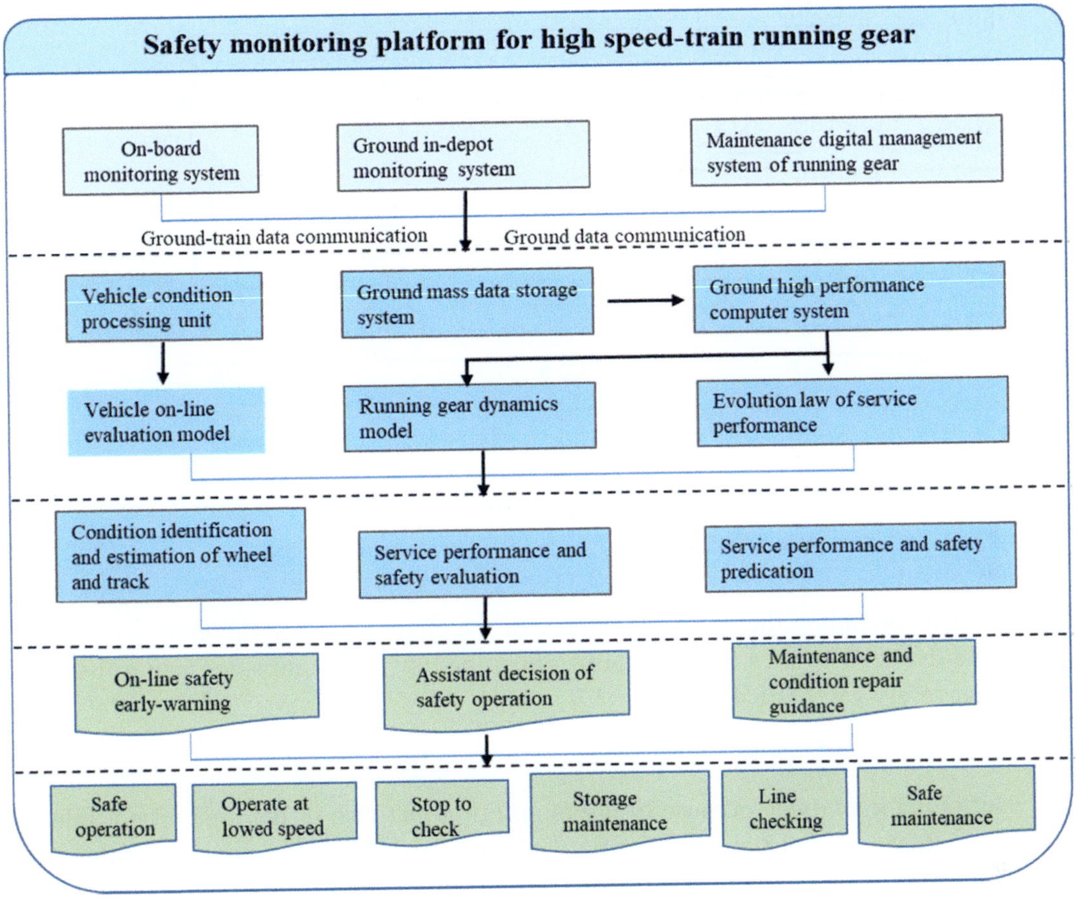

Figure 8.21

Framework of the safety monitoring platform for the high-speed train running gear.

nonlinear, and non-gaussian) system, resulting in the dynamic safety monitoring data being disturbed or submerged by strong noises. Furthermore, the diagnosis and assessment of the condition of the running gear becomes more difficult for the processing and analysis of mass data. First, modern signal processing methods should be conducted to denoise, purify, extend, decompose, and reconstruct various condition signals. Combining the dynamics theories and the simulations of the high-speed train coupled systems with the bench test and tracking test of the running gear, the variation of state parameters along with the service time, can be investigated. The evolution rule of service performance of the running gear can be analyzed for every component and for the entire system. Then, a highly efficient identification, diagnosis, and assessment model of the condition and performance evolution trend of the vehicle and track is built up. Further, a comprehensive safety assessment model of the running gear is structured according to the failure effect model and contributions from every component to the

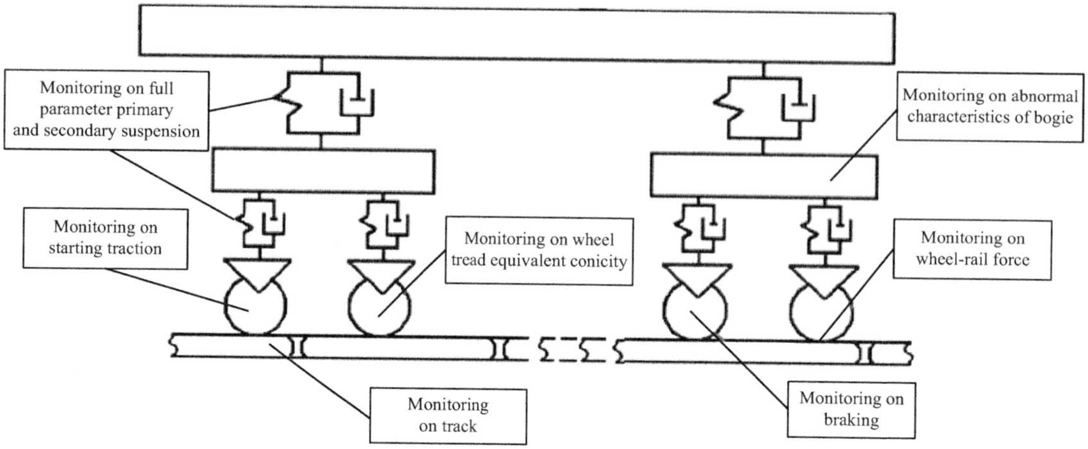

Figure 8.22
Monitoring diagram of high-speed train running gear.

system security, including the deviation degree of the characteristics, the gradient of damage, and the development trend of fault. Thus, the diagnosis and assessment of the running gear safety can be realized.

(3) Safety control

Safety control is the ultimate purpose of the monitoring platform of the high-speed train running gear. Based on real-time monitoring, diagnosis, and assessment of the running safety of the running gear, online assistive decision-making and warning information are provided for safety operations, such as deceleration, stop, on-site inspection, and depot maintenance. The performance and evolution trend of the running gear system, together with all its components, can be controlled within the safety domain of the coupled system dynamics of the high-speed train. Thus, the running gear is guaranteed to run in a condition of no faults or minor faults, while the safety of the vehicle is monitored systematically in real-time.

8.2.2 Onboard safety monitoring and detection technology

8.2.2.1 Framework of the onboard safety monitoring system

The parameters of the high-speed train running gear will be slowly varied. These parameters include spring stiffness and wheel tread, along with increase in service time. Concurrently, there will also be occurrences of shape and position error of some components, such as wheel out-of-roundness and flange corrosion. Furthermore, some transient damages, such as wheel flat, steel rail fracture, and air spring leakage, may occur under special excitation or special physicochemical environments. The slow varying process or trend of the running gear parameters cannot be sensed in real time or be predicted by means of regular inspection, decomposition, and maintenance. Thus, it is nearly impossible to assess the safety performance of the running gear online. The onboard safety monitoring system (shown in Fig. 8.23) is capable of online detection,

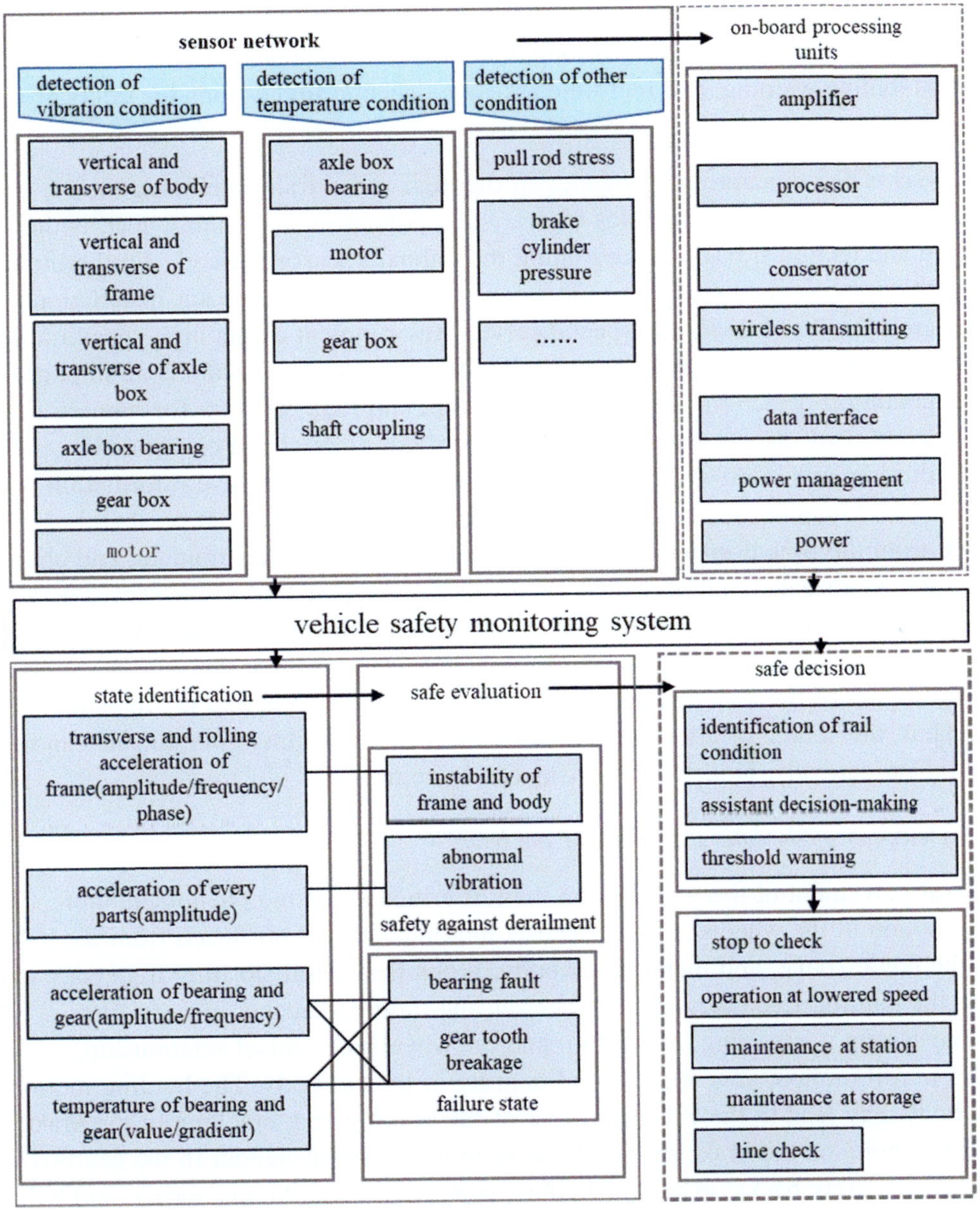

Figure 8.23

Framework of onboard safety monitoring system of high-speed train running gear.

tracking, identification, diagnosis, and prediction of the condition or performance of the running gear in real time, and is also capable of highlighting the state difference of the same train and the same position by means of comparison in both time and space domains. These techniques can detect potential risk and early fault timely, and also track the slow varying process of the running gear parameters to prevent accidents.

The onboard safety monitoring system is primarily composed of the onboard sensor network and state processing units. Through onboard condition detection, the on-road assessment, online warning, and real-time assistive decision suggestions for safety are realized.

The analysis of the dynamic performance of the high-speed train, including the mechanics and motion characteristics of every component in the running gear, is the theoretical and technical basis for designing the onboard sensor network. Furthermore, the adaptation requirements for dynamic safety detection and the sensor installation methods are considered under the special service environment of the high-speed train. The reverse deduction of dynamic performance from the dynamic state is an important basis for condition detection, damage identification, and fault location for running gear safety. The onboard state processing unit will carry out hierarchical classification, storage, filtering, and multi-sensor information integration for massive information with several physical quantities. The effective features of the state information shall be extracted from the time domain, frequency domain, time-frequency domain, and chaos domain, and abnormal signals detected via the system response from the sensor network. The identification of the safety state, damage, and failure location is the reverse deduction process of the dynamic performance of the running gear, i.e., the occurrence and transfer path of the abnormal state are analyzed with reference to the dynamics model. Thus, the source of abnormal vibration or heat, and failing components can be detected. Consequently, hidden safety risks can be removed.

8.2.2.2 Detection of the lateral stability of the bogie

The lateral movement of the wheelset on the rail brings about the phenomenon of hunting motion in the wheelset. The wheelset hunting can be transferred from the axle box to the bogie frame, and then the car body. When the lateral sloshing frequency is equal to the hunting frequency of the wheelset, the train becomes a self-exciting resonance system comprising of the train and the wheel-rail contact relationship, resulting in roll motion, yaw motion, and hunting of the car body. The hunting motion of the car body may lead to the eccentric wear of the braking pad, cracking of the brake disc, loss of bolts on the brake disc, bolts loosening off the suspender in the gearbox, rupture of the safety backstop of the gearbox, rupture of the axle box spring, and bulge of the rubber bushing of the anti-roll bar, etc.

When the total damping in the lateral direction and the speed reach a certain matching, zero-attenuation vibration (sustained oscillation) may occur. This kind of oscillation can be attenuated by change in speed conditions and may become more violent with increase in speed. Therefore, in the high-speed train industry, the sustained vibration is monitored to detect the lateral stability of the bogie [1-3]. The wheelset and the bogie frame have a similar form of lateral motions, a quasi-sustained oscillation [3] shown in Fig. 8.24. When hunting of the bogie frame occurs, the self-excited vibration is the main vibration mode in the lateral direction. Therefore, it can be detected by monitoring the lateral acceleration of the bogie frame [4]. Detection of the hunting motion of the bogie can be conducted based on TB/T3188-2007. For different train types, the installation position of the sensor is approximately consistent with the similarity in the bogie structures (shown in Fig. 8.25).

Hunting of the bogie is primarily caused by the conicity of the wheel tread and other factors, including the positioning stiffness of the rocker arm joint and characteristics of the anti-yaw damper. With increased operational mileage under various working conditions, the wear of these elements is intensified, and the performance declines. Therefore,

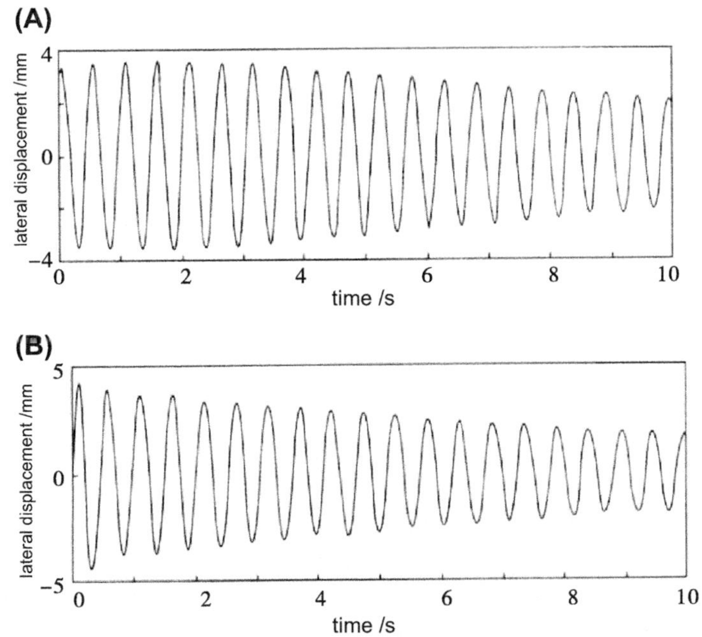

Figure 8.24
Diagrams of hunting motion of the wheelset and bogie frame. (A) Lateral instability of the wheelset (B) Lateral instability of the frame.

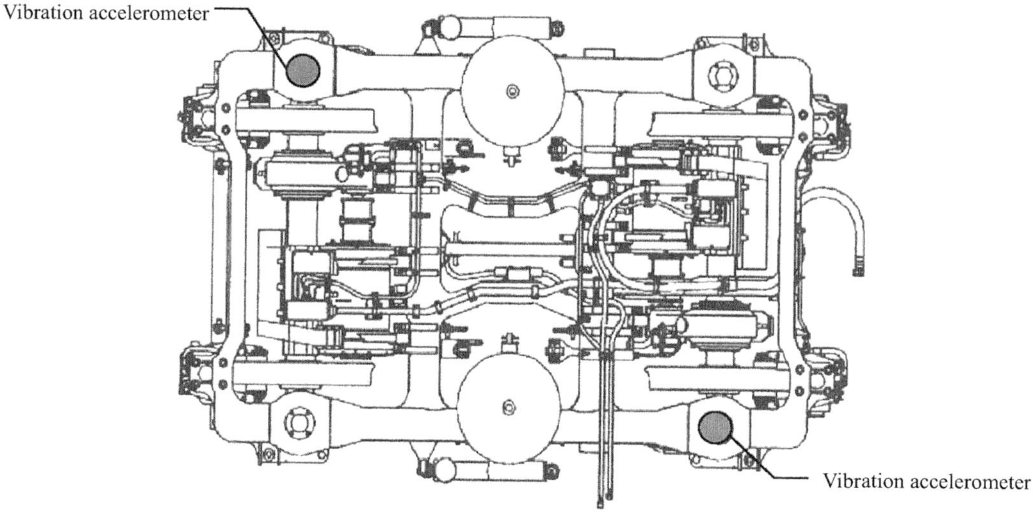

Figure 8.25
Layout of installation position for detecting hunting motion of the bogie.

condition-based maintenance of the bogie can be realized through lateral stability monitoring of the bogie.

8.2.2.3 Detection of abnormal vibration of the running gear

Abnormal vibration of the running gear refers to vibration that has a significant influence on running safety and requires emergency shutdown and troubleshooting [5]. The monitoring of abnormal vibration is primarily targeted at the axle box, bogie frame, gearbox, motor, car body, and other key parts (Fig. 8.26).

With the significant impact of coupled forces on the dimensional properties of the high-speed train and random factors within the complex running gear, it is difficult to detect specific sources causing abnormal vibration, vehicle faults, or excitation from track irregularities. Applying a uniform threshold is not the best scientific approach. Instead, a comprehensive diagnosis, that includes relative as well as absolute judgment, should be applied. The principle of absolute judgment should be adopted for measuring the dynamic performance of the running gear with an existing absolute boundary value. For the performance, which is hard to determine, the boundary of a statistical feature value and the principle of relative judgment, are used to classify the performance dynamically. The principle of relative judgment is based on the distribution characteristics of the particular dynamic performance of the same bogie type under the same railway conditions.

8.2.2.4 State detection of rotating parts

Rotating parts are vulnerable and have tremendous effects on the running safety of the high-speed train. Examples are the axle box bearing, motor bearing, and the gear. Due to the

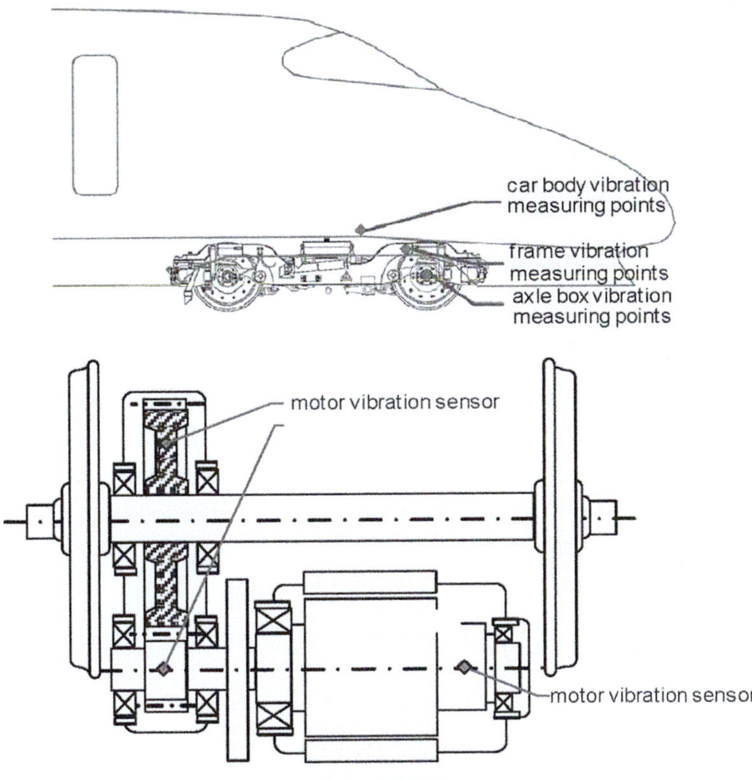

Figure 8.26
Layout diagram of the monitoring point for abnormal vibration of running gear.

influence of the structure design, installation technology, application condition, and discontinuous load, various defects easily occur in the rotating parts after a period of running. Examples are cracks, pitting corrosion, abrasions, pocks, stripping, and wear. The extension of these defects will deteriorate performance under special service environments (high speed, vibration, coupled effects, and long-term time variation. Furthermore, these defects may cause a hot axle, heated box, cut axle, or/and breaking of gear tooth, within a short time. These may result in serious accidents, such as overturning of the train. Presently, periodical maintenance is conducted to control the safety status of rotating parts, but this method may lead to over maintenance A more undesirable outcome is that it may result in insufficient maintenance, such that there may be hidden safety hazards that cannot be detected or eliminated in time. Therefore, onboard real-time monitoring of the rotating parts needs to be conducted for early detection of faults using the evolution rule of the fault rate, to remove hidden safety hazards in the high-speed train running gear.

The monitoring technology for the rotating parts of the running gear is based primarily on the temperature rise mechanism caused by the strong frictional heat accompanying failure.

This mechanism is effective for detecting moderate to severe faults or damage. However, it is not sensitive to tiny faults in the early period and is easily influenced by the specific heat of the lubricating oil, which may trigger a false positive or false negative. The damage and fault of rotating parts can intensify vibration, and change in vibration characteristics is synchronous to the fault process, from occurrence to development. Furthermore, the vibration signal has good sensitivity to early tiny faults. Therefore, omnidirectional monitoring of the state of the rotating parts can be realized by introducing the measurement of vibration signals to the existing temperature monitoring systems for rotating parts.

An abnormal running state of the rotating parts generates additional impact vibration, which is random vibration consisting of multiple frequency components due to the irregular shape of the damage or fault. The natural frequency and damping of the system structure determines the excited high-frequency response. Comprehensive detection for rotating parts, including the fault mode and degree, can be attained by extracting additional impact vibration signals from the system response, combined with the vibration characteristics and the dynamics model of the rotating parts.

8.2.2.5 State detection of suspension parts

Failure of a primary suspension element, such as spring fracture and damper failure, will cause the transfer of vertical high-frequency vibration to the bogie frame, resulting in violent vibration and fatigue damage at the end of the side beam of the bogie frame, and consequently, running safety is influenced [5]. A fault in secondary suspension element, such as air spring leakage, spring fracture, and damper failure, cause aggravation of the vibration on the car body, resulting in poor riding comfort. Furthermore, the rate of wheel load reduction increases when passing through a transition curve, further impacting running safety [5].

State detection of suspension parts is realized by analyzing the relations between the damage characteristics and the vibration of the running gear or car body based on the dynamic model, and then designing a state identification algorithm from the vibration signals of other parts. The monitoring vibration of the axle box and bogie frame can be used to detect the state of the primary suspension element, and that of the bogie frame or car body can be used for the secondary suspension element. Generally, parameter estimation of the suspension parts is carried out based on the dynamics model of the running gear system by applying a Kalman filter or particle filter [6-8], while the adaptive dynamics model is established based on the normal running condition of the vehicle. When one suspension component breaks down, the vibration characteristics of the vehicle are inconsistent with normal conditions. Linear characteristics are assumed to approximately describe the suspension elements and the vertical vibration is taken as an example: Under normal conditions, the pitch motion and vertical motion of the car body are independent of each other. However, this symmetry of vibration is damaged when one vertical suspension part breaks down, and the pitch motion and vertical motion of the car body are no longer independent, resulting in coupled vibration [9]. Therefore, dynamics modeling of the running gear system under fault conditions must be incorporated into the state detection algorithm for the suspension parts, rather than limited only to normal conditions.

8.2.3 Ground safety monitoring and detection technology

8.2.3.1 Framework of the Ground safety monitoring system

Common internal faults in the running gear of the high-speed train occur primarily at the wheelset (flange wearing, tread circle wearing, tread stripping, wheel crack), rolling bearing (fatigue damage, cage crack), frame (structural fracture), axle (axle crack), and the pull rod (pin fracture). Being limited by a complex and severe detection environment, and the onboard information processing capability of the running gear, the onboard safety monitoring system cannot cover all detection elements of the running gear. The performance state of the running gear can be more comprehensively mastered by the implementation of online and offline three-dimensional monitoring of the running gear, by structuring a ground safety monitoring system (Fig. 8.27), carrying out ground warehousing static detection, building a digitized management system for state information, and combining the maintenance file records of the running gear.

Fundamental aspects of the ground safety monitoring system for the running gear includes mainly static state detection and state information digital management. The ground static state detection considers mainly warehousing conditions, wheel lathing conditions, and other maintenance conditions. The main detection objects are the wheelset, axle, and other parts, the detection of which is hard to perform via onboard dynamic detection. The state information digital management aims at implementing systematic, integrated management of dynamic and static state data, the model base, experience base, records of maintenance, equipment, and personnel for the running gear. It is also the background processing platform for massive state information, performing time-space chain analysis by integrating historical data and real-time data. Furthermore, it performs further feature extraction on state information utilizing the major system theory of the high-speed train. It identifies track, environment, and vehicle states; explores the evolution rule of service performance for every component and system; and performs forecasting for the running stationarity, stability, and safety of the vehicle running gear, so that safety diagnosis and assessment of the running gear are implemented, and the assistive safety decision suggestions and warning information are presented.

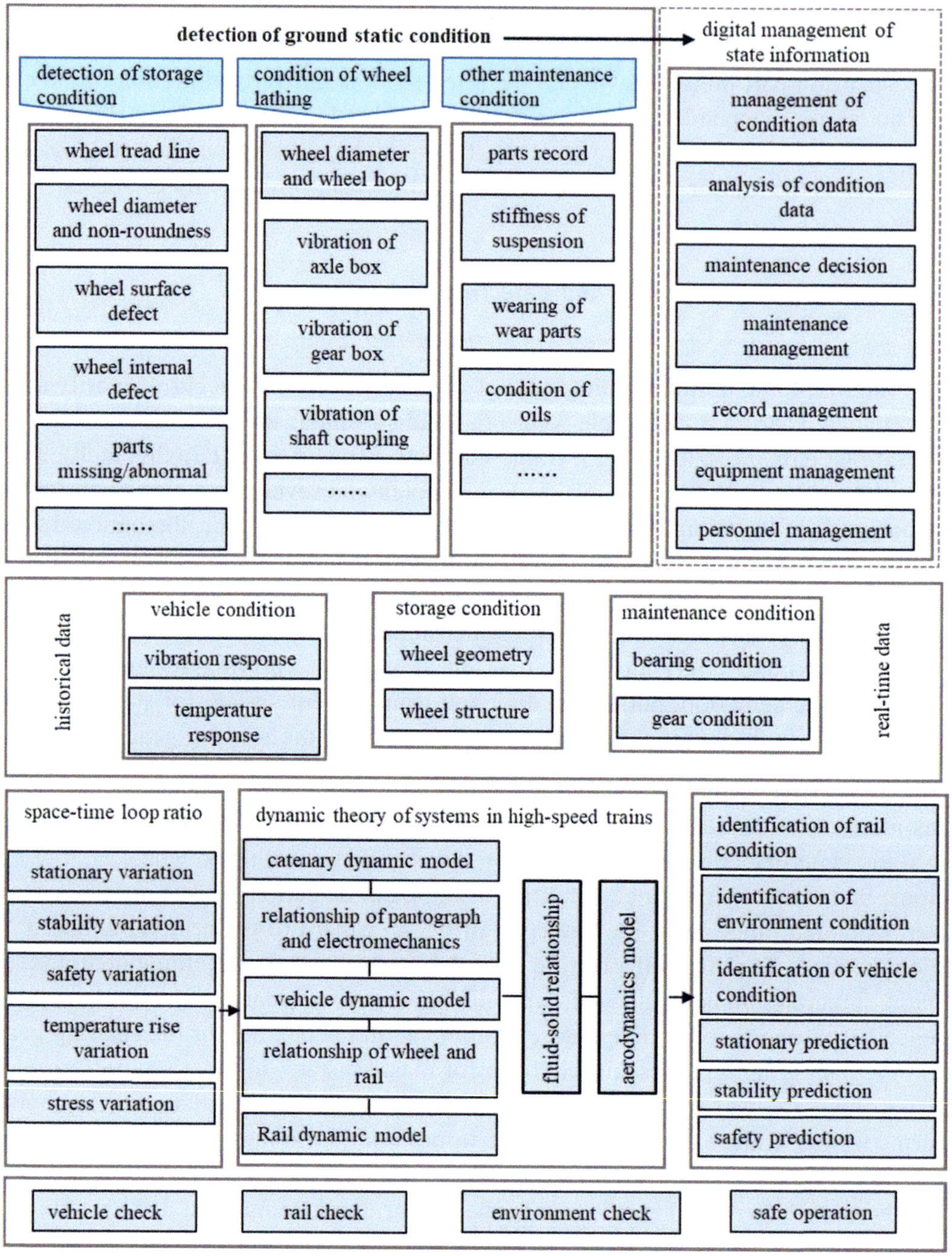

Figure 8.27
Framework of ground safety monitoring system for the running gear of the high-speed train.

8.2.3.2 Flaw detection technology for the wheel axle

The rail vehicle axle is a component that is vital to driving safety, and its quality requirements are very strict, both at home and abroad. A nondestructive detection using ultrasound waves or rays is conducted on the wheelset immediately after production. For rail vehicles that are operational, quality detection for the wheel axle is generally carried out during medium-term repairs or an overhaul. The wheels have to be dismounted and moved to a quality detection device during the maintenance. However, local detection can be carried out only roughly and inefficiently at ordinary times [10]. Wheel axle damage generally happens before the expiration of a class of repairs. Consequently, flaw detection via the warehousing detection method is generated gradually. German, French, and other countries with advanced railways have designed and developed advanced ultrasound wheel flaw detection devices. China has also independently developed a mobile rim spoke flaw detection system. The mobile rim spoke flaw detection system combines phase controlled ultrasonic flaw detection and regular flaw detection technology and achieves online flaw detection of the rim spoke and detection of fatigue defect and material defect [11]. The through-type wheelset flaw detection device researched in China (Fig. 8.28) applies a combined technical method of big-angle transverse wave probe, longitudinal wave double crystal confocal probe, and monocrystal (small angle) longitudinal wave probe, and is able to automatically and effectively detect surface cracks, rim depth cracks, and web plate cracks. The detection accuracy is as follows:

(1) Near-surface stripping and crack of tread (within 10 mm beneath tread): equivalent crack of radial 30 mm (length) x 3mm (depth).
(2) rim inner crack (within 90 mm beneath tread): equivalent defect of Φ3 horizontal hole at 30 mm beneath tread.

Figure 8.28
Through-type wheelset flaw detection device.

8.2.3.3 Detection technology for overall dimension of wheelset

Wheelset dimension measurement is directly linked to vehicle maintenance. The geometrical state of the wheel (especially the change in the equivalent taper caused by wear of the tread, the occurrence of wheel-to-left, and right wheel diameter difference caused by wear) contributes directly to degradation of the stability of the vehicle, while abrasion and wear of the wheel causes out-of-roundness, which directly affects ride comfort. Furthermore, shape damage to the wheel also becomes a source of additional excitation for the running vehicle, which makes the wheel track system generate strong impact vibrations, degrading the running environment and accelerating the wear of the components.

The warehousing detection device for the overall dimensions of the wheelset has been widely applied to railway vehicle repair, home and abroad (Fig. 8.29). It achieves parameter detection of the basic overall dimension of the wheelset, including flange height, flange thickness, rim thickness, wheelset inner-side distance, wheel diameter, flange value (flange slope), flange angle. However, the detection device is unable to detect strain capacity of vehicle speed, actual wheel diameter, deformation of rail, and others. Furthermore, it is unable to precisely position the wheel center. The detection precision is low, mostly for applying the first-generation detection technology based on pure photography. The second -generation detection technology is based on single laser line and single photography methods. The third-generation wheelset overall dimension detection technology is based on laser detection technology and the multi-time photography method. It can precisely ascertain the wheel center, and its detection precision is improved to 0.25 mm. However, information from massive warehousing detection cannot be used effectively because the warehousing detection technology fails to combine with the

Figure 8.29
Wheelset overall dimension detection device.

real-time operating state of the vehicle. Furthermore, the technology is not free from function indicators and technical judgment indicators of wheelset overall dimension change in response to vehicle equivalent conicity, critical velocity, vertical vibration characteristics, and others. Therefore, the wheelset overall dimension detection technology is required to effectively link with the onboard monitoring system to further improve the detection precision, and thus provide a direct and effective basis for vehicle safety operation and maintenance.

8.2.3.4 Wheel tread detection technology

Long-term operational service conditions always include abnormal actions such as emergency braking, or external environmental forces like extreme weather causing wheel lock, slip, or impact vibration. Surface abrading, surface hardening, exfoliated stripping, or hot cracks may result in the reduction of the service life of the vehicle and line. Such hidden vulnerabilities could become safety risks. To prevent accidents, it is necessary to detect wheel or tread damage early and oversee wheel grinding or lathing.

The detection of wheel tread, home and abroad, consists of static detection and dynamic detection. The process of static detection is complex because it requires disassembling the vehicle, and is mainly performed during maintenance processes. Dynamic detection has an onboard and ground type. The onboard type of wheel tread detection technology is not widely applied and is also challenging because of high application cost, limited detection precision, and other problems. A domestically developed ground type wheel tread detection device (Fig. 8.30), complete with electromagnetic ultrasonic flaw system, has a high degree of automation, and the wheel tread detection can be completed during the vehicle running process by adopting the passing type detection method. However, the wheel brings heavy impact and vibration to the detection system during the vehicle

Figure 8.30
Wheelset tread detection device.

running process, such that the load bearing is easily and instantly isolated from the flange, measuring datum is lost, and missed measurements or measurement errors occur. Therefore, the detection system must effectively reduce the wheel impact and vibration, and the measurement accuracy can be guaranteed only by eliminating detection error from the lateral sway, as shown in Fig 8.29. The detection precision of the detection device for wheel tread is 0.2 mm.

8.3 Prediction and threshold of service performance of the high-speed train

8.3.1 Influence and threshold of wheel tread wear on dynamic performance

Wheel wear of the high-speed train is unavoidable during the running process (Fig 8.31). It may cause change in the tread shape and lead to change in equivalent conicity and wheel diameter difference. Consequently, the dynamic performance of EMUs is influenced [12,13].

8.3.1.1 Influence of equivalent conicity

To investigate the influence of equivalent conicity on vehicle stability and to roughly consider the shape of the wheels after wear, different equivalent conicities can be acquired

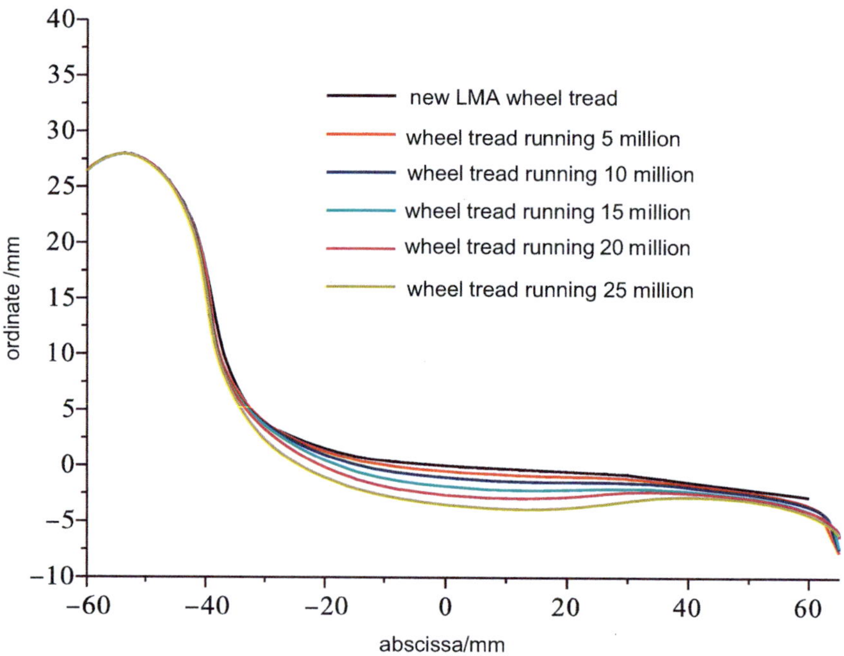

Figure 8.31
4-times tracking measurement results of tread wear.

Table 8.4: Influence of tread equivalent conicity on critical velocity.

Equivalent conicity	0.05	0.10	0.15	0.20	0.25	0.30	0.35	0.40	0.45	0.50
Critical velocity Vcr (km/h)	> 550	545	513	491	478	456	407	351	283	186

through fine-tuning of the wheel gauge based on actual measurements of tread, and then brought into the train system to analyze the critical velocity. The calculated results are presented in Table 8.4. Based on these results, the critical velocity of the train system decreases gradually with increase in the tread equivalent conicity.

Considering a 30% safety allowance for critical velocity, the tread equivalent conicity should be controlled to within 0.3, specific to EMUs with a 350 km/h speed grade; the tread equivalent conicity should be controlled to within 0.4, specific to EMUs with a 250 km/h speed grade.

8.3.1.2 Influence of the wheel diameter difference

For a bogie, wheel diameter difference of four wheels can be expressed in four typical wheel diameter difference combination methods, shown in Fig 8.32. The diameter of the wheels on the same side of the front and back wheelsets of the bogie shown in Fig 8.32(A) are equal to each other, so the size of the wheel diameter difference of the front and back wheelsets of the bogie are equal, the positive and negative signs are the same, and the difference is referred to as an equivalent same-phase wheel diameter difference. The wheel diameters on the diagonal of the bogie shown in Fig 8.32(B) are equal, so the wheel diameter difference of the front and back wheelsets of the bogie are equal, the positive and negative signs are opposite, and thus the difference is referred to as an equivalent reverse-phase wheel diameter difference. Fig 8.32(C) shows only the rear wheelset of the bogie with wheel diameter difference. Fig 8.32(D) shows only the front wheelset of the bogie with wheel diameter difference.

The degree of influence on the dynamic performance of the vehicle is different for the various wheel diameter difference expression forms of the bogie [12]. From Fig. 8.33 to Fig. 8.35, it can be seen that the influence on safety is small when the front and back wheelsets of the bogie have the same-phase wheel diameter difference. Conversely, the influence on safety is huge when the front and back wheelsets have the reverse-phase wheel diameter difference. The influence of other situations of wheel diameter difference on safety is between the degree of influence of the same-phase wheel diameter difference and the reverse-phase wheel diameter difference of the front and back wheelsets. When the reverse-phase wheel diameter difference of the bogie is more than 1 mm, the safety of the vehicle system is apparently compromised.

To guarantee driving safety and riding comfort of the high-speed train, it is recommended that the wheel diameter difference (especially, the reverse-phase wheel diameter difference

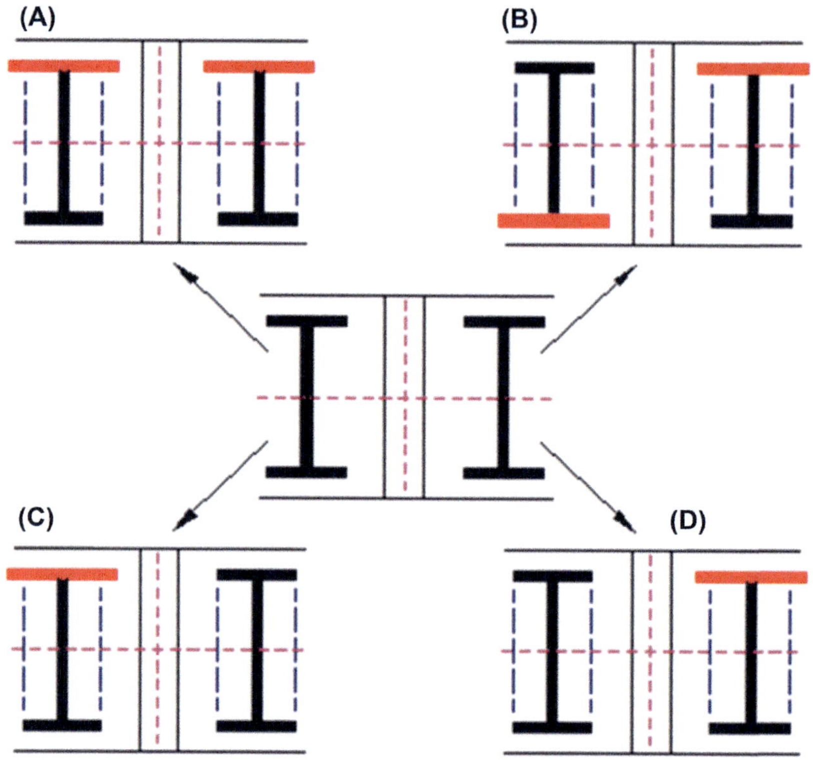

Figure 8.32
Diagram of wheel diameter difference of the bogie.

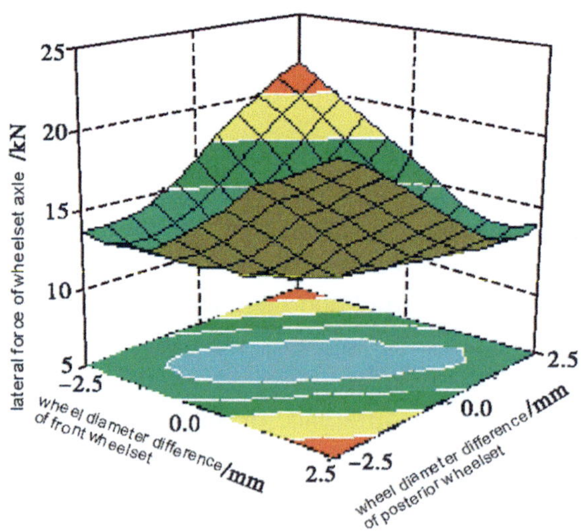

Figure 8.33
Influence of wheel diameter difference on the lateral force of the axle.

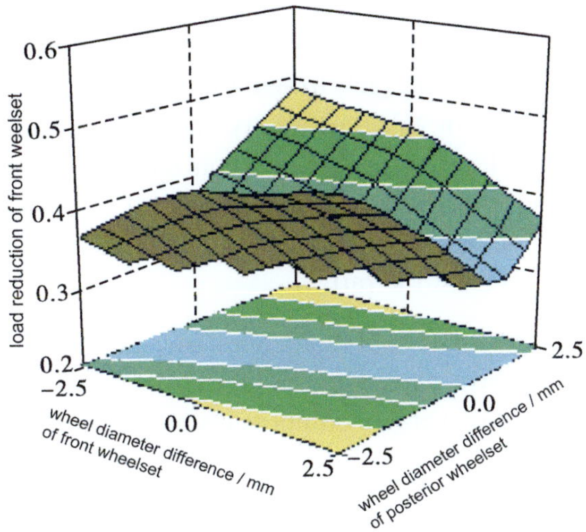

Figure 8.34
Influence of wheel diameter difference on the derailment coefficient.

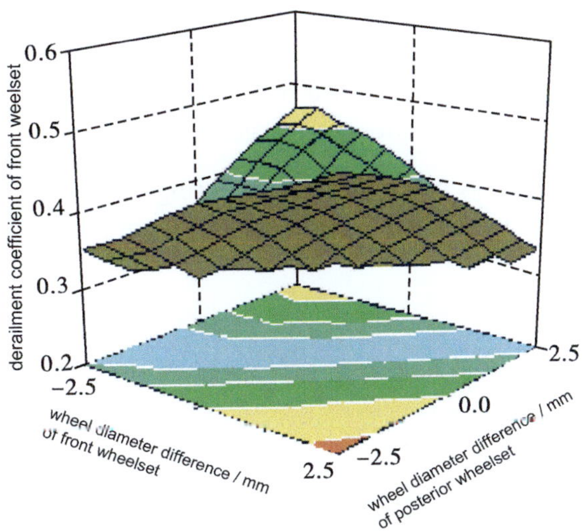

Figure 8.35
Influence of wheel diameter difference on rate of wheel load reduction.

of the front and back wheelsets of the bogie) be controlled to within the scope of 1 mm while the train is running.

8.3.1.3 Influence of wheel polygonization

When the railway vehicle operates at high speed, the wheels will experience the out-of-roundness phenomenon, exhibiting a polygonal state known as wheel polygonization, due to large change in the wheel-rail normal force caused by unavoidable mass eccentricity or geometric eccentricity in the wheel manufacture and repair process, excitation by vehicle vibration, wheelset bending vibration, track irregularity, etc. Small amplitude wheel polygonization leads to a decline in vehicle stability, increase in running noise, and other adverse factors. When the amplitude of wheel polygonization is large enough, the rail jump phenomenon may occur. Every wheel jump on the rail will inevitable have considerable impact on the rail and vehicle. Thus it poses a big threat to the safe running of the vehicle and the service life of the rail.

Research shows that low orders wheel polygonization have a significant influence on the vehicle system. For the same wave depth, the influence of 1st to 4th order wheel polygonization on the vibration of vehicle components are as below:

It can be seen from figures 8.36–8.38 that the order of the wheel polygonization has great influence on the vertical vibration of the vehicle system but little influence on the lateral vibration generally. The 3rd order wheel polygonization has the greatest influence on the vehicle vibration. Under this order, the influence of different wave depths on vehicle vibration is as below:

Figure 8.39–8.41 shows both lateral and vertical vibration gradually intensifies with the increase of wave depth. The relative influence on vertical vibration is larger, indicating that the influence of wheel polygonization on the vehicle system is mainly vertical.

Vehicle vibration puts more stress on comfort level during the vehicle running process and negatively influences operational safety. To analyze this from the perspective of physics, it

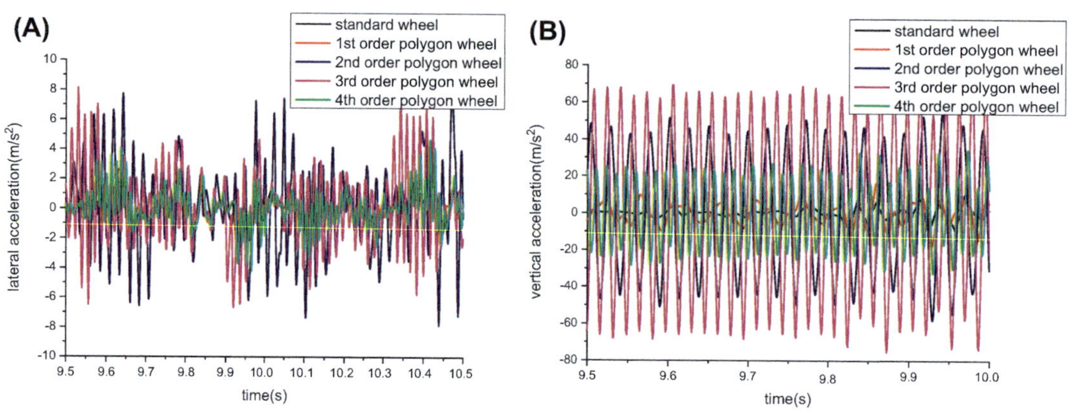

Figure 8.36
Influence of the order of wheel out-of-roundness on axle box vibration. (A) Lateral vibration (B) Vertical vibration.

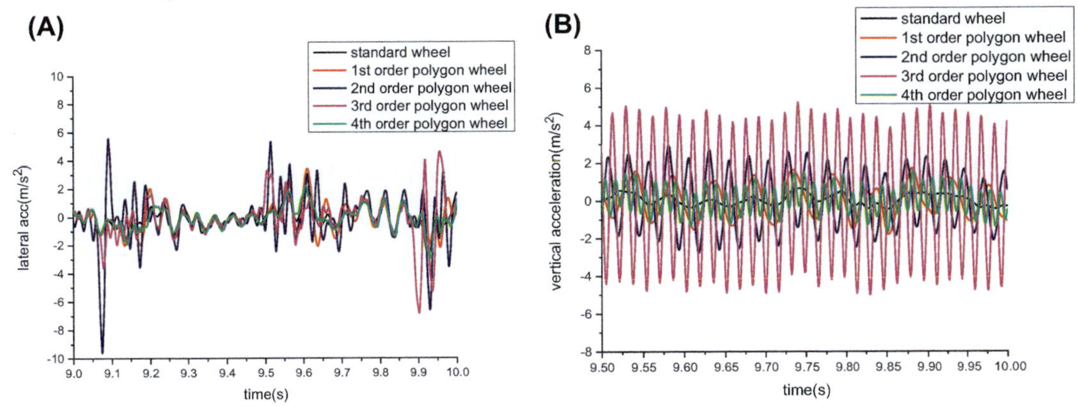

Figure 8.37

Influence of the order of wheel out-of-roundness on frame vibration. (A) Lateral vibration (B) Vertical vibration.

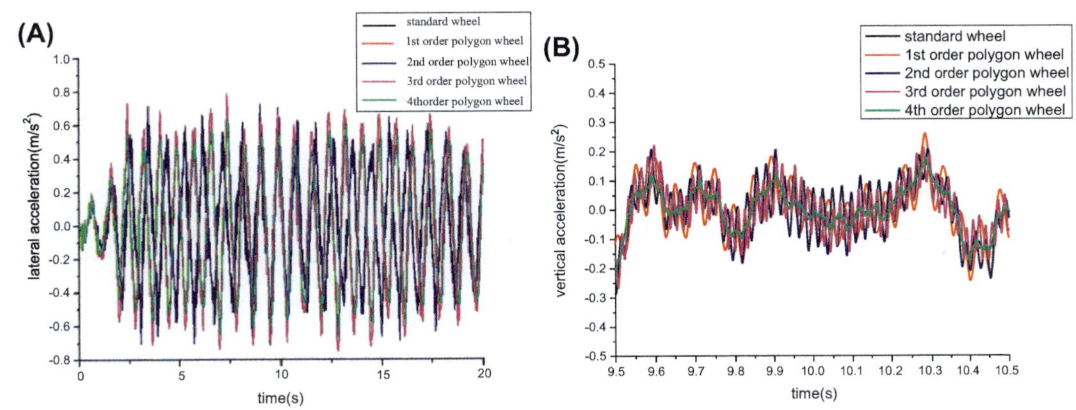

Figure 8.38

Influence of the order of wheel out-of-roundness on car body vibration. (A) Lateral vibration (B) Vertical vibration.

is necessary to look at the influence of different orders of wheel polygonization, and wheel polygonization with different wave depths, on vertical wheel-rail force, as presented below:

Figures 8.42 and 8.43 show that the trend for wheel-rail vertical force is consistent with that for vertical vibration. The influence of wheel polygonization on wheel-rail vertical force is very significant: the degree of this influence varies with the order of the wheel polygonization, and is most pronounced on the triangular wheel. Furthermore, the peak

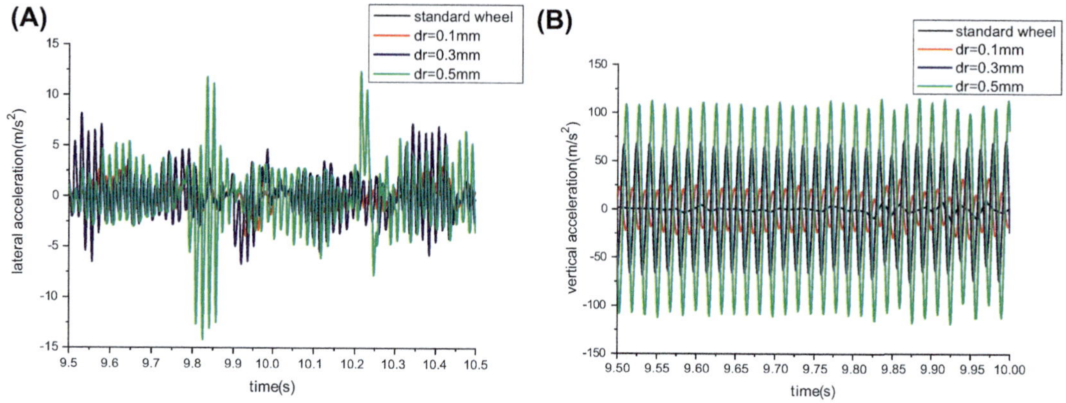

Figure 8.39
Influence of different amplitudes of 3rd order wheel out-of-roundness on axle box vibration. (A) Lateral vibration (B) Vertical vibration.

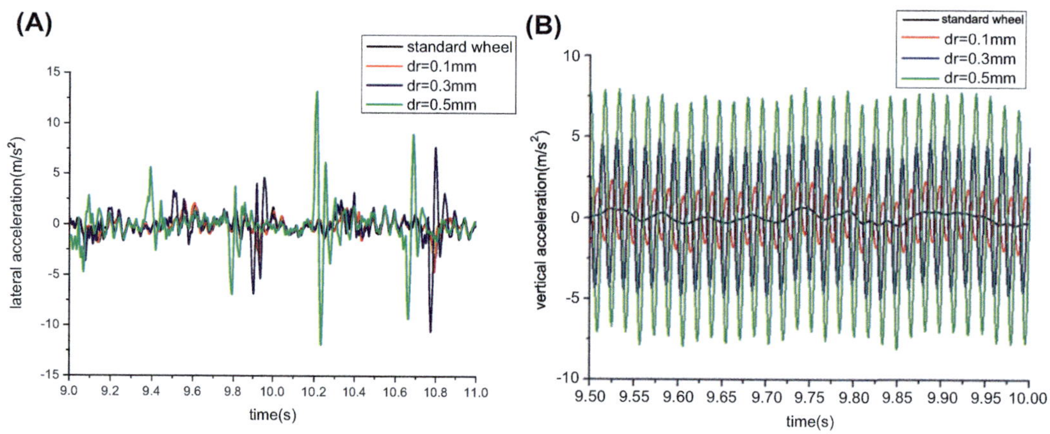

Figure 8.40
Influence of different amplitudes under 3rd order wheel out-of-roundness on frame vibration. (A) Lateral vibration (B) Vertical vibration.

value of the vertical force increases with increase in wave depth under the same order of harmonic excitation.

According to the *Specifications for Vehicle Test of High-speed EMUs*, the peak value of the wheel-rail vertical force of the high-speed train should not exceed 170 kN. Therefore, wave depth thresholds for different orders of wheel polygonization can be acquired based on the wheel-rail force limits.

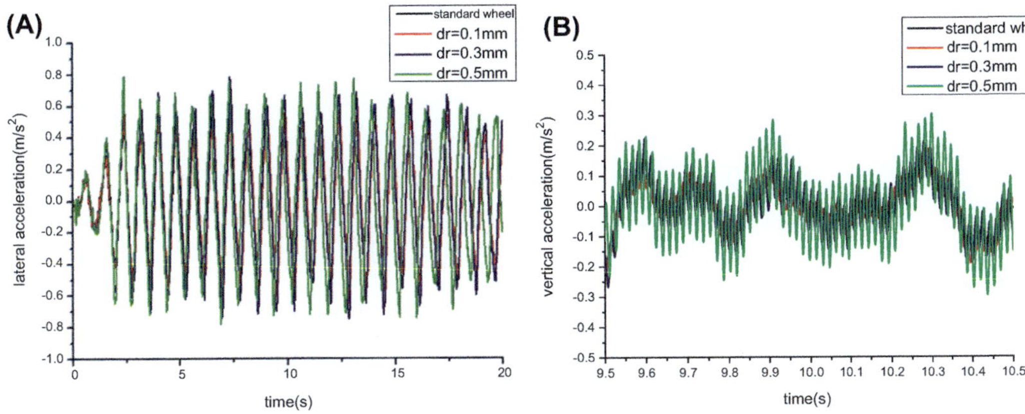

Figure 8.41
Influence of different amplitudes under 3rd order wheel out-of-roundness on car body vibration. (A) Lateral vibration (B) Vertical vibration.

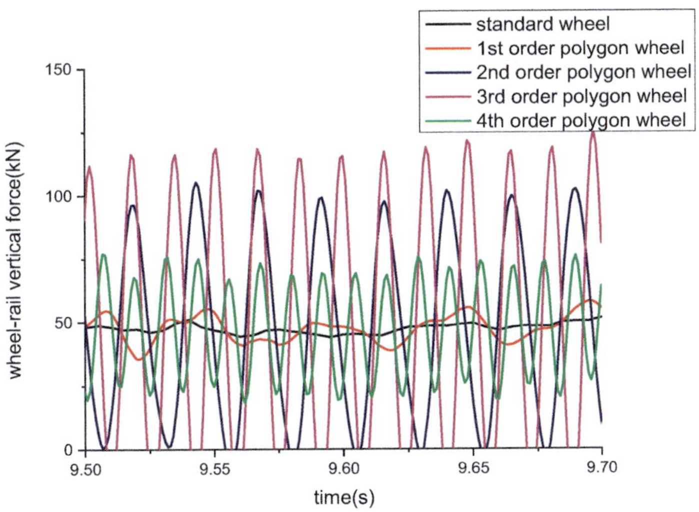

Figure 8.42
Influence of different orders of wheel out-of-roundness on wheel-rail vertical force.

8.3.2 Influence of dimension error on dynamic performance and its threshold

8.3.2.1 Shape and position errors of wheelset installation

The ideal standard bogie (free from shape and position errors) has the features of equal wheelbase of the left and right sides of the front and back axles, and equal distance between the diagonal lines of the front and back axles. The inequality of the wheelbases of the left and right sides of the front and back axles is generally referred to as a parallelism

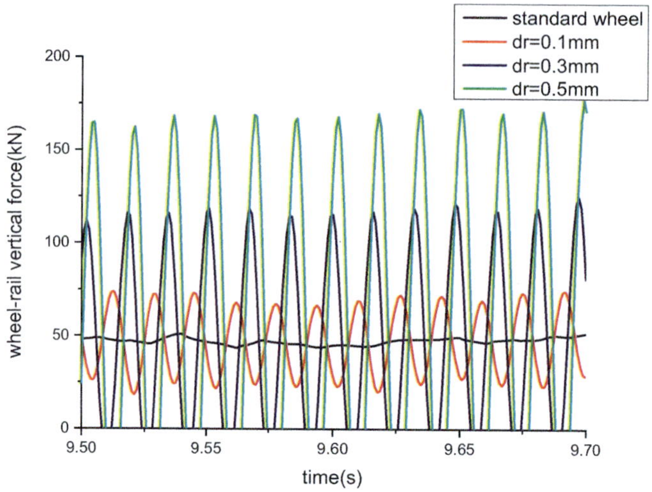

Figure 8.43
Influence of different wave depths of wheel out-of-roundness on wheel-rail vertical force.

error, while inequality of the diagonal lines of the front and back axles is generally referred to as diagonal error of the bogie. These two errors are referred to as installing shape and position errors of the bogie (or, installing shape and position errors of the wheelset). Such errors may occur during manufacture, operation, and maintenance processes, and they directly influence the dynamic performance of trains.

Fig. 8.44 presents the four most basic shape and position errors of the wheelset. The front and back wheelsets of the bogie shown in Fig 8.44(A) are simultaneously deflected at the

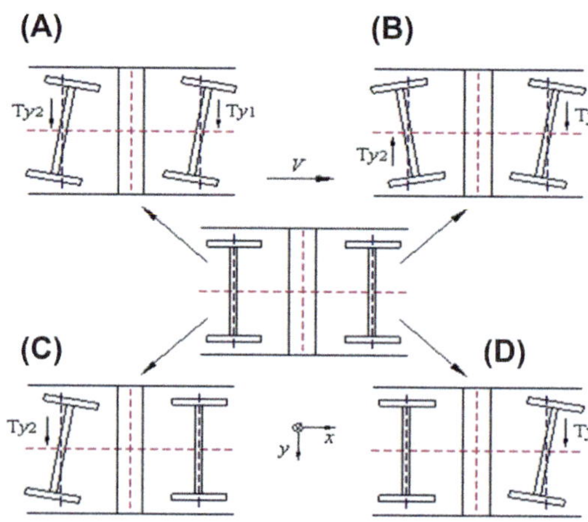

Figure 8.44
Schematic diagram of installing shape and position errors of the wheelset.

same angle in the same direction. Although the wheelbase of the left and right sides of the front and back axles are still equal to each other at the moment, the distance between the diagonal lines of the front and back axles is not equal. Therefore, the bogie has a diagonal error. The front and back wheelsets of the bogie shown in Fig. 8.44(B) are deflected at the same angle towards the opposite direction simultaneously. Although the length of the diagonal lines of the front and back axles are equal, the wheelbases of the left and right sides of the front and back axles are not equal to each other. Therefore, the bogie has a parallelism error. Under the conditions that deflected the back wheelset of the bogie shown in Fig. 8.44(C) and deflected the front wheelset of the bogie in Fig 8.44(D), the length of the diagonal lines of the front and back axles are not equal, and the front and back axles are not parallel to each other. Therefore, the parallelism error and the diagonal error exist simultaneously. There are many expression forms of shape and position errors of the wheelset, and these forms can be acquired by combining the most basic four forms.

The degree of influence of different wheelset installation shape and position errors on vehicle dynamic performance are different [13]. From Fig 8.45–Fig 8.47, it can be seen that the influence on safety is small when the installation deflection angle of the front and back wheelsets of the bogie changes in the same phase. Conversely, the influence on safety is larger when the deflection angle of the front and back wheelsets of the bogie changes in the reverse phase. The influence of other shape and position errors of the wheelset on safety is between the same phase deflection and reverse phase deflection of the front and back wheelsets of the bogie. When the parallelism error of the front and back wheelsets of the bogie is more than 0.5 mrad, the safety index of vehicle system is apparently

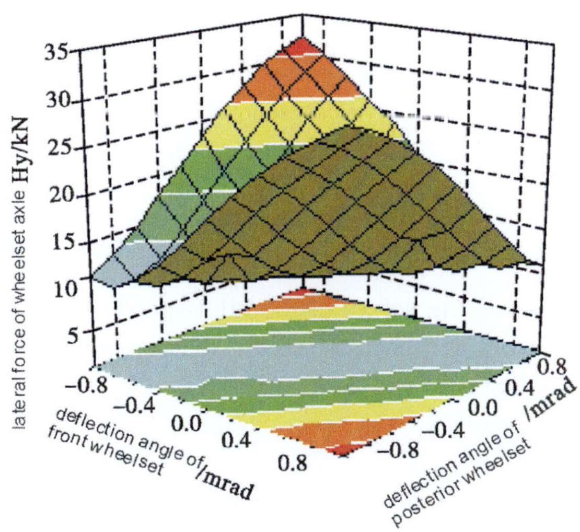

Figure 8.45
Influence of shape and position errors on the lateral force of the axle.

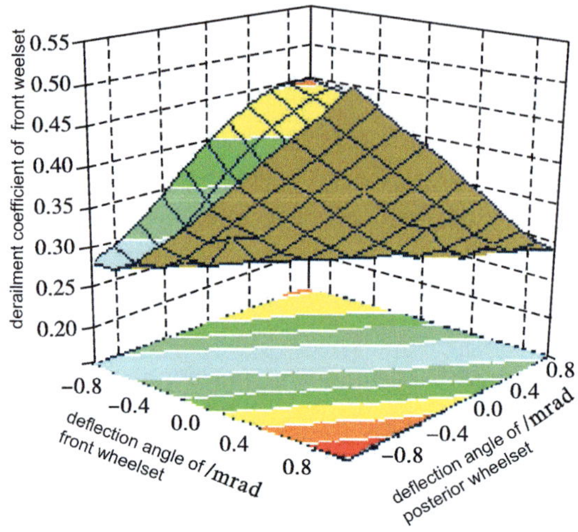

Figure 8.46

Influence chart of shape and position errors on derailment coefficient.

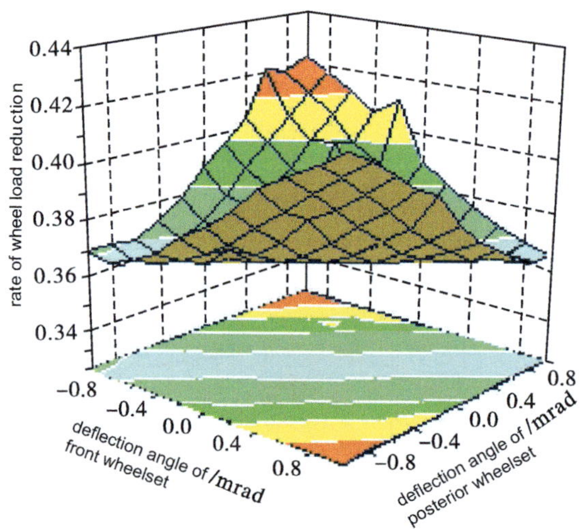

Figure 8.47

Influence of shape and position errors on the rate of wheel load reduction.

worsened. In order to ensure driving safety of the high-speed train, it is recommended that the parallelism error of the front and back wheelsets of bogie be controlled to within the scope of 0.5 mrad during operation.

8.3.2.2 Wheel-weight difference

The function of the high-speed train is to transport passengers. Movement of passengers inside the vehicle may cause an unbalanced load and even cause wheel-weight difference. This assumes that the center of gravity of the unbalanced car body is at the geometric center of the car body. The center of gravity of the unbalanced car body changes in the shadow zone around the geometric center of the car body shown in Fig. 8.48. There are two basic forms of partial load: the first one is longitudinal partial load (the center of gravity of the car body shown on the left side in Fig. 8.48 changes only in x direction and will change the wheelbase of the rotary arm positioning the bogie); the second one is lateral leaning load (the center of gravity of the car body shown on the right side in Fig. 8.48 changes only in y direction and will change the parallelism of the front and back axles of the rotary arm positioning the bogie). The center of gravity of the car body with leading load may be deflected in both longitudinal and lateral directions relative to the geometric center (the shadow zone in Fig. 8.48 is the variation range of the center of gravity of the unbalanced vehicle). However, various forms of partial load can be acquired through the combination of longitudinal and lateral unbalance load, which are the two most basic forms of partial load.

From Fig. 8.49 to Fig. 8.51, it can be seen that the vehicle running safety index is gradually worsened with increase in leaning load. However, different leaning load forms have different degrees of impact on driving safety [14]. The longitudinal leaning load has a small influence on vehicle running safety, while lateral leaning load has huge influence

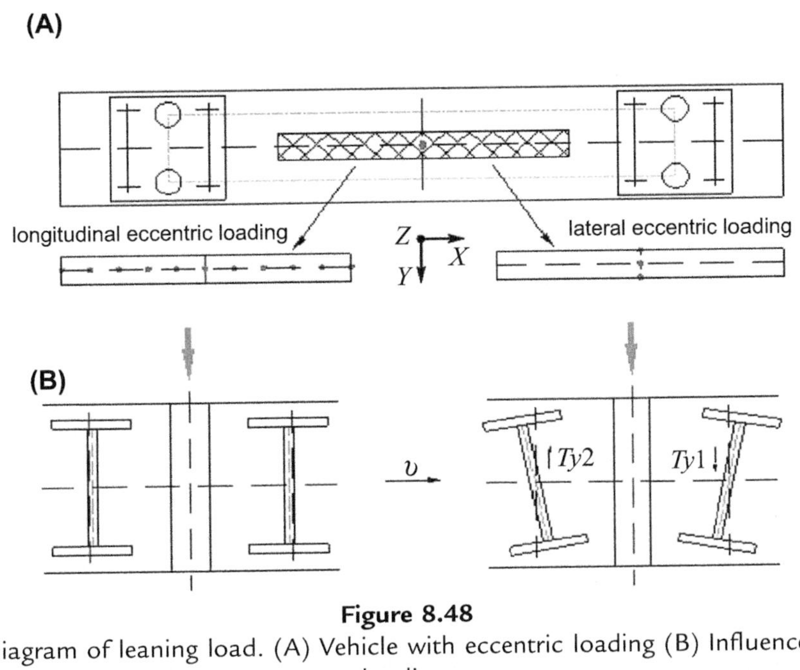

Figure 8.48
Schematic diagram of leaning load. (A) Vehicle with eccentric loading (B) Influence of eccentric loading.

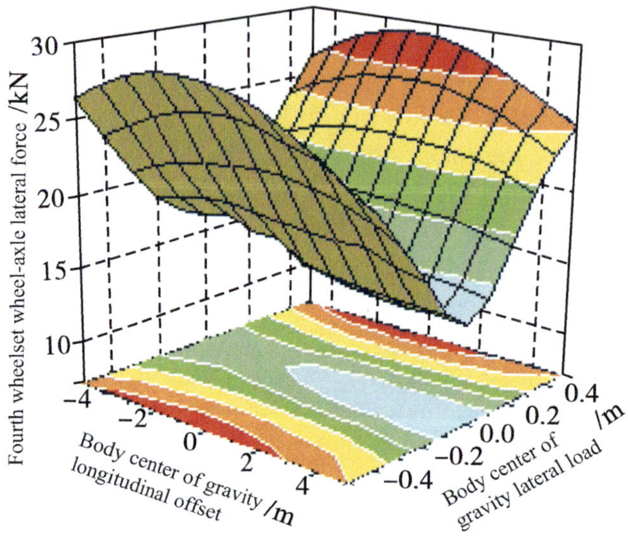

Figure 8.49
Influence of leaning load on lateral force of the axle.

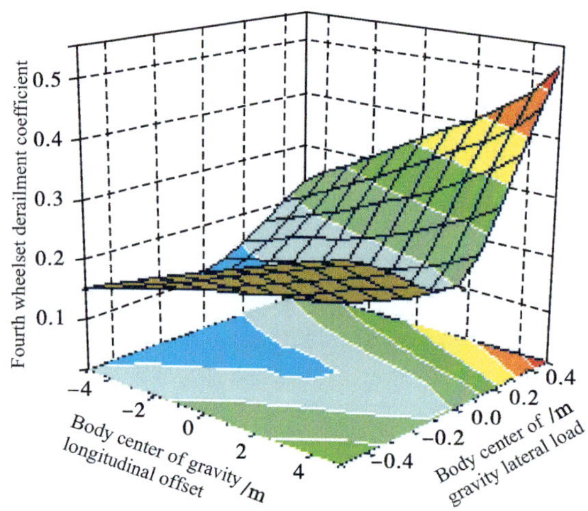

Figure 8.50
Influence of leaning load on derailment coefficient.

on vehicle running safety. The main reason is that the impact of longitudinal leaning load on the incidence angle of the front and back wheelsets of the bogie is small, but the lateral leaning load can cause a large additional incidence angle to the front and back wheelsets of the bogie. The Table 8.5 shows that the influence on safety is not too large when the lateral leaning load is controlled to within 0.25 m (the wheel-weight difference is 10%).

Table 8.5: Contrast table of leaning load and wheel-weight difference of the car body.

Lateral gravity offset of the car body (m)	0	0.05	0.1	0.15	0.2	0.25	0.3	0.35	0.4
Wheel-weight difference (%)	0	2%	4%	6%	8%	10%	12%	14%	16%

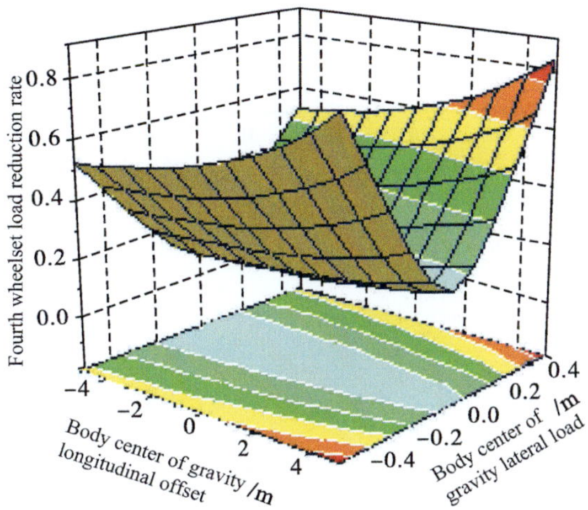

Figure 8.51
Influence of leaning load on the rate of wheel load reduction.

In conclusion, it is recommended that the wheel-weight difference be controlled to within 10% during operation to guarantee driving safety of the high-speed EMUs.

8.3.3 Influence of suspension parameters on dynamic performance and its threshold

8.3.3.1 Control threshold of primary positioning stiffness

Longitudinal and lateral positioning stiffness of the positioning device of the axle box of the wheelset have significant influence on the critical velocity of bogie (especially longitudinal locating stiffness). When the rubber of the positioning joint of the axle box of the wheelset is aged, and the stiffness has changed, the critical velocity of the vehicle is reduced. The degree of this influence varies for different vehicles. For CRH EMUs, the influence of the primary longitudinal positioning stiffness on critical velocity is shown in Fig. 8.52 and Fig. 8.53, and is such that it cannot meet the normal operating requirement. Therefore, it is essential to set a reasonable elasticity position for the wheelset axle box to guarantee high critical velocity of the bogie. The primary longitudinal stiffness should generally be controlled to within 10–20 MN/m, and the lateral stiffness should generally be controlled to within 5–10 MN/m. When the bogie has sufficiently high critical velocity, longitudinal and lateral positioning stiffness should be appropriate, and consequently, the

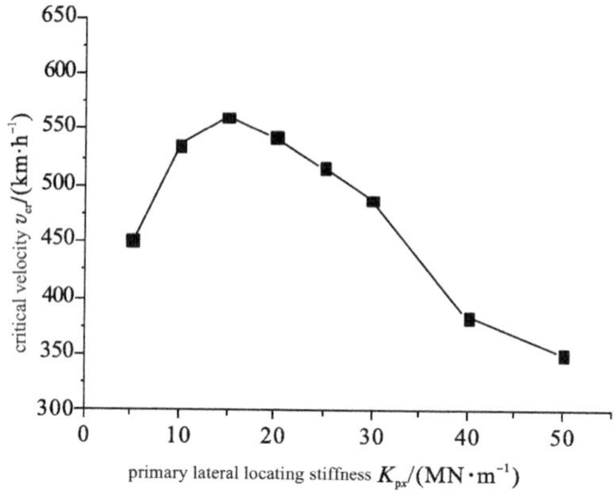

Figure 8.52
Influence of primary longitudinal locating stiffness on critical velocity.

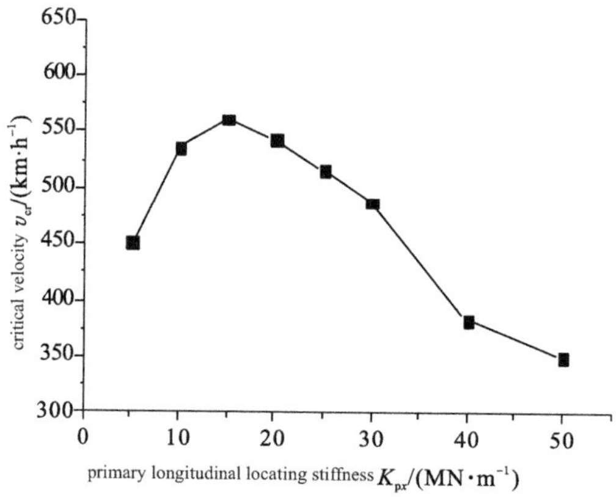

Figure 8.53
Influence of primary lateral locating stiffness on critical velocity.

curving performance can be improved simultaneously while the wheel-rail force and wheel-rail wear can be reduced.

To guarantee dynamic performance of the vehicle system, it is recommended that the actual primary locating stiffness of the bogie during operation should not exceed $\pm$ 20% of the theoretical design value.

8.3.3.2 Control threshold value of the yaw damper

The yaw damper is very important in the high-speed train. It is able to effectively inhibit hunting of the bogie to improve stability and riding comfort.

The damping of the yaw damper of CRH EMUs is about 2500 kN.s/m (the unloading force is 7.36 kN, with an unloading velocity of 0.003 m/s); Fig. 8.54 and Fig. 8.55 present the calculated results of varying equivalent damping of the yaw damper by adjusting the

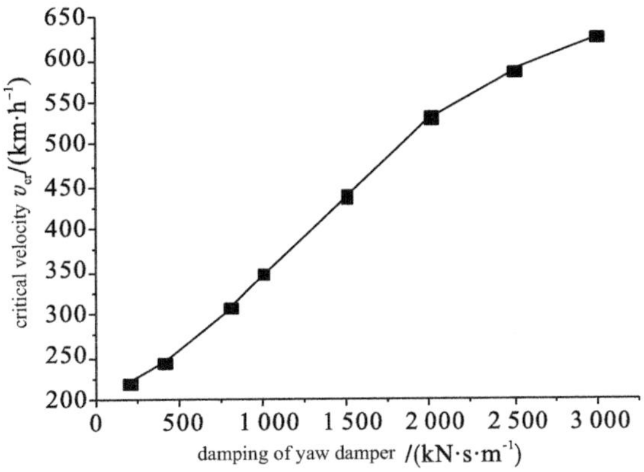

Figure 8.54
Influence of damping of the yaw damper on critical velocity.

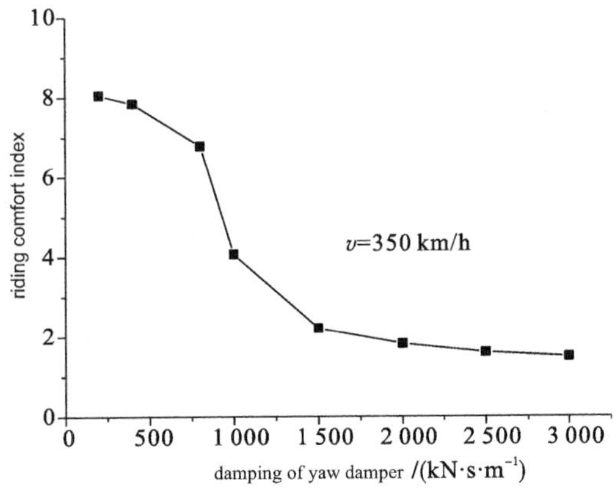

Figure 8.55
Influence of damping of the yaw damper on ride comfort.

unloading velocity while keeping the unloading force unchanged. When the equivalent damping of the yaw damper approaches the theoretical value, the stability allowance of the EMU unit is sufficient, and the riding comfort is excellent. However, when the equivalent damping of the yaw damper is reduced to 80% of the theoretical value (2000 kN.s/m) or lower, the critical velocity of the EMU units is greatly reduced, and the riding comfort seriously diminished.

The equivalent damping of the yaw damper may be reduced to compensate for oil leakage during the operating process. To maintain good dynamic performance of the EMU units, it is recommended that the reduction amount of the equivalent damping value of the yaw damper should not exceed 20% of the theoretical design value.

8.4 Control of service performance of the high-speed train

8.4.1 Control strategy of tolerance and deviation from parameter design

Dynamic performance of the railway locomotive is determined by structure parameters, suspending stiffness, damping, and other dynamic parameters. This section primarily discusses the control strategy of reliability of tolerance and deviation from parameter design with reference to the bogie structure.

There are many impact parameters on dynamic performance of the locomotive, such as wheel-rail tread shape, suspension stiffness, and damping. These parameters depend on and support each other. Therefore, the safety and reliability of the high-speed train are guaranteed through the reliability of these parameters.

8.4.1.1 Reliability design of the wheelset tread

(1) Design objective for the wheelset tread

There are three reliability objectives of the wheelset tread: 1) Critical velocity of hunting instability of a wheelset running in a straight line has a high degree of reliability. 2) The lateral force between wheel rails when passing a curve is sufficiently small that the degree of its reliability reaches distribution reliability. 3) Contact pressure between the wheels and rails is as small as possible.

Considering the standardized design of the wheel-rail system, the tread is designed independently. There are TB, LM, and LMA types of treads in China, and these are adapted to low speed, quasi-high speed, and high speed, respectively. For stability, control of the equivalent conicity of the tread is essential.

(2) Randomness of equivalent conicity parameters

The wheel-rail geometrical surface is composed of several arcs with different radiuses or straight lines. The contact point of the wheel and rail in this model can change from one arc to another arc with a different radius during the rolling contact process. The contact point can be jumped, and the actual wheel-rail extrusion deformation can remit the jump and mutation. The equivalent conicity is characterized by randomness because it is determined by the tread shape and influenced by wheel-rail extrusion deformation. Therefore, the equivalent conicity is deemed a random variable. When there is shortage of statistical data, it can be assumed that the equivalent conicity complies with lognormal distribution, because the equivalent conicity can be influenced by climate, environment, vehicle parameters, and other factors.

The motion stability of the high-speed train is determined by vehicle structure parameters and suspension parameters. The tolerances of these parameters are allowed for in actual manufacture, and these parameters may be changed in application. Thus, we can acquire the distribution rule of the parameter by testing parameters of the actual product.

Fig. 8.56 shows the equivalent conicity data calculated according to measurement and simulation of CRH EMU units running on the track. The distribution law of tread

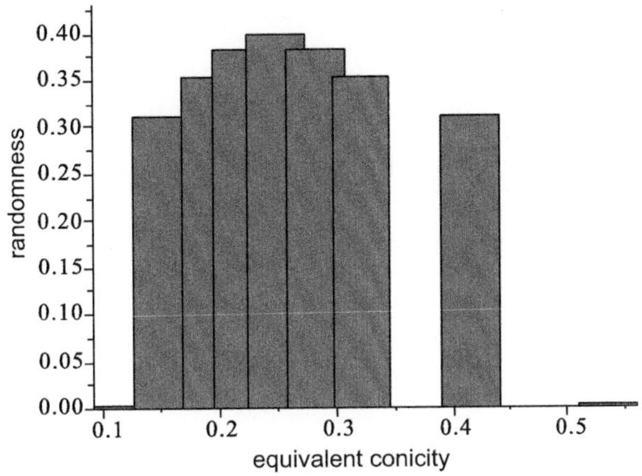

Figure 8.56
Randomness of tread equivalent conicity.

equivalent conicity in the 200,000 km service process is acquired and fit by lognormal distribution. The distribution function is expressed as Equation (8.1), where the mean value u_λ, is lg 0.25, and the mean square deviation σ_λ, is 0.11.

$$\lg\lambda \sim N\left(u_\lambda, \sigma_\lambda^2\right) = N\left(\lg 0.25, 0.11^2\right) \tag{8.1}$$

(3) Relationship between equivalent conicity and line critical velocity

Fig. 8.57 represents the critical velocity under different equivalent conicities, which is calculated by adopting equivalent linearization based on CRH EMUs.

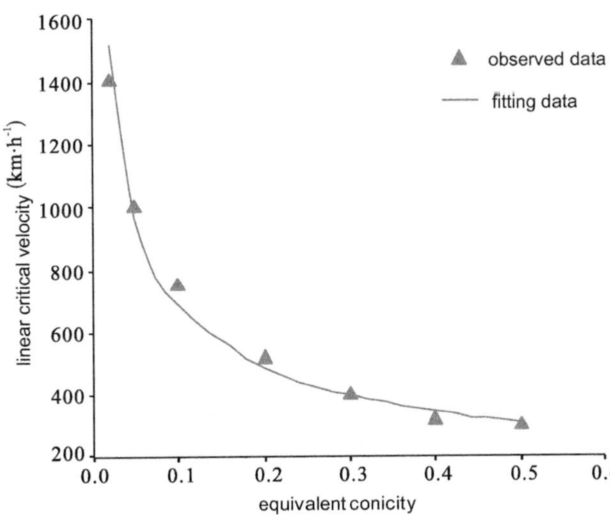

Figure 8.57
Fitting function curve.

A function is applied to fit the relationship of tread equivalent conicity and critical velocity of the CRH EMUs:

$$\lg v = \lg a - b\lg\lambda = \lg 219.821 - 0.4944 \times \lg\lambda \tag{8.2}$$

where λ represents tread equivalent conicity and v represents critical velocity.

(4) Forecasting of reliability of motion stability

When the allowable critical velocity of the high-speed train is set as $[v]$, and the critical velocity $v \geq [v]$, the train operation is safe, and its degree of reliability is:

$$R = P\{v \geq [v]\} \tag{8.3}$$

Equation (8.2) on the relationship between tread equivalent conicity and critical velocity is substituted into Equation (8.3) to calculate the reliability:

$$\begin{aligned} R &= P\{v \geq [v]\} \\ &= P\{\lg v \geq \lg[v]\} \\ &= P\{\lg a - b\lg\lambda - \lg[v] \geq 0\} \\ &= P\left\{\lg\lambda \geq \frac{\lg a - \lg[v]}{b} = \lg[\lambda]\right\} \\ &= \Phi\left(\frac{(\lg a - \lg[v])/b - u_\lambda}{\sigma_\lambda}\right) \end{aligned} \tag{8.4}$$

Equation (8.4) gives the degree of reliability of tread equivalent conicity. From Equation (8.4), the degree of reliability of tread equivalent conicity is also related to the mean value u_λ and mean square deviation σ_λ of $\lg\lambda$ of the tread equivalent conicity, but not to the allowable critical velocity [v]. When the distribution of the tread equivalent conicity is presented as Equation (8.1), and the relationship between equivalent conicity and critical velocity is presented as Equation (8.2), different values of allowable critical velocity [v] have varying influence on the degree of reliability, based on Equation (8.4):

The allowable critical velocity is set as 250 km/h, 275 km/h, 300 km/h, 350 km/h, and 400 km/h. The degree of reliability of tread equivalent conicity is calculated, and the results presented in Table 8.6. The relationship between the degree of reliability of the allowable critical velocity and the tread equivalent conicity is shown in Fig 8.58.

From Fig 8.58, it can be seen that the train movement stability varies with the degree of reliability under different allowable critical velocities for the discrete distribution property of equivalent conicity. The degree of reliability of stability is decreases with increase in running speed, and this indicates that the rate of occurrence of hunting instability is improved.

Table 8.6: Degree of reliability of tread equivalent conicity.

Allowable critical velocity [v]	degree of reliability R
250	0.999996
275	0.999886
300	0.998605
350	0.960708
400	0.755727

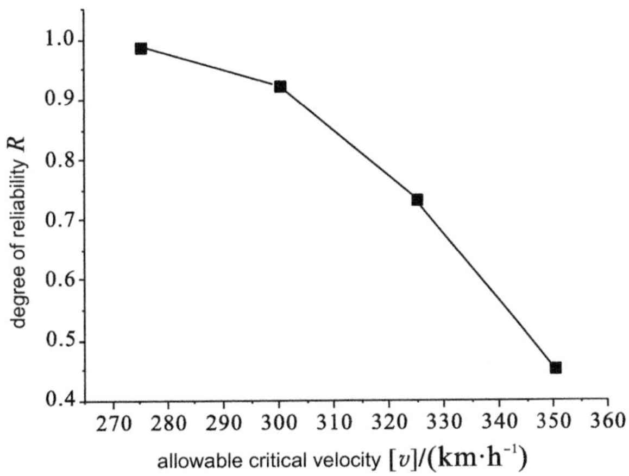

Figure 8.58
Relationship between the degree of reliability of the allowable critical velocity and tread equivalent conicity.

Influence of the dispersion σ_λ of different equivalent conicities on the degree of reliability:

The mean square deviations σ_λ of logarithms of tread equivalent conicity is set as 0.100715, 0.141001, 0.20143, 0.251788, 0.302145, and 0.40286. The calculated degree of reliability of the corresponding tread equivalent conicity are presented in Table 8.7. The relationship between the two variables is shown in Fig 8.53.

Table 8.7: Relationship between mean square deviation and the degree of reliability of tread equivalent conicity.

σ_λ	Degree of reliability of logarithm of tread equivalent conicity R
0.100715	0.997832
0.141001	0.979207
0.20143	0.923115
0.251788	0.873081
0.302145	0.829171
0.40286	0.76213

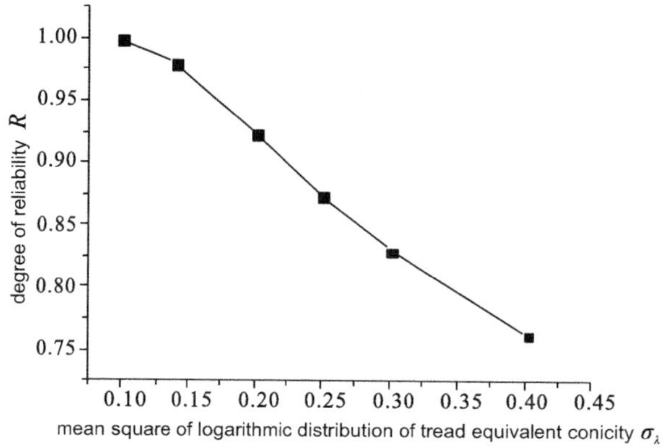

Figure 8.59
Relationship between mean square deviation of logarithm and the degree of reliability of tread equivalent conicity.

From Fig. 8.59, it can be seen that even though the allowable critical velocity [v] and the mean value of logarithm of tread equivalent conicity u_λ is unchanged, the degree of reliability R of safety of tread equivalent conicity can also be reduced along with the increase in mean square deviation of the logarithmic distribution of tread equivalent conicity.

(5) Relationship between the degree of reliability $[R]$ and design tread equivalent conicity

Because the mean square deviation of the tread equivalent conicity is caused by many factors such as the application environment and material defect and is hard to control, the mean value of the tread equivalent conicity is determined by the design of the tread profile. The degree of reliability of the critical velocity of the running vehicle is guaranteed by the mean value of the tread equivalent conicity. Based on Equation (8.4), we obtain:

$$\lg[\lambda] = z_{[R]} \cdot \sigma_\lambda + u_\lambda$$
$$u_\lambda = \lg[\lambda] - z_{[R]} \cdot \sigma_\lambda \tag{8.5}$$

where $[\lambda]$ is the allowable equivalent conicity for guaranteeing critical velocity.

The degree of reliability is set as 0.5, 0.9, 0.95, 0.99, and 0.999 and substituted into Equation (8.5), and compare with the design mean value of corresponding equivalent conicity. The calculated results are presented in Table 8.8:

From the Table, the mean value of tread equivalent conicity decreases with increase in the degree of reliability to reach the allowable degree of reliability.

Table 8.8: Calculated results for tread equivalent conicity.

R	lg[λ]	Quantile	Mean square deviation	u_λ	Mean value of equivalent conicity
0.999	-0.62898	3.090232	0.11	-0.96891	0.107422
0.99	-0.62898	2.326348	0.11	-0.88488	0.130353
0.95	-0.62898	1.644854	0.11	-0.80991	0.154912
0.9	-0.62898	1.281552	0.11	-0.76995	0.169844
0.5	-0.62898	-1.4E-16	0.11	-0.62898	0.234974

8.4.1.2 Reliability design of primary location system

There are many structure forms of locating the axle box, such as the guide frame type, guide column type, pull plate type, pull rod type, swing arm type, and location method by rubber element. The rotary arm positioning and multiple rubber positioning are two typical location methods for the axle box.

(1) Design objective of reliability of the primary location system

The design objective of reliability of the primary location system is: 1) the critical velocity of hunting instability of the wheelset operated on a straight line has a high degree of reliability; 2) the index of stationarity of the wheelset operated on the straight line meets the high degree of reliability for standard requirements; 3) high degree of reliability of the lateral force of the axle meets the standard requirements; 4) primary locating stiffness has high reliability for curving performance.

The primary location system has stiffness in three directions, and the design of its reliability is primarily the control of stiffness.

(2) Randomness of primary location system

Taking the rubber location system as an example: its performance is influenced by environmental temperature and discrete characteristics. Its stiffness is a random variable, and the statistics show that the stiffness of the primary location system always follows the normal distribution, namely, $k \sim N(u_k, \sigma_k)$.

(3) Relationship between stiffness and vehicle dynamic response

Taking CRH EMUs as an example: if the vertical stiffness of the primary location system is changed, the vehicle dynamic response are presented in Table 8.9.

The relationship curve for primary vertical stiffness k and derailment coefficient is fitted and presented as Equation (8.6) and Fig 8.60.

$$y_1 = 0.21716 + 1.91327 \times 10^{-5} x \tag{8.6}$$

Table 8.9: Relationship between primary locating stiffness and vehicle dynamic response.

Vertical stiffness of primary suspension MN·m^{-1}	Wheel-rail lateral force (kN)	Wheel-rail vertical force (kN)	Lateral force of axle (KN)		Derailment coefficient	Stationarity index	
			Maximum value	Mean value		Lateral acceleration	Vertical acceleration
0.9996	14.5219	131.0050	24.9782	−1.1327	0.2351	2.3774	1.6451
1.0584	15.1623	131.2080	26.1923	−1.0857	0.2370	2.3803	1.6454
1.1172	15.7095	134.3670	26.7673	−1.0430	0.2401	2.3810	1.6456
1.176	16.1626	134.3130	27.8062	−1.0047	0.2411	2.3791	1.6458
1.2348	15.9705	134.1740	27.4727	−0.9685	0.2399	2.3851	1.6459
1.2936	16.6729	134.1210	28.4413	−0.9371	0.2422	2.3862	1.6460
1.3524	16.8000	134.2080	28.6462	−0.9071	0.2422	2.3835	1.6460

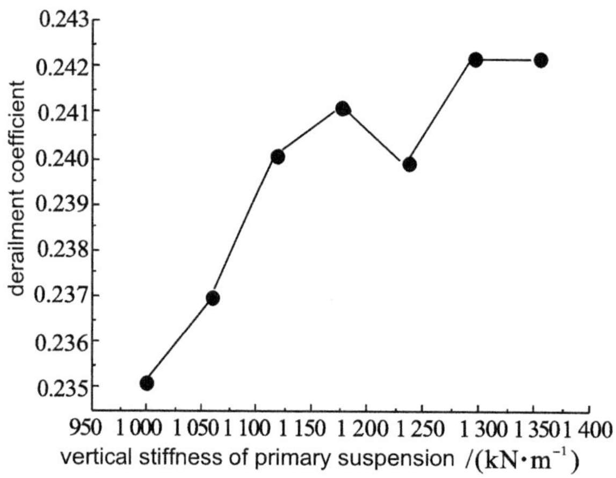

Figure 8.60
Relationship between vertical stiffness and derailment coefficient.

The curve of the relationship between primary vertical stiffness k and wheel-rail lateral force is fitted and presented as Equation (8.7):

$$y_2 = 15.61 + 0.009k \tag{8.7}$$

(4) Reliability analysis of primary locating stiffness

The evaluation index of the derailment coefficient of running stability of domestic trains is presented in Table 8.10.

Thus, the degree of reliability of a good running vehicle is R1:

$$R_1 = P\{y_1 \leq 0.8\} \tag{8.8}$$

Table 8.10: Safety limit of Chinese derailment coefficient.

Index	GB5599-1985		TB/T 2360-1993		95J 01-L(M)	Complete vehicle test specifications for high-speed EMU units
Derailment coefficient	First limit ≤ 1.2	Second limit ≤ 1.0	Good 0.8	Qualified 0.9	≤ 0.8	≤ 0.8

Substitute Equation (8.6) into Equation (8.8). Thus:

$$\begin{aligned} R_1 &= P\{a + bx \leq 0.8\} \\ &= P\{x \leq (0.8 - a)/b\} \\ &= \Phi\left(\frac{(0.8 - a)/b - \mu_k}{\sigma_k}\right) \end{aligned} \tag{8.9}$$

Substitute Equation (8.6) and $u_k = 1176kN, \sigma_k = 117.6$ into Equation (8.9), and obtain $R_1 = 1$. Substitute Equation (8.6) and $u_k = 1176kN, \sigma_k = 78.4$ into Equation (8.9), and obtain $R_1 = 1$.

Degree of reliability of qualified running vehicle is set as R_2:

$$R_2 = P\{y_1 \leq 0.9\} \tag{8.10}$$

Substitute Equation (8.6) into Equation (8.10). Thus:

$$\begin{aligned} R_1 &= P\{a + bx \leq 0.9\} \\ &= P\{x \leq (0.9 - a)/b\} \\ &= \Phi\left(\frac{(0.9 - a)/b - \mu_k}{\sigma_k}\right) \end{aligned} \tag{8.11}$$

Substitute Equation (8.6) and $u_k = 1176kN, \sigma_k = 117.6$ into Equation (8.11), and obtain $R_2 = 1$; substitute Equation (8.6) and $u_k = 1176kN, \sigma_k = 78.4$ into Equation (8.11) and obtain $R_2 = 1$.

Based on $R_1 = 1$ and $R_2 = 1$, the primary retaining spring can reach the degree of reliability of a qualified running vehicle as well as reach the degree of reliability of a good running vehicle.

Complete vehicle test specifications for the high-speed EMUs stipulate that the assessment limit value of the lateral force of the axle should adopt the standard below:

Maximum value: $H \leq (10 + P_0/3)$ (P_0 represents axle net weight)

Mean value: $H_{ave} \leq 20kN$

Thus, the degree of reliability of the assessment value of the lateral force of the axle of running vehicle $R_2 = P\{y_2 \leq 0.9\}$ is set as R3:

$$R_3 = P\{y_2 \leq 10 + P_0/3\} \tag{8.12}$$

Substitute Equation (8.7) into Equation (8.12). Thus:

$$\begin{aligned} R_3 &= P\{b_0 + b_1 k \leq 10 + P_0/3\} \\ &= P\left\{k \leq \frac{10 + P_0/3 - b_0}{b_1}\right\} \\ &= \Phi\left(\frac{(10 + P_0/3 - b_0)/b_1 - u_k}{\sigma_k}\right) \end{aligned} \tag{8.13}$$

where $b_0 = 15.61, b_1 = 0.009, p_0 = 51.012kN$, the degree of reliability of the lateral force of the axle is 0.7780 when $u_k = 1176kN, \sigma_k = 117.6$; the degree of reliability of the lateral force of the axle is 0.8745 when $u_k = 1176kN, \sigma_k = 78.4$; the degree of reliability of the lateral force of the axle is 0.9892 when $u_k = 1176kN, \sigma_k = 39.2$.

By exploring R_1, R_2, R_3, the degree of reliability of primary locating stiffness is $R = \min\{R_1, R_2, R_3\}$.

(5) Stiffness design of the primary retaining spring with given degree of reliability

Analysis (4) shows that the multi-objective design of the degree of reliability can be carried out on the lowest state. Therefore, the primary retaining spring can have a design spring stiffness reliability up to the degree of reliability of a good running vehicle. u_k is designed, given degree of reliability R_1, and $k_1 = 0.151408$, $k_2 = 2.287154$, and $\sigma_k = 0.4$. Based on Equation (8.9), the given degree of reliability is solved using the numerical approximation method. When $R_1 = 0.99, u_k = 1.325MN \cdot m$; if $\sigma_k = 0.2$, the given degree of reliability $R_1 = 0.99, u_k = 1.815MN \cdot m$.

8.4.2 Deterioration law of service reliability of the high-speed train

8.4.2.1 Reliability index

Generally, the service life of a product (or system) is described by a nonnegative random variable X. The distribution function of X is:

$$F(t) = P\{X \leq t\}, t \geq 0 \tag{8.14}$$

Thus, the probability that the system worked normally before moment t is:

$$R(t) = P\{X > t\} = 1 - F(t) \tag{8.15}$$

$R(t)$ is referred to as the reliability function or reliability of the system. $R(t)$ is the probability that the system will not fail within time $[0, t]$.

For a repairable system, its normal and failure running course appear alternately. X_i and Y_i distributions are recorded as the starting time and downtime of the cycle I, and $i = 1, 2, \ldots$. Generally, $X_1, X_2, \ldots$ or $Y_1, Y_2, \ldots$ are different distributions. The main indexes for describing the system reliability includes:

(1) Average time before first failure

$$MTTFF = EX_1 = \int_0^{+\infty} t dF_1(t) \tag{8.16}$$

where $F_1(t) = P\{X_1 \le t\}$ is the distribution of time X_1 before the first failure of the system.

(2) Availability

Set

$$X(t) = \begin{cases} 1 & \text{if the system is normal at the moment } t \\ 0 & \text{if the system is failure at the moment } t \end{cases} \tag{8.17}$$

Then the definition of instantaneous availability of the system at the moment t is:

$$A(t) = P\{X(t) = 1\} \tag{8.18}$$

The average availability of the system within $[0, t]$ is:

$$\widetilde{A}(t) = \frac{1}{t} \int_0^t A(u) du \tag{8.19}$$

If the limit

$$\widetilde{A} = \lim_{t \to \infty} \widetilde{A}(t) \tag{8.20}$$

exists, $\widetilde{A}$ is referred to as steady-state average availability.

If the limit

$$A = \lim_{t \to \infty} A(t) \tag{8.21}$$

exists, it can be referred to as the steady-state availability of system.

(3) Steady-state failure frequency

for a repairable system, $N(t)$ is set as the failure number of the system within $[0, t]$. Thus the failure number distribution of the system within [0, t] is:

$$P_k(t) = P\{N(t) = k\}, k = 0, 1, 2, \ldots \tag{8.22}$$

The average failure number of the system within [0, t] is:

$$M(t) = EN(t) = \sum_{k=1}^{+\infty} kP_k(t) \tag{8.23}$$

The steady-state failure frequency of the system is defined as:

$$M = \lim_{t \to \infty} \frac{M(t)}{t} \tag{8.24}$$

8.4.2.2 Deterioration law of the reliability of key components

The operation and maintenance reliability centered maintenance (RCM) database on the high-speed train is the foundation of studying the deterioration law of reliability of key components. Massive, multisource, and heterogeneous service data accumulated from massive and long-term service of the high-speed train, which is main source of the RCM database, a valuable resource and treasure, and possessed of important value for improving the service reliability of the high-speed train. Furthermore, perfect vehicle and component record keeping is an important basis for revealing the deterioration law of reliability of key components, such as vehicle operation route and mileage, wheel reprofiling record, and replacement records of key components.

In considering big differences in vehicle condition during the service of a high-speed train, and the existence of attachment roll-in/roll-out, the deterioration law of reliability is studied by adopting the survival analysis method. The splattering point of the degree of reliability of key components along with change of operation mileage is calculated by fully considering the influence of deleted data. Thus the splattering point of the degree of reliability is further fit by empirical distribution to obtain a parameterized deterioration law model of the degree of reliability of the key components.

Fig. 8.64 takes wheel material defect found through rim and spoke flaw detection of one high-speed train as an example, applies survival analysis to calculate the deterioration law of the degree of reliability along the operation mileage, and adopts Weibull distribution to carry out parameter fitting inspection of goodness of fit. Its shape parameter is 0.9471, the dimension parameter is 3.9330×10^8, and the goodness of fit is 0.9930.

This method of analysis can also be applied to the analysis of the deterioration model of the overall reliability of key components; namely, multiple failure models can be fit according to reason for failure, failure influence, etc. Similarly, the deterioration law of the overall reliability of key components can be studied by adopting this method. As shown in Fig. 8.65, all bogie repair failures, including wheelset replacement, wheel reprofiling, air spring replacement, and others, are integrated. The fitting result of the Weibull distribution

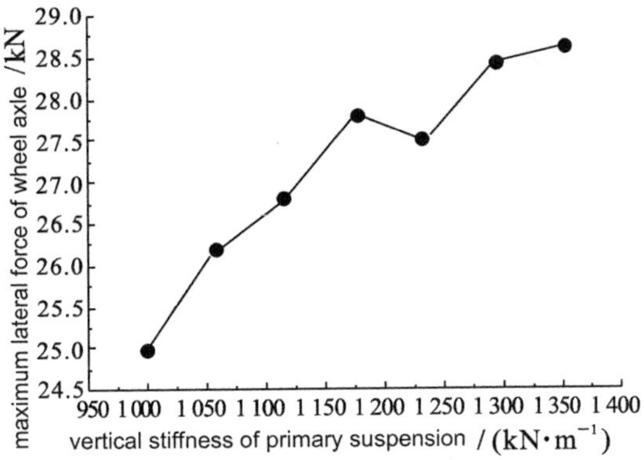

Figure 8.61

Relationship between vertical stiffness and maximum lateral force.

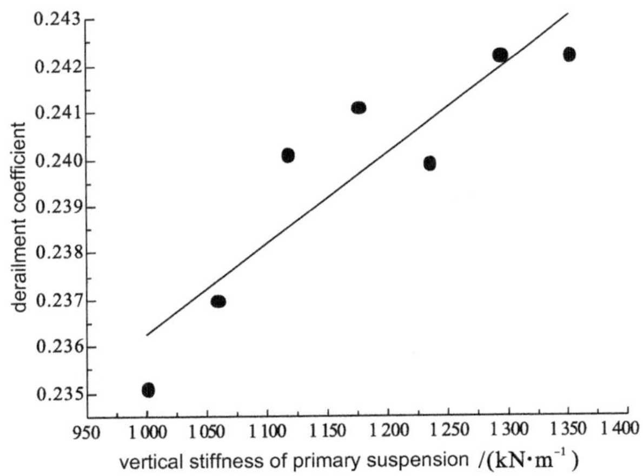

Figure 8.62

Curve fitting for the relationship between vertical stiffness and derailment coefficient.

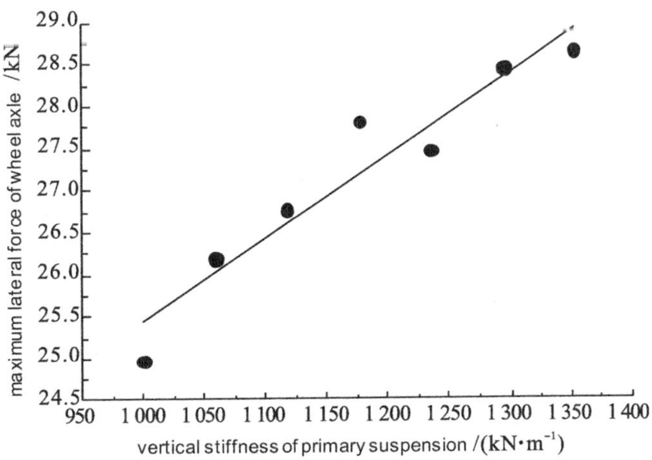

Figure 8.63

Curve fitting for the relationship between vertical stiffness k and maximum value of wheel-rail lateral force.

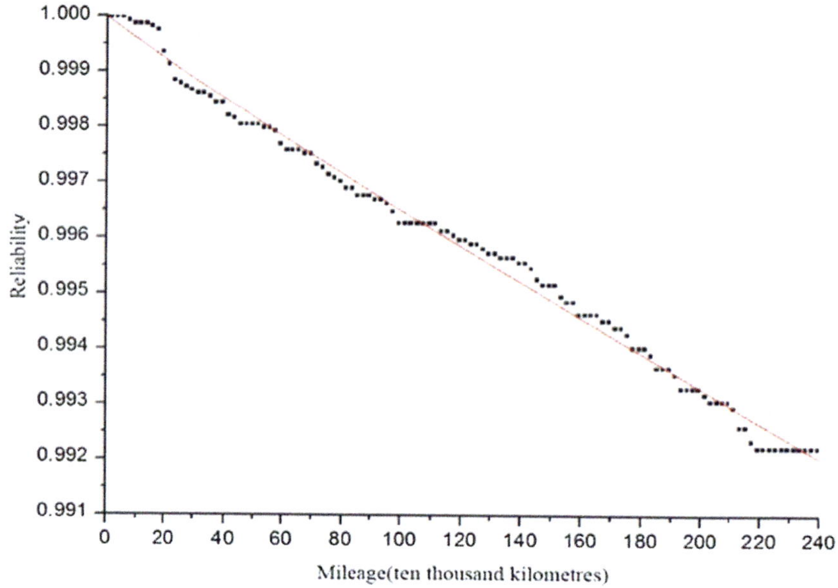

Figure 8.64
Model of deterioration law of the degree of reliability of material wheel defect.

of the degree of reliability and main failure model ratio can be acquired by adopting the above method. The sample volume of analysis method by overall reliability is larger, and the goodness of fit is higher. Thus, the failure rule of different failure modes can be compared.

8.4.2.3 Structural importance of technical parameters and vibration parameters

According to actual experience, parts in the system are not equally important: failure of some parts may cause system failure, and some may not. The concept [15,16] of structural importance of system components is given below.

It assumes that the system is composed of n parts. x_i represents the state of part i; $x_i = 1$ represents that the part i is normal; $x_i = 0$ represents that the part i has failed. The subscript sets of parts are set as $N = \{1, 2, \ldots, n\}$, vector valued as 0,1, only $x = (x_1, \ldots, x_n)$ of all components is recorded as the state vector of parts, and $(\cdot_i, x) = (x_1, \ldots, x_{i-1}, \cdot, x_{i+1}, \ldots, x_n)$. It assumes that the system only has normal and failure states which are represented by 1 and 0 respectively. The system state of the given state vector x is expressed by $\phi(x)$. It is a function on $\{0, 1\} \rightarrow \{0, 1\}$ and referred to as the structural function of the system. If $\phi(x) = 1$, x is referred to as the path vector of ϕ.

Let J be any part of the system; if one $(\cdot_j, x)$ satisfies:

$$\phi(1_j, x) - \phi(0_j, x) = 1 \tag{8.25}$$

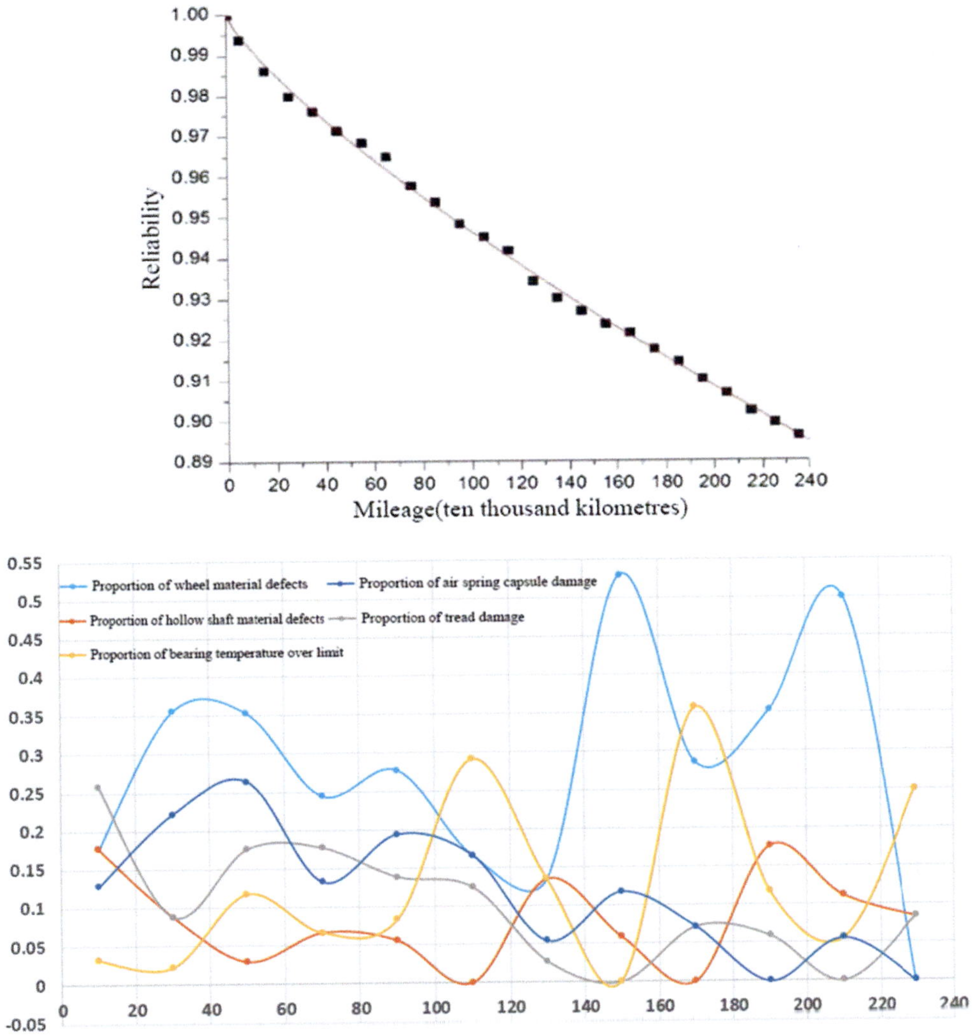

Figure 8.65
Deterioration law model of the degree of overall reliability of bogie.

it indicates that j is a key part under the situation $(\cdot_j, x)$. Because equation (8.25) is equivalent to

$$\phi(0_j, x) = 0, \phi(1_j, x) = 1 \tag{8.26}$$

it indicates that the system is normal when the part j is normal and failed when the j is failed. $(1_j, x)$ is referred to as a key path vector of j. Record

$$n_\phi(j) = \sum_{\{xx_j=1\}} \{\phi(1_j, x) - \phi(0_j, x)\} \tag{8.27}$$

Apparently, $n_\phi(j)$ is the total number of key path vectors of j. When $x_j = 1$, the state vector $(1_j, x)$ has 2^{n-1} different kinds of results. Therefore, define

$$I_\phi(j) = \frac{1}{2^{n-1}} n_\phi(j) \tag{8.28}$$

as the structural importance of part j. It indicates the ratio of the number of key path vectors of j in all 2^{n-1} kinds of possible situations. Therefore, parts with any structure can be sequenced according to their structural importance.

8.4.3 Control strategy for service reliability of the high-speed train

8.4.3.1 Service evaluation parameters of the high-speed train

The service reliability evaluation parameters of the high-speed train include the parameter distribution function describing the distribution characteristics of parameters; parameter reliability function and parameter failure rate function; parametric mean time to first failure, parametric availability, and parametric fault frequency; parametric structural importance for describing parameter sensitivity.

Reliability analysis of disturbance factors in high-speed train service is carried out; the influence of the disturbance factors' degeneration on critical velocity, derailment coefficient, rate of wheel load reduction, overturning coefficient and other safety indexes are analyzed; the hazardous sources are identified; safety risk management follows the principle of "*reducing to reasonable and feasible*" (ALARP) stipulated in EN50126.

Fault tree analysis (FTA) is necessary for high-risk events identified in security analysis, and corresponding preventive and corrective measures should be applied to reduce the risk.

8.4.3.2 Disturbance factors affecting service safety and comfort of the high-speed train

When the high-speed train runs along the track, various vertical and lateral forces are generated, and various vibrations of the vehicle system are caused due to the interaction between wheel and rail. The causes of vehicle vibration include excitations from the vehicle and disturbances from the track. The excitation from the vehicle comprises wheel wear, wheel eccentricity, wheel uneven weight, wheel tread abrasion, hunting motion of tapered tread wheelset. The disturbance factor from the rail comprises wheel-rail impact at rail joints, vertical deformation of the rail, local track irregularity, random track irregularity, etc. The excitations are shown in Fig. 8.66.

8.4.3.3 Mapping relation between excitation factors and running safety of the high-speed train

In the deterioration law of excitation, factors that vary with increase in mileage are considered. Because the excitation factors are characterized by dispersity, the degradation

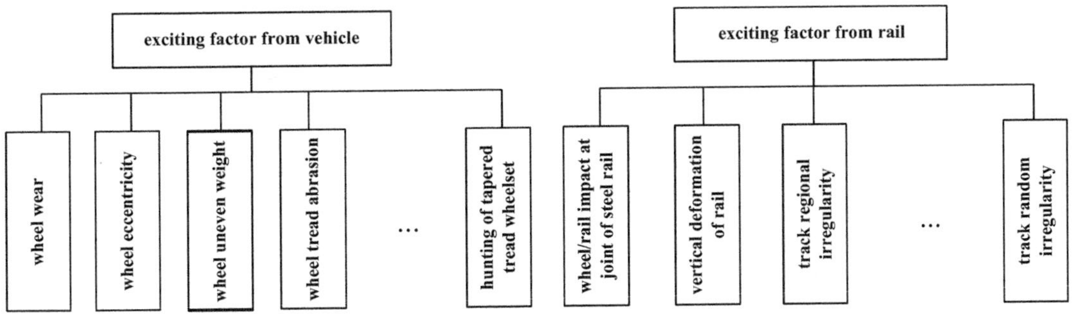

Figure 8.66
Exciting factor causing vibration of high-speed train.

law includes the degradation law of the mean value of excitation factors with increase in mileage and the distribution characteristics of excitation factors at the same mileage.

Fig. 8.67 indicates that the mean value of the excitation factors deteriorates with mileage based on the black solid line. At any mileage, the deterioration has dispersity and a certain distribution.

Taking wheel wear as an example, the relationship between wheel diameter wear and operation mileage is analyzed. Wheels are gradually worn in a wheel-lathing period. Tracking test is performed on profile and wheel diameter of eight wheels on the same car regularly five times based on the operation mileage, and five groups of measured profile and wheel diameter data are obtained and integrated into five wear conditions. Thus the

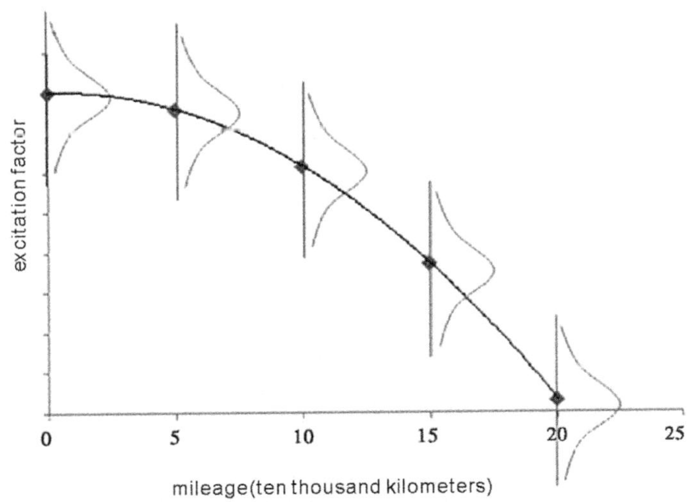

Figure 8.67
Deterioration feature of probability of excitation factor plotted against mileage.

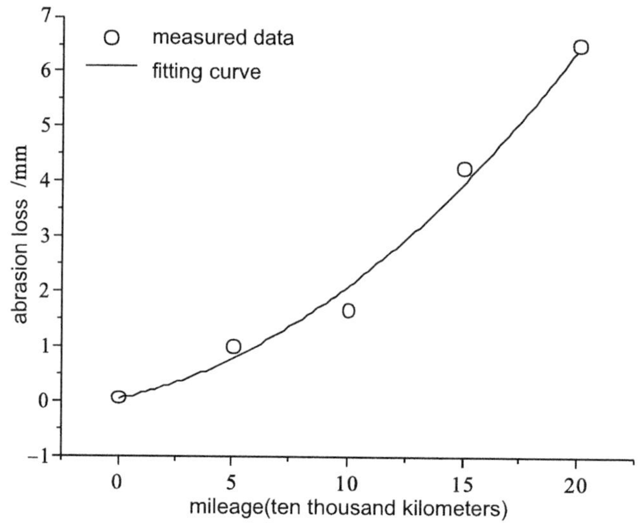

Figure 8.68
Relationship between operation mileage and abrasion loss.

relationship between the wheel diameter abrasion loss and the operation mileage is obtained and shown in Fig. 8.68.

Fitting the relationship between wheel diameter wear and operation mileage using quadratic function, the fitting function and goodness of fit are presented as Equation (8.29):

$$\begin{aligned} y = f_1(x) &= 0.0457 + 0.0917x + 0.0113x^2 \\ R^2 &= 0.97706 \end{aligned} \tag{8.29}$$

where x represents operation mileage, and y represents wheel diameter abrasion loss. The goodness of fit is 0.97706.

Distribution characteristics of wheel wear: Except for relevance to operation mileage, the wheel wear is also influenced by rail face state, vehicle load, climatic environment, and other factors. Therefore, the wear is often under normal distribution. The statistics of wheel diameter abrasion loss from track test show that the deviation of wheel diameter abrasion loss of a certain type of CRH EMU is about 0.04 mm.

There are many excitation factors affecting the safe operation of the high-speed train, and there are complex coupling characteristics among various factors. It is difficult to achieve an accurate mathematical model description for the mapping relation between the excitation factors and high-speed train operational safety. Neural network technology, based on the working model of the human brain, studies adaptive and non-programmed information processing methods that can be used to describe the intelligent behavior of cognition, decision-making, and control. This section discusses the feasibility of describing the mapping relation between excitation factors and train running safety using a back propagation (BP) neural network, which has the widest application at present.

The BP neural network was first put forward by a PDP (parallel distributed processing) scientist research group represented by Americans, Rumelhart and McClelland. It is a multilayer feedforward network trained on the basis of an error back-propagation algorithm. Fig. 8.69 presents the topological structure of a 3-layer BP neural network. It includes an input layer, a hidden layer, and an output layer. The number of neurons in each layer are n, m, and l, respectively. $w_{ij}(1 \leq i \leq n,\ 1 \leq j \leq m)$ is the connection weight between input layer and hidden layer; $v_{jk}(1 \leq j \leq m,\ 1 \leq k \leq l)$ is the connection weight between hidden layer and output layer; $f(x)$ and $g(x)$ are transfer functions from input layer to hidden layer and from hidden layer to output layer. The calculation formulas of front and back values of the neuron node are presented as Equations (8.30) and (8.31):

$$\text{Hidden layer}: \begin{cases} c_{1j} = \sum_{i=1}^{n} w_{ij}x_i - \theta_j \\ c_{2j} = g(c_{1j}) \end{cases} \tag{8.30}$$

$$\text{Output layer}: \begin{cases} y_{1k} = \sum_{j=1}^{n} v_{jk}c_{2j} - t_k \\ y_{2k} = f(y_{1k}) \end{cases} \tag{8.31}$$

where the subscript 1 represents the input side and 2 represents the output side; θ_j and t_k represent the output thresholds of every neuron. The transfer function applied to the traditional BP neural network model is the sigmoid function:

$$f(x) = g(x) = \frac{1}{1 + e^{-x}} \quad (-\infty < x < \infty) \tag{8.32}$$

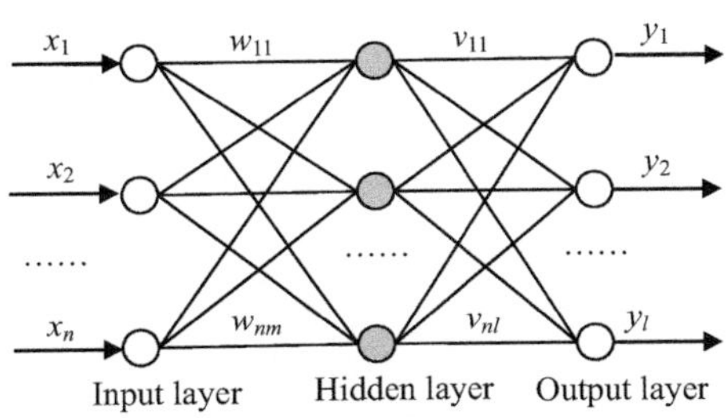

Figure 8.69
Topological graph of a three-layer BP neural network.

The formula of the global root-mean-square error between output value and expected value in the training stage model is:

$$E = 0.5\sum_{t=1}^{p} E_t = 0.5\sum_{t=1}^{p}\sum_{k=1}^{l}\left(e_k^t\right)^2 = 0.5\sum_{t=1}^{p}\sum_{k=1}^{l}\left(y_{2k}^t - y_{0k}^t\right)^2 \tag{8.33}$$

where y_{2k}^t and y_{0k}^t, respectively, represent the output value and expected value corresponding to the tth training sample; and p is the number of the training sample.

The forecasting process for the target object of the BP neural network can be summarized as: First, perform nonlinear processing and forecasting on the input sample data by simulating the brain neural network system using Equations (8.29)–(8.31). The error between the forecasting value and the input expected value is calculated based on Equation (8.32), and assess whether the value meets the preset goal; if not, adjust the weights and thresholds of the network through back propagation based on the learning rules of the steepest error descent method or the gradient descent method, and the process is repeated until the root mean square error meets the requirements. Thus the optimal neural network model is trained. Finally, the predicted data are calculated based on the trained neural network model.

Taking the lateral stability of one high-speed train as an example, and using a BP neural network to construct the mapping relation model with vehicle disturbances such as suspension parameters and wheel parameters: First, the key factors influencing the EMUs are determined using the single parameter impact analysis of the vehicle-track coupling system dynamics model. Then set parameters of multiple groups of key influencing factors are generated randomly, and the parameters are possible values during the operation period. The critical velocity of the lateral stability of trains corresponding to the combination of parameters is calculated using the simulation model. Taking simulation results as training samples and setting different BP neural network parameters, optimal parameter values of the sample are compared and screened, and the empirical model of the mapping relation is generated through sample training. Finally, the forecasting of the lateral stability of trains is realized by inputting the state excitation factors based on the generated model. A specific flow chart is presented in Fig 8.70.

Based on the analysis of single parameter influence, the key influencing factor sets of the suspension parameters and wheel parameters of this type of EMUs are confirmed as: damping of yaw damper, damping of secondary lateral damper, primary longitudinal locating stiffness, coaxial wheel diameter difference, wheel diameter difference of the same bogie, tread equivalent conicity, and order and depth of wheel polygonization. Using the data generated by simulation as training samples, the mapping relation model of the excitation parameters, and the critical velocity of lateral stability is obtained. The contrast curve between the forecast value and the actual value (expected value) of the model in

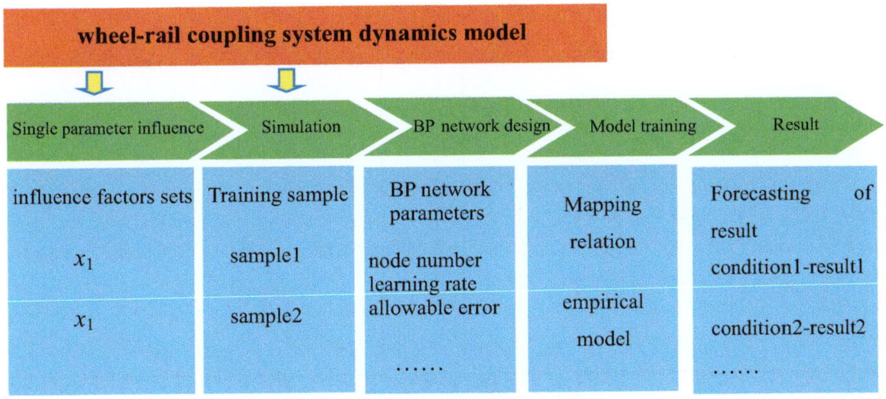

Figure 8.70
Modeling flow of the mapping relation model.

training stage and forecasting stage is shown in Fig. 8.71. Fig. 8.72 presents the result of the analysis of the linear correlation of forecast value and expected value.

From Fig. 8.72, it is apparent that the output values and expected values of the BP neural network have good coincidence. The goodness of fit of the linear correlation in the training stage and forecast stage are 0.97 and 0.93, respectively. There is a strong linear correlation between them. The results show that the BP neural network model built in this section can well represent the mapping relation between excitation parameters and lateral stability. The forecasting model of vehicle lateral stability based on the mapping relation is effective.

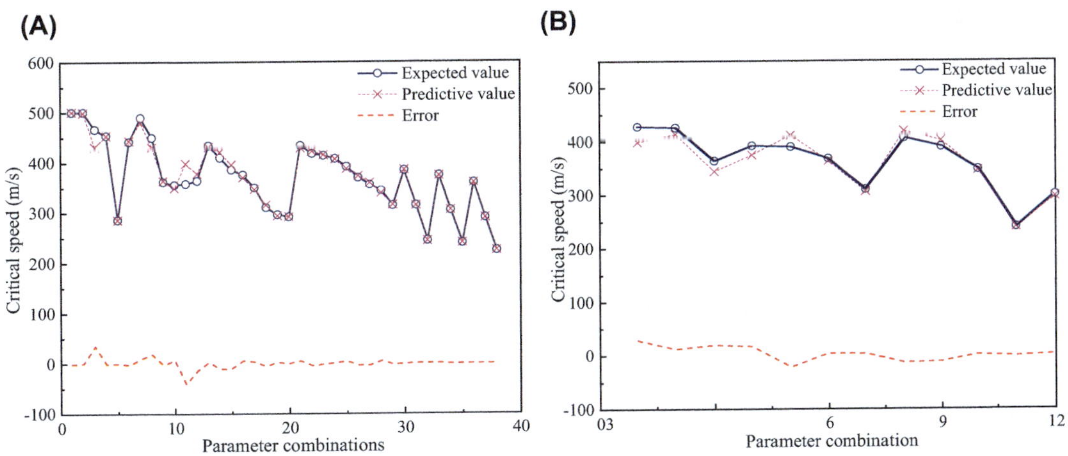

Figure 8.71
Contrast curve between forecast value and expected value of BP neural network. (A) Training stage (B) Forecast stage.

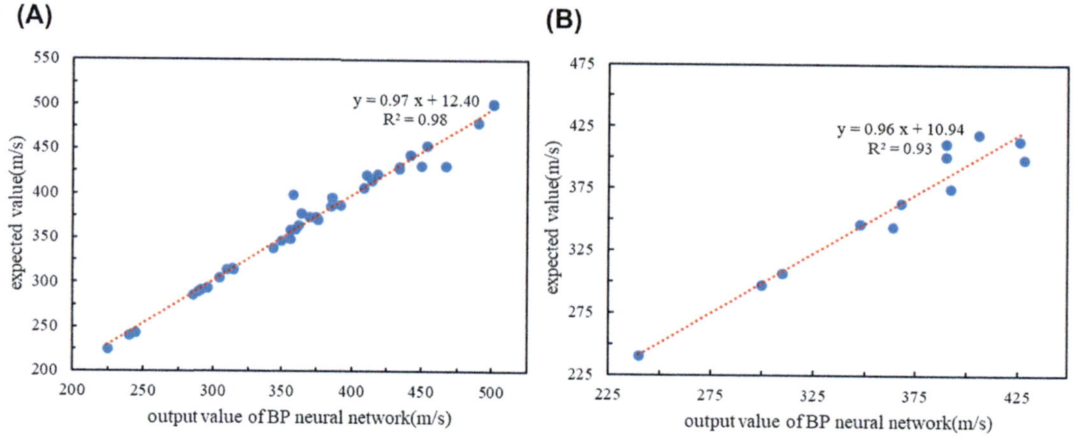

Figure 8.72
Relevance of output value and expected value of the BP neural network. (A) Training stage (B) Forecast stage.

8.4.3.4 Relationship between excitation factors and service safety

The wheel wear may cause an increase in equivalent conicity and further reduce the critical velocity. Thus, its influence on the safety and reliability of train service.

(1) Wheel diameter abrasion loss and equivalent conicity

With increase in operation mileage, the wheel profile and diameter change, which makes the wheelset traverse and shake. Great offset occurs in the new balancing position of the wheel-rail matching, inevitably leading to change in the wheel-rail contact relationship. The taper curve is significantly different from the original design curve.

Based on the simulation calculation, the relationship between wheel abrasion loss and equivalent conicity of a CRH high-speed train is obtained. The relationship is presented in Fig. 8.73.

Fitting the relationship between equivalent conicity and wheel diameter abrasion loss: Using a linear function to fit the relationship between wheel diameter abrasion loss and operation mileage, the fitting function and the goodness of fit are presented as the Equation (8.34).

$$\begin{aligned} \lambda &= f_2(y) = 0.067y \\ R^2 &= 0.730 \end{aligned} \tag{8.34}$$

where λ represents equivalent conicity, and y represents wheel diameter abrasion loss. The goodness of fit is 0.730.

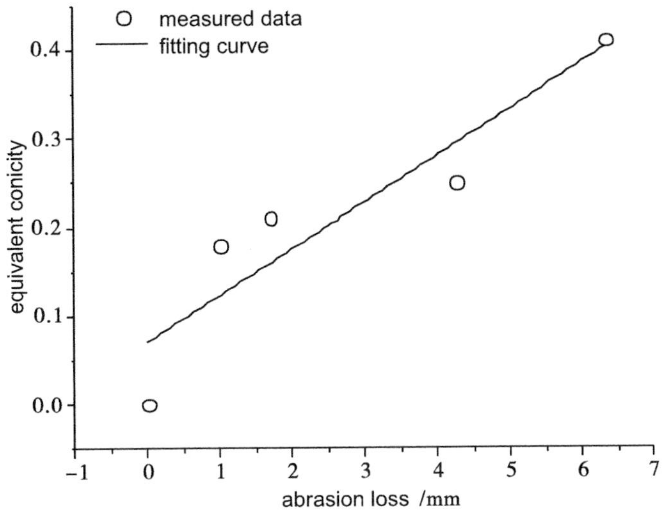

Figure 8.73
Relationship between abrasion loss and equivalent conicity.

(2) Equivalent conicity and critical velocity

With increase in operation mileage, the wheel gradually wears, and the tread equivalent conicity changes. These lead to change in critical velocity. Based on the analytical calculated results of Table 8.4, the relationship between the equivalent conicity and critical velocity of a certain type of EMU is shown in Fig. 8.74.

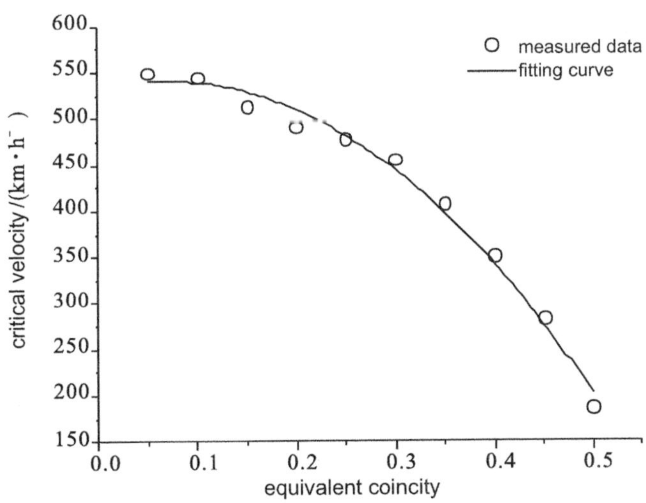

Figure 8.74
Relationship between equivalent conicity and critical velocity.

The relationship between equivalent conicity λ and critical velocity v is fitted and presented as Equation (8.35):

$$v = f_3(\lambda) = 530.633 + 267.545\lambda - 1851.51\lambda^2$$
$$R^2 = 0.9860 \tag{8.35}$$

8.4.3.5 Maintenance decisions centered on reliability

(1) Optimization of maintenance period based on safety margin

Based on the wheel wear rule, equivalent conicity evolution rule and its influence on critical velocity: Based on the allowable critical speed of EMUs, being about 130% of the operating speed, if the running speed is 300 km/h, the allowance critical velocity [v] is 390 km/h, and when the running speed is 350 km/h, the allowance critical velocity [v] is 455 km/h. If EMUs run x 0,000km, the meaning of R degree of reliability of its critical velocity can be expressed by a safety margin as Equation (8.36):

$$R = P([v] - f_3(f_2(f_1(x))) > 0) \tag{8.36}$$

Through inverse solution of Equation (8.6), the wear lifetime capable of meeting R degree of reliability of wheels with allowable critical velocity of 390 km/h is solved, and the calculated results are presented in Table 8.11.

(2) Optimization of maintenance period based on risk threshold

Based on the degree of damage caused by different failure modes of components, including safety, usability, economy, failure, deterioration, speed, redundancy, and other indexes, the acceptable risk threshold value F_c of planned preventive maintenance with fixed interval can be determined. To determine the maintenance period of each failure mode, combine the degradation model of the degree of reliability of every failure mode obtained in 8.4.2.2. The maintenance period of every failure mode is further packaged and integrated:

$$max\left\{\frac{R(T_i) - R(T_{i+1})}{R(T_i)}\right\} \leq F_c \tag{8.37}$$

Table 8.11: Wheel wear lifetime with given degree of reliability.

Degree of reliability	Wheel wear lifetime (10, 000 km)
0.5	17.9464
0.9	17.84307
0.95	17.81369
0.99	17.75848

Based on the overall reliability degradation law of components and the proportion of failure modes obtained in 8.4.2.2, the corresponding risk changes for prolonging maintenance period can be obtained. For example, if the 3-class repair period of the bogie is extended to 1.4 million kilometers, the risk of car-smashing maintenance failure will increase by 12.9%, and the risk of wheel material defect, axle material defect, and overheated rolling bearing will also increase.

(3) Optimization of overall maintenance period based on the system reliability model

To fully consider the mutual influence and combined effect of multiple factor reliability in the vehicle system on the overall reliability of the system, it is necessary to set up a system-level reliability model with further optimization of maintenance period based on the failure law model of each component. This section takes the bogie stability system as an example and introduces the reliability modeling method based on orthogonal tests and range analysis. Then carries out overhaul cycle optimization of the bogie by combining failure rules of relevant components.

The stability of the high-speed train is based on a multi-factor coupling function system: any parameter failure in the system may cause lateral instability to a certain degree. Thus the logic relationship of the system can be described by the equivalent series model as shown in Fig. 8.75. The system reliability analytical model is set up as Equation (8.38) by further considering the number, importance, and working time of every unit:

$$R_s(t_s) = \prod [1 - E_i \cdot (1 - R_i(t_i))] \tag{8.38}$$

where $R_s(t_s)$ represents the degree of reliability of the system worked for t_s, R_i represents the degree of reliability of the i unit worked for t_i, N_i represents number of the i unit in the equivalent serial system, E_i represents importance of the i unit.

Determination of the importance of each element based on orthogonal tests and range analysis: The vehicle critical velocity under various combination conditions can be determined using orthogonal test method based on the parameter distribution of failure parts in the service process of the high-speed train and vehicle bench test or digital

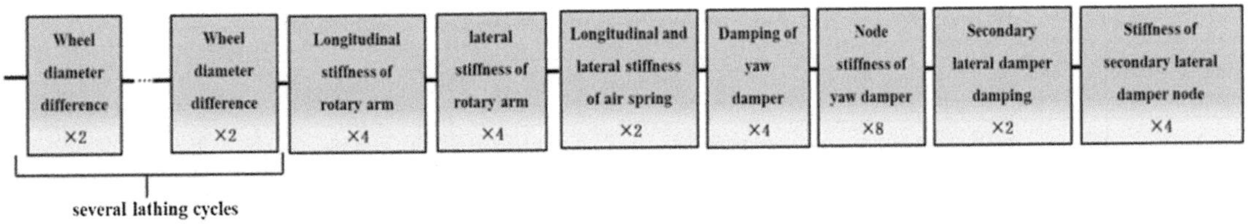

Figure 8.75
Logic relationship of the reliability of the stability system of the high-speed train.

simulation, and the use of range analysis to further determine the importance of each factor.

$$E_i = \frac{R_i}{\overline{y_{max}} - \overline{y_{min}}} = \frac{max\left[\overline{y_{i1}}, \overline{y_{i2}}\ldots\right] - min\left[\overline{y_{i1}}, \overline{y_{i2}}\ldots\right]}{\overline{y_{max}} - \overline{y_{min}}} \tag{8.39}$$

where $\overline{y_{ik}}$ represents the mean value of all j test index results corresponding to the k level of the factor i; $\overline{y_{max}}$, $\overline{y_{min}}$ represents the mean value from previous j to the following j of all test index results sequenced from large to small.

The importance of every parameter is calculated using the method above and is presented in the Table 8.12.

Furthermore, based on the system reliability model and the reliability degradation law of each parameter, the reliability-centered optimization of the advanced repair cycle of the bogie is carried out. The integral multiple relationship between the advanced repair cycle T_s and wheel lathing cycle T_w of the bogie is maintained. The optimization algorithm of the system maintenance period T_s and wheel lathing cycle T_w are:

$$\begin{cases} T_s = k \cdot T_w \quad k = 1, 2, \ldots \\ R_w = 1 - F_w(T_w) \\ R_i = 1 - F_i(T_s) \\ R_s \leq \prod [1 - E_w \cdot (1 - R_w)] \prod [1 - E_i \cdot (1 - R_i)] \end{cases} \tag{8.40}$$

The recommended bogie 3-class repair cycle and wheel lathing cycle are: extending the wheel lathing cycle to 240000–250000 km and prolonging the bogie 3-class repair cycle to 1,44–1.5 million kilometers. The reliability of the vehicle stability can be guaranteed to meet the requirements.

Table 8.12: Calculated results of the importance of every parameter.

Parameters	**Wheel diameter difference**	**Longitudinal locating stiffness of rotary arm**	**Lateral locating stiffness of rotary arm**	**Longitudinal and lateral stiffness of air spring**
Importance	0.078204	0.468585	0.179304	0.145406
Parameters	Damping of yaw damper	Node stiffness of yaw damper	Secondary lateral damper damping	Stiffness of secondary lateral damper node
Importance	0.405611	0.482516	0.109426	0.081772

References

[1] Translated by Zeng Minshi. Technical condition of networking high-speed train of European high-speed railway. Standard Measurement Institute of Ministry of Railway; 1994.
[2] TB/T3188-2007. Technical specification for railway car safety monitor and diagnosis system [S].
[3] Liu Feng, Yan Jiulei, Dong Xiaoqin, Zhang Ruifang. Lateral instability real-time online monitoring devices for bogies. Railway locomotive 2010;30(2):5–7.
[4] Wang Futian. Dynamics of vehicle system. Beijing: China Railway Publishing House; 1994.
[5] Zhang Bin. Theory and Analysis Research of Monitoring and Detecting the Key Components of Train. Chengdu: Southwest Jiaotong University; 2007. p. 10.
[6] Simani Silvio, CesareFantuzzi, Ron J. Patton.Model-based Fault Diagnosis in Dynamic Systems Using Identification Techniques. London: Spriner-Verlag; 2003.
[7] GodaKenjiro Goodall Roger. Fault-detection-and-isolation system for a railway vehicle bogie. Vehicle System Dynamics Supplyment 2004;41:468–76.
[8] Bruni Stefano, Goodall Roger, X Mei T, Tsumashina Hitoshi. Control and monitoring for railway vehicle dynamics. Vehicle System Dynamics 2007;45(7):743–79.
[9] Li Haitao, Wang Chengguo. Secondary vertical suspension element state monitoring based on frequency domain model of railway vehicle. Railway locomotive 2008;28(2):1–5.
[10] Jiang Changhong. Study on failure detection system of railway vehicle axle. Jilin University; 2006.
[11] Wang Zhen. Study on defect automatic locating algorithm of wheelset ultrasound flaw detection system. Chengdu: Southwest Jiaotong University; 2011.
[12] Maoru Chi, Zhang Weihua, Zeng Jing, Jin Xuesong, Zhu Minhao. Influence of wheel diameter difference on running safety. Journal of Traffic and Transportation Engineering 2008;8(5):19–22.
[13] Maoru Chi, Zhang Weihua, Zeng Jing, Jin Xuesong, Zhu Minhao. Influence of wheelset installing error on railway vehicle running safety. Journal of Southwest Jiaotong University 2010;45(1):12–6.
[14] Maoru Chi, Zhang Weihua, Zeng Jing, Jin Xuesong, Zhu Minhao. Influence of deflection load on running safety of rotary arm locating bogie. China Railway Science 2009;30(3):81–5.
[15] Cao Jinhua, Cheng Kai. Introduction to reliability mathematics. Beijing: Higher Education Press; 2006.
[16] Song Baowei. System reliability design and analysis. Xi'an: Northwestern Polytechnical University Press; 2008.

Index

Note: 'Page numbers followed by "f" indicate figures and "t" indicate tables'.

D

T

W

Y

Z

CPI Antony Rowe
Eastbourne, UK
November 20, 2019